ksf
3/26/97

Practical Engineering Perspectives

Plant Operation and Optimization

Gail F. Nalven, Editor

American Institute of Chemical Engineers
345 East 47th Street • New York, New York 10017-2395

Practical Engineering Perspectives Plant Operation and Optimization
ISBN 0-8169-0710-2
American Institute of Chemical Engineers
345 East 47th Street
New York, New York 10017-2395

Library of Congress Catalog Card Number: 96-24176

cover design: Joseph Roseti

PREFACE

Keeping a plant running right requires grappling with a host of issues. These include ones related to the actual hardware in which the chemical feeds and products are handled, the devices and electronics that monitor and adjust those processing operations, and how the various elements in a plant can be tied together most effectively and efficiently. This book, aimed at chemical engineers and other technical professionals in the chemical process industries who are involved in the running of process plants and improving their operation, is a compendium of recent articles of interest to practitioners.

This book addresses seven key issues, each in a separate section with at least five articles. The volume begins with a section on fluid flow. This particularly looks at hardware issues that can be crucial for reliable plant operation. Articles cover valves, piping, and centrifugal pumps, as well as compressors and dryers for plant air.

Energy frequently comprises the greatest cost in plant operations. So, the role of heat transfer in a process receives considerable attention and, indeed, forms the largest section in this volume. Design and selection information is provided on such topics as heat-transfer fluids, heat tracing, reboilers, spiral heat exchangers, and immersion heaters. In addition, this section includes insights on making the most efficient use of energy.

To understand how a plant is performing, accurate data are necessary. Thus, a section focuses on choosing the right measurement device and avoiding data errors. Particular attention is paid to the measurement of flow, one of the most common needs in any plant and one with a wide variety of different measurement technologies available, and to pH, one of the most troublesome.

Data must be analyzed correctly and used properly to control the process. The next section covers these issues, with emphasis on control valves and the process control system.

The next section focuses on mixing of materials. It looks at design issues, such as the mechanical design of agitators, but also provides guidance, for instance, on selecting agitators for viscous liquids, when to consider small, magnetically driven units, and how to assess heat transfer.

This is followed by a section on reactors that has a wide breadth: particulates reaction engineering; sizing of relief systems; scale up and optimization of fluidized-bed reactors; and using reaction calorimetry for scale up.

The overall aim for any plant is to achieve the most efficient operation. So, appropriately, the final section of this volume focuses on issues related to optimization, and addresses a diverse range of issues, including maintenance, engineering design, equipment selection, and pinch technology.

In this compendium, articles have been selected based on their usefulness to practitioners, and their emphasis on practical and generic guidance. The roster of authors includes many recognized authorities—and represents a mix of individuals who work for operating, engineering/construction, and equipment companies, as well as research organizations, and academia.

All the articles have been published by the American Institute of Chemical Engineers over the past three years. Two articles were drawn from the Institute's ongoing Symposium Series. All the others appeared originally in *Chemical Engineering Progress*, the monthly magazine of AIChE.

Mark D. Rosenzweig
Editor-in-Chief
Chemical Engineering Progress
May 1996

ACKNOWLEDGMENTS

Creating a new publishing program is a daunting task that can only be made possible with the help of a group of talented publishing professionals who share their expertise.

Thanks go to Mark D. Rosenzweig, Cynthia Fabian Mascone, and Rich Greene for their guidance in developing the text content, and to Stephen R. Smith for his vision in creating this program. Special appreciation goes to Chung S. Lam whose assistance in handling the "details" is really what made this series possible!

Gail F. Nalven
June 1996

CONTENTS

Mixing

Reactors

Plant Optimization

CONTRIBUTORS

Peter R. Baen, Manager
Marketing Support, Thermon Industries, San Marcos, TX

Kevin M. Bailey, PE, Product Manager
Alfa Laval Thermal, Inc., Richmond, VA

Roy E. Barth, Vice President
Research and Development, Thermon Industries, San Marcos, TX

André Bakker, Ph.D., Principal Research Engineer
Chemineer, Inc., Dayton, OH

Thore Berntsson, Ph.D., Professor
Chalmers University of Technology, Göteborg, Sweden

Mohan A. Bhalodia, Ph.D., Staff Engineer
Exxon Research and Engineering Co., Florham Park, NJ

Donna G. Blackmond, Ph.D., Associate Director
Merck & Co., Inc., Whitehouse Station, NJ

David L. Bradford, Technical Manager
Chemelex Division, Raychem Corp., Cincinnati, OH

Gerald R. Brown, PE, Project Engineer
Bechtel Corp., Houston, TX

Jeffrey A. Caldwell, Doctoral Candidate
University of Louisville, Louisville, KY

John W. Carson, Ph.D., President
Jenike & Johanson, Inc., North Billerica, MA

Annika Carlsson, Doctoral Candidate
Chalmers University of Technology, Göteborg, Sweden

Joel B. Christian, PE, Advanced Research Engineer
Osram Sylvania, Inc., Towanda, PA

William J. Colt, Staff Project Engineer
Morrison Knudsen Corp., Cleveland, OH

John Cuthbert, Development Leader
Thermal Fluids Technical Service and Development Laboratory
The Dow Chemical Co., Midland, MI

Pradeep B. Deshpande, Professor
Chemical Engineering Department, University of Louisville, Louisville, KY

Alexandre Dimian, Ph.D., Associate Professor
University of Amsterdam, Amsterdam, The Netherlands

John W. Dolenc, PE, Chemical Industry Manager
Fisher-Rosemount, Inc., Eden Prairie, MN

Timothy E. Dowdle, General Manager
Quantum Chemical Co., Morris and Tuscola, IL

L.-S. Fan, Ph.D., Professor
Ohio State University, Columbus, OH

Julian B. Fasano, PE, Director
Parts and Field Service, Chemineer, Inc., Dayton, OH

James D. Fleshman, PE, Chief Process Engineer
Foster Wheeler USA, Perryville, NJ

Per-Åke Franck, Ph.D., Managing Director
CIT - Energy Technology Analysis, Göteborg, Sweden

V. Ganapathy, Heat-Transfer Specialist
ABCO Industries, Abilene, TX

Manfred Gans, President
Technology Evaluation & Development Associates, Inc., Hoboken, NJ

Lewis E. Gates, PE, Vice President
Engineering and Development, Chemineer, Inc., Dayton, OH

Peter Glavic, Ph.D., Chair
Chemical Engineering Department, University of Maribor, Maribor, Slovenia

A. Ghosh-Dastidar, Ph.D., Post-Doctoral Research Associate
Ohio State University, Columbus, OH

Brian R. Goedke, Plant Manager
Quantum Chemical Company, Tuscola, IL

Jerry Graczyk, Product Manager
ITT Engineered Valves, Lancaster, PA

Bill Hannon, Marketing and Product Manager
ITT Engineered Valves, Lancaster, PA

Russell E. Hannula, Ph.D., Manager
Total Quality Systems Department, Rohm and Haas Kentucky, Inc., Louisville, KY

Clay W. Hansen, Applications Engineer
Exxon Chemical Co., Baton Rouge, LA

James W. Hayes, Marketing Consultant
Waukesha, WI

John L. Hedrick, President
Automation & Control Technologies, Inc., St. Louis, MO

Behzad Jazayeri, Process Principal Engineer
Fluor Daniel, Inc., Irvine, CA

Michael F. Jeglic, Product Development Manager
Dow Plastic-Lined Products, Bay City, MI

Pankaj K. Khandelwal, Senior Process Engineer
Uhde India Ltd., Bombay, India

Shoichi Kimura, Ph.D., Associate Professor
Oregon State University, Corvallis, OR

Robert C. Klein, Product Manager
Watlow Electric Manufacturing Co., Hannibal, MO

George E. Klinzing, Ph.D., Professor
University of Pittsburgh, Pittsburgh, PA

Ted M. Knowlton, Ph.D., Associate Director
Fluidization Research, Institute of Gas Technology, Chicago, IL

Hemant Kumar
APV Baker Therm. Div., Goldsboro, NC

Ralph N. Landau, Ph.D.
Technical Operations Fellow, Merck & Co., Inc., Whitehouse Station, NJ

Nancy L. Lindley, Marketing and Customer Service Manager
Dow Plastic-Lined Products, Bay City, MI

Bodo Linnhoff, Ph.D., Head, Department of Process Integration
University of Manchester Institute of Science and Technology, Manchester, UK
and Chairman, Linnhoff March International, Houston, TX

Norman J. Little, PE, Process Engineering Consultant
Kilborn, Inc., Sarnia, Ontario, Canada

Edward H. Livingston, Vice President and General Manager
Burton Corblin North American, Inc., Horsham, PA

Joseph T. Lonsdale, Technical Marketing Specialist
Chemelex Division, Raychem Corp., Menlo Park, CA

Vincent M. McCalley, Chief Engineer
Pressure Products Industries, Inc., Warminster, PA

Alfred J. McCarthy, Senior Engineering Consultant
The M. W. Kellogg Co., Houston, TX

Gregory K. McMillan, Fellow
Integrated Manufacturing Control Group, Monsanto Co., St. Louis, MO

Jerry L. Miller, PE, Principal Mechanical Engineer
Chemineer, Inc., Dayton, OH

Zorka Novak, Graduate Student
Chemical Engineering Department, University of Maribor, Maribor, Slovenia

James Y. Oldshue
Oldshue Technologies International, Fairport, NY

Burton Palmer, Industrial Analytical Operations Consultant
Tucson, AZ

Gianni Papa, Process Engineer
Snamprogetti, SpA, Milan, Italy

Andrew Parr, Divisional Process Engineer
Humberside Refinery of Conoco, Ltd., Grimsby, UK

Lee R. Partin, PE, Research Associate
Eastman Chemical Co., Kingsport, TN

Steve A. Pasley, Manager
Product Engineering, Chemineer, Inc., Dayton, OH

J. Rathman, Ph.D., Assistant Professor
Ohio State University, Columbus, OH

Siegfried H. Richter, PE, Retired, Principal Process Engineer
John Brown, Bridgewater, NJ

Tim Robinson, Executive Vice President
Asahi/American, Malden, MA

Charlie Rose, Senior Product Manager
Fischer & Porter Co., Warminster, PA

Fred F. Sanders, Engineering Specialist
Monsanto Chemical Co., Pensacola, FL

Brian R. Smith, Engineer
The M.W. Kellogg Co., Houston, TX

Frederick M. Sitter, Chief Application Engineer
Van Air Systems, Inc., Lake City, PA

Ming T. Tham, Ph.D., Lecturer
University of Newcastle upon Tyne, Newcastle upon Tyne, UK

Frank Turner, Principal Process Engineer
John Brown, Bridgewater, NJ

Hank Van Ormer, Founder
Air Power U.S.A., Inc., Pickerington, OH

Gabor E. Vass, Applications Manager
Fischer & Porter Co., Warminster, PA

Michael T. Walsh, Product Manager
Pressure Products Industries, Inc., Warminster, PA

Alan W. Weimer, Ph.D., Research Associate
Dow Chemical, Midland, MI

Barry A. White, PE, Sales and Marketing Manager
Star Systems, Valtek International, Provo, UT

Owen Woodside, PE, Consultant
Woodside Associates, Arcadia, CA

Wen-Ching Yang, Ph.D.
Advisory Engineering, Science and Technology Center
Westinghouse Electric Corp., Pittsburgh, PA

Srinivas S. Yerrapragada, Ph.D., Post-Doctoral Research Associate
University of Louisville, Louisville, KY

Frederick A. Zenz, Ph.D., Director
PEMM-Corp., Cold Spring, NY

The Editor

Gail F. Nalven, Acquisitions Consultant
American Institute of Chemical Engineers, and Technical Editor, New York, NY

INTRODUCTION

The concept of plants and chemical engineering is a historic mixture which remains contemporary. It was the need for a special discipline to design and understand chemical plants that provided the basis for the formation of the American Institute of Chemical Engineers and subsequently the underlying body of science and technology from which Chemical Engineering education evolved. Chemical Engineering distinguished itself from other disciplines by concept of unit operations. Conceptual decomposition of the plant into intellectual "bite size" problems that were solvable, first by slide rule then by computer, facilitates understanding and progress. While the increased capability of our solving tools now increases the size of the problem that we can deal with, we continue to think about the plant in separate unit operations terms. In fact, this approach may continue to be the most efficient for some time.

Plants are complex entities that convert feeds into products. They are parts of commercial enterprises that, as a whole, exist to provide necessary, desirable, and beneficial commodities. They must function amidst a conglomeration of constraints to be successful. These constraints include: profitability, management goals, government regulation, and the human ability of engineers and operators, to name a few.

If the plant is analogous to Humpty Dumpty whose fragments are fluid flow, heat transfer, process measurements, mixing, plant optimization, process analysis and control and reactor characteristics, etc., then it is up to the chemical engineer to accept the critical role of "all the King's horses and all the King's men" by putting it all together again. Many would argue that it is this "systems mentality" that distinguishes chemical engineers from others. In theory, it is possible to quantify the effect of every variable that influences any unit operation on the overall net present value of the plant objective. In practice, this approach would lead to extensive study and very little action. Instead, chemical engineers, by intuition or necessity or both, chose the path for plant functioning based on their arsenal of knowledge. Choosing the "better" path depends on their breadth of experience and education. One of the goals of this book is to complement your arsenal.

The challenges for operating, modifying and designing chemical plants and their sibling refineries are both great and exciting. Agents that cause change are also changing. Markets change to create new demand for some products while shrinking others. Uses for products change, causing changes in specifications. Hardware for making things happen—pumps, exchangers, tower trays, mixers, etc.—evolve to accomplish better general or specific jobs. Organizations change. Technical knowledge changes. For example, knowledge about the underlying principles of chemistry, and molecular transport expands in fits and jerks, while the capabilities for more closely modeling the intricacies of chemical reactors follows this roller coaster path. More and more of the plant is squeezed inside the control loop. On line optimization self optimizes through almost, but never quite, adequate process models. It's like the grasshopper and the wall. We can hop halfway there each jump, but we will never get there.

Who are the creators of agents of change? Well, for one, we are—at least to a major extent. It is our insatiable drive to compete and to create newness—to somehow sense achievement. By creating new methods, new processes, new ways to look at processes and underlying new knowledge, we create new challenges for continued improvement.

How do we start to meet these challenges? Well, one way is to learn about what others think and are doing about things. That's where this book can help. It is a set of snapshots of technology and its application to plants. It is a collection of articles that were published in *CEP* over the past three years. The articles are organized generally by subject. It is not all encompassing in any one subject (but the references help those who want more). It doesn't replace your texts, but it's content may be included in their next update. If you take the time to familiarize yourself with it, you may find that it helps you in your change quest more than you expect.

Jerry L. Robertson
May 1996

Modernize Valve Selection

Process automation and demands to reduce emissions are two of several key factors that increasingly must be weighed in selecting valves.

Jerry Graczyk and Bill Hannon,
ITT Engineered Valves

In recent years, as environmental regulations have tightened and community awareness of environmental issues increased, the effect on both the chemical process industries (CPI) and valve manufacturers has been profound.

With the Clean Air Act Amendments of 1990, "emissions control" quickly became one of the most important purchasing criteria used by the CPI when selecting a valve. The 1990 amendments also led valve manufacturers to reexamine the design criteria of their valves, and to focus their efforts on providing valves with higher performance at a competitive price.

At about the same time that emissions control became a driving force in valve selection, the CPI also saw a tremendous increase in process automation and its complexity. While this trend in automation greatly increased production levels, it also placed much greater demand on valves, particularly in the areas of shutoff and controllability. Today's valves no longer simply open or close, they must communicate with switches for opening, closing, and throttling instructions. Systems have become dependent on feedback to a computer or process controller from both valves and switches.

■ *Figure 1. Quarter turn valve in acid service.*

When choosing the proper valve for their application, CPI engineers are often confronted with a wide array of performance requirements. There are many factors that must be considered, including:

• *Temperature.* The materials that the valve is produced from must be able to handle the temperature requirements of the process materials. Abrasive or corrosive processes tend to be more detrimental to the valve as the temperature increases, so simply specifying a valve based solely on the process temperature may result in an inadequate selection. Selection must be based on all factors of the process.

• *Pressure.* Valves are designed and manufactured based on the pressure in which they will operate. Typically, chemical processes are operated in the 150-600 ANSI class range.

• *Shutoff.* There are various ratings of shutoff on valves depending on style and design. Typically nonresilient seated control valves will follow ANSI B16.104 test standards. Under that standard, Class I leakage is equivalent to Class II, III, or IV, but by agreement between user and supplier no compliance test is required. Class VI is the most stringent of ANSI B16.104 standards, but still allows for some leakage (stated in bubbles per min) through the valve (see Table 1).

Resilient (polymeric or elastomeric) seated valves, such as ball and plug valves, refer to Manufacturers Standardization Society Standard Practice (MSS SP) and American Petroleum Institute (API) for leak testing methods and rates. MSS SP-72 and API 598 involve two standards testing procedures for ball valves with flanged or butt weld ends. Section 7.2.4 states that there shall be no visible leakage, as defined by MSS SP-82, past the seat for the duration of the test for valves with resilient seats. This is more commonly known as a bubbletight shutoff.

• *Throttling.* Some valves are designed specifically for throttling. It's important to have the right valve for throttling, because those that are not designed to handle the stresses accompanied with throttling will often self-destruct under throttling pressures. Globe valves are specifically designed for throttling applications while floating type ball valves are better suited for open/close shutoff applications.

• *Size of pipe.* The size of the pipeline, as well as operating flow coefficient, C_v, are critical elements when specifying a valve. Typical piping applications can afford to have some pressure drop throughout the piping system. Regular ported (reduced port) valves work well in most applications where pressure drop is not a concern, and are less expensive than a full-ported valve. For those rare applications where pressure drop and low C_v are detrimental to the process, the expense of a full ported valve may be necessary. C_v calculations will determine if the piping system and pumps can handle the pressure drop of the selected valves.

• *Nature of process fluids.* Valves are designed to work in different process conditions, and each application should be checked throughly to determine if the selected valve will work in the applied process. For instance, a ball valve with a cavity area will have difficulty operating in a high slurry/solids application, while the straightway diaphragm valve is designed to effectively handle a high solid/abrasive slurry process.

• *Velocity.* One of the most critical factors in valve selection, velocity can wear down a valve as easily as solids can, especially if there is excessive velocity through the system. High velocity should be avoided when handling abrasive solids.

• *Concentration.* Chemical concentration is a very important factor when selecting materials for the wetted surfaces, as well as those parts of the valve or piping system that are in contact with the process fluid. The impact of concentrations of materials vary, because certain chemicals are more aggressive at diluted concentration levels than at higher levels. High concentrations of sulfuric acid, for example, can be handled by ordinary steel pipe while the lower levels of sulfuric acid are much more aggressive, and would require high-alloy metal, plastic, or plastic-lined piping. The level of chemical aggression of a specific chemical at a specific concentration can vary between different piping materials such as metal, plastic or elastomer.

• *Firesafe.* Some applications require the valve to be firesafe rated when processing flammable or explosive media. The most recognized and used firesafe rating in the CPI is the American Petroleum Institute's (API) 607 rating. Others include Factory Mutual (FM), United Laboratories (UL), and Exxon testing.

• *Hazardous material.* While some valves are designed for general purpose use, there are others that are specifically designed to handle more serious processes which must be contained in the pipeline to avoid injury to plant personnel. These valves are typically manufactured with redundant stem packing designs, or in some cases use a bellows stem seal design. Leak detection ports, which can indicate primary stem seal failure, are common in this type of application. Some lethal applications are classified by ANSI B31.3-1987 as a "Category M" service. (See box on "Hazardous or Lethal Material".)

Taking these factors into consideration before purchasing a valve can save time and money in the long run. While choosing the right valve and the correct materials may cost a little more upfront, it is important to consider the complete cost of valve ownership. Valves that operate smoothly with little or no downtime will more than return a higher initial investment than a valve that cannot handle the specific requirements of your application.

There are six types of valves typically used in CPI applications. The following is a brief description of each valve, with specific benefits and limitations inherent in each design.

Hazardous or Lethal Material

According to ANSI, fluid services are classified as Category M when:

• the fluid is toxic;

• a single exposure (breathing or bodily contact), caused by leakage, to a very small quantity of the fluid can cause serious and irreversible harm, even when prompt restorative measures are taken; and

• considering experience, service conditions and location involved will not sufficiently protect personnel from exposure to very small quantities of fluid in the environment.

ANSI makes clear that valve manufacturers cannot categorize the fluid service. Paragraph 300 (b) (1) of B31.3 says that "The owner (of a piping system) is responsible for identifying those fluid services which are in Categories D and M..."

Specific requirements for valves under Category M are as follows:

• Valves having threaded bonnet joints (other than union joints) shall not be used.

• Only metallic valves conforming to the following requirements may be used:

1. Special consideration shall be given to valve design to prevent stem leakage to the environment.

2. Bonnet or cover plate closures shall be flanged, secured by at least four bolts with gasketing conforming to paragraph 308.4; or proprietary, attached by bolts, lugs, or other substantial means, and having a gasket design that increases gasket compression as fluid pressure increases; or secured by straight thread sufficient for mechanical strength, a metal-to-metal seat, and seat weld made in accordance with paragraph M3111, all acting in a series.

Globe valve

A globe valve is a linear, rising stem valve that has a plug and is generally used for throttling applications. This valve has limited shutoff capabilities and generally does not offer bubble-tight capabilities.

They are available with standard stem packing or a bellows configuration to prevent fugitive emissions. A potential drawback to the linear bellows design is their limited cycle life. Improved packing materials, such as perfluoroelastomer, have helped reduce or eliminate stem leakage without the expense of a bellows. Typical applications include steam, chemical process controlling, and gas service.

Gate valve

Because of their solid, dependable nature, gate valves are the most commonly used valve in the CPI. This "traditional" valve has been around for many years, is relatively inexpensive, and offers a long service life. Like globe valves, these valves offer high temperature and high pressure capabilities. Because these valves are metal-seated, they cannot consistently offer bubble-tight shutoff and have lots of cavities, which makes them unsuitable for handling solids. Although a top entry valve, they are difficult to maintain, and their long stroke of a linear stem packing makes it difficult for the valve to maintain a leak free, emission free service environment. Indeed, due to its outdated performance, in recent years the gate valve has been targeted by manufacturers as the valve to be replaced by newer, more performance-oriented models. Subsequently, the number of gate valves in use today, while still greater than other valve types, has fallen from the levels of only five or ten years ago. Typical applications include crude oil processing, butadiene, and steam.

Diaphragm valves

Diaphragm valves, which can be constructed from basic ferrous metals through high performance plastics, have earned a reputation as being extremely adaptable. Excellent for handling corrosive chemicals and solids from trace quantities to heavy slurries, they are typically available in two different configurations—weir pattern and straight-through pattern. The weir is like a dam in the valve body and is what the diaphragm seats against. In the straight-through pattern, the diaphragm seats against the body, enhancing performance in abrasive type applications which need minimal obstruction in the valve. Because of their top-entry nature and simplicity of design, both weir and straight-through designs lend themselves to ease of repair/service. Because diaphragm valves are packless, they have seen increasing duty in areas of the CPI affected by the increasing emissions standard set by the Clean Air Act Amendments of 1990.

The diaphragm valve is soft-seated, which assures bubble-tight shutoff. They are often found in applications where automation (remote operation) is necessary because of the valve's implicit dependability. Actuated diaphragm valves have only one moving part—the spindle which attaches the valve diaphragm to the actuator diaphragm. A limitation to the diaphragm valve design is that the main sealing member—the diaphragm—is manufactured from a synthetic material, so maximum pressures are restricted. Maximum recommended pressure is 200 psi; in contrast, the maximum pressure for gate valves can exceed 3,500 psi. Typical applications include sulfuric acid, methylene chloride, and chlorinated brine.

Plug valves

Although there are three types of plug valves, lubricated, nonlubricated, and eccentric, CPI applications generally call for nonlubricated plug valves, of which there are two designs, fully lined and sleeved. Fully lined plug valves are manufactured using fluoropolymer derivative material and are used primarily for corrosive chemical applications. The unlined, or sleeve, design is available in a wide range of materials, from ductile iron through high alloys, and is typically used for handling corrosive and toxic materials. A limitation

Table 1. Leakage guidelines for valve users *(1)*.

Leakage Class	Maximum Seat Leakage
Class I	A modification of any class II, III, or IV valve where design intent is the same as the basic class, but by agreement between user and supplier, no test is required.
Class II	0.5% of rated valve capacity
Class III	0.1% of rated valve capacity
Class IV	0.01% of rated valve capacity
Class V	5×10^{-4} ml per minute of water per in. or orifice diameter per psi differential
Class VI	(See Table 2)

Table 2. Class VI valve maximum valve seat leakage.

Nominal Port Diameter			
Millimeters	In.	ml per min	Bubbles* per min
25	1	0.15	1
38	1½	0.30	2
51	2	0.45	3
64	2½	0.60	4
76	3	0.90	6
102	4	1.70	11
152	6	4.00	27
203	8	6.75	45

* Bubbles per min as tabulated are a suggested alternative based on a suitable calibrated measuring device, in this case a 1/4 in. outside diameter by 0.032 in. wall tube submerged in water to a depth of from 1/8 to 1/4 in. The tube shall be cut square and smooth with no chamfers or burrs and the tube axis shall be perpendicular to the surface of the water.

of the plug valve is the inherent high turning torque. Typical applications include anhydrous hydrogen fluoride (HF), water, and liquid sulfur.

Butterfly valves

There are two basic types of butterfly valves—elastomer seated and high performance. The elastomer seated type has a liner that is manufactured from a material such as Buna or ethylene-propylene-diene monomer (EPDM). In this design, a disk, which is mounted through a spindle in the center of the body, rotates in and out of the seat to affect a 360° seal around the outside diameter of the disk. Quality is especially critical in this type of valve since sealing is dependent on precise fit of the disk against the liner. High-performance butterfly valves are typically built to ANSI Class 150 and 300 specifications. They are used where precise throttling or bubble tight shutoff is required. The high-performance butterfly valve uses advanced stem packing and corrosion resistance seats such as perfluoroalkoxy fluoropolymer backed up by a spring mechanism to assure a bubble-tight shutoff. Although the butterfly valve is cost-effective, compact solution for lighter duty process applications, this design makes it unsuitable for abrasive fluids. Typical applications include various chemicals (dependent on compatibility to the seat and body materials), steam, water, and chlorinated water.

Ball valves

Ball valves, which use a ball with a "drilled-out" hole to allow flow passage, have traditionally been one of the most commonly used valves in the CPI. They are available in four different body designs, with ANSI ratings of 150, 300, and 600 (higher ratings, while possible, are rarely used in CPI applications):

- end entry;
- two-piece;
- three-piece; and
- top entry.

The top entry design allows for multiple end configurations such as flanged, butt weld, socket weld, and threaded. It also allows for easy maintenance without the need to remove the valve for the pipeline. Full port ball valves have a higher C_v value but are considerably more expensive than reduced port valves. Calculations of C_v values should determine which valve is needed in your application. Ball valves are excellent in on/off applications and are available in various seat options and body materials to handle the most demanding applications. The ball valve's quarter-turn motion enhances its ability to quickly move from one position to another with a clearly marked position indicator. Some ball valves are designed for positioning and control by means of a specialized ball configuration. Typical applications include chlorine, hydrogen fluoride (HF), and phosgene.

Bellows stem seals

Bellows stem seals are available on some quarter-turn ball and plug valves, and most linear motion globe valves. Bellows are useful in fugitive emissions applications where a leakless stem seal is required. The decision to use a bellows must take into consideration the type of valve being used and in what applications. Typically, a high quality ball valve with a low operating torque and live loaded packing would not justify the expense of using a bellows since its leak rate would be far below the Clean Air Act's permissible level of 500 ppm when properly maintained. Valves that have high operating torques tend to exert an unacceptable amount of lateral loads on the stem and stem seals causing wear and distortion eventually leading to stem leakage. On these types of valves a bellows stem would be beneficial.

This is also true for linear motion valves that have a stem that moves in an upward and downward motion along a static stem seal. This motion puts linear friction and wear on the stem seal to cause a premature seal failure. The use of a bellows on a globe type valve is beneficial in eliminating stem leakage. New advancements in stem seal materials like perfluoroelastomers have greatly improved the sealing ability of linear motion valves, so the cost of a bellows vs. any improved stem seal materials should be taken into consideration when choosing a valve. CEP

Literature Cited

1. **ANSI B16.104-1976,** American National Standards Institute, New York (1976)

Further Reading

"Control Valve Handbook," 2nd ed., Fisher Controls Co. Marshalltown, IA (1977).

Merrick, Ronald C., "Valve Selection and Specification Guide," Van Nostrand Reinhold, New York (1991).

J. GRACZYK is diaphragm valve product manager at ITT Engineered Valves, Lancaster, PA (717/291-1901; Fax: 717/291-2025). In 1982, he began his employment with ITT Engineered Valves as a district sales manager in St. Louis. After six years there, he moved to ITT Engineered Valves' world headquarters in Lancaster, PA. He began his career in the sale of process valves in 1974. He graduated with a bachelor's degree in science from Loyola University Chicago.

B. HANNON is chemical process industries marketing manager and quarter-turn valve product manager at ITT Engineered Valves, Lancaster, PA (717/291-1901; Fax: 717/291-2025). In 1988 he joined ITT Engineered Valves as a district sales representative in southern Louisiana. After a brief stay in the field, he then moved to ITT Engineered Valves world headquarters in Lancaster, PA. He began his career in the chemical process industries in the sale of process instrumentation equipment. He graduated with a bachelor's degree in marketing/business administration from the University of New Orleans.

Compare Nonmetal and Metal Valves

With careful consideration, you can successfully select the right valve materials.

■ *Figure 1. Metallic and nonmetallic valves and piping in vertical support racks.*

Tim Robinson,
Asahi America

The annual sale rate in the thermoplastic valve market is currently $75 million, about 3% of the total industrial valve market. But it is growing at twice the rate of the larger market; 4-8%/yr during this recession with expectations of a return to 10% plus in the near future. The many manufacturers, distributors, and major customers of pipe believe its potential is far greater than that.

Inertia, outdated codes, and lack of experience with new thermoplastic materials are the reasons metal valves continue to be specified in applications where thermoplastic models offer significant cost and performance advantages.

There are six primary criteria to weigh for selecting thermoplastics vs. metal for valves or pipe:

- corrosion resistance;
- abrasion resistance;
- freeze resistance;
- antileaching properties;
- structural flexibility; and
- cost.

As a rule, thermoplastic valves should not be used for temperatures above 250°F or for pressures in excess of 150 psi. It is estimated that 75% of all process lines, 90% of all utility lines, and 85% of all drainage lines fall below these levels, which make thermoplastics a safe, cost-effective alternative to metal.

Table 1. Physical properties of thermoplastics.

Properties	Unit	PVC	CPVC	PP	PPG	PVDF	Test Method
Specific Gravity	—	1.43	1.45	0.92	1.03	1.76	ASTM D792
Tensile Strength	psi	7,000-7,800	8,500-9,200	4,300-5,000	9,950-11,000	7,000-7,800	ASTM D638
Elongation	%	60-120	20-40	150-200	3-5	30-50	ASTM D638
Tensile Modulus	10^3 psi	455-485	485-510	130-170	12,800-14,220	170-200	ASTM D638
Flexural Strength	psi	14,220-15,560	14,220 17,060	7,820-9,240		13,510-14,930	ASTM D790
Flexural Modulus	10^3 psi	384-412	426-455	213-228	540-570	227-256	ASTM D790
Compressive Strength	psi	12,800-14,220	14,220-15,560	8,530-9,950		12,800-14,220	ASTM D695
Compression Modulus	10^3 psi	242-256	256-284	120-156		14-199	ASTM D695
Poisson's Ratio	—	0.37	0.35	0.44		0.28	ASTM D638 ASTM D790
Hardness (Rockwell R)	degree	115	118	95	107	110	ASTM D785
Impact Strength (Izod) w/V-notch	kg-cm/cm²	4-5	4-5	4-7	6-8	10-12	ASTM D256

Resins

Over the past 50 years, thermoplastic materials have been refined and have evolved into sophisticated polymers, each engineered with specific characteristics for specific needs. A general knowledge of the various types of thermoplastics and their features is as important to a specification engineer or buyer as is a knowledge of metals. The greater the understanding, the better the specification in terms of both performance and economy.

There are five main thermoplastic materials used in the injection molding production of solid resin valves. Other materials are used in small specialty valves; however, production is often limited, and the products are generally fabricated from extruded stock. An understanding of the five primary materials should be sufficient, as more than 98% of the solid resin valves produced are of these materials.

Polyvinyl chloride (PVC). This material has excellent chemical resistance, strength, rigidity, and modulus of elasticity. Its temperature range is 32° to 140°F. PVC resists attack by most acids and strong alkalis, as well as gasoline, kerosene, aliphatic alcohols and hydrocarbons, and salt solutions. Chemicals not recommended for use with PVC are aromatics, chlorinated organic compounds, and lacquer solvents.

The highest quality PVC valves are molded with "unplasticized" PVC resin. Unplasticized means that the resin is pure PVC and free of additives or fillers. Although unplasticized PVC is difficult to mold, the end product offers much better weathering properties, chemical resistance, impact resistance, and other physical properties compared to a valve molded of plasticized PVC. The low cost and overall balance of properties of PVC make it the material best suited to the widest number of corrosive applications.

Chlorinated polyvinyl chloride (CPVC). The properties of CPVC and its advantages are very similar to those of PVC; however, its maximum working temperature is higher than that of PVC: 200°F. It should be specified in certain instances where hot corrosive liquids are being handled or an extra margin of safety is required.

Polypropylene (PP). A member of the polyolefin family, PP is one of the lightest plastics known. It is highly crystalline and thus, strong and hard. It possesses excellent chemical resistance to many acids, alkalis, and organic solvents. PP is not recommended for use with chlorinated hydrocarbons and aromatics. Its maximum temperature limit is slightly lower than that of CPVC. By adding glass fibers to PP during the polymerization process, higher temperatures limits, 220°F, and greater strength can be achieved.

Polyvinylidene fluoride (PVDF). This material is a high molecular weight fluorocarbon that has superior abrasion resistance, chemical resistance, dielectric properties, and mechanical strength. The PVDF fluorocarbon resin family includes difluoride, trifluoride, and tetrafluoride chemical combinations. Though having the best chemical and heat resistance, tetrafluoride resin (PTFE) is relatively soft. It is suitable for the diaphragm membranes and seats rather than bodies of solid valves.

Trifluoride resin also has excellent chemical and heat resistance,

Properties	Unit	PVC	CPVC	PP	PPG	PVDF	Test Method
Resistance to heat (continuous)	°F	140	194	194	221	248	—
Deflection Temperature (at 66 psi)	°F	165	250	230	302	302	ASTM D648
Thermal Expansion	10^{-5}/°F	43-46.5	43-46.5	52-54	39-41	52-54	ASTM D696
Thermal Conductivity	kcal/mh°C	0.13	0.12	0.08		0.11	ASTM D177
Dielectric Strength	kV/m	0.90	0.90	1.02	1.18	1.18	ASTM D149
Volume Resistivity	Ohm-in.	2.0×10^{15}	2.3×10^{15}	1.9×10^{15}	0.4×10^{15}	1.95×10^{15}	ASTM D257
Dielectric Constant							
10 Hz		2.8-3.0					
60 Hz		3.15	2.93	2.42		9.8	
103 Hz		3.14	2.92	2.41		9.5	ASTM D150
106 Hz		2.85	2.69	2.41	2.2	7.5	
Dissipation Factor							
60 Hz	10^{-2}	1.18	1.09			0.05	
10^3 Hz	10^{-2}	1.91	1.10	0.044		0.048	ASTM D150
10^6 Hz	10^{-2}	1.72	0.92	0.063	0.02	0.160	
Water absorption 24 h, ⅛ in. thickness	%	0.07	0.15	0.01	0.02	0.03	ASTM D570

but it is very difficult to mold and is very high in cost. As a thermoplastic, it is hard and fragile.

Almost all solid PVDF valves are molded of difluoride resin. This material is relatively easy to mold, has strong resistance to stress cracking caused by halogens, and is fairly low in price compared to trifluoride and tetrafluoride resins. PVDF retains its structural characteristics over a temperature range of -40° to 250°F. It is highly resistant to wet or dry chlorine, bromine and other halogens, most strong acids and bases, aliphatics, aromatics, alcohols, and chlorinated solvents. It is not recommended for ketones or esters.

Ethylene chlorotrifluoroethylene (ECTFE). A durable polymer of ethylene and chlorotrifluoroethylene, ECTFE is well suited for harsh chemical applications and severe environments that exceed the capabilities of other materials. It has excellent resistance to solvents, caustics, chlorides, and most corrosive chemicals, but is not recommended for amines. Its cost is lower than that of trifluoride or perfluoroalkoxy (PFA) resins.

Table 1 compares some physical properties of PVC, CPVC, PP, PPG (glass-filled polypropylene), PVDF and ECTFE.

Pressure and temperature

Compared to thermoplastics, the physical properties of steel are

Table 2. Pressure limits (psi) for PVC valves

Size, in.	70°F	150°F	120°F
¾	150	150	100
1	150	150	100
1½	150	150	100
2	150	150	100
2½	150	100	100
3	150	100	100
4	100	70	70
5	100	70	70
6	100	70	60
8	70	40	40

significantly higher in terms of tensile, compression, and bending strength. Consequently, metal valves are capable of handling temperatures above 500°F and pressures above 1,000 psi.

The maximum ratings of thermoplastic valves are based on a correlation among temperature, pressure, and valve size. These ratings vary from one manufacturer to another depending on their own designs. Basically, as the temperature increases, the pressure rating decreases; as the valve size increases, generally above 4 in., the pressure capabilities decrease.

Generally, the highest pressure rating for a thermoplastic valve is 150 psi at ambient temperature. PVDF valves have the highest correlation between temperature and pressure of the thermoplastic valves. They are pressure-rated up to 250°F. CPVC and PP are pressure-rated up to 200 and 180°F, respectively.

An example of the temperature, pressure, and size correlation is shown in Table 2 for a solid PVC thermoplastic horizontal swing check valve with rubber seats.

Thermoplastic valves from reputable manufacturers usually have a body burst rating ten times the recommended working pressure. For example, the body and bonnet of a 1 in. PVC diaphragm valve rated 150 psi at ambient temperature will burst at approximately 1,500 psi. However, leakage through seals to the atmosphere will occur at about 500 psi. The safety factor for most thermoplastic valves is three to four times the maximum rated working pressure for breakage.

Corrosion resistance

Since the development of thermoplastic resins in the early 1930s, researchers have explored the chemical resistance properties unique to each resin and projected the use of these resins as raw materials for corrosive fluids handling equipment. As predicted, solid thermoplastic valves are providing a dependable and economical way to handle corrosive chemicals such as sulfuric acid, nitric acid, and other oxidizing chemicals, as well as caustics, solvents, halogens, and other problem materials.

Corrosion almost exclusively applies to metals. In addition, because thermoplastics are dielectric, they are practically immune from corrosion in the conventional sense. If plastic is used in an installation where grounding is important, the user can buy plastic that has special conductive material added. This is available from some manufacturers as a special order item. There are some chemicals that can cause thermoplastics to deteriorate. For this reason, most manufacturers publish a chemical resistance guide to assist in specifications and to ensure that the resin selected will work for a particular application.

Industry replaces millions of dollars' worth of valves each year due to corrosion. It is estimated that one-half of these valves fail due to external corrosion and deterioration of stems, and critical sealing surfaces. Acknowledging this problem, most metal valve manufacturers offer protective coatings and paintings to temporarily delay the inevitable corrosion, since coatings are susceptible to chipping and pinhole voids.

Plastics cannot be used in extremely high temperature or pressure situations. They should also not be used where the pipe itself is under high external stress, such as in high-vibration areas. In these three cases, many applications still need the corrosion resistance plastic provides, and also the extra strength metal provides. The answer has been to use coated metal pipe. In some cases this coating is painted on, and in some it is baked on. It is never as corrosion resistant as solid plastic as it is susceptible to wear (especially in the case of thin painted on or dipped materials, and especially at maximum friction points) and to chipping. These coated metal pipes and valves are generally more expensive than solid metal or solid plastic pipes. But in some cases they are the only answer.

Solid thermoplastic valves give the same excellent corrosion resistance on the outside as on the inside. Whether the process liquid or the environment is corrosive, thermoplastic valves are designed to give complete corrosion resistance service.

Table 3 is a general guide to the chemical resistance of the five thermoplastic resins used in the production of valves.

Table 3. Chemical resistance.

	Weak Acids	Strong Acids	Oxidizing Acids	Mixed Acids	Weak Alkalis	Strong Alkalis	Organic Solvent 100%	Organic Solvent	Aggressive Gas (Cl_2, SO_2, and so on)
PVC	✔	✔	✔	✔	✔	✔	X	✔	✔
PP	✔	✔	X	*	✔	✔	*	✔	X-✔
CPVC	✔	✔	✔	✔	✔	✔	X	✔	✔
PPG	✔	✔	X	✔	✔	*	*	✔	X-✔
PVDF	✔	✔	✔	✔	✔	*	✔	✔	X-✔

Ratings: ✔ = good (no effect)
* = fair (some effect with possible impairment of serviceability)
X = not recommended

Abrasion resistance and flow coefficient (C_v)

Resistance to flow is lower in thermoplastic piping systems and valving than with metallic systems. This flow advantage is significant at the initial installation and becomes greater with time. The interior walls of thermoplastic pipe, fittings, and valves are either molded or extruded with extremely smooth surfaces which, unlike metal products, will never rust or form scale that will reduce flow.

Since piping systems are designed to deliver a specific flow with minimum pump sizing and energy consumption, consideration must be given to the reduced service a metal system will yield as the system ages. The ultimate criteria for sizing metal pipe are the expected C_v values the pipe will yield over time. Thermoplastic systems can, in most cases, be designed one size smaller than metal systems and still deliver the required flow with maximum energy efficiency.

Thermoplastic valves perform better and, in most cases, can at least double the life of a metal valve in abrasion service where flow does not exceed 12 ft/s and particle size does not exceed 300 mesh. Abrasive applications where thermoplastics are successfully being used include glass etching, pulp and paper production, phosphate, uranium, coal and copper mining, river and harbor dredging, drilling mud production, and the manufacturing of polishes.

PVDF offers the best abrasion resistance, followed by PVC, CPVC, and PP. Thermoplastic diaphragm valves should be used in relatively low flow velocity conditions (less than 7 ft/s). For flow velocities less than 12 ft/s, thermoplastic butterfly valves are recommended; that is, butterfly valves using a full resilient seat on the body rather than a small O-ring on the disk.

To realize optimum life from a thermoplastic valve for severe abrasion service, specifications should require PVDF for either the body material or the diaphragm valve or the disk material of the butterfly valve. Diaphragms or butterfly valve seats should be specified as natural rubber with a durometer rating of approximately 50. If the natural rubber will not withstand the chemical attack of the application's solution, an elastomer seal such as ethylene-propylene-diene-monomer (EPDM), nitrile, perfluoroelastomer, polytetrafluoroethylene (PTFE), or silicone should be used.

Extensive testing by manufacturers (one by Asahi/America available upon request) and trade associations have been conducted comparing the abrasion resistance of different valve materials. One such test which typifies the superior properties of thermoplastic valves for abrasion service is summarized as:

• valve type	2 in. diaphragm
• body material	PVC, PP and cast iron (lined with natural rubber)
• diaphragm	natural rubber (50 durometer)
• slurry	manganese dioxide and water
• particle size	200 mesh
• hardness	Vicker's — above 600 deg.
• content	200 g/L
• temperature	100°F
• pressure	15 psi

Of the body materials tested, PVC showed the best results. After 108 days only the natural diaphragm perforated; (see Table 4). The body of the natural rubber-lined cast iron valve (⅛ in. lining) perforated on the 45th day of the test, long before the anticipated life of the diaphragm.

Table 4. Degree of abrasion per day.

Material	Abrasion wt. (g)	volume (cm^3)
PVC	0.10	0.07
PP	0.13	0.14
Cast Iron	7.85	1.12

Perforation occurred past the liner and completely through the cast-iron wall. The lined cast-iron valve wore 16 times faster each day compared to the PVC valve, and the cast-iron valve wore 78 times faster. It should be noted, although the PVDF was not tested, it has 25% better abrasion-resistance than PVC, according to the Akron Abrasion Tests.

Freeze resistance

Although PVC, CPVC, and PP are rated for temperatures above freezing, they can and are successfully used in temperatures below freezing. These three thermoplastic materials become brittle at subfreezing temperatures and, if struck with an object, are more susceptible to breakage. PVDF, on the other hand, is pressure rated at -40°F. No thermoplastic valve is rated for temperatures below -40°F.

When liquids freeze inside thermoplastic valves, those valves will crack just like metal valves. However, two factors make thermoplastic valves the better choice in outdoor areas where an overnight freezing may occur.

First, thermoplastics are 1/250 in degree of heat conductivity (or heat loss) compared with metals. So, if the atmospheric temperature drops below freezing, the liquid in an all-thermoplastic system will freeze 250 times more slowly than the liquid in a metal system, giving plant personnel more time to react to the situation. Second, thermoplastic valves have a greater thermal expansion rate and a greater degree of elasticity. Once the liquid inside a thermoplastic valve freezes, expansion in the valve will compensate for some of the expanded frozen liquid. Not only does it take a liquid longer to freeze inside a thermoplastic system, but once frozen, the length of time between freezing and valve breakage is far longer than with metal valves.

Tests by a leading manufacturer (done by Asahi Yukizia Kogyo Co., Ltd., a supplier of Asahi/America, fully documented and available) show that after freezing occurred in 2 in. sample diaphragm valves made of PVC, PP, PVDF, and cast-iron, the cast-iron valve burst 40 min. after freezing while all three thermoplastic valves continued to function 20 h after freezing. Even after thawing, all three thermoplastic valves passed a 215 psi hydrostatic pressure test with no leakage.

Anti-leaching qualities

During the last decade many technology-based industries made advances requiring purity levels in the subpart-per-billion (ppb) range for high purity water and chemicals. In the very near future that range will increase to parts-per-trillion (ppt). The high purity distribution system was once dominated by materials such as pure tin, stainless steel, titanium, and PVC. These are now being replaced with state-of-the-art polymers such as PVDF, PTFE, PP, polyphenylene sulfide (PPS) and polyetheretherketone (PEEK). These polymers maintain point-of-use purity through very low or virtually no leachability, excellent interior surface finishes, and relatively clean jointing methods.

Flexibility

Generally, the wall thickness of a thermoplastic valve body or bonnet is several times the thickness of a steel or cast-iron valve and thus can support a cantilevered weight. When solid thermoplastic valves are used in either an all-plastic piping system or as a component in a metal piping system, the higher flexibility of the plastic valves can be advantageous. The valves will absorb more vibration and more bending caused by earthquakes, ground settling, misaligned piping, and thermal expansion than metal valves. Comparing a 4 in. cast iron gate valve with a 4 in PVC gate valve, the cast-iron valve cracked at a 2.18 in. deflection angle while the PVC valve continued to function under pressure at a 7.5 in. deflection angle.

The disadvantages of plastics are that they lack the strength for applications with pressures greater than 150 psi (if plastic valves are involved—plastic pipe used with metal valves can go to as much as 850 psi for ½ in. PVC Schedule 80 at ambient temperatures), and temperatures greater than 250°F. They should not be used in conditions where there is unusual external abuse, such as high vibration or in locations where the pipe might be damaged by machinery or equipment. It is estimated that these three cases are represented in 15 to 20% of pipe and valve installations.

Economy

The real cost of a valve is determined by adding the purchase price, freight, and installation and maintenance costs, then dividing the total by the life of the valve. The cost of thermoplastic valves is usually much less than for metal valves which can handle the same chemicals and offer equal performance. Cast-iron valves and small brass valves are priced lower than thermoplastic valves, but adding in installation and life expectancy can make them more expensive. Stainless-steel trimmed valves and high alloys valves are generally much more expensive in terms of both initial and installed cost.

Thermoplastic valves generally range from one-half to one-third the weight of an identical size and type metal valve. Because of this significant difference one person can easily install a 6 in. gate valve, for example, where it would take two people and a hoist to install the same valve in cast iron or steel. Therefore, plastics can reduce installed cost. Other installation considerations such as extra valve supports, coatings, and groundings are not necessary with thermoplastic valves.

Table 5. Industrial thermoplastic process valves.

Valve Design	Materials	Size Range (in.)
Ball (union design)	PVC, CPVC, PP, PVDF	¼-4
Ball (compact design)	PVC, CPVC	½-3
Ball, multiport	PVC, CPVC, PP, PVDF	½-3
Diaphragm	PVC, CPVC, PP, PVDF	½-10
Butterfly	PVC, CPVC, PP, PVDF	1½-24
Globe	PVC, CPVC, PP	½-4
Gate	PVC	1½-14
Ball check	PVC, CPVC, PP, PVDF	1-4
Swing check	PVC, PP, PVDF	¾-8
Labcock	PVC	¼
Foot	PVC, CPVC, PVDF	½-4
Pressure relief	PVC, CPVC, PP	½-4
Solenoid	PVC, CPVC, PP	½-1

Availability

Today most of the standard valve designs manufactured in metal are available in an injection molded thermoplastic in sizes to 24 in. These valves are offered with the same optional equipment as metal valves, including electric and pneumatic actuators and control devices, stem extensions, square nuts, "T" nuts for buried service, chain wheel operators for overhead piping, and many others. Standard "off-the-shelf" thermoplastic valves available through industrial distributors are shown in Table 5.

Thermoplastic valves are here to stay

The growth of the thermoplastic valve industry has been phenomenal considering its relatively short history, and the potential for above-market growth is massive. Thermoplastic valves today account for 3% of the industrial valve market with a potential of handling more than 80% of all industrial valving applications.

For chemical applications to 250°F and pressures to 150 psi, thermoplastic valves outperform metal valves in terms of corrosion, abrasion and freeze resistance, and with respect to a generally lower total installed cost.

Metal valves are best applied where pressures are higher than 150 psi and where temperatures are either above 250°F or lower than -40°F. Understanding the advantages and limitations of both thermoplastics and metals as valve materials is essential to making the right valve choice in terms of both performance and real cost. CEP

T. ROBINSON, is executive vice president of Asahi/America, Malden, MA (617/321-5409 or 800/343-3618; Fax: 800/426-7058.) Mr. Robinson has worked in the piping industry since 1973. He joined Asahi/America shortly after it was founded 16 years ago. He is a graduate of the University of Louisville.

Reduce Flanged Connections in Plastic-Lined Pipe

New fabrication methods and longer spools can cut installation and maintenance costs, and help meet upcoming fugitive-emission regulations.

Michael F. Jeglic and Nancy L. Lindley
The Dow Chemical Company

Plastic-lined metal pipe has been used widely for nearly 50 years in the chemical process industries (CPI) to solve corrosion problems. Such plastic-lined pipe (PLP) is a cost-effective alternative for handling a wide variety of acids, bases, and high-purity fluids. It withstands a wide range of operating conditions, with temperatures ranging from -20 to +450°F, and pressures from full vacuum to 475 psi.

The most common plastic liners are polytetrafluoroethylene, polyvinylidene fluoride, polypropylene, and polyvinylidene chloride. Liner selection primarily depends upon the chemical service and anticipated operating temperature (*1*).

Traditionally, each PLP joint is assembled using an American National Standards Institute (ANSI) flanged connection (FC). PLP systems typically require a large number of connections because: pipe spools are relatively short; flanged fittings are needed for directional changes in the pipe; and flanges have been the only way to mate PLP with other components, such as valves, pumps, and instruments. Bolted, flanged joints have been the only connection technique available, because it is impossible to weld PLP without damaging the liner, or tap the steel housing without sacrificing the liner's integrity.

The liner forms a plastic gasket surface at the flange face of the PLP spool or fitting. Variability in sealing force, stress relaxation, and process conditions can require careful maintenance to keep flanged connections from leaking (*2*).

Leaks and fugitive emissions

Several simple preventative measures can help prevent flange leaks (*3*). These include performing piping stress analysis, using proper bolting materials, employing trained PLP fabricators, using torque wrenches for tightening, following the manufacturer's recommended bolt-torquing instructions, and periodically retorquing. Surprisingly, despite today's increased scrutiny of connectors, many simple preventative measures often are ignored, resulting in emissions and even leaks at FCs.

Complicating this picture is a new, stricter fugitive-emission regulation, proposed by the U.S. Environmental Protection Agency (EPA) for the synthetic organic chemical manufacturing industry. This rule is expected to become law this year. The proposed regulation underscores connections as a primary potential source of fugitive emissions.

■ *Figure 1 (left). Using long lengths of precision-bent PLP and other design methods reduced flanged connections in this chlor-pyridine plant.*

The proposed regulation lowers the threshold amount for identified organics within a process stream. Thus, more process lines will become subject to regulation than are monitored now. Leak definition criteria will become more stringent, dropping from 10,000 ppm, detected visually, to a maximum of 500 ppm, detected by an organic vapor analyzer.

Strict record keeping will be required and poor performance will be penalized by requiring more-frequent monitoring. Conversely, good performance will be rewarded by decreased monitoring. EPA's method for calculating the percentage of leaking connections favors reduction in the total number of FCs (*4*).

Reducing the number of FCs has a benefit beyond the regulatory one, however. Fewer FCs mean speedier and less expensive installation, and lower weight, which decreases the number of supports and hangers. In addition, reduced retorquing requirements (*5*) cut maintenance costs.

Assess the use of FCs

PLP systems contain a high number of FCs for various reasons. Historically, PLP has been limited to 20-ft or even 10-ft lengths. Often, lines are field-engineered, resulting in the use of many field-fabricated spools. Tapered spacers are used to impart slope to PLP to assist in drainage. Yet, use of a spacer doubles the number of sealing surfaces at a FC. Flanged fittings, like 90° and 45° elbows, are used, often in combination, for directional changes. Standard fittings are sometimes stacked together for line size reductions. Valves and pumps are integrated into the line with a FC. Since PLP cannot be tapped, tees are sometimes installed with the outlet blinded with a spacer and flange, if retrofit is anticipated.

Careful design and use of recent PLP developments can reduce the number of FCs by 50–75%. This reduction is possible through using:

• the longest length piping available;

• precision bent piping for large and small directional changes; and

• custom fabricated flangeless assemblies.

Figure 1 shows PLP that minimizes the number of flanges, due to using such methods.

Use the longest length piping available

The maximum available lengths of different PLPs vary. Length limitations usually depend on liner manufacturing technique, or the ability to pass various American Society for Testing and Materials (ASTM) thermal-cycling tests (*6*), not on shipping and handling considerations. Thus, different suppliers offer different maximum spool lengths. Not every maker offers all liners or diameters in 20-ft lengths, the traditional maximum. Some quote two 10-ft spools at the price of a competitor's 20-ft spool. The engineer supervising the installation often is unaware of the difference until the shorter pipe arrives.

Also, some manufacturers may offer 20-ft pipe from their factories but limit field-fabricated pipe to 15- or even 10-ft maximums. Often, this shorter "field-fab" pipe is used by local distributors, which can fabricate quickly custom lengths near the job site. Some CPI plants maintain in-house PLP fabrication shops and specify a supplier that offers only shorter pipe for field fabrication. Sometimes the physical constraints of the pipe shop may limit the length of the pipe stocked.

Even if the local shop handles only 10-ft pipe, select the longest length possible for capital projects. These larger orders are usually factory-fabricated with only a few "field-verified" or closure spools fabricated either onsite or by a local distributor. If a 20-ft factory-fabricated spool needs to be replaced, two 10-ft shop-fabricated pieces can be used if 10 ft is the maximum locally available.

One manufacturer now offers 40-ft PLP in some common diameters and liners. These 40-ft lengths are well suited for eliminating FCs along transfer lines and other straight runs.

Some CPI firms design and install PLP in standard lengths, using combinations of pipe that are 6-, 8-, and 10-in., and 1-, 2-, 5-, and 10-ft long. These firms maintain a large inventory of these fabricated standard-length spools for ready replacement of failed ones. With all the FCs used with these standard lengths, it is no mystery why these companies need pipe for quick and frequent replacement.

The only standard length, fabricated spool offered from manufacturers is their longest length spool, usually 20 ft. All others are custom-fabricated either at the factory or by a local distributor. All of the major PLP manufacturers have extensive networks of trained, stocking distributor-fabricators, so most failed PLP can be replaced in a matter of hours. If this is not acceptable, then some of the manufacturers can train and certify a plant's shop personnel to fabricate. Thus, only maximum-length spools need to be stocked.

■ ***Figure 2. In this HCl line, the right side of this photo uses precision-bent PLP, while the left has conventional, multiflanged PLP.***

In addition to being the longest pipe available, factory-fabricated PLP usually has the best sealing characteristics since it is made under controlled conditions by skilled technicians who specialize in PLP fabrication. This quality is difficult to duplicate when PLP is field-routed and fabricated in uncontrolled (often outdoor) environments by contractors or plant personnel who occasionally handle PLP, in addition to other piping products.

Plan before field routing

Good upfront engineering and design can reduce the number of FCs over and above that attainable by minimizing the need for field routing and fabrication of PLP. For instance, consider the following options:

Eliminate tapered spacers. Often it is necessary to impart a slight angle to a pipeline for gravity-assisted drainage. This usually is done by installing a tapered spacer between flange faces, doubling the number of sealing surfaces at a FC. Metal pipe frequently is bent when a drainage slope is needed.

One PLP manufacturer has developed precision-bending technology to create small angle bends in its factory-fabricated pipe. This eliminates the use of tapered spacers. However, PLP cannot be bent by the end-user or contractor.

Eliminate flanged elbows. This same precision-bent PLP can be supplied with bend angles from 0° to 90°. The standard bend radius is 3 diameters (3*D*), which is the tightest bend possible, given the mechanical properties of the steel and liner.

Standard flanged PLP elbows have a 1.5*D* bend radius, so directional changes should be examined to determine if the longer bent-pipe face-to-center dimensional requirements can be met within the initial design constraints. Sometimes, minor design adjustments, such as moving a valve downstream by a few inches, permits the use of precision bent pipe, thereby eliminating another FC.

Take the most direct route. Often flanged elbows are "stacked" for complex directional changes. Since precision-bent pipe is not limited to 45° or 90° bends, the designer can simplify complex routings and plan the most direct route.

Precision-bent PLP can be supplied with multiple bends in one piece. For example, expansion loops can be incorporated with a single piece of bent PLP. The greatest reduction in FCs is possible if the use of bent PLP is considered early in the design phase so that directional changes are not rigidly designed using the traditional 45° and 90° limitations. Figure 2 shows how multiple-bend precision-bent pipe eliminates flanged connections.

Contractors and end-users should not bend PLP. Use of special bending equipment and proprietary manufacturing methods minimize liner distortion on both the inner and outer radii during bending, resulting in a high-quality product that meets ASTM PLP performance requirements.

Reduce the use of fittings

Two steps apply here:

1. *Plan line-size reductions.* Most manufacturers offer reducing tees, elbows, and instrument tees with various sized outlets. These create line-size reductions in place of combinations of standard tees, elbows, and reducers. For example, use a tee with a reduced size outlet to replace a standard tee that is combined with a concentric reducer or reducing filler flange.

2. *Eliminate flanged fittings.* Although PLP has been joined by FCs, flangeless joint technology has been developed by one manufacturer. This method overcomes difficulties that prohibited welding of PLP. Because the high heat generated by welding the metal shell can damage the plastic liner, PLP manufacturers and ASTM recommend against any welding on PLP. However, the new flangeless assembly process consists of both a structurally sound metal weld and a high-integrity plastic weld that, together, eliminate flanged connections while preserving the structural integrity of the PLP.

The greatest opportunity to reduce FCs is at branched connections. A special plastic-lined flangeless tee is used in these custom assemblies. The tee has the annulus and stub required for flangeless joining. Other fittings, such as concentric reducers, have been developed and tested.

Although most of the tools and procedures used to make flangeless joints were developed for ultimate use in the field, current limitations require that all joints be shop-fabricated. Thus all flangeless joining is done on custom assemblies of PLP made at the factory.

Start a design review early

To minimize the number of FCs, early in the design stage assess the piping design, the engineering firm's piping plan, fabricating abilities of the end-user, and the stock offered by the PLP supplier. Review the applicability of each of the three joint reduction techniques, considering the need for adequate clearance around surrounding equipment, existing piping, and structural members (*7*).

Areas for joint-reduction include:

Straight runs. Compared to conventional shorter lengths of swaged PLP, 40-ft spools cut the number of flanges by up to 75%. Consider using 40-ft pipes for transfer lines and other long, straight runs where there is sufficient room to maneuver them into pipe racks.

Elbows. Most 1- to 4-in. elbows are candidates for potential flange elimination. When considering precision-bent pipe, carefully examine directional changes to determine whether or not the face-to-center dimensional requirements of bent pipe can be met within the parameters of the initial design. If not, design adjustments may be required.

Tees. Branched fittings with their multiple connections present a significant opportunity to reduce the number of joints. Although a flanged connection may be required on at least one branch for maintenance purposes, many flanged tees can be replaced with special flangeless ones. These can be mated to any other flangeless pipe end. Special flangeless tees can be made with short stubs so that one end can be flanged for connection to valves, pumps, or nozzles. CEP

Literature Cited

1. **Lednicky, R., and N. L. Lindley,** "Reduce Leaks and Corrosion with Plastic-Lined Piping," *Chem. Eng. Progress*, **87** (3), pp. 45–48 (March 1991).
2. **Henthorn, G. V., and R. Lednicky,** "Sealing Tips for Plastic-lined Piping," *Chem. Eng.*, pp. 142–146 (November 1992).
3. "Guidelines for Selection, Installation and Operation of Plastic Lined Pipe and Fittings Systems," Amer. Pipe Fittings Assn., Springfield, VA (1990).
4. "National Emission Standards for Hazardous Air Pollutants; Announcement of Negotiated Regulation for Equipment Leaks," U.S. Environmental Protection Agency, 40 CFR, Ch. 1, *Federal Register*, **56** (44) (Mar. 6, 1991).
5. **Jeglic, M. F.,** "Reducing Flanged Connections with Plastic-Lined Piping Systems," First International Symposium on Process Industry Piping, Mat. Tech. Inst. and NACE International, pp. 41/1–41/17 (Dec. 1993).
6. "Plastic Pipe and Building Products," **08.04**, in "1993 Annual Book of ASTM Standards," ASTM, Philadelphia, PA (1993).
7. "New Joint Reduction Technologies," Form No. 178-00311, The Dow Chemical Co., Midland, MI (1993).

M. F. JEGLIC is the product development manager for Dow Plastic-Lined Piping Products, Bay City, MI (517/638-2641; Fax: 517/638-2697). During his 14 years with The Dow Chemical Co., he has worked in various technical positions within Dow's mining chemicals, engineering ceramics, and plastic-lined piping products businesses. He earned a BS degree in mining engineering from Virginia Polytechnic Institute and State Univ. He is an active member of ASTM's subcommittee F17.11 on composite piping.

N. L. LINDLEY is marketing and customer service manager of Dow Plastic-Lined Piping Products, Bay City, MI (517/638-2630; Fax: 517/638-0522). She has been with The Dow Chemical Co. for 15 years in technical service, sales, and marketing. She is a member of Tau Beta Pi, and Pi Tau Sigma. She earned a BS degree in mechanical engineering from the Univ. of Houston.

Build Your Working Knowledge of Process Compressors

How good are you at selecting medium-power reciprocating compressors?

Edward H. Livingston,
Burton Corblin
North America, Inc.

The role of compressors in the chemical process industries (CPI) is critical since they are used to circulate gas through a process, enhance conditions for chemical reactions, provide inert gas for safety or control systems, recover and recompress process gas, and maintain correct pressure levels by either adding and removing gas or vapors from a process system.

The chemical engineer involved with process design and equipment selection must have a working knowledge of compressors, since they are the most mechanically complex machinery used in the CPI. This working knowledge must not only be related to the thermodynamics of the gas being compressed, but also to the type of compressor to be used for a particular process. The latter has become more important as the result of the passage of the Clean Air Act Amendments of 1990, legislation which places limitations on emissions from process equipment. Fugitive emissions from such sources as compressors, pumps, valves, and piping systems and connections must be reduced over the coming years in order to comply with EPA Equipment Leak Regulations.

Brief background

The principal types of compressors found in the CPI are reciprocating, turbo (centrifugal and axial), and rotary flow designs. Within some of the types are variations. Nevertheless, in all cases, compressors are used to convert energy from one form to another.

1. Reciprocating machines are those in which gas is moved by the linear motion of a piston within a confining cylinder. The work done enhances the pressure and density. Flow through the cylinder is controlled by valve actions. Examples of reciprocating machines include piston compressors, lubricated and nonlubricated, and metal diaphragm compressors.

2. Turbomachinery, or dynamic compressors, are those in which a dynamic head is imparted to the gas by means of high speed impellers rotating in a confining case. This category includes axial-flow, radial, centrifugal and fan-blower compressors.

3. Rotary machines are those in which gas is moved by the positive displacement of two rotating lobes or by oscillating vanes confined in an eccentric cylinder.

4. Ejector machines are those in which gas is moved by kinetic energy induced through high-velocity nozzles.

This article will primarily deal with reciprocating, positive displacement compressors with emphasis placed on applications and machines having installed power of 200 kW or less. However, some comments with respect to larger compressors will be made due to their importance in process applications.

Before we discuss specific compressor types in detail, let's look at typical applications of the units.

Oil refinery processes

Both positive displacement and dynamic compressors are used in the refining of crude oil. Crude oil feed-stock contains polluting compounds such as sulfur, chlorides and salts. Refining processes extract these pollutants and convert them into needed byproducts, thus reducing emissions into the atmosphere. Hydrogen sulfide and carbon dioxide are being treated by chemical absorption systems such as methyldiethanolamine (MDEA) with a high absorption efficiency.

Hydrogenation processes occur in the initial distillation column, as well as hydrotreating, hydrocracking, and catalytic reforming units.

Hydrotreating

Hydrotreating removes objectionable elements such as sulfur, nitrogen, oxygen, and halides from feedstock by reacting them catalytically with hydrogen. In this application, three compressors are frequently used: the recycle compressor which takes hydrogen-rich gas from the hydrogen separator and recycles it to the process front end, the make-up compressor that adds hydrogen to the process at the front end, and the vent gas compressor which handles hydrocarbon gas mixtures of molecular weights from 20 to 40.

The size of the recycle and make-up compressors depends on the hydrotreater capacity. In general, recycle compressor flow rates vary from 25,500 to 127,500 m^3/h and reciprocating or centrifugal compressors may be used since compression ratios are low. The make-up compressor flow rates vary from 5,950 to 29,750 m^3/h, and while the flow rates can support centrifugal design compressors, the compression ratios are too high for centrifugal design and thus reciprocating designs are used.

Hydrocracking

Hydrocracking produces gasoline from heavy feedstocks. In many cases, hydrocracking and cat reforming units work in unison. Hydrocracking takes place at higher pressures, 6.9 MPa to 13.8 MPa and has a high concentration of hydrogen with a fixed-bed catalyst. Two compressors are used for this application: the first is the recycle compressor which takes hydrogen-rich gas from the separator and recycles that gas to the process front end where it mixes with the liquid feedstock, and the second is the make-up compressor which adds hydrogen to the process after the separation and before the recycle compressor. The recycle compressor is usually a barrel-type centrifugal compressor since the compression ratio is low, while the make-up compressor is a multistage balanced opposed reciprocating compressor because the compressor must boost the gas from 1.4 MPa to 13.8 MPa.

To reduce emissions within the refinery, hydrocarbon gases are collected, recompressed and used elsewhere. Hydrogen collected from the reciprocating compressor packing group and from the barrel-type compressor mechanical seal are often recycled back into the process by small, positive displacement, reciprocating compressors.

Polymerization

Compressors are required to feed gases at elevated pressures into reactors or compress gases to a pressure that will permit liquefaction after which the liquid is pumped directly into the reactor. Typical gases compressed include ethylene, hydrogen, hydrogen chloride, methyl chloride, phosgene, propane, and butane. Depending on the polymer and the process used, pressures can range from 0.1 MPa to 380 MPa.

Higher flow rate applications require centrifugal or axial flow machines.

The original low density polyethylene process required gas pressures of 200 MPa to 320 MPa. Three positive displacement compressors were used in series to compress ethylene from 0.5 Pa to 240 MPa. The combined power requirements of these three compressors exceeded 7,500 kW.

The operating pressures of the linear linear low density process have been reduced to 1 to 2.1 MPa, however, compression equipment is required to keep the gas circulating in the reactor and fluidize the polymer particles.

Some specialized co-polymers still use pressures of 200 to 320 MPa. The flow rates are small in comparison to standard, full-size plants. Feed compressors for additives operate to 320 MPa, and power is in the range of 75 kW.

High density polyethylene is often co-produced with polypropylene in a pressure range of 1 to 3.5 MPa. The catalyst is fed to the reactor with the gas is flashed, separated, and recycled.

Electronics and semiconductors

Gases are produced for the manufacture of electronic components and semiconductors. Purity of the gases is vital and ultrapure systems are used in all phases of manufacturing. Adding to the purity requirements is the handling problem associated with strong oxidizers, flammable, pyrophoric, and highly toxic compounds.

Oxidation processes are used for the formation of protective silicon dioxide coating on wafer surfaces. This is accomplished in a diffusion furnace in an oxygen atmosphere.

Protective atmosphere doping uses nitrogen in the purity range of of 99.9999% for the manufacture of microelectronic components to protect the material as well as being a carrier gas for dopants. *Chip manufacturing* involves ultrahigh-purity argon for silicon crystal growing, oxide removal (etching) and doping of wafers for desired chemical composition.

Manufacturing of integrated circuits and semiconductors use ultrahigh purity gaseous chemicals for doping, etching, epitaxy, and ion implantation. Gaseous chemicals such as arsine, phosphine, silane or chlorosilanes, diborane, halocarbons, hydrogen selenide, hydrogen sulfide, and sulfur hexafluoride are used in a mixture of diluent gases like argon, helium, nitrogen, and hydrogen.

Hydrogen recovery

Even though the cost of hydrogen is relatively low, hydrogen is recovered for safety reasons. The hydrogen vapor from liquid storage tanks is recovered by a compressor rather than venting it to the atmosphere. This reduces the possibilities of auto-ignition, a situation that could occur when vapor is vented by a relief device.

High-pressure piston compressors use packing vents to dispose of effluent leaks. Small compressors recover, recompress, and inject the gas into the process stream.

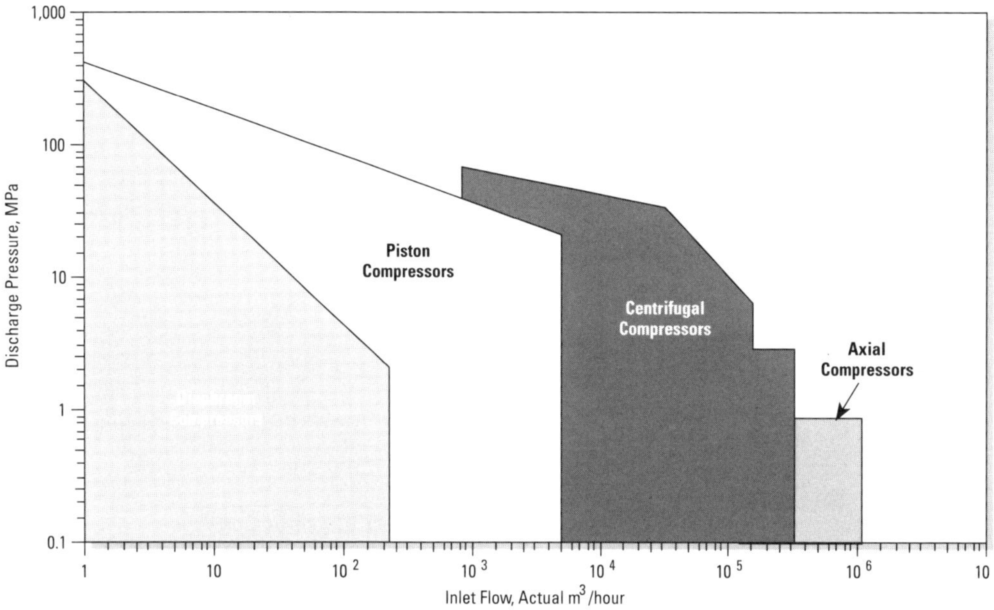

■ *Figure 1. Typical performance range of compressors used in chemical processes.*

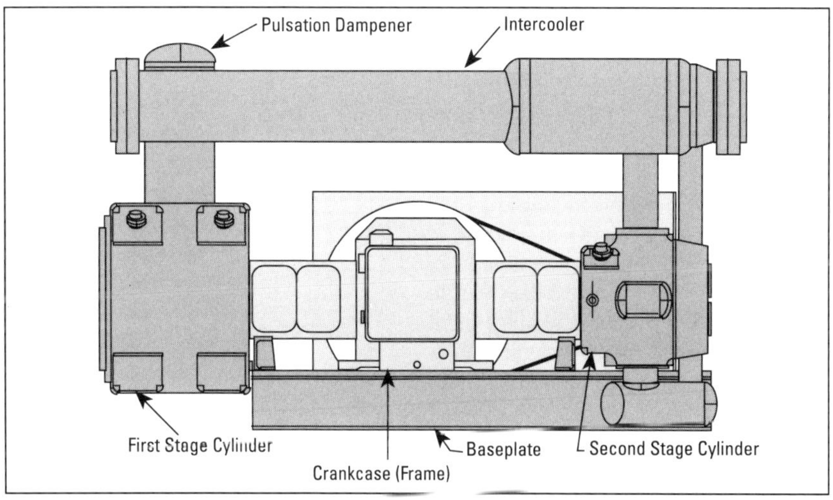

■ *Figure 2. Piston design.*

Reactor gas feeds

Hydrogen. Hydrogen is required for full or partial hydrogenation of fats and oils by converting unsaturated radicals of fatty glycerides to highly or completely saturated glycerides, pharmaceutical intermediates, and final products. Compressors are used for initial hydrogen feed and recycle of hydrogen in the process.

Hydrogen from either gas plants or tube trailers is boosted by positive displacement piston and diaphragm compressors to the reaction feed pressures required.

Carbon dioxide is a solvent at supercritical conditions. High purity carbon dioxide is compressed to the critical pressure conditions and is then recycled after separation.

Nitrogen is sparged into reactors to reduce dissolved oxygen, blankets sensitize compounds against oxidation and contamination, and purges process reactors and piping systems at shutdown and startup.

Refrigerant gases such as chlorodifluoromethane are fed to reactors for manufacture of fluoropolymers such as polytetrafluoroethylene (PTFE) .

Gas separation

Cryogenic separation of atmospheric gases results in the highest purity levels in comparison to membrane separation or Pressure Swing Adsorption (PSA). If the final product can tolerate lower gas purity levels, PSA production of nitrogen and oxygen can result in cost reductions of 20-60%.

Conventional air compressors feed the PSA unit. Depending on the pressure requirements, reciprocating piston compressor can be installed to boost the PSA outlet pressure to that required by the plant. These compressors are sized to meet peak demands for the gas (gases) and are frequently specified with a standby unit of 50 or 100% capacity.

Membrane separation can produce higher purity gas streams and are used in upgrading hydrocarbon gases. Processing pressure drops may require the treated gas to be compressed for recycle. In most cases centrifugal compressors are applicable because of the low compression ratio.

Nitrogen boosting

Under certain conditions, nitrogen is required to maintain flow of liquid from storage tanks or tank cars. Typically, bone dry nitrogen is introduced under pressure to induce flow. Chlorine tank cars frequently use this method during cold weather and the importance of using an oil-free, nonlubricated compressor is apparent. Hydrocarbons from lubricants could carryover and react with the chlorine, while water vapor (150 ppm or more) would cause a highly corrosive condition.

Nitrogen also acts as a blanketing gas to prevent fire or explosion conditions and creates an oxygen free environment to enhance long-term storage, especially for perishable products.

Nitrogen is used to balance pressures in high-speed mechanical seals to prevent leakage of gas to the environment. The system also permits monitoring of seal performance by pressure decay in the seal. Small, positive displacement, reciprocating compressors are used to pressurize the seal and maintain purge flow.

Compressor types

Compressor selection is based on the process operating variables and how those variables fit the design ranges of the types of available compressors. Figure 1 illustrates the range of compressors used in the CPI.

Positive displacement piston compressors are normally selected for applications where the inlet flow rate is no greater that 6,800 m^3/h. Design discharge pressures can range from 0.5 MPa to 380 MPa. The latter for small displacement machines assure low density polyethylene process. Diaphragm compressors, a specialized version of positive displacement piston compressors, are limited to single cylinder inlet flows up to 204 m3/h.

Centrifugal compressors range from inlet flows of 850 m^3/h to 340,000 m^3/h with case pressures to 70 MPa for small centrifugal units and considerably lower pressure for large units, and axial flow compressors range from 34,000 m^3/h to 1,020,000 m^3/h with case pressures generally limited to 1 to 2 MPa for all sizes.

Engineering considerations and economics

For the higher flow rate applications, the process engineer has few alternatives, with selections being limited to either centrifugal or axial flow machines. However, in the lower flow range, several choices are available: positive displacement piston and diaphragm compressors, or hybrid machines that combine piston and diaphragm cylinders on one crankcase. General arrangements of these compressors are shown by Figures 2 through 4.

Piston compressors

All piston compressors have a crankcase that converts rotary motion to linear motion: a crosshead for guiding the motion of the piston, a piston fitted with seal rings, a cylinder in which gas compression takes place, and one or more suction valves and one or more discharge valves that regulate the flow of gas into and out of the compression cylinder. A simplified diagram of these elements is shown in Figure 5.

Air compressors should not be considered as process gas compressors. There are significant design differences inherent in the air compressors and the improper use of an air compressor in a process gas application could have severe consequences,

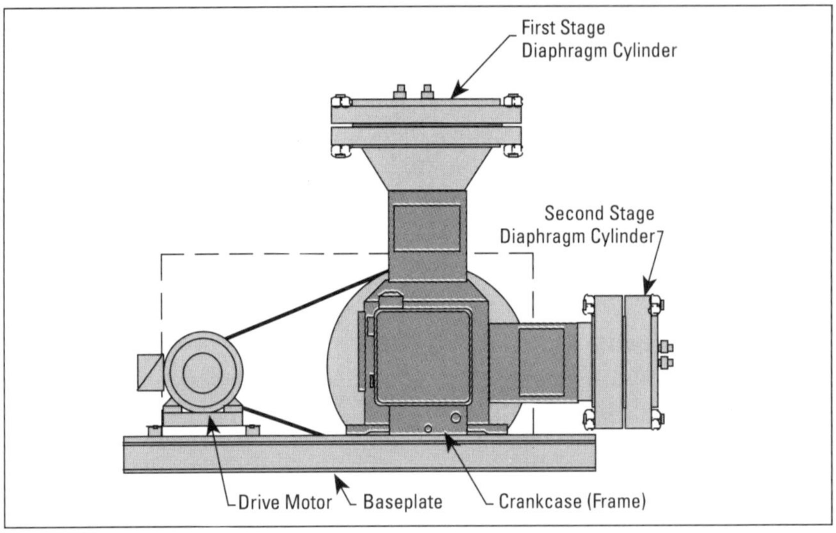

■ *Figure 3. Diaphragm design.*

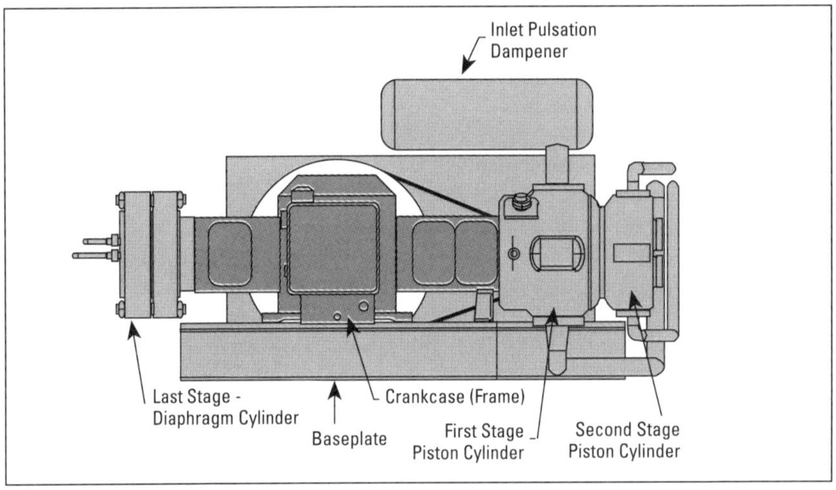

■ *Figure 4. Hybrid design.*

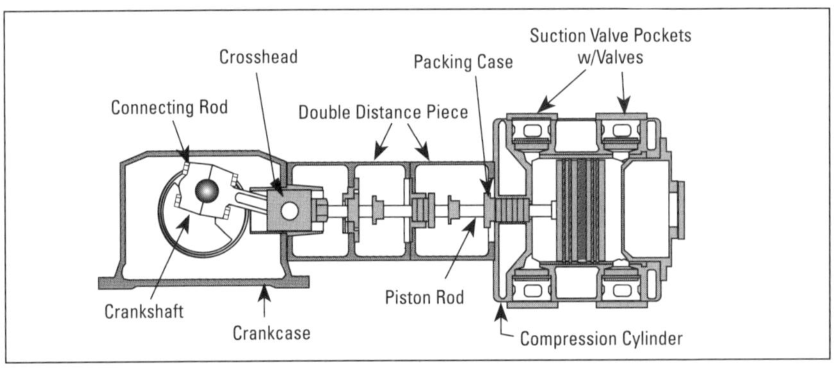

■ *Figure 5. A simplification of compressor crankcase elements.*

such as excessive leakage, fire, or explosion. For example, the air compressor piston is sealed from the crankcase by piston rings, usually of cast iron. Since gas can leak into the crankcase, the simple air compressor should be limited to air or nitrogen even if the crankcase is pressurized.

Process-gas piston compressors are designed with a distance piece to eliminate gas leakage into the atmosphere and crankcase. The piston is mounted on a piston rod which is connected to a crosshead. The piston rod is relatively small in diameter, and therefore can be sealed within a packing case with one or more sets of packing. Figure 6 illustrates a typical packing case. Special purged packing cases and distances pieces are fitted on these compressors to further minimize gas leakage.

In the lubricated process-gas compressor fluid is injected into the compression cylinder to provide lubrication for the piston rings and the compressor valves. Packing is both lubricated and cooled by injecting the same lubricant into the packing set(s). The lubrication is provided by either a separate, crankshaft driven lubrication pump, or an auxiliary motor driven lubricator.

Nonlubricated compressor operation prevents the entrainment of oil in the gas being compressed. Types of service for nonlubricated compressors include oxygen compression, food processing, container manufacturing, breweries, chemical, and specialty gas plants where oil contamination of the end product cannot be tolerated. Nonlubricated reciprocating compressors vary in design, but fundamentally the piston is driven through a crosshead from the crankshaft. The piston rod is sealed by a stuffing box with packing rings. The compression cylinders are mounted to distance pieces which isolate them from the crankcase thus preventing oil carryover. The cylinders may be mounted horizontally, vertically, at an angle, or in combinations of these on multistage, multithrow crankcases. Nonlubricated cylinders are normally limited to pressures of 25 to 41 MPa. A typical non-lubricated piston cylinder is shown in Figure 7.

For the successful operation of nonlubricated piston compressors, nonmetallic piston rings are required. Materials such as PTFE with fillers have proved to be the most efficient.

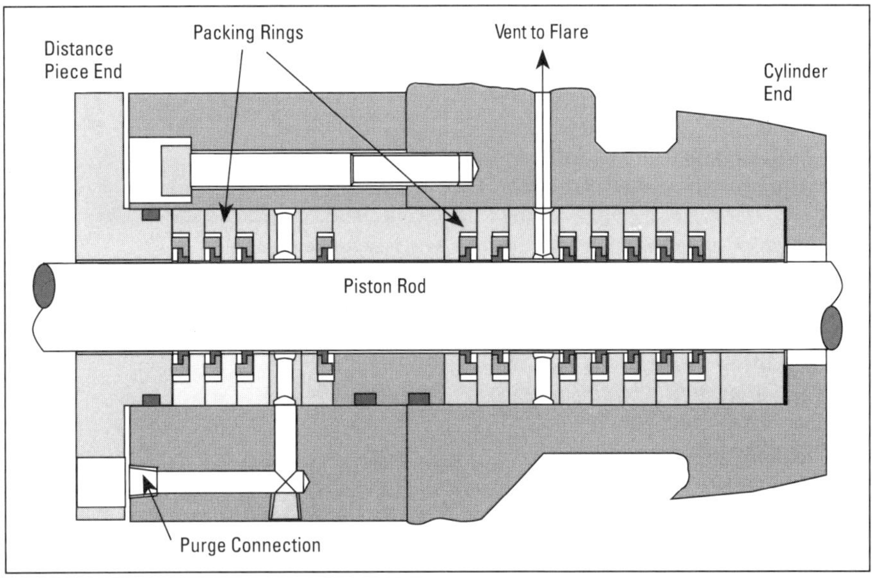

■ *Figure 6. A typical packing case.*

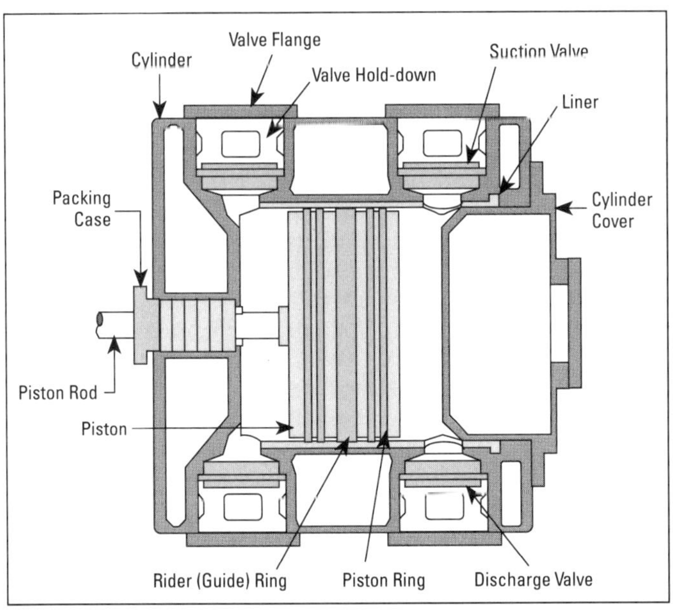

■ *Figure 7. A typical lubricated piston cylinder.*

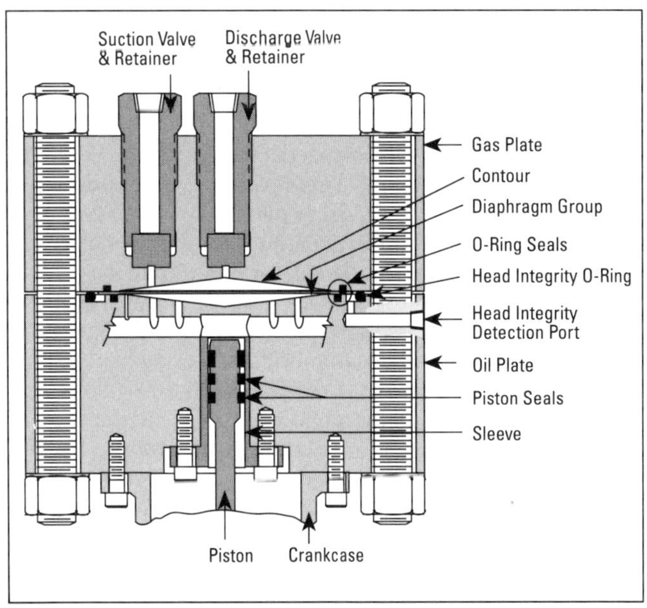

■ ***Figure 8. Motion of the displacing element causes the diaphragm to move into the compression chamber to reduce volume and thereby increase gas pressure.***

Not only are the wear rates extremely low on the rings made of these materials, but cylinder wear is also reduced. It is vital that these materials be uniform throughout the entire cross section so that wear rates and sealing properties are continuous for the life of the ring.

Another type of oil-free piston compressor is one in which a labyrinth profile is used. Normally, the leakage volume past the labyrinth does not exceed 5% of the rated displacement. A two-compartment distance piece for oil wiping and for holding piston alignment is required.

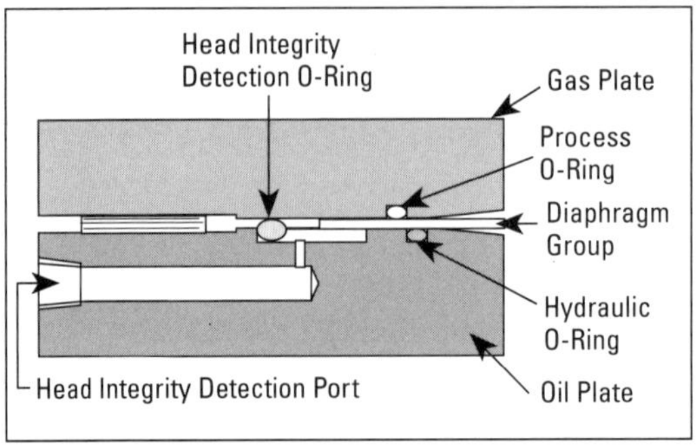

■ *Figure 9. Typical leak detection arrangement for a diaphragm compressor.*

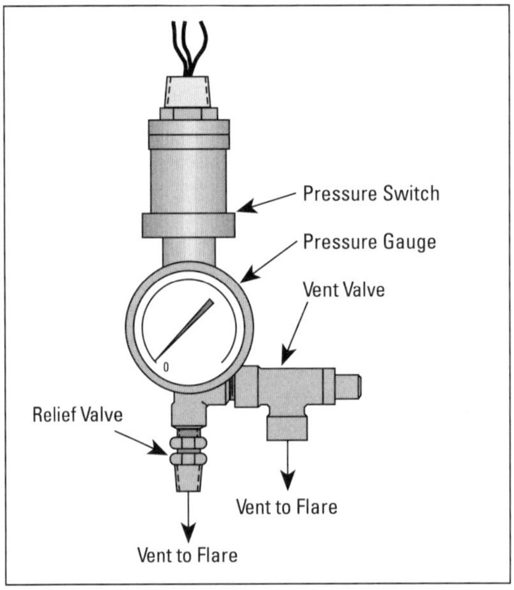

■ *Figure 10. Typical leakage monitoring system.*

Diaphragm compressors

Piston compressors and diaphragm compressors share many of the same components: a crankcase, crankshaft, connecting rod(s), and piston.

The main difference between piston and diaphragm compressors is how the gas is compressed. Unlike other types of reciprocating piston compressors in which the primary displacing element, a piston, contacts the gas, the metal diaphragm compressor completely isolates the gas from the displacing element during the entire work cycle. The motion of the displacing element is transmitted to a hydraulic fluid, and the hydraulic fluid transmits its motion to one or more thin, flexible metal discs called "diaphragms." This motion causes the diaphragm to move into the compression chamber, reducing the volume and thereby increasing the gas pressure. See Figure 8.

Because the diaphragms isolate the gas from the compressor lubricants, the discharged gas is as pure as the gas entering the compression head. The gas only contacts clean, dry metallic surfaces and static elastomer or metallic seals. With improved diaphragm materials and contour configurations, reliability of these machines has been demonstrated by years of service in critical applications. With proper installation and maintenance, the likelihood of premature diaphragm failure is minimal. As a result, this equipment has become widely accepted for all types of contamination-free applications in laboratory, pilot, and plant operations.

Corrosive gases can be handled in these compressors because the compression cylinder can be manufactured from virtually any machinable material. Certain limitations do apply and these limitations are related to the diaphragm material. Materials of construction commonly used for gas contacting parts include 17-4ph, 17-7ph, 304SS, 316SS, 400SS, 20Cb, nickel, carbon and low alloy steels. Designs to handle H_2S and conforming to National Association of Corrosion Engineers (NACE) MR-01-75 can be produced.

The benefit of a no-leakage design is evident and is of greater importance with the advent of the Clean Air Act. Synthetic Organic Chemical Manufacturing Industry (SOCMI) factors have been developed by the EPA for all equipment, including compressors. For compressor seals, the emission factor is 0.228 kg/h/source. Using standard design and construction methods, leak rates from a diaphragm compressor are in the order of 1×10^{-7} standard cc/s. When extremely low leak rates are required, in the order of 1×10^{-8} standard cc/s or less, the diaphragm can be sealed by metallic "O" rings or it can be seal-welded to the gas head.

In the event the integrity of the diaphragm or seal is breached, effluent process gas is retained in the head assembly detection system. With a relatively inexpensive monitoring system, an anomaly can be detected and corrected before there is a discharge to the atmosphere. Figure 9 illustrates a typical arrangement for diaphragm compressors and Figure 10 illustrates a monitoring system

Hybrid compressors

Hybrid compressors are unique since they combine nonlubricated piston technology with diaphragm technology on one reciprocating frame (crankcase). These compressors find application when inlet pressures are low, the gas flow rates are relatively high, and the gas must be compressed to high pressure. Depending on the gas flow rate, multiple two or three stage diaphragm compressors, or a five stage piston compressor may otherwise be required. To avoid such a situation, two or three stages of nonlubricated piston cylinders are used with a final stage diaphragm cylinder. In a single machine, the large capacity of a piston compressor is combined with the high pressure and leak-tight performance of a diaphragm cylinder.

Piston and diaphragm compresssor efficiency

The volumetric efficiency of a positive displacement compressor is the ratio of the gas handled to the compressor displacement including gas compressibility. Several factors influence volumetric efficiency:

- compression ratio;
- compressibility factors of the gas at suction and discharge conditions;
- cylinder clearance volume;
- valve action (losses);
- piston ring leakage (piston compressors);
- adiabatic or polytropic exponent; and
- water vapor.

Compressor manufacturers can control clearance volume, valve action (losses), piston ring leakage, and the compression ratio (by multistaging). Of great concern to the compressor manufacturer is the control of clearance volumes at high-compression ratios and when gases have a low specific heat ratio.

Piston compressor clearance volumes can range from 7% to 22% with fixed valve pockets; diaphragm compressors are normally designed with clearance volumes between 4% and 7% depending on the size of the diaphragm cylinder. For pressure applications to 300 MPa diaphragm compressor clearance volumes may be as high as 10% to 12%, due to practical manufacturing tolerance limits, particularly in the valve pocket area.

The effect of clearance volume is illustrated in Figure 11. Clearance volumes of 5%, 10%, and 15% are given for illustrative purposes. The 5% compression slope ABC will attain P2 quicker than the compression slopes of 10%. Likewise, upon re-expansion at the end of the compression stroke DEF, the slope is steeper and therefore allows the gas to enter the cylinder sooner during the suction cycle.

Compression efficiency is controlled by the valves and valve pocket design. For example, the compressor designer may concentrate on the clearance volume reduction to improve volumetric efficiency and valve losses could be high, thus affecting compression efficiency. A decrease in compression efficiency leads to increased power requirements.

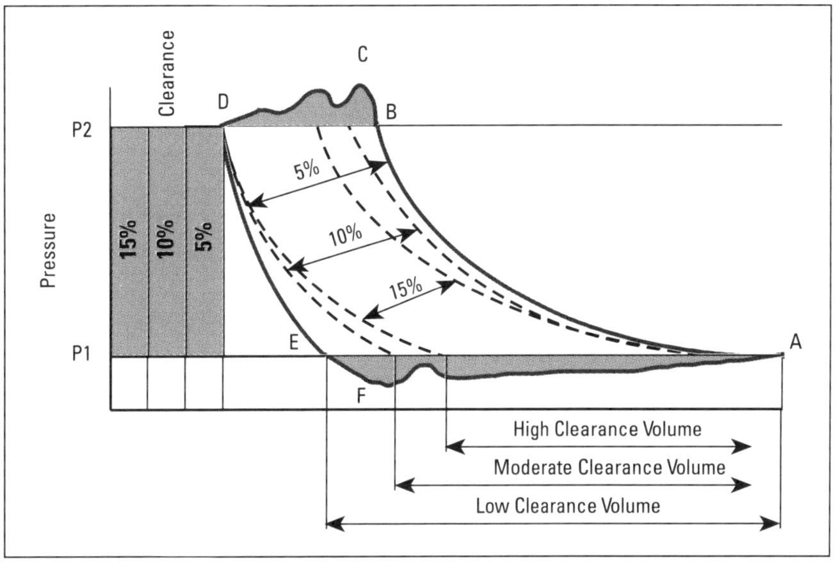

■ *Figure 11. Compressor efficiency (valve design effect).*

Compression efficiency (valve design effect) is also illustrated in Figure 11. The pressure increases along curve ABC, and when the cylinder pressure exceeds the line discharge pressure P2 at point B, system energy unseats the discharge valve. Pressure spikes to point C, and the gas is released into the discharge piping. The discharge event is a series of pressure waves that will degenerate until the piston reaches the end of its stroke at point D. From top dead center, unexpelled gas pushes on the piston and expands along curve DEF. Slightly below point E at P1, the pressure is further reduced to point F. Here a sufficient differential pressure exists to unseat the suction valve. Gas is drawn in by the piston for the remainder of the stroke until the piston reaches bottom dead center.

Volumetic efficiency increases with a decrease in clearance volume and a decrease in compression ratio. While the other factors do influence volumetric efficiency, clearance volume has the most pronounced effect. In the end, a compromise position must be taken to balance volumetric and compression efficiency. In general, the design compromises are related to compression ratio. For high compression ratios (6 to 15), clearance volume is the principal factor and valves are secondary. For the intermediate compression ratios (3 to 6), clearance volume and valve design should be balanced. For low compression ratios (less than 3), valve design is primary.

For a given operating condition where either a piston or diaphragm compressor can be considered, diaphragm compressors will have a higher volumetric efficiency since clearance volumes are less and ring leakage is not a factor. This will also permit compression ratios of 15:1 across one cylinder. Most piston compressors limit compression ratios to 5:1 or less in nonlubricated designs due to the inherent problem of heat removal and the effect of high temperatures on piston rings.

Isothermal efficiency is usually higher in the diaphragm compressor. This is due in part to the large, flat surface area of the cylinder, the proximity of the cooling passages to the compressor valves, and the recirculation of the hydraulic fluid on the back side

Table 1. Relative cost factors.

Installed Power	Lubricated	Nonlubricated	Diaphragm
To 30 kW	1.0	1.3	1.5
31 to 50 kW	1.0	1.3	1.9
51 to 100 kW	1.0	1.4	1.7
101 to 200 kW	1.0	1.4	1.6

Table 2. Typical selection guidelines.

Process and Operating Requirements	Compressor Type
Contamination during compression	Lube or NL piston
Limited contamination during compression	NL piston or diaphragm
Medium corrosive gas	All
Highly corrosive gas	Diaphragm
Hazardous gas (toxic, flammable)	Diaphragm or NL piston
Leaktight	Diaphragm
Low suction pressure, low compression ratio	Lube or NL piston
Low suction pressure, high compression ratio	Hybrid
High discharge pressure (to 25 MPa)	NL piston or diaphragm
High discharge pressure (> 25 MPa)	Lube piston or diaphragm

of the diaphragm group. However, isothermal efficiency of any positive displacement compressors is less than a dynamic compressor. Hence, a comparison of isothermal efficiency between compressor types is a false comparison. Overall efficiency, volumetric and compression, will normally favor diaphragm types.

Engineering economics

For comparison purposes, capital cost factors for lubricated, non-lubricated and diaphragm compressors are given in Table 1. These values are relative and are based on manufacturers' standard materials of construction and exclude drivers.

Performance cost factors are usually based on such ratios as cost/kW, cost/m^3/h, and cost/Nm^3/h. Whatever measurement is used, the piston compressor has the most favorable ratio within the context of its application. However, relative cost factors should not be the sole criteria for compressor selection assuming that any of the types discussed have the pressure and displacement capabilities. Process or environmental constraints can raise the cost factors of lubricated piston compressors above other types. Cleanup and disposal equipment and monitoring instruments for hazardous gases will rapidly escalate the ratios. This points out the need to match application. In Table 2, some typical selection guidelines are given.

Applications requiring compression equipment to boost pressure in the range of 101 kPa to 21 MPa and as high as 40 MPa should consider the use of hybrid compressors. These machines combine the high flow of the nonlubricated piston compressor with the leaktight, high-pressure capabilities of diaphragm compressors. The benefit to the engineer is that the total package is on one compressor and driver, not multiple compressors and drivers. A limitation of such equipment is the diaphragm cylinder operating speed which must be in the range of 400 to 450 rpm.

Process and equipment considerations

Process contacting materials of construction. Table 3 lists typical materials of construction for lubricated and non-lubricated piston compressor cylinders.

For temperatures of -25°C to -200°C, type 304SS low carbon, alloyed cast iron, or nickel alloy steel should be used. These materials will provide low temperature impact strength and thus avoid brittle failure of critical components.

Liners are recommended for cylinders since they can be replaced easily in comparison to complete cylinder. Two types are used—a wet liner or a dry liner. Wet liners form the inside pressure boundary of the cylinder and also the inside of the water cooling jacket. Because it is a pressure boundary, the liner wall must be designed to withstand the internal gas pressure and the external cooling water pressure. The liners are either pressed or shrunk into place. Dry liners are not pressure boundaries. They are thinner in section and are pressed into the cylinder. Liner materials include steel, stainless steel, nodular iron or gray iron. The liner material selection should be reviewed with the manufacturer since there may be limitations relating to pressure and piston ring combinations.

Table 4 lists typical materials of construction for diaphragm compressor cylinders.

Diaphragm compressor gas plates can be produced from any machinable material. Some limitations do apply because of the special requirements placed on diaphragm material by the compressor manufacturers. Certain alloys are not readily available, or are available only with a tremendous cost impact. Such materials include high nickel alloys, and so forth.

The diaphragm compressor cylin-

Table 3. Typical piston compressor materials of construction.

Component	Material	Remarks
Cylinders and Heads	Gray Iron	Pressures to 10 MPa
	Ductile Cast Iron (DCI)	Pressure to 16 MPa
	Steel	Pressure to 400 MPa
	Stainless Steel	Pressure to 25 MPa
Piston	Aluminum	Large cylinders, low inertia
	Cast Iron, Steel	High pressures, chlorides
	Stainless Steel	Corrosive conditions
Liners	Gray Iron, DCI	Most common
	Ni-resist	Low temperature, corrosion
	Stainless Steel	Corrosive conditions
Packing	PTFE-Filled	Pressures to 28 MPa
	Metal	Pressures > 28 MPa

Table 4. Typical diaphragm compressor process materials of construction.

Component	Material	Remarks
Gas Plate	Carbon Steel	Pressure to 63 MPa
	Low Alloy Steel	Pressure to 200 MPa
	304 SS, 316 SS	Pressure to 63 MPa
	17-4 PH, A286	Pressure to 200 MPa
	High Nickel Alloys	Pressure to 63 MPa
	20Cb-3	Pressure to 63 MPa
Diaphragms	301 SS, 316 SS	Standard for most service
	Ni-Cu alloy	Oxidizer service

Table 5. Typical compressor valve materials of construction.

Component	Material	Remarks
Valve Seat/Guards	Steel	Standard, medium pressure
	400 SS	Standard, all pressures
	17-4 PH	Optional, corrosion resistance
Valve Discs	400 SS	Standard
	17-4 PH/17-7 PH	Corrosion resistant
	316 SS	Corrosion resistant
	Plastics	Corrosion resistant/valve action
Springs	17-7 PH	Standard
	Nickel superalloy	Corrosion resistant
	302 SS, 316 SS	Corrosion resistant

der can be designed for structural temperatures from -150°C to 317°C. Special construction methods and hydraulic fluids with acceptable viscosity at these temperature extremes are required for successful operation of the compressor at low or high temperatures. Occasionally, the diaphragm cylinder is located off the crankcase with an extended distance piece and piston rod so that the temperature extremes do not affect the reciprocating frame. Separate hydraulic systems, one for frame lubrication and the other for diaphragm pulsing can be employed.

Valves are common to all reciprocating compressors. Table 5 lists typical materials of construction for compressor valves.

Certain gases or gas mixtures are corrosive, flammable, or explosive. When pressures and temperatures are increased, or when water vapor is present, the gas can be more difficult to handle. Process contacting materials of construction should be specified by the user because the user is most familiar with the process. However, the experience of the compressor manufacturer permits solutions to be offered, including material of construction choices. In Table 6, guidelines are given for some of the common gases used in the CPI.

Energy use

In today's energy conscious environment, users require the correct solution for gas compression applications. Energy costs can amount to 80% of a compressor's operating cost. It is important to use only the energy required for the process.

For constant speed compressors, several methods can be used to control capacity. They are suction control, bypass control, and suction valve clearance control. All of these methods can be applied to piston compressors. Diaphragm compressors are limited to suction control and bypass control.

Suction control limits the gas pressure at the suction valve. Because piston and diaphragm compressors are positive displacement, the inlet flow varies directly with the suction pressure and provides an infinite number of steps between fully opened and fully closed. For example, a 10% decrease in suction pressure will result in a minimum flow decrease of 10% since both the gas density and the volumetric efficiency are decreased. Some power savings can be expected, but compression ratios and discharge temperatures will be higher affecting compressor performance.

Bypass control (capacity bypass) requires additional external piping, a control valve, and instrumentation. This provides an infinite number of steps, but the compressor operates at full discharge pressure and capacity all the time. Compressed gas is recycled back to suction but it must be cooled to the normal suction pressure or else higher discharge temperatures will result. Power savings are non-existent.

Suction valve unloading controls capacity by maintaining one or more suction check valves in a partially or fully opened position. This is a step control and capacities of 0%, 25%, 50%, 75%, and 100% can be achieved. A variation of this control method is the progressive capacity control. The action of the suction valve is delayed and the compression diagram is altered. This permits capacity fine tuning from 40% to 100%. No-load power is 15% to 20% of the full-load power.

Clearance control can be either fixed or variable, the latter preferred for fine control. This control method permits the capacity to vary by changing clearance volumes which, in turn, affects the volumetric efficiency. Like suction valve unloading, it is efficient and results in power savings almost proportional to the capacity with no-load and full-load power.

Power savings can be realized by using variable-speed control. Capacity and power vary linearly with speed. A speed decrease of 10% will result in a power decrease of 10%. However, variable-speed drives have a high initial cost and they must be carefully sized to meet the torque requirements of the compressor which generally limits the practical turndown to 50% to 80%.

Multistage compressors

Multistage compressors are selected when the single stage compressor design limitations are reached. Generally, these limitations are:

- compression ratio (clearance volume);
- pressure differential;
- discharge temperature; and
- power savings.

Practical, maximum single-stage

E. H. LIVINGSTON is vice president and general manager of Burton Corblin North America, Inc., Horsham, PA (215/443-5600; Fax: 215/443-5609). His responsibilities include management of overall operations, application and design engineering of piston and diaphragm type reciprocating compressors. He has spent 30 years associated with chemical processing equipment, such as compressors, pumps, and high pressure vessels and piping. The author of several articles on diaphragm compressors and high pressure equipment, Mr. Livingston has also lectured in courses on compressors and high pressure equipment for applications in the chemical and petrochemical industries. He received his engineering degree in chemical engineering from Drexel University. He is an active member of AIChE, American Society for Testing and Materials (ASTM), ASM International, and National Association of Corrosion Engineers (NACE).

compression ratios for piston diaphragm compressors are 5:1, and for diaphragm compressors 8:1 to 10:1. These limitations are based on the clearance volumes and the heat characteristics of each compressor type.

High-pressure differentials can create higher loads (stresses) on the mechanical parts of the compressor. In most cases, this can be solved by multistaging.

High discharge temperatures are to be avoided to prevent the deterioration of piston rings, packing, and cylinders. Gas discharge temperatures should be limited to 200°C. Diaphragm compressors can tolerate higher gas discharge temperatures because of the limited use of nonmetallics and a higher rate of heat rejection.

Power savings from multistaging can range from 5% to 25% depending on the number of stages. Actual savings will be a function of the compressor load factor and compressor size.

Table 6. Material and construction guidelines.

Gas	Type	Remarks
Acetylene	Explosive	Temperature limit, 55°C, low gas velocity, no copper/copper alloys
Ammonia	Corrosive	No copper/copper alloys
Carbon Dioxide	Corrosive	Corrosive when wet, use 316 SS
Carbon Monoxide	Toxic	Temperature limit 150°C, carbon steel or low nickel alloys
Chlorine	Toxic	Temperature limit 125°C, no hydrocarbon greases/oils, corrosive when wet
Chlorofluorocarbons	Environmental hazard	Temperature limit 110°C, leaktight construction
Fluorine	Corrosive	Fluorinated hydraulic fluids, 316 SS or high nickel alloys, degrease all components
Hydrogen	Explosive	Leaktight construction, potential embrittlement above 30 MPa
Hydrogen Sulfide	Corrosive	Leaktightness, toxic gas, material hardness limitations, 21 Rc per NACE publications
Oxygen	Flammable	Fluorinated hydraulic fluids, 316 SS or high nickel alloys, degrease all components, low gas velocity, limit compression ratios
Vinyl Chloride	Flammable	Temperature limit 90°C, will polymerize or liquify

In conclusion

The compressor types available to the process engineer are many and the proper selection can be quite difficult. More users are relying on the compressor manufacturers to develop the complete gas handling system for a particular process. Still, the process engineer must evaluate and select the type of compression equipment to be used. During this phase of the project, several important points must be considered:

1. Are the process conditions accurately stated and are there any contingencies relating to flow rate and pressure identified?
2. Will the process be scaled-up in the future and, if so, should the compressor specifications reflect immediate needs, future needs or both?
3. Are standard specifications such as API 618 relevant to the type of equipment to be specified? Should a stand-alone specification be written?
4. Can lubrication be tolerated during compression and if so, must lubricants be removed from the gas?
5. Can leakage be tolerated?
6. Can neither lubrication or leakage be tolerated?
7. Is there an actual or potential corrosion problem?
8. Can the manufacturer service the equipment?

The responses to these questions are very important and they will help in the selection process. If there are questions relating to the type of compressor, controls, process changes, and the like, discuss these issues with a manufacturer or another user. If these steps are not taken, the engineer is faced with difficult equipment and cost comparisons. CEP

Choose the Proper Compressed Air Drying Equipment

Evaluate the application to determine the required dew point for the compressed air. This guides the selection of the dryer.

Frederick M. Sitter,
Van Air Systems, Inc.

Compressed air has two major uses in chemical process industries (CPI) plants: as a power source or utility, and as a process ingredient. Because compressed air is almost always produced on-site, it is up to the user to treat it as needed to make the process work most efficiently. How dry and clean the air must be to provide trouble-free operation depends on the intended application.

This article provides some background on how compressed air becomes contaminated and the problems this contamination causes, outlines the considerations for determining the level of treatment needed, and offers guidance on choosing the best equipment to provide that treatment. The economics of various treatment alternatives is also discussed briefly.

Contamination problems

First, let's look at how contaminants enter the air system. As a compressor takes in ambient air, it also takes in all of the moisture present in that air. At atmospheric pressure, this is usually not a problem. Each cubic foot of atmospheric air will contain a specific amount of moisture, based on its temperature and relative humidity.

During the compression process, 7.8 ft^3 of atmospheric air becomes 1 ft^3 of compressed air at 100 psig, and it contains all of the moisture that was in the original 7.8 ft^3 of air. The result is a cubic foot of air with a high water content and dew point. (Note that even though the compressor has an inlet filter, it removes only a portion of the atmospheric particles, the exact amount depending on the filter's efficiency.) The amount of water as water vapor in the air is proportional to the compression ratio — the higher the pressure, the more water that is squeezed into each cubic foot of compressed air.

In order for the air to hold this moisture, it must remain hot. But as the air proceeds through the system piping, it cools, and the moisture condenses.

The condensed moisture causes numerous problems. It floats away lubricating oils, causes corrosion in air piping, and contaminates materials. It can also result in complete blockage of air lines in freezing temperatures.

Evaluating your needs

Determining the dew point inside the compressed air system (without any drying equipment installed) involves some relatively simple calculations and a dew point moisture chart (Table 1 contains excerpts from such a chart):

1. Read the moisture content of the atmospheric air (in gr/ft^3 [one grain equals about two small drops]) from the chart based on the ambient temperature and relative humidity (RH).

2. Multiply the atmospheric air's

moisture content by the compression ratio to get the moisture at the higher pressure.

3. Read from the chart the temperature at which 1 ft^3 of air containing that amount of moisture is saturated (that is, when its relative humidity is 100%). That is the dew point of the system.

For example, if the ambient air at the compressor intake is at 40°F with relative humidity of 70%, the dew point in a system at 100 psig is 92°F. Thus, if piping is exposed to ambient temperatures less than 92°F, moisture will condense in the system.

On the average, an aftercooler placed before the dryer will remove about 70% of the moisture that entered the system. However, as the air exits the aftercooler, it is still 100% saturated and any further reduction in temperature will still result in condensation.

The dryer can remove approximately 45–99% of the 100% RH saturated air left after the aftercooler. In most applications, the dryer must only reduce the relative humidity of the compressed air to a level such that no moisture will condense within the air system.

To determine what type of equipment best suits an application, first examine the location of the air system and all related piping and components, and then consider the use of the compressed air.

Looking at the existing air system, locate all air piping and note the temperatures to which it is exposed. This will help in determining the required dew point. To ensure that no moisture condenses in the system, specify a dew point at least 10°C lower than the lowest ambient temperature that the air piping runs through. (Instrument Society of America [ISA] Standard S7.3 for instrument air requires a dew point 10°C (18°F) lower than the lowest ambient temperature at any given time.)

In conducting this analysis, be sure to take into consideration piping that runs outdoors, that may be exposed to winter subfreezing temperatures, or that runs through air-conditioned areas. This is a major factor in many process applications.

Then review the intended application(s) of the compressed air. It may be used to operate pneumatic instruments and controls, to convey dry ingredients, to operate dust collectors, or to package chemicals. This is important, because different degrees of dryness are needed for different applications. For example, air used for cleaning parts need not be as dry as air used for running critical pneumatic instrumentation.

Finally, cost is another important factor. A lower dew point costs more to achieve and maintain.

There are three basic types of compressed air drying systems:

- a central air drying system for the entire facility;
- point-of-use dryers for particular applications; and
- one central dryer to provide a level of drying suitable for most applications combined with smaller point-of-use dryers for a few critical applications.

Installing the correct central dryer for all applications will ensure that all the air is sufficiently dry, which is particularly useful if it is not clear where all the piping runs. If the plant has only a few critical applications, point-of-use dryers placed upstream of these applications may be more efficient and cost-effective. If most of the facility needs moderately dry air but a few applications require drier air, a central dryer may be aided by a few point-of-use dryers.

The main dryer types are distinguished by dew point performance, location where the dryer can be installed (indoors or outdoors, and conditions such as dusty or hazardous), initial cost, and operating costs.

Table 2 compares the three basic dryer designs: single-tower (deliquescent), refrigerated, and regenerative (twin-tower).

Single-tower dryers use no electricity for operation, which makes them ideal for applications in hazardous or corrosive environments. They have a very low pressure drop (which means more energy stays in the system). And they have low capital and operating costs.

Single-tower dryers use absorbent consumable desiccants as the drying medium. In general, these desiccants are stable, inorganic materials that are hygroscopic, deliquescent (that is, tending to dissolve upon the absorp-

Table 1. Dew point moisture chart — grains of water vapor in 1 ft^3 of air.

	Degree of Saturation (RH)			
Temperature, °F	70%	80%	90%	100%
38	1.86	2.126	2.391	2.657
39	1.931	2.207	2.483	2.759
40	2.003	2.290	2.576	2.862
41	2.079	2.376	2.673	2.970
42	2.157	2.466	2.774	3.082
...				
90	10.470	11.966	13.462	14.958
91	10.789	12.330	13.871	15.413
92	11.107	12.694	14.281	15.868
93	11.440	13.075	14.709	16.344
94	11.774	13.456	15.138	16.820

Table 2. Comparison of the three major types of compressed air dryers.

	 Single-Tower (Deliquescent)	 Refrigerated	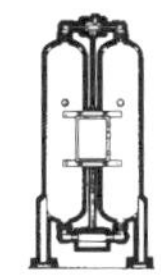 Twin-Tower (Regenerative)
How Dryer Removes Moisture	Absorptive desiccant absorbs moisture and dissolves slowly.	Cools air to condense moisture.	Adsorptive desiccant attracts and holds moisture.
Dryer Design	Single upright pressure vessel. Grid located above inlet supports desiccant bed. Rated capacities to 16,000 scfm for a single unit;. manifolding can be added for higher flows.	Two independent systems — air and refrigerant. Air system consists of two heat exchangers in series with separator and/or automatic drain. Refrigerant system includes compressor, condenser, and regulating valves. Capacities to 7,500 scfm, or higher with special units.	Two upright vessels interconnected by piping and equipped with valves to control direction of air flow. Capacities to 10,000 scfm or larger.
Principles of Operation	Air enters lower central area of vessel, where larger droplets and solid particles are separated mechanically. As the air moves upward through the desiccant bed, moisture is absorbed by the desiccant tablets. The water and dissolved desiccant drop to the claim area, which can be drained manually or automatically.	Entering air is precooled in the first heat exchanger before entering the second for final cooling. Heat from the air is transferred to the cool refrigerant. Condensed moisture is trapped and discharged through an automatic drain. Exit air is reheated by the first heat exchanger.	One tower receives wet air, which flows through desiccant bed, and moisture collects on the desiccant. Simultaneously desiccant in the other tower is being regenerated. Collected moisture is driven off by electric or steam heat and/or a purge of dry air from the operating tower.
Desiccant Required	Treated chemical tablets.	None.	Activated alumina, molecular sieves, or silica gel.
Energy Requirements	None.	Electricity (all models). Water for water-cooled condensers where specified.	Electricity (all models). Power for regeneration; may be electric or steam heat or compressed air purge.
Dew Point at 100 psig	Compensating 20°F dew point suppression from inlet air temperature. (Greater suppressions, up to 63°F, are available.)	+38°F or +50°F; constant.	–40°F to –100°F; constant.
Oil Removal Ability	Will remove some larger liquid-phase oil particles.	Will condense some oil droplets during cooling process.	None.
Pressure Drop	Less than 1% of operating pressure.	2–5 psid at rated capacity.	2–5 psid.
Inlet Air Temperature	Maximum: 100°F; Minimum: None.	Maximum: 110°F; Minimum: 40°F.	Maximum: 120°F; Minimum: Varies.
Ambient Air Temperature	Maximum: 100°F; Minimum: None.	Maximum: 110°F; Minimum: 40°F.	Maximum: 120°F; Minimum: Varies.
Location	Indoors or outdoors.	Indoors.	Indoors.
Prefilter Required?	Recommended where heavy amounts of oil are present to prevent coating of desiccant tablets.	Recommended; prevents oil from fouling heat exchangers.	Yes; prevents oil and liquid water from damaging or overloading desiccant bed.
Afterfilter Required?	Yes; prevents particle migration from desiccant and removes the fine oil aerosol.	Not required, but optional for removal of fine oil aerosol in cricital applications.	Yes; prevents particle migration from desiccant.
Maintenance Requirements	Drain condensate once per shift, either manually or with motorized drain valve. Add desiccant every 3–4 mo.	Blow dust from condenser fins regularly. Check and clean automatic drain.	Replace desiccant every 2–3 yr. Adjust controls. Clean or replace filter elements.
Repair Requirements	None; no moving parts to repair.	Requires qualified service personnel.	Requires qualified service personnel.

Dispelling common misconceptions

• Running piping into warm areas will not raise the dew point inside an air line. It will raise the temperature inside the air line, but the dew point — the temperature at which moisture will begin to condense — will not change.

• The lowest possible dew point is not necessarily the most practical or economic choice. While an extremely low dew point will prevent any moisture problems, it is not cost-effective to spend the extra money to produce a dew point lower than what is actually needed for the application.

• Filters and separators do not provide dry air. They can only remove oil, particulates, and liquids — not water vapor. Condensation of moisture can only be prevented by lowering the dew point with a dryer.

• Price alone should not dictate the drying equipment purchased. Even if the drying equipment is a "good buy," money spent on equipment that will not do the job is money wasted. Money spent on the right equipment will yield a multiple return in terms of less maintenance, less downtime, and better product quality.

tion of moisture), and 100% soluble. It is necessary to add more desiccant just a few times per year; however, the desiccant does not have to be removed or changed.

Most single-tower installations also require an aftercooler, which will be discussed later. When used with an aftercooler, such a dryer provides an approximate 20°F to 50°F dew point suppression — that is, it reduces the dew point to about 20–50°F below the dryer inlet temperature, depending on the desiccant used. When installed outdoors, the dryer automatically adjusts its dew point, providing a moderate dew point in the summer and a lower dew point, to protect against freeze-ups, in the winter.

Refrigerated dryers use mechanical cooling to condense entrained moisture. They are designed for the typical indoor industrial atmosphere and provide dew points from about 35°F to 50°F. They do not provide protection against freeze-ups. They are also simple to operate and require minimal maintenance.

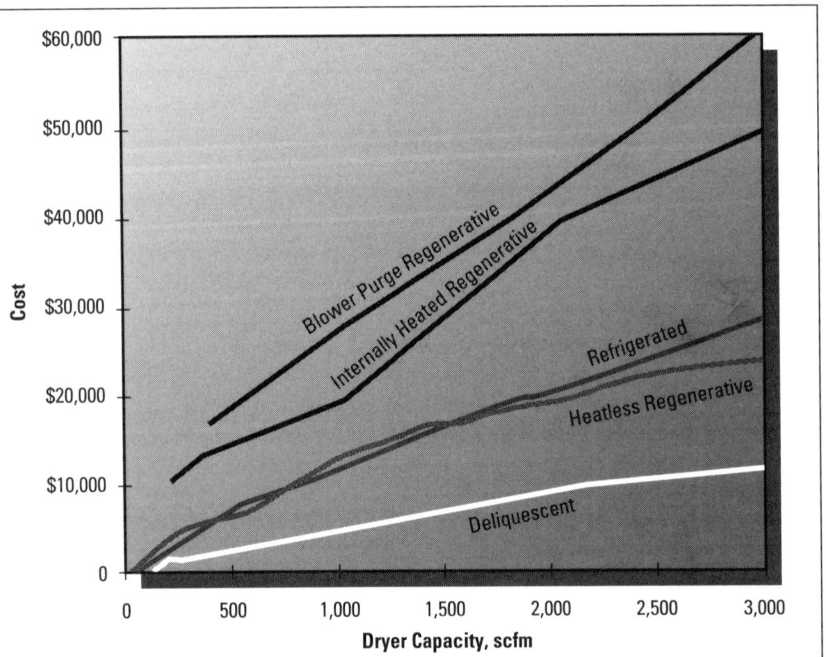

■ *Figure 1. Average capital investment as a function of dryer capacity.*

Two types of refrigerated dryers are cycling and noncycling. In a cycling dryer, the compressor shuts down during low-demand periods, thus saving money. However, according to refrigerant compressor manufacturers, compressors in noncycling dryers typically have a longer life because they are not subjected to starting and stopping. Noncycling units also ensure a reliably consistent dew point by responding more readily to changes in load conditions.

Some larger non-cycling units can be provided with unloaders for compressor capacity control, which reduce the electrical consumption of the compressor. This lowers operating costs without subjecting the compressor to frequent starting and stopping or temperature changes.

Regenerative dryers use an adsorbent desiccant that attracts moisture to its surface. Tower operation is cycled — that is, while one tower is on-stream drying the compressed air, the other tower is being regenerated. Regeneration can be accomplished with a purge of extremely dry compressed air from the on-stream tower (heatless), with internal tower heaters and purge air, or with atmospheric air heated by an external blower.

Heatless dryers are the simplest and safest of the regenerative types. They may be used when the extra capacity for the purge, usually about 15% of the operating capacity, is available; this usually limits their use to systems less than 1,000 cfm.

Internally heated dryers use less purge air — about 4.5% of operating capacity — but they require additional electricity for the heaters. These are typically used for dryers in the 500–3,000 cfm range.

Externally heated blower units use no purge air for regeneration. They are most commonly used for large systems, of 3,000 cfm and larger, because of the costs of purge air for heatless systems of these sizes and the difficulty in providing even internal heating in large desiccant beds.

Table 3. Typical costs for compressed air dryers*.

	Single-Tower	Refrigerated	Heatless Regenerative	Blower Purge Regenerative	Heated Regenerative
Initial Cost ($)					
Dryer Only	3,700	9,000	8,400†	22,000	17,500
Full System‡	7,800	12,100	12,600	27,200	22,600
Annual Operating Costs ($/yr)					
Dryer Only	2,440	5,913	7,884	5,490	10,748
Full System†	2,980	6,453	8,645	6,528	11,786

* Basis: 600 scfm, 125 psig line pressure, 180°F compressor outlet temperature and 100% RH, 80°F ambient air temperature, oil carryover of 10 ppm maximum, electrical costs of $0.075/kWh, and continuous operation for 8,760 h/yr; the dry air must be free of condensed water and oil contamination is limited to 1 ppm.

† Not including the energy saving option, which would also reduce the operating costs.

‡ "Full system" refers to the dryers plus standard accessories, which, depending on the type of dryer, include air-cooled aftercoolers, separators, afterfilters, prefilters, adsorbent or absorbent, and so on.

The operating costs for regenerative dryers are higher than for other types — generally 50% to 150% higher. But they provide extremely low dew points — usually -40°F to -100°F — for uninterrupted operation of critical processes. Energy-saving options on regenerative dryers can reduce the number of times the towers switch so that the desiccant is regenerated by dew point demand rather than at a pre-timed rate. (For heatless dryers this is typically a 10-min cycle, with each tower on-stream for 5 min and regenerating for 5 min; heated dryers usually have an 8-h cycle, on-stream for 4 h and regenerating for 4 h.)

Selecting a dryer

One of the most important factors in choosing a dryer is often cost. The single-tower dryer has a low initial investment and low operating costs, requiring only adding desiccant a few times per year. It may also require an aftercooler (which requires some electricity), compressed air, or water, depending on the type. Refrigerated dryers need electric power. Heatless regenerative dryers require both electricity and a compressed air purge, the latter of which requires sufficient compressor capacity to allow for this lost purge air (about 14–20% of the dryer's inlet capacity, depending on the inlet line pressure). Other types of regenerative dryers will operate without purge air, but would have a higher initial investment and would require more electricity or steam.

Table 3 compares the initial investment and operating costs for the different dryer types. These data are based on the following conditions: 600 scfm, 125 psig line pressure, 180°F compressor outlet temperature and 100% RH, 80°F ambient air temperature, oil carryover of 10 ppm maximum, electrical costs of $0.075/kWh, and continuous operation for 8,760 h/yr; the dry air must be free of condensed water and oil contamination is limited to 1 ppm.

Figures 1 and 2 illustrate the capital and operating costs of each dryer type (for the dryer only, not the complete system). These figures were developed from actual measured costs, and for convenience, curve smoothing was performed. In reality, however, the relationships are not strictly linear. For example, if some of the same components are used in systems of different sizes, the corresponding capital and operating costs for the higher capacity systems do not increase linearly as a function of capacity (scfm).

Auxiliary equipment

In addition to the dryers, other components are needed to make the drying system complete.

Aftercoolers placed after the compressor but before the dryer can remove a large portion — up to 75%

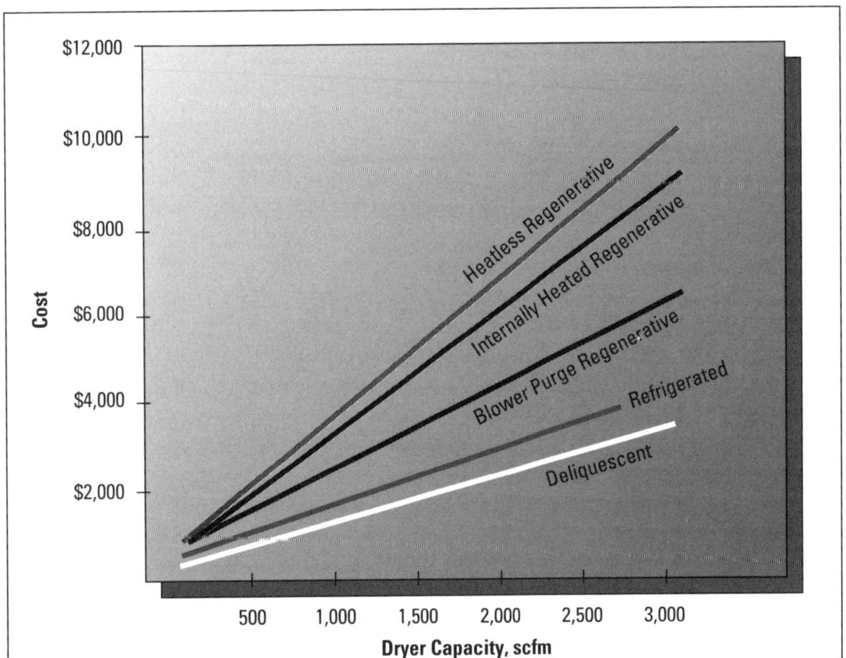

■ *Figure 2. Estimated annual operating costs as a function of dryer capacity.*

An alternative application

A variation of the single-tower dryer can be used for low-pressure (2–15-psig blower operation) applications — for blanketing and for preventing moisture contamination of bulk chemicals. With the installation depicted here, the typical drying efficiency would be a humidity reduction to 12% RH or 1% RH at atmospheric pressure, depending on the type of desiccant used.

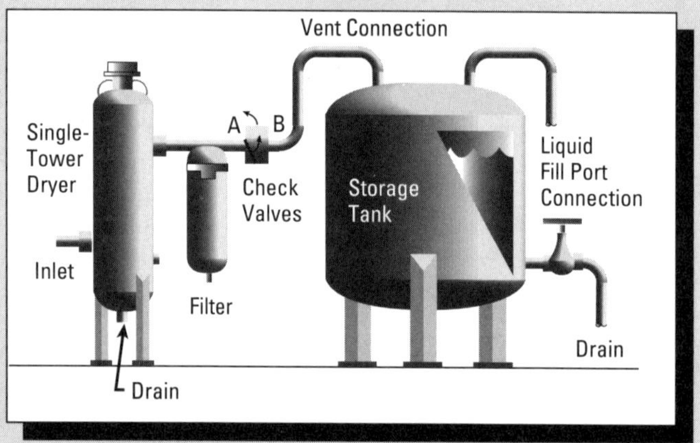

The following factors should be considered for this type of application:

• the type of chemical to be protected — as measured by the water vapor equilibrium, to help determine the amount of dryness needed to prevent moisture contamination of the chemical;

• the maximum flow (in gpm) from the bulk holding tank — converted to scfm, this helps determine the size of dryer needed; and

• the volume (in gal/mo or gal/yr) drawn from the holding tank — to determine the approximate desiccant usage rate and how often desiccant will need to be added.

To size this type of dryer, the air flow through the dryer is calculated indirectly based on the volumetric flow rate from the bulk tank.

— of the moisture for more efficient dryer operation. But having an aftercooler does not eliminate the need for a dryer because the air exiting the aftercooler is still saturated and moisture will condense if the downstream temperature is reduced further.

Two types of aftercoolers are available: air-cooled and water-cooled. Air-cooled units may be either wind-cooled, with finned tubes to increase the surface area and using no electricity, or fan-cooled, with fans blowing ambient air over the heat exchangers.

A good aftercooler should provide a low approach, of about 5–10°F, to remove the greatest amount of moisture. An air-cooled aftercooler installed outdoors or in a cool area achieves the greatest amount of cooling. More technical performance and sizing information is available from aftercooler manufacturers.

Separators and drain valves are important to expel the condensed liquids from the air system and prevent flooding of the dryer and air system. Many types of drain valves are available. Stainless steel ball valves are good for corrosive conditions and high pressures up to about 2,000 psi. Bronze ball valves are slightly less expensive and are rated up to 720 psi at 100°F. Solenoid valves are best for filters and refrigerated dryers. Float drains are acceptable for most clean water condensate, but electric ball valve drains should be used for more contaminated condensate and to ensure against clogging. For natural gas applications, pneumatic dump systems are available, which prevent gas from escaping into the atmosphere as condensate is expelled. Solenoid valves should not be used for any thick or highly contaminated solutions.

Further Reading

"How to Select a Compressed Air Dryer," Van Air Systems, Inc., Lake City, PA (1993).

Rollins, J. P., ed., "Compressed Air and Gas Handbook," 5th ed., Compressed Air and Gas Institute, Cleveland, OH (1989).

A small but valuable accessory is a high condensate level alarm, which alerts maintenance personnel of an increase in condensate level so that any problems can be corrected to prevent the dryer and downstream equipment from flooding. It is installed in the air receiver; in single-tower dryer systems, it can be installed directly in the dryer.

Filters are necessary to remove particle and oil contamination from the compressed air in order to protect the dryer and downstream equipment. Many styles are available with different micron and oil-removal ratings. Figure 3 summarizes the filters suggested for each dryer type and the air quality that can be achieved with the equipment arrangements shown.

Filters should be checked before installation to make sure the correct filter element has been selected. A differential pressure indicator should be added to show when the element should be changed to maintain system efficiency; some indicators are available with alarm contacts. Filters need to be drained daily or after each shift. For this reason, electric drains are often used to automate draining.

The benefits of proper drying

Treating compressed air as outlined here helps keep many pro-

F. M. SITTER is chief application engineer with Van Air Systems, Inc., Lake City, PA (814/774-2631; Fax: 814/774-3482). He has held a variety of positions in his 33 years with the company, and is currently responsible for standard design applications of dryers, aftercoolers, filters, and absorbent and adsorbent desiccants. He holds two filter and drain patents, and has authored several articles. He received a BS in chemistry with a minor in math from Gannon Univ. (Erie, PA), and has been active in the American Chemical Society.

cesses working efficiently and without interruption. Often the drying equipment pays for itself within the first year because it eliminates downtime caused by frozen or blocked air lines and the need for extra compressors to make up for lost pressure. It also prevents contamination of moisture-sensitive materials and eliminates other moisture-related downtime and maintenance.

All of these benefits translate into substantial savings of both time and money. Taking the time to analyze your particular conditions and determine what you actually need can truly pay off. CEP

Typical Uses	Suggested Filter/Dryer Installations*	Nominal Air Quality[†]		
		Oil, ppm	Solids, microns	Dew Point
■ Instrument Air[‡] ■ Blanketing and Padding ■ Nitrogen Replacement ■ Pipeline Purging	Compressor → After-cooler → Separator → General Purpose Coalescing Filter[§] → High-Efficiency Coalescing Filter → Heatless Regenerative Dryer → General Purpose Particulate Filter	0.008	1.0	–40°F
	or Vapor Adsorbing Filter	0.004	0.01	–40°F
■ Pneumatic Hand Tools ■ Robotics and Machine Tools ■ Spray Painting ■ HVAC Temperature Controls	Compressor → After-cooler → Separator → General Purpose Coalescing Filter → Refrigerated Dryer → High-Efficiency Coalescing Filter	0.008	0.01	38°F
■ Shot/Sand Blast ■ Dust Collectors ■ Spray Painting ■ Sawmills and Lumber Yards ■ Natural Gas ■ Low Pressure Air/Gas	Compressor → After-cooler → Separator → Coarse Coalescing Filter → Single-Tower (Deliquescent) Dryer → General Purpose Coalescing Filter	0.78	1.0	20–55°F Suppression
Point-Of-Use				
■ Bulk Contaminant Removal	→ Coarse Coalescing Filter →	3.9	5.0	N/A
■ Instrumentation ■ Air Cylinders and Motors ■ Pneumatic Conveyors	→ General Purpose Coalescing Filter → High-Efficiency Coalescing Filter →	0.008	0.01	N/A
■ Food/Beverages/Dairy ■ Pharmaceutical ■ Clean Rooms ■ Critical Instrumentation ■ Blow Molding	→ General Purpose Coalescing Filter → High-Efficiency Coalescing Filter → Vapor Adsorbing Filter →	0.004	0.01	N/A

* Mechanical separation device should be installed prior to filters to remove slugs of liquid oil and/or water.

† Assumes inlet contaminants to filter at less than 50 ppm liquid oil and water-saturated air at 100°F. Actual air quality may vary with operating conditions unique to each installation. Critical applications may require additional filtration.

‡ ISA S7.3, Quality Standard for Instrument Air, requires maximum particle size of 3 mm and oil content of no more than 1 ppm.

§ General Purpose Coalescing Filter is optional for heatless regenerative dryers.

Figure 3. Recommended filter combinations for various compressed air dryers.

Protect Centrifugal Pumps From Low Flows

Guidelines are given on when protection is needed and the types of systems that provide it.

Owen Woodside,
Consultant

Operation of centrifugal pumps below their minimum flow requirements is the primary cause of premature pump failure. Hydraulic instability occurs at low flows, causing cavitation, surging, and excessive vibration in the pump. This instability is more restrictive in setting the minimum allowable flow than using the temperature rise, which was used in the past. If you are the system design engineer, you should consider providing a protective system during the preparation of the piping and instrumentation diagrams (P&IDs). Not all pumps require a protective system. Your problem is to determine which pumps do and which do not, and decide the type of system to employ. This decision must be made as early as possible in the development of the P&IDs to avoid costly rework. This article gives guidelines on when to provide a protective system and offers examples of appropriate designs.

Why provide protection?

Protection against operation below the safe low-flow limits is required for the following reasons:

1. To prevent damage due to overheating;

2. To prevent cavitation and vibration due to excessive internal recirculation;

3. To prevent damage due to excessive radial reaction;

4. To prevent damage due to cycling between two points if the performance curve is humped; and

5. To prevent overloading of drives for certain types of pumps (axial flow, mixed flow, and regenerative turbine) which have horsepower curves that rise toward shutoff.

The extent and type of damage to a pump depends on how long low flow persists and on the magnitude of the forces and vibrations that are generated. Failure of the mechanical seals is the most common problem, followed by failure of the bearings. Loss of efficiency due to wear of the impellers by cavitation is usually a slower process.

The cost of pump maintenance amounts to about 40% of the total for many process plants *(1)*. If the addition of a low-flow protection system can reduce these costs and also reduce plant upsets, then the cost of the system is more than justified.

When to provide protection

Every pump should be examined to determine whether or not low-flow protection is required. To make the decision, you must be familiar with the modes of operation of the plant and the effect the control systems may have on the pump operation. The following operating modes should be examined:

1. Is the plant designed for stable operation during prolonged turndown conditions, *e.g.,* 50% of capacity?

2. Is the plant to be started up slowly at reduced flow rates?

3. Is the flow reduced when there is a plant upset?

4. Is the pump flow controlled by a level controller (or a flow controller reset by level)?

5. Can the flow rate be cut back or stopped by a downstream user?

Table 1. Guidelines to determine when a low-flow protection system is needed.

Type of Service	Recommended Minimum Flow, % of Design	$NPSH_a/NPSH_r$
Hydrocarbons, single- and double-suction	25–50	1.1
Water and water solutions, single-suction	40–70	2.0
Water and water solutions, double-suction*	70-85	2.0
Pumps with domed head/flow curves	Below shutoff head	1.1–2.0

* Generally above 1,000 gpm.

6. Is the pump in critical service?

7. Are two pumps with different head/capacity curves operating in parallel in the same system with a common control?

8. Is the fluid at its bubble point or does it contain dissolved gas?

If the anticipated flow rate for any of these modes of operation is less than the minimum required for stable operation of the pump, then a low-flow protective system is warranted. Capital and operating costs, reliability, and maintainability are factors to consider.

Applications that frequently are provided with low-flow protective systems are:

• Feed pumps;

• Product pumps;

• Intermediate transfer pumps on level control;

• Boiler feedwater pumps on level control;

• Steam condensate pumps on level control;

• Firewater pumps;

• Utility water pumps;

• Drinking water pumps; and

• Pumps operating in parallel with different head/capacity characteristics.

Applications that normally do not require minimum flow protection are:

• Circulating cooling-water systems;

• Circulating heating or cooling medium systems; and

• Any system with a fixed flow rate.

Setting minimum flow rate

In the past, temperature rise in the pump was considered to be the limiting factor setting the minimum flow requirements *(2)*. The minimum flow was set at 10–15% of design based on a temperature rise of about 20°F. This resulted in few pumps being provided with minimum flow protection systems and many pump failures. Due to research over the past 30 years *(3–6)*, it has been realized that the minimum flow requirements can be as high as 70% to prevent excessive internal recirculation causing vibration, cavitation, and surging. This can occur even with considerable excess available net positive suction head (NPSH). Pump vendors now specify the low-flow limit for both thermal limitations and hydraulic stability.

Since pump vendors usually have not been selected during the initial phase of P&ID development, the specified minimum safe flow is unknown. It is impossible to determine the low-flow limit for hydraulic stability without having specific information on the internal design of the pump. The guidelines in Table 1 may be useful in deciding whether or not to provide a minimum flow protection system. These guidelines are based on articles by Taylor *(7)* and Makay *(8)*. If the recommended minimum flow limit is above any of the anticipated low-flow modes, then the addition of low-flow protection may be justified.

The percentages in Table 1 are based on the design flow (normal flow) of the pump, which are usually 5% below the flow at maximum efficiency (best efficiency point or BEP). The BEP is normally halfway between the design and rated flows. It is common practice to specify the rated capacity as 110% of design. In some cases the rated flow may be 125% of the design flow. In these instances the minimum flow listed in Table 1 should be increased proportionally. Pump vendors normally quote the minimum flow based on the BEP. Unstable flow conditions can also occur if the pump is operated much above its rated capacity, that is, >110% of design. If pre-selection of the pumps for a project can be made, a better estimate of the low-flow limit may be obtained.

The recommended minimum ratio of $NPSH_a$ to $NPSH_r$ is also given in Table 1. During the hydraulic calculations for the P&IDs, you must set the heights of the suction vessels to provide adequate NPSH for the pumps. As indicated, it is general practice to

Nomenclature

ARC	=	automatic recirculation valve
FC	=	flow controller
FI	=	flow indicator
HLL	=	high liquid level
LC	=	level controller
LLL	=	low liquid level
NLL	=	normal liquid level
$NPSH_a$	=	net positive suction head available, ft
$NPSH_r$	=	net positive suction head required, ft
PC	=	pressure controller
PT	=	pressure transmitter
RO	=	restriction orifice
SV	=	solenoid valve

Table 2. Guidelines for using low-flow protection systems.

Protection System	High Power Cost Pump Size, hp	Low Power Cost Pump Size, hp
Continuous bypass	<15	<30
On/off bypass	10–15	20–30
Automatic recirculation valve	>15	>35
Modulating control valve	>15	>30

provide a safety factor for the $NPSH_a$ which is based on experience. It is normally not necessary nor economical to provide more than the recommended minimum.

Figure 1 shows typical operating ranges for centrifugal pumps in different services. It should be noted that there is a limit for operation at too low a flow as well as too high a flow. In general, the safe operating range for high-capacity (over 1,000 gpm), high-head (over 250 psi), high-horsepower (over 300 hp) pumps is less than for small, low-head pumps.

The required minimum turndown for a pump may be specified to the pump vendors. Generally, it is normally not economical nor possible to purchase a specially designed pump that can cover the required range of conditions. However, there are some special cases where this has been justified for improved safety and reliability.

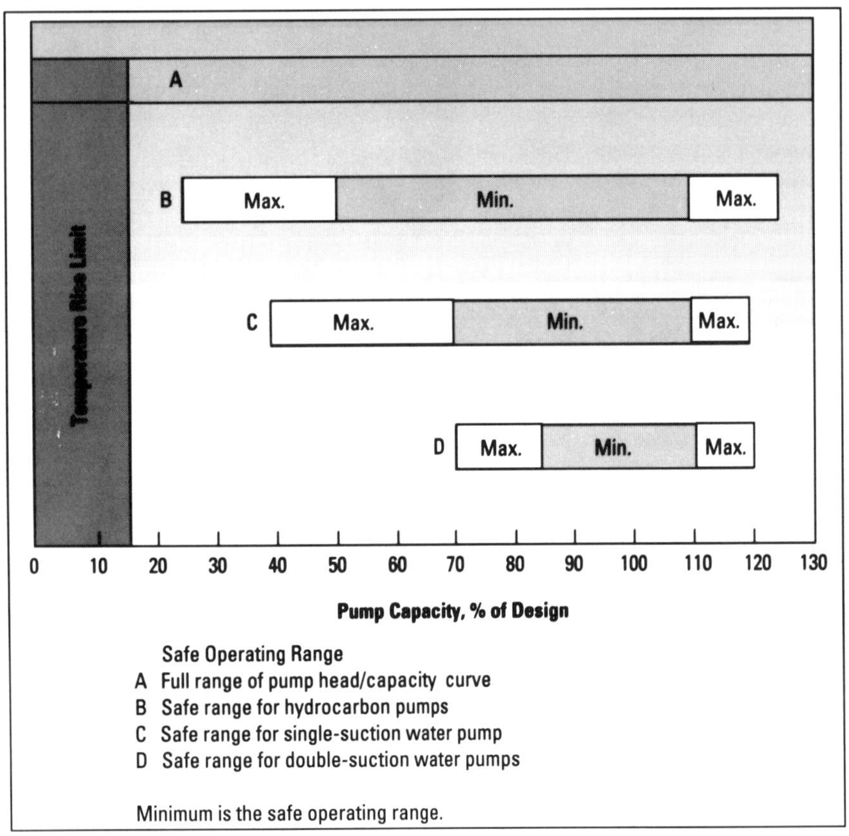

Figure 1. Safe operating range.

At the time the pumps are purchased and the manufacturer's recommended minimum flow rates are received, the low-flow requirements should be reviewed. This is normally before the issue of the P&IDs but after many of the piping drawings have been prepared. It is cost-effective to provide the protective systems on the P&IDs as early as possible — that is, do it right the first time.

Finally, during the process hazards review of the P&IDs, the need for protective systems for the pumps, and the design of the protective systems should be analyzed. Lack of protective systems or faulty designs have resulted in many disruptions to operations.

Recirculation systems

There are three basic types of bypass systems for low-flow protection: continuous, on/off, and variable flow. Which type to select depends on the service, capital, and power costs. Table 2 gives guidelines for each application. In the table, high power costs are 8¢/kWh and low costs are 3 ¢/kWh.

The recycle line should return to the vapor space of the suction vessel. If this is not possible, then recycle through a cooler and return to the suction line. If this alternative is not feasible and the recycle is short-term, that is, of a few minutes' duration, then return the recycle to the suction line as far upstream as practical. The design of these different protective systems is now considered.

Continuous recirculation (Figure 2) — This is the least expensive low-flow protection system. A fixed orifice in the bypass line controls the minimum flow. The continuous bypass must be added to the normal flow to set the design capacity of the pump. Thus there is a continuous waste of energy when the pump operates above the minimum required flow.

Recirculation with an on/off valve (Figure 3) — This is the next least expensive system. The on/off valve in the recirculation line opens wide whenever the flow element in the main line drops below the safe minimum flow. A fixed orifice controls the recirculating flow. A disadvantage of this system is that when the on/off valve opens or closes it can cause sufficient hydraulic shock to upset sensitive systems. This type of system is not generally recommended. It should only be used if the minimum recommended flow is less than 30% of the design flow.

Some experts will disagree with providing a single recycle system, as shown in Figures 3 and 5, for a pump and its spare. When a single recycle system is provided, it is incumbent on the operators to be aware that only one pump at a time should be operating. This is accepted as the normal practice in many oil refineries and chemical plants.

A check valve is provided in the main process line downstream of the bypass branch. This check valve is required to prevent reverse flow from any downstream source returning to the suction drum via the recycle line in case of pump failure. There is a check valve at the discharge of the pumps for added insurance against reverse flow through the pump.

Automatic recirculation valve (Figure 4) — The automatic recirculation (ARC) valve has been used for many years to control the minimum flow for pumps in clean service, such as for boiler feedwater and steam condensate. More recently, it has been used in many chemical and hy-

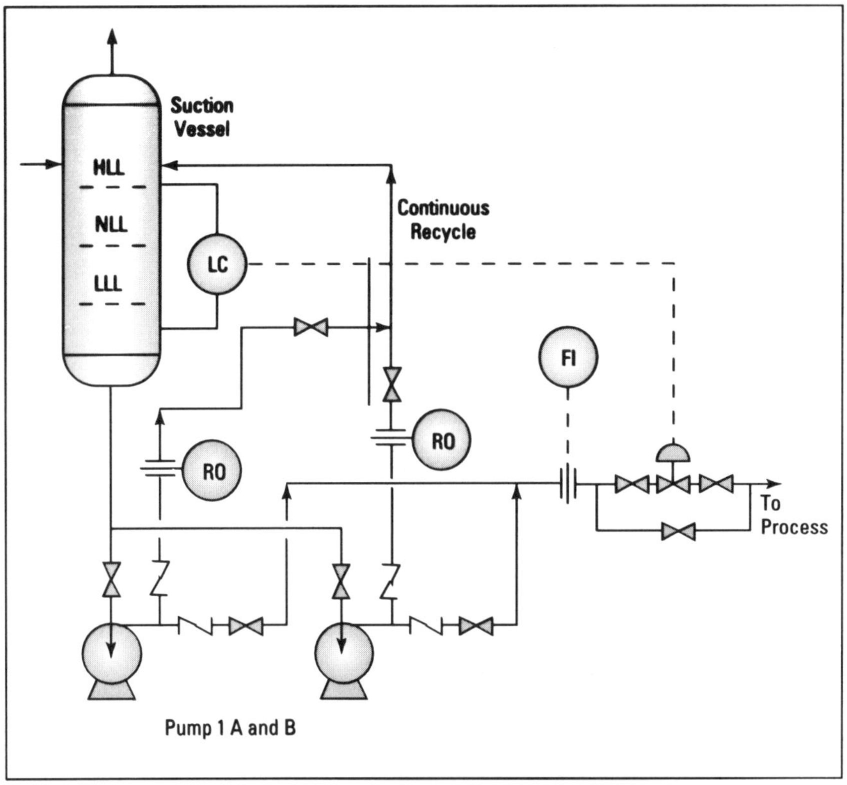

Figure 2. Continuous recycle system.

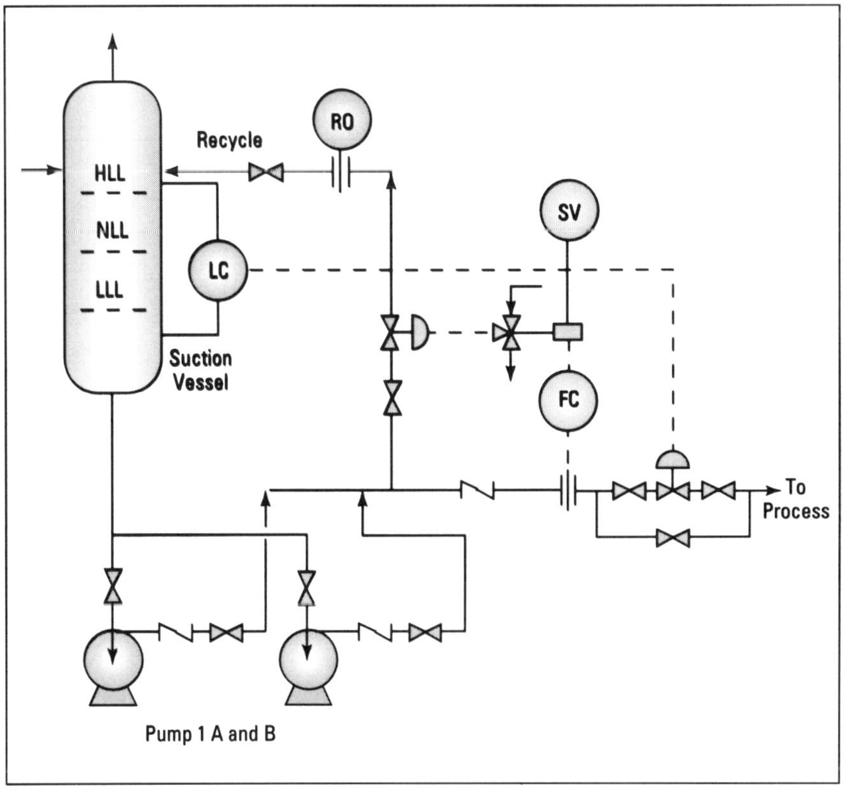

Figure 3. On/off recycle system.

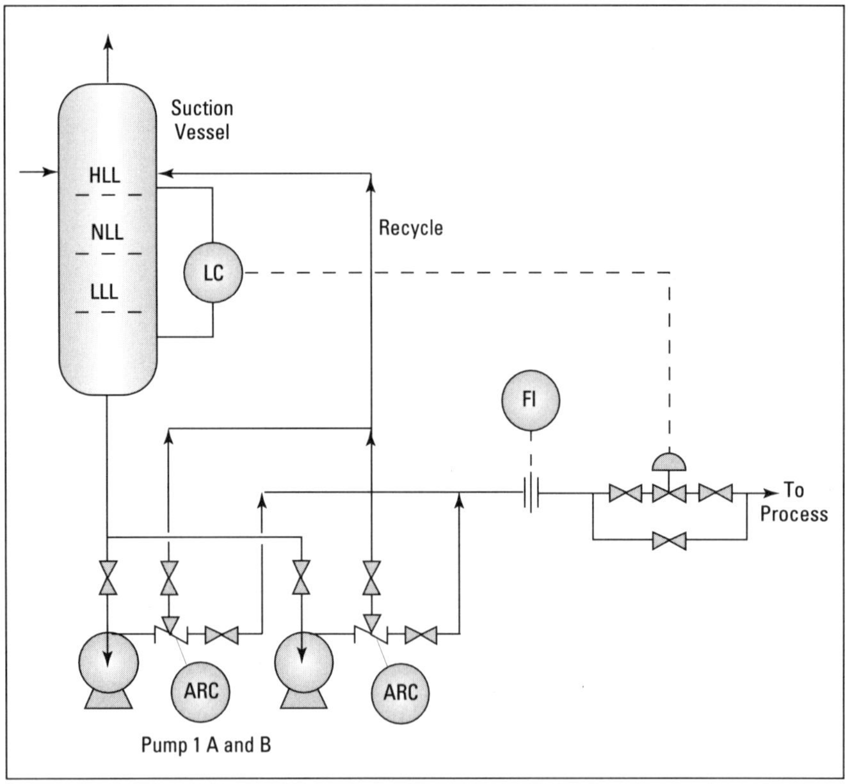

Figure 4. Automatic recirculation system with ARC valve.

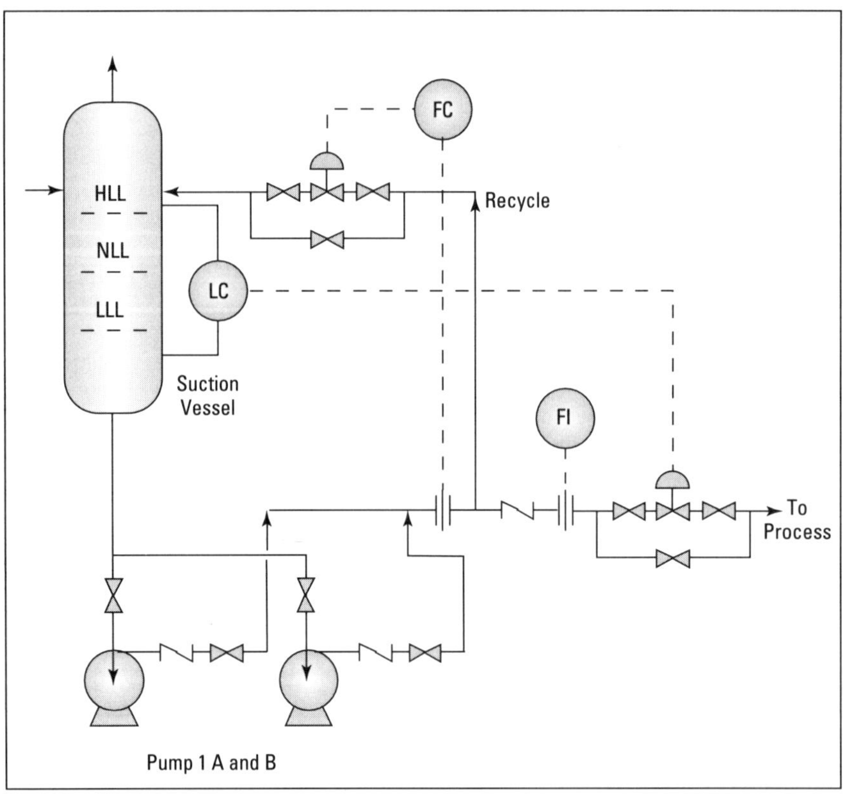

Figure 5. Flow-controlled recycle system.

drocarbon services. It is a combination of a check valve and modulating control valve. The spring-loaded check valve acts as a flow-sensing element. Its position controls the opening of the bypass port. It is factory set to begin opening the bypass port when the flow drops to the minimum recommended rate.

The bypass is fully open when the forward flow stops and the check valve is in the closed position. An orifice in the bypass port limits the bypass flow to that required for safe operation. The ARC valve allows smooth operation over the full capacity range of the pump.

The total cost of an ARC valve installation to protect one pump and its spare is about the same as for a single control valve recycle system up to 3 in. For 4 in. and over, the latter system is less expensive. ARC valves can also be purchased with the dual function of protecting against low flow and excessively high flows. This system is particularly useful when two or more pumps subject to low flow are operated in parallel.

Modulating control valve (Figure 5) — This is the most commonly used system in process plants. It is used in services where the presence of scale or solid particles makes the use of ARC valves unsuitable. The recycle valve is frequently controlled by a flow element on the discharge of the pump. Other methods are also used, such as pressure control, differential temperature across the pump, and motor amperage. Figure 6 shows an example of a pressure controlled system. The controller is set to begin opening at a pressure corresponding to the minimum flow requirement. In general, pressure control of the bypass is not as good as flow control because it is not as sensitive to flow changes.

The system shown in Figure 5 is for the case where there are two pumps, one of which is spare. The flow controller is set to begin opening when the discharge flow falls below the minimum recommended rate. If both pumps are operating, the minimum

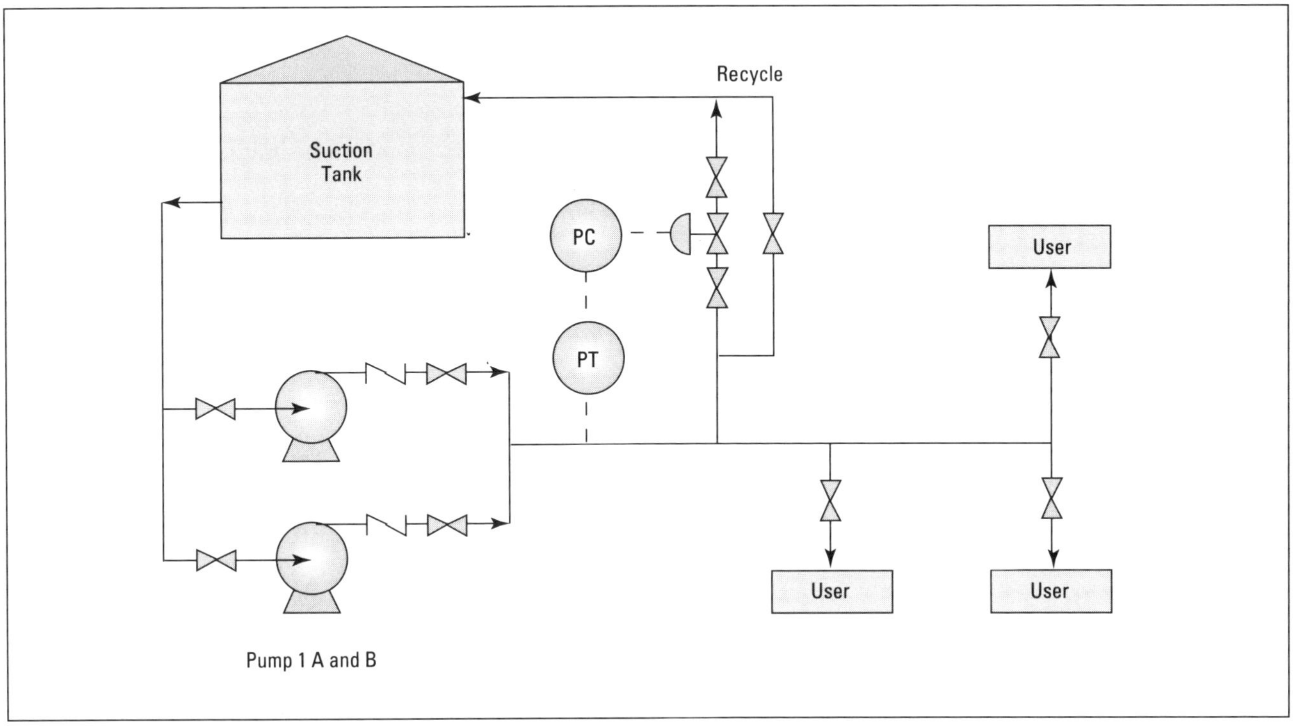

■ *Figure 6. Pressure-controlled recycle system.*

flow setting should be doubled or the control system changed to protect both pumps, for example, change to pressure control if applicable.

Existing units

If a pump in your processing plant has excessive failures, it may be the consequence of periodic operation outside the safe limits for low or high flow. Find out from the operating staff if there are any periods of low or high flow, either short- or long-term. If this is the potential cause of the failures, consider increasing the NPSH available, modifying the operating procedures or providing a low- or high-flow protection system. Check the pump specification for the recommended range for stable flow. If it is not listed, ask the manufacturer for its recommendation. CEP

Literature Cited

1. **Clark, E., and D. Littlefield**, "Maximize Centrifugal Pump Reliability," *Chem. Eng.*, pp. 90–96 (Feb. 94).
2. **Panesar, K. S.**, "Consider Suction Specific Speed," *Hydrocarb. Proc.*, pp. 107–108 (June 1978).
3. **Heald, C. C, and R. Palgrave**, "Backflow Control Improves Pump Performance," *Oil & Gas J.*, pp. 96–105 (Feb. 25, 1985).
4. **Rayner, R. E.**, "Watch Suction Specific Speed," *Chem. Eng. Progress*, **89** (3), pp. 79–84 (Mar. 1993).
5. **Ross, R. R., and P. H. Falbeck**, "Flexible Inducer Pumps Can Have Wide Operating Range," *Oil & Gas J.*, pp. 96–100 (Aug 5, 1985).
6. **Fraser, W .H.**, "Recirculation In Centrifugal Pumps," Worthington Group, paper presented at American Society of Mechanical Engineers Winter Meeting, ASME, New York (Nov. 1981).
7. **Taylor, I.**, "Pump Bypasses — Now More Important," *Chem. Eng.*, pp. 53–57 (May 11, 1987).
8. **Makay, E.**, "How to Avoid Field Problems With Boiler Feed Pumps," *Hydrocarb. Proc.*, pp. 79–84 (Dec. 1976).

O. W. WOODSIDE is a consultant, residing in Arcadia, CA (818/355-7372). Previously, he was project chemical engineer and project manager with C.F. Braun/Brown & Root, from 1965 to 1993, and process technologist and plant manager with Shell Oil Companies, from 1947 to 1965. He has been involved with the operation, design, and project management of oil refineries and chemical plants, particularly ethylene plants, in the U.S., Canada, U.K., Germany, Japan, and Trinidad. Woodside is a PE in California, a member of AIChE, and holds a BApSc from the University of British Columbia, Vancouver.

Acknowledgments

The author wishes to thank S. Uchizamada of Ingersoll-Dresser Pumps for supplying material for this article and D. West (machinery specialist) and H. Heidsman (systems specialist) for taking the time to review it.

Design Better Heat Exchanger Network Retrofits

This new approach to heat exchanger network retrofitting considers the relative importance of various technical and economic parameters.

Annika Carlsson
Per-Åke Franck
Thore Berntsson
Chalmers Univ. of Technology

Several methods and approaches have been developed for the design of heat exchanger networks (HENs) in grassroot systems. However, these methods are not directly applicable in retrofit situations. In retrofitting of HENs, the relative importance of various parameters is different than in grassroot design, so other approaches are necessary. The retrofit approaches presented previously in the literature do not take into account all parameters relevant to the cost equation for a match, but rather focus mainly on minimizing the total exchanger area.

This article presents a new method, based on pinch technology, for use in retrofit situations. It enables the engineer to consider many different parameters that affect network cost. [The basics of the method were presented at last year's Institution of Chemical Engineers' Research Event in Manchester, U.K., *(1)*.] It is assumed that the reader is familiar with pinch technology *(2,3)*.

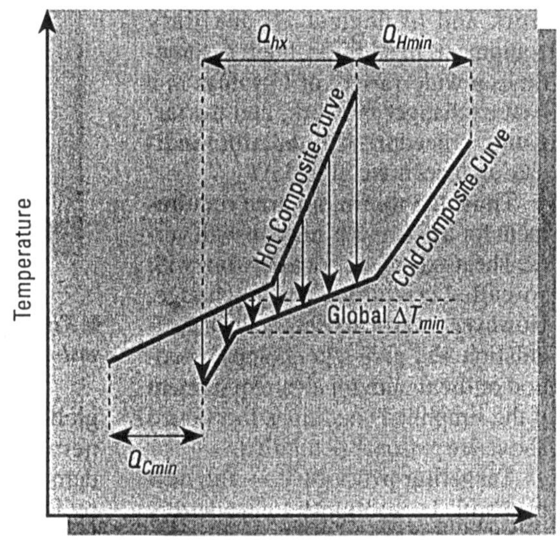

■ ***Figure 1.*** ***The composite curves show the targets for the hot and cold utilities, the energy recovery, and the location of the pinch.***

Defining the problem

In a retrofit situation, improvements can be made by introducing new heat exchangers and/or by using existing ones more efficiently. The latter can be achieved by either moving the exchangers temperature-wise or by rearranging them to other streams.

The goal of current retrofit methods is to minimize the use of heat exchanger area in the network. However, in order to find a cost-optimal solution, a number of other parameters must also be taken into account, such as:

- the physical distance between streams;
- material requirements;
- the types of heat exchangers;
- auxiliary equipment (valves);
- heat-transfer coefficients;
- space requirements;
- annual pressure-drop (*i.e.*, pumping) costs;
- annual maintenance costs; and
- fouling.

Targeting procedure — existing methods. Existing retrofit methods are aimed at minimizing the total heat exchanger area in the network and minimizing the number of units. The relationship between heat exchanger area and utility consumption is used to perform a targeting procedure for the minimum area prior to design. For a given global ΔT_{min}, (defined as shown in Figure 1), the corresponding utility demands, Q_{Hmin} and Q_{Cmin}, and the minimum area requirement based on vertical heat exchange between the composite curves are calculated.

However, the result of this targeting procedure is correct only when the same cost equation for the area can be applied to all units in the network and no other costs (such as piping costs) are considered. In addition, a uniform U-value must be used for all units in the network. In reality, it is usually impossible to assign the same cost equation to all matches in a network and to neglect all auxiliary equipment. In addition, there is sometimes a wide variety of U-values in a heat exchanger network, and the targeting procedure will therefore indicate an inaccurate global ΔT_{min}.

Thus, the minimum area requirement for a network is not one with vertical heat exchange, but a network with so-called crisscross heat exchange. However, a vertical heat exchange calculation will give a good approximation of the minimum area requirement if the simplified conditions mentioned above are reasonably fulfilled.

Targeting procedure — the new method. Including the parameters mentioned above in a retrofit situation requires a new approach to the targeting procedure. When these parameters are considered, the most economically correct network is usually not one with vertical heat exchange, but a crisscross network, as illustrated in Figure 2. This, of course, is because the aim of the retrofit is not to minimize the total area of the network but rather the total cost of the network. (The issue of targeting is discussed in more detail later.)

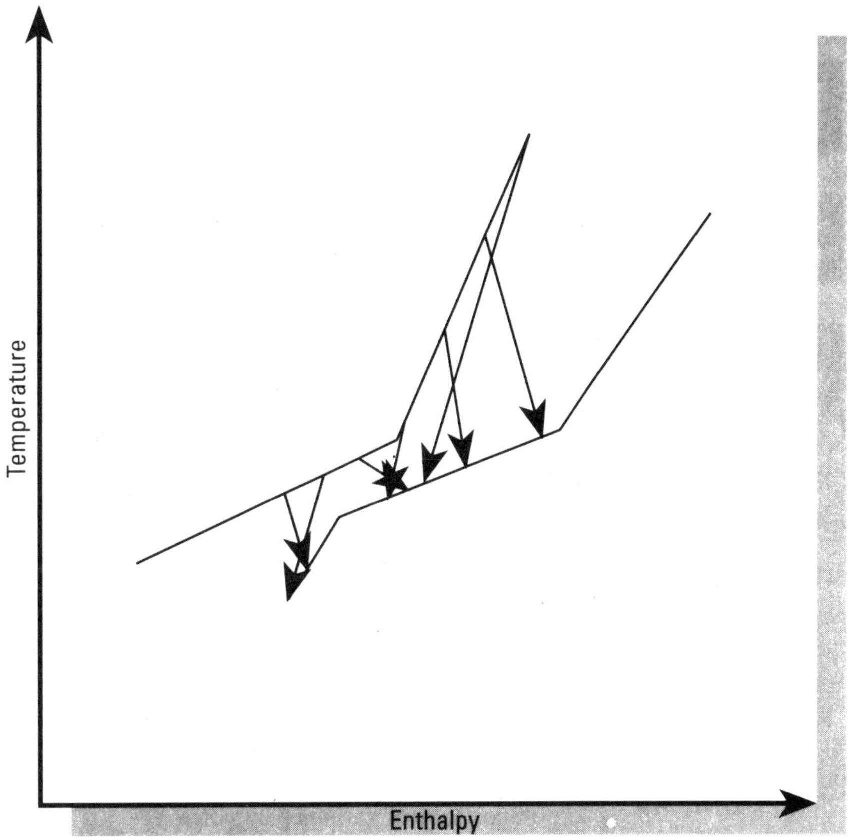

■ ***Figure 2. Crisscross heat exchange between the composite curves will result in a cost-optimum solution.***

Earlier retrofit methods

One method for the retrofit of heat exchanger networks based on pinch technology was presented by Tjoe *(4)*. The method is mainly concerned with minimizing the total required heat exchanger area, and a targeting procedure for the most cost-effective global ΔT_{min} is carried out before the network is constructed. The utility demands and area requirement (based on vertical heat exchange) are calculated for each global ΔT_{min} less than the existing one. The results of these calculations can be compared to the existing utility demands and area, and then the savings in utility costs and the cost for additional area (assuming a uniform price for all units) can then be calculated. These calculations will result in an investment/savings plot.

An acceptable payback period or investment is specified, which will indicate a global ΔT_{min} in accordance with these economic criteria. The selection of a global ΔT_{min} is also guided by a value of the total area efficiency, α. This is the ratio of the installed area to the minimum area requirement for the existing global ΔT_{min}, and it indicates how well the existing area is installed. In the design stage, the main objective is to make use of the existing area and new heat exchangers as efficiently as possible through vertical heat exchange. The area efficiency of the resulting design should not be lower than the original area efficiency.

The initial targeting procedure for a suitable global ΔT_{min} was improved by Polley and Panjeh Shahi *(5)* by including a relationship between pressure drop and heat-transfer coefficients in order to obtain a more accurate area calculation. This method is mainly applicable for uniform processes where area is the dominating cost and heat exchangers can be relocated easily between different positions in the network. The main drawback is that the temperature driving forces between the streams (*i.e.*, the heat exchanger areas) are considered as the main factors influencing the economy. This means that most of the other parameters mentioned above cannot be taken into account. And, in many cases, these parameters may have a greater influence on the economics of the HEN than the heat exchanger area.

Linear programming. Some other methods for the retrofit of heat exchanger networks *(6, 7)* employ various linear programming models. Yee and Grossmann *(7)* proposed a two-stage mixed integer nonlinear programming (MINLP) method that includes a prescreening stage and an optimization stage. In the prescreening stage, an economic feasibility study is performed to evaluate the potential for a more rigorous retrofit study. In the optimization stage, a so-called superstructure that includes all interesting retrofit designs is constructed. To find the most economic retrofit option embedded in the superstructure, a MINLP model is developed and solved.

According to the objective function used, five different costs are considered: cost for additional area, fixed charges for new units and new segments of piping, and costs for hot and cold utilities. However, one uses the same costs everywhere in the network rather than specifying costs for individual matches. Other drawbacks of MINLP-based methods are that significant computational time is required in order to solve the MINLP model, and that they provide no opportunity for interaction during the optimization. This means that the economic impact of choosing a nonoptimal solution for certain matches due to practical or flexibility reasons (among others) is complicated.

The new method

Our aim in developing the method presented here was to create a computer-based model for finding near-optimum HENs in retrofit situations with the following criteria:

• The model shall be able to take into account all relevant parameters discussed previously.

• It shall be interactive with the user so that practical constraints and "process peculiarities" that are not programmable can be considered early in the selection process.

• It shall allow the user to conduct a sensitivity analysis, that is, to study and quantify the penalty associated with a chosen solution caused by operating flexibility needs and constraints.

Targeting procedure. Since the degree of crisscross corresponding to an optimum solution is unknown for a HEN in which many different costs are considered, a targeting procedure for a most economical global ΔT_{min} prior to design cannot be carried out. This is because even though some of the parameters included can be linked to the energy recovery (such as area), others (such as length of piping) are linked only weakly or not at all. In addition, individual cost equations are used for all matches in the network and different combinations of matches at the same global ΔT_{min} may result in large differences in cost.

Instead of a targeting procedure, solutions are produced for a number of global ΔT_{min}s. The global ΔT_{min}s to be studied can be selected by means of a Q_{Hmin} = f(global ΔT_{min}) plot, from which a number of interesting global ΔT_{min}s are chosen.

For each global ΔT_{min} less than the existing one, the utility demands, Q_{Hmin} and Q_{Cmin}, are determined. The savings in annual utility costs can be evaluated for these global ΔT_{min}s compared to the existing one.

Based on experience in designing a network, some simplifying assumptions can be made as a starting point for a design.

In addition, the energy recovery, Q_{hx}, is fixed at each global ΔT_{min}. Hence, the optimal retrofit design is the one with minimum total annual costs for changes in the existing HEN to satisfy Q_{hx}.

This design consists of a combination of matches, each of which contributes an individual annual cost, E_{match}, to the total cost of retrofitting the heat exchanger network, E_{HEN}. The combination of matches that yields the lowest E_{HEN} is the solution to the retrofit problem at each global ΔT_{min}.

By selecting matches, it is possible to find optimal or near-optimal HENs. An appropriate global ΔT_{min} for the whole retrofit project can be selected by comparing payback periods or annual saving for the best solution at each global ΔT_{min}.

Since it is not possible to perform a targeting procedure, the degree of nonoptimality is not known. However, when the method has been applied to examples found in literature in which simplified assumptions regarding cost equations and U-values have been used, the network solutions found have been at least as good as thoses previously published.

Ticking off matches. This method is based on pinch technology. The network is split into two parts, above and below the pinch point, which are treated separately. The design of the HEN, that is, the selection of matches, starts at the pinch and continues outward. If possible, matches are placed so that all of the heat in one of the streams involved in the match is completely transferred; this stream is said to be ticked off (or checked off).

However, the main objective of the method is to limit the cost of the retrofit project — and this is done by relaxation of pinch theory when necessary. The two most important types of relaxation are to allow a lower temperature difference in the heat exchangers than the global ΔT_{min} and to use alternative tick off points. (This will be clarified later in the article with the example, specifically matrix (a) above the pinch.)

Based on experience in designing a network, some simplifying assumptions can be made as a starting point for a design. In most situations, the economically correct way of designing a network is to accept new matches and extend existing matches so that at least one of the streams in the match is ticked off. If accomplished, this will lead to the minimum number of units in the network, according to Euler's General Network Theorem.

Situations when a "non-tick-off" match is either necessary or more economical can also be identified, and actions can be taken to terminate a match before tick off, thus including one extra unit each time a non-tick-off match is accepted. However, one should recognize that even if adding one extra unit seems more economi-

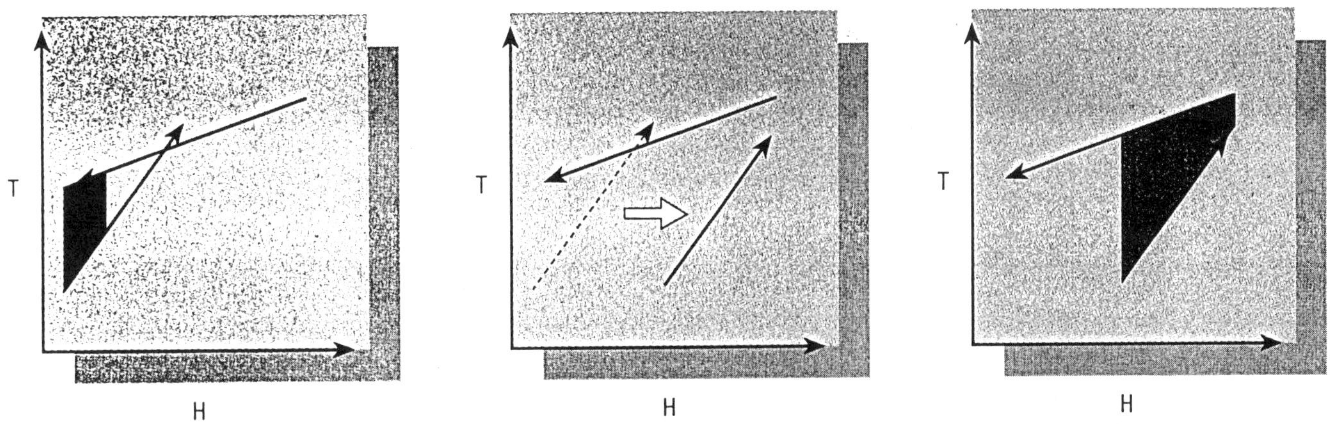

■ *Figure 3. A match that was terminated when a specific ΔT within the heat exchanger was reached can be shifted into a "shift-tick-off" match.*

cal, it will also increase the complexity of the network.

Stream splitting is presently not considered as an option unless necessary in finding a solution, but stream split options in order to reduce the total cost of the network will be included later.

One characteristic of this method is the simple way of dealing with existing and new matches. Existing matches are regarded as new installations but are given a relevant value by means of the cost equation (as discussed below). If an existing heat exchanger must be extended by another heat exchanger in order to tick off one stream, the investment cost of matching the two streams is only the cost of the new heat exchanger and repiping (assuming that existing pumps and piping can be used). Maintenance and pressure drop costs are considered for the two heat exchangers on the streams. If a possible new heat exchanger coincides with an existing one, no investment cost at all is attached to the match, and the total cost for the match consists of pressure drop costs and maintenance costs.

Because the method is interactive, one can change the conditions for a match during the evolution of the design. This is valuable when dealing with more dynamic parameters, such as fouling.

Restrictions. By including basic pinch technology in the method, constraints are automatically placed on transferring heat across the pinch point. Even so, networks transferring heat across the pinch point can be considered if the amount of heat transferred from below to above the pinch is equal to the heat transferred from above to below the pinch. This is essentially allowing crisscross in the pinch region.

This extension to the method has not been fully worked out and for now no heat transfer is allowed across the pinch. However, this is a very important aspect of the retrofit problem because it opens the possibility of reducing the total number of units in the network. It will be included in the future. In regions adjacent to the pinch point, crisscross solutions are considered.

As in any pinch analysis, it is necessary to have relevant thermodynamic data for all streams.

The cost equation. An individual annual cost can be assigned to each match in the network. The total annual cost for one match is divided into four parts:

$$E_{match} = E_{HE} + E_{PAE} + E_{\Delta P} + E_M$$

where: E_{HE} is the cost for heat exchanger area; E_{PAE} is the cost for piping and auxiliary equipment such as pipes, valves, supports, and so on; $E_{\Delta P}$ is the cost for pumps and fans to overcome pressure drops in heat exchangers and pipes as well as power costs for pumps and fans; and E_M is the maintenance cost. Annual costs are obtained by including capital recovery factors in the equations.

It is sometimes difficult to obtain rigorous data for the cost equations. It is possible, however, to use costs generated by computer programs, and it is also possible to assign standard U-values to the heat exchangers in the network. Once more rigorous data are found, the cost equations can easily be modified. The degree of precision of the cost equations is decided by the user and the availability of data.

It is possible to generate optimal U-values for all matches in the network for different types of heat exchangers. This will give an indication for possible heat exchanger types and configurations in the particular match being studied. Using the results from these calculations the pressure drop in the match can be calculated, and from this the pressure drop costs.

Calculation of the match matrices. When applying the method to a heat exchanger network for retrofit, the network is split into two parts, above and below the pinch. Starting at the pinch point, the individual cost equations are applied to each thermodynamically possible match, including crisscross matches. For each match, the maximum possible heat load between the hot stream and the cold stream in the match is calculated. The termination of the match occurs in three different cases — the hot stream is ticked off, the cold stream is ticked off, or the match is terminated when a ΔT, specified by the user, is reached within the heat exchanger.

However, a non-tick-off match of the third type is usually not very valuable. Instead, this type of match can sometimes be modified into a "shift-tick-off" match, as shown in Figure 3. In addition, matches that are thermodynamically impossible to start at the pinch point can be modified into this type of tick off.

When the maximum possible heat load between two streams has been calculated, the corresponding temperatures are known, and it is then possible to calculate heat exchanger area using the Bath formula *(8)* and the cost of the match. The cost calculation for each match includes the cost of heat exchanger area, piping and pumps, and so on.

The result of this calculation is placed in a cost matrix. The left side of Figure 4 illustrates such a cost matrix for the region below the pinch. Hot streams are arranged in rows and cold streams in columns. Thus, the value in box 8-1 is the cost (in thousands of dollars) of exchanging heat between hot stream 8 and cold stream 1 starting at the pinch point, or as close to the pinch point as possible, until the maximum heat load is exchanged.

The matrix starting from the pinch point is the first in a series of matrices that are calculated during the evolution of a design. To solve the retrofit problem, the user manually selects matches from these matrices until the heat recovery, Q_{hx}, corresponding to the global ΔT_{min} chosen, is accomplished. Each time a match has been selected, a new matrix is calculated to account for effects on the remaining matches from the previous choice, as shown in Figure 4. By adding the costs from the accepted matches, the total cost is obtained:

$$E_{HEN} = \Sigma E_{match}$$

All the matches in the matrix are classified according to their type and marked with an exponent, as outlined in Table 1.

Hot \ Cold Streams	1	4	5	6
1	2	44	43	18
7	49^{H}	44^{C2}	0^{C1}	25^{H}
8	36^{C3}	44^{C3}	54^{C3}	9^{H} (italic)
9	x	x	x	x

36^{C3} →

Hot \ Cold Streams	4	5	6
1	44	43	18
7	44^{C2}	0^{C1}	25^{H}
8	44^{C3}	54^{C3}	2^{H}
9	x	x	x

■ *Figure 4. When a match is selected, a new cost matrix is calculated.*

Applying the method

As in any pinch analysis, it is necessary to have relevant thermodynamic data for all streams. Using these data, a traditional pinch stream network is constructed and energy targets are calculated. Some additional data are necessary to account for match costs independent of thermodynamic properties. These include information about existing heat exchangers, piping, pumps, distances between streams, and material requirements for specific streams. All of these data are used to create reliable cost equations for all matches in

Table 1. Matches are classified by type and designated by an exponent.

Above the pinch

C = The cold stream is ticked off, Figure a

H1 = The hot stream is ticked off, and the residue of the cold stream cannot be used immediately for another match, Figure b

H2 = The hot stream is ticked off, and the residue of the cold stream can be used immediately for another match to tick off the cold stream, Figure c

H3 = The hot stream is ticked off, and the residue of the cold stream can be used immediately to tick off at least one other hot stream, Figure d; this type is referred to as a multiple tick off match

a.

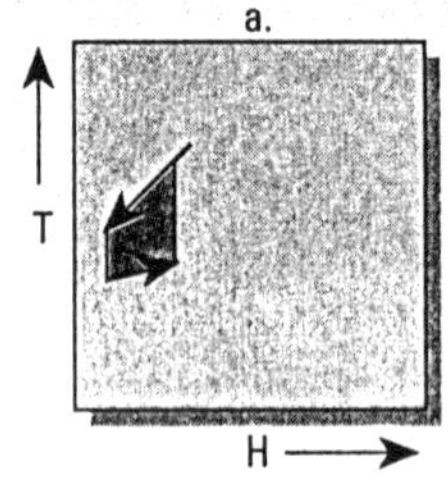

b.

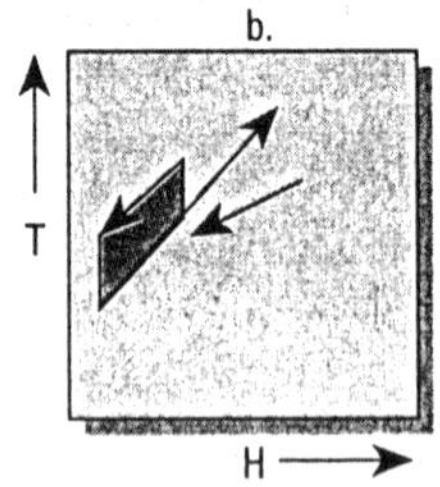

Below the pinch

H = The hot stream is ticked off

C1 = The cold stream is ticked off, and the residue of the hot stream cannot be used immediately for another match

C2 = The cold stream is ticked off, and the residue of the hot stream can be used immediately for another match to tick off the hot stream

C3 = The cold stream is ticked off, and the residue of the hot stream can be used immediately to tick off at least one other cold stream (multiple tick off match)

c.

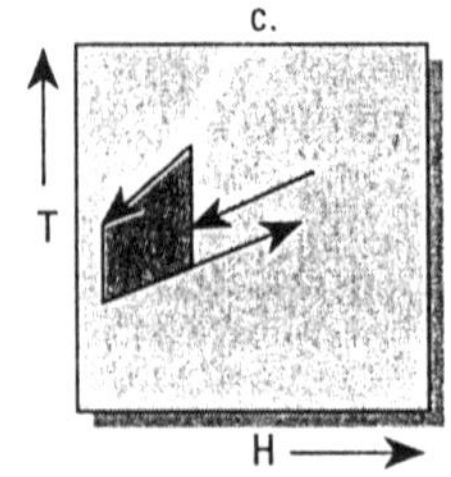

d.

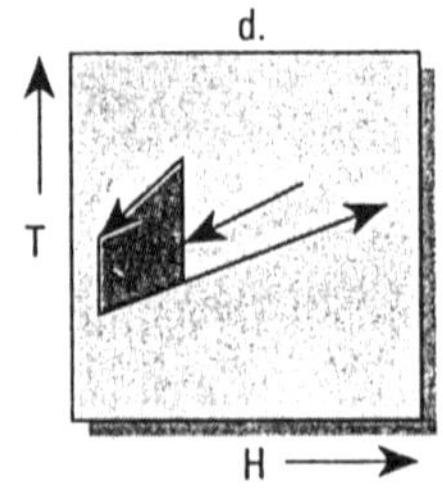

Other notations

– = The match is forbidden

x = The match is not thermodynamically possible, that is, the cold stream is warmer than the hot stream

italics = The match includes an alternative tick off point (as illustrated in the example)

NOMENCLATURE

E = annual cost, \$/yr
H = heat content, kW
h = heat-transfer coefficient, kW/m²•°C
Q = heat content, kW
Q_{Cmin} = minimum cold utility, kW
Q_{Hmin} = minimum hot utility, kW
Q_{hx} = internal heat recovery, kW
U = overall heat-transfer coefficient, kW/m²•°C
T = temperature, °C

LITERATURE CITED

1. **Carlsson, A., *et al.*,** "Retrofit of Heat Exchanger Networks," Institution of Chemical Engineers' Research Event, Manchester, U.K. (1992).
2. **Linnhoff, B., *et al.*,** "A User Guide on Process Integration for the Efficient Use of Energy," Institution of Chemical Engineers, Rugby, U.K. (1982).
3. **Linnhoff, B., and D. Vredeveld,** "Pinch Technology Has Come of Age," *Chem. Eng. Progress,* **80**(7), pp. 30–40 (July 1984).
4. **Tjoe, T. N.,** "Retrofit of Heat Exchanger Networks," PhD Thesis, Univ. of Manchester (UMIST), U.K. (1986).
5. **Polley, G. T., and M. H. Panjeh Shahi,** "Interfacing Heat Exchanger Network Synthesis and Detailed Heat Exchanger Design," *Chem. Eng. Res. Des.*, **69**, pp. 445–457 (1991).
6. **Ciric, A. R., and C. A. Floudas,** "A Mixed Integer Nonlinear Programming Model for Retrofitting Heat Exchanger Networks," *Ind. Eng. Chem. Res.*, **29**, pp. 239–251 (1990).
7. **Yee, T. F. and I. E. Grossmann,** "A Screening and Optimization Approach For the Retrofit of Heat Exchanger Networks," *Ind. Eng. Chem. Res.*, **30**, pp. 146–162 (1991).
8. **Townsend, D. W., and B. Linnhoff,** "Surface Area and Capital Cost Targets for Heat Exchanger Networks," IChemE Annual Research Meeting, Bath, U.K. (1984)

the network. If it is difficult to collect these data, one can perform a sensitivity analysis to identify which parameters are most relevant to consider in detail.

From the matrices, one can easily obtain much information about the nature of the network, such as:

(a)

Hot \ Cold	1	2	3
1	0^{H3}	x	x
2	7^{H3}	x	x
3	695^{H1}	325^{C}	2

(b)

Hot \ Cold	1	2	3
1	2^{H3}	82^{H2}	x
2	2^{H3}	94^{H2}	x
3	695^{H1}	325^{C}	2

■ *Figure 5. The initial matrix at the global ΔTmin = 10°C (a), and at the new global ΔTmin = 25°C (b).*

1. Identification of which streams are difficult to match and how different streams interact with each other. (This will be examined in the example.)

2. Identification of matches that are economically feasible to terminate before heat load tick off. (This, too, will be considered in the example.)

3. Information about the need for stream splitting and the possibility of creating a network with the minimum number of units. Figure 5a is a matrix for a small network at ΔT_{min} = 10°C. Two H3-type matches are indicated. A design with the minimum number of units must start with either match 1-1 or match 2-1, since cold stream 1 must be used immediately after the first match for another match. In addition, cold stream 1 must be used for both hot stream 1 and hot stream 2. In this case, this is not possible. Thus, the indication H3 on the two matches refers to match 3-1, that is, the hot stream that can be ticked off immediately after hot stream 1 or 2 has been ticked off is hot stream 3. In order to find a solution in this case, cold stream 1 must be split. Even if this issue is a conventional pinch problem that can be dealt with using basic pinch rules, this type of representation is an improvement in the assessment of the need to split streams.

4. The influence of the global ΔT_{min} on the complexity of the network, concerning, for example, the need for stream splitting and extra units. Let us now consider the same network as in Figure 5a, but this time at the global ΔT_{min} of 25°C, as in Figure 5b. At global ΔT_{min} = 25°C, it is possible to use cold stream 2 for matches in the pinch region. By choosing a new global ΔT_{min}, a stream split is avoided.

5. Identification of the parameters having the greatest impact on the solution (that is, heat exchanger costs, piping costs, pressure drop costs, and so on). This involves a separate study where one specific cost is considered at a time. If it is difficult to find the cost data necessary for the matches in the network, a sensitivity analysis can be made to examine the impact of the different parameters. The collection of necessary cost data can then be minimized and the important parameters concentrated on.

Finding a solution. To obtain a solution to a network when no heat transfer is allowed across the pinch point, two essential criteria must be satisfied: (1) all hot streams above the pinch must be used for internal heat exchange above the pinch, and (2) all cold streams below the pinch must be used for internal heat exchange below the pinch. Thus, the aim in finding a solution in the matrices is to find the most economic combination of matches (minimum E_{HEN}) while obeying these criteria.

Since the strategy is identical above and below the pinch, from here on the method is described only for the evolution of a design above the pinch.

Choosing matches from the matrices to find a near optimum solution involves the following steps:

1. Accept matches where the hot stream can be matched in only one way.

2. Investigate possible matches to be terminated before heat load tick off (in the example, match 3-5 above the pinch).

3. Select matches in order of cost, beginning with the most economic

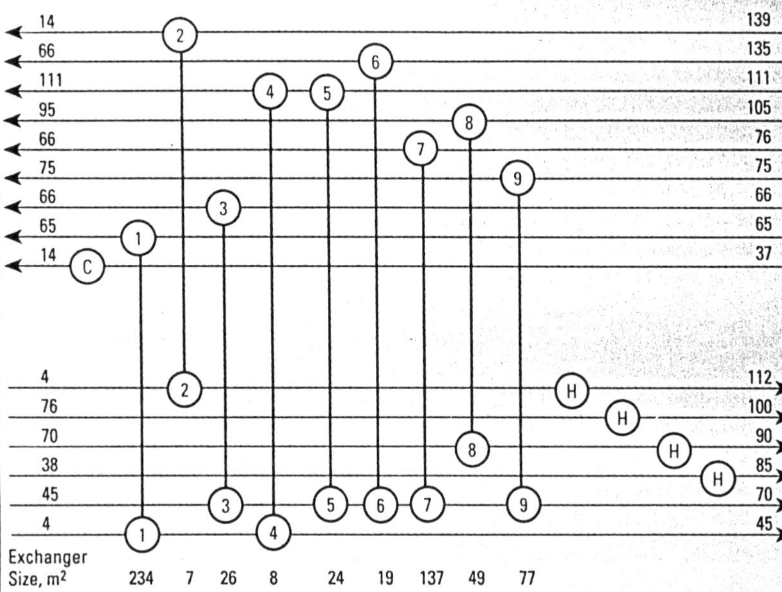

Stream	T_{supply} (°C)	T_{target} (°C)	FC_p (kW/°C)	Q (kW)	h (kW/m²•°C)
Hot 1	137	14	5.3		4.0
Hot 2	135	66	16.8		4.0
Hot 3	111	111		3,327	3.1
Hot 4	105	95	135.4		1.4
Hot 5	76	66	134.6		2.0
Hot 6	75	75		928	3.0
Hot 7	66	66		840	3.0
Hot 8	65	65		12,000	2.0
Hot 9	37	14	324.0		4.0
Cold 1	4	112	56.6		4.0
Cold 2	76	100	108.1		0.9
Cold 3	70	90	205.0		4.0
Cold 4	38	85	18.4		4.0
Cold 5	45	70	280.1		4.0
Cold 6	4	45	314.0		4.0

Hot Stream	Cold Stream	Fixed Cost ($)	Hot Stream	Cold Stream	Fixed Cost ($)	Hot Stream	Cold Stream	Fixed Cost ($)
1	1	2,000	4	1	170,000	7	1	42,000
1	2	42,000	4	2	17,000	7	2	17,000
1	3	170,000	4	3	2,000	7	3	85,000
1	4	42,000	4	4	85,000	7	4	42,000
1	5	42,000	4	5	17,000	7	5	2,000
1	6	17,000	4	6	42,000	7	6	17,000
2	1	170,000	5	1	17,000	8	1	17,000
2	2	85,000	5	2	42,000	8	2	17,000
2	3	42,000	5	3	170,000	8	3	85,000
2	4	170,000	5	4	42,000	8	4	42,000
2	5	2,000	5	5	2,000	8	5	42,000
2	6	42,000	5	6	42,000	8	6	2,000
3	1	170,000	6	1	42,000	9	1	170,000
3	2	42,000	6	2	17,000	9	2	170,000
3	3	42,000	6	3	85,000	9	3	17,000
3	4	170,000	6	4	42,000	9	4	170,000
3	5	2,000	6	5	2,000	9	5	42,000
3	6	2,000	6	6	17,000	9	6	17,000

Note: For all matches, the area cost is $350/m².

■ *Figure 6. The original network for the example.*

match. If two matches have approximately the same economic value, the order of preference regarding type of match is: H3, H2, H1, and C (below the pinch the order of preference is C3, C2, C1, and H). Do not choose the most economic match if it is evident that the design cannot be pursued any further if the match is selected (this will be illustrated in the example by matrix (b) below the pinch).

4. When a combination of matches has been found that corresponds to the global ΔT_{min} studied, the user is aware of the costs contributed by the individual matches. Thus, the expensive matches are known and these matches are often the ones selected late in the selection process when no alternative matches remain for the hot streams involved. If a certain match or matches represent a large part of the total cost, a new combination should be investigated to avoid these matches or to select them earlier in the process. When a new combination of matches is considered, the hot streams involved in the expensive matches are given priority to be chosen earlier than the cost of matching them suggests, usually after the matches in step 2 have been selected. Consequently, an iterative procedure is sometimes necessary, but usually only a few runs are required to find an "optimal" solution.

5. If all existing heat exchangers have not been used, deduct the cost associated with these heat exchangers from the total cost if it is possible to use these units as parts of the resulting network.

	1	2	3	4	5
1	2^{H3}	x	x	44^{H2}	44^{H3}
2	176^{H3}	x	x	176^{C}	2^{H3}
3	201^{H1}	106^{C}	63^{H2}	173^{C}	5^{H3}
4	182^{H2}	66^{H2}	0^{H2}	93^{C}	26^{H3}
5	54^{H3}	x	x	47	2^{H3}
6	53^{H3}	x	200^{H3}	47	2^{H3}

Matrix (a)

	1	2	3	4	5
1	2^{H3}	x	x	44^{H2}	44^{H3}
2	176^{H3}	x	x	176^{C}	0^{H3}
3	173^{H3}	57^{H3}	47^{H3}	173^{C}	5^{H3}
4	182^{H3}	66^{H3}	0^{H3}	93^{C}	28^{H3}
5	54^{H3}	x	x	47	2^{H3}
6	53^{H3}	x	200^{H3}	47	2^{H3}

Matrix (b)

	1	2	3	4	5
3	174^{H1}	57^{H1}	48^{H1}	173^{C}	3^{C}

Matrix (c)

	1	2	3	4
3	172^{H1}	52^{H1}	46^{H1}	173^{H1}

Matrix (d)

■ *Figure 7. The matrices above the pinch for the example.*

Example

The problem. Figure 6 represents a typical HEN in the pulp and paper industry. The existing hot and cold utility demands are 11.9 MW and 7.5 MW, respectively, corresponding to a global ΔT_{min} of 27°C.

A study is made at a new global ΔT_{min} = 18°C. The pinch is at 66°C for the hot streams and at 48°C for the cold streams. The corresponding hot and cold utility requirements are 8.6 MW and 4.2 MW, respectively.

The investment cost for each match is split into one area-dependent part and one fixed part, which consists of investment costs for piping and auxiliary equipment. For simplicity, maintenance and pressure drop costs are not included. The values of the cost equations have been chosen to emphasize the impact of variations in cost between different matches and should not be taken as exact values. Since only investment costs are considered, no capital recovery factors are included in the cost calculation (however, the results are the same if capital recovery factors are included).

The emphasis of the example is on internal heat exchangers (that is, heaters and coolers have not been included). It is possible, however, to include utility streams in the matrices.

Above the pinch. Six hot and five cold streams are located above the pinch. Three hot streams and three cold streams reach into the pinch region. Six existing heat exchangers are situated above the pinch point: heat exchangers 2, 5, 6, 7, 8 and 9. In order to reach a solution above the pinch, all hot streams must be used for internal heat exchange. One match, 3-5, has an alternative tick off point, so a study taking alternative tick offs into consideration will be made.

The initial matrix is presented in Figure 7a. At the global ΔT_{min} = 18°C, none of the heat exchangers placed above the pinch on cold stream 5 transfers heat across the pinch. Therefore, it should be investigated if the existing arrangement can be kept for the heat exchangers placed on this stream.

Starting from the pinch point, the first match on this stream is match 3-5. This match has an alternative tick off point, shown in Figure 8, which will be investigated. The alternative tick

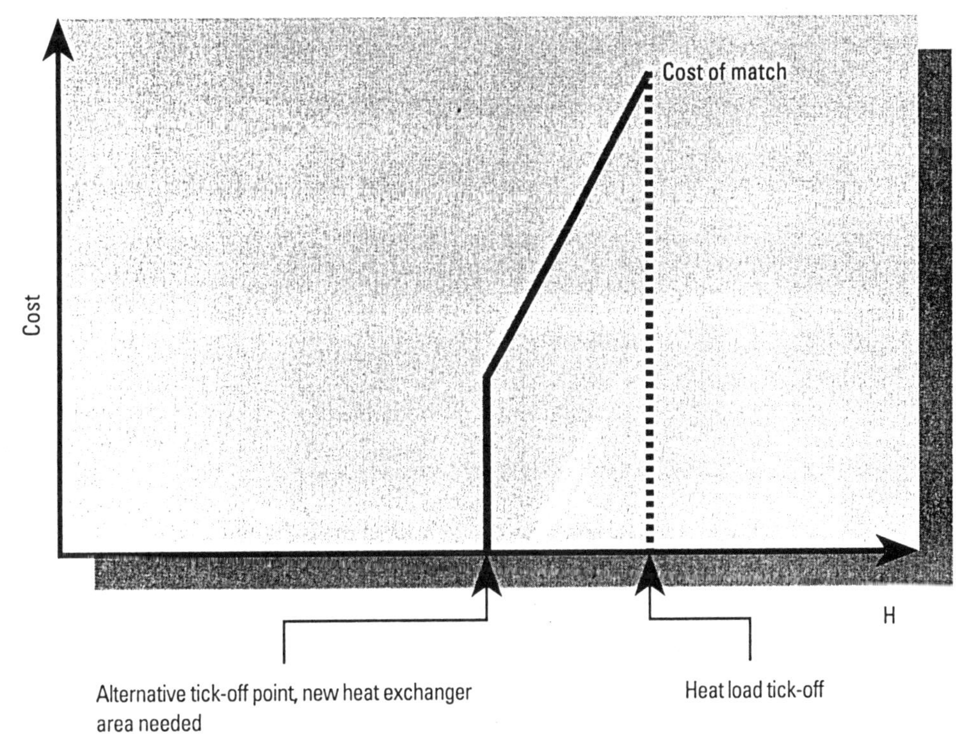

■ *Figure 8. Match 3-5 has an alternative tick off point.*

	1	4	5	6
1	2	44	43	18
7	49^{H}	44^{C2}	0^{C1}	25^{H}
8	36^{C3}	44^{C3}	54^{C3}	9^{H}
9	X	X	X	X

Matrix (a)

	1	4	6
1	2	44	18
8	36^{C3}	44^{C3}	9^{H}
9	X	X	X

Matrix (b)

	4	6
1	44	18
8	44^{C2}	2^{H}
9	X	X

Matrix (c)

■ *Figure 9. The matrices below the pinch for the example.*

off point indicates that there is an economic possibility to terminate the match before one of the streams involved in the match is ticked off by means of the heat load tick off. In this case, the alternative tick off point exists because the existing area in the match is smaller than the area corresponding to a heat load tick off. Hence, if a match is placed between the two streams, it should be either a heat load tick off match or a match that makes use of all the existing area. Figure 8 shows that no cost is involved with this match until the alternative tick off point is reached. The alternative match 3-5 is selected, that is, the existing area is kept, but the heat exchanger is not extended to tick off one of the streams. Thus, no stream is ticked off when this match is selected.

The resulting matrix when match 3-5 has been selected is Figure 7b. The next match on cold stream 5 is match 2-5. This match is selected and hot stream 2 is ticked off.

The resulting matrix after matches 3-5, 2-5, 5-5, 6-5, 4-3 and 1-1 have been selected is Figure 7c. These matches were selected because of the low cost they contribute to the total cost. Match 3-3 is the cheapest match to tick off hot stream 3. But is it possible to reduce the cost of any of the matches that will tick off hot stream 3 by including match 3-5, even though this will lead to a loop in the network? Figure 7d shows what happens if match 3-5 is selected. From this matrix it is clear that there is an improvement in match 3-3, but this improvement is less than the extra cost of including match 3-5 in the network. Thus, only match 3-3 should be used to tick off hot stream 3.

Below the pinch. Four hot and four cold streams are located below the pinch. Three cold streams and three hot streams reach into the pinch point region. Two existing heat exchangers, heat exchanger 1 and heat exchanger 3, are located

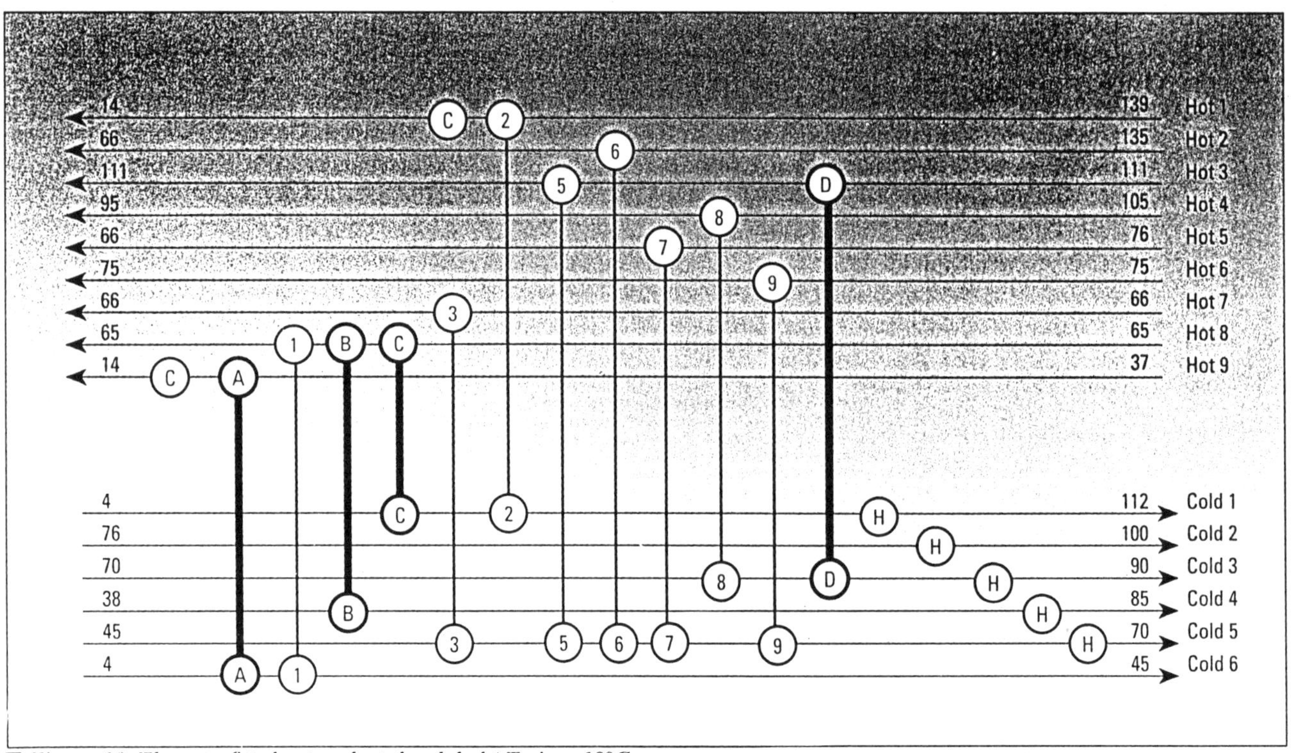

■ *Figure 10. The retrofitted network at the global ΔTmin = 18°C.*

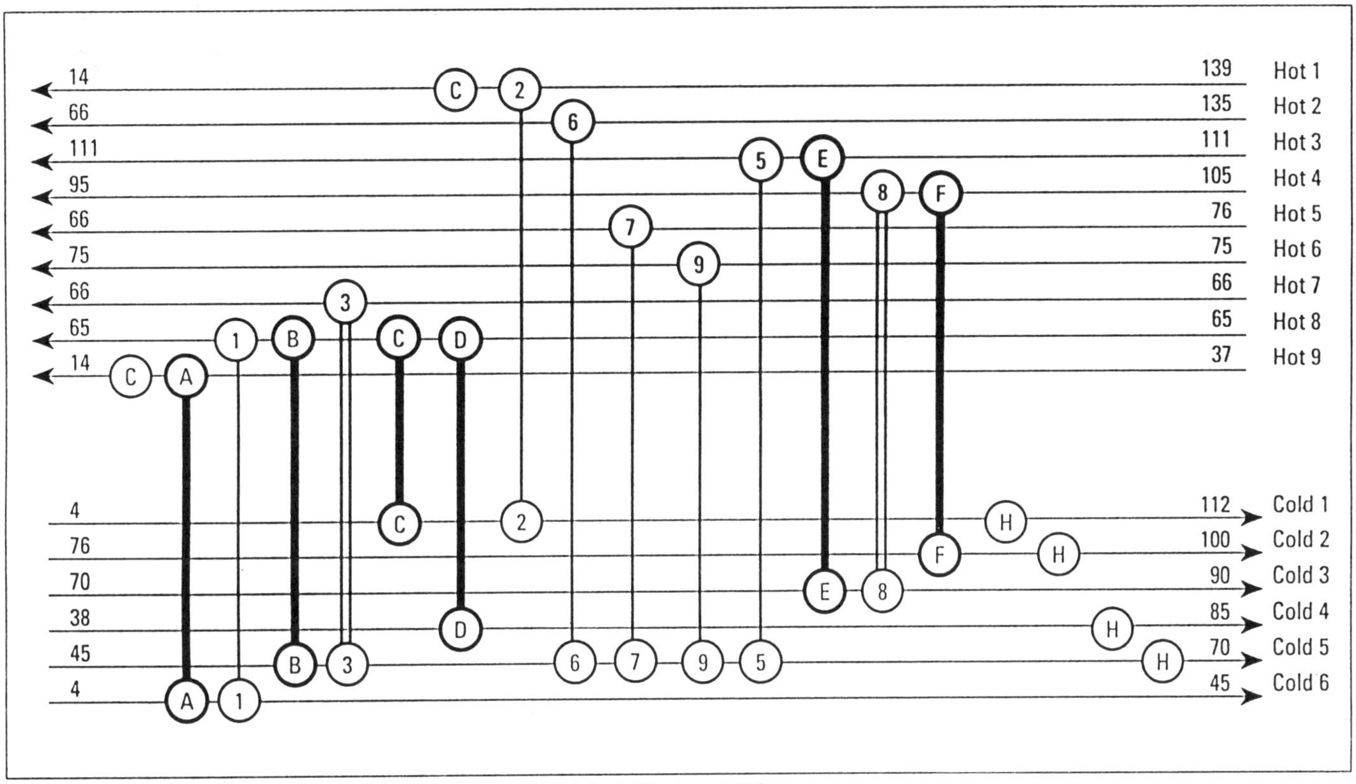

Figure 11. The alternative retrofitted network at the global $\Delta Tmin$ = 10°C.

below the pinch. Below the pinch all cold streams must be used for internal heat exchange.

The initial matrix is depicted in Figure 9a. Match 7-5, which is identical to the existing match, is selected, and hot stream 7 and cold stream 5 are ticked off.

The resulting matrix after match 7-5 has been selected is Figure 9b. It is necessary to match all remaining cold streams to hot stream 8, since this is the only hot stream that reaches into the pinch point and the only one that will result in tick off matches. It is not possible to start with match 8-6, since this will tick off hot stream 8 leaving no alternatives for cold stream 1 and 4. Instead, match 8-1 or match 8-4 must be selected. Match 8-1 is selected first, and it is found that the cost of match 8-4 does not worsen due to this choice. Match 8-1 is accepted and cold stream 1 is ticked off, as shown in Figure 9c.

Match 8-4 is then selected. This is the only match that can be selected and at the same time make it possible to match another cold stream with hot stream 8. Cold stream 4 is ticked off.

A. CARLSSON is a doctoral student in the dept. of heat and power technology at Chalmers Univ. of Technology, Göteborg, Sweden (011-46-31-772-1000; Fax: 011-46-31-811-928). She holds an MS in chemical engineering from Chalmers.

P-Å FRANCK is managing director at CIT — Energy Technology Analysis, where he is active in process integration projects. He holds master's and doctoral degrees in chemical engineering from Chalmers.

T. BERNTSSON is professor at the dept. of heat and power technology at Chalmers Univ. His research areas are process integration and industrial energy technologies, and he has been responsible for several national demonstration projects on process integration in Sweden. He received his undergraduate and doctoral degrees in chemical engineering from Chalmers.

Finally, match 8-6 and match 9-6 are selected to complete the retrofit HEN.

Results. The investment cost for the retrofitted network at the global ΔT_{min} = 18°C, Figure 10, is \$176,000. Approximately 32% of the total cost represents new heat exchanger area. Assuming a steam cost of \$11/MWh, no cost for cooling water, and an operating schedule of 8,000 h/yr, the payback period for the network at the global ΔT_{min} = 18°C is approximately seven months.

An alternative scenario, at the global ΔT_{min} = 10°C, was analyzed using the same cost equations for heat exchangers and fuel. This network is shown in Figure 11. It has an investment cost of \$385,000, of which 52% is attributable to the cost for new heat exchanger area. This design has a payback period of nine months.

It is interesting to note the small difference in payback periods between the two global ΔT_{min}s. Thus, it is necessary to study a network at different global ΔT_{min}s to get an understanding of the impact of choosing a specific global ΔT_{min}. If a capital recovery factor of 0.30 is included in the calculations, the annual saving at ΔT_{min} = 18°C is \$237,600 /yr, whereas at ΔT_{min} = 10°C it is \$421,300/yr. CEP

Completely Analyze Energy-Integrated Processes

With the inclusion of the Extended Grand Composite Curve, pinch technology attains a new level of usefulness.

Peter Glavic and Zorka Novak,
University of Maribor

Pinch technology has established itself as a key tool to analyze and synthesize or retrofit energy-important parts of processes, for instance, heat exchanger networks (HENs) *(1)*, heat engines *(2)*, separators *(3)*, reactors *(4)*, and heat pumps *(5)*. For HENs, a grand composite curve (GCC) simultaneously represents heat sources like steam, and heat sinks like cooling water *(6)*. The GCC also has been used for other pinch analyses, but all the energy-active process units have not been accounted for on a single GCC. There have been two reasons for this:

1. Units that are inappropriately placed against the pinch temperature cancel other streams in the GCC, thus giving a false picture of the energy cascading; and

2. Many units bypass the HEN and thus do not participate in the usual GCC — that is, energy is cascaded from energy sources directly to energy sinks, for instance, from a reactor to a boiler.

We can overcome these two limitations by first eliminating all inappropriately placed process units from the GCC and then placing them in front of the GCC as additional composite curves together with adequate process utilities *(7)*. We call this presentation the Extended Grand Composite Curve (EGCC). The EGCC thus is composed of a GCC that is extended with utility composite curves (UCCs). While the GCC represents indirect heat exchange (energy flows from hot utilities through the HEN to cold utilities), UCCs graphically reproduce direct heat exchange between energy sources and energy sinks. In such a manner, a simplified exergy (energy availability) analysis of the whole process is possible, as we will describe. The EGCC has the same advantages over traditional composite curves (CCs) that the GCC has.

The extended pinch design method will be illustrated by the design of an ammonia plant *(8)*, whose simplified process scheme is shown in Figure 1.

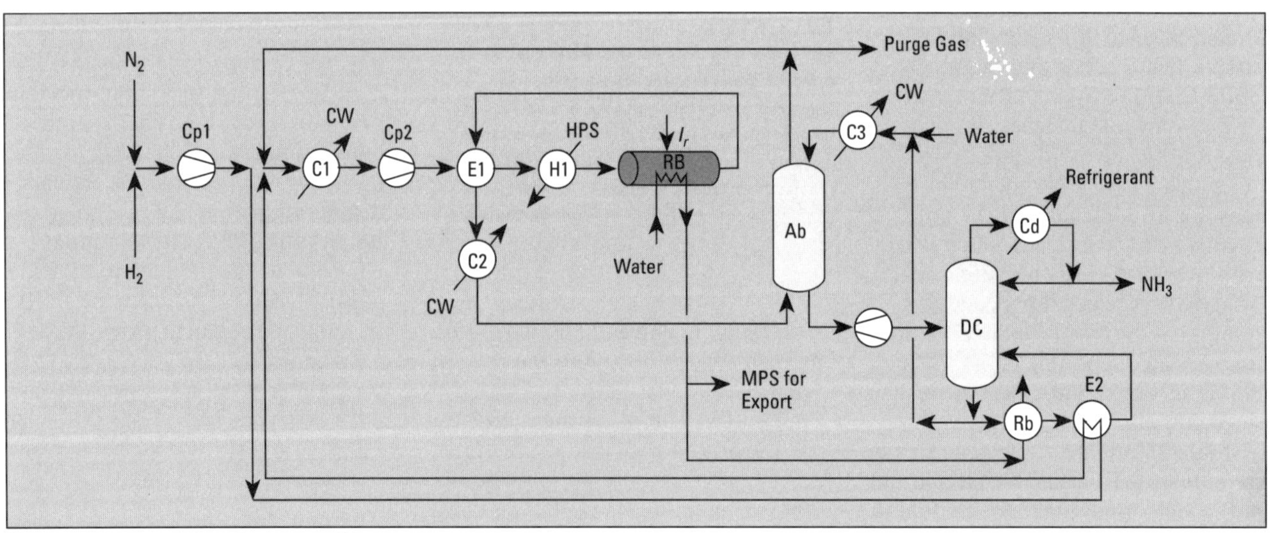

■ *Figure 1. Simplified flowsheet of base process.*

Ammonia is produced by a reversible, exothermic reaction:

$$\div H_2 + \tfrac{1}{2} N_2 \Leftrightarrow NH_3 \quad (1)$$

with a reaction enthalpy ΔH_r of -46.2 kJ/mol.

The reaction mixture is compressed to 70 bar (266°C) cooled in heat exchanger E2 and water cooler C1, and then compressed again to 210 bar (335°C). It is then heated first by reactor effluent stream to 408°C in heat exchanger E1, and finally with high-pressure steam (HPS) in heater H1 to 430°C. The exothermic reaction is carried out isothermally, and the heat evolved is absorbed by a built-in boiler (RB). In the boiler, cold water is heated up to 187°C and evaporated at 11.4 bar. The reactor effluent stream is cooled by its inlet stream in E1 and by cooling water in C2. The NH_3 containing product is contacted with water in absorber Ab; gaseous reactants are recovered and fed back to the compressor, while unwanted inerts are exhausted with purge gas. The product is separated from the solvent in distillation column DC using propane as a refrigerant. The process was simulated rigorously with a commercial simulator Design II using Peng-Robinson thermodynamics. Important HEN streams are reproduced in Table 1, together with their supply and target temperatures, their heat capacity flow rates, and their individual temperature contributions.

Table 1. Heat exchanger network, energor and utility streams of an ammonia plant.

Process unit*	T_s,°C	T_t °C	*CF*, kW/K	ΔT_c, K	*I*, kW
HEN hot streams					
E1	430	131	229	5.0	68,500
E2	266	158	55	5.0	6,000
C1	158	129	61	20.0	1,800
C2	131	42	234	8.0	20,900
C3	147	40	25	6.0	2,700
Cd	2	-33	679	4.0	23,800
HEN cold streams					
E1	35	408	939	5.0	68,500
E2	147	147	6,000	5.0	6,000
H1	408	430	239	5.0	5,300
Rb	147	147	17,100	5.0	17,100
Utilities					
HPS	470	470	5,300	5.0	5,300
R	430	430	42,800	20.0	42,800
MPS	187	187	17,100	5.0	17,100
PH	27	187	68	3.0	10,900
B	187	187	31,800	3.0	31,800
CW	29	49	1,757	3.0	34,300
Refrig.	-41	-41	23,700	3.0	23,700

*Abbreviations are defined in the nomenclature box.

The pinch design approach

In the pinch method, all HEN cold streams are summed up in a cold composite curve (CC_c), which is represented in a temperature/enthalpy flow rate (*T/I*) diagram; we are using enthalpy flow rates instead of the heat flow rates because they can also represent energy flow of material in reactors, open-cycle heat pumps, absorbers, and so on. The same procedure is used for hot process streams. Reboiler and condenser streams often are included in the cold and hot CCs, respectively. Both CCs are separated by a minimum temperature difference (ΔT_{min}). Figure 2a shows CCs for our base process; the gap between the curves at the top represents the hot utility wanted (22.4 MW) and the gap at the bottom indicates the cold utility needed (58.0 MW). At the pinch temperature (T_p = 152°C), the vertical difference between the two CCs, ΔT_{min} = 10 K. When the cold CC is moved to the right, ΔT_{min} increases, and both utilities increase, too. Moving cold CC to the left cuts the use of utilities.

While the utilities are well quantified with CCs, their temperature levels are not so clearly seen. We need 22.4 MW of hot utility, for example, but we do not know whether steam or flue gas is used as a heat source and what the temperature of the utility streams is. Therefore, we compose a grand composite curve. It can be obtained when the hot CC is lowered by contribution temperatures of all the hot streams (5 K at the pinch point in our case) and the hot CC is increased by them (5 K again at T_p), yielding interval temperatures as shown in Figure 2b.

In our case study, starting stream temperature contributions in Table 1 were determined according to well-known rules: 20 K for gaseous streams, 10 K for liquid streams, 5 K for boiling and condensing streams, and direct contacting streams having ΔT_c = 0 K. Contributions were lowered in those cases when they overpassed their heat source or heat sink temperature, for instance, from 10 to 6 K in water cooler C3. These changes automatically provide the necessary driving forces for heat flow. Utility levels can now be clearly seen — we have two levels of steam (HPS, and medium-pressure steam, MPS), as well as two different coolants (cooling water and refrigerant) in our case. We can also predict their mini mum enthalpy flow rates as shown in the Figure 2b.

Even with the GCC, however, we cannot evaluate the most energy-intensive process unit — the reactor with the built-in boiler.

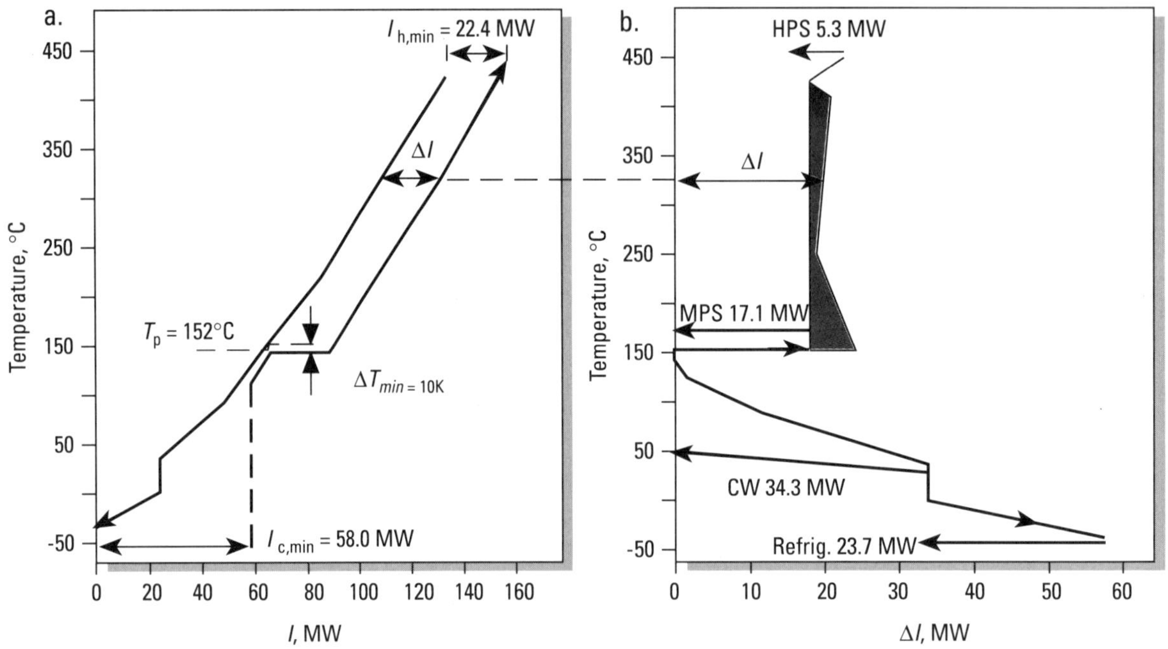

■ ***Figure 2. Composite curves (a) and grand composite curve (b) of the base process.***

Energors

Heat exchangers transmit energy from one stream to another. Utilities supply heat deficit to a process or absorb heat surplus from it. In contrast, a process unit's primary task is to fulfill some process requirement (such as compression, reaction, or separation) that is normally connected with energy changes. As an example, energy that is evolved from an exothermal reaction can be used as a heat source just like steam energy. There is actually no difference between the chemical utility (reaction) and the thermal one (steam) as far as energy serviceability is concerned. We, therefore, call all sources of energy *donors* and all energy sinks *acceptors*. A common name for donors and acceptors is *energors*. Besides steam, flue gas, and heat engine exhaust, typical energy donors are: exothermal reactors, furnaces, and compressors. The most frequent process energy acceptors besides coolants and refrigerants are boilers, evaporators, air dryers, endothermal reactors, and expansion engines.

In a conventional pinch analysis, energor streams are not directly taken into account although their effects may be factored into downstream heat exchangers, heaters, or coolers. But all the energors are vital parts of process energetics, and thus they should be included in the thermodynamic analysis, that is, in our EGCC. For that reason, Table 1 lists the streams passing the reactor/boiler (reactor R, preheater PH and boiler B) in addition to all the HEN and utility streams of our example process.

In a conventional pinch analysis, energor streams are not directly taken into account.

Thermal separators (distillation columns, multistage evaporators, dryers, and so forth) are often shown in a form of EGCC (see the distillation column DC in Figure 3). They should be separated from the GCC and be presented outside it whenever we want to study multiple separator stacking by changing pressures in the separators. This is not necessary, however, when their structure is fixed — in this case, the streams passing separators are included in the GCC to study their integration with the HEN, as shown in Table 1 and in Figure 3. Note that by excluding the distillation column from the process, the pinch temperature of the remaining HEN moved from 152°C in Figure 2 to 428°C in Figure 3! We shall discuss this phenomenon later.

Energors always can be shown as a part of the process EGCC — their streams are included in the EGCC: as a part of the HEN, as a utility above or below the GCC, or as UCCs in front of the GCC.

Energy mixers

Now, let us consider energy mixers. They use temperature differences and heat flows to enable executing separation — and include stream mixers, expansion valves, and direct heating, cooling, or quenching towers. Their energy can be integrated only outside them, provided that sufficient temperature driving forces are available. So, energy mixers usually are not shown as a part of the EGCC; in our example, energy mixers have not been considered.

Distillation columns and absorbers, although energy mixers, are sometimes included in the analysis. Absorbers often are connected with chemical reactions that supply enthalpy to or remove it from process

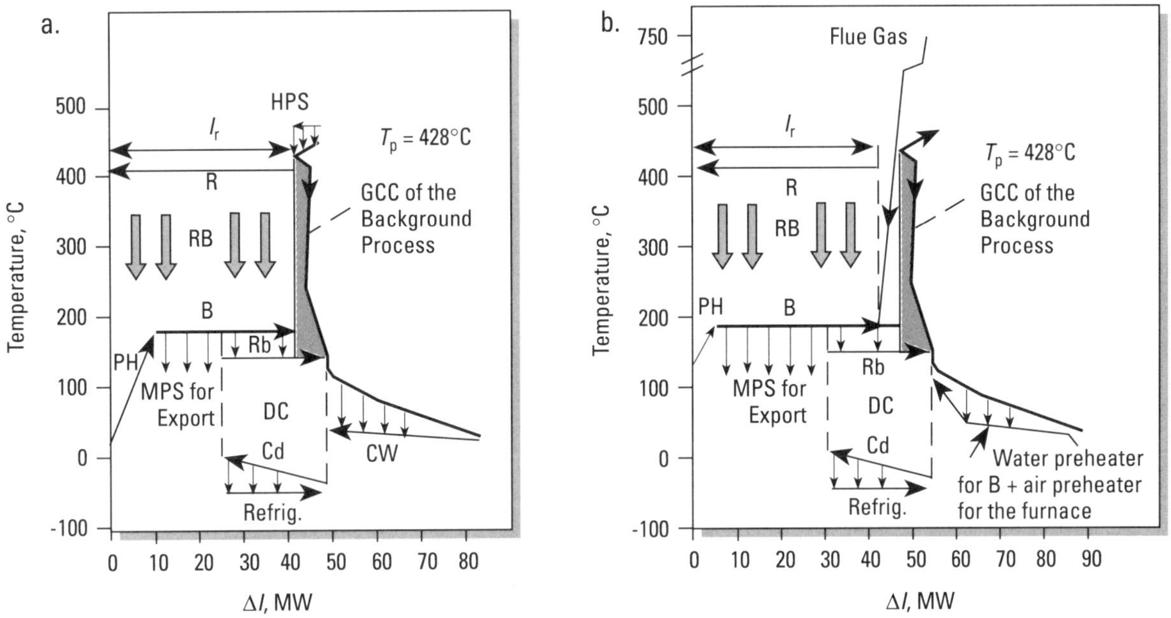

■ ***Figure 3. Extended grand composite curve of: nonintegrated process (a), and energy integrated process (b).***

streams; an example was shown with sulfuric acid optimization *(9)*. In stagewise separators, each interstage of a distillation column, for example, can be provided with a side-reboiler or a side-condenser *(10)*. Their integration potentials can be studied by including the separators into the analysis *(11)*. In our study, only condensers and reboilers of distillation columns were included in the analysis; the column itself was excluded.

The Extended Grand Composite Curve

Figure 3a presents an EGCC of the example process *(12)*. The reaction enthalpy flow rate (I_r) of the reactor (R) is used to preheat (PH) and evaporate (B) boiler feedwater. Compressors are left out of the EGCC, but we could place them in the UCCs with steam or electrical energy as their hot utility. Some of the medium-pressure steam (MPS) produced by the boiler is used for reboiling the distillation column and the rest is exported from the process. An appropriately placed heat source (HPS) is situated above the GCC to drive the process by cascading heat flow through the HEN to process sinks. When energors are inappropriately placed (for instance, our boiler as an acceptor is located above the pinch temperature, 152°C

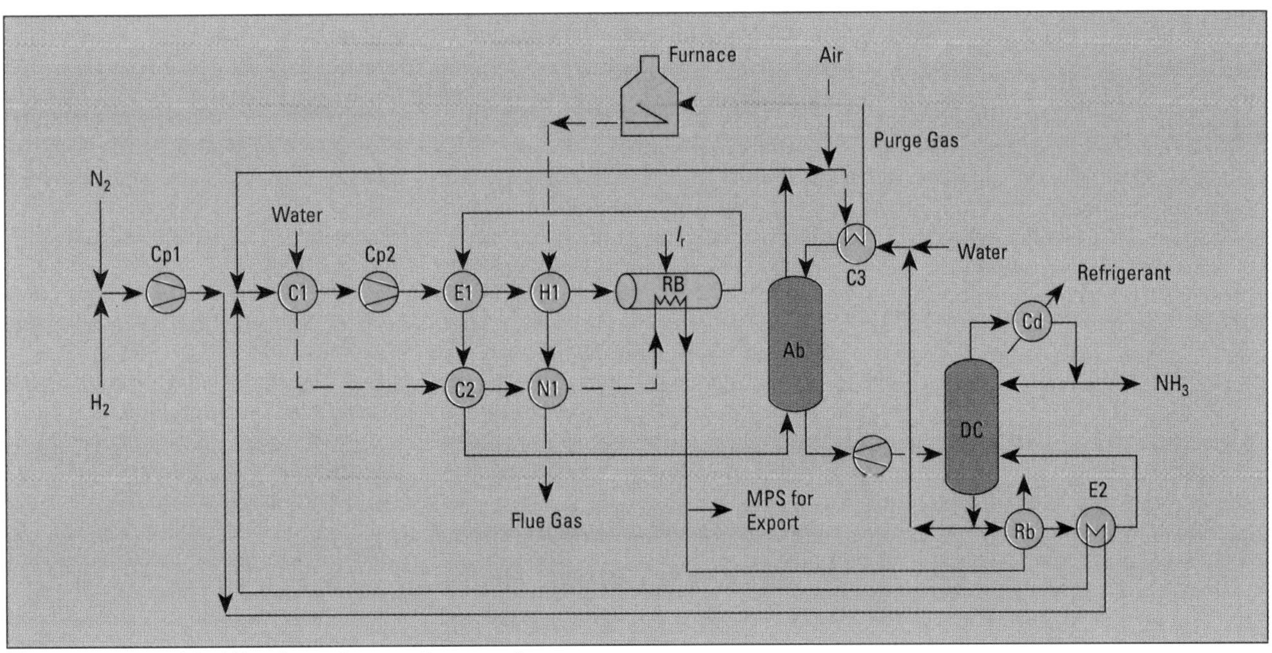

■ ***Figure 4. Simplified flowsheet of energy-integrated process.***

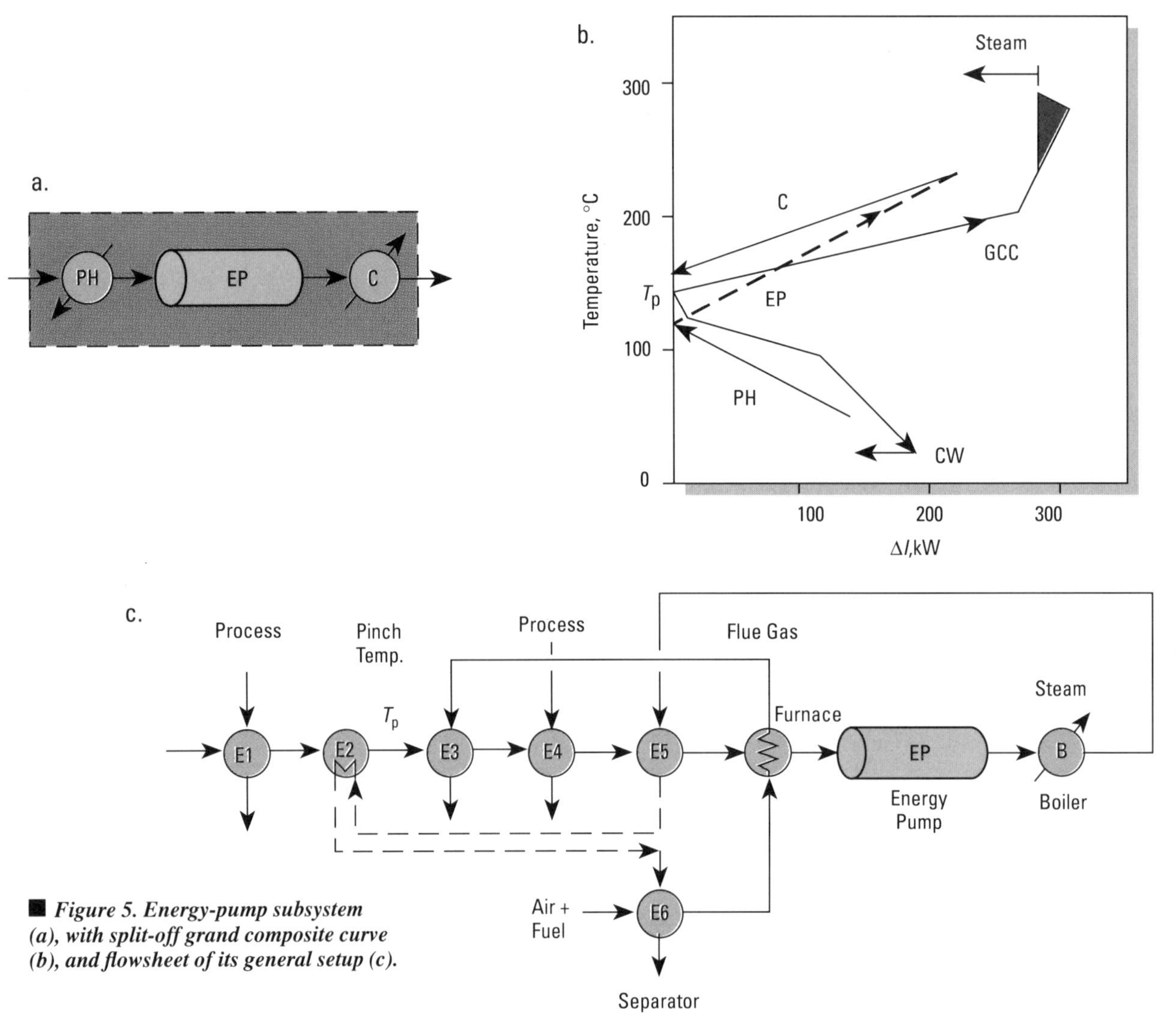

■ ***Figure 5. Energy-pump subsystem (a), with split-off grand composite curve (b), and flowsheet of its general setup (c).***

or directly transmit enthalpy flow from hot utility to cold utility outside the HEN (for example, from our reactor to the boiler), they are shown as pairs of UCCs outside the GCC.

The distillation column that is crossing the pinch temperature (T_p = 152°C) can be heated by the reactor outlet stream and by a process stream that may be taken out of the GCC and put into a UCC to visualize exergy losses *(12)*. The EGCC offers great flexibility in the use of different streams, but their interpretation needs careful consideration to make sure that they reflect the real world.

While a large temperature difference is used for preheating boiler feedwater, the GCC streams deliver heat flow into cooling water. This clearly is not favorable as we could preheat feed water with the HEN below the pinch temperature. Also, purge gas could be burned to heat the HEN above the pinch temperature instead of using HPS, and also to heat the reboiler instead of MPS. This is shown in Figure 3b: furnace air, purge gas, and boiler feed water now are preheated under the pinch by the HEN, thereby reducing cold water usage and more than doubling the boilup of MPS for export at the same time. The complete energy situation is visible clearly. Improvements easily can be realized in a new process flowsheet (Figure 4): one additional heat exchanger (N1) is needed in addition to the furnace.

Lift energy up the heat cascade, through the pinch.

For the case study, we have calculated the following potential annual savings:

- burning of purge gas instead of using HPS, \$2.58 million/yr;
- additional steam produced by the reactor, \$0.79 million/yr;
- preheating boiler feed water with the HEN, \$0.16 million/yr; and
- preheating furnace air and purge gas with the HEN, \$0.04 million/yr.

Altogether, the net annual effect of the improvement is a \$3,570,000 savings.

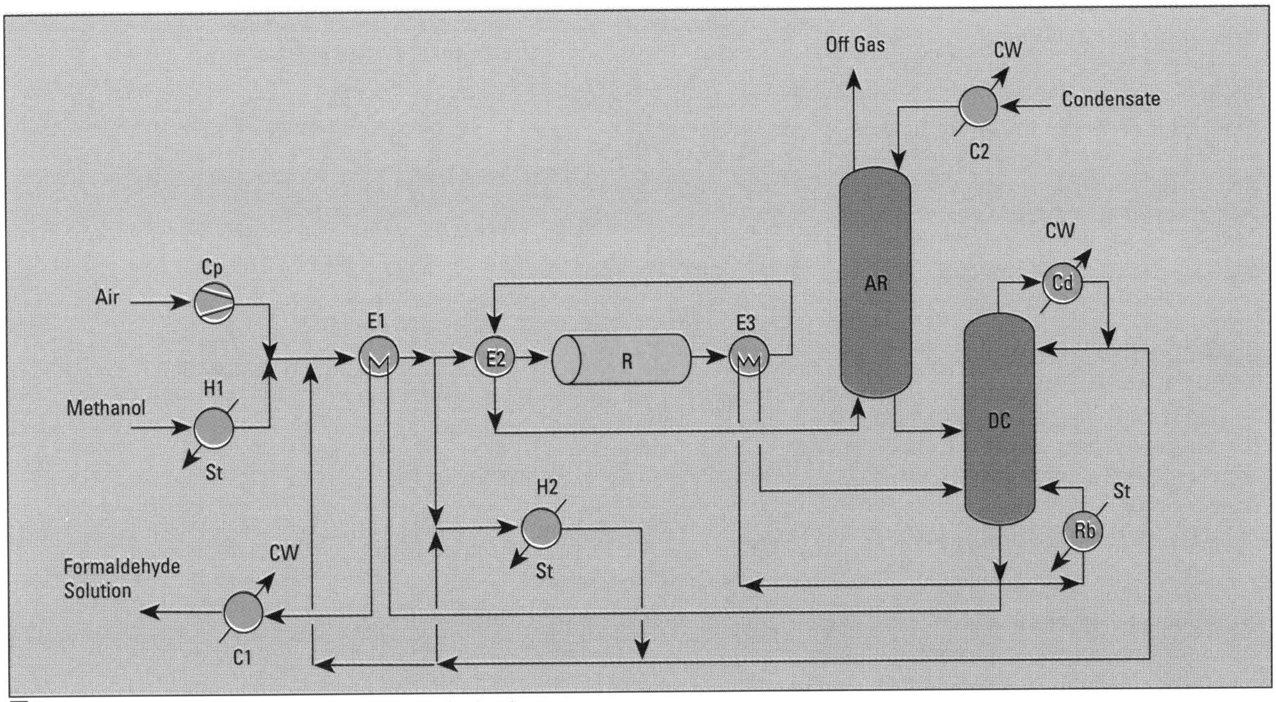

■ *Figure 6. Flowsheet of a modern formaldehyde plant.*

Energy pumps

A heat cascade devalues energy so as to enable a process to go on. The pinch point is a bottleneck in the heat cascade like a valve is a bottleneck in emptying a reservoir *(13)*. Normally we try to reduce energy flow downwards through the pinch — but a much more profitable action is to lift energy up the heat cascade, through the pinch. This can be accomplished by energy pumps that consume surplus heat from the process under the pinch temperature, lift it across the bottleneck by enriching it with chemical, electrical, or thermal energy, and deliver the consumed heat together with the enriched energy afterwards to the process above the pinch, as shown in Figure 5 *(4)*. Exothermal reactors and furnaces, motor or steam driven compressors, and mechanical heat pumps can be used for this purpose. To check and optimize the energy pump's placement, its preheaters and aftercoolers (that is, the energy-pump subsystem in Figure 5a) should be split off the GCC to be superimposed onto it afterwards (Figure 5b). Splitting often uncovers a different pinch temperature of the background process and only then can appropriate integration of the energy pump really be carried out.

We have seen that several pinch temperatures exist in each process. If energors are not included into the GCC, a traditional HEN pinch temperature is obtained but, in energy-rich processes, often a threshold problem results. Besides the normal HEN pinch, a process pinch temperature can be obtained when energors are included in the HEN analysis *(4)*. Streams passing the energors are included in the analysis as normal streams — examples are the reactor (R) and the boiler (PH, B) streams in Figure 3 passing our reactor/boiler. Utility pinches have been described already *(6)*. Each energy pump can have a different temperature of the background process pinch and it can be obtained if the energy-pump subsystem is split off the GCC *(4,5)*. The same is true when taking out our distillation column — the process pinch temperature, which was originally at 152°C (Figure 2), changes to the background one of 428°C (Figure 3). Whenever we introduce or eliminate a stream that crosses the pinch, the pinch temperature will change. Pinch temperatures can be moved not only by omitting or including streams that cross the pinch temperature but also by changing the ΔT contributions of the streams, that is, the enthalpy flow rates through the pinch! The process pinch is an overall one, while all the others (including the HEN pinch) are of local importance.

An efficient energy-pump subsystem should absorb heat from the process with its preheaters below the background process pinch temperature (E1 and E2 in Figure 5c). In that case, the most efficient energy-pump inlet temperature is at or above the pinch temperature. If the pump has to work far above the pinch temperature, as reactors and furnaces often do, the stream to be lifted can be heated further above the pinch temperature, for example, by: flue gases or exhaust streams (E3), process streams (E4), and the energy-pump outlet stream (E5). Final heating of reactants above the E5 outlet temperature can be carried out by a furnace.

The energy-pump outlet stream should be cooled down, for example, in turn by: producing steam in a boiler (B), or by heating up the inlet stream of the energy pump itself (E5) and other process streams. It should leave the process just above the pinch temperature except when it is used to heat inlet streams of an energy pump below the pinch temperature as shown by a dashed line in E2 of Figure 5c. Also, the outlet stream can be used for heating furnace air and fuel, or boiler feed water (E6). Some of the above-mentioned energy-exchange units can be left out, as it is often the case with heat pumps. Their appropriate deployment depends on local temperature potentials, and heat-capacity flow rates.

Splitting the Grand Composite Curve

Our example process is not suitable for illustrating energy-pump synthesis — its reactor is appropriately placed, far above the pinch, and so are the compressors, with inlets just under the pinch temperature. Therefore, we will look at another example to show an energy-pump subsystem with its preheaters and aftercoolers, and to present consecutive splitting of energy pumps as well as a separator from the GCC. We shall demonstrate our approach on an existing formaldehyde plant, which is shown in Figure 6. Important HEN, energor, and utility streams appear in Table 2.

NOMENCLATURE

Abbreviations in figures and tables

Ab = absorber
AR = absorber/reactor
B = boiler
C = cooler
CC = composite curve
Cd = condenser
Cp = compressor
CW = cooling water
DC = distillation column
DSD = direct search discretized algorithm
E = heat exchanger
EGCC = extended grand composite curve
EP = energy pump
GCC = grand composite curve
H = heater
HEN = heat exchanger network
HPS = high-pressure steam
MINLP = mixed integer nonlinear programming
MPS = medium-pressure steam
N = new heat exchanger
PH = preheater
R = reactor
RB = reactor/boiler
Rb = reboiler
Refrig. = refrigerant
SGCC = split-off grand composite curve
St = steam
UCC = utility composite curve

Notation

CF = heat capacity flow rate, W/K
I = enthalpy flow rate, W
$I_{C,min}$ = minimum cooling utility requirement, W
$I_{h,min}$ = minimum heating utility requirement, W
I_r = reaction enthalpy flow rate, W
T = temperature,
T_p = pinch temperature,
T_s = supply temperature,
T_t = target temperature,
ΔH_r = reaction enthalpy, J/mol
ΔI = enthalpy flow rate difference, W
ΔI_{cp} = compressor enthalpy flow rate difference, W
ΔT_c = stream temperature contributions, K
ΔT_{min} = exchanger minimum approach temperature, K
ΔT_{opt} = exchanger optimum approach temperature, K

Table 2. Heat exchanger network, energor and utility streams of a formaldehyde plant.

Process unit*	T_s,°C	T_t,°C	CF, kW/K	ΔT_c, K	I, kW
HEN hot streams					
E1	94	67	10.0	10.0	270
E2	150	115	4.6	15.0	160
E3	580	150	4.6	20.0	1,980
C1	67	41	6.3	10.0	160
C2	40	31	3.7	2.0	30
Cd	52	40	311	4.0	3,730
HEN cold streams					
E1	50	52	135	5.0	270
E2	52	85	4.9	15.0	160
E3	94	94	1,980	4.0	1,980
H1	15	54	1.9	10.0	70
H2	53	55	72	10.0	140
Rb	94	94	2,170	4.0	2,170
Reactor and compressor cold streams					
R	85	580	4.9		2,420
Cp	6	45	1.2		50
Utilities					
Steam	158	137	1.2	0.5	25
Steam	137	137	2,090	0.5	2,090
Steam	137	40	4.1	0.5	400
CW	28	40	452	0.5	5,420

*Abbreviations are defined in the nomenclature box.

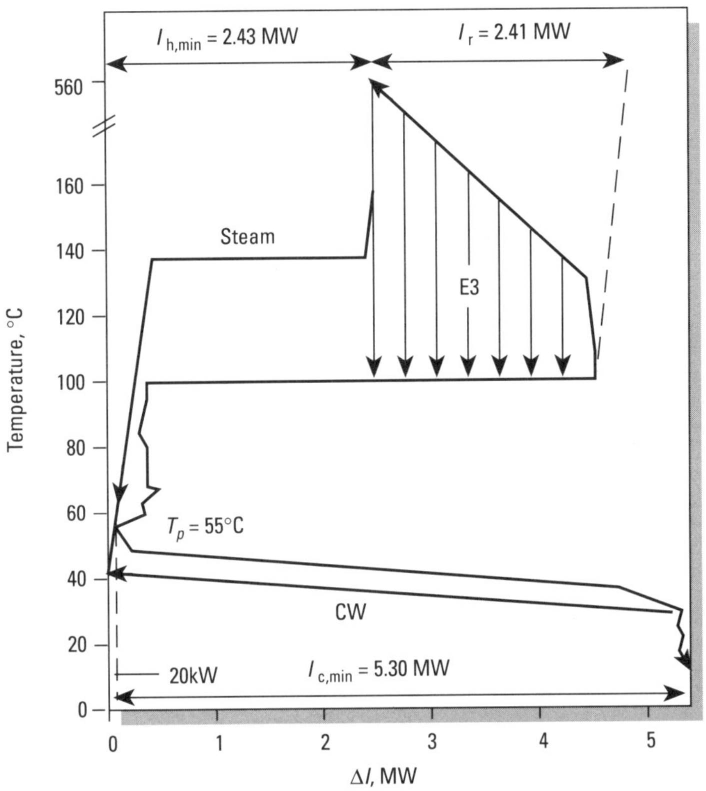

■ *Figure 7. Extended grand composite curve of the formaldehyde plant.*

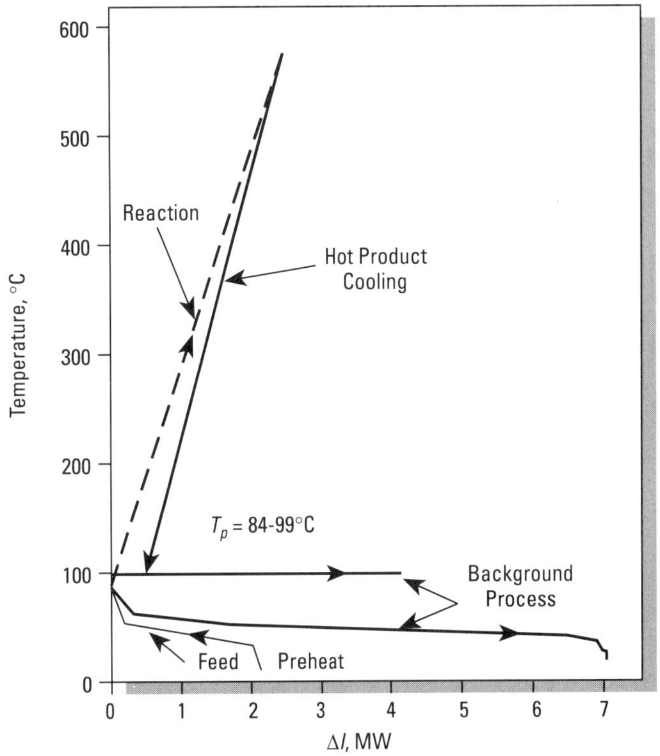

■ *Figure 8. Splitting reactor subsystem off formaldehyde plant grand composite curve.*

Methanol is evaporated in H1 and air is compressed into a mixing vessel. Reaction mixture is heated by product in E1 and recycle streams in E2 before entering the reactor (R). Silver catalyst is used to oxidize methanol to formaldehyde with air:

$$CH_3OH + \tfrac{1}{2}O_2 = HCHO + H_2O \quad (2)$$

with ΔH_r = -159 kJ/mol.

This and seven other complex reactions were modeled as a series of tubular reactors, each with an independent reaction, using Design II as a process simulator *(13)*. The extents of the reactions were adjusted according to the actual composition of the real reactor outflow. The reactor outlet stream is cooled with the feed stream in E2. In the absorption/reaction column (AR), a reaction between formaldehyde and steam condensate takes place to form methylene glycol. The liquid is fed to a distillation column (DC), where unreacted mixture is stripped out, heated with steam in H2 and recycled to the reactor inlet stream. Bottoms product is cooled by reactants in E1 and by cooling water in C1.

The new method consists of two steps — a conceptual and a rigorous one.

The first analysis step is to construct an EGCC of the present plant. Figure 7 comprises all the HEN streams, together with the reboiler and the condenser streams of the distillation column. The upper part of the GCC reveals a pocket (arrowed surface); 1.98 MW of enthalpy flow is appropriately transferred from the reactor to the distillation column by the heat exchanger E3. We have excluded the cold stream passing the reactor from the process streams. When we include it into the process streams, the upper gap of the hot utility widens from 2.43 to 4.84 MW (dashed line), because it includes the summed reaction enthalpies (2.41 MW). Hot reactor utility supplies the heat flow for the arrowed part of the reboiler stream — the effect is the same but presentations differ. The last presentation would reveal the correct process pinch temperature of 55°C when excluding the reboiler and condenser streams of the distillation column from the GCC (as evidenced by the GCC of Figure 9).

Integrated utilities (steam and cooling water) are shown as thick lines. Steam is desuperheated and then condensed at an interval temperature of 137°C; the condensate subsequently delivers heat to the process. As the condensate transfers enthalpy flow downwards through the pinch (in C2), 20 kW of the hot utility are indirectly wasted in the cooling water. Steam condensate, which is used as process water, is inappropriately cooled across the pinch temperature (55°C)! Utilities can now be modified. The heat loss incurred can be reduced by returning the condensate to the boiler just above the pinch temperature — using fresh process water in the absorption column saves cooling water. The savings amount to \$5,700/yr.

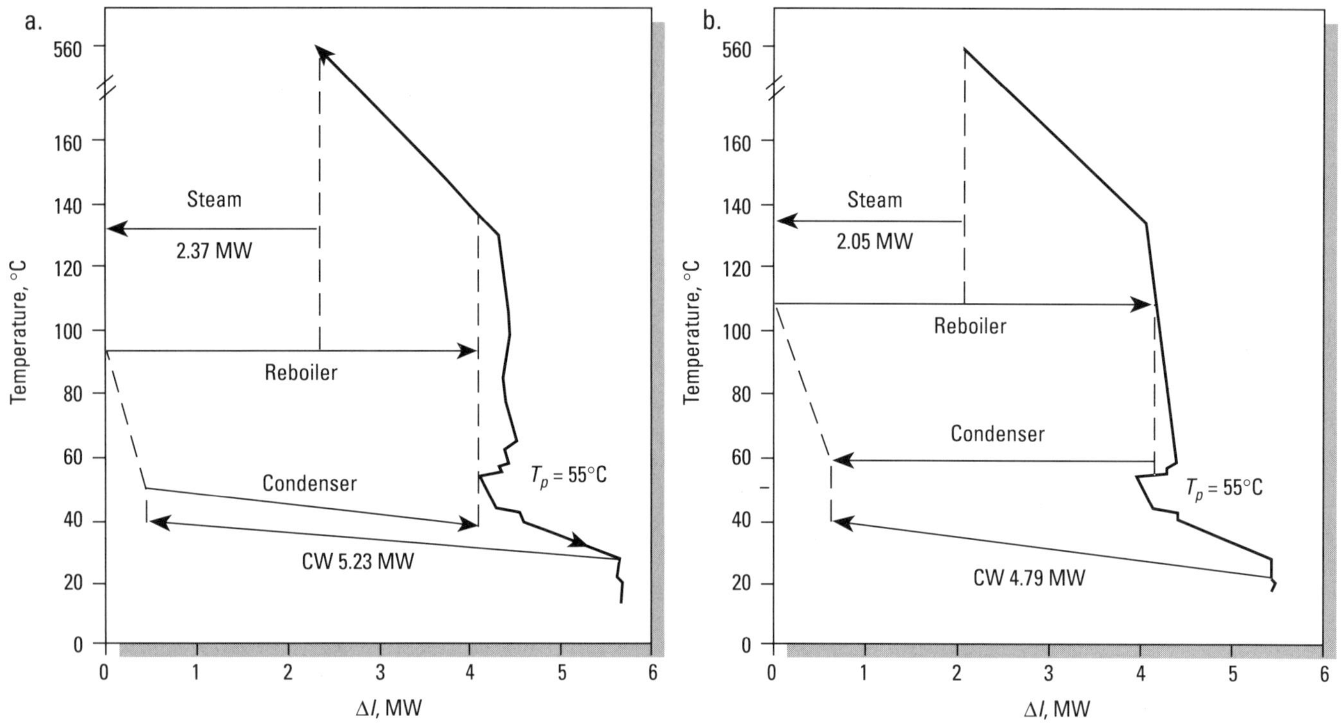

■ *Figure 9. Splitting distillation column off formaldehyde plant grand composite curve (a) and raising it above the pinch temperature (b).*

After the utilities have been selected, we proceed by splitting off energy-active units: reactors, separators, heat engines, heat pumps, and flashes. We take each energy-active unit or a group of similar units (for instance, all distillation columns) out of the GCC and check in turn their placement against the GCC; we name this GCC modification a split-off GCC (SGCC). Placement of all the units against the background process pinch temperature can be adjusted one by one.

Splitting of the reactor. Energy-intensive process units, such as the reactor, can be studied the same way that we have studied energy pumps in the GCC. To be more precise: the original reactor subsystem containing the reactor R, its feed preheaters E1,

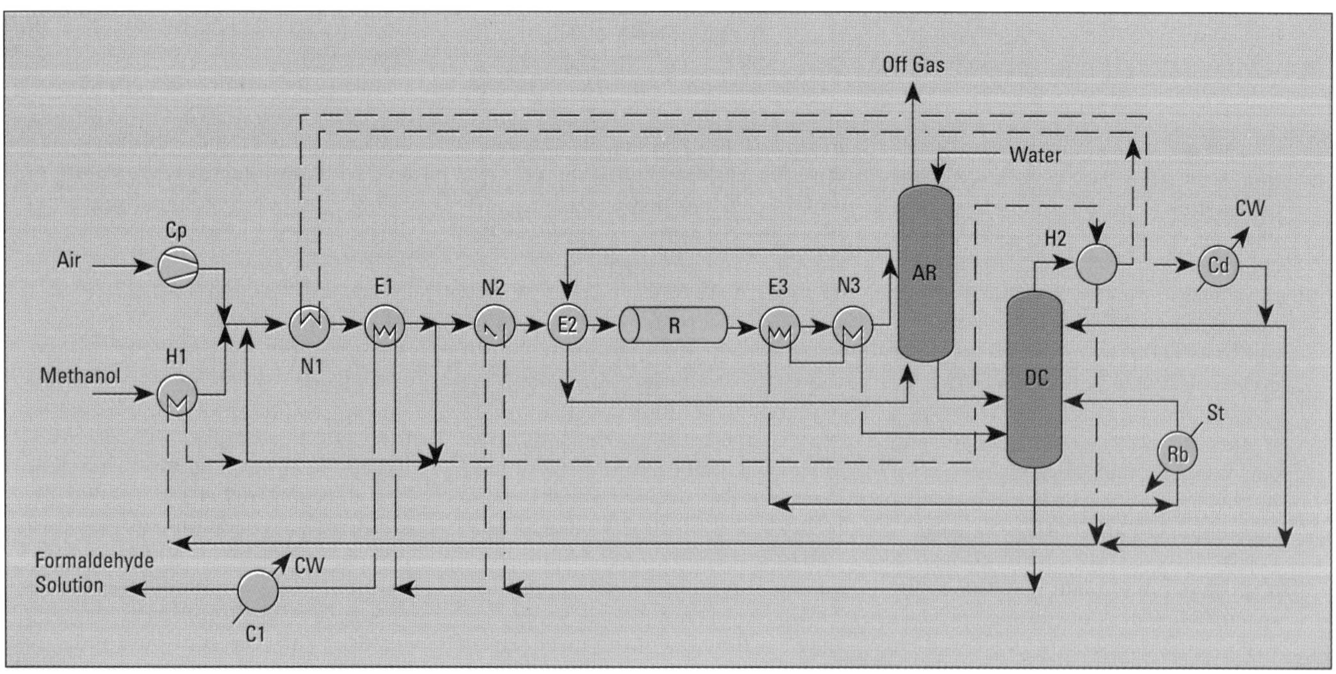

■ *Figure 10. Energy-integrated formaldehyde plant.*

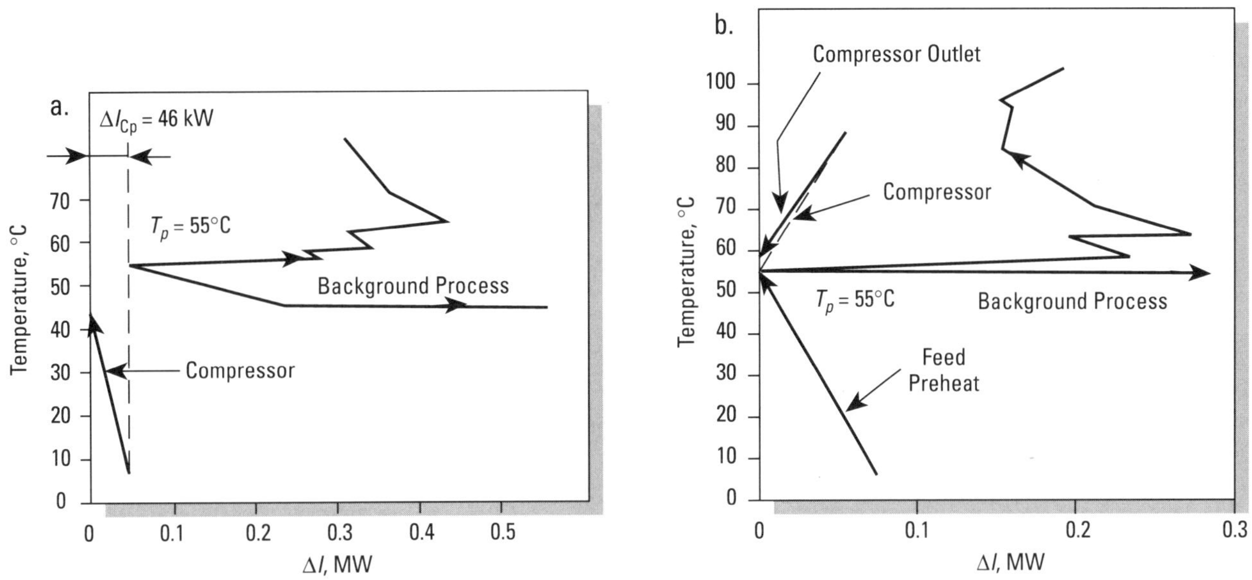

■ ***Figure 11. Splitting off the compressor (a) and shifting it above the pinch temperature (b).***

E2, and its product aftercoolers E3, E2 is split off. The reactor subsystem is superimposed on the GCC; the reboiler and the condenser streams of the distillation column, of course, remain included in the GCC of the remaining process (Figure 8). Note that by splitting off the preheaters and the aftercoolers, the pinch temperature moved again from 55°C to an interval of 84–99°C. As the heaters are situated below the background pinch and below the background GCC and as the coolers are above them, the reactor subsystem is appropriately placed. Now, we can return our preheaters and aftercoolers streams back into the GCC without any worries!

Splitting off the distillation column. This operation can be considered next. To study our distillation column, we, of course, should split its reboiler and condenser streams off the GCC and present them outside it as an EGCC (Figure 9a). The difference between the enthalpy flow rates of the reboiler and the condenser results from the enthalpy flow rate difference between the feed and the effluents of the column. Exclusion of constant temperature reboiler and condenser streams does not influence the pinch temperature; variable ones, however, could change it if they traversed the pinch temperature. We immediately see that our column is inappropriately placed because it stretches across the pinch *(3)*. We can improve the design by shifting the column above the pinch. This can be accomplished by raising its pressure from 0.4 to 0.9 bar. The GCC is very steep, though; therefore only 0.32 MW (2.37 – 2.05 MW) is saved (Figure 9b).

This design method has been successfully applied to several other processes.

Additional energy is now transferred from the process to the column reboiler in a new heat exchanger N3 that is heated by reactor outlet gases (Figure 10). An additional 110 kW of heat flow from the column bottoms can be used to preheat reactor inlet flow in a new exchanger N2. The condenser now gives up additional heat flow to the process: 140 kW of steam can be saved in the condenser H2, 70 kW of steam can be economized in H1, and 110 kW can be transferred from the condensing methanol to the fresh methanol in a new heat exchanger N1. These changes can save $80,600/yr of steam and $7,600/yr of cooling water. The total savings amount to $94,000/yr, and require only $15,000 of differential investment costs for the alternative utilities and distillation column retrofit; so, the payback period is only two months.

Splitting of the compressor. This is shown in Figure 11a. Since the compressor transforms electrical energy into heat, it absorbs hot utility under the pinch and is inappropriately placed. Instead, it should be located just above the pinch temperature. Its effect can be compared to that of the exothermal reactor. Therefore, to appropriately place it, a process preheater and an aftercooler can be added to the subsystem (Figure 11b).

Evaluation of the new process scheme in Figure 11b, together with the shifting of the distillation column, results in a new process flowsheet with two additional heat exchangers that are very large because of the gas/liquid contact and a small ΔT_{min} *(13)*. With the hotter outlet of the compressor, we could reduce the consumption of steam and we could eliminate one exchanger. Yet, the potential savings of $32,000/yr are small in comparison with the investment costs ($102,800). Because the payback period is too long (3.2 yr), this alternative structure is rejected without the need to first do a rigorous design!

The proposed method shows that splitting of energy donors and energy acceptors can be carried out in series to obtain better designs. In our case, the distillation column and the compressor are inappropriately placed while the reactor placement is correct. Shifting the column above the pinch temperature is very profitable, but raising the compressor is too expensive. We can evaluate placement of each unit in turn, change the design and then evaluate its effects. Thus, we can determine the most promising alternatives by evaluating incremental changes.

The potential savings in the formaldehyde plant under study are small; this evidently reflects that the design of the plant, the 16th of its kind, is well down the learning curve.

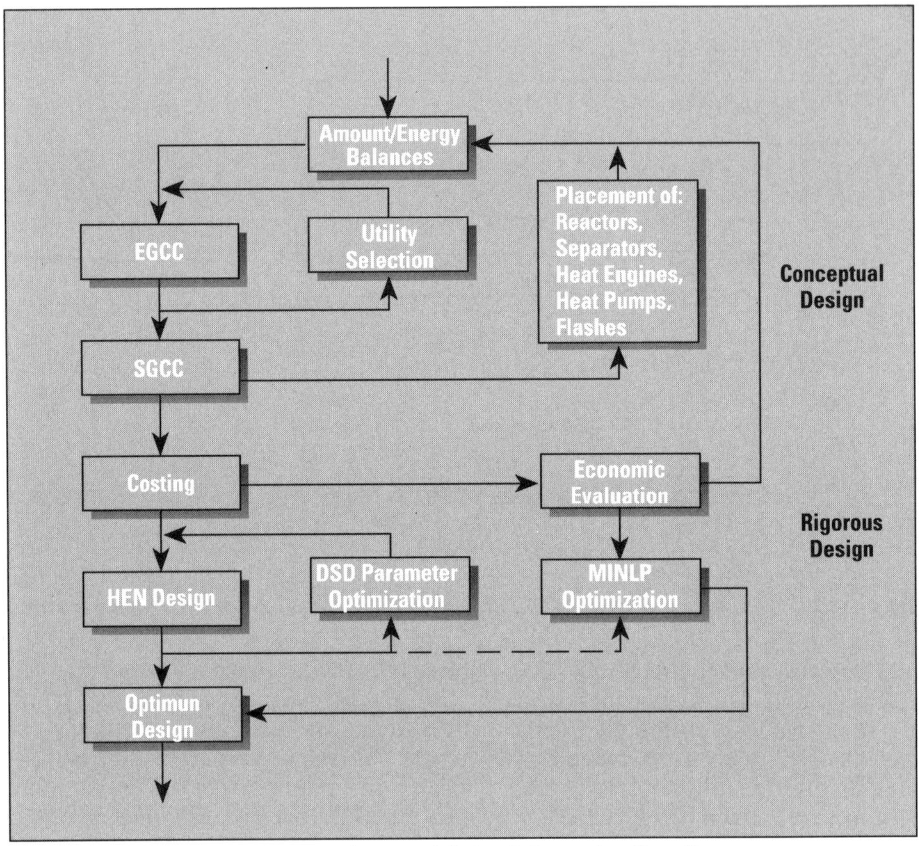

■ *Figure 12. Process analysis, synthesis and design information flow diagram.*

Modified pinch rules

Integration of energy pumps suggests to us the following modifications of traditional rules *(6)* for the pinch method:

• Do not transfer heat across the pinch down the heat cascade.

• Lift heat across the pinch up the cascade with energy pumps.

• Use only profitable acceptors above the pinch.

• Do not use donors below the pinch.

Design procedure

The general applicability of the thermodynamic pinch method to process design was first demonstrated by B. Linnhoff and coworkers. He detailed a procedure to bridge the process/utility interface *(14)* and a "supertargeting method" for optimal synthesis of energy systems *(15)*. In this article, an improvement over the supertargeting method is introduced.

The new method, shown as a block diagram in Figure 12, consists of two steps — a conceptual and a rigorous one. Conceptual design includes synthesis of utility system and appropriate placement of energy-important process units, energors. This step is interactive and driven by chemical engineers. It enables new ideas to be found and significantly reduces the number of possible alternatives.

The proposed conceptual design methodology includes integral analysis of energors and of the HEN in the EGCC with overall pinch temperature, as well as differential analysis of individual energor subsystems around the local pinch temperature in the SGCC. The integration procedure may differ substantially for energy-poor and energy-abundant processes. Splitting can be carried out in turn for each energor together with its heaters and coolers; the subsystem is checked for placement against its background pinch temperature; and the structure of the inappropriately placed subsystem or parameters of the units are changed, evaluated, and accepted or rejected.

When this conceptual synthesis step is finished and a few promising structures have been found, the second step, a rigorous design starts. It is computer aided and proceeds according to the well-known thermodynamic or algorithmic approach. Thermodynamic design closely follows the supertargeting method *(15)*. Design begins at the pinch and proceeds away from it *(1)*. Area and energy targets may be optimized through ΔT_{opt} at the pinch; then, heat exchanger network design can follow. Sequential structural heat integration and optimization, however, does not necessarily provide the true optimum design. So, when the structure is optimized, we can continue with parameter optimization. For rigorously modeled flowsheets with a small number of decision variables, we can use a direct-search discretized (DSD) optimization algorithm having individual stream cost and film coefficients *(16,17)*.

Another approach uses simultaneous heat integration of process streams with total flowsheet optimization *(18)*. Optimum process structure and optimum parameters of their units are determined by a Mixed Integer Nonlinear Programming algorithm (MINLP). It uses simultaneous structural and parameter energy integration and yields the true optimum. Unfortunately, it is not yet

LITERATURE CITED

1. **Linnhoff, B., and E. Hindmarsh,** "The Pinch Design Method for Heat Exchanger Networks," *Chem. Eng. Sci.*, **38** (5), p. 745 (1983).
2. **Townsend, W., and B. Linnhoff,** "Heat and Power Networks in Process Design," *AIChE J.*, **29** (5), p. 742 (1983).
3. **Smith, R., and B. Linnhoff,** "The Design of Separators in the Context of Overall Process," *Chem. Eng. Res. Des.*, **66** (3), p. 195 (1988).
4. **GlaviW, P., Z. Kravanja, and M. HomEak,** "Putting the Pinch on Reactors," *Chem. Eng.*, **97** (6), p. 106 (1991).
5. **Krajnc, M., and P. GlaviW,** "Energy Integration of Mechanical Heat Pumps with Process Fluid as Working Fluid," *Chem. Eng. Res. Des.*, 70 (A4), p. 407 (July 1992).
6. **Linnhoff, B., et al.,** "A User Guide to Process Integration for the Efficient Use of Energy," IChemE, Rugby, U.K. (1982).
7. **HomEak, M. and P. GlaviW,** "Thermodynamic Analysis of Inappropriately Placed Energetic Units," in "Computer-Oriented Process Engineering," L. Puigjaner, and A. Espuña, eds., Elsevier, Amsterdam, p. 363 (1991).
8. **Bike, S. G.,** "Design of Ammonia Synthesis Plant," case study, CACHE Austin, TX (1985).
9. **Kravanja, Z., and P.GlaviW,** "Heat Integration of Reactors — II, Total Flowsheet Integration," *Chem. Eng. Sci.*, **44** (11), p. 2,667 (1989).
10. **Terranova, B. E., and A. W. Westerberg,** "Temperature-Heat Diagrams for Complex Columns," *Ind. Eng. Chem. Res.*, **28** (9), p. 1,374 (1989).
11. **Dhole, V.R., and B. Linnhoff,** "Setting Targets for Distillation," *Proc. Eng.*, Pt. I, p. 33 (June 1992), Pt. II, p. 37 (July 1992), Pt. III, p. 38 (Aug. 1992).
12. **GlaviW, P., Z. Novak, and N. Malek,** "Conceptual Design of Energy Integrated Processes," 1992 Research Event, IChemE, Rugby, U.K., p. 51 (1992).
13. **GlaviW, P., and Z.Novak,** "Improved Pinch Technics for Process Design," *Vestn. Slov. Kem. Drus.*, **38** (3), p. 271 (1991).
14. **Linnhoff, B.,** "Pinch Technology for the Synthesis of Optimal Heat and Power Systems," *J. Energy Resources Technol.*, **111**, p. 137 (Sept. 1989).
15. **Linnhoff, B., and S. Ahmad,** "Supertargeting: Optimum Synthesis of Energy Management Systems," *J. Energy Resources Technol.*, **111**, p. 121 (Sept. 1989).
16. **Kravanja, Z., and P. GlaviW,** "Cost Targeting for HEN through a Simultaneous Optimization Approach," *Computers Chem. Eng.* (accepted for publication);
17. **Kravanja, Z.,** "Heat Integration of Chemical Reactors through Flexible Energy Integration of Process Flowsheet," PhD Thesis, Univ. of Maribor (1990).
18. **Lang, Y. D., L. T. Biegler, and I. E.Grossmann,** "Simultaneous Optimization and Heat Integration with Process Simulators," *Computers Chem. Eng.*, **12** (4), p. 311 (1988).
19. **KovaW, A.,** "Retrofit of Energy Integrated Processes," MS Thesis, Univ. of Maribor (1992).
20. **GorEek, A., and P. GlaviW,** "Optimal Process Design for Specialty Products," *Computers Chem. Eng.*, **16** Suppl., p. S321 (1992).

developed enough to be able to handle more complex processes. Therefore, MINLP should not be applied before the conceptual design described above is carried out and the solution space is reduced. Preliminary results indicate that even very complex processes can be analyzed and optimized by the combined conceptual design and MINLP optimization approach *(19)*.

This design method has been successfully applied to several other processes: production of acetaldehyde, benzene, methanol, methyl ethyl ketone, nitric acid, and sulfuric acid, and also to continuous production of a specialty chemical *(20)*.

The benefits

To sum up, design with the improved method:

- simplifies process utility and exergy analysis using its EGCC and SGCC steps; and
- produces new ideas for synthesizing better design.

The technique enables the process to be seen from different perspectives, each view giving a better understanding of the system. EGCC provides an overall insight into the process energy flows and temperature potential devaluation. SGCC allows each individual energor subsystem to be taken out of the total process system in turn, and it thereby yields an insight into the remaining background process to improve their mutual fitting. The method winnows potential structures and MINLP enables simultaneous energy integration and optimization of promising structural alternatives and their parameters. CEP

holds the chair in chemical engineering at the University of Maribor, Maribor, Slovenia (011-38-62-25-461; Fax: 011-38-62-212-013). He has nine years of experience in the chemical process industries, and 12 years in research and application of process design methods. Author of numerous papers, he is a founder and president of the Slovenian Association of Chemical Engineers. He received a BS in chemical technology, and MS and PhD degrees in chemistry, all from the University of Ljubljana, Slovenia.

is a graduate student in chemical engineering at the University of Maribor, from which she received a BS degree in chemical engineering. She is currently involved in research on process synthesis and optimization.

Choose the Right Heat-Transfer Fluid

Heat-transfer fluid selection involves balancing economic and engineering objectives.

John Cuthbert,
The Dow Chemical Co.

Choosing the best heat-transfer fluid is critical for many chemical processes. A mismatch between the fluid and the process can result in excessive fluid costs due to rapid degradation, lost production time, and damage to equipment.

Selecting a fluid is a complex task, based on a number of interrelated economic and technical considerations. While the price per gallon of each fluid is important, the actual cost of operation, including capital, operating, and maintenance costs, will be determined by many other factors that should not be ignored. For example, the fluid's physical properties will influence heat exchanger sizes, pump horsepower requirements, and pressure vessel ratings. In addition, regulatory issues may dictate (or significantly influence) fluid choice.

The most commonly used heat-transfer fluids, aside from water, are synthetic organic fluids, silicone fluids, inhibited glycol-based fluids, and paraffinic and/or mineral oils. (Although outside the scope of this article, molten salts can be used for very high process temperatures [above 750°F].) These chemistries offer a wide range of performance characteristics, service life, and price (Table 1). Selecting the appropriate fluid for an application requires careful consideration of all of these factors. The priority of each of these factors will vary from one application to the next, and even from one installation to the next within the same industry.

This article discusses synthetic organic, silicone, and glycol-based heat-transfer fluids in detail and briefly touches on water and paraffinic and mineral oils. It addresses the impact of fluid chemistry on system design and provides guidance on selecting the most appropriate fluid.

Synthetic organic fluids

Synthetic organic fluids, which include alkylated aromatics, aromatic ethers, and alkylated diphenyls, offer exceptional thermal stability for efficient heat transfer, long fluid life, optimum operating economics, and a broad temperature range of –100°F to 750°F. Some of these fluids can be used in the vapor phase as well as in the liquid phase.

Synthetic organic fluids also provide reasonably good heat transfer in the –100°F to 0°F temperature range. Because they have good physical properties at temperatures down to –100°F, synthetic organic fluids can be used in systems where both low- and high-temperature capabilities are required, such as single-fluid batch processing. In systems where low temperature heat transfer is not required, using an organic fluid that remains pumpable at low temperatures can eliminate the need for costly steam tracing.

Synthetic organic fluids are especially effective in systems that require both a broad temperature range and low vapor pressure. Most synthetic heat-transfer fluids, even at 600°F, have vapor pressures below 180 psig. This makes them a good choice for high temperature applications.

Synthetic organic fluids also provide

high thermal stability. This enables engineers to avoid most problems associated with localized or temporary temperature increases beyond the upper process temperature limit.

To maximize fluid life, engineers should select a synthetic organic fluid that offers sufficient thermal stability to accommodate the system's maximum planned operating temperature and allows for unplanned temperature increases. Choosing a synthetic organic fluid with a maximum recommended use above the highest anticipated operating temperature will provide not only longer fluid life, but also optimum heat-transfer efficiency and operating economy.

Thermal stability is an important consideration in the selection of a heat-transfer fluid. Thermal stability is a function of the fluid's chemistry, the temperature to which it is exposed, and the duration of the exposure. Over time, at a given temperature, the fluid will begin to break down or degrade. Degradation usually produces "low boilers," which have boiling points lower than the remaining fluid (and are also known as noncondensables), and "high boilers," which have higher boiling points than the fluid. Low boilers are vented from the system; high boilers remain in the fluid, and their concentration determines when the fluid has reached the end of its useful life. The rate at which this breakdown occurs is called the degradation rate.

A fluid's degradation rate is a very strong function of temperature, typically following an Arrhenius type of relationship (Figure 1). A general rule with synthetic organic fluids is that the degradation rate doubles for every 18–21°F increase in fluid temperature. In addition to temperature, fluid chemistry also has a dramatic impact on fluid degradation (Table 2).

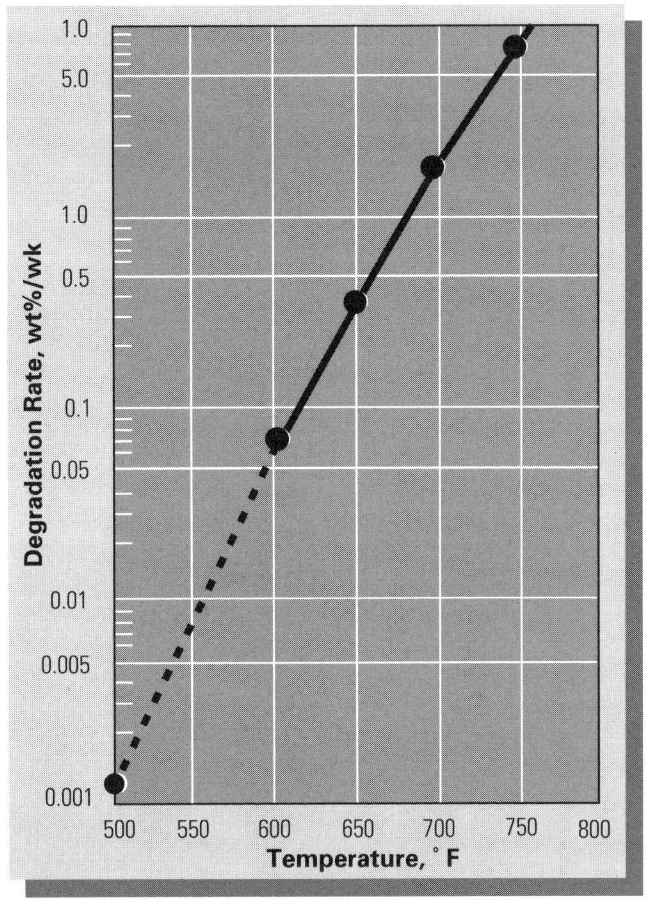

■ *Figure 1. Degradation rate for a mixture of di- and triaryl ethers.*

As an example of the impact of fluid stability on fluid economics, consider the hypothetical system in Table 3. For each fluid, the actual degradation rate at each location in the heat-transfer system is determined by multiplying the degradation rate for the fluid at the given temperature by the fraction of the time the fluid is exposed to that temperature (represented by percent of total system divided by 100). These individual degradation rates are then added to give each fluid's overall degradation rate — 0.0312 wt%/wk for Fluid A and 0.623 wt%/wk for Fluid B. Thus, Fluid A is approximately 20 times more stable than Fluid B. Although Fluid A will typically have a higher price per pound than Fluid B, this premium can usually be justified based on Fluid A's superior performance.

Silicone fluids

Polydimethyl siloxane fluids, also known as silicones, offer excellent heat-transfer performance, and provide exceptionally long service life without periodic reclaiming. They are essentially odorless and low in acute oral toxicity. Their broad temperature range, –100°F to 750°F, makes them ideal for use in petrochemical, pharmaceutical, and fine chemical processing.

Due to their chemistry, silicone fluids behave somewhat differently than other heat-transfer fluids. This difference affects their physical properties and thermal stability.

When heated above 500°F, silicones undergo a molecular rearrangement whereby the higher molecular weight compounds rearrange to form lower molecular weight compounds. This rearrangement is an equilibrium process — at a given operating temperature, the fluid com position will change until the equilibrium composition is reached. Thereafter, the composition will remain constant, unless fluid is lost from the system or the operating temperature changes. As the composition of the fluid changes, so will the fluid's physical properties. The lower molecular weight compounds will increase the vapor pressure, reduce the flash point, and reduce the fluid's viscosity.

When comparing the thermal stability of silicone fluids to synthetic organic fluids, it is important to keep in mind the significant differences between the two chemistries. The degradation of a synthetic organic fluid is nonreversible. When heated to sufficiently high temperatures, these fluids will begin to break down and

Table 1. Properties of major heat-transfer fluids.

Fluid Composition*	Maximum Usage Temperature, °F†	Minimum Pumping Temperature, °F	Film Coefficient at 600°F, Btu/h•ft2•°F		Pressure Drop at 600°F, psi/100 ft	
			at 5 ft/s	at 7 ft/s	at 3 ft/s	at 7 ft/s
Synthetic Organic Fluids						
Mixture of diphenyl and diphenyl oxide (A)	750	53.6	464	607	1.3	6.9
Mixture of diphenyl and diphenyl oxide (B)	750	53.6	449	588	1.3	7.0
Mixture of di- and triaryl ethers	700	–4	387	506	1.5	7.6
Hydrogenated terphenyl (A)	650	31	353	462	1.4	7.2
Hydrogenated terphenyl (B)	650	30	319	417	1.4	7.3
Aromatic blend	650	–50	384	503	1.3	6.9
Alkylated aromatic	600	<–100	522	684	0.9	4.9
Polyaromatic	600	–50	355	465	1.4	7.0
Synthetic hydrocarbon	600	4	302	395	1.2	6.2
Isomeric dibenzyl benzenes	662	–18	366	479	1.4	7.4
Isomeric dimethyl diphenyl oxide	626	–46	359	469	1.4	7.1
Alkyl diphenyl	707	–22	350	458	1.4	6.9
Alkyl benzene	590	–49	330	433	1.3	6.4
Silicones						
Dimethyl siloxane	750	<–50	239	313	1.2	6.2
Polydimethyl siloxane‡	500	<–148	214	280	0.97	5.02
Inhibited Glycols						
50% Inhibited EG/H_2O‡	350	–60	1,358	1,778	1.60	8.47
50% Inhibited PG/H_2O‡	325	–50	1,030	1,348	1.65	8.42
Paraffinic and Mineral Oils						
Paraffinic oil (A)	600	<15§	268	351	1.3	6.5
Paraffinic oil (B)	600	<40§	280	367	1.3	6.4
Mineral oil (A)	600	<35	319	417	1.4	7.0
Mineral oil (B)	600	<25§	310	406	1.3	6.3
Mineral oil (C)	600	20	298	390	1.3	6.3

*(A), (B), and (C) denote different product formulations.
†Maximum use temperature recommended by fluid manufacturer.
‡Properaties are evaluated at the maximum use temperature.
§Reported pour-point temperature.

will continue to do so until all the parent molecules (the starting fluids) are gone. As synthetic organic fluids degrade, periodic venting of the system and fluid change-out is required. Since silicone fluids achieve an equilibrium, periodic venting and fluid replacement are not required. The low molecular weight compounds are intended to remain with the fluid.

In terms of initial cost, silicone fluids may be more expensive than synthetic organic or glycol-based fluids — roughly 10–30% on a price-per-pound basis. However, use of silicone fluids can result in lower long-term expenses, since they do not degrade in the same manner as other fluids or require top-off refills. At their maximum use temperatures, silicones can last as long as ten years.

Glycol-based fluids

Inhibited glycol-based fluids are frequently used in the 0°F to 350°F temperature range. These fluids are formulated products, containing either ethylene glycol or propylene glycol and a proprietary corrosion inhibitor package. When used in heat-transfer systems, glycol-based fluids are diluted with water.

Table 2. Relative stabilities of several common heat-transfer fluids (at 600°F).

Fluid Composition	Relative Stability
Mixture of diphenyl and diphenyl oxide	0.0024
Mixture of di- and triarylethers	1.0
Akylated aromatic	1.5
Hydrogenated terphenyl	1.6
Aromatic blend	2.2

Table 3. Degradation rate depends on the fluid's temperature, exposure time, and chemistry.

Fluid Location	Temperature Within System, °F	Percent of Total System†	Degradation of Fluid A,* wt%/wk: At Given Temperature	Degradation of Fluid A,* wt%/wk: Within System	Degradation of Fluid B,* wt%/wk: At Given Temperature	Degradation of Fluid B,* wt%/wk: Within System
Film in heater	650	1.0	0.396	0.0040	7.92	0.079
Heater discharge	600	34.0	0.0696	0.0237	1.39	0.473
Return loop	550	34.0	0.0103	0.0035	0.206	0.070
Expansion tank	450	31.0	0.0001	0.0000	0.002	0.001
Total degradation				0.0312		0.623

*Fluid A is a mixture of di- and triaryl ethers, Fluid B is a petroleum oil.
†"Percent of Total System" is an estimate of the time the fluid is exposed to a given temperature.

Glycol-based fluids have several advantages over synthetic organics, silicones, hot oils, and water. Of the fluids covered in this article, ethylene glycol and propylene glycol based fluids have the lowest cost (with the exception of water).

The specific heat and thermal conductivity of aqueous glycol solutions are much higher than those for synthetic organics, silicones, and hot oils at similar temperatures. Since they have better physical properties, they are more effective heat-transfer fluids.

The addition of glycol to water lowers the freezing point of water, which allows the solution to be used to transfer heat down to 0°F. It should be noted that the addition of glycol to water negatively impacts the physical properties of water. The more glycol added, the greater the impact. This is why careful consideration should be given to choosing a glycol concentration. In fact, ethylene glycol and water have a eutectic composition at approximately 60 wt% ethylene glycol. At glycol concentrations above this, the freezing point of the solution increases.

Solutions of glycol and water can remain pumpable down to –40°F. This low temperature pumpability eliminates the need for heat tracing of glycol-water systems.

Another valuable property of glycol-water solutions is that they can provide burst protection for piping and equipment. Water expands upon freezing with a force sufficient to rupture piping. A solution of ethylene glycol or propylene glycol will prevent this. As the temperature drops below the solution's freezing point, ice crystals begin to form. Because water in the solution freezes first, the remaining glycol solution becomes further concentrated and remains fluid. The combination of ice crystals and fluid results in a flowable slush. Fluid volume increases as this slush forms, with the extra volume flowing into the available expansion volume in the system. Ethylene glycol concentrations greater than 30 wt% and propylene glycol concentrations greater than 35 wt% will provide burst protection down to –60°F.

The choice of glycol concentration will vary depending on the type of protection, (that is, freeze or burst) that is desired. Burst protection will typically require less glycol than freeze protection. The lower glycol requirements result in better fluid transport properties as well as lower cost. When choosing a glycol concentration, keep in mind that burst protection will allow some freezing in the system. Once a glycol slush has formed, it is highly unlikely that circulation through the system can be maintained as long as the solution is below its freezing point. A glycol concentration for burst protection should be chosen for systems that have seasonal uses, such as building heating, ventilation, and air conditioning (HVAC) systems, or that are used intermittently.

Like synthetic organics and silicones, the concept of thermal stability for glycol solutions is intimately connected with the fluid's chemistry. An issue for glycol solutions that is of little or no concern for synthetic organics or silicones is corrosion. Uninhibited solutions of ethylene glycol and propylene glycol are quite corrosive to steel, the most common material of construction for heat-transfer systems in the chemical process industries. Glycol solutions become corrosive as the glycols oxidize to form organic acids. The addition of a corrosion inhibitor package substantially reduces corrosion by neutralizing the organic acids as they form (Table 4).

The degradation of glycol solutions is a strong function of temperature. Both reserve alkalinity and pH indicate degradation. The reserve alkalinity is a measure of the fluid's ability to buffer the solution against a change in pH. As the glycol degrades, the reserve alkalinity will decrease,

Table 4. Corrosion test results (weight loss, in mg) for glycol solutions.

	Water	Ethylene Glycol	Propylene Glycol	Inhibited Ethylene Glycol	Inhibited Propylene Glycol
Copper	2	4	4	3	1
Solder	99	1,780	1,096	4	2
Brass	5	11	5	3	2
Mild Steel	212	974	214	1	1
Cast Iron	450	1,190	345	3	1
Aluminum	110	165	15	4	*

Notes:
Tests were conducted over a 2-week period according to ASTM D1384, at 190°F and 30%(vol) glycol. Results represent weight loss in mg/2 wk.
*Weight gain of 3 mg.

usually with a slight decline in pH. Once the inhibitor is depleted, the pH will drop and the fluid will become corrosive. Regular fluid analysis will detect inhibitor depletion long before a corrosive condition can develop. Most industrially inhibited glycols are easily reinhibited.

Fluid life will vary according to use temperature, but fluid life at maximum use temperatures is typically four to five years. For fluids used at temperatures below 180°F, lives of 20 years or more have been documented.

Note that good quality water for dilution is critical. Deionized or distilled water is best. Water containing excessive hardness or chlorides can result in premature inhibitor depletion and increased corrosion rates.

Water

Water is the most widely used heat-transfer fluid. Over the temperature range of 32°F to 350°F, it is typically the fluid of choice.

However, it has several disadvantages that make other fluids more desirable for many applications. It freezes at 32°F and has a very high vapor pressure at elevated temperatures (1,543 psia at 600°F vs. 45 psia for fluids based on diphenyl oxide and biphenyl at the same temperature). Water may react with certain process fluids and typically requires corrosion inhibition.

Paraffinic and mineral oils

Paraffinic and mineral oils are an economical choice for systems operating in the 300°F to 600°F temperature range. They offer very good thermal stability (at 500°F, a hot oil will last for 10–15 years) and acceptable physical properties. They cannot, however, be pumped at low temperatures and must therefore be heat traced.

At lower usage temperatures (below about 550°F), the lower cost of the hot oils — about \$1.50/gal to \$8/gal, vs. \$15/gal to \$30/gal for synthetic organics and silicones — makes oils economical. But when temperatures exceed about 550–600°F, or when other properties such as good low-temperature heat transfer are desired, the premium priced fluids become good, economical choices.

Fluid chemistry and system design

The choice of fluid chemistry can dramatically influence the design of the heat-transfer system. There are several basic equations an engineer can use to size a system's heat exchanger, furnace, pump, and piping. The goal is to optimize all of these units to meet the necessary heating and cooling requirements of the process, while minimizing the overall cost of the heat-transfer system. The physical properties of the heat-transfer fluid impact all of these calculations in a very significant way.

Nomenclature

- A = area, ft²
- bhp = brake horsepower requirements for pump, hp
- c_p = specific heat, Btu/h•°F
- d = diameter, in.
- f = friction factor
- h = film coefficient, Btu/h•ft²•°F
- h_i = film coefficient inside tubes, Btu/h•ft²•°F
- h_o = film coefficient outside tubes, Btu/h•ft²•°F
- k = thermal conductivity, Btu/h•ft•°F
- L = length, ft
- m = mass-flow rate, lb/h
- Nu = Nusselt number
- ΔP = pressure drop, psi
- Pr = Prandtl number = $c_p\mu/k$
- Q_h = heat-transfer rate, Btu/h
- Q_v = volumetric flow rate, gpm
- r_f = fouling resistance, h•ft²•°F/Btu
- r_w = tube resistance to conductive heat transfer, h•ft²•°F/Btu
- Re = Reynolds number = $dv\rho/\mu$
- sg = specific gravity
- ΔT = temperature difference, °F
- ΔT_{lm} = log mean temperature difference, °F
- TDH = total dynamic head, ft
- U = overall heat-transfer coefficient, Btu/h•ft²•°F
- v = fluid velocity, ft/s

Greek letters

- ε = surface roughness
- ε_p = pump efficiency
- μ = viscosity, cP
- μ_w = viscosity at wall temperature, cP
- ρ = density, lb/ft³

The engineer must first consider the heat balance equation. For a single phase fluid:

$$Q_h = mc_p\Delta T \quad (1)$$

Table 5. Effects of fluid chemistry on physical properties and heat-transfer system design.

	50% Ethylene Glycol/Water	Alkylated Aromatic	Polydimethyl Siloxane
Mass Flow,* lb/h	53,705	87,873	102,040
Flow Rate, gpm	896	1,918	2,227
h_i,† Btu/h•ft²•°F	1,998	522	364
ΔP, psi/100 ft	5.84	9.37	9.13
Vapor Pressure, psia	102	13.06	15.23

*Based on 1 million Btu/h duty and 20°F ΔT.
†Based on fluid velocity of 7 ft/s in a ¾-in., 14-BWG tube.

Q_h is determined by the process, and to a large extent so is ΔT. The specific heat is an inherent property of the fluid. Obviously, the higher the specific heat, the lower the mass-flow rate required to accomplish the desired heat-transfer duty.

Once the duty is determined, the engineer can begin to size heat exchangers. The basic equation used for this is:

$$Q_h = UA\Delta T_{lm} \qquad (2)$$

The overall heat-transfer coefficient, U, combines heat transfer by conduction and convection. It is a function of the film coefficient of the process fluid, the film coefficient of the thermal fluid, the conductivity of the tube wall (or resistance), and a fouling factor (corrections for geometry have been omitted):

$$1/U = 1/h_i + 1/h_o + r_w + r_f \qquad (3)$$

The transport properties of the thermal fluid will determine one of the film coefficients. The thermal stability of the fluid at the operating temperatures will influence the value of the fouling factor.

The film coefficient in a forced convection system is a function of the Reynolds number (Re), the Prandtl number (Pr), and the Nusselt number (Nu). For fully developed turbulent flow in tubes, the Sieder-Tate equation applies:

$$Nu = hd/k$$

$$= 0.027Re^{0.8}Pr^{1/3}(\mu/\mu_w)^{0.14} \qquad (4)$$

The transport properties of the heat-transfer fluid at the temperature of interest completely determine the Prandtl number and substantially determine the Reynolds number. In the case of the Reynolds number, the flow rate and tube diameters can be varied to some small extent, although the ability to do this is somewhat limited when other factors, such as pressure drop, are taken into consideration.

In general, fluids with low viscosity, high specific heat, and high density will have higher film coefficients.

It should be noted that the film coefficient for the process side also has a significant impact on U. If the value of $h_{process}$ is much less than h_{fluid}, or if r_f or r_w is very large, then the impact of the thermal fluid on the overall size of the heat exchanger will be less.

In a liquid phase system, the thermal fluid must be pumped around the system, typically by centrifugal pumps. Once the required duty of the system and ΔT are determined, the physical properties of the chosen heat-transfer fluid will determine the pumping requirements. Pump brake horsepower requirements are calculated from:

$$bhp = [Q_v(TDH)(sg)]/3{,}960\varepsilon_p \qquad (5)$$

Q_v is fixed by the heat balance (Eq. 1). Once the duty and ΔT are fixed, the choice of fluid determines the volumetric flow rate. The choice of fluid is also a factor in the total dynamic head; TDH is comprised of several factors, of which fractional pressure drop is key. The pressure drop of a single phase fluid flowing in a pipe can be estimated using the Darcy equation:

$$\Delta P = 0.001294(fL\rho v^2/d) \qquad (6)$$

In this equation, f is the friction factor, which is an implicit function of the Reynolds number. Using the Colebrook Equation:

$$\frac{1}{\sqrt{f}} = -2.0 \log\left(\frac{\varepsilon/d}{3.7} + \frac{2.51}{Re\sqrt{f}}\right) \qquad (7)$$

Since Reynolds number is a strong function of the fluid's physical properties, the friction factor will be, too.

Looking at pump sizing overall, flow rate and pressure drop are two key elements in determining the pump size. The physical properties of the thermal fluid determine, to a large extent, the flow rate and the pressure drop.

J. CUTHBERT is a development leader in the Thermal Fluids Technical Service and Development (TS&D) Laboratory at The Dow Chemical Co., Midland, MI (517/636-3251; Fax: 617/638-9241). His areas of expertise include the use of silicone and synthetic organic heat-transfer fluids in the chemical, fiber, and pharmaceutical manufacturing industries; he has also worked with glycol-based fluids in heating, ventilation, and air conditioning (HVAC) and thermal energy storage applications. He has two patents in the field of heat-transfer fluids. He received a BS in chemical engineering from Rensselaer Polytechnic Institute.

Further Reading

Aerstin, F., and G. Street, "Applied Chemical Process Design," Plenum Press, New York (1982).

Holman, J. P., "Heat Transfer," McGraw-Hill, New York (1976).

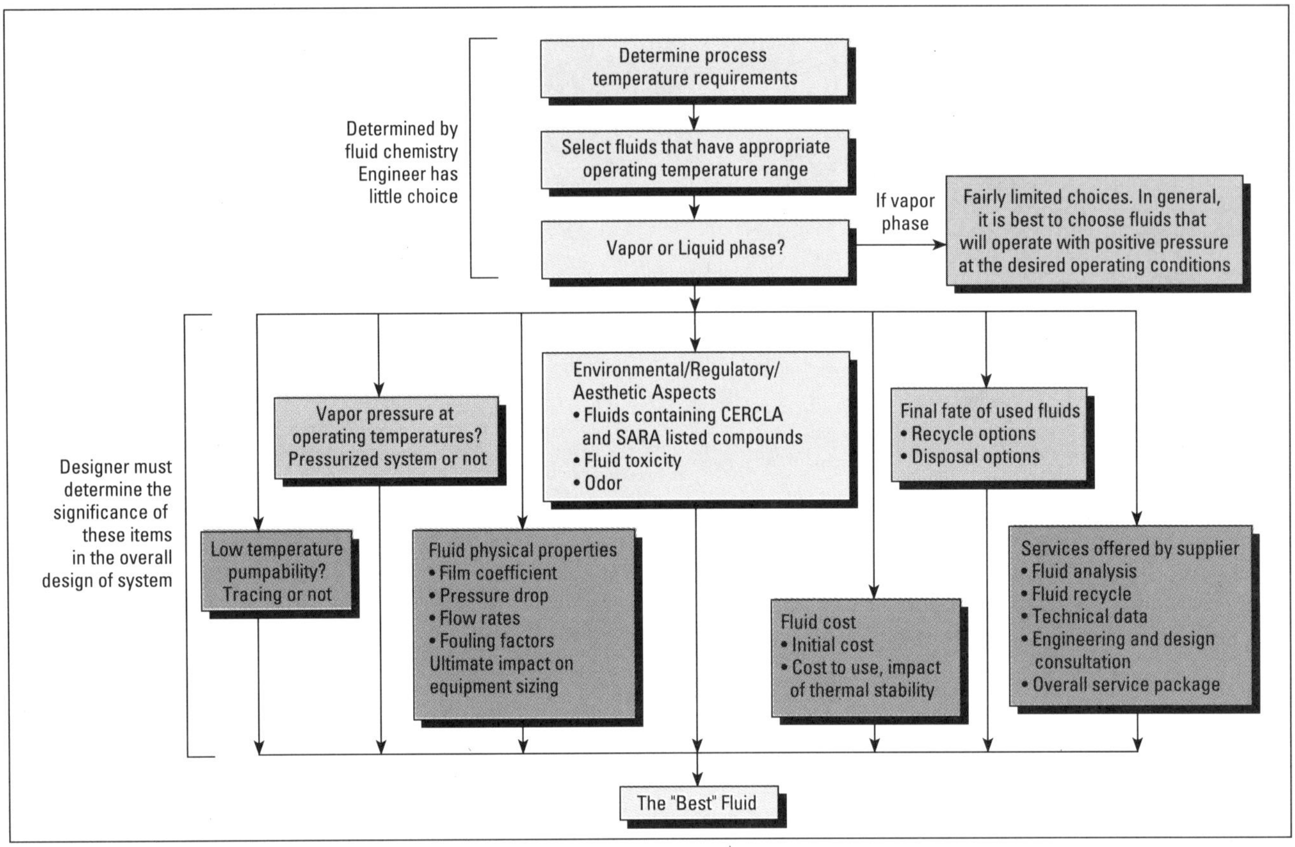

■ ***Figure 2. Factors to consider when choosing a thermal fluid.***

It should be noted that pump horsepower requirements translate to energy consumption. Choosing a fluid that requires a higher pump horsepower may be more expensive in terms of energy consumption than other options.

The vapor pressure of the fluid, which is a function of the fluid's chemistry, is also a factor in the design of a heat-transfer fluid system. The net positive suction head for the system's pump must consider the vapor pressure of the fluid at the pump inlet conditions. The vapor pressure of the fluid will also impact the design and operation of the system's expansion tanks and pressure relief valves.

Table 5 illustrates the impact of fluid chemistry on physical properties at a given condition. The fluids are a solution of 50% ethylene glycol and water, an alkylated aromatic, and a polydimethyl siloxane fluid; the temperature is 350°F and the flow rate is 7 ft/s in a ¾-in., 14-BWG tube. This example is extreme, of course, as it is attempting to compare all of the general types of chemistry that have been discussed.

Fluid maintenance

Even in the most carefully controlled system, all fluids will eventually degrade. Poor flame patterns, flame impingement, and high heat fluxes within a heat-transfer system, if not corrected, will cause a fluid to degrade prematurely. Degradation accelerates as temperatures reach the fluid's rated capacity, but varies from one type of fluid to another.

Several heat-transfer fluid manufacturers offer periodic fluid evaluation as part of the initial cost of a heat-transfer fluid. These services monitor the condition of the fluid, its pH, degradation products, and residue. Many manufacturers will reprocess degraded fluid for a service charge. Some suppliers credit customers for the value of the reclaimed amount, which can significantly reduce the life-cycle cost of the material.

Independent reprocessors also recondition degraded or contaminated fluid. Reprocessing options have increased in recent years as the service has grown in economic and environmental importance.

Health and safety considerations

Health, employee safety, and environmental acceptability have also assumed primary importance in the selection of heat-transfer fluids and the calculation of their costs. Most commercially available heat-transfer fluids are moderate to low in oral toxicity and present little risk when operating personnel handle them according to the manufacturer's instructions. This includes the use of protective gear, such as gloves, goggles or safety glasses, and protective clothing, to help prevent skin or eye irritation that

Table 6. Selecting a heat-transfer fluid involves evaluating the fluid's temperature limits plus a variety of physical properties.

Fluid Composition	Temperature Use Range, °F Liquid	Temperature Use Range, °F Vapor	Minimum Pumping Temperature, °F*	Flash Point, °F†	Fire Point, °F†	Autoignition Temperature, °F‡	Vapor Pressure, psig§	Thermal Conductivity, Btu/h•ft^2•°F	Specific Heat, Btu/lb•°F	Density, lb/ft^3	Viscosity, cP
Synthetic Organic Fluids											
Mixtures of diphenyl	60/750	495/750	53.6	255	275	(1,139)	30.6	0.061	0.579	49.29	0.19
and diphenyl oxide	60/750	495/750	53.6	240	260	1,150	31.0	0.056	0.564	49.78	0.183
Mixture of di- and triarylethers	20/700	—	–4	285	295	(1,083)	9.5	0.059	0.540	53.91	0.30
Hydrogenated	25/650	—	31	355	375	(662)	(434)	0.062	0.630	49.55	0.38
terphenyls	15/650	—	30	350	380	705	(434)	0.0545	0.628	50.30	0.40
Aromatic blend	–40/650	—	–50	275	295	(932)	17.3	0.055	0.590	48.67	0.24
Alklylated aromatic	–100/600	358/600	<–100	145	155	(788)	159.8	0.0625	0.720	35.46	0.11
Polyaromatic	–60/600	—	–50	310	320	835	6.9	0.0540	0.622	49.05	0.30
Synthetic hydrocarbon	0/600	—	4	350	410	675	(341)	0.0535	0.699	42.71	0.36
Isomeric dibenzyl benzenes	–4/662	—	–18	374	N/A	932	(134)	0.0644	0.608	50.85	0.38
Isomeric dimethyl diphenyl oxide	–4/626	—	–46	275	N/A	1,013	15.3	0.0534	0.552	49.79	0.273
Alkyl diphenyl	–39/707	—	–22	266	N/A	N/A	11.3	0.0569	0.625	47.96	0.322
Alkyl benzene	32/590	—	–49	358	367	N/A	(310)	0.0609	0.645	43.70	0.35
Silicones											
Dimethyl siloxane	–40/750	—	<–50	320	350	725	50.0	0.052	0.49	42.00	0.43
Polydimethyl siloxane#	–100/500	—	<–148	130	155	662	61.2	0.0295	0.541	35.02	0.19
Inhibited Glycols											
50% Inhibited ethylene glycol#	–20/350	—	–60	N/A	N/A	N/A	88.2	0.238	0.931	59.93	0.27
50% Inhibited propylene glycol#	0/325	—	–50	N/A	N/A	N/A	67.6	0.210	0.976	57.06	0.39
Paraffinic Oils											
Paraffinic oils	15/600	—	<15	390	430	670	(110)	0.046	0.70	44.42	0.40
	40/600	—	<40	430	N/A	675	(150)	0.064	0.70	42.4	0.54
Mineral oils	5/600	—	<35	350	N/A	N/A	(210)	0.060	0.70	47.30	0.45
	25/600	—	<25	380	N/A	N/A	(160)	0.065	0.67	42.43	0.43
	5/600	—	20	400	N/A	N/A	(120)	0.065	0.670	42.43	0.47

Notes:

Data supplied by fluid manufacturers. Values reported for physical properties at 600°F unless otherwise noted. N/A = data not available.

** For mixtures of diphenyl and diphenyl oxide, this is the freezing point; for other fluids, this is the temperature at which the fluid viscosity is 1,000 cP.*

† Measured by the Cleveland Open Cup (COC) method.

‡ Values in parentheses are determined by ASTM Method E-669-78; other values are determined by ASTM D-2155.

§ Values in parentheses are in mm Hg rather than psig.

 # Properties are evaluated at the maximum use temperature.

may occur when the fluids are not handled properly.

Since the fluids are normally used in closed systems, they do not ordinarily enter the atmosphere or pose the likelihood for incidental contact. Should the material accidentally contaminate the air, water, or soil, or contact skin or eyes, material safety data sheets (MSDSs) are the best source of information regarding followup procedures.

Selecting a fluid

Choosing a heat-transfer fluid for a given process can involve many variables, as illustrated in Figure 2 and Table 6. It is up to the engineer to carefully consider the weight that should be given to each.

There are two hard-and-fast rules of fluid selection:

- Do not use a fluid above the manufacturer's maximum recommended use temperature; and
- If the process requires a vapor phase fluid, only those that are suitable for use in the vapor phase may be chosen.

After the process temperature requirements are established and the decision to use liquid or vapor phase heat transfer has been made, a large array of other variables comes into play. ("Vapor phase heat transfer" is actually a misnomer, though, since the heat transfer is accomplished by condensation of the vapor.)

In Figure 2, these variables are represented in parallel. There is usually no "best" choice. What will ensue is a series of decisions that evaluate the various attributes of the many heat-transfer fluids. These decisions vary from the strictly technical, such as a comparison of the fluid's engineering properties, to the more aesthetic, such as the desire for a fluid with little or no odor.

Process temperature requirements are the most significant criteria for selecting a heat-transfer fluid. As noted, the manufacturer's recommended maximum use temperature should not be exceeded, as the fluid will tend to degrade more rapidly and give generally unsatisfactory performance. The fluid selected should be able to operate at temperatures that extend beyond those of the process. Engineers should choose a fluid with temperature capabilities at least 50°F higher than required at the high end of the range to safeguard the system against unexpected overheating and fluid degradation. A similar margin is not necessary at the low end of the range, as long as the lowest anticipated process temperature is considered.

Engineers should also consider a fluid's vapor pressure and the effect it will have on the heat-transfer system. The expansion tank should be capable of handling the fluid's vapor pressure at planned operating temperatures, as well as the additional pressure created by fluid thermal degradation, water, or other types of system contamination. The expansion tank's capabilities should be able to withstand pressure at least 15 to 20 psi above the highest anticipated fluid vapor pressure.

Many decisions will be based on economics. Decisions to use the lowest priced fluid should be carefully considered, however. The lowest priced fluid may not be the most cost-effective when equipment sizing and operating economics are factored in (although there are many times when the lowest priced fluid is the best).

In summary, selecting a fluid that best meets the operation's required temperature range and offers high thermal stability, along with some safety margin, will extend the fluid's service life. Estimating the return on investment often helps the engineer determine whether the anticipated overall cost is reasonable based on the fluid's expected service life. CEP

Changing fluids in an existing system

Heat-transfer systems are designed around the physical properties of a particular type of heat-transfer fluid, which influence the sizing of heat exchangers, pumps, pressure relief systems, and pressure vessels. In general, once a system is built, the same type of fluid is used for the life of the system.

However, there may be occasions where it is necessary or desirable to change heat-transfer fluids in an existing system. The temperature requirements of the process may change, thus requiring a fluid with different temperature capabilities. The wrong fluid may have been chosen initially and excessive degradation has made continued operation of the system impractical. Or the fluid may be subjected to increasing regulatory pressure (such as chlorofluorocarbon refrigerants).

Changing from one type of heat-transfer fluid to another can be simple or quite complicated — the degree of difficulty depends entirely on the system under consideration. Conversion can be as easy as removing one fluid and replacing it with another, or it can require extensive redesign and equipment changes.

The following factors should be considered when converting from one type of heat-transfer fluid to another:

1. Is the system vapor phase? If so, conversion options are very limited. Converting from vapor phase to liquid phase is extraordinarily complicated and expensive. In these cases, "converting" is frequently a euphemism for rebuilding.

2. Can the new fluid be used over the required temperature range?

3. What is the vapor phase of the new fluid at the maximum operating temperature? The vapor pressure of the new fluid may be substantially higher, possibly exceeding equipment pressure limitations.

4. Are the pressure relief valves sized properly for the new fluid? Are the set pressures appropriate?

5. How will the new fluid impact the operation of the process? If the new fluid has a lower heat-transfer film coefficient, will the performance of heat exchangers, furnaces, or chillers be adversely impacted? If the fluid has a lower specific heat, is there enough spare capacity to compensate?

6. How will the new fluid impact the system hydraulics? If the new fluid has a higher pressure drop, or if the new fluid requires a higher circulation rate, will the system's existing pumps be able to meet the required increase? If the new fluid has a higher vapor pressure, will the pumps have sufficient net positive suction head (NPSH)? How will the new fluid affect system start-up?

7. Will the new fluid be compatible with the system's gasketing and sealing materials? For example, gaskets compatible with glycol-water solutions may not be suitable for use with silicones.

8. Is the expansion tank design appropriate for the new fluid?

9. Are there any concerns about the chemical reactivity of the new fluid? Are any modifications to the site's environmental permits needed?

10. How thoroughly should the system be cleaned prior to conversion? Will it be as simple as draining the old fluid and replacing it with the new, or should the system be chemically cleaned?

Simulation Aids Cogeneration System Analysis

Simulating the performance of heat-recovery steam generators provides valuable information for system design, as well as for the operation of an existing system.

V. Ganapathy,
ABCO Industries

Cogeneration systems using gas turbines and heat-recovery steam generators (HRSGs) are widely used in chemical process industries (CPI) plants. Because these plants are quite expensive and the HRSG is an important part of the system, it is prudent to analyze the heat-recovery system or simulate its performance well in advance of finalizing plant specifications. Simulation is a method of predicting the performance of the HRSG under different operating modes and gas and steam conditions without physically designing the equipment.

Such a study will provide the engineer with valuable information about the HRSG and its performance capabilities. The simulation results could influence the choice of steam system parameters and the selection of the steam or gas turbine. In addition, one may also obtain information about the performance of the HRSG and the cogeneration system.

This article explains what HRSG simulation is and the basic methodology. Its applications are then illustrated through several examples.

What is HRSG simulation?

It is not necessary to design an HRSG in terms of surface area, tube size, fin configuration, and so on in order to evaluate its performance under different gas and steam conditions.

HRSG designers arrive at the geometry of the HRSG by computing the overall heat-transfer coefficient U and then the surface area S from the equation

$$S = Q/U\Delta T$$

The simulation process does not require the computation of U, which requires information on tube size, pitch, geometry, fin configuration, length, and so on. In simulation analysis, the term $US = Q/\Delta T$ is computed for each surface and used in the prediction of off-design performance. The HRSG's mechanical details are not important at this stage.

Thus, engineers can use the simulation methodology to obtain valuable information about the thermal performance of the HRSG before developing its specifications or committing to a particular design or supplier.

Two limitations apply to the simulation technique discussed here:

• The HRSG should be of the convective type. Nearly 85%–95% of the HRSGs for gas turbines fall into this category.

• The gas and the steam sides should be clean. Gas turbine exhaust, even with oil firing, may be considered clean compared to municipal solid waste or similar applications.

Why simulate?

One might wonder why simulation is necessary when the HRSG supplier can provide information on the thermal performance of the HRSG. The supplier may not have the time to analyze hundreds of operating conditions during the early stages of a study or development of a cogeneration project, particularly if the steam system has multiple pressure levels and is complex. The engineer who is more familiar with the plant requirements can perform the simulation and conduct several "What if" studies to arrive at firmer data on steam parameters and pressure levels and to optimize the gas/steam temperature profiles based on the gas turbines being considered. Such a study may also eliminate certain gas or steam turbines from consideration due to incompatibilities between actual and required steam parameters. The HRSG designer also benefits by having fewer options to work on. And, since the plant owner will know what to expect in terms of steam-side performance, less time is spent on evaluating or second-guessing the HRSG design.

Some specific applications of HRSG simulation are as follows:

• The exhaust gas flow rate, temperature, and gas analysis of a gas turbine vary with ambient conditions, load, fuel used, and NO_x control requirements. Using simulation analysis, one can see how a given HRSG behaves when gas-side or steam-side parameters such as steam pressure, feed water temperature, or steam temperature change.

• Energy recovery can be optimized by placing appropriate heat-recovery surfaces one behind the other or by relocating heat-transfer surfaces, particularly in multiple-pressure HRSGs, such that the gas and steam temperature profiles are closely matched. One may also consider adding secondary recovery surfaces, such as condensate heaters, deaerators, or fluid heaters, to maximize energy recovery. Simulation helps in analyzing gas/steam temperature profiles in order to optimize energy recovery.

• Plant engineers may use simulation to predict the performance of their existing units and study the effects of variations in steam and gas parameters. Such a study can also predict fouling of heat-transfer surfaces. If the actual duty or energy transferred is less than what is simulated under clean conditions, then one can infer that fouling has occurred.

Engineers often compute steam generation based on a certain exit gas temperature without checking to see if it is even thermodynamically feasible.

The simulation methodology

The entire theory and procedure for performing simulation analysis for a complex HRSG is outlined in *(1)*. Here we will briefly describe the procedure for simulating the design and off-design performance of a boiler evaporator, such as the unit in Figure 1.

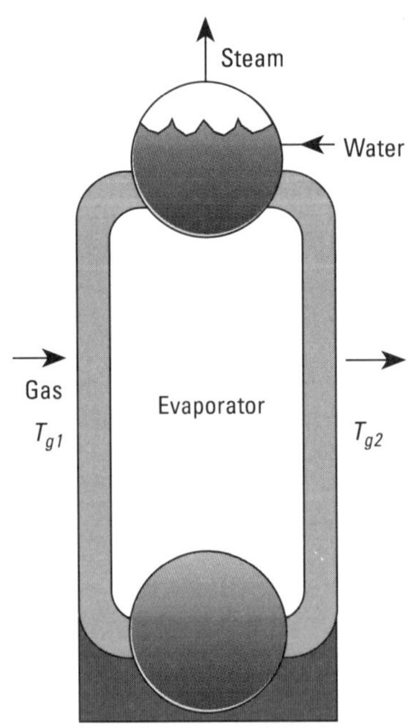

■ ***Figure 1. Evaporator for a heat-recovery boiler.***

1. Given the gas flow W_g, inlet gas temperature T_{g1}, and steam pressure, let us assume that the evaporator has been designed to cool the gases to a temperature T_{g2} and thus transfer a duty Q. The relationship among the variables (neglecting heat losses for the sake of simplicity) can be described by the following equation, which we call the design case:

$$Q = (US)_d \Delta T = W_g C_p (T_{g1} - T_{g2}) = W_s \Delta H \quad (1)$$

The steam flow W_s is obtained by computing the enthalpy added to steam, ΔH, and solving Eq. 1. We now have all of the process information about the evaporator except the surface area or the geometry.

One of the objectives of the simulation process is to see how the above evaporator performs under different conditions of gas flow, temperature, and steam conditions. This is outlined below.

2. From Eq. 1, compute the term

$$(US)_d = Q/\Delta T \quad (2)$$

3. Under the new conditions of gas flow and temperature, the heat-transfer coefficient will be different. This is accounted for by correcting the term $(US)_d$ for the new conditions as follows:

$$(US)_p = (US)_d (W_{gp}/W_g)^{0.65} F(t) \quad (3)$$

where W_{gp} is the gas flow under the new performance conditions and $F(t)$ corrects for the different gas temperature in the off-design mode *(1)*. The gas-side heat transfer governs the overall heat-transfer coefficient; hence, the design heat-transfer coefficient may be corrected for the off-design case by using the factor $(W_{gp}/W_g)^{0.65}$ shown in Eq. 2. [A convective HRSG has been assumed in which the gas-side heat-transfer coefficient and hence the overall heat-transfer coefficient is proportional to the gas flow raised to the power of 0.65. *(2)*].

4. The new duty is

$$Q_p = (US)_p \Delta T = W_{gp} C_p (T_{g1p} - T_{g2p}) \quad (4)$$

Equation 4 can easily be solved by assuming the exit gas temperature T_{g2p}, computing ΔT, and then calculating Q_p. One or two iterations may be required. Steam flow is obtained by simply dividing Q_p by ΔH. Thus, we can arrive at the new gas exit temperature, steam generation, and heat-transfer duty without computing the surface area *S per se.*

In other words, we have simulated the off-design performance of a boiler.

The complete simulation procedure is quite involved and requires several trial-and-error calculations and iterations, particularly for multi-pressure, complex HRSGs with auxiliary firing. In such cases, commercial software developed for this purpose (such as *(3)*) may be used. [For information on general simulation packages and heat-transfer software, see *CEP's Software Directory,* the next issue of which is scheduled to be published in December. — Editor]

Examples

Example 1. An engineer is developing specifications for a process plant and is considering three gas turbines (designated GT1, GT2, and GT3). The major gas/steam data are shown in Table 1. The HRSG operates in unfired and fired modes. Simulate the design with 15°F pinch and approach points; obtain the firing temperature and fuel consumption required to generate 40,000 lb/h of saturated steam in the off-design mode using natural gas containing 95% methane and 5% ethane.

The HRSG consists of an evaporator and economizer. "Pinch point" in this case refers to the difference between the temperature of the gas leaving the evaporator and the steam

Table 1. Data on gas turbines and steam parameters (Example 1).

	GT1	GT2	GT3
Gas Flow, lb/h	165,000	137,000	165,000
Exhaust Temperature, °F	900	1,040	950
Steam Pressure, psig	200	200	200
Feed Water Temperature, °F	230	230	230
Required Steam, lb/h	40,000	40,000	40,000

Exhaust Gas Analysis (volume %):
CO_2 = 3%, H_2O = 7%, N_2 = 75%, O_2 = 15%,
Heat Loss = 1% and Blowdown = 3%

Table 2. Design and off-design performance of gas turbines (Example 1).

	GT1		GT2		GT3	
	Unfired	Fired	Unfired	Fired	Unfired	Fired
Exhaust Temperature, °F	900	900	1,040	1,040	950	950
$t_{o,HRSG}$, °F	900	1,200	1,040	1,347	950	1,194
$t_{o,Evaporator}$, °F	403	409	403	408	403	408
$t_{o,Economizer}$, °F	309	293	281	265	299	285
Steam, lb/h	25,450	40,000	27,300	40,000	28,000	40,000
Pinch, °F	15	22	15	20	15	20
Approach, °F	15	42	15	41	15	37
Fuel, (MM Bth/h*)	0	14.41	0	12.63	0	11.76

*Based on the lower heating value.
t_o = outlet temperature of the gas from the ...

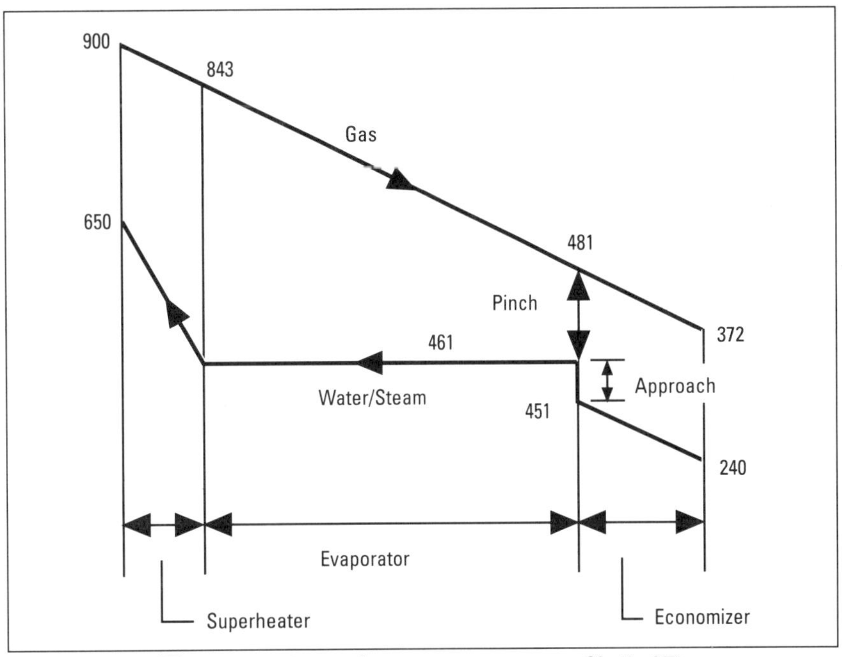

■ *Figure 2. Gas and steam temperature profile (in °F) for a single-pressure turbine (Example 2).*

saturation temperature; "approach point" refers to the difference between the saturation temperature and feed water temperature entering the evaporator. (Reference 1 provides for more information on the selection of these values and their influence on HRSG size.)

Using a simulation program, the design and off-design performances for the three machines were evaluated. The results are summarized in Table 2.

It can be seen that GT2 generates more steam than GT1 in the unfired mode (27,300 lb/h vs. 25,450 lb/h). GT3 requires the least fuel (11.76 MM Btu/h) to generate 40,000 lb/h steam. The HRSG for GT3 will be larger and, hence, slightly more expensive than the HRSG for GT2. If fuel costs are not high, then GT2 may be a better choice if the other aspects, such as electrical power from the turbines and the cost of the turbines, are favorable. The purpose of simulation is only to see how a HRSG would perform and what range of steam-side performance one can expect.

Example 2. While developing specifications for the HRSG, engineers often assume that incorporating large surface areas can lower the exit gas temperature to any desired level. This may not always be thermodynamically feasible, as this example illustrates.

Consider the case where 150,000 lb/h of turbine exhaust gases at 900°F generate superheated steam at 450 psig and 650°F with feed water at 240°F. Is a 300°F exit gas temperature feasible for this single-pressure system?

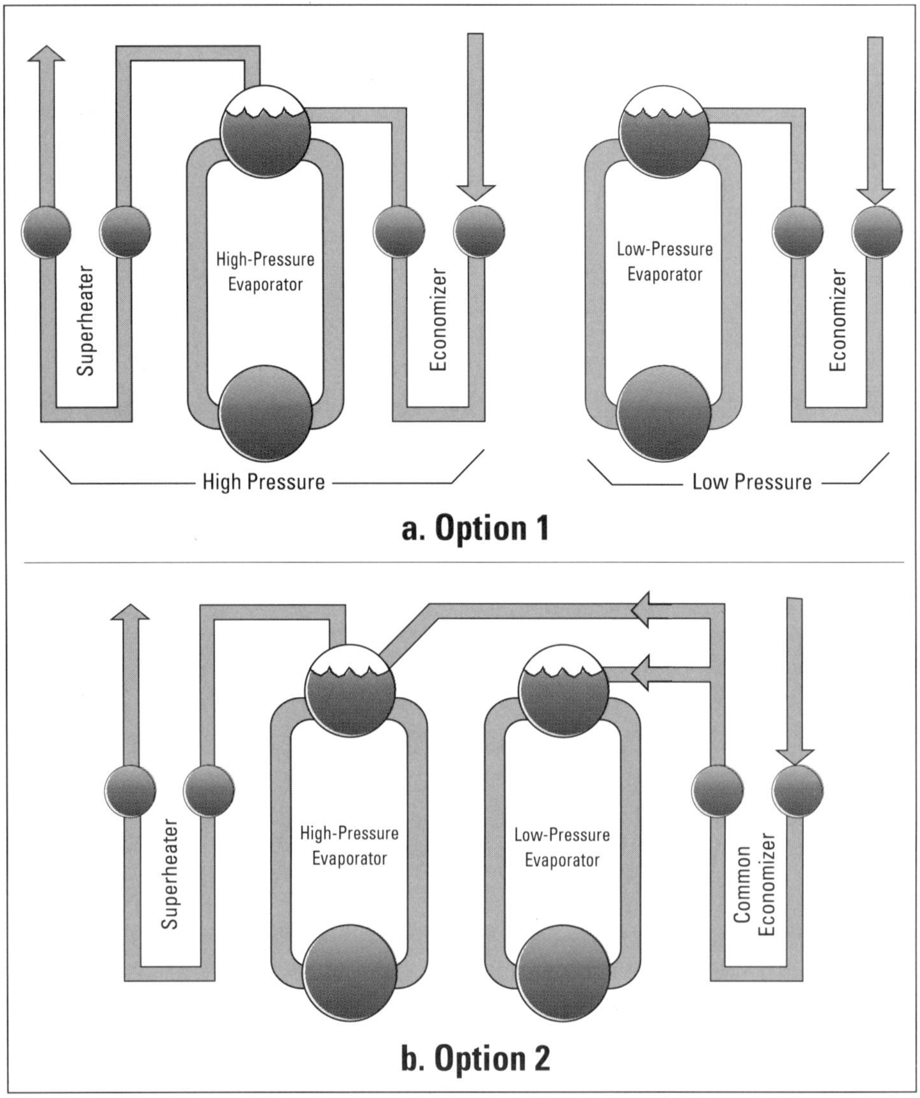

■ *Figure 3. Two configurations for a dual-pressure gas turbine HRSG (Example 3).*

One need not physically design an HRSG in order to obtain information about its performance.

Through simulation, the gas and steam temperatures were obtained using a 20°F pinch and a 10°F approach, as shown in Figure 2. It can be seen that the exit gas temperature is 372°F, nowhere near 300°F. Even with a pinch and an approach of 0 (which, of course, is physically not possible as it would require infinite surface area), the exit gas temperature can be lowered to only 342°F. To lower the temperature to 300°F, secondary heat recovery or a multiple-pressure HRSG is required.

Engineers often compute steam generation based on a certain exit gas temperature without checking to see if it is even thermodynamically feasible. Simulation can help them in analyzing the gas and steam temperature profiles and make sensible demands on HRSG performance, thus saving valuable time when interacting with HRSG suppliers.

Example 3. It was mentioned earlier that temperature profiles could be maximized by relocating heat-transfer surfaces. This example shows how this is done for a multiple-pressure HRSG.

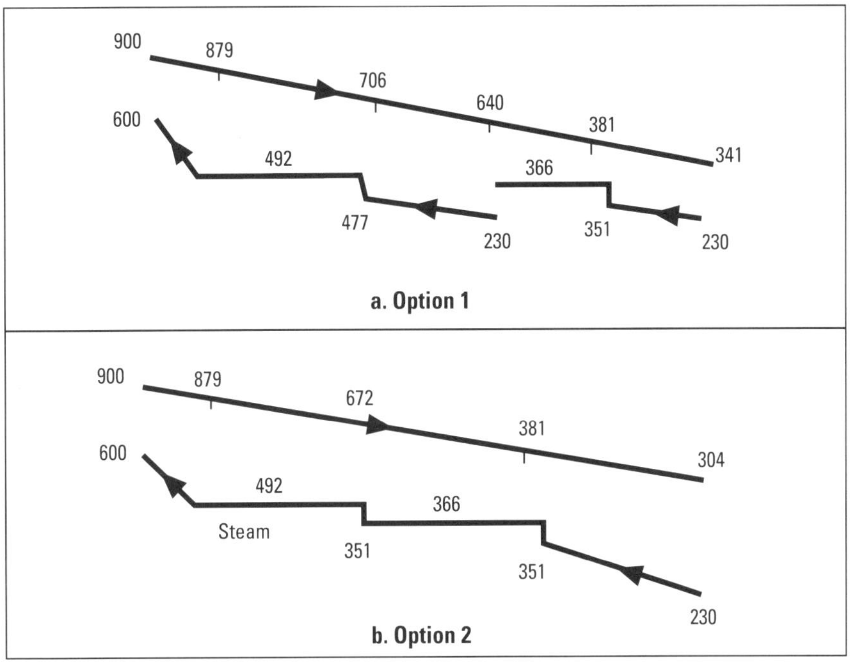

■ *Figure 4. Gas and steam temperature profiles (in °F) for dual-pressure gas turbine HRSG (Example 3).*

Table 3. Data for multipressure HRSG analysis (Example 3).

Gas Flow	500,000 lb/h	
Exhaust Temperature	900°F	
Feed Water Temperature	230°F	
HP Steam Pressure	600 psig	
HP Steam Required	31,500 lb/h at 600°F	
LP Steam Pressure	150 psig	
LP Steam Required	Maximize	
	Option 1 (Figures 3a, 4a)	**Option 2 (Figures 3b, 4b)**
HP Steam, lb/h	31,500	31,500
HP Steam Pressure, psig	600	600
HP Steam Temperature, °F	600	600
LP Steam Pressure, psig	150 (sat)	150 (sat)
LP Steam Flow, lb/h	38,500	43,400

A dual-pressure gas turbine HRSG is to be simulated using the data shown in Table 3. The high-pressure (HP) steam flow required is 31,500 lb/h and the low-pressure (LP) steam is to be maximized.

Figure 3a shows an HRSG configuration consisting of a HP superheater, evaporator, and economizer followed by a LP evaporator and economizer. This design will get the job done, but it does not maximize the energy recovery.

Figure 3b shows another configuration, which generates the same HP steam but more LP steam. This is achieved by relocating the economizer, thus making the heat sink behind the LP evaporator larger. The exit gas temperature is now lower. Note that this system has four modules, while the system in Figure 3a has five modules.

Table 3 shows the data and the results. Figure 4 shows the gas and steam temperature profiles obtained after performing the simulation.

In summary

HRSG simulation is a valuable tool and has several applications. One need not physically design an HRSG in order to obtain information about its performance. Such analysis can help consultants, plant engineers, and HRSG designers to learn more about the HRSG and the steam system and, hence, arrive at optimum HRSG configurations. CEP

Literature Cited

1. **Ganapathy, V.,** "Waste Heat Recovery Deskbook," Fairmont Press, Atlanta, GA (1991).
2. **Ganapathy, V.,** "How Important is Surface Area?" *Chem. Eng. Progress*, **88**(10), pp. 70–76 (Oct. 1992).
3. **Ganapathy, V.,** "HRSGS — Software for Simulation of HRSGs," available from the author (1992).

Further Reading

Ganapathy, V ., "Heat-Recovery Boilers: The Options," *Chem. Eng. Progress*, **88**(2), pp. 59–64 (Feb. 1992).

Ganapathy, V., "Effective Use of Heat-Recovery Steam Generators," *Chem. Eng.*, **100**(1), pp. 102–106 (Jan. 1993).

Ganapathy, V., "Recovering Heat When Generating Power," *Chem. Eng.*, **100**(2), pp. 94–98 (Feb. 1993).

V. GANAPATHY is a heat-transfer specialist with ABCO Industries, Abilene, TX (915/677-2011; Fax: 915/677-1420), where he is engaged in the engineering of packaged steam generators and waste heat recovery boilers. He holds a B. Tech. degree in mechanical engineering from Indian Institute of Technology, Madras, India, and a master's degree in boiler technology from Madras Univ. He has authored over 175 articles and papers on boilers, heat recovery, and steam plant systems, as well as five books.

Heat Tracing: Steam or Electric?

Use these guidelines to evaluate the relative economics of steam and electric heat tracing and to select the appropriate system.

Joseph T. Lonsdale,
Raychem Corp.
Norman J. Little,
Kilborn Inc.
David L. Bradford,
Raychem Corp.

Choosing the proper heat-tracing system requires an understanding of the strengths and weaknesses of the available systems and a method of estimating their relative installed and operating costs. This article provides some basic information about steam tracing and electric tracing, and presents some simple decision rules for choosing the most economic system for most types of chemical process industries (CPI) applications.

Heat tracing is used to prevent pipes from cooling to ambient temperature when they are in the no-flow condition. This is usually accomplished by placing a source of heat on the pipe under thermal insulation. Preventing pipes containing water from freezing and preventing fuel oil pipes from dropping below the pour point of the fuel oil are two common applications for heat tracing.

The two most common methods of heat tracing are steam tracing and electric tracing. In the 1960s, over 90% of industrial heat tracing was steam tracing. Today, the market share of electric tracing has increased to over 30% due to advances in electric tracing technology and increases in energy costs.

Steam tracing

In a steam tracing system, a tube (usually copper, although stainless steel and more exotic metals are sometimes used) is placed against the pipe to be traced before the thermal insulation is applied. Steam, usually at low pressure (25–150 psig), is introduced into this tube, which is called a steam tracer. The steam tracing system consists of steam supply piping from the plant's existing steam mains to the traced pipe, the steam tracing tubing, steam traps, and piping from the traps to the existing condensate return system.

Steam tracing has several advantages:

• *High heat output.* A steam tracing system provides a large quantity of heat to the pipe. There is a very high heat-transfer rate *(1)* between the metallic tracer and a metallic pipe. Even with damage to the insulation system, there is little chance of a low-temperature failure with a steam tracing system.

• *High reliability.* While many things can go wrong with a steam tracing system, such as leaks and failed steam traps, very few of the potential problems lead to a low-temperature failure.

• *Safety.* Although steam burns are fairly common, there are generally fewer safety concerns with steam tracing than with electric tracing.

• *Use of waste steam.* If excess low-pressure steam is available, it can be used for steam tracing. This improves the economics of steam tracing (discussed later). However, just because "free" steam is available does not always mean that steam tracing is the correct choice for a particular application.

But it also has several weaknesses:

• *Energy inefficiency.* The total energy consumed by a steam tracing system is often more than 20 times the actual energy required to keep the traced pipe at the required temperature. The steam tracer itself puts out significantly more energy than required. Steam traps, steam leaks, and the

Self-regulating electric heat tracers

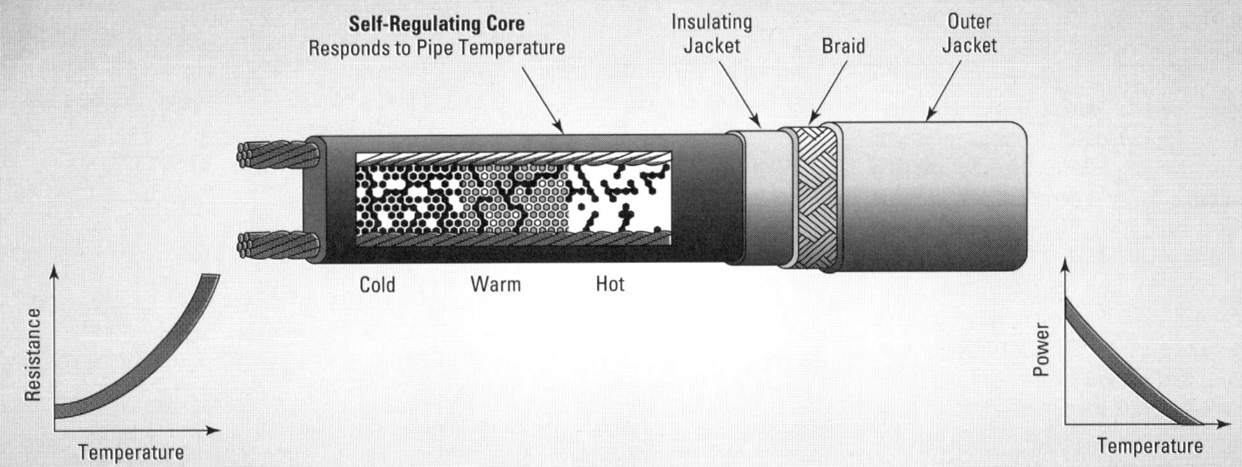

A self-regulating electric heat tracer consists of a polymer heating element between two parallel wires. As the temperature of the polymer increases, its electrical resistance also increases, causing the heat output to decrease. This effect occurs independently at each point on the heater. The self-regulating feature prevents the heater from generating temperatures that can destroy the heater. Eliminating tracer failure by overheating (burnout) significantly increased the reliability of electric tracing systems.

supply and return system all waste large quantities of energy.

• *Poor temperature control.* A steam tracing system offers little temperature control capability. The pipe reaches an equilibrium temperature somewhere between the steam temperature and the ambient temperature. However, the area of pipe in contact with the steam tracer will effectively be at the steam temperature.

• *High installed costs.* The incremental piping required for the steam supply and condensate return systems must be installed and insulated, and additional steam traps are often required in the steam supply system.

• *High maintenance costs.* Leaks must be repaired and steam traps must be checked and replaced when they fail.

Electric tracing

An electric tracing system consists of a power supply system, an electric tracing cable placed against the pipe under the thermal insulation, and any control or monitoring system that may be used (optional). The power supply system usually consists of an electrical distribution panel and electrical wiring in conduit or cable trays. Depending on the capacity of the plant's existing electrical system, an additional transformer may be required.

The strengths of electric tracing are:

• *Lower installed and operating costs.* Many studies have shown that electric tracing is less expensive to install and operate.

• *Reliability.* In the past, electric heat tracing had a well-deserved reputation for poor reliability. However, since the introduction of self-regulating heaters in 1971 (see above sidebar), the reliability of electric heat tracing has improved dramatically. Self-regulating heaters cannot destroy themselves with their own heat output. This eliminates the most common failure mode of polymer-insulated constant-wattage heaters. Also, advances in the technology used to manufacture mineral-insulated cables has significantly improved the reliability of high-temperature electrical tracing systems.

• *Temperature control.* Even without a thermostat or any control system, an electric tracing system usually provides better temperature control than a steam tracing system. With thermostatic or electronic control, very accurate temperature control can be achieved.

• *Energy efficiency.* Electric heat tracing can accurately provide the heating required for each application without the excessive energy used by a steam system.

• *Safety.* Self-regulating heaters and ground-leakage circuit breakers are the answer to the safety concerns of many engineers considering electric tracing. Self-regulating heaters eliminate potential problems from high-temperature failures, and ground-leakage circuit breakers minimize the danger of an electrical fault to ground.

• *Monitoring capability.* One question often asked about any heat-tracing system is, "How do I know it is working?" Electric tracing now has available almost any level of monitoring desired. The temperature at any point can be monitored with both high- and low-alarm capability. This has allowed many users to switch to electric tracing with a high degree of confidence.

The weaknesses of electric heat tracing are:

• *Poor reputation of old technology.* In the past, electric tracing systems were less than reliable. Due to past failures, some operating personnel remain unwilling to take a chance on any electric tracing, despite improvements over the last twenty years.

- *Design requirements.* A slightly higher level of design expertise is required for electric tracing than for steam tracing. User-friendly design programs have recently made electric tracing design significantly quicker and easier.
- *Lower power output.* Since electric tracing does not provide a large amount of excess energy (beyond what is required), it is less forgiving to problems such as damaged insulation or below-design ambient temperatures. Most designers include a 10%–20% safety factor in the heat loss calculation to cover these potential problems.

Table 1. Ranges of variables evaluated.

Variable	Base Value	Low Value	High Value	Other Values Considered
Maintenance Temperature, °F	150	50	250	400
Supply/Return Length, ft	100	40	300	
Labor Cost, $/h	25	15	40	
Circuit Length, ft	100	50	400	
Pipe Size, in.	4	2	8	24
Energy Cost				
Electricity, $/kWh	0.07	0.04	0.10	0.07
Steam, $/1,000 lb	4	3	6	0 (Excess/ Free Steam)

Additional information on base case:
The pipe to be traced is standard steel pipe, 400 ft long. The minimum ambient temperature is −20°F. The application is located outdoors in an electrically hazardous area. A condensate return system equal to the steam supply system is assumed. No cost is assigned to steam tracing system for boiler or main steam header. The electric tracing system is charged for panel and transformer capacity. All electrical supply is assumed to be run in rigid conduit; if existing cable trays are available, the electric supply would be significantly cheaper.

Choosing without an economic analysis

Some applications require either steam tracing or electric tracing regardless of the relative economics. For example, a large line that is regularly allowed to cool and needs to be quickly heated would require steam tracing because of its much higher heat output capability.

Applications that require accurate temperature control are generally limited to electric tracing. For example, chocolate lines cannot be exposed to steam temperatures or the product will degrade, and if caustic soda is heated above 150°F, it becomes extremely corrosive to carbon steel pipes.

It is considerably more expensive to supply steam from a long distance than to provide electricity. Unless steam is available close to the pipes being traced, electric tracing is usually the correct choice.

For most applications, particularly in chemical processing plants, either steam tracing or electric tracing can be used. In these cases, the best choice will depend on the installed costs and the operating costs of the two types of systems.

Economic decision rules

The question of the economics of various tracing systems has been examined in many articles and papers [(2–5), for example]. All of these publications have concluded that electric tracing is generally less expensive to install and significantly less expensive to operate. However, no one has suggested that electric tracing is more economic in all applications, and no effort has been made to determine the types of applications in which steam tracing is more economic.

Reference 5 was the most detailed study of real systems. In it, two applications in chemical plants were examined in detail by a major chemical company's engineering group. In each case, the costs of an existing tracing system were determined (one steam and one electric); then the alternative system was designed in detail and its costs were estimated using standard cost estimating techniques.

A computer program, the Auto-Trace Steam Analysis Program (ASAP), for estimating the costs of steam tracing and electric tracing systems was also used to estimate the costs of each tracing system for both applications. The results of the computer program compared well with the engineering estimates. Thus, we have a high degree of confidence that our analysis using the ASAP program will be reasonably accurate.

The program was used to study several key variables (supply line length, labor costs, circuit length, pipe size, and energy costs) at three different

Literature Cited

1. **Kohli, I. P.,** "Steam Tracing of Pipelines," *Chem. Eng.,* **86**(7), pp. 156–163 (Mar. 26, 1979).
2. **Lam, V., and C. Sandberg,.** "Choose the Right Heat-Tracing System," *Chem. Eng. Progress,* **88**(3), pp. 63–67 (Mar. 1992).
3. **Lonsdale, J. T., and J. E. Mundy,** "Estimating Pipe Heat-Tracing Costs," *Chem. Eng.,* **89**(24), pp. 89–93 (Nov. 29, 1982).
4. **Sandberg, C.,** "Heat Tracing: Steam or Electric," *Hydrocarbon Processing,* **68**(3), pp. 107–110 (Mar. 1989).
5. **Erickson, C. J.,** "A Study of Steam vs. Electrical Pipeline Heating Costs on a Typical Petrochemical Plant Project." IEEE paper No. PCIC-90-02, Presented at the IEEE Petroleum and Chemical Industry Conference, Houston, TX (Sept. 30, 1990).
6. **Gupta, D. C., and N. R. Rafferty,** "Proven Tools and Techniques for Heat-Tracing, " *Chem. Eng.,* **102**(5), pp. 104–107 (May 1995).

Table 2. Economic comparison for base case conditions*.

	Steam Tracing			Electric Tracing			
Installed Costs							
	Material	Labor	Total	Material	Labor	Total	Ratio, Stm/Elec
50°F	$5,186	$17,079	$22,265	$5,161	$2,572	$7,733	2.88
150°F	$5,186	$17,079	$22,265	$10,021	$3,092	$13,113	1.70
250°F	$5,363	$17,444	$22,807	$14,162	$3,462	$17,624	1.29
400°F	$6,299	$20,625	$26,924	$10,120	$3,936	$14,056	1.92
Operating Costs							
	Energy	Maintenance	Total	Energy	Maintenance	Total	Ratio, Stm/Elec
50°F	$1,378	$293	$1,671	$151	$183	$334	5.00
150°F	$4,063	$293	$4,356	$1,709	$183	$1,892	2.30
250°F	$5,055	$293	$5,348	$1,931	$183	$2,114	2.53
400°F	$6,431	$293	$6,724	$3,759	$183	$3,942	1.71

* Base case conditions are outlined in Table 1.

Table 3. Effects of supply line length on relative economics.

	Total Installed Costs			Total Operating Costs		
	Steam	Electric	Ratio, Stm/Elec	Steam	Electric	Ratio, Stm/Elec
Temperature = 50°F						
40 ft	$13,169	$6,387	2.06	$1,232	$334	3.68
100 ft	$22,265	$7,733	2.88	$1,671	$334	5.00
300 ft	$56,376	$12,477	4.52	$4,098	$334	12.3
Temperature = 150°F						
40 ft	$13,169	$11,565	1.14	$2,908	$1,892	1.54
100 ft	$22.265	$13,113	1.70	$4,356	$1,892	2.30
300 ft	$56,376	$19,512	2.89	$10,694	$1,892	5.65
Temperature = 250°F						
40 ft	$13,571	$16,077	0.84	$3,226	$2,114	1.53
100 ft	$22,807	$17,624	1.29	$5,348	$2,114	2.53
300 ft	$57,406	$24,023	2.39	$13,943	$2.114	6.60

temperatures (50°F, 150°F, and 250°F). This information is the basis for the general rules on the economic tradeoffs between steam tracing and electric tracing presented below.

The ranges for each of the variables are shown in Table 1. These ranges were chosen to cover most typical heat-tracing applications. Table 2 summarizes the economics of the base case. Tables 3–5 compare the effects of those variables that have the most significant impacts on relative economics (namely, supply line length, circuit length, and energy costs).

Although the choice between steam and electric tracing is complicated, most decisions can be made correctly by considering the desired pipe temperature and the length of supply lines (that is, from the power or energy source to the beginning of the tracer) required for the steam and the electric systems. The other key variables are unlikely to have a major effect on the correct choice of tracing system, unless one (or more) of them is well outside of the usual range (see Table 1).

1. For low-temperature applications — 50–150°F — electric tracing is usually much more economic (installed costs are less than half and operating costs are less than one-fourth those of steam tracing) if electricity is available. Even electric supply lines three times the length of the steam supply lines will usually not change the economics.

2. For medium-temperature applications — 150–250°F — electric tracing is usually somewhat more economic (30–50% lower installed costs and 50% lower operating costs) unless there is something unusual about the application. Long electric supply lines can make electric tracing more expensive (see Table 3) and larger process pipes that require more than one tracer may also make electric tracing less economic.

3. For high-temperatures — above 250°F — electric tracing is somewhat more economic (20–40% lower installed costs and 20–30% lower operating costs). The length of supply lines can easily change the economics. High-temperature applications are difficult for steam tracing because high-pressure steam must be used. They are difficult for electric tracing because the high-temperature tracers are less convenient than the lower-temperature tracers.

Circuit length considerations

Circuit length is a key determinant of total heat-tracing system costs *(6)*. This is because most of the costs of both steam tracing systems and electric tracing systems are related to the number of circuits. These costs are for supply lines (for both steam and electric systems), steam traps, condensate return lines, electric components such as power connections and end seals, and space in electric panels.

Circuit length is defined as the length of pipe that can be traced with one steam tracer (without additional traps or steam supplies) or one electric tracer (without an additional source of electricity). The circuit length may be determined by the geometry of the system being traced (many short, independent pipes) or the capabilities of the tracing system being used (for example, if the pipe length exceeds the heater's circuit length capability). Steam tracing (½-in. tubing) is generally limited to 150 ft per circuit (60 ft for ⅜-in. tubing). Electric tracing can generally handle circuit lengths of at least 300 ft, up to 700 ft for some standard heaters.

In addition to shorter circuit length capabilities, steam tracing also has higher per-circuit costs. The costs of the steam supply lines, steam traps, and condensate returns are usually higher than the electric supply and electrical connection costs. Because electric tracing has longer circuit length capabilities, electric designers are better able to minimize the number of circuits by maximizing circuit length.

Recent advances in electric tracing design have concentrated on using different heater control techniques to minimize the number of circuits and thus reduce total system costs. In some cases, total costs of electric tracing can be reduced by as much as 50%.

This effect can be seen by comparing the 50°F and 150°F cases in Table 4. In each of these cases, the total installed costs for 400 ft of electric tracing using 400-ft circuit lengths are less than half the costs of the same tracing using 50-ft circuit lengths. This effect is not seen at 250°F, as high-powered, high-temperature electric heaters have shorter circuit length capabilities.

Table 4. Effects of circuit length on relative economics.

	Total Installed Costs			Total Operating Costs		
	Steam	Electric	Ratio, Stm/Elec	Steam •	Electric	Ratio, Stm/Elec
Temperature = 50°F						
50 ft	$42,492	$12,236	3.47	$2,858	$477	5.99
100 ft	$22,265	$7,733	2.88	$1,671	$334	5.00
400 ft	$17,209	$4,354	3.95	$1,341	$156	8.60
Temperature = 150°F						
50 ft	$42,492	$16,459	2.58	$7,438	$1,609	4.62
100 ft	$22,265	$13,113	1.70	$4,356	$1,892	2.30
400 ft	$17,209	$7,250	2.37	$3,477	$1,334	2.61
Temperature = 250°F						
50 ft	$43,590	$22,108	1.97	$9,645	$2,297	4.20
100 ft	$22,807	$17,624	1.29	$5,348	$2,114	2.53
400 ft	$17,612	$17,432	1.01	$4,273	$2,047	2.09

Table 5. Effects of energy cost on relative economics.

	Total Installed Costs			Total Operating Costs		
	Steam	Electric	Ratio, Stm/Elec	Steam	Electric	Ratio, Stm/Elec
Temperature = 50°F						
0/0.07*	$22,265	$7,733	2.88	$293	$335	0.87
3/0.04	$22,265	$7,733	2.88	$1,327	$270	4.91
4/0.07	$22,265	$7,733	2.88	$1,671	$334	5.00
6/0.10	$22,265	$7,733	2.88	$2,360	$400	5.90
Temperature = 150°F						
0/0.07	$22,265	$13,113	1.70	$293	$1,892	0.15
3/0.04	$22,265	$13,113	1.70	$3,341	$1,160	2.88
4/0.07	$22,265	$13,113	1.70	$4,357	$1,892	2.30
6/0.10	$22,265	$13,113	1.70	$6,389	$2,625	2.43
Temperature = 250°F						
0/0.07	$22,807	$17,624	1.29	$293	$2,114	0.14
3/0.04	$22,807	$17,624	1.29	$4,084	$1,286	3.18
4/0.07	$22,807	$17,624	1.29	$5,348	$2,114	2.53
6/0.10	$22,807	$17,624	1.29	$7,875	$2,941	2.68

*Costs are given in $/1,000 lb for steam (first number) and $/kWh for electricity (second number). For example, 0/0.07 refers to free (excess) steam and electricity at $0.07/kWh.

J. T. LONSDALE is a technical marketing specialist for the Chemelex Div. of Raychem Corp., Menlo Park, CA (415/361-6717; Fax: 415/361-2389; e-mail: jlonsdal@raychem.com). He has worked in the field of heat tracing for Raychem for 15 years, and previously he was a production manager in a chemical plant for Kaiser Chemicals. He has a BS from Northeastern Univ. and an MS from Purdue Univ., both in chemical engineering, and an MBA from Harvard Univ. He is a member of AIChE, IEEE, and ISA. He has authored many articles on heat tracing, including the heat tracing chapter in the "McGraw-Hill Piping Handbook."

N. J. LITTLE is a process engineering consultant with Kilborn Inc., Sarnia, Ontario (519/336-0201; Fax: 519/336-0209). He has over 30 years of experience in process design, plant optimization, and energy conservation in process industries. He has a BASc in chemical engineering from the Univ. of Waterloo, is a registered Professional Engineer in Ontario, and is a member of the Instrument Society of America and the Canadian Institute of Electronics.

D. L. BRADFORD is a technical manager for the Chemelex Division of Raychem Corp., Cincinnati, OH (513/762-7665; Fax: 513/721-4628), where he has worked since 1980, mostly in the field of heat tracing. Previously, he was a product research engineer for Procter & Gamble Co. He has a BS in chemical engineering from the Univ. of Maryland and an MBA from the Univ. of Virginia, and is a member of AIChE.

Insulate Heat Tracing Systems Correctly

The right insulation for heat tracing is critical for proper temperature maintenance, economic operation, and long equipment life.

Peter R. Baen and Roy E. Barth,
Thermon Industries

Many chemical processes involve the storage and transfer of products in pipelines. Often these products can freeze, become extremely viscous, or condense if allowed to cool to ambient temperatures. While insulation and an adequate flow rate can sometimes prevent problems during processing, it may not be possible to prevent cool down during extended nonprocessing times. In these cases, heat must be added to the piping and equipment to offset the heat losses.

A common way to provide this heat is to install electric or fluid heat tracing along the pipe (Figure 1). A critical component of the heat tracing system is an effective and efficient insulation.

This article outlines the factors that need to be considered in choosing the insulation and optimizing tracer design. The selection process involves:

1. Choosing the type of insulation that best fits the application.

2. Determining the thickness of insulation based on an economic analysis.

3. Selecting a weather barrier to help maintain system integrity after installation and extend insulation service life.

4. Matching the heat tracing system to the application.

Choosing the type of insulation

When selecting an insulation type for a specific heating application, the engineer should consider the following characteristics:

Thermal conductivity. Insulating materials can be described as a matrix of gas cells, fibers or polymers, and binders. The fiber or polymer type, the density of the gas cells, and the binder together determine the insulation's thermal conductivity. Depending on the type of insulation, the dry thermal conductivity typically varies from 0.18 to 0.45 Btu•in./h•ft^2•°F.

Maximum and minimum service temperatures. Some insulants, such as fiberglass, have binders that volatilize at elevated temperatures, which can lead to degradation of mechanical strength. Other insulants, such as polyurethane, can char or burn if their maximum service temperature is exceeded.

Some insulants that are highly susceptible to moisture absorption also have minimum service temperatures. Below these temperatures, their use is discouraged due to the likelihood of condensation being present.

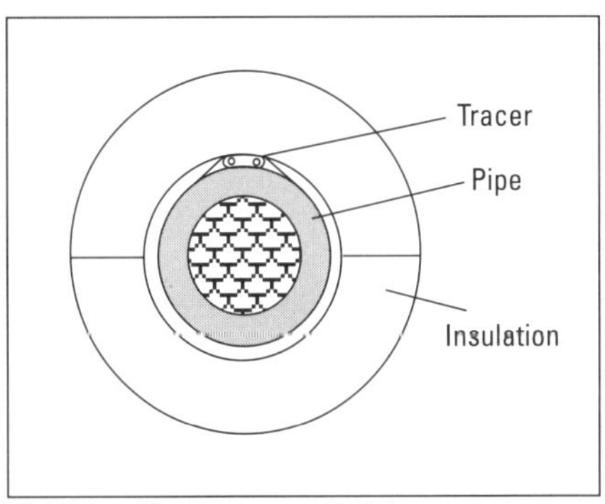

■ ***Figure 1. Heat traced piping requires proper insulation.***

	Fiber Glass	Mineral Wool	Cellular Glass	Polyiso-cyanurate	Rubber Foam	Expanded Perlite	Calcium Silicate
Temperature Range, °F*	Ambient to 800	Ambient to 1,200	–320 to 900†	–320 to 250	–40 to 220	Ambient to 1,200	Ambient to 1,500
Thermal Conductivity, Btu•in./h•ft²•°F‡	0.25	0.27	0.36	0.18	0.29	0.42	0.28
Average Density, lb/ft³ §	7	8	8.5	2	5.5	12	15
Vibration Resistance	Good	Good	Fair	Good	Good	Fair	Fair
Combustibility and Flame Propagation	Yes, Binders	Yes, Binders	No	Yes	Yes	No	No
Dimensional Stability#	Good	Good	Excellent	Excellent	Good	Excellent	Excellent
Compressive Strength, psi¶	Low	Low	100	20	Low	100	160
Water Absorption, vol%**	50–90	Nil–90††	0.2	1.6	1	2–90††	90

Table 1. Properties of common insulations used with heat tracing systems.

* Temperature ratings can vary from product to product within a generic class.
† Special expansion joint construction required above 400°F.
‡ Typical values at 100°F mean insulation temperature.
§ Density can vary with insulation size and thickness.
# Nonrigid insulants are given a lower rating due to their inherent field compressibility.
¶ Ratings can improve as density increases.
** Water absorption data are based on information in *(4)* and *(5)*.
†† Waterproofing agents may be destroyed when exposed to temperatures above 250°F.

Vibration resistance. Almost any insulation used in an industrial application will be subjected to some vibration due to compressors, pumps, or the motion of the fluids passing through pipelines.

Under extreme vibration conditions, cellular glass is abrasive to the pipe and heat tracing. Rigid insulations, such as expanded perlite, calcium silicate, or cellular glass, can be susceptible to cracking.

Chemical reactivity. Some materials being processed may react if absorbed by the insulation. One such example is the combination of calcium silicate and ethylene oxide.

Combustibility. Some insulants, if exposed to extreme temperatures, can be combustible. Others may not autoignite but will support and propagate a flame.

Corrosion to piping. Certain combinations of insulants and metals may result in corrosion. For example, any insulant having free chlorides can cause stress corrosion cracking of austenitic stainless steel.

Water absorption. In heat tracing applications, one of the most common insulation problems is moisture intrusion. An insulant gains its effectiveness (and its low thermal conductivity) from the tiny gas cells (or pockets) that are present throughout. If moisture replaces the gas in these pockets, the thermal conductivity of the insulant increases drastically. If an insulant's cell structure is "closed cell," it will be inherently resistant to moisture absorption; if the material is "open cell," the opposite is true.

Capillarity. This refers to the insulation's tendency to wick water or other liquid. In a high-capillarity insulation, damage caused by water or product entry into the insulation spreads quickly. This wicking can damage significant lengths of pipe insulation, even though only a small portion was directly exposed to moisture.

Good design practice requires the review of many properties when selecting an insulation or a weather barrier.

Compressive strength. Where the insulation must support a load such as a pipe or occasional foot traffic, a high compressive strength is required. Conversely, where insulation is used as an expansion joint, a low compressive strength may be required to allow for insulation movement during pipe or vessel temperature changes.

Dimensional stability. Many insulants, unlike piping, contract when heated. Dimensional stability data indicate the amount of dimensional change that can be expected when an insulation is used at high temperatures. Dimensional stability is commonly the basis used by manufacturers in establishing the maximum temperature rating of an insulant.

Nonrigid insulation materials can experience dimensional changes due to compression during installation. This can also impact the insulant's overall resistance to heat loss.

Flexural strength. This affects the ease with which the insulation can be installed. A material with low flexural strength usually indicates that breakage and extra waste during installation can be expected.

Any one (or more) of these insulation properties may influence the selection process for heat tracing applications. Good design practice requires a review of all of these in order to pinpoint any properties that may be critical to the application under consideration. Table 1 compares some key characteristics of several common insulation materials.

Choosing the weather barrier

Properly designed and installed weather barriers play a vital role in ensuring the effectiveness of insulation, especially for insulants that can absorb water.

Some of the characteristics that should be considered in the selection of the weather barrier are:

Mechanical strength. Various types of mechanical forces are likely to be exerted on the insulation system, including:

- forces due to different thermal expansion (or contraction) rates of the pipe, insulation, and weather barrier materials;
- compressive forces due to pipe hangers or plant personnel walking on the insulated pipes; and
- shearing forces due to wind.

Maximum and minimum service temperatures. Temperature limitations are more commonly a concern with mastic type coatings and nonmetallic weather barriers. These can become brittle and crack at subfreezing temperatures. At high temperature extremes they can quickly deteriorate, melt, or crack.

Corrosion resistance. Many plant environments contain chemicals and vapors that are corrosive. In coastal regions, salt-laden air contributes to corrosion. Sunlight (ultraviolet radiation) can degrade mastic and nonmetallic weather barriers. In addition, the insulation may contain moisture and contaminants that could attack the weather barrier on its inner surface.

Fire resistance. If the insulant is combustible and can propagate a flame under the weather barrier, a barrier that is *weak* in fire resistance — permitting fire hoses to get to the burning insulant — is appropriate.

Chemical stability. Some chemicals, when accidentally spilled on a weather barrier, have been known to destroy the barrier's integrity in a matter of minutes. This most often results in moisture getting to the insulation, ruining its efficiency.

Sealing capability. Regular shaped objects such as pipes can be sealed by standard jacketing. Irregular objects, such as valves, pumps, and instruments, are often more reliably sealed by a mastic weather barrier. In either case, assuring that weather barriers are properly overlapped and joints are sealed is critical to keeping the insulation dry over its lifetime.

Any of these characteristics can be critical to the successful performance of a weather barrier. Remember that a poorly designed or installed weather barrier can damage the thermal insulation, which can, in turn, render the heat tracing system ineffective. Table 2 summarizes some of the key characteristics of insulation weather barriers.

Evaluating insulation products

Once the desired properties of an insulation system are determined, the next step is to evaluate the commercially available insulating materials and weather barriers. Manufacturers supply a wealth of technical data concerning their products.

Most of the physical characteristics are determined through standard tests developed by the American Society for Testing and Materials (ASTM). Interpretation of the results, however, requires some knowledge of

Table 2. Relative characteristics of insulation weather barriers.

	Polyvinyl Chloride	Fiberglass Reinforced Plastic	Aluminum*	Stainless Steel*	Paper†	Mastic‡
Temperature Range, °F	–20 to 150	–40 to 180	–60 to 500	–60 to 500	50 to 200	–40 to 200
Flexural Strength	Excellent	Excellent	Good§	Good§	Good	Good
Corrosion Resistance#	Good	Excellent	Good	Excellent	Poor	Fair
Fire Resistance	Low	Low	High	High	Low	Low
Chemical Stability	Good	Excellent	Good	Excellent	Poor	Fair
Sealability¶	Good	Good	Good	Good	Poor	Excellent

* Temperature ratings of metal jackets are limited by the maximum temperature range of the coolants.
† Paper jackets are intended for indoor service and, hence, do not exhibit good weatherability.
‡ Ratings can vary depending on the specific mastic formulation.
§ Depending on the thickness and material type, metallic jackets can permanently distort when used in combination with non-rigid insulation.
\# Corrosion resistance to direct exposure to chemicals should be evaluated on a case-by-case basis. These ratings are based on general outdoor plant environment experience.
¶ With the exception of mastics and paper, it is assumed that longitudinal joints are properly overlapped and proper sealants are used at all circumferential joints.

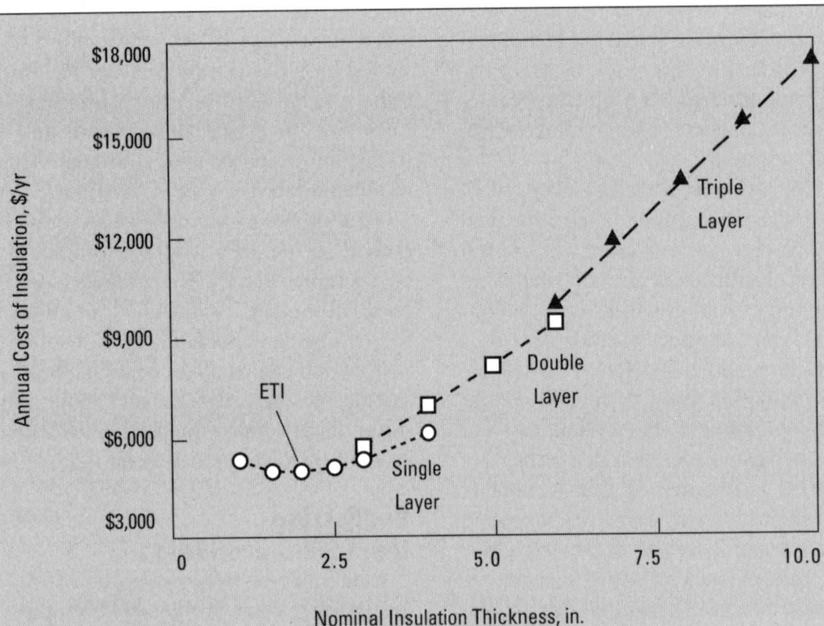

■ *Figure 2. ETI is the thickness corresponding to the lowest annual cost of insulation.*

Results for Single-Layer Insulation Thickness*

Insulation Thickness, in.	Total Annual Cost, $/yr	First-Year Estimated Insulation Cost, $	After-Tax Energy Savings, $	Insulation Payback Period, yr	Estimated Return on Investment, $
1.0	$5,426.39	$9,422.50	$23,965.27	0.3971	$251.82
1.5	4,978.27	11,220.50	24,402.75	0.4644	215.33
2.0	5,032.31	13,768.50	25,650.47	0.5641	177.26
2.5	5,110.85	15,776.00	24,826.21	0.6418	155.81
3.0	5,356.29	17,924.00	24,927.11	0.7262	137.69
4.0	6,080.27	23,020.00	25,089.95	0.9267	107.91

*For simplicity, only the results for single-layer insulation are tabulated here. Results for double- and triple-layer insulation are presented in the curves above.

Data:

Temperature: Process = 366°F Maintenance = 40°F
Hours of Operation: Process = 7,960 h/yr Maintenance = 780 h/yr
Ambient Conditions: Temperature = 55.6°F Wind = 6.6 mph
Insulation: Type = 7PCF Fiberglass
Size = 4.0 in. Length = 1,000 ft Complexity = 1 Fittings = 1
Unit Price for 2-in.-by-2-in. Pipe Insulation = $4.00
Insulation Project Life = 10 yr
Energy Cost: $4.00/million Btu
Local Labor Rate = $30/h (Location = West Virginia)
Cost to Finance Project = 15% Corporate Tax Rate = 46%

the test procedures *(1)*. In some cases, the tests may not be pertinent to the specific application.

For example, ASTM C 335-89, "Steady State Heat Transfer Properties of Horizontal Pipe Insulation," outlines the test method commonly used to determine the thermal conductivity of an insulation. The testing is based on bone-dry material. Because moisture affects different insulation materials in different ways, the conductivities published are not always accurate under actual service conditions. The conductivities of insulants relative to each other may provide a good rough comparison, but the effects of water or condensate should be considered, as they will affect overall performance. Therefore, the selection process should also consider the intended service and not focus on conductivity alone.

As another case in point, ASTM C 165-92, "Standard Test Method for Measuring Compressive Properties of Thermal Insulation," describes the test method for determining the compressive properties of thermal insulation. The test is carried out at room temperature after the insulant has been dried in an oven at 215–250°F and held in a desiccator prior to testing. Extrapolation of these test results for comparing the compressive strengths of the two insulations in a higher temperature application is difficult, if not impossible.

In situations where the test method data do not appear to confirm the applicability of an insulation material or weather barrier, the designer must make an engineering judgment based on performance data from previous installations and/or recommendations from manufacturers, contractors, or consultants. Additional information such as that given in *(2)* may also be useful.

Determining insulation thickness

The proper thickness of piping insulation should be based on simple economics. The addition of each half-inch of insulation will reduce heat losses and energy costs. However, the additional insulation costs more for the initial installation, and thus tends to diminish any operating cost savings. Multiple layers of insulation are sometimes also considered, but this adds significantly to the labor costs.

The insulation thickness that minimizes energy costs can be determined as follows: First add the annual cost of energy lost through the insulation and the amortized cost of installing and maintaining the insulation system for each thickness of insulation. Then compare these costs and select the thickness corresponding to the lowest total annual energy costs. This thickness is commonly referred to as the Economic Thickness of Insulation (ETI).

Although calculating the ETI is quite a lengthy exercise when done by hand, computers have greatly simplified the iterative procedure. Various ETI programs are available from insulation manufacturers, contractors, and consultants, as well as from the North American Insulation Manufacturers Association (NAIMA), Alexandria, VA. [*CEP*'s Annual Software Directory also lists such programs. — *Editor*] Some companies also publish recommended insulation thickness tables for virtually any application. A computer analysis of an ETI calculation is shown in Figure 2.

More information on economic insulation thickness appears in *(3)*.

Matching heat tracing to the application

The heat tracing may be either a tube that carries heated fluids, such as steam, hot water, ethylene glycol and water, or heat-transfer oils, or may be an electric heating cable. Fluid tracers can be installed with heat-transfer cements, banding, or tie wire; or, if only a small amount of heat transfer is needed, the tracer can be isolated from the pipe using insulating spacing material. Electric heating cables are available in a variety of types, including self-regulating, power-limiting, parallel constant-wattage, and series-resistance. These cables may be installed with a parallel coverage of aluminum tape and circumferential bands of polyester tape, polyester tape alone, or in some cases banding.

The tracing systems may operate without temperature control devices, be controlled by an ambient sensing control unit, or may be controlled by a pipe-wall-sensing temperature-control unit. Clearly, there are many possible combinations of tracers, insulation, and temperature control. Selecting the right combination involves evaluating the following tracing system characteristics.

Temperature maintained. Under equilibrium conditions, the heat delivered will exactly equal the heat lost through the insulation. For a fluid tracing system, the temperature maintenance capability is a function of the resistance to heat loss represented by the insulation, the conductance of heat from the heat tracer to the pipe, and the temperature of the heat-transfer medium. For an electric heat tracing system, the temperature maintenance capability is a function of the resistance to heat loss represented by the insulation and the power output of the heating cable (measured in W/ft or W/m).

The proper insulation thickness is determined based on an economic analysis.

Tracer operating temperature. This refers to the sheath (or tube surface) temperature at which a tracer must operate to achieve the desired pipe temperatures. For a fluid tracer, this temperature is essentially that of the tube through which the heating medium flows, and is one of the primary criteria upon which tracer material selection is based. For an electric tracer, this temperature is a function of the watt density at which the cable is operated, the method of cable attachment, and the pipe temperature being maintained. For an electric tracer, the tracer operating temperature determines the type of electrical insulation required on the heating cable.

Maximum and minimum tracer temperature. Electric heat tracing cables with plastic or polymeric type materials in their construction have lower temperature limits than those with mineral insulation and metal sheathed materials. Upper temperature ratings for plastic/polymeric insulation materials range from 200°F to 500°F. Ratings for mineral insulated cables may range as high as 1,000°F. Depending on the pressure, copper fluid tracers are suitable for use up to 406°F and stainless steel tracers up to 1,000°F.

Temperature swings. Electric tracing systems are generally thermostatically controlled. Assuming a pipe or vessel wall sensing controller, temperature swings as high as ±20°F or as low as ± 2°F can occur, depending on the precision and accuracy of the temperature controller and the dynamics of the pipe and tracing system. If an ambient sensing controller is used, temperature swings as high as the differential between the design maintenance temperature and the minimum ambient temperature can be expected.

Fluid tracing systems are generally not thermostatically controlled. Pipe or vessel temperature swings vary with ambient changes as well as with the inherent overdesign in the tracing system.

Corrosion resistance. A fluid heat tracing system's ability to withstand corrosive effects depends on the fluid medium being used, the product or process being maintained, and the plant environment. For example, steel tubing is seldom recommended for use with steam tracing due to the internal rusting problems that can occur (especially in intermittent service). Copper tubing is seldom specified in ammonia plants because copper deteriorates rapidly in the presence of ammonia vapors. Certain grades of stainless steel tubing can be affected by stress corrosion caused by the free chlorides present in insulation such as calcium silicate.

An electric heat tracing system can also be affected by corrosion. Grounding braids are often specified to be nickel-plated, and they can also be covered with a plastic or polymeric overjacket. Terminations must be well sealed to prevent corrosion at normal or elevated operating temperatures. Additionally, the plastic or polymeric materials used as the primary heater cable dielectric insulation must be compatible with the environment.

Mechanical and flexural strength. Heat tracing may be exposed to compressive forces if personnel walk on the piping (and hence the tracer), bending forces if the tracer must be twisted around many irregular surfaces, and vibrational forces due to pulsating flows, vibrating pumps, and the like. Tubing or electric heat tracing cables generally have minimum specified bend radii that should be adhered to under all installation temperature conditions. Fluid tracing is also exposed to the internal pressure of the heating medium being contained.

Water absorption. While electric heat tracing cables are generally designed to be quite resistant to moisture intrusion, the end terminations and power connections, which are field fabricated, can sometimes be vulnerable to moisture intrusion if improperly constructed. Among the large group of dielectric materials (insulants) currently in use, some, such as magnesium oxide (MgO), tend to be more hygroscopic. Hence, these terminations are preferably constructed at the factory under controlled conditions.

Hazardous location characteristics. The suitability of electric tracing for use in hazardous (classified) areas is commonly based on an evaluation of the tracer's maximum heater sheath temperature under the conditions specified in the 1993 National Electric Code (NEC), Article 427 and Article 500. Additional requirements on the tracing system design can be imposed by approval agencies such as Factory Mutual (FM), the Canadian Standards Association (CSA), and Underwriters Laboratories (UL), as well as the Institute of Electrical and Electronics Engineers (IEEE) Standard 515-1989, "IEEE Recommended Practice for the Testing, Design, Installation, and Maintenance of Electrical Resistance Heat Tracing for Industrial Applications."

Fluid tracing systems do not fall under the jurisdiction of agencies or industry standards and generally enjoy unrestricted use in electrically classified areas (providing local standards are met).

Stagnant pipe heat-up. A pipe heat-up application generally requires larger heat outputs than required for temperature maintenance. Most heat tracing applications are for temperature maintenance only (that is, preventing the product from cooling below a given temperature), so the pipe or vessel is presumed to be held at a given temperature. However, since heat loss calculations assume a safety factor and the lowest expected ambient temperatures, heat-up can usually be accomplished by a "maintain system" if sufficient time is allowed.

Wrapping up

In closing, heat tracing performance is critically dependent on the thermal insulation system. Considering not only the thermal conductance of the insulation and tracer heat delivery, but also reviewing all other anticipated operating conditions before the design and installation phases will help achieve a properly operating system and a long service life. A maintenance program that includes routine inspections and audits of the thermal insulation and heat tracing system's performance as well as reviews of any design or operational changes should also be considered to assure system longevity. CEP

Literature Cited

1. **American Society for Testing and Materials,** "Annual Book of ASTM Standards, Vol. 04.06: Thermal Insulation; Environmental Accoustics," ASTM, Philadelphia, PA (1993).
2. **National Insulation Contractors Association,** "National Commercial and Industrial Insulation Standards," NICA, Omaha, NE (1988). [NICA became the National Insulation and Abatement Contractors Association (NIACA) in 1989 and is now headquartered in Alexandria, VA.]
3. **Irwin, W.,** "Insulate Intelligently," *Chem. Eng. Progress,* **87**(5), pp. 51–55 (May 1991).
4. **Turner, W. C., and J. F. Malloy,** "Thermal Insulation Handbook," Krieger/McGraw-Hill, New York, NY (1981).
5. "FOAMGLAS Insulation Systems," Brochure FI-201A, Pittsburgh Corning Corp., Pittsburgh, PA (1992).

P. R. BAEN is currently manager of marketing support for Thermon Industries, San Marcos, TX (512/396-5801; Fax: 512/396-3627). During his 17 years with Thermon he has held various positions in sales and marketing, as well as in research and development. He holds bachelor's degrees in physics and mathematics and a master's degree in mathematical analysis from Southwest Texas State Univ. He has authored numerous papers and technical articles on heat tracing, and has been involved in the development of three industrial heat tracing standards and practices.

R. E. BARTH is vice president of research and development for Thermon. He has over eleven years of engineering experience in the design and installation of industrial heat tracing systems, plus seven years of research experience in the areas of liquid metal boiling, arctic deicing and anti-icing, instrument heating, and commercial heat tracing systems. He has directed new product development projects in the areas of insulation and electric and steam tracing, and has been involved in the design and development of computer-aided design tools for the heat tracing industry. He holds a BS and an MS from the Univ. of Texas at Austin, both in mechanical engineering. He contributed to the "Thermal Insulation Handbook" as well at portions of IEEE Standard 515-1989. He is a member of the American Society of Mechanical Engineers and an associate member of the IEEE, and is a registered professional engineer in Texas.

Reboiler System Design: The Tricks of the Trade

Paying attention to the little details can help prevent big problems. Here's how to make plants safer, less expensive, more efficient, and easier to maintain.

Alfred J. McCarthy and Brian R. Smith,
The M. W. Kellogg Co.

Reboiler systems have been an important part of process plant design since the beginning of the chemical process industries (CPI). However, some of the details that go into the design of a reboiler system are unappreciated by many chemical engineers, and "attention to detail is of utmost importance" *(1)*.

This article reviews the details of reboiler system design and how they affect the system from maintainability, operability, and safety standpoints.

MAINTAINABILITY

Maintainability should be one of the first issues addressed during reboiler selection.

All other things being equal, various physical parameters can be used to select the proper reboiler system configuration, such as the selection guide presented in *(2)*. When using charts of this sort, the engineer must first examine all the premises upon which the chart was developed and be sure that these premises fit the situation at hand.

One of the most important considerations in determining system configuration is the fouling potential of the individual streams. Where one of the streams, either the process reboiled stream or the heating medium, is judged to have a significantly higher fouling potential, that stream should be assigned to the tube side of the reboiler. Such an arrangement is better in terms of maintainability, because it is far easier to clean the tube side of a reboiler than the shell side.

A situation from some 40 years ago illustrates this point. A mixed C_4 fractionation column had a kettle-type reboiler with the process stream on the shell side. The accepted cleaning process was to pull the bundle and place it in an incinerator to burn off the rubber-like deposits. This practice would be questionable in today's environmentally driven plant design atmosphere.

There are also economic reasons for assigning the more fouling fluid to the tube side. If a fouling fluid is allocated to the shell side, a square-pitch tube layout is a virtual necessity to enable cleaning. Because this generally results in a larger tube bundle than a conventional triangular pitch, the square-pitch exchanger will be more expensive per unit of heat-transfer area.

Either the process stream or the heating stream can be fouling. Unfortunately, sometimes both are. When this situation arises, the selection of sides can be determined by the type of cleaning that may be effective. For example, if one side could be cleaned by circulating a solvent (water, hydrocarbon solvent, or chemicals) while the other requires reaming, scraping, or hydrojetting, the fluid needing the physical method should be assigned to the tube side.

If all else fails and a fouling fluid must be on the shell side, one must then fall back on exchanger design types that will allow shell-side cleaning, such as pullable bundles and square-pitch layout. Figure 1 illustrates the designations used by the Tubular Exchangers Manufacturers Association (TEMA).

The inlet head (which will also be the outlet in even multipass designs) should be a TEMA A or C design with removable end covers. Type B, which has an integral cover (or bonnet), requires that the process connections be disassembled for cover removal. This results in a heavier load for

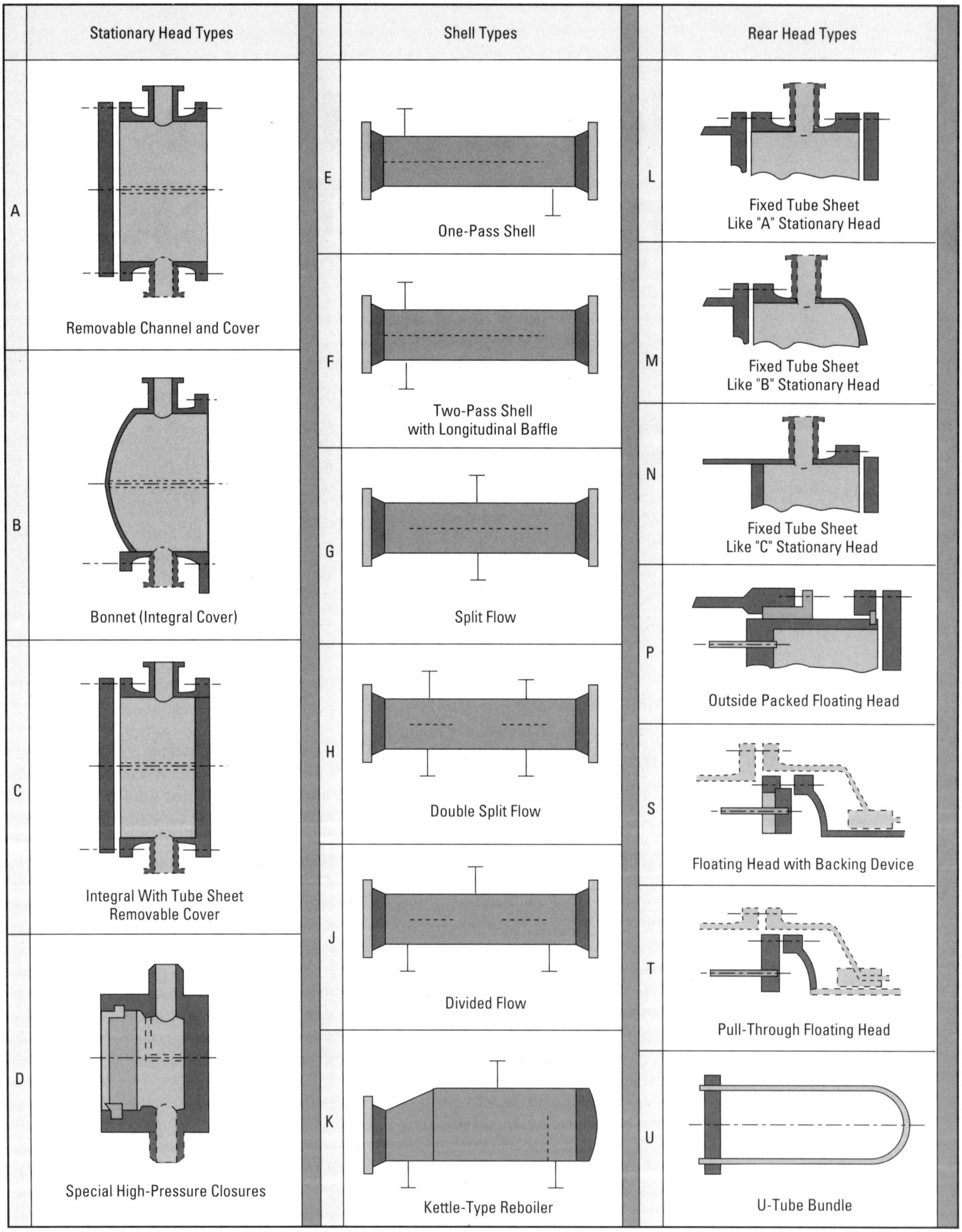

■ ***Figure 1. TEMA designations for shell-and-tube exchangers.***

maintenance removal and subjects the flanges on both the head and the piping to potential damage during cover removal and replacement.

The rear head selection is subject to the same considerations. If the number of tube passes is odd, a type L or type N design (with removable covers) is preferable. For an even number of passes, a type M is preferable over type L because it has no major process connections and it has only one full body flange instead of two.

If a U-tube design is desired and tube cleaning will be necessary, types L, M, and N, as well as types S and T (which are used when differential expansion is a concern), are preferred over a true U-tube bundle. If reaming is used in a U-tube exchanger, the return bend may be worn away. If hydrojetting is used, any dislodged material, as well as the hydrojet stream, would be directed back at the hydrojet operator.

Once the fluids have been assigned to exchanger sides, the exchanger orientation is generally fixed, since orientation is usually a function of the process fluid side assignment. When the process fluid is on the shell side, horizontal thermosyphons or kettles are the normal choice, whereas vertical thermosyphons are used when the process fluid is on the tube side.

Once-through reboiler configurations can be either way (shell-side/horizontal or tube-side/vertical). However, this is not a hard-and-fast rule. On a recent project involving a regenerative acid-gas stripper utilizing diethanolamine (DEA), a once-through design was selected, to minimize the DEA contact with hot surfaces. The process fluid was assigned to the tube side because DEA is fouling, and a horizontal orientation was chosen to minimize the pressure at the process side inlet. The designers recommended a horizontal CEN exchanger (TEMA type C inlet head, type E shell, and type N rear head) with two outlets on the rear head, a vertical up for the vapor and entrained liquid, and a smaller vertical down for any liquid separated in the head, both discharging to the reboiler return section of the tower.

OPERABILITY

After the fluids have been assigned to the exchanger sides based on maintainability considerations, there are a number of things that can be done to enhance operability.

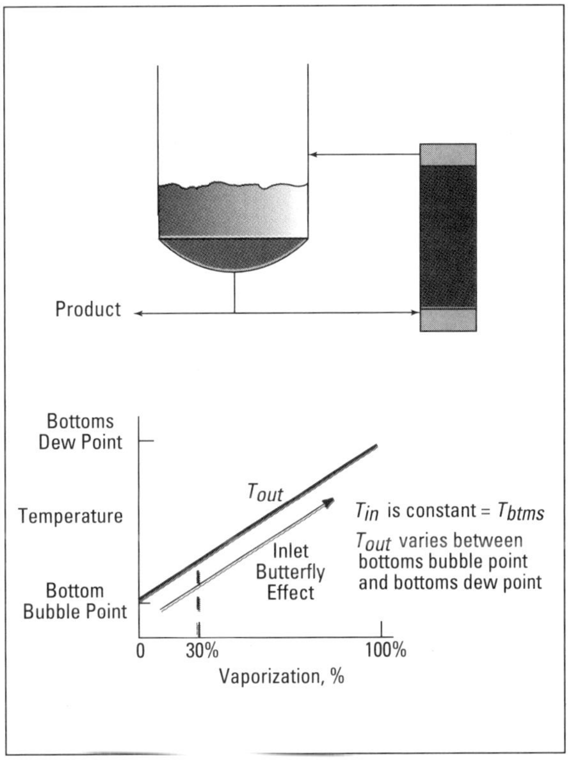

■ *Figure 2. Open base column.*

Exchanger configuration

Most reboilers are on towers that have, or may have, a wide boiling range mixture as the bottoms product and steam as the reboil medium. Though some of these may be in light hydrocarbon service where the bottoms product is clean and stable (not likely to polymerize or otherwise degrade), the steam is usually on the shell side because it is almost always the cleaner of the two fluids. Because this rules out both a kettle and a horizontal thermosyphon, the choices left are a vertical once-through or recirculation thermosyphon.

Note that the selection guide in *(2)* rules out a vertical thermosyphon unless the vaporization is less than 30%. However, the text points out that this restriction is for once-through (trap-out) vertical thermosyphons. A recirculation vertical thermosyphon can be designed so that the recommended vaporization restriction can be followed even if the tower design requires a higher liquid vaporization.

Having determined the requirement for a recirculation vertical thermosyphon, there are still two options for this configuration (although they refer to the tower bottom internals rather than reboiler) — the standard open base configuration, and the partition plate (or baffled bottom) arrangement. There are distinct benefits to a baffled bottom that should be considered during the selection process, and in many instances these benefits far outweigh the added cost. To understand the differences between the two arrangements from both a process and reboiler thermal rating basis, refer to Figures 2 and 3.

Open base. Figure 2 shows the open base design, as well as the temperature of the process fluid through the exchanger as a function of percent vaporization of that fluid (on a simplified straight-line basis). The process fluid entering the reboiler must be equivalent to the bottoms product in quality and, therefore, temperature. The outlet temperature is, thus, a function of the percent vaporization and can theoretically range to the dew point of the tower bottoms (100% vaporization).

For the simplest of cases, a pure component bottoms product, this presents no problems for a thermal design, since the reboiler outlet is essentially constant regardless of the percent vaporization chosen for design. (Note that the curve of percent vaporization vs. temperature is developed from a series of equilibrium flashes of the bottoms product, and is not a bottoms distillation curve.)

For wide-boiling multicomponent mixtures, however, the change in temperature with percent vaporization can be significant. This relationship must be available to the exchanger rating engineer, who will generally select the design percent vaporization to satisfy other thermal rating criteria. If this is not done and a percent vaporization that is different from the dew-point vaporization/temperature is specified, the exchanger may not have enough surface area.

Very often a temperature indicator is specified at the reboiler outlet. (In some instances this has even been specified as the control point for the reboiler.) For a wide-boiling mixture, a temperature indicator at this point, in conjunction with other data (as discussed below), can be used as an indicator of the "health" of the reboiler — that is, the progression of fouling. However, it is not a suitable control point for this type of reboiler configuration (although it could be a suitable control point for a once-through vertical thermosyphon).

All exchangers are prudently designed with fouling factors. These fouling factors are usually based on the design or operating company's experience, or they can be obtained from generic sources such as TEMA. For services like reboilers, which generally have excellent heat-transfer coefficients on both sides (boiling hydrocarbon on one side, usually condensing heat-transfer fluid on the other), even modest fouling factors result in significant exchanger oversurface in a new and clean state — 50% oversurface is not uncommon for reboilers.

Fouling factors are selected based on the propensity of the fluids (both sides) to foul and the expected campaign time of the plant. In some instances, due to a very high fouling potential of one side, it is common to provide a 100% spare reboiler or three 50% units. This allows the fouled exchangers to be cleaned while the plant is in operation, thus maintaining the desired campaign time.

With the open bottom configuration, it is generally accepted practice to design the hydraulic system based upon the liquid level in the tower bottom at the bottom tangent line (BTL), or at most at the bottom tap of the level instrument (generally about 6 in. above the BTL). It is also often common to design considerable surge volume into the bottom of the tower. Depending on the quantity of the bottoms product and the specified surge time, this can result in significant "normal" level ranges, sometimes 10 ft or more.

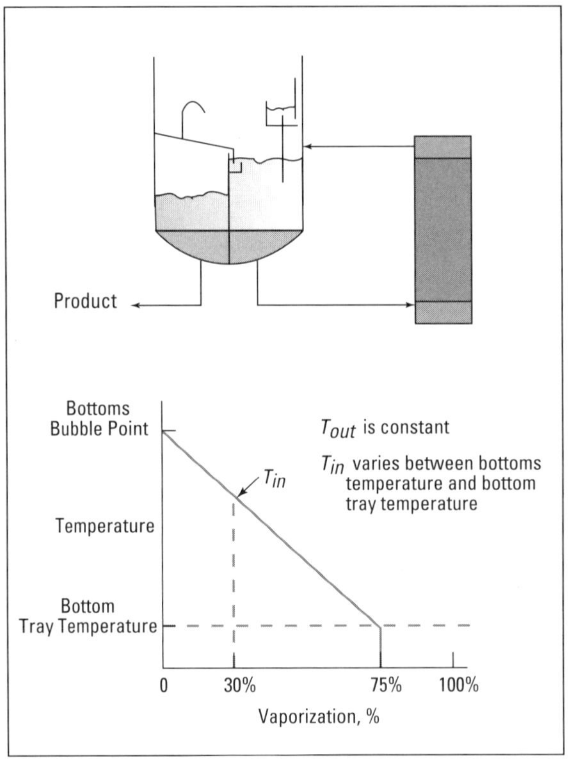

■ ***Figure 3. Partition plate, or baffled bottom, base.***

When all of these considerations are taken into account, along with the options available for heat-transfer fluid control (discussed later), operators have experienced difficulty in stabilizing reboiler operation, primarily during startup. These problems have led to the inclusion of a flow restriction valve in the inlet to the reboiler. This is a common requirement, even written into design standards, of many companies. The problem is that it caters only to one specific mode of operation, and for normal operation it can be a curse rather than a blessing. The flow restriction valve can be of assistance in stabilizing flow to the reboiler during startup when the bottom of the tower is filled to its maximum level and the other factors noted above conspire to give an initial reboiler operation similar to a percolating coffee pot.

However, once in operation at rates high enough to satisfy tower tray stability requirements (about 50% load), the valve is not necessary, and in fact starts to cause problems just by being there. Even wide open, it imposes additional pressure drop and, therefore, lower circulation rate, and with this configuration gives a higher process fluid outlet temperature for any given load.

These valves always have local manual operation, which introduces the possibility for operator error. Failure to open the valve fully after startup imposes an even higher pressure drop, and higher process fluid temperature, during normal operation. In effect, it steals process-side fouling factor.

Thus, as recommended in *(1)*, these valves should not be provided. Yet, it is not practical for all companies to change their design specifications, so a more pragmatic approach is needed.

The problem arises with wide-boiling tower bottoms, and is especially critical with fouling or reactive materials. In these instances, rather than rewrite specifications, it is probably more practical to ensure, either during design or when writing operating manuals or special operating procedures (SOPs), that the valve be returned to its 100% open position after use. The criticality of the service could be used as an indicator of whether an administrative control (an SOP) or an engineered control (say, spring return to open, timer return to open, position switch with alarm, and so on) is selected.

Baffled bottom. Consider now the baffled bottom reboiler configuration, shown in Figure 3. The tower bottom is first divided by a vertical baffle, with the reboiler feed on one side and product accumulation on the other side.

The top of this partition plate is fitted with a trough having unequal-height sides (the lower side is on the reboiler feed side), which serve as overflow weirs. A "shed" baffle is provided over the product side with a lip positioned in the trough at the top of the partition plate, below the lower trough weir. The shed baffle ensures that liquid return from the reboiler and any bottom tray weeping are preferentially returned to the reboiler feed side. Because this effectively seals the two compartments from one another, the shed baffle must have a vent to allow the product side to "breathe" as the product level fluctuates under control of the product offtake instrumentation.

The bottom tray downcomer seal pan is fitted with downcomer pipes sealed in the liquid on the reboiler feed side. The pipes are positioned several inches below the partition plate so that the liquid must pass through the reboiler. They are sized to pass the design quantity of bottom tray liquid, so that the liquid does not overflow the downcomer seal pan weir, splash directly to the surface of the liquid in the reboiler feed side, and potentially contaminate the bottoms quality liquid overflowing into the product side of the partition. These downcomer pipes are usually extended up into the tray seal pan to provide the bottom tray downcomer seal. If they are located flush with the bottom of the seal pan, a downcomer seal weir is needed between the downcomer and the pipes.

The reboiler return is generally oriented so that the return vapor and liquid do not impinge on the downcomer pipes. For multipass trays, the pipes may have to be kinked because the bottom tray seal pans cannot be located directly above the reboiler feed partition.

One would expect there to be some return for the extra cost associated with this design. Indeed, there are several advantages, and they should be weighed against the added cost, though in some cases it will be difficult to put a price on the advantage.

The first advantage relates to the process fluid temperature through the reboiler. Unlike the open bottom design, this configuration must operate at a constant outlet temperature (the bottoms product temperature), because a portion of the reboiler return liquid (that not overflowing to the reboiler feed side) overflows to the product side of the partition baffle. Something must change as percent vaporization varies, and the changed variable is the reboiler inlet temperature (see Figure 3). The curve (here of inlet not outlet temperature) is shown for a theoretical 75% vaporization required of the bottom tray liquid (again simplified as a straight line). Unlike the open bottom configuration, this assumption is not critical to the design, so there is less benefit to determining the actual shape of this curve. The important thing to note is that for any given percent vaporization, the maximum bulk fluid temperature is the bottoms product bubble point. This is a definite positive aspect for this arrangement with fouling (polymerizable) or reactive tower bottoms streams.

> ***The flow restriction valve, which causes problems with an open bottom, could be advantageous with a baffled bottom.***

Another interesting aspect of the design is that as the reboiler fouls on the tube side and the circulation rate starts to decrease, the inlet temperature goes down, marginally increasing the exchanger log mean temperature difference (LMTD). This is the exact opposite of what happens with the open bottom design. A further benefit from this phenomenon is that, with any consistent heating medium control scheme, the exchanger's base LMTD is better than the LMTD in the open bottom design, sometimes dramatically so. This results in lower exchanger costs, as well as lower exchanger hydrocarbon inventories, which is a safety advantage.

Of course, where the required percent vaporization of the bottom tray liquid is less than 30%, a once-through vertical thermosyphon should be considered as the minimum-cost operable design.

From a process design standpoint, the baffled bottom reboiler system acts as a theoretical tray, always delivering a reboiler vapor stream in equilibrium with the bottoms product, derived from the net input of the bottom tray liquid. From a calculational point of view, the excess circulated liquid, of bottoms quality, just "goes along for the ride" through the reboiler. Again, compared to the open bottom design, this results in one, possibly two, fewer actual trays being needed in the tower. From a hydraulic design standpoint, a further benefit accrues because the liquid head driving the thermosyphon is now at a fixed elevation, the elevation of the partition baffle overflow weir to the product side. This allows more precise design of the thermosyphon circuit, with no perturbations of the reboiler operation due to product withdrawal. And, both the reboiler feed and return lines could be reduced by one or more line sizes (compared to the open bottom design).

It is interesting to note that the flow restriction valve, which causes problems with the open bottom design, could actually be advantageous with the baffled bottom design. By throttling this valve, the reboiler could be driven to a higher percent vaporization, resulting in a lower temperature feed to the reboiler. Obviously there are limits to this strategy — one being the percent vaporization required of the bottom tray liquid, and another being the maximum percent vaporization that can be achieved before risking dry-out of the exchanger tubes near the top of the bundle (in the range of 50–60% vaporization). If this strategy is to be tried, temperature indicators in the reboiler feed liquid and the bottom tray liquid should be installed to provide guidance to the operator.

Another benefit comes as a result of maintainability. If there are no extraneous constraints (such as a bottoms product pump or bottoms product flow orifice), maintainability of the exchanger should set the tower/reboiler height. The bottom flange cover of the vertical reboiler should be approximately 7 ft above grade for head clearance during operation and accessibility during cleaning. Therefore, the baffled bottom configuration, due to the elevation of the fluid inside the tower, could save six or more feet of tower skirt height compared to the open bottom design, whose hydraulics are set by the bottom tangent of the tower. Note that this advantage may be lost if pumps must be provided for reboiling medium removal (discussed below).

Though the baffled bottom tower configuration can yield the best results in a highly fouling or potentially reacting tower bottoms environment, from a minimum temperature standpoint, even this temperature reduction may not always be enough. In this case, the tower design premise must be questioned. Lowered pressures, by means of refrigerated overhead condensers or vacuum operation (or both), will yield less reactive bottoms conditions (lower temperatures) as well as an inherently safer design, but this alternative is beyond the scope of this paper. (See *(3)* for further information.)

Reboiling medium configuration

This discussion will focus on a steam-heated reboiler in a vertical thermosyphon with the process fluid on the tube side. With some minor modifications, it is equally applicable to any other condensing heat-transfer fluid, and is generally applicable to tube-side condensing reboiling media in such services as kettles or horizontal thermosyphons.

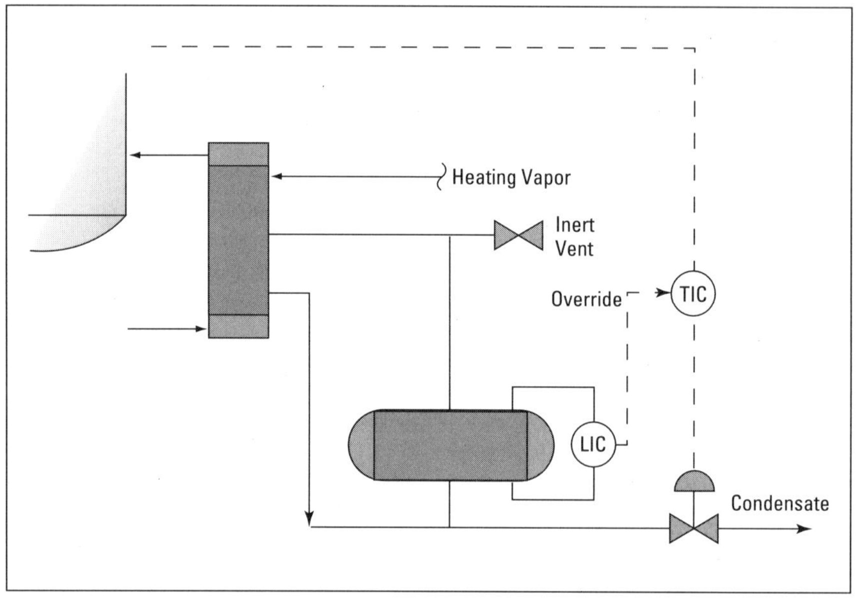

■ ***Figure 4. Bundle flooding control.***

It is most common to control the heat input to a reboiler based upon achieving a specified process parameter that indicates the degree of fractionation desired in the tower. Fixed heat input is another option for control and will generally respond similarly, as described below. The most common parameter chosen for this control configuration is a tray temperature in the stripping section of the tower. (The selection of this control point is beyond the scope of this article, which is to show how the reboiler controls react in response to that control input and how these reboiler controls affect the process environment of the reboiler.)

Heat transfer in the reboiler is governed by the classical equation:

$$Q = U\,A\,LMTD \qquad (1)$$

where Q is the heat transferred, U is the overall heat-transfer coefficient, A is the active heat-transfer area of the exchanger, and $LMTD$ is the log mean temperature difference between the hot and cold sides. U is made up of the film coefficients for heat transfer of both the hot and cold fluids, the resistance of the tube wall to heat transfer (usually negligible compared to the other components), and the fouling factors assigned to each fluid side.

In order to vary Q, the control signal must act on one or more of the entities on the righthand-side of the equation. U is somewhat immutable, changing only gradually over time as fouling occurs; it is more susceptible to change as the percent vaporization through the reboiler on the process side changes, which is a function of both the heat input and the hydraulic head established in the bottom of the column. Therefore, input heat control is accomplished by varying either the active heat-transfer area or the LMTD of the fluids.

Flooding the bundle. Varying the active heat-transfer area is usually referred to as flooding the bundle. It is limited to those cases where a condensing fluid is used as the reboil medium — generally steam.

Figure 4 illustrates a basic control system commonly employed to accomplish this. The only control element required is the outlet condensate control valve. The control parameter adjusts this valve opening to pass slightly more or less condensate, which lowers or raises the condensate level in the exchanger, increasing or decreasing the area available for condensing heat transfer (the bulk of the heat transferred).

A necessary adjunct to this is a level pot, which can be located below the

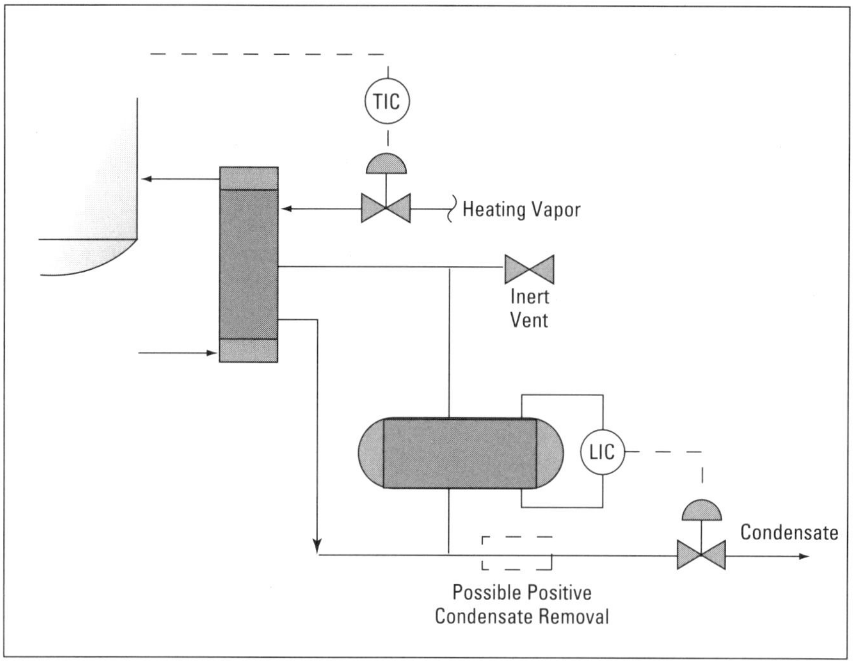

■ *Figure 5. Vapor throttling control.*

condensate outlet nozzle of the exchanger. This pot indicates impending vapor blow-through through the condensate control valve, which would result in loss of pressure and, generally, most of the duty in the reboiler. The pot would be configured to detect a liquid/vapor interface and would either alarm or automatically constrain the signal from the control element that is stroking the valve (or both). This detection means that, for whatever reason, there is no further heat-transfer capability left in the exchanger, and corrective action must be taken (discussed later).

This control scheme has several advantages. First, it is fairly simple — a single control valve with a process signal input. And, condensate removal is generally sure and secure and does not tend to interact with the process-side controls. (Note that this control scheme is equally applicable to a horizontal exchanger configuration with the steam on the tube side.)

The main disadvantage of the scheme is seen when it is used with a fouling or reactive process fluid. In a new and clean condition, the tube-wall temperature on the process side will be very close to the condensing temperature of the medium (with 50-psig steam, about 295°F). This could cause accelerated fouling or unwanted side reactions (degradation) in the wall film of the process fluid. At a minimum, this results in an early expenditure of the process-side fouling factor. At the worst, if used with an open bottom thermosyphon configuration, it could result in lowered circulation, a higher process-side bulk fluid temperature, even faster fouling, or, even worse, accelerated reaction or degradation.

Throttling the vapor. The other method of heat-transfer control is LMTD control, commonly called throttling the vapor. The good and bad points of this method are essentially the opposite of those of area control. However, it is significantly more expensive and complex. Thus, it should only be considered for extremely fouling or reactive process fluids. For the same reason, it is the ideal companion to a baffled bottom tower configuration.

In this scheme, illustrated in Figure 5, there is still a condensate control valve and an associated level pot, but the control valve is now under control of the level instrument on the pot (the intent being to maintain the full area of the heat exchanger for condensing). The inlet vapor line has a control valve, generally driven by the same process input as would be used for the condensate control valve in area control. Because this new control valve limits the heating vapor to the amount required to provide the controlled amount of heat, the condensing-side pressure decreases with a concomitant decrease in condensing temperature until the heat-transfer equation is again satisfied.

Note that when vapor throttling is used with low-pressure heating vapors, the condensing pressure can easily fall below the condensate return pressure. In fact, with low-pressure steam and a low-temperature process side, initial operation could even be at a slight vacuum. If this is so, a positive means of condensate removal must be provided, such as condensate pumps or pumping traps. Otherwise, the control system will convert itself to a hybrid flooding/throttling system, building level in the condenser (even with a wide-open condensate control valve), until the combination of higher liquid head and higher vapor-side pressure (due to decreased area) combine to flush the condensate. Without positive condensate removal, this system will be extremely erratic until it fouls enough to allow condensate removal by hydraulics only.

Vapor throttling is most suitable for fouling services but can be used in other configurations. If this is done, and positive condensate removal is required, one must keep in mind the maintainability of the design. It can come as a rude shock, when everything was done correctly to minimize tower skirt height and locate reboiler heads at an elevation that facilitates cleaning, to find that the tower skirt is now 10 to 15 ft higher and the reboiler is out of reach because a pump was added to remove condensate. Any pump chosen for this service should have a low enough net positive suction head requirement ($NPSH_R$) so that it does not destroy an otherwise well-planned design. If a pump with a suitable $NPSH_R$ cannot be found, a pumping trap is a viable alternative. Pumping traps do not have an $NPSH_R$; rather, they only need a suitable hydraulic design on the inlet side, from the condensate source to the trap, to fill by gravity. An exchanger elevated for maintainability, as described above, generally has sufficient elevation to allow use of these devices.

Condensing-side concerns

The first issue to be addressed on the condensing medium side is vapor superheat. Although it may be a concern with other heat-transfer media, it is usually of most concern with steam, simply because of the way an overall steam system is designed.

A low-pressure header is generally supplied with steam from back-pressure turbine exhausts or from letdown stations with probably only a small percentage of saturated steam input. The turbine exhausts and letdown stations may or may not have desuperheaters. If desuperheaters are present, they will be designed so that the low-pressure header will have sufficient superheat to avoid massive condensation in the header system.

The result is that 50-psig steam headers, for example, generally run in a 350°F to 450°F temperature range (50–150°F of superheat). Condensing steam exchangers can tolerate moderate amounts of superheat (40–60°F) and operate up to this range in a purely condensing mode with a film of condensate on the tubes. Above this range, the higher superheat may dry this film, and a portion of the bundle operates as a desuperheater. Though this yields a higher LMTD for heat transfer, the vapor film coefficient is much lower than the condensing coefficient, so more condenser area may be needed. In a fouling or reactive service, this has even worse implications, since the tube-wall temperature in this desuperheating area can be considerably higher than in the condensing area. For these reasons, a desuperheater is generally specified for any condensing service where the fluid has excessive superheat.

Generally, the first thing that comes to mind when desuperheater is mentioned is the classical spray-type desuperheater. This consists of a spray nozzle inserted in the steam line through which high-pressure, high-quality water is sprayed into the steam. High-pressure boiler feedwater (BFW) is the most common water source if convenient to the unit. Pumped condensate may be a convenient alternative, if the pumped pressure is high enough to generate a reasonable spray.

The spray nozzle assembly is sometimes fitted with a higher pressure steam assist to help produce a fine mist spray. This device relies on evaporation of the water utilizing the sensible heat of the superheated steam. The process is governed by complex heat-transfer and mass-transfer phenomena, but in simple terms the drops (spray) should be fine enough to completely evaporate before settling out of the steam flow.

> ***The sparge desuperheater is simpler than a spray-type desuperheater, and provides a more positive superheating mechanism.***

Important design details include:

- locating the temperature sensing point far enough downstream of the injection point for the heat and mass transfer to occur (generally 25 to 30 pipe diameters);
- a fine spray nozzle;
- the nozzle oriented cocurrent with the steam flow (to avoid particle agglomeration or impingement on the nozzle);
- a reasonable straight run upstream and downstream of the nozzle (approximately 10 and 15 pipe diameters, respectively) to avoid, as much as practicable, impingement of droplets on the pipe wall; and
- a line boot-and-trap assembly in the vicinity of the temperature point (or just after a change of direction, if earlier) to remove excess spray water from the steam line.

Despite following prudent installation details, several things can go wrong with this system. The most common problem occurs as a result of an incorrect temperature control setting. Because this type of desuperheating depends on heat and mass transfer, one cannot achieve 100% desuperheat. If the control temperature is set too close to saturation, the control system will indiscriminately flood the system to the hydraulic capability of the desuperheating water system, which may be more than the boot-and-trap installation can handle. Excess water would then tend to accumulate in pockets, causing slug flow in the piping and possibly overtaxing the condensate removal capacity of the exchanger itself. A reasonable control temperature is 15–20°F above saturation (which achieves the target from the exchanger standpoint).

These systems do not have good turndown. At low steam flow rates, more water droplets will settle from the steam before full vaporization, and at low water flow rates, atomization in the spray nozzle is much less efficient. This could cause the system to cycle, with a high water rate giving a good spray but desuperheating too much, followed by reduction of the water, followed by poor atomization, followed by too high a temperature, then back to a high water flow rate — and the cycle begins again.

The system also introduces four more components that are prone to failure: the spray nozzle (which can plug), the water control valve, the temperature sensor, and the boot-and-trap assembly.

A further complication arises when this system is used in conjunction with steam throttling control on the exchanger. If the desuperheater is installed before the steam throttle valve and a target 15–20°F above saturation is met, the steam could still be significantly superheated after letdown through the duty controlling steam throttle control valve. Assuming a process-side reboiler temperature of 180°F utilizing 50-psig steam as the heating medium (condensing at approximately

300°F), the LMTD at the end of the campaign is approximately 120°F.

During startup and initial operation, the required duty could be as low as 50% of design. Because the exchanger is new and clean, there is no fouling to inhibit heat transfer, so there could be another 50% extra surface that the control system must account for. In an area control system, the steam side would be roughly 65% flooded; in an LMTD control system, the required LMTD would be about one-third the design LMTD (in this case, about 40°F with steam condensing at about 220°F and 2.5 psig).

Under these conditions, the Joule-Thompson (isenthalpic) expansion would cool the steam about 20°F, but the steam would still have approximately 80°F of superheat, which may be judged excessive if it dries out the top portion of the tubes and exposes the process fluid to a temperature of about 300°F. Obviously, this is no higher a temperature than would be seen on the steam side when the exchanger has reached the end of its campaign time, though the fouling at that time would limit the temperature of the hydrocarbon tube to a somewhat lower value. With a clean exchanger, it may just accelerate the tube-side fouling, shortening the projected campaign time or, in some cases, accelerating an unwanted degradation reaction and possibly raising safety concerns.

It would seem that a potential solution to this problem would be to desuperheat the steam, in the conventional manner, downstream of the throttling valve. However, this poses several problems.

In a piping layout, it is generally good practice to keep the control valve near the exchanger. This would then require extra, convoluted piping runs to meet the straight run requirements of the desuperheating station, adding congestion to a potentially high-maintenance-activity area.

The more significant problem is the temperature setting selected for the desuperheater. In the example discussed earlier, the actual condensing temperature of the steam will vary with time (due to fouling) from 220°F to 300°F and, therefore, the desuperheat target will vary from 240°F to 320°F. A control algorithm could be developed that would input the throttled pressure, convert it to a condensing temperature, and add 20°F to the result as the input to the desuperheating temperature controller. However, this approach involves more instrumentation, which in turn means more sources of possible system failure and more ongoing instrumentation maintenance.

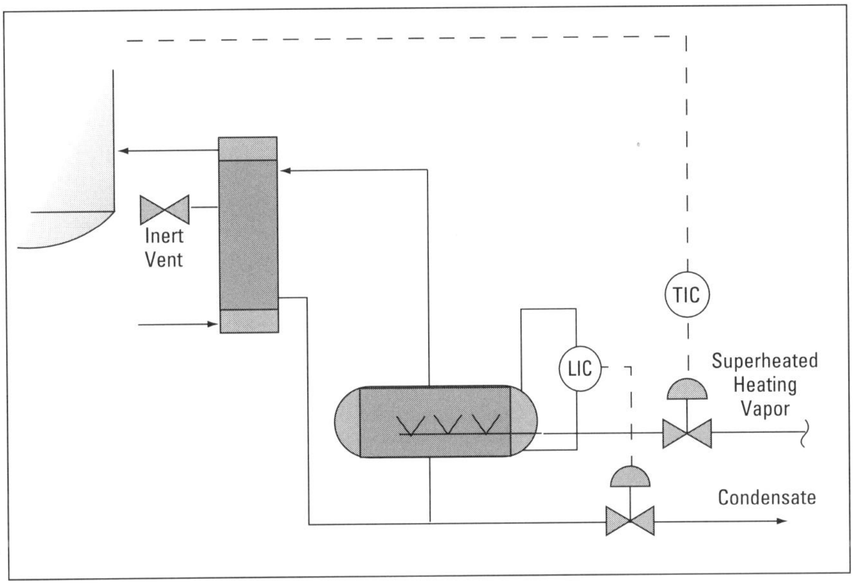

■ *Figure 6. Sparge desuperheater.*

A much simpler and much more positive desuperheating system exists, but unfortunately it is seldom seen in this service — the sparge desuperheater, shown in Figure 6. In this configuration, the inlet steam to the reboiler is first intimately contacted with the reboiler outlet condensate through the sparger dispersion head, which is generally just a submerged header, with or without branches, that has tangential slots cut in the bottom all along its length. The contact of finely divided vapor bubbles rising through equilibrium condensate is a much more positive desuperheating mechanism, yielding almost complete desuperheating. In throttling control, the drum and reboiler operate at the throttled pressure of the reboil vapor. The sparge drum also acts as the condensate level control pot. The advantage of this system is the lack of controls other than the reboiler duty controls (the condensate control valve and the vapor throttle).

Note that the sparge desuperheater does not rely on an extraneous supply of desuperheating water, and the water used is clean condensate. The steam (or other condensing vapor) is always within a few degrees of condensing temperature no matter what that temperature may be or how it varies with load or fouling. Thus, there is no concern from an exchanger design standpoint or from a safety standpoint that the exchanger tubes will dry out locally.

This type of desuperheater can also be used with a flooded bundle control system with a minor modification to the condensate level control pot. With this type of control system, as noted above, the pot is always full of liquid (except at the end of the run). For the sparger to work, there must be a measurable liquid level in the pot. This can be accommodated by placing a level control valve on the liquid (condensate) inlet line to control the pot level. In so doing, however, the utility of the pot as a blow-through indicator is lost. So, consideration should be given to adding another pot for this purpose, or alternatively a level indicator across the exchanger (which can have another beneficial use, as described below).

The most probable reason for not using a sparge desuperheater is cost: the full drum required will typically be more expensive than a condensate pot made from pipe and piping components. This generalization takes into account the excess instrumentation cost and the cost to make the traditional system viable, but not the ongoing maintenance and testing that is required for this more complex instrumentation system (and the drain on instrument technicians' valuable time).

Note that, unlike almost all other drums installed for liquid/vapor separation, the separation here is not critical. One could design a much tighter drum and accept a modicum of liquid entrainment with the vapor to the reboiler, and design the condensate removal system (from the exchanger to the drum) for a moderately higher load. (This is what is done within any distillation tower when designing the tray downcomer loading.) This will reduce the installed cost differential of the two systems even more. Care should be taken during equipment layout to ensure that this larger piece of equipment does not destroy the maintainability of the system by driving the exchanger, and tower, to a higher elevation. A low-$NPSH_R$ condensate removal device, if one is needed, is even more critical in this scheme.

Inerts removal

All completely condensing services require a means for removing inerts from the condensing chamber. There will always be inerts (or light ends in a hydrocarbon stream) that will exceed the capacity for removal with the condensed stream. These inerts accumulate in the condensing chamber and eventually exert their own partial pressure on the condensing chest. This means that, at any given condensing pressure and temperature of the heating medium, the total pressure must be increased to accommodate this partial pressure effect. With LMTD control, the inlet control valve (throttle) would have to open wider; with flooding control, more exchanger surface would have to be uncovered to accommodate the lower condensing temperature.

Inerts accumulation will be especially noticeable during startup of a unit because of the difficulty and time required to completely replace an inerted system volume with the required process fluid. Any remaining inerts will eventually show up in these fully condensing services, as well as any inerts that may be continuously present in these vapor streams, even at otherwise negligible levels.

It is impossible to distinguish between inerts accumulation and fouling — both steal either surface or LMTD.

There is nothing inherently unsafe in this accumulation of inerts in an exchanger. However, inerts have the same impact on reboiler active surface that fouling does: they steal either surface or LMTD.

It is impossible to tell the difference between inerts accumulation and fouling. There have been several instances where a spare reboiler was provided in a fouling service and when the primary reboiler was unable to provide the required duty, it was assumed to be fouled. The reboilers were switched, and when the fouled exchanger was opened for cleaning, it was found to be free of any deposits. Thus, a manpower-intensive operation was undertaken when simply opening a small vent valve would have sufficed. (The vent valves were provided, they just were not used.)

As stated above, vents must be provided. But it should also become part of operators' routine walk-around to exercise these vents regularly on all condensing services.

There are some disagreements as to where the vent should be located — at the high point of the condensing space or at the liquid/vapor interface. The vent will work in either location, as long as it is attached to the body of the exchanger and preferably oriented at least 90° around from the inlet vapor nozzle (if it is near the top). Because the condensate pot for condensate level control must be vented back to the exchanger for a proper hydraulic design, it is quite common to place the vent valve on this pot equilibrating line. With flooding control, the equilibrating line will be near the top of the exchanger; with LMTD control, it can be near the bottom and still be in the vapor space of the exchanger. This operating inerts vent valve will not serve its purpose if located on the inlet heating medium line — all that will be vented from this point, when the exchanger is in operation, is inlet quality vapor, not accumulated inerts.

A further caution applies to the throttled vapor reboiler control scheme. With services where the cold-side temperature is below the normal boiling point of the hot-side fluid, there is the possibility that the condensing side could be (or should be, if it is clean) operating in a mild vacuum. If this is the case, consideration should be given to providing an eductor on the vent. The motive vapor for this eductor can generally be the reboil vapor taken from upstream of the vapor throttle valve. The eductor fluid and educted inerts should be directed to a safe location away from where the operator manipulating the valve can be expected to stand, and, if the fluid is process hydrocarbon, to a contained system (back to the process at another location, or to the hydrocarbon disposal system).

In fouling services, it is reasonable to monitor the state of the reboiler to have an indication of the rate of progression of fouling and to predict when the reboiler must be taken off-line for cleaning. The monitoring instrument obviously depends on the type of control chosen. For flooding control, the height of liquid on the condensing side is used; for throttling control, the pressure in the reboiler condensing side gives the proper indication.

Because inerts accumulation on the condensing side forces the identical control movements of the heat duty controller, comparative monitoring points for the trend are best taken after a routine venting of the exchanger. As noted earlier, much more inerts venting is required immediately after an initial startup than when well into the run length of the exchanger. If it is necessary to vent frequently for extended periods of time, one should suspect an exchanger leak (if the process side is at a higher pressure than the reboil fluid side, as is generally the case). The capability to collect a sample of vented gas from the normal vent connection would be useful in establishing this fact.

SAFETY

Underlying both maintainability and operability are a number of safety-related issues.

Safety during maintainance

Safety during maintenance is a prime thrust of designing for maintainability. The issues addressed previously concerning exchanger types and grade clearances are general maintainability concerns.

In addition, if an exchanger is to be serviced while a unit is in operation, consideration should be given during basic design to the types of equipment that will be used to accomplish this mission. If a spare exchanger is provided, it is obvious that maintenance (cleaning) will be performed while unit operations continue. Even if this is not the case, it is possible that an exchanger will become fully fouled before the planned campaign time of the plant, necessitating cleaning while the balance of the plant remains on-line. It is hoped that in situations of this sort, a quick cleaning of the tube side of the exchanger will be all that is necessary (even if the shell side is potentially fouling also).

Equipment spacing criteria abound for operational access, maintenance access, fire-fighting access, and so on, and it is not the intent of this article to review these. It is assumed that the layout will conform to these criteria.

It is common to design for maintenance access and manipulation by mobile equipment. In the case of on-line maintenance, vehicle access may be prohibited. Also, though it may be acceptable to scaffold for access during scheduled maintenance, scaffolding takes time that may not be available during on-line maintenance. Access platforms should be considered during the design, by extending either vessel platforms or nearby structures.

Partial disassembly of the exchanger is necessary, and cleaning equipment must have the necessary headroom and may need support assistance. Davits on horizontal exchangers, similar to those usually provided for vessel accessways, provide an easy manual method for swinging tube-side covers out of the way. For vertical exchangers, covers should have lifting lugs and overhead steel structures should have hitch points for chain falls, to allow a bottom cover to be lowered to a handcart at grade or a top cover to be lowered to the access platform by a davit or traveling beam. Connections to the covers should be minimized, and those absolutely necessary should be easily disassembled from a fixed point outside the necessary maintenance access radius.

For planned on-line maintenance, positive isolation from the live system is necessary. This requires valving on the inlet and outlet of both sides and blinding points on the exchanger side of the valves (or alternatively double blocks with bleeds). Access to these blinding points and the chain-fall hitch points should be provided.

Safety and operability

Ease of operation (operability) enhances safety. Generally, the less complicated the instrumentation around a given operation, the more likely it can be operated safely.

Sophisticated instrumentation schemes have blossomed with the advent of electronic control systems, in many instances seemingly only because the control system has the ability to manipulate complex control algorithms. Besides the added costs for the extra instrumentation, the more complex programming, and the increased workload for instrumentation calibration and maintenance, another negative effect of this trend is that the true control objectives may be masked from the operator.

When everything is operating satisfactorily, the operator has little to do, which further divorces him or her from the control system. However, the systems are seldom able to start up automatically — this function, it is generally agreed, is still within the operator's scope, with the instrumentation on manual. It should come as no surprise, then, that approximately half the incidents and accidents in modern process plants occur during startup.

As so aptly stated in *(4)*: "Human error often figures as a major factor in analyses of incidents. When a failure occurs, there is frequently an administrative requirement to determine the cause, and allocation to human error is notorious ... Often the error is more truly that of another party who is responsible for some aspect of the work situation. This point is important, because whereas assignment to human error suggests little can be done, recognition of the effect of the work situation does tend to indicate possibilities for improvement."

Liquid inventory minimization. Within a tower bottoms and reboiler system, the major concern that can be addressed from a safety standpoint is liquid inventory and its minimization. Both the American Petroleum Institute's (API) Recommended Practice 750, "Management of Process Hazards," *(5)* and the Occupational Safety and Health Administration's (OSHA) "Process Safety Management of Highly Hazardous Chemicals" *(6)* apply to processes having a total unit inventory of flammable materials of 10,000 lb or more. For highly hazardous, toxic, or reactive chemicals, OSHA's inventory cutoff is even lower. The OSHA regulation also requires that information on the maximum intended inventory of the process be available as information pertaining to the process. And, indications are that one of the questions that will be asked during an OSHA compliance audit is what has the facility done to minimize inventory.

Liquid inventory is potentially more hazardous than vapor, partly because of the ability to accumulate more mass per unit volume than vapor. But in addition, for any loss-of-containment scenario postulated, the rate at which liquid escapes from containment is approximately an order of magnitude higher than the escape rate of vapor at the same pressure *(7)*. This is because a liquid at its bubble point at elevated pressure (such as tower bottoms), once released from containment, flashes isenthalpically to form a vapor cloud and entrains a portion of the liquid from the flash as an aerosol. Work continues on defining this entrained quantity, but a good rule of thumb is to assume that the entrained liquid equals the flash vapor *(7)*. Thus, a liquid flash of 50% or more ends up completely in the loss-of-containment vapor cloud, with potentially disastrous consequences.

There are three areas in a reboiler system where reducing liquid inventory can be addressed:

- the interconnecting piping;
- the reboiler; and
- the tower bottoms.

Piping. Piping will generally contribute the least to the overall inventory assessment, but a few observations are in order. The piping of concern for liquid inventory is the reboiler feed piping. The vapor return piping from a kettle obviously does not contribute to liquid inventory. And the actual volume of liquid in the return line for a two-phase liquid/vapor return system is also generally low — for example, in a system designed for 30 wt.% vaporization (70 wt.% liquid in the return), only 10–15 vol.% will be liquid.

For the feed piping, in the size range of 6–12-in., for example, a reduction of one line size will reduce liquid inventory by about one-third. Smaller line sizes, for a given percentage vaporization, can only be justified by higher liquid heads, which can be achieved with trap-out reboilers or thermosyphon reboilers with partition plate tower internals.

For open tower bottoms, one might consider doing the hydraulic analysis from the normal liquid level rather than the bottom tangent line of the tower. If this is done, however, one should check that an inoperable situation does not occur at low liquid level. (If 30% vaporization were the base at normal liquid level, a higher percent vaporization would be obtained at lower levels at the design heat input. If this higher vaporization percentage is close to the point of exchanger tube dry-out, the design could be inoperable. The effect of a higher outlet temperature at the higher vaporization rate must also be acceptable.)

The most fertile area for inventory reduction in a reboiler system is the tower bottoms.

Line length is also a factor, since inventory increases in direct proportion to length, and, if too long, a size increase may also be necessary. Close coupling the exchanger and tower nozzles, with little or no interconnecting piping, should always be the first layout consideration.

Reboiler. Once an exchanger's heat duty parameters have been set, there is little that can be done to minimize liquid inventory. For vertical reboilers, the main source of inventory is in the inlet head (or barrel). The barrel diameter is a function of the exchanger diameter, while the length is generally a function of the inlet line size. The diameter can be minimized by increasing the length of the tubes (within the constraints of the hydraulic design), and the length by minimizing the inlet line size.

However, these minimal changes can be far surpassed by reconsidering those elements of the basic design that may have resulted in a potentially far larger exchanger than is really needed. As noted above, high fouling factors can increase surface requirements in a service that has inherently high heat-transfer coefficients (like reboilers) far more than one would expect. Adding a process safety factor, in the form of an extra duty capability (a nominal 10–15%) or extra surface, compounds the problem.

These aspects contribute inordinately to exchanger size, which means extra inventory and extra investment. Safety, when addressed at the right time, can result in an investment savings rather than an increased cost *(8)*.

Tower bottoms. The most fertile area for inventory reduction in a reboiler system, however, is the tower bottoms itself. Generally the bottom of the tower is designed to perform two functions: to provide a liquid inventory to drive the reboiler and to provide surge time for the tower bottoms product. If these two concepts can be divorced in the mind of the tower designer, then some progress can be made in reducing inventory. There are two generic types of reboiler systems where these concepts are already separated, a trap-out reboiler system and a kettle reboiler.

In both of these types, the liquid inventory to drive the reboiler is provided by a trap-out pan located just below the bottom tray of the tower. All liquid accumulated in the trap-out pan is fed to the reboiler, and no liquid out of the reboiler is returned to the trap-out pan. Trap-out pans are generally relatively small, representing only seconds of inventory for feeding the reboiler, yet these reboiler systems perform well.

Bottoms product accumulation (surge) is provided in the bottom of the tower in the case of a once-through reboiler, or in the back head (beyond an overflow baffle) in a kettle. The volume of surge provided is estimated by the rules of thumb developed 40 or 50 years ago — generally, 3–5 min of product volume, which, though valid when developed, is probably higher than necessary given the need to minimize liquid inventory.

Separation of the two concepts is not quite so evident with thermosyphon reboilers. In an open bottom configuration, the same space in the tower bottoms is allocated to reboiler feed and product accumulation.

A recent project involved a distillation tower of significant size (10 ft in diameter), with a bottoms section allocation of 10 ft in height to maximum liquid level. When full, this would represent approximately 800 ft^3 of liquid, or over 20 tons of a gasoline blending compound held at its boiling point. This inventory could be reduced by an order of magnitude by lowering the maximum liquid level to 1 ft, at the same time saving 9 ft of 10-ft-dia. pressure shell. Unfortunately, this would have reduced the span of the bottoms level controller used for bottoms product offtake control to a potentially unworkable height. The alternative, which accomplishes the same goal, is to immediately swage the bottom of the tower (below the reboiler return nozzle) down to a 3-ft diameter and then provide the same 10 ft of height for the level instrument in the smaller diameter section. The savings in investment would not be quite as much, but inventory would be reduced by the same order of magnitude, while retaining a viable instrumentation system.

A similar arrangement can be considered for a baffled bottom tower. The normal baffled bottom has a vertical baffle dividing the bottom into two compartments with combined volumes similar to the volume of the open bottom design. The reboiler feed section is maintained at a constant height (by design) and the bottoms product side will vary.

If instead of a vertical baffle a horizontal baffle is used for the reboiler feed (similar to a trap-out tray design but with the other features of the baffled bottom), then the bottom section could again be immediately swaged down to a significantly smaller bottoms diameter (in this instance below the trap-out pan and shed baffle) and individual and nonrelated holding volumes assigned to the two sections. The total volume of liquid inventory in this arrangement will be slightly larger than that of the open bottom design, basically by the amount of liquid held on the trap-out pan. The discrepancy could be reduced, however, by reducing the height of the swaged boot.

The positive impact on safety associated with reducing inventory at the early design stage is generally accompanied by a decrease in investment — it is a win-win situation. If this is so, why is it not done regularly?

> ***The need for surge and the need to "give the operator time to do something" are relics of the past and need to be re-examined.***

Two reasons are often given for needing significant inventory at every step of the process: the need for surge, and the need to "give the operator time to do something." These two concepts are relics of the past and need to be reexamined in light of several major changes that have taken place over the last 30 years *(8)*:

- the escalation of plant sizes (throughput);
- the conversion of plant control systems from pneumatic to electronic (with computer-based control systems);
- the "leaning" of operations staff, especially outside operators; and
- the relocation of the control house to the periphery of the plant (if not to a more remote and safer location).

The escalation of plant size is the major factor related to the need for surge. The provision of extra volume has never been very scientific. Generally, it is the equivalent volume of the total inflow or outflow of 3–5 min (and possibly more) of the material being accumulated. What would have been a nominal volume 30 or more years ago is now a significant hazardous material inventory.

Surge was originally provided to enable the plant operators to change the throughput of the plant by changing the plant feed rate, and then following the change through the plant, sequentially trimming the controls on equipment as the feed change impacted it. Today's electronic control systems do this job much more quickly and economically than any operator could ever have hoped to do.

Surge is necessary, but its main benefit is in temporarily isolating an operating unit from the vagaries of other operating units that are producing feed or consuming product. Surge can be, and properly should be, limited to offplot tankage (or pressurized storage), the status of which can then be used to dictate necessary changes to unit throughput.

The other side of this coin, however, is that the instrument technician may now be of more importance than the hands-on operator of the past. In order for the instrumentation system to do the job it is capable of, it must have accurate information on which to base control changes. Field instrumentation should be subject to a planned continual program of inspection, testing, and calibration.

In order to keep the information manageable, though, there should be a clear distinction between necessary information that must be known by the control system and information that is "nice to know." Information that may be needed to close the material balance around a piece of equipment for troubleshooting purposes (and therefore may never really be needed) should be given a low priority, possibly even to the extent of providing only primary instrument connections (a valve for a pressure gage, a thermowell with no thermocouple, or an orifice flange and plate set) that will provide access to the information by using the appropriate test equipment if and when needed. Old strategies, such as cascading level control to flow control, were developed because of the practice of providing surge at every point in the unit. If surge volume is seriously curtailed, these strategies may hinder more than help the control system (in addition to requiring more instrument technician attention, adding one or more items in the control loop that can go wrong, and generally adding to the cost of the unit).

The need to give the operator time to do something must also be seriously challenged. For example, consider a tower bottoms system with a product pump and a classical 5-min holding time in the tower bottoms — how much time does the operator really have should something malfunction?

The system should have a minimum flow bypass from the pump back to the tower, since the control system is level control (even if actually configured as level resetting flow), which can be expected to stop bottoms product. The primary malfunctions that would be expected are the bottoms product control valve fails closed, the pump stops, and the minimum flow control valve fails open. All three would cause the level to rise — the first two quickly, the third more gradually. Assuming the level is held at the midpoint of the normal span (2½ min to high or low level), the operator has two and a half minutes before the high level is reached (and generally an alarm is sounded) *only* if he happens to be scanning that particular system at that time and notes that the level has started to rise with the product valve closed, the kickback valve open, or the pump stopped.

If this serendipitous situation does not prevail, then the operator will be made aware of the situation when the high-level alarm sounds, 2½ min after the event, rather than much sooner if the holding time were significantly less. Thus, the operator will have only the time available from the high or low level alarm to a system trip (if provided), which is generally on the order of 15–30 s, and *not* the 2½–5 min built into the system at significant cost and potential hazard.

Final thoughts

By understanding the basic principles of a reboiler system design and paying special attention to the little details (which can cause big problems if unaddressed) during the design process, CPI facilities can be made safer, less expensive, more efficient, and easier to maintain. In addition, step-change improvements in inherent safety can be achieved by substantially reducing liquid inventories.

To take this step, the engineer must examine common paradigms, which if one is aware can be detected while discussing the system. Statements such as "We need surge" or "Our standards require ..." or "That's the way we've always done it" are definite tip-offs. If you cannot answer the question "Why?" after such a statement, it will be difficult to take the next step. CEP

Literature Cited

1. **Viera, G. A., *et al.*,** "Lessons Learned from the Ethylene Oxide Explosion at Seadrift, Texas," *Chem. Eng. Progress,* **89**(8), pp. 66–75 (Aug. 1993).
2. **Love, D. L.,** "No Hassle Reboiler Selection," *Hydrocarbon Processing,* **71**(10), pp. 41–47 (Oct. 1992).
3. **AIChE Center for Chemical Process Safety,** "Guidelines for Engineering Design for Process Safety," Chapter 2, "Inherently Safer Plants," American Institute of Chemical Engineers, New York, NY, pp. 5–52 (1993).
4. **Lees, F. P.,** "Loss Prevention in the Process Industries," Butterworths, London (1980).
5. **American Petroleum Institute,** "Management of Process Hazards," API Recommended Practice 750, First Edition, API Production and Refining Departments, Washington, DC (Jan. 1990).
6. "Process Safety Management of Highly Hazardous Chemicals," U.S. Occupational Safety and Health Administration, Washington, DC, 29 CFR 1910.119 (Feb. 24, 1992).
7. **AIChE Center for Chemical Process Safety,** "Guidelines for Chemical Process Quantitative Risk Analysis," American Institute of Chemical Engineers, New York, NY, pp. 71, 77 (1989).
8. **Frey, R. C., and A. J. McCarthy,** "History Revisited — Do We Learn From Or Accept The Past?," presented at the 1992 Process Plant Safety Symposium, AIChE South Texas Section (Feb. 1992).

A. J. McCARTHY is a senior engineering consultant for The M. W. Kellogg Co., Houston, TX (713/753-2000; Fax: 713/753-5353), where he is responsible for safety and operability reviews of plant design at various stages of projects, ensuring that each project conforms to industry and company standards of safety, operability, and quality. Over his 33-yr career with Kellogg, he has served as process engineer, process manager, engineering manager, project manager, and, in 1989, was assigned to his current position in the chief engineer's office. He holds a BS from the Univ. of Massachusetts and an MS from the City Univ. of New York, both in chemical engineering. He is a member of AIChE and serves on the Engineering Practices Subcommittee of the Center for Chemical Process Safety, and is a member of Sigma Xi. He has written several technical papers and is a lecturer for an AIChE continuing education course.

B. R. SMITH is an engineer in the process systems dept. of The M. W. Kellogg Co., Houston, TX (713/753-2000; Fax: 713/753-5353), where he is responsible for the development of piping and instrumentation diagrams and utility/offsites systems for integration with the process unit design, and production of overall plant and hydraulic design criteria. In his five years with Kellogg, he has worked on grassroots MTBE projects, MTBE/ETBE expansions, LNG offsites/utility front-end engineering, and specialty chemical facilities in various stages of project execution. He holds a BS in engineering from Cornell Univ. and is near completion of an MBA in finance from the Univ. of Houston.

Understand Spiral Heat Exchangers

A spiral exchanger can be an economic alternative for difficult heat-transfer applications.

Kevin M. Bailey,
Alfa Laval Thermal Inc.

Shell-and-tube heat exchangers are standard in most chemical process industries (CPI) applications. However, they do have limitations related to thermal efficiency, mechanical design, and maintenance requirements that will not allow the standard straight-tube fixed-tubesheet shell-and-tube (S&T) heat exchanger to work properly in certain applications. When constrained by these limitations, S&T exchangers often require complex and expensive design features, such as floating heads, removable tube bundles, and welded tubesheets, to enable them to operate in a particular service. These design features allow the S&T exchanger to operate, although often in a limited capacity and with drawbacks.

It is in these problem areas that spiral heat exchangers (SHEs) have been used successfully worldwide for over 60 years. The SHE can be a viable alternative to the complex and often expensive shell-and-tube heat exchanger. The SHEs' unique spiral countercurrent monochannel design gives them exceptionally high heat-transfer rates and low fouling tendencies. The mechanical configuration of the SHE also allows full access to all heat-transfer surfaces for simplified inspection, maintenance, and cleaning.

This article describes how SHEs operate, discusses their advantages in terms of thermal efficiency, fouling, mechanical design, and maintenance characteristics, and provides guidance on choosing between spiral and tubular exchangers.

SHE configuration

The SHE (Figure 1,) is constructed by rolling two relatively long seamless metal strips around a mandrel to form two concentric spiral channels. The channels are alternately welded on opposite ends to form a hot channel and a cold channel, each accessible from its respective head. The welding of the channels eliminates the potential for cross-contamination of the fluids and is analogous to a welded tube-to-tubesheet joint in a shell-and-tube heat exchanger.

On one side, the hot fluid enters the center nozzle of its head and flows outward in a spiral to a nozzle on a peripheral header. The cold fluid simultaneously enters a peripheral header and flows countercurrently to the hot fluid to the center nozzle on the cold side head. Removable cover heads with full face gaskets are used to seal the open end of the channels and prevent bypassing of a fluid from the peripheral header to the center nozzles. The heads are easily removed to allow access to all heat-transfer surfaces.

SHEs are fabricated in accordance with ASME Section 8 Division 1 Pressure Vessel Codes. They can be made of carbon steel, stainless steel, titanium, or virtually any other metal that can be cold formed, rolled, and welded.

Heat-transfer rates of up to 2,000,000 Btu/h can be transferred in single-shell units with surface areas up to 4,000 ft^2. Flow rates up to 3,000 gal/min are feasible in single shells. If larger units are required, multiple SHEs may be placed in series or parallel to accomplish the required heat-transfer duty.

SHEs can typically be designed with design pressures up to 350 psi and design temperatures up to 1,500°F, depending on size. Large SHEs with high design pressures and temperatures may be uneconomical due to their high capital cost.

With these design capabilities though, SHEs can handle a large majority of the applications in the CPI.

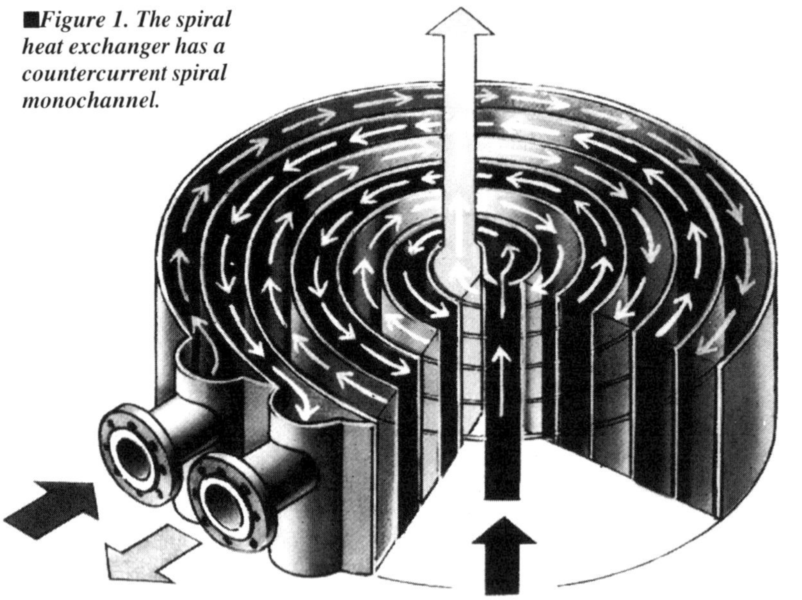

■*Figure 1. The spiral heat exchanger has a countercurrent spiral monochannel.*

Basic heat exchanger design

Understanding basic heat-transfer theory is the basis for any good heat exchanger design.

The ability of any heat exchanger to transfer heat can be represented by

$$Q = UAfLMTD \qquad (1)$$

The economical design of a heat exchanger results from maximizing the overall heat-transfer coefficient, U, for a given LMTD and allowable pressure drop in order to minimize the heat-transfer area. It must be remembered that the capital cost of the heat exchanger is directly related to the amount of heat-transfer area.

Fouling, which directly affects the value of U through the fouling factor, R_f, must also be carefully considered when designing a heat exchanger because as R_f, increases, so does the required heat-transfer area, A. Fouling also has to be considered from a maintenance and operational cost standpoint, which can often outweigh the capital cost of a heat exchanger.

Thermal efficiency

SHEs work very well in high thermal efficiency duties, such as energy recovery applications, where temperature crosses and very close terminal temperature approaches exist. Using the nomenclature shown in Figure 2, temperature crosses occur when $T_{ho} < T_{co}$, and close temperature approaches occur when $T_{ho} \approx T_{ci}$ or $T_{co} \approx T_{hi}$.

In these applications, the LMTD is low — typically <20°F — so it becomes important to maintain the LMTD by having a countercurrent heat exchanger. The spiral exchanger, being fully countercurrent, has an LMTD correction factor of 1.0 and, thus, does not have an LMTD penalty. For S&T heat exchangers, the LMTD correction factor becomes very small (<<1) when there are temperature crosses and close temperature approaches, and the LMTD is effectively lowered to what is commonly known as the corrected LMTD.

The LMTD correction factor corrects for the fact that a shell-and-tube heat exchanger is essentially a crossflow heat exchanger, where the shell side flow is forced by a series of baffles in the shell to cross over the tubes. Thus, for any given application, the SHE will have a higher LMTD than an S&T heat exchanger and will require less heat-transfer area for a given heat-transfer rate.

Very often a single SHE can transfer as much heat as several S&T exchangers in series because the shell - and - tube exchanger's lower effective LMTD requires more area. The SHE can also handle temperature approaches down to 5°F in a single shell, again because no correction is necessary to the LMTD.

Another reason why SHEs work well in high efficiency applications is because they generate U

Nomenclature

A	= heat-transfer surface area, ft²
f	= LMTD correction factor to account for noncountercurrent flow arrangement (f is always less than 1.0, except in true countercurrent applications)
h_c	= cold side convection heat-transfer coefficient, Btu/ft²•°F•h
h_h	= hot side convection heat-transfer coefficient, Btu/ft²•°F•h
k	= conductivity of heat-transfer surface, Btu/ft²•°F•h
$LMTD$	= log mean temperaure difference across heat exchanger, °F
Q	= heat-transfer rate in the heat exchanger, Btu/h
R_f	= fouling factor, ft²•h•°F/Btu
t	= thickness of heat-transfer surface, ft
T_{ci}	= cold fluid inlet temperature, °F
T_{co}	= cold fluid outlet temperature, °F
T_{hi}	= hot fluid inlet temperature, °F
T_{ho}	= hot fluid outlet temperature, °F
U	= overall heat-transfer coefficient, Btu/h•°F•ft² = $1/[1/h_h + t/k + 1/h_c + R_f]$

values 120–150% higher than those of an S&T unit for the same pressure drop. For water-to-water heat transfer, SHEs can generate U values in the range of 400 Btu/h•°F•ft^2. These high U values are achieved by obtaining very high convection heat-transfer coefficients on both the hot- and cold-side fluids and minimizing the potential for fouling.

As shown in Figure 3, the constant spiraling flow channel induces tremendous turbulence and secondary flow effects (eddy currents and vortices) that result in excellent convective heat transfer. At a velocity that would be marginal and approaching laminar flow in straight tubes, good turbulence is obtained in the spiral heat exchanger, owing to secondary flow effects caused by the continuously curving channel. Heavily solid-laden fluids are easily handled at velocities of 3 ft/s.

Fouling considerations

Fouling describes the buildup of material over time that retards heat transfer by forming an insulating layer on the heat-transfer surface. Whether it is caused by biological materials, organic substances, scaling, or plating, fouling reduces the heat-transfer coefficient, U.

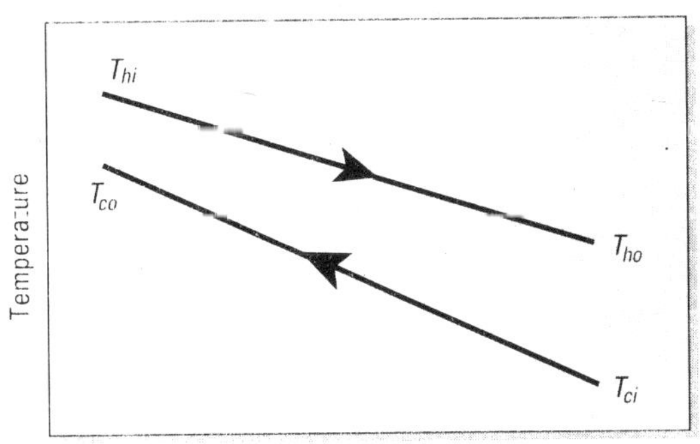

■*Figure 2. Temperature profile within the exchanger.*

To compensate for fouling, a margin of safety known as the fouling factor, R_f, is used. This fouling factor essentially lowers the clean U value, typically by 10% to 40%, to a service U value, which is used to calculate the required heat-transfer area under the fouled conditions. Fouling factors are derived mainly from experience with S&T heat exchangers, which tend to foul relatively easily due to their configuration and, thus, have high fouling factors — in the 0.001 ft^2•°F•h/Btu range and higher.

One reason why shell-and-tube exchangers foul easily is because of low turbulence and shear rates in the tubes, dead spots on the shell side, and multiple parallel flow channels. Shell-and-tube exchangers very often have hundreds of parallel tubes, which can develop pressure gradients through each tube. If a tube begins to foul or plug, the pressure drop in that tube will increase. The flow through this plugged tube will decrease as flow is bypassed into the tubes that are not plugged. With reduced flow, a partially plugged tube will become further fouled due to the lower velocity and will ultimately plug completely. The velocity in the other tubes will then increase until a steady-state condition occurs and little additional pluggage will occur. Unfortunately, a percentage of the heat-transfer area has already been bypassed at this point. A successful heat exchanger design has estimated this percentage of decrease in heat-transfer area correctly so that the heat exchanger will transfer the required amount of heat in the fouled condition.

A SHE is unique in that it has a monochannel flow passage. There is no bypassing of fluids due to pressure gradients. The highly turbulent flow and constant impingement on the heat-transfer surface keeps scale and fouling from building up. If a pluggage does initiate, the velocity in the flow passage will increase at the point of pluggage until the surface is cleaned.

With this configuration, a SHE can use fouling factors as low as one-third those of S&T heat exchangers, thus increasing the overall service U value. The fouling factor for SHEs are in the range of 0.0003 to 0.001 ft^2•°F•h/Btu. Experience has shown that even with these lower fouling factors, SHEs will typically operate two to three times longer than an S&T exchanger before cleaning is needed. Thus, SHEs work exceptionally well in fouling, viscous, slurry, and fibrous applications where fouling is expected or known to occur.

In services where mechanical cleaning is not practical and chemical cleaning must be used, the SHE's monochannel works very well. The monchannel ensures that the cleaning fluid will reach all the heat-transfer surfaces. In an S&T with many tubes that may be plugged, chemical cleaning is not as effective, since the chemical cleaning solution will only pass through the unplugged tubes.

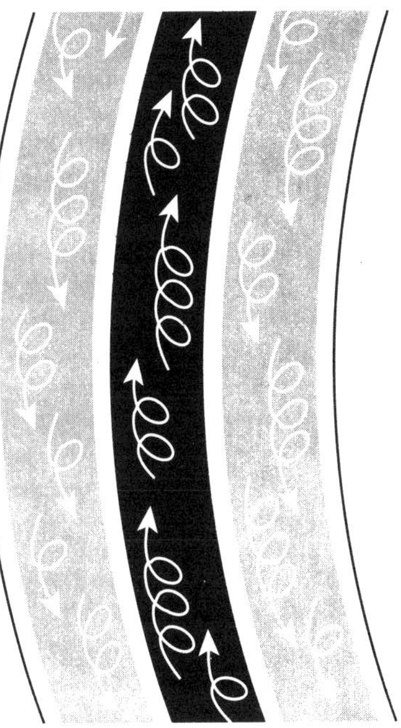

■*Figure 3. High turbulence results in high convective heat-transfer coefficients.*

Mechanical design and maintenance

While thermal design is the primary consideration in heat exchanger design, it is also very important to consider the long-term implications of a design, such as installation, operation, maintenance, and repairs, all of which figure into the cost of transferring heat. A heat exchanger must be designed not only to minimize capital cost, but also to minimize operating and maintenance costs (including production losses) as well.

SHEs offer the advantage of fully accessible heat-transfer surfaces for simplified maintenance, inspection, and cleaning. Both the hot and cold surfaces are accessible (simultaneously) by removing the respective heads. The surfaces can then be inspected and, if necessary, cleaned. Cleaning is typically done with high pressure water and can normally be completed in one 8-h maintenance shift, or about one-half to one-third the time needed for other types of heat exchangers.

Shell-and-tube heat exchangers typically do not have access to the shell side unless removable tube bundles are used. However, removable tube bundles make the shell and the tubes much more expensive, as well as difficult to maintain because of the need to remove the large and cumbersome tube bundle. In services that require frequent cleaning of the shell and/or tubes, the feasibility of using shell-and-tubes exchangers greatly diminishes.

The configuration of a spiral heat exchanger also allows for thermal expansion without generating high mechanical stresses that can result in mechanical failure. The SHE is always manufactured from a single metal, so there is no potential for the high stresses from expansion that occurs with bimetallic construction. The SHE is free to float and can absorb thermal expansion by coiling and uncoiling. This allows the SHE to be used with hot and cold fluids that have very large temperature differences.

Bimetallic shell-and-tube exchangers, which typically have a carbon steel shell and stainless steel tubes, are cost-effective from a material standpoint. However, they experience high mechanical stresses that can lead to premature failures in tubesheet joints or shell expansion joints. These are some of the most common types of failures in S&T exchangers, especially at elevated temperatures.

The SHE's fully welded heavy-wall construction also makes it extremely rugged and reliable. SHEs are typically fabricated with wall thicknesses from 12 gage (0.105 in.) to 3 gage (0.25 in.). These sizes are much thicker than typical tube wall thicknesses of 20 gage (0.035 in.) to 16 gage (0.065 in.). Such thin tubes often fail at baffles due to tube vibration and cause cross-contamination of the hot and cold fluids. The heavy walls in a SHE provide a margin of safety against cross-contamination, erosion, and corrosion without retarding heat transfer to any great extent. (For highly conductive metallic heat exchangers, *U* values are much more influenced by convective coefficients than the conductivity and thickness of typical heat-transfer surfaces.) In addition, the welded seals on the SHE provide a seal against cross-contamination of high integrity, comparable to welded tube-to-tubesheet joints.

Selection guidance

Although SHEs can operate in most CPI applications, their relatively high capital cost often limits their use to only the most difficult services. Many of the

Table 1. Selection guide for spiral and shell-and-tube heat exchangers.

	Spiral	Shell-and-tube
Operational Requirements		
Design Temperature Limits, °F	–30 to 1,500	–100 to 2,000
Design Pressure Limits, psia*	0 to 350	0 to 3,000
Maximum Flow Rate per Shell, gal/min	3,000	35,000+
Heat-Transfer Area, ft^2	4,000	20,000+
Fouling Factor, °F•h•ft^2/Btu	0.0003 to 0.001	0.001 to 0.01
Heat-Transfer Requirements		
Liquid-to-Liquid	Good	Good
Gas-to-Liquid	Good	Good
Gas-to-Gas†	Fair	Excellent
Condensation	Excellent	Excellent
Vaporization	Good	Excellent
Temperature Cross	Excellent	Fair
Close Temperature Approach (LMTD<20°F)	Good	Fair
Slurries, Fibrous, Viscous Fluids	Excellent	Fair
Highly Fouling Services	Excellent	Fair
Mechanical Requirements		
Cleanability on One Side	Excellent	Excellent
Cleanability on Both Sides	Excellent	Fair
Chemical Cleaning	Excellent	Fair
High Thermal Differential	Good	Good
Cyclic Conditions	Fair	Fair
Repairability‡	Fair	Good

* Depending on the size of the unit.
† Assuming an enhanced-surface shell-and-tube exchanger.
‡ Depends on the nature of the repairs.

inherent features of the SHE are not economically justified when other types of heat exchangers will suffice.

Table 1 compares SHEs and S&T exchangers in various types of applications. As can be seen, SHEs are typically reserved for fouling, fibrous, energy recovery, interchanging, or wastewater applications, where other heat exchangers have difficulties.

Specifying a SHE requires only completing a typical heat exchanger data sheet (Table 2), showing fluid properties, flow rates, and temperatures, as well as mechanical information such as design pressure, temperature, and materials of construction. A typical Tubular Exchanger Manufacturers Association (TEMA) specification sheet can also be used by simply disregarding the shell-and-tube data.

SHE manufacturers typically complete U value, pressure drop, and ASME calculations based on in-house software programs. If necessary, these designs can be verified by using the correlations in Reference *(1)*. It is recommended that a SHE manufacturer be consulted on size and mechanical limitations as final specifications are being developed. CEP

Table 2. Spiral heat exchanger specifications sheet.

	Job No.	
	Reference No.	
	Inquiry No.	
Plant Location	Date	
Service of Unit	Item No.	
Type — No. of Exch. Req'd	Connected In	
Total Surface ft.²	Surface/Exch. ft.²	
Guaranteed Performance	**Hot side**	**Cold Side**
Fluid Circulated		
Total Fluid Entering		
Vapor		
Liquid		
Non-Condensables		
Fluid Vaporized or Condensed*		
Physical Property Data Temp °F		
Specific Heat BTU/lb. °F		
Specific Gravity		
Thermal Conductivity BTU/lb. ft. °F		
Viscosity Cp		
Latent Heat vapors BTU/lb.		
Non-Newtonian k/n		
Molecular Weight		
Temperature In °F		
Temperature Out °F		
Operating Pressure Psig		
Max Allow. Pressure Drop Psig		
Thermal Margin %		
Heat Exchanged	BTU/Hr.	LMTD: °F
*For two phase duties, also provided either condensing curve or vapor pressure data.		
All units are built and stamped to ASME Section 8 Division 1 Pressure Vessel Code		
Construction		
	Test Pressure: Per ASME	
Design Pressure Psig		
Design Temperature °F		
Spiral Material:		
Gasket Material		
Remarks:		

K. M. BAILEY is the spiral heat exchanger product manager with Alfa Laval Thermal Inc., Richmond, VA (804/236-1342; Fax: 804/236-1360). He has over 10 years of engineering experience in the utilities, shipbuilding, and CPI. His work has included project, maintenance, and reliability engineering positions, and he has considerable experience with pressure vessels, piping, and heat exchangers. He holds a BS in mechanical engineering from Virginia Polytechnic Institute and State University and an MBA from the Univ. of Richmond. He is a registered professional engineer in the state of Virginia.

Literature Cited

1. **Minton, P. E.**, "Designing Spiral Plate Heat Exchangers, *Chem. Eng.*, **77**(10), pp. 103–112 (May 4, 1970).

Further Reading

Beckman, A. T., et al., "Applications of Spiral Heat Exchangers," *Chem. Eng. Progress*, **63**(7), pp. 62–67 (July 1967).

Devore, A., et al., "Specifying and Selecting Heat Exchangers," *Chem. Eng.*, **87**(20), pp. 133–148 (Oct. 6, 1980).

Fanaritis , J. P., and J. W. Bevevino, "How to Select Optimum Shell and Tube Heat Exchangers," *Chem. Eng.*, **83**(14), pp. 62–71 (July 5, 1976).

Employ Immersion Heaters Properly

Follow these guidelines for selecting, sizing, installing, and maintaining electric immersion heaters.

Robert C. Klein,
Watlow Electric
Manufacturing Co.

Immersion heaters have a wide variety of applications in the chemical process industries (CPI). Knowing which ones to specify and how to properly install them can make a manufacturing process more cost-efficient. This article describes the different types of immersion heaters and their typical applications, explains how to select, size, and specify immersion heaters, and provides guidance on installing and using the heaters.

Immersion heaters, as the name implies, are directly immersed in water, oils, viscous materials, solvents, process solutions, molten materials, and gases. By generating all the heat within the liquid or process, these heaters are virtually 100% energy efficient.

Many designs are available from heater suppliers as stock items and are offered in numerous sizes, kilowatt ratings, and voltages, and with a variety of termination connections, sheath materials, and accessories. Or, a manufacturer will custom engineer an immersion heater or system for a specific application.

The basic types of immersion heaters are the screw plug, flange, pipe insert or bayonet, circulation or in-line, booster, over-the-side, and vertical loop. They are usually available in either a round tubular design or a flat tubular design. Flat tubular immersion heaters can typically run at a higher watt density — 30 W/in.^{2} compared to 23 W/in.^{2} for a round tubular heater in a typical application such as heating low viscosity lubricating oil — while not having a higher sheath temperature. Heaters are also grouped into two categories — pressurized (closed) systems and nonpressurized (open tank) systems.

■ *Figure 1. Square flange immersion heaters.*

Pressurized systems

The **square flange immersion heater** (Figure 1) is used in industrial water boilers and storage tanks holding degreasing solvents, fuel oils, heat-transfer fluids, and caustic solutions. The assembly consists of either a round or flat tubular heater brazed or welded to a four- or six-bolt flange with screw lug or threaded stud terminals for wiring connections. These heaters bolt directly to a mating companion flange that is welded to a tank wall or nozzle. Assembly change is as easy as unbolting the flange and replacing it with another heater, which minimizes extensive equipment downtime.

The **screw plug heater** (Figure 2) is

inserted into a threaded opening in a tank wall or into a mating full or half coupling. Screw plug immersion heaters are available in a variety of National Pipe Thread (NPT) sizes, materials, wattages, and voltages, and with various sheath materials and thermostats. Applications include deionized or demineralized water, oils, hydraulic and crude oil, caustic cleaners, chemical baths, antifreeze (glycol) solutions, liquid paraffin, process water, and industrial and clean-water rinse tanks.

ANSI flange heaters are through-the-side heaters for liquid immersion applications requiring high wattage in large tanks — up to 500 kW or higher. Applications are similar to those of screw plug heaters, but ANSI flange heaters are used in higher pressure applications (up to 3,000 psi), such as tanks of superheated steam, compressed gases, or liquids.

Pipe insert, or bayonet, heaters (Figure 3) are used for heating liquids in huge (millions of gallons) storage tanks. The heater is mounted inside a pressure-tight bayonet pipe that mates to a flange connection on the side of a storage tank, thus supplying the pressure boundary. The heater is then inserted into the open end of the bayonet, allowing for removal of the heater without draining the tank.

Circulation, or in-line, heaters (Figure 4) are all-in-one units with the heater mounted inside its own insulated tank. The heater has inlet and outlet piping and the liquid or gas flows through the tank. By the time the material comes out the end, it is heated to the proper temperature. This design has a fast response and an even heat distribution. Heaters can be as small as a 1¼-in. NPT screw plug size to as large as 14 in. in diameter; custom units have been made up to 36 in. nominal pipe size.

Booster heaters are a type of circulation heater. They are ideal for applications using lower wattage, including in-line operations or engine preheating. Booster heaters with copper and steel sheaths are ideal for heating water and oils.

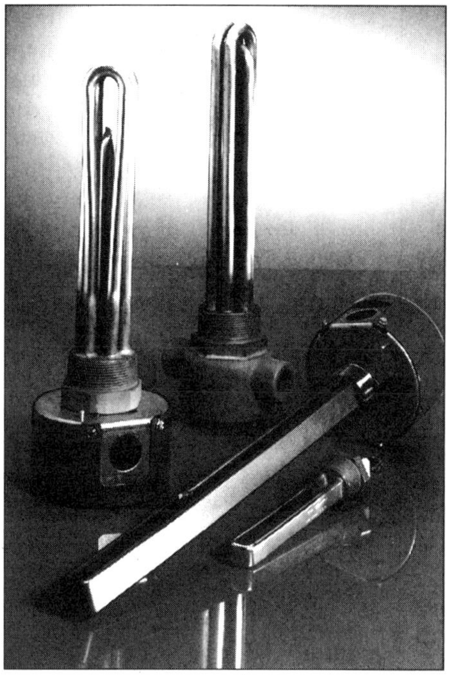

■ *Figure 2. Screw plug immersion heaters.*

■ *Figure 3. Pipe insert immersion heater.*

Nonpressurized systems

Over-the-side heaters (Figure 5) are formed into L and O shapes and are installed in the top of a tank, with the heated portion directly immersed along the side or at the bottom. Over-the-side heaters evenly distribute heat to liquids and viscous solutions. They are portable, easily removed for cleaning of heaters and tanks, and provide ample working area inside the tank. A variety of optional sheath materials, kilowatt ratings, terminal enclosures, and mounting methods are available.

Over-the-side heaters are ideal for heating small quantities of water, oils, solvents, salts, and acids. They are often used for freeze protection.

The thin-profile **vertical loop heater** (Figure 6) is available in a round tubular heater design and hangs over the side of an open tank. Another over-the-side type is the **drum heater** (Figure 7), which easily fits into the bung hole of a 55-gal

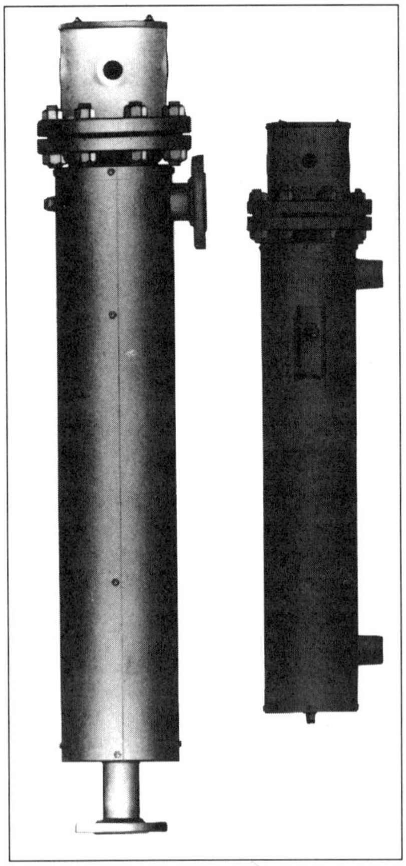

■ *Figure 4. Circulation heaters.*

drum. It typically melts heat-sensitive materials, such as wax, lard, grease, and coconut oil. A pre-wired thermostat protects the material from being overheated.

Cost comparisons

Many heater choices are limited by specific characteristics or requirements of the application. However, if there is a choice, the following cost relationships generally hold true — a screw-plug heater is the least costly, a flange heater somewhat more costly, and an over-the-side heater can cost the most.

Selecting a heater

Most electrical heating problems can be solved by determining the heat required to do the job. The heat requirement is converted to electrical power, and the most practical heater can then be selected for the job. Whether the problem is heating solids, liquids, or gases, the method for determining the power requirement is the same.

In defining the problem, the following factors should be considered:

1. Properties of the material to be heated. It is very important to know the type and quality of the fluid being heated.

For example, if the fluid is rinse water for parts, is the water clean or is it contaminated with traces of acids or alkalis, which are often left behind when rinsing parts? Acids cause corrosion and buildup on the sheath of the heater, which acts as an insulator and causes the heater coil to fail prematurely due to overheating.

If the fluid is an oil, what type of oil is it? For instance, a crude oil is very thick and viscous and requires a very low watt density, whereas a very light oil such as vegetable oil could use up to 30–40 W/in.[2] The watt density depends on the viscosity, specific heat, and thermal conductivity of the oil (Table 1). Choosing the proper watt density ensures that coking does not occur.

2. Startup and maximum operating temperatures. The expected temperatures are needed for determining the fluid's temperature change, which is needed for calculating the kilowatt requirements.

3. Maximum flow rate of the material being heated. This is needed to determine the wattage requirements. The minimum flow rate may also be required to help determine the watt density requirements. If the flow rate is too low and the watt density too high, excessive coking can occur in oils or excessive sheath temperatures can occur in air and gases.

4. Required time for startup heating and process cycle times. The longer the startup time allowed, the lower the kilowatt requirement. This is because the kilowatt requirement is inversely related to the time needed to heat up the medium to operating temperature.

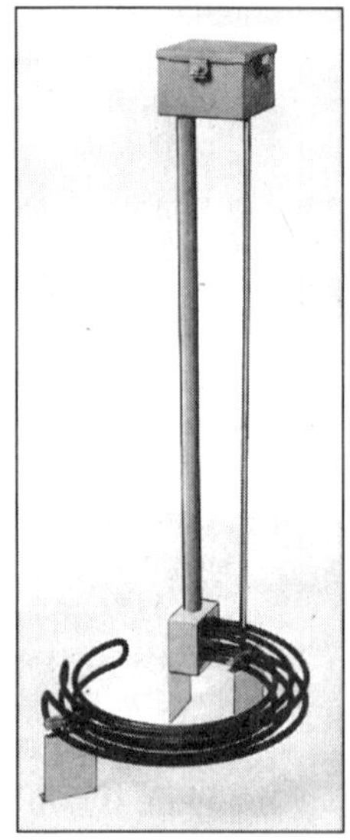

■ *Figure 5. O-shaped over-the-side immersion heater.*

5. Volume or weight of the heated material. These are needed to determine kilowatt requirements for startup.

6. Characteristics of the containing vessel. The weight is used to determine the kilowatt requirement for startup. The dimensions of the vessel are required to determine heat

Table 1. Watt density depends on the oil being heated.

Material	Maximum Operating Temperature, °F	Specific Heat*, Btu/lb•°F	Thermal Conductivity*, Btu•in./h•ft²•°F	Maximum Watt Density†, W/in²
Fuel Oils				
Grades 1 and 2 Distillate	200	0.45	0.98	23
Grades 4 and 5 Residual	200	0.41	0.875	12
Grades 6 and Bunker C	160	0.405	0.832	8
Machine Lubrication Oils				
SAE 10–30	250	0.43		16
SAE 40, 50	160	0.43		13

* Average values
† Noncirculating

losses in the initial startup equations and to determine kilowatt requirements for maintaining the operating temperature. The vessel's material of construction could affect the type of heater chosen and the way it is supported in the tank, especially if it is a plastic container. Factors involved in material selection include the threat of galvanic corrosion, wattage requirements, and structural support. Also, whether the vessel has an open top or closed top will greatly affect heat loss — a closed top will significantly reduce the kilowatt requirement to heat up and maintain the process.

7. Vessel insulation. If any insulation is present, its thickness and thermal properties will affect the heat loss from the vessel. Heat loss on connecting piping is normally compensated for by alternative heating, such as mineral insulated cable or heat-tracing cable.

8. Temperature monitoring and control. Sensing and control methods and locations vary greatly depending upon the precision requirements for the process and heater sheath temperatures. For example, a simple freeze protection application may only require the use of a mechanical bulb and capillary type thermostat, which is most economical, to monitor the process. For more precise measurement and control, a thermocouple or resistance temperature detector (RTD) sensor may be used in conjunction with a microprocessor-based controller. A high-limit sensor located on the sheath prevents overheating, which could lead to premature failure or accelerated buildup of contaminants.

The temperature sensor should be located at the point where the process temperature is most critical.

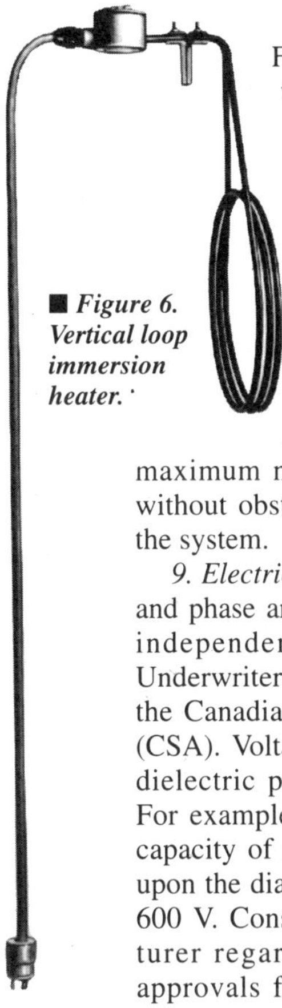

■ *Figure 6. Vertical loop immersion heater.*

For instance, in a circulation heater application, the sensor should be located in or nearest to the outlet nozzle of the vessel. In an open tank, the sensor should be positioned high enough to avoid contamination from sludge and low enough to receive maximum natural fluid convection without obstructing the operation of the system.

9. Electrical requirements. Voltage and phase are governed primarily by independent agencies, such as Underwriters Laboratories (UL) and the Canadian Standards Association (CSA). Voltage is limited due to the dielectric properties of the heater. For example, the maximum voltage capacity of most heaters, depending upon the diameter, would not exceed 600 V. Consult the heater manufacturer regarding the agency approvals for heater voltage and diameter limitations. Phase is not limited by anything other than certain heater designs.

10. Electrical limitations. The biggest limitation is a maximum of 600 V because of dielectric capabilities of the heater. Resistance limitations are also encountered when there are voltage and wattage extremes, either too low or too high. For example, if the voltage is too high and wattage too low, the resistance on the heater coil would be so high that the thin-gauge wire typically used would be too fine. The reverse, high wattage and low voltage, creates a need for a wire of such heavy gauge that it is impractical to manufacture the heater. Consult suppliers for their various manufacturing capabilities. Always consider UL and CSA agency approvals.

11. Environmental conditions. The ambient temperature and wind conditions can affect heat loss and should be taken into consideration when calculating kilowatt requirements.

Hazardous environments, such as corrosive and explosive situations, are also important considerations. For example, a stainless steel enclosure is resistant to corrosive processes. In explosive atmospheres, a NEMA 7 explosion-resistant electrical enclosure must be used. NEMA 4 ratings are for moisture resistance and may be needed in outdoor or wash- or rinse-down cleaning. Often, a combination NEMA 4 and 7 rating is required. General-purpose NEMA 1 enclosures are typically used when environmental conditions pose no problem.

12. Contingencies. Because the thermal system design may not take into account all the possible or unforeseen heating requirements, a safety or contingency factor that increases heater capacity beyond calculated requirements is applied. A factor of 10% is typically used. However, when there are many variables and some unknowns, safety factors up to 20% may be considered.

Sizing the heater

The basic steps in sizing an immersion heater involve calculating the following:

1. Power required for initial heating of the fluid and the tank. Use the basic heat-transfer equation (heat equals mass times heat capacity times temperature change: $Q = WC_p\Delta T$) to calculate the fluid heating requirement.

2. Power required to heat the fluid during the operating cycle.

3. Heat required to melt or vaporize materials during initial heating.

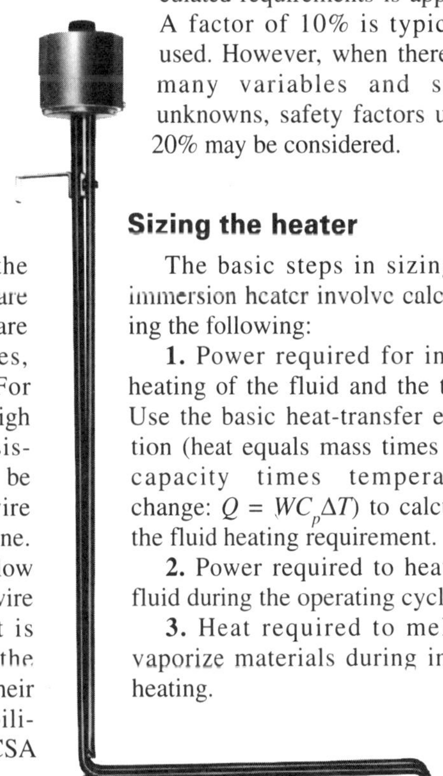

■ *Figure 7. Over-the-side drum heater.*

4. Heat required to melt or vaporize materials during operating cycle.

5. Thermal system heat losses. The heat losses are calculated by multiplying the exposed surface area, the startup time, and a surface loss factor ($Q_L = AtF_{SL}$).

6. Total startup power requirements. The results of Steps 1 and 3 are added together and an appropriate safety factor (typically 10%) is applied.

7. Total operating power requirements. The results of Steps 2, 4, and 5 are added and the safety factor applied.

8. Watt density. The total wattage is divided by the active heater surface area. The latter is calculated based on the length of heater element immersed in the fluid, the surface area per linear inch, and the total number of heater element lengths (two lengths per element multiplied by the number of elements).

Table 2. Wattage ratings for heating water.

Amount of Liquid, gal	Temperature Rise, °F 20°	40°	60°	80°	100°	120°	140°
5	0.3	0.5	0.8	1.1	1.3	1.6	1.9
10	0.5	1.1	1.6	2.1	2.7	3.2	3.7
13	0.8	1.6	2.4	3.2	4.0	4.8	5.6
20	1.1	2.2	3.2	4.3	5.3	6.4	7.5
25	1.3	2.7	4.0	5.3	6.7	8.0	9.3
30	1.6	3.2	4.8	6.4	8.0	9.6	12.0
40	2.1	4.0	6.4	8.5	11.0	13.0	15.0
50	2.7	5.4	8.0	10.7	13.0	16.0	19.0
60	3.3	6.4	9.6	12.8	16.0	19.0	22.0
70	3.7	7.5	11.2	15.0	19.0	22.0	26.0
80	4.3	8.5	13.0	17.0	21.0	26.0	30.0
90	5.0	10.0	14.5	19.0	24.0	29.0	34.0
100	5.5	11.0	16.0	21.0	27.0	32.0	37.0
125	7.0	13.0	20.0	27.0	33.0	40.0	47.0
150	8.0	16.0	24.0	32.0	40.0	48.0	56.0
175	9.0	18.0	28.0	37.0	47.0	56.0	65.0
200	11.0	21.0	32.0	43.0	53.0	64.0	75.0
250	13.0	27.0	40.0	53.0	67.0	80.0	93.0
300	16.0	32.0	47.0	64.0	80.0	96.0	112.0
400	21.0	43.0	64.0	85.0	107.0	128.0	149.0
500	27.0	53.0	80.0	107.0	133.0	160.0	187.0

Heating water — a shortcut

For many immersion heating applications, some relatively simple formulas are available for quickly estimating kilowatt requirements.

Table 2 can be used for water heating applications. To use the chart, find the amount of liquid in gallons on the left and the desired temperature rise at the top. Read from the chart the wattage needed to heat the water in one hour.

Alternatively, one of the following equations could be used, where Q_g = flow rate in gal/min, ΔT_F = temperature rise in °F, Q_L = flow rate in L/min, ΔT_C = temperature rise in °C, and t_h is the heating time in h.

For heating flowing water:

$$\text{kW} = Q_g \times \Delta T_F \times 0.16 \quad (1a)$$

$$\text{kW} = Q_L \times \Delta T_C \times 0.076 \quad (1b)$$

For heating water in tanks:

$$\text{kW} = (Q_g \times \Delta T_F)/(375 \times t_h) \quad (2a)$$

$$\text{kW} = (Q_L \times \Delta T_C)/(790 \times t_h) \quad (2b)$$

Make certain the sheath material and watt density ratings are compatible with the liquid being heated.

Installation tips

Each heater's current should be checked before installation, since during shipping and storage moisture can enter the heater element insulation and affect the heater's performance. The same problem may occur if the heater has been idle for a week or more.

Each circuit should be checked using a 500-V d.c. megohm meter, and the reading should be at least 10 MΩ. Lower values may be acceptable, but the supplier should be consulted for more information.

A low current reading does not mean the heater is bad and has to be returned. There are several ways to increase the megohm level. One is to put the heater in an oven at 200°F–300°F and leave it overnight or until the megohm readings are acceptable. The second way is to energize the heater at no greater than 50% of the rated voltage until the megohm reading reaches its proper specification. Consult the heater manufacturer for more details.

The proper temperature rating of the wire coming into the heater is also important. A minimum of 200°C wire for process heaters is recommended, although higher-rated wire may be required for some applications. It is best to consult the supplier for the proper wiring information.

Power feeder line connections must be compatible with the heater and meet National Electric Code (NEC) specifications. All installation wiring should be done in accordance with the NEC and other state and local codes.

Immersion heaters used in tanks should be mounted horizontally near the tank bottom to allow convective circulation. They must be located high enough to be above any scale or sludge buildup on the bottom of the tank.

The entire heated length of the heater should be immersed at all times. Do not locate the heater in a restricted space where free boiling or a steam buildup could occur. Low-level shutoff switches can be installed to avoid heater failure should the liquid level drop too low.

Maintenance to maximize performance

To increase the life and performance of an immersion heater, follow these tips:

Make sure the sheath material and watt density ratings are compatible with the liquid being heated. Application and specification guides supplied by manufacturers provide a complete listing of materials along with maximum temperatures and watt density recommendations.

For circulation or in-line heaters, make sure there is adequate flow. If the flow rate is too low, the heater will overheat and fail prematurely. A flow switch can be used to monitor or shut off the heater if a blockage occurs in the system. Alarms are often used in conjunction with them.

Corrosion of the heater can lead to problems ranging from equipment downtime to serious safety hazards. Because sheath temperature plays such an important role in the corrosion process, it is important to accurately monitor the heater during operation. Place temperature sensors on the areas of the sheath where the highest temperatures are expected — on the top of the heater bundle in an open tank with a horizontally mounted heater or nearest the vessel outlet in a circulation heater.

Also check inside the terminal housing for corrosion due to ambient conditions or loose line connections. If oxidation is present on the line connections, clean and retighten them. If moisture or fumes are present, a different terminal housing may be required. Once the maintenance is complete, thoroughly blow clean with dry, oil-free air.

Scale buildup on the sheath and sludge on the bottom of the tank must be minimized. If not controlled, they will inhibit heat transfer to the liquid and possibly cause overheating and failure.

The flat tubular heater design prevents the buildup of scale in water immersion applications. Because of its unique geometry the heater "breathes," breaking scale and deposits off its sheath. If scale buildup is discovered on other tubular elements, it is important to clean the units as required.

There are various brands of cleaning chemicals that can remove scale buildup. Water treatment companies are a good source for this information.

Another way to remove scale is to periodically clean the heater. Brush the scale off with a wire brush or clean the heater element in a mild, caustic chemical solution using a brush and chemical that will not harm the heater sheath. A mild sandblasting of virtually any type of sheath is often very effective, although one must take great care to not damage the heater sheath.

Coking is another problem that can lead to early heater failure. It often occurs in oil or other viscous products and increases as sheath temperatures increase. A flat tubular element's sheath runs cooler than that of a round tubular element when operated at the same watt density, so the flat element has a lower potential for coking. The degree of coking varies greatly, depending upon the maximum operating temperature of the oil being heated.

Extreme caution should be taken not to get silicone lubricant on the heated section of the heater. The silicone will prevent "wetting" of the sheath by the liquid and act as an insulator, possibly causing the heater to fail.

In addition to coking and scaling, poor wiring connections account for a large percentage of problems in the field. Make sure that the electrical connections are tight. This should be done on a regular basis because process temperature as well as the amperage going through the terminal area create heat in the terminal enclosure. When the process heats up and cools down, the connections can be loosened, which eventually leads to heater failure. A torque of 20 in.•lb on each heater stud is recommended. In addition, the connections should be free of oxide, dust, and dirt buildup.

Make sure the interior of the terminal enclosure is clean and dry, and free of dirt, dust, oil, and rust.

Thermal cycling may also cause sealed joints, such as flange mounting bolts, to relax, resulting in leaks. Tighten threads and flange bolts.

Periodically check the sensing probes (thermostat or thermocouple) to make sure they are operating properly and that the connections are all good. Check proper grounding for safety.

Make sure the power is turned off before doing any maintenance procedures. CEP

R. C. KLEIN is product manager with Watlow Electric Manufacturing Co., Hannibal, MO (314/221-2816; Fax: 314/221-3723). He has worked for Watlow for 17 years in production and design and has assisted a variety of customers with the proper application and troubleshooting of industrial heaters. He holds a BS in industrial occupations from Northeast Missouri State Univ.

Evaporation in Plate Heat Exchangers

Hemant Kumar

APV Baker Therm. Div., 1200 W. Ash St., Goldsboro, NC 27533-2046

Plate heat exchangers are employed for a wide range of evaporation duties. In this article, heat transfer coefficients and pressure drops are discussed relative to the geometry of the plate channels which varies from essentially flat rectangular ducts to highly turbulent cross-corrugated channels. Differences between upward and downward performance are also outlined.

Whereas plate heat exchangers (PHE's) are becoming common place for liquid to liquid heat transfer, their use in phase change thermal transfer is relatively unknown to most users. In fact, plate evaporators have been used to concentrate liquid foods for at least forty years.

Frequently, this has involved transfer of heat with boiling suppressed until the outlet whereafter flash evaporation takes place. This is a straightforward heat exchange duty but with obvious limitations on single pass vapourising capacity. Except to state that this technique is often combined with recirculation to achieve required evaporation rates, further discussion in this article is limited to boiling within the PHE itself.

GENERAL CONSTRUCTION

A PHE consists of an array of rectangular plate elements sealed by elastomeric gaskets with suitable headers such that process and service media flow in alternate channels. The entire pack of elements is compressed between two rigid members by a number of peripheral bolts to a predetermined dimension. See Figure 1. For completeness, below are some important variations to the preceding description:

In small PHE's, the entire plate pack is sometimes copper brazed, finding widespread usage in small freon evaporators.

Of similar size, a novel plate and shell design has recently appeared on the market, where a pack of fully welded circular plates is housed within a shell. See Figure 2. In common with the conventional plate heat exchanger and unlike the shell and tube, concurrent or counter flow is maintained. Avoidance of braze material allows its use in a wider variety of chemical and refrigerant evaporator applications.

Large PHE's constructed from welded plate pairs such that the process medium flows within welded channels, passing through headers bounded by trapped ring gaskets. See Figure 3. Primary usage is in refrigerant evaporators and for evaporation of aggressive chemical media.

Small to mid-size PHE's where dissimilar metals may be employed without welding for the two heat exchanger media of different corrosive tendencies.

HEAT TRANSFER SURFACE AND FLOW GEOMETRY

For evaporation, two geometries are discussed in this article. The first consists of familiar elements with closely spaced corrugations inclined to the general flow direction alternating with similar elements such that their forms cross and abut at a large number of points. See Figure 4. In some special variants, not covered here, the corrugations on adjoining plates become aligned essentially in the vertical direction.

APV Baker Goldsboro, North Carolina U.S.A.
Kumar, Hemant

In the second kind of plate evaporator geometry for low pressure evaporation, corrugations are kept to a minimum such that the process side heat transfer surfaces are essentially flat with just sufficient dimples to maintain the flow spacing. See Figure 5. The service channel in which steam is used as the heating medium is also a plain rectangular duct with intermittent baffles to lift the condensate from the surface and to thin the non-condensible layer which may build up on wall.

PLATE EVAPORATOR: CORRUGATED CHANNELS

Primarily developed for liquid/liquid duties, the usual advantages of the PHE relative to other heat exchangers also apply to their use in evaporation.

However, the turbulating channels do have a more pronounced effect on pressure drop, vapour pressure and therefore the boiling temperature and ΔT. Whereas this is not a problem for high molecular weight organic vapours (seals permitting) or lighter vapours under pressure (pressurised ammonia chillers, superatmospheric water evaporation), wider gap channels have to be used for the common case of boiling under vacuum. In such instances, header losses at the exit from the evaporator may also force selection of a larger ported PHE. Sometimes, for example in compact plants on board ships producing drinking water by desalination, water is evaporated in PHE's at 40°C and pressure drops are minimised by transverse flow, bypassing the limiting plate ports and reducing the channel path length at the same time. See Figure 6.

HEAT TRANSFER: WATER AND AQUEOUS SOLUTIONS

In a wide ranging study, PHE's were found to exhibit some interesting characteristics. Figure 7 is a reproduction from APV's evaporation of water against counter-current steam flow as the heating medium. Some basic conclusions are evident as follows:

Since $Q'' = U \cdot \Delta T$
and approximatly $Q = M_e \cdot \lambda$

It follows that,
for a given surface area, $M_e \propto U \cdot \Delta T$

Since U was shown to be invariant with M_e,
we have the following approximate proportionality.

$M_e \propto \Delta T$

Other work under similar conditions has confirmed this conclusion for a range of exit qualities which occur in practice, with burnout only apparent at very high vapour fractions.

The second conclusion discussed here is the difference in htc's for two different boiling temperatures. Clearly in vacuum evaporation, the overall htc's are significantly lower because the contributing fluid properties are worse for heat transfer not only in the boiling channel but in the adjacent condensing passage too.

The final conclusion is that overall htc's attained were good.

For the same experimental setup discussed above, Figure 8 is reproduced. The difference this time is that instead of pure water evaporation, a high viscosity glucose syrup with a feed strength of 60% solids is being boiled to a higher concentration. This is a common PHE application in the sugar industry.

Another interesting observation from Figure 7 & 8 is that the reduction in overall heat transfer coefficient between the cases of 212°F and 160°F (100°C and 71°C) exit temperatures is in each case roughly one-third. This can only happen if the htc drop is only a function of thermodynamic and transport properties and not a result of non-condensible enrichment at the wall. In other words, in corrugated channel PHE evaporators where the heat source is condensing steam, small mass fractions of non-condensibles are not a problem. Naturally, one would propose that the turbulating geometry encountered by the fluid keeps the non-condensibles reasonably well-mixed in the bulk. (It also raises the obvious question whether in multi-component two-phase heat transfer, models of mass transfer built for regular surfaces such as plain tubes or flat plates will be invalid for a PHE).

Further conclusions about the behaviour of the PHE channels may be understood as follows. There is a drop of some 50% in overall htc relative to water primarily because the viscosity is orders of magnitude higher. An htc of 500 Btu/h deg F·sq. ft. (2.8 kW/sq. m·K) is still quite high for an exit temperature of 212°F (100°C). The true htc is in fact higher because the analysis in constructing Figure 8 ignored the boiling point elevation of the syrup relative to the exit vapour temperature. Hence the true ΔT across the plate was lower. Furthermore, the apparent ΔT here has been calculated based on end saturation temperatures and therefore takes into account the effect of pressure drop on ΔT. (A greater pressure drop occurred in boiling 60% syrup compared with pure water). Which brings us to the shape of the htc/evaporation rate curve showing a weak decline with increasing quality. Had the two reasons given above been analysed differently a greater invariance between M_e and U would have been found. In other words, the conclusions reached in this regard for pure water boiling are not negated by Figure 8.

Finally, a note about falling film evaporation in corrugated PHE channels. Experiments show that whereas reasonable htc's are still attainable, they are lower than in the rising mode. The opposite is usually true for the flat plate evaporator discussed later.

HEAT TRANSFER: REFRIGERANT EVAPORATION

The reader is referred to Reference [2] for a more detailed discussion on this topic. Heat transfer correlations take the conventional single phase form, Reference [3], modified by an enhancement factor which is a function of the geometry itself and also of the pressure drop characteristics. Added to this is a nucleate boiling component with a suppression factor.

With significant investments in commercial R&D in this field, both in creating experimental correlations and developing large welded plate pairs for added security, the use of PHE's in evaporation of refrigerants is growing rapidly.

PLATE EVAPORATOR: FLAT PLATES

Whereas the corrugated PHE is reasonably well known in the industry, little is known of flat plate evaporators which operate at modest pressures or under vacuum, and are prevalent in concentration processes for a number of liquid foods. Examples are milk and fruit juices where the low residence time in plate evaporators yields a better quality concentrate.

Such evaporators can be rising or falling film and usually the complete plant consists of multiple effects such that a unit mass of steam can evaporate several units of water. Since the evaporating medium is heat sensitive, low temperature differences, e.g. 3 to 10 deg C, are used.

Figure 9 shows a combined climbing/falling film scheme showing a one up/one down arrangement. In a plate pack, the feed is distributed into sets of plates, each set with one upward and one downward boiling channel. Unlike corrugated PHE's, wettability of the surface is very important in these evaporators. For example, in falling film, adequate wetting rate may be maintained by adopting a 2 up/1 down scheme. A special distribution system is used to ensure that the falling pass is wetted correctly. In Figure 9 steam enters the upper left hand corner and zigzags down the plate. Baffles with closely spaced holes are employed which serve two purposes. Firstly, they lift the condensate off the plate so that fresh surface becomes available for high steam coefficients to be generated down the plate, not only at the top. Secondly, the baffles allow a higher steam velocity to aid mixing of non-condensible air into the flow.

Figure 10 shows a comparison of upward and downward evaporating coefficients for boiling water at 120°F (49°C), Reference [4]. It can be seen that up to a point, falling film htc's are higher. The fall-off at high qualities is attributed to the poor wettability of water and breakdown of the liquid film into rivulets especially when the feed rate is low.

The differences in up and down flow are less dramatic at higher boiling temperatures.

Figure 5 shows a falling film plate arrangement which has a special flash chamber and distributor to ensure complete and even wetting. The plate is symmetrical about its vertical centre line so that two stage operation can be performed for high concentration ratios desired. For example, concentrated product and vapour exit from the bottom left rectangular port and the concentrate is fed back at the top right hand edge for further concentration.

Figure 11 plots overall htc for this plate boiling a 1% caustic solution (for better wettability) at 180°F (82°C). Reference [5].

PRESSURE DROP

The pressure drop during evaporation in corrugated channels has been modelled employing a type of correlation which uses a two-phase multiplier on liquid alone pressure drop. This multiplier is, among other things, dependent on the channel geometry.

In corrugated passages, pressure drop can be quite significant and for low ΔT operations, crucial in determining the model which can be chosen. In upflow, at low vapour rates there is significant static head effect in aqueous evaporation under vacuum, whereas for refrigerant boiling under pressure, this has a relatively minor effect in the design of the evaporator.

In flat plate evaporation, the designs have been tailored to give very little pressure drop and these usually only become deleterious when evaporating water at low temperatures, e.g. 140°F (60°) or below.

Figure 12 shows pressure drops for upward boiling in the corrugated channels of a very small PHE with a small flow cross-sectional area. Reference [6]. Figure 13, on the other hand, plots the pressure drop in a large falling film evaporator. Reference [7].

CONCLUSIONS

Depending on available pressure drop and seal suitability, PHE's can be employed for a wide range of evaporation applications.

ACKNOWLEDGMENTS

The author wishes to thank the Directors of APV for permission to publish this paper.

NOTATION

- Q Heat Load (kW)
- Q'' Heat Flux (kW/sq m)
- U Overall heat transfer coefficient (kW/sq m·K)
- ΔT Temperature difference between the media (deg C)
- λ Latent heat of boiling (kJ/kg)
- M_e Evaporation rate (kg/s)

REFERENCES

1. APV: Internal evaporator R&D report E263

2. Kumar, H. The Design of Plate Heat Exchangers for Refrigerants, Proc. Inst. R. 1991-92. 5-1

3. Heavner, R. et al, Performance of an Industrial Plate Heat Exchanger: Effect of Chevron Angle, Submitted for 1993 National Heat Transfer Conference, Atlanta Georgia

4. APV: Internal R&D Evaporator Report E237

5. APV: Internal R&D Evaporator Report E290

6. APV: Private Communications, M. Layko, 1993

7. APV: Private Communications, J.G. Shaw, 1991

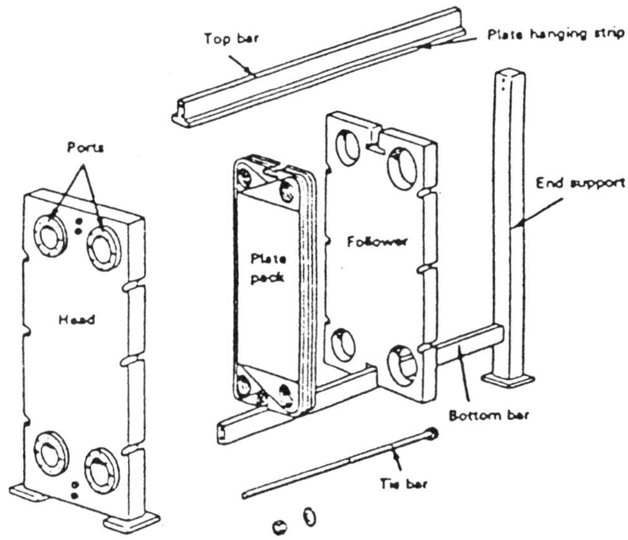

Figure 1 Exploded view of a P.H.E.

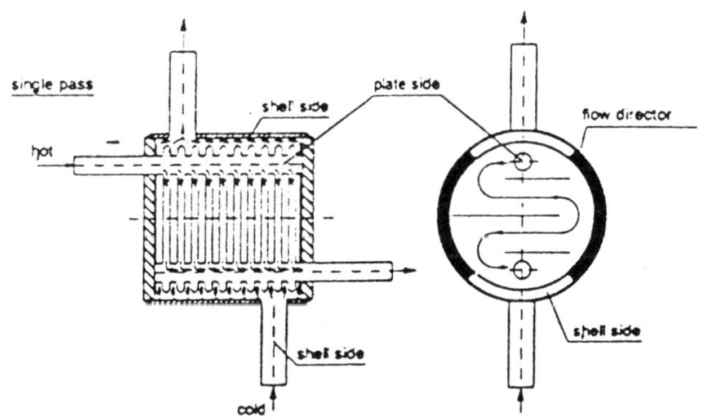

Figure 2 Welded Plates in a shell

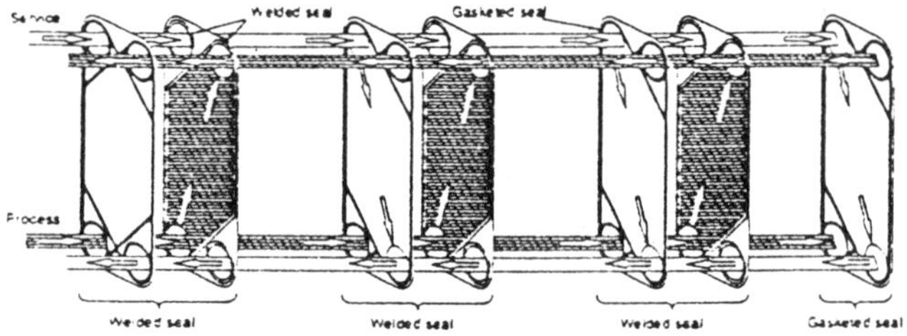

Figure 3 Welded Pair P.H.E.

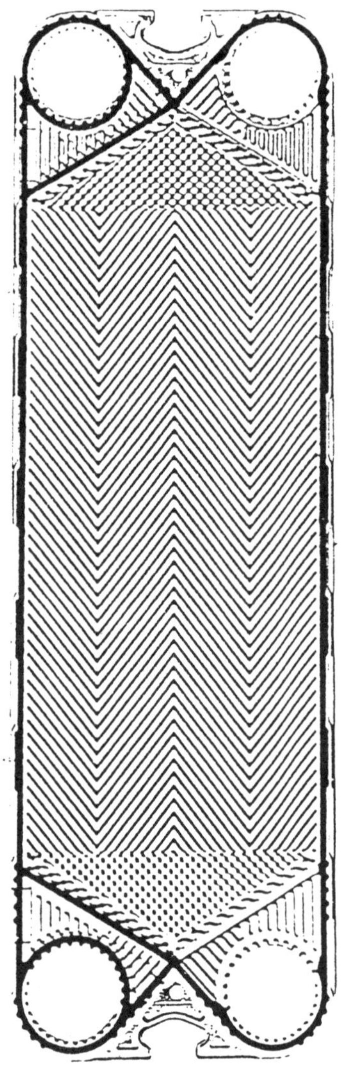

Figure 4 A Corrugated Plate

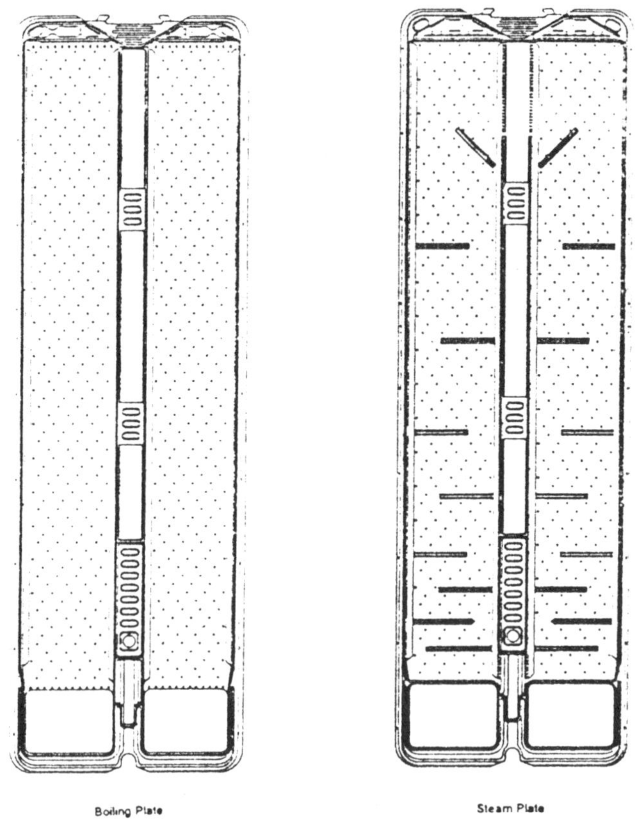

Figure 5 Flat Plate Evaporator

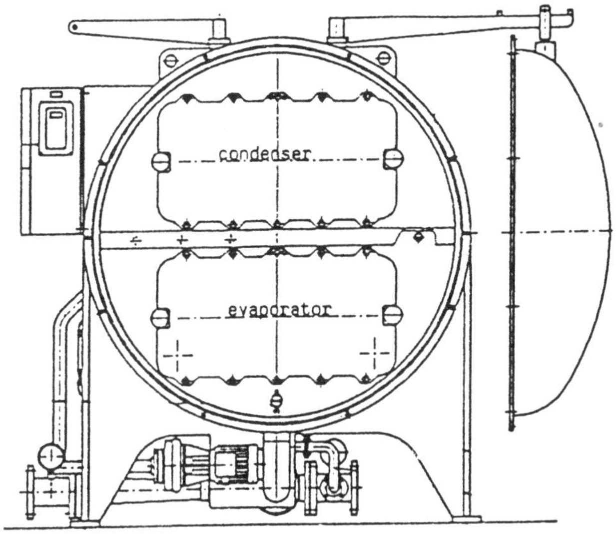

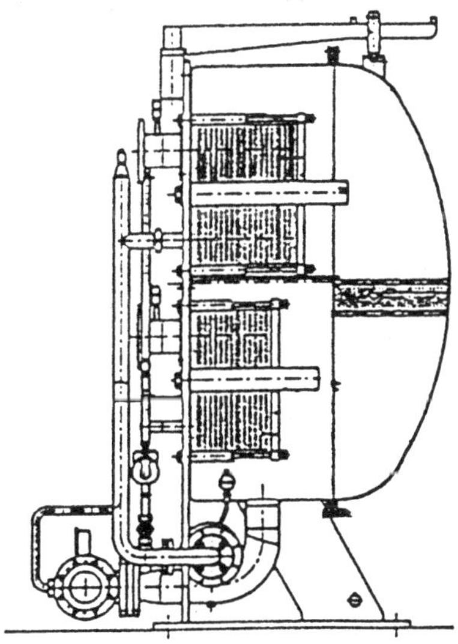

Figure 6 Cross-flow Evaporator & Condenser Shipboard Desalination

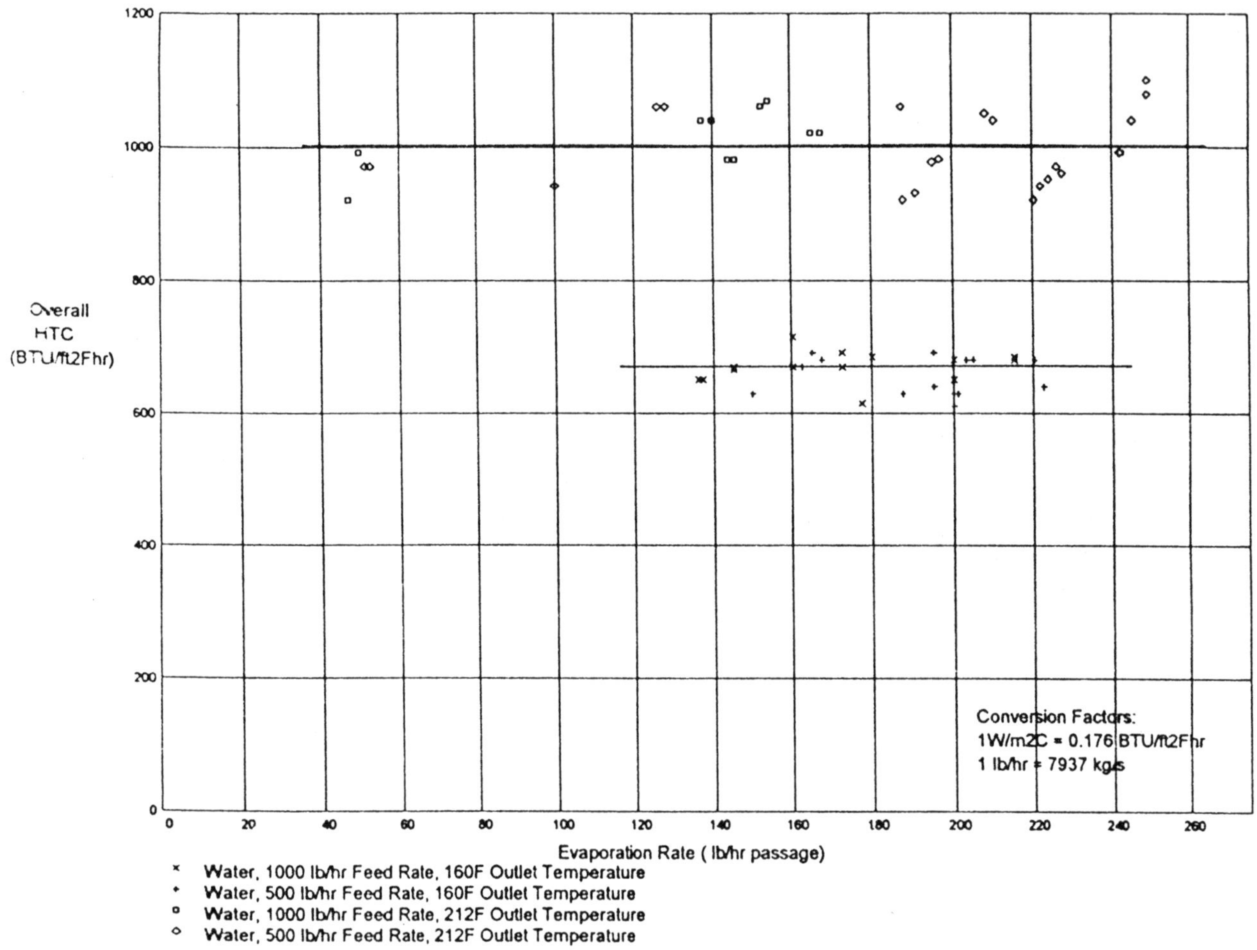

Figure 7 Climbing Film Water Boiling in a Medium Sized Corrugated Channel

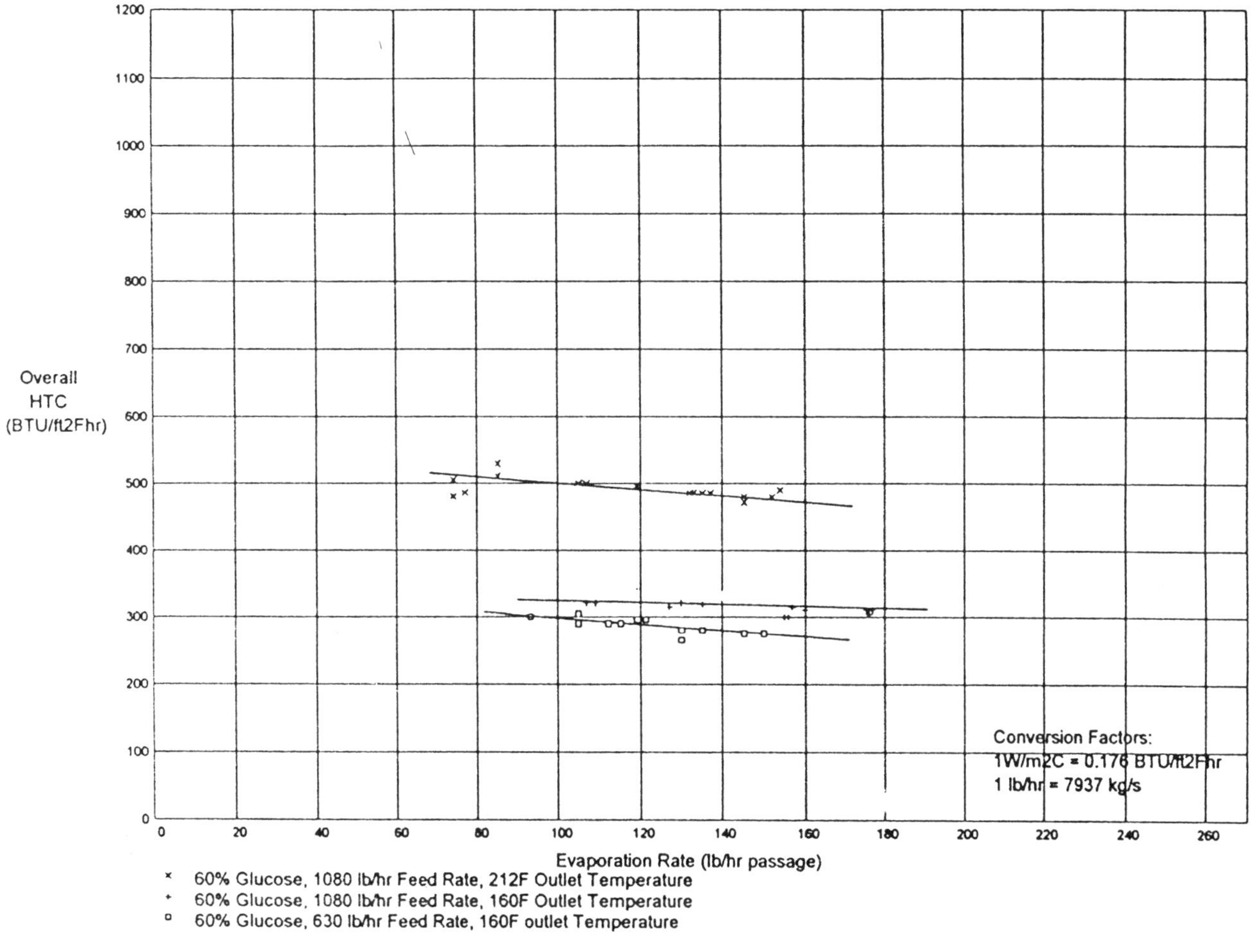

Figure 8 Climbing Film Boiling of 60% Glucose Feed Medium Sized Corrugated Channel

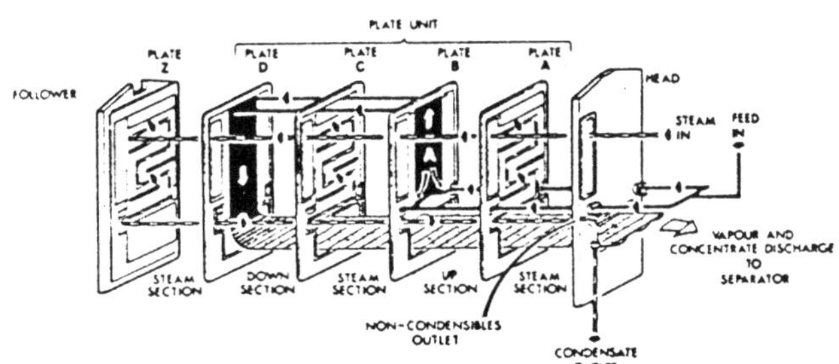

Figure 9 Climbing / Falling Flat Plate Evaporator

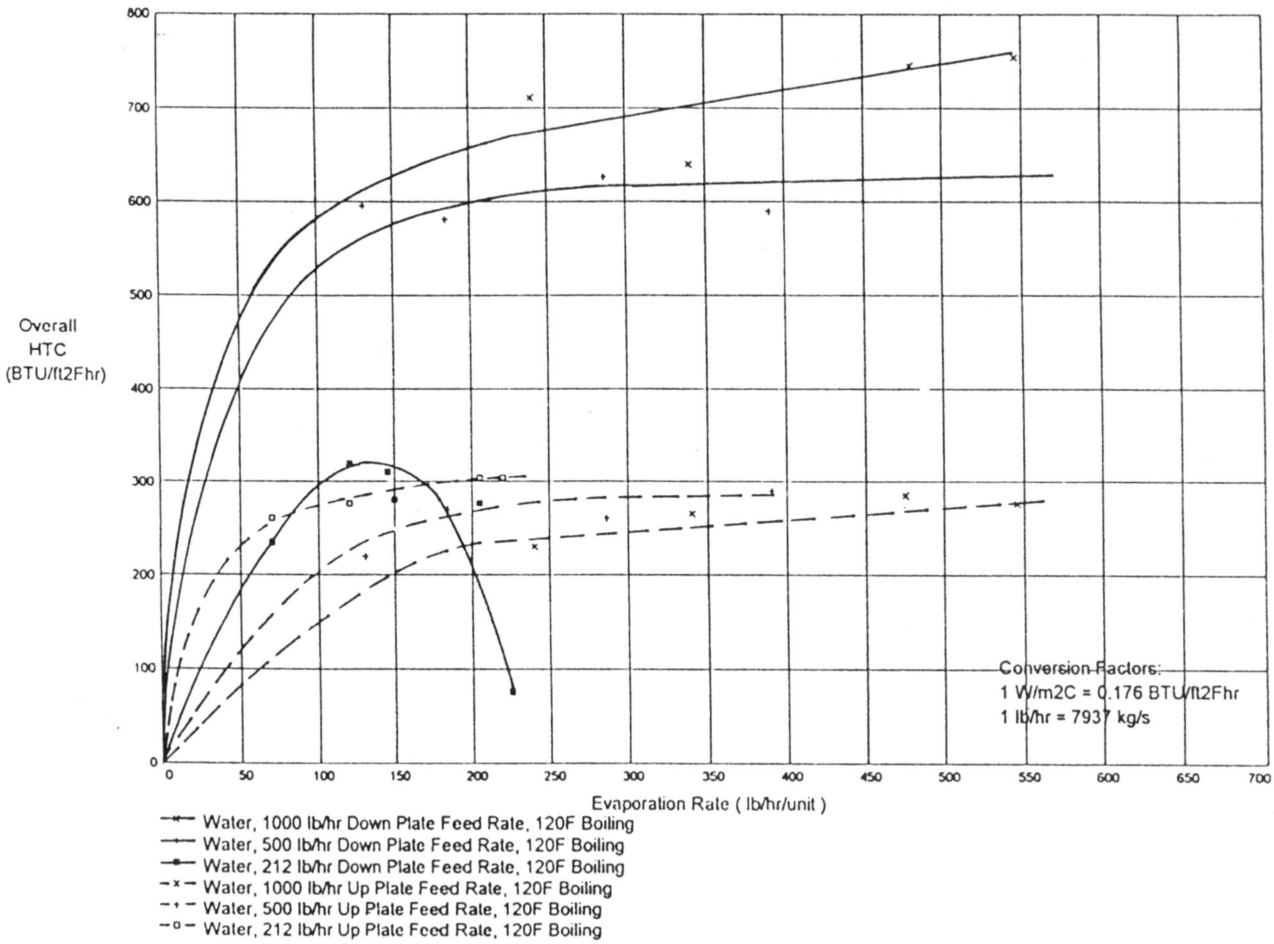

Figure 10 HTC's in Up and Down Boiling—Climbing and Falling Flat Plate Evaporator

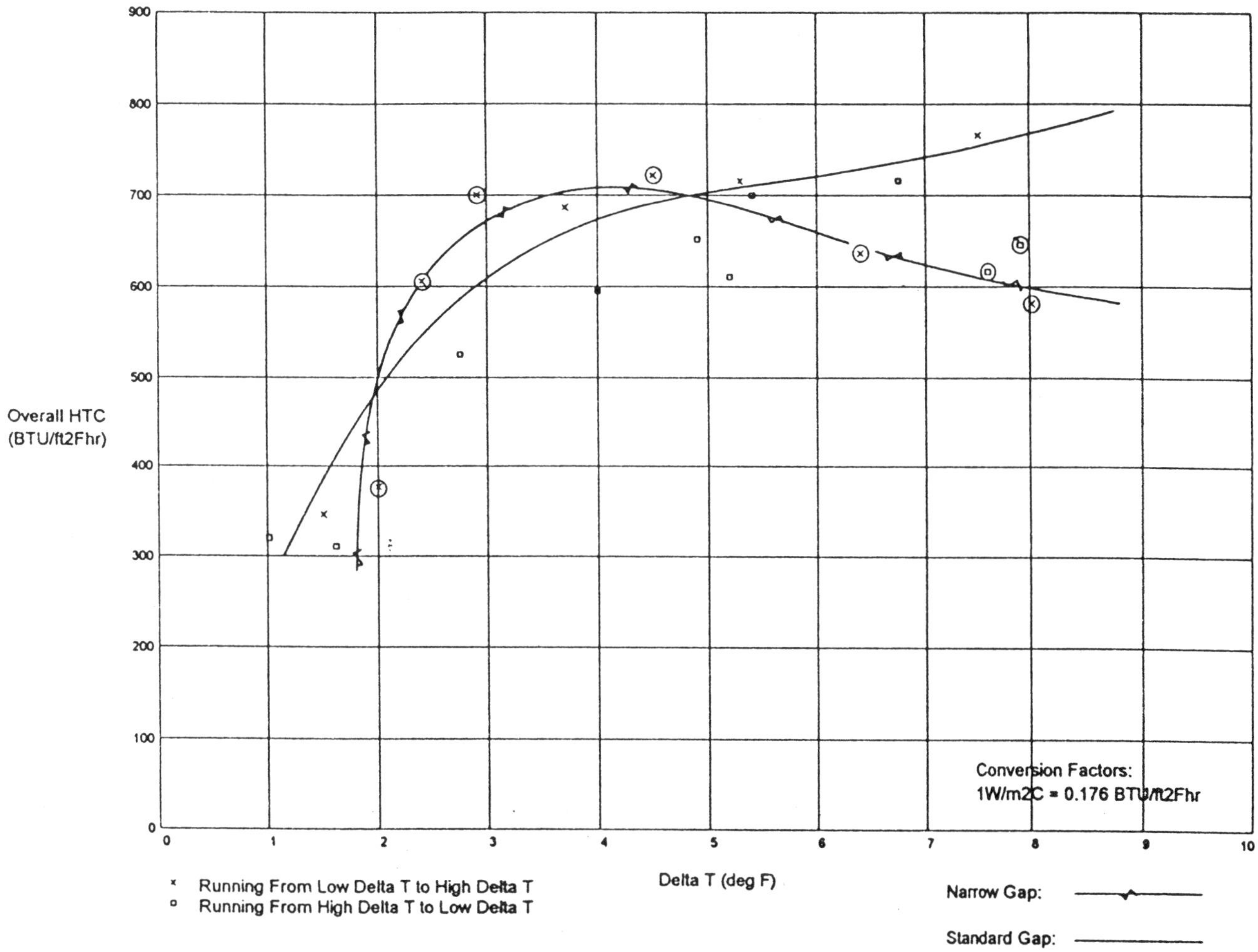

Figure 11 Falling Film Flat Plate Evaporator Overall HTC's, 1% Caustic Solution Boiling at 180F.

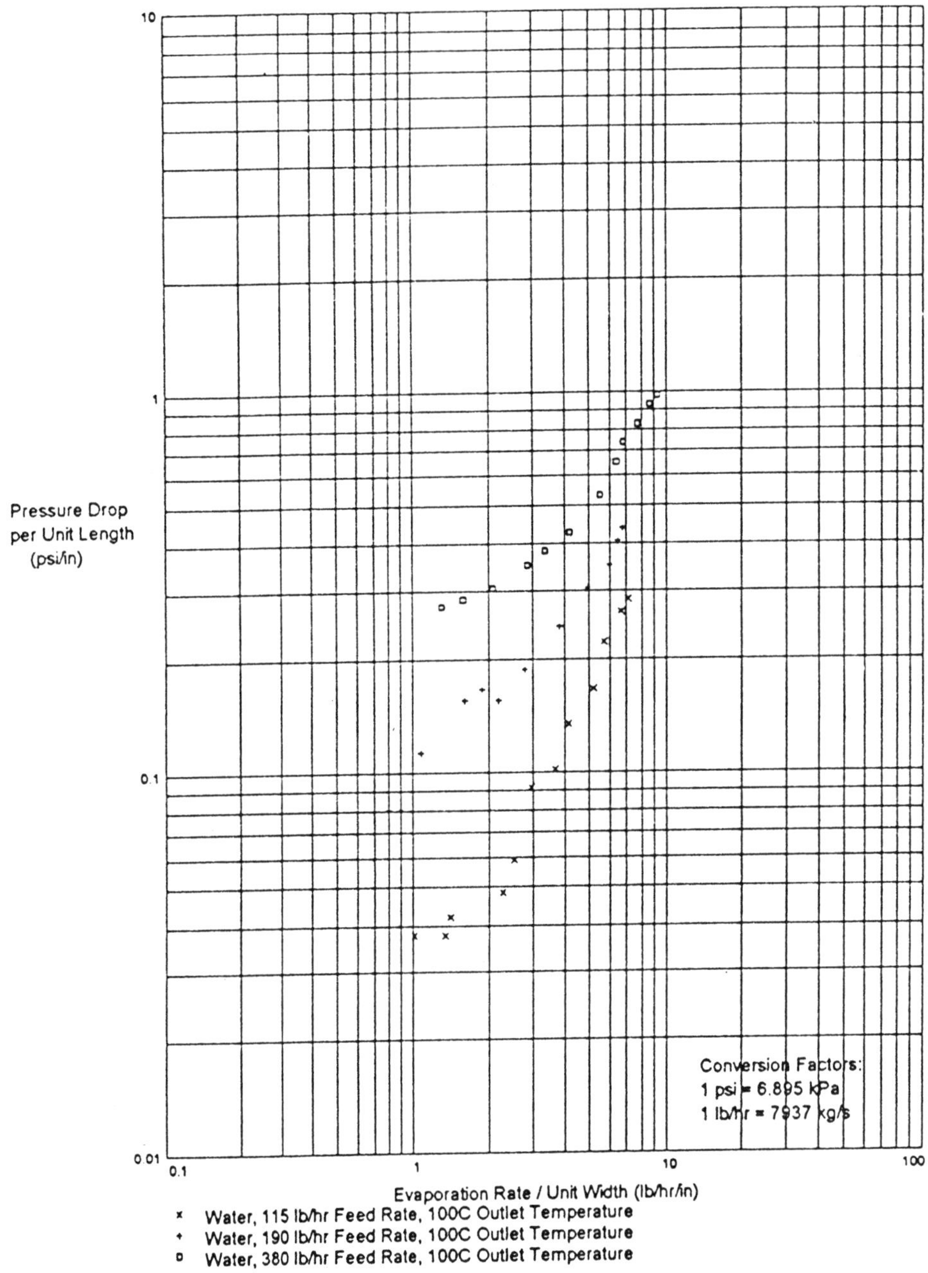

Figure 12 Small Narrow Gap Corrugated Channel Pressure Drop / Unit Length v Evaporation Rate / Unit Width

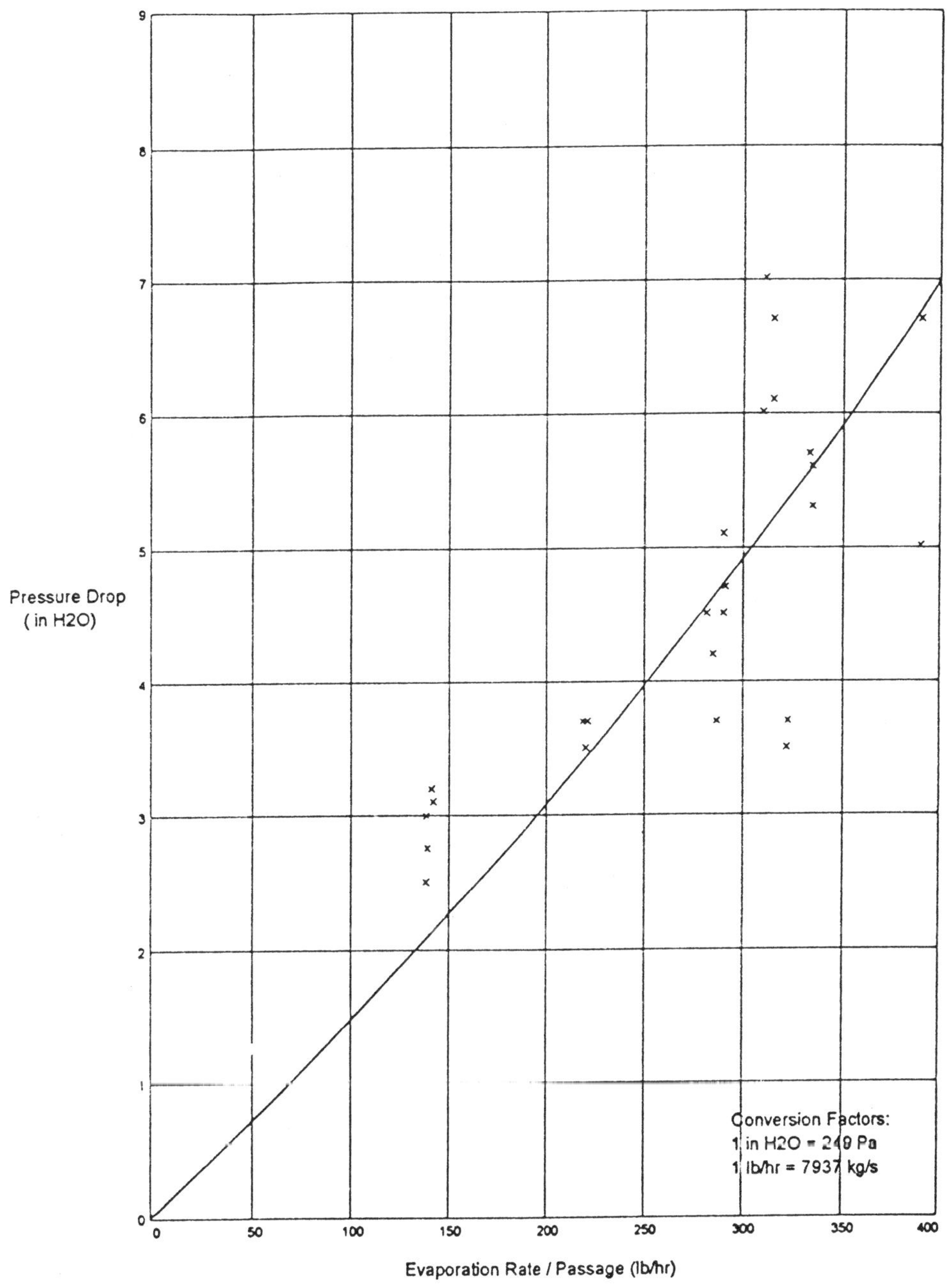

Figure 13 Pressure Drop in a Large Plate Evaporator, Water boiling at 140F

Choose the Right Flow Meter

Here's a guide to properly sorting through the wide variety of devices.

John W. Dolenc,
Fisher—Rosemount Inc.

Flow measurement is distinguished from other measurements by the large number of viable technologies available. Many flow meter selection guides have been published, all of which come to the same basic conclusions: choice depends upon the purpose of the measurement, the value of the fluid being measured, its physical characteristics, and installation constraints associated with the location of the meter.

When selecting a flow measurement technology for any application, seven key questions must be answered:

1. What is being measured?

2. Why is the particular measurement being made?

3. What is the value of the fluid?

4. What are the physical properties of the fluid at the point of measurement?

5. What materials of construction are acceptable?

6. What are the installation constraints? and

7. What are the track records of the various technologies in this application?

Finally, after these questions are answered, performance vs. cost can be judged. This judgment can be made via a formal procedure with written documentation or, for an experienced engineer, based on an informal analysis utilizing accumulated knowledge or preferences.

Start with the fluid

The type of fluid to be measured determines available options. Let's classify the fluid types commonly measured in the chemical process industries (CPI) as:

- clean liquid;
- dirty liquid;
- slurry containing particles;
- slurry having fibrous solids;
- corrosive liquid;
- cryogenic liquid;
- clean gas;
- dirty gas; and
- steam.

Certainly, fluids can be categorized in other ways and, indeed, have been in other selection guides. Regardless, the point of any such classification is to provide a coherent basis for winnowing options. Because most flow meters cannot measure all types of fluids, determining the fluid type begins the process of eliminating some flow measurement devices.

Ultimately, of course, all key physical characteristics of the fluid must be taken into account. Another consideration is the physical effect the fluid can have on any intrusive parts of the flow meter. Will metallic parts corrode? Will entrapped solids, dirty fluids, or slurries cause erosion? With certain differential-pressure primary elements, it's important to evaluate the potential buildup of solids on the upstream side of an orifice plate, or the plugging of small orifices in an averaging pitot device. With fibrous slurries, you should consider the possibility of hangup in the measuring element or any other internal parts such as the flow splitters in certain Coriolis mass flow meters.

A matrix of fluid type vs. flow measurement technology can be an excellent aid in bracketing viable options. Table 1 provides such a matrix for the categories that we have chosen. Others have been published in handbooks and by manufacturers of instruments (for instance, *1,2*). Care needs to be exercised when using any

Table 1. Applicability of the main types of flow meters.

	Fluid								
Technology	**Clean Liquid**	**Dirty Liquid**	**Slurry, Particles**	**Slurry, Fibrous**	**Corrosive Liquid**	**Cryogenic Liquid**	**Clean Gas**	**Dirty Gas**	**Steam**
Differential Pressure									
Averaging Pitot	++++	–	–	–	+	+	++++	+	++++
Nozzle	++++	+	–	–	++	+	++++	++	++++
Orifice	++++	+	–	–	+	+	++++	++	++++
Venturi	++++	++	–	–	++	+	++++	++	++
Target	++++	++++	+	+	+	+	++++	++++	+
Velocity									
Electromagnetic	++++	++++	++++	++++	++++	–	–	–	–
Turbine	++++	–	–	–	–	++++	++	–	–
Ultrasonic Doppler	–	++++	++	+	+	–	–	++	–
Ultrasonic Transit Time	++++	–	–	–	+	+	++++	–	+
Vortex	++++	++	+	–	+	++++	++++	++	++++
Mass Flow									
Coriolis	++++	++++	+	+	++++	++++	+	+	+
Thermal Mass									
Heated Element	–	–	–	–	–	–	++++	++++	–
Temp. Rise Flow-Through	++++	++++	++	++	++	–	+	+	+
Temp. Rise Bypass	–	–	–	–	–	–	++++	–	–
Volumetric									
Positive Displacement	++++	+	+	–	++	–	–	–	–
Other									
Variable Area	++++	–	–	–	++	+	++++	–	+

Note: (++++) = acceptable; (++) = normally acceptable; (+) = acceptable under certain conditions; and (–) = not acceptable.

flow selection guide, however, as some technologies can be borderline. In these cases, nothing surpasses past experience as the deciding factor. If a certain flow measurement technology for a specific fluid type has not been used in-house, then ask the manufacturer about its applicability — and for references about such applications.

Understand the role of the measurement

The measurement's purpose is the next item to consider. Is the flow rate value to be used for critical control, noncritical control, monitoring, or local indication? This determines the needed degree of accuracy — which ranges from the extreme accuracy required for critical control to the rough measurement that suffices for simply indicating a minimum flow rate condition.

Critical control applications are those that directly and decisively affect the successful operation of a process. Flow measurements that can play such a key role include the totalization of feedstocks into a batch reactor, and custody transfer, among many others.

Noncritical control includes all other flow measurements used as process variables of a flow control loop — such as solvent flow rates into reactors (especially if a separation step to remove the solvent is downstream of the addition), and feed and reflux flow rates in distillation columns. In these, absolute accuracy is not necessary. In essence, the point of diminishing return is reached with accuracy of approximately ± 1%; no real benefits are realized with more accurate measurements. In many cases, repeatability is more important than extreme accuracy.

Flow measurements that are made for monitoring purposes are not tied to control loops. Such measurements are taken, for instance, to ensure that flow rates are in the desired range, to totalize the flow of utility fluids for inventory purposes, to use in troubleshooting, or to confirm heat and material balances. For example, one of the key reasons to monitor noncontrolled flow rates is to enable comparison of operating parameters between times of good and bad plant performance. Again, repeatability often is more important than accuracy. For plants that are not very large or for pipes that are easily reached, it may not be necessary to transmit flow rate measurements to a centralized control room; local indication may suffice.

When indication of flow is the sole purpose of a measurement, accuracy plays almost no role at all. In such situations, it only is important to know that there is actual flow in the pipe, and the general range of the flow. Examples are setting the flow rate to the seals of a liquid-ring vacuum pump, and ensuring that the flow rate of a scrubbing tower remains in the range of 2,000 gal/min, not 500 gal/min. In many cases, such measurements are used in interlocks to ensure that a flow is actually occurring so that plant operation can continue. Pressure or flow switches may even be used instead of meters.

Accurately assess accuracy

Unfortunately, flow accuracy is defined in different ways. Accuracy statements may be expressed as percent of actual flow rate, percent of fullscale flow, or percent of maximum differential pressure. Always keep this in mind when comparing different flow meters or manufacturers.

Also, some flow measurement accuracy standards are for combinations of more than one device. For example, a primary element and a transmitter each may have stated accuracies. The actual flow measurement accuracy is a combination of these two. The combined accuracy, however, is not additive because the possibility of all errors having the same sign (positive or negative) can't be assumed. As long as none of the associated possible errors is extremely large relative to the others (say, ten times greater), the total accuracy can be reasonably approximated by:

$$Total\ Accuracy = (Error_1^2 + Error_2^2 + \ldots + Error_n^2)^{1/2} \quad (1)$$

where n is the total number of errors *(2)*.

The accuracy that makes economic sense largely depends upon the value of the fluid. With expensive fluids, the ability to operate closer to specification, or more accurately measure the amount of fluid sold, decreases the payback period of the initial purchase — thus making it easier to justify a higher cost, but more accurate meter.

Regardless of the accuracy needed, reliable operation of the meter is a must. Even if the purpose is simply indication of flow, proper operation of the process depends upon the instrument performing its specified function reliably.

Consider the fluid's physical properties

A number of physical properties significantly influence meter selection.

Fluid temperature and pressure at the point of measurement are used in several ways. First, they enable identification of whether the fluid is a liquid or gas. If the fluid is a liquid, its vapor pressure and the expected pressure drop through the meter should be checked to see if flashing would occur. Also, of course, the temperature and pressure limits of each meter under consideration should be compared to the process conditions.

Flow rate of the fluid also must be considered. Normally, three flow rates must be specified: the normal flow rate passing by the measurement device, as well as the maximum and minimum flows to be measured. The ratio of the maximum to the minimum system flow rate defines the required turndown. An application with a maximum flow rate of 100 gal/min and a minimum flow rate of 4 gal/min has a turndown of 25:1. The allowable turndown for different flow technologies varies greatly — from 4:1 for orifice plates to 100:1 for certain Coriolis mass flow meters.

Fluid viscosity plays an important role in determining the effectiveness of certain flow technologies. A higher fluid viscosity can affect accuracy or greatly increase the pressure drop expected through the instrument. Viscosity, along with the specific gravity, flow velocity, and pipe diameter are used to calculate the Reynolds number through the measurement device. The Reynolds number defines the kinetic energy in the flowing stream, and is related to the ratio between the internal and viscous forces. Certain devices are very sensitive to the Reynolds number. For example, vortex meters depend upon a minimum Reynolds number in the fluid to provide enough energy to cause vortices to form when a shedding bar is placed in the flow stream. If the Reynolds number falls below the minimum (between 10,000 and 20,000), then vortices are not formed and the meter does not work.

Lastly, *fluid conductivity* has to be considered when evaluating an electromagnetic flow meter. The conductivity must be above a minimum value, approximately 5 micromhos/cm (5 microSiemens/cm), for magnetic flow meters to operate.

Investigate installation constraints

An equally detailed analysis of installation conditions must be performed. Consider such factors as pipe size, and available length of straight pipe on both sides of the measurement point. Straight-pipe-length requirements vary greatly among flow technologies, ranging from up to the equivalent of 40 pipe diameters for turbine meters or orifice plates to none for Coriolis instruments.

Electrical classifications of the measurement location must match the device classification. Manufacturers meet code requirements in different ways. Some choose to build devices that are intrinsically safe, while others provide the option of explosionproof housings. Some manufacturers can provide a positive pressure purge of their equipment housing to meet the extremely difficult requirements of NEMA Class I, Division 1, Group A.

With some instruments such as electromagnetic flow meters, it inherently is impossible to meet the most stringent electrical-code requirements. If allowable, it may be best to consider placing the measurement device outside the Division 1 region. In this way, a desirable flow measurement technology may still be usable.

Is the fluid hazardous? Is it on the OSHA list of chemicals with threshold limits that require mandatory monitoring? Or does the fluid pose particular safety hazards? In any of these cases, carefully consider the way the flow measurement device is connected to the process. Flanged connections may be used with some technologies, such as electromagnetic, Coriolis, and some styles of vortex and turbine meters, while other technologies require screwed connections, impulse lines, or rotating shaft seals.

Is pipe vibration likely? Historically, vibration has been a concern for Coriolis mass-flow and vortex technologies. This, however, does not mean that all vibration must be eliminated for their use. Undesirable vibration conditions are specific to the meter and the frequency of the vibrations. Recent advances in sensor technologies and microprocessor-based electronics almost have eliminated this problem in certain manufacturers' meter designs. Consult the manufacturers if doubts exist about vibration.

Also, for outdoor installations, consider ambient temperature conditions. Subjecting measurement device electronics to wide temperature swings, say from hot (100+°F) during the day to cold (32°F) at night, is the main concern. Again, consult manufacturers for information about specific instruments.

The measurement device itself poses some installation issues. Wetted parts, of course, should be constructed of materials that are compatible with the fluid at process conditions, especially the temperature and fluid velocity through or past the meter. When evaluating lined meters in vacuum service, check the type of liner and its method of attachment to make sure that it won't collapse during use.

For food or pharmaceutical applications, a sanitary design may be necessary. This implies eliminating minute cracks, pockets, and crevices where fluid might be trapped and allowed to collect. A sanitary meter must be easy to clean and inspect, which normally is done by means of special connections to the process piping. The installation also must allow the meter to be self-draining. These requirements prevent contamination of the process fluid and eliminate the risk of bacterial growth within the device.

Flow measurement inaccuracies often stem from failing to correct for large differences between the assumed design conditions and the actual flow.

The spectrum of flow meter technologies

Most flow meters used in the CPI can be grouped into four categories: differential pressure, velocity, mass, and volumetric.

- *Differential pressure meters* restrict fluid flow to create a differential pressure that provides an indirect measurement.
- *Velocity flow meters* use a primary device to generate a signal linear with the fluid velocity that then is multiplied by the cross-sectional area of the meter to obtain a volumetric flow rate.
- *Mass flow meters* directly measure the mass flow rate of the fluid. So, no value for density need be assumed, as it must for other types of flow meters.
- *Volumetric flow meters* divide the flow into discreet, known volumes that then are counted to provide the total flow.

Now, let's look at each of these categories in more detail.

Differential pressure meters

Differential pressure devices represent the largest segment of the flow metering market. They are based upon the conservation of mass and energy as given by Bernoulli's Equation. When a restriction, commonly referred to as a primary element, is placed in a flow stream, the velocity of the fluid flowing through the restricted opening increases, causing a corresponding decrease in the static pressure. The difference between the upstream pressure and the downstream pressure due to the primary element restriction is related to the square of the fluid velocity, as well as to the ratio of the diameter of the restriction bore to the pipe diameter, known as the β ratio, and the density of the fluid.

The differential pressure is measured by a pressure sensing device referred to as a secondary element. Commonly, the secondary element is a transmitter that provides an electronic output signal proportional to either the differential pressure or, if a square root extraction is done, to the fluid velocity. The volumetric flow rate is determined by multiplying the fluid velocity by the known cross-sectional area.

The true flow rate, however, frequently is lower than the theoretical flow rate. Therefore, a factor called the discharge coefficient is applied to the measurement to correct for such items as the velocity profile, fluid contraction geometry, and the location of pressure taps. A thermal expansion factor also is applied. This factor corrects for changes in bore and pipe diameters with temperature, and is factored into all β-dependent terms.

When the measured fluid is a gas, other corrections also must be made. The pressure of the gas drops while passing through the reduced area of the primary element bore, thus expanding the gas. This expansion is corrected by use of a gas expansion factor, which is a function of the gas pressure, the β ratio, the measured differential pressure, and the isentropic exponent for the gas.

Always keep in mind that flow measurement inaccuracies often stem from failing to correct for large differences between the assumed design conditions and the actual flow conditions.

Advances in microprocessor design have allowed manufacturers to increase the functionality of the secondary element, the transmitter, to include the appropriate flow equation. (Details on a number of flow equations

Figure 1. Multivariable transmitters provide pressure and temperature compensation for orifice plate measurements.

that have been developed are given in Ref. 2.) Transmitters like the one shown in Figure 1 take into account such factors as the pressure and temperature of the fluid at the measurement point, and adjust the corrections accordingly for more accurate measurements. This is of special interest in natural gas transmission, where AGA equations can be programmed into a microprocessor in the transmitter.

Primary element designs vary. The most common types are the orifice, venturi, nozzle, and static pitot tube (see Figure 2).

The *orifice* is the most popular. It simply is a piece of flat metal with an orifice bore, normally in the center. The upstream edge of the bore usually is cut sharp with a beveled edge on the downstream side. This design works well with clean liquids and gases. Slurries or dirty fluids can erode the sharp leading edge and cause the flow measurement to drift. Eccentric orifices also are used. For example, the bore can be positioned at the top of the plate to allow passage of gases when measuring liquids, or at the bottom to allow suspended solids to pass. Half circle bores often are used with light slurries or dirty gases.

Orifice meters are popular in the CPI in gas flow applications, and for liquid flows in pipe sizes of six inches or larger, because of the relatively consistent cost regardless of pipe size. More information on these devices is given in Ref. 3.

A *venturi tube* is used as a primary element when it is important to limit the permanent pressure loss across the flow measuring device. It consists of a gradually narrowing cone creating a smooth entry, and another cone on the discharge side creating a smooth exit. This smooth design makes it a good primary element for dirty fluids and light slurries. The venturi tube also is employed in high flow rate applications, such as main steam lines. The high cost of manufacturing venturi tubes, however, restricts their application.

A *flow nozzle* is used as a lower cost alternative to a venturi tube. Its design provides for a smooth entry but a sharp exit. Flow nozzle performance falls between that of the orifice plate and the venturi tube. Flow nozzles are dimensionally more stable than orifice plates at high temperatures and high flow rates, thus making them a better choice for main steam lines. Nozzles also can be used in fluids that contain solids because their design limits solids buildup.

A *pitot tube* is a point-velocity measuring device — meaning it measures the velocity of the fluid at a specific point in the pipe. The design consists of a small-diameter impact port positioned normal to flow in the pipe. As the fluid hits the port, the velocity is reduced to zero and the resulting pressure is measured. This pressure is known as the total pressure. Static pressure at the impact point also is measured, and the difference between total pressure and static pressure is called velocity pressure. Because the flow velocity profile across a pipe diameter is not linear, to be accurate the pitot tube must be positioned to mea-

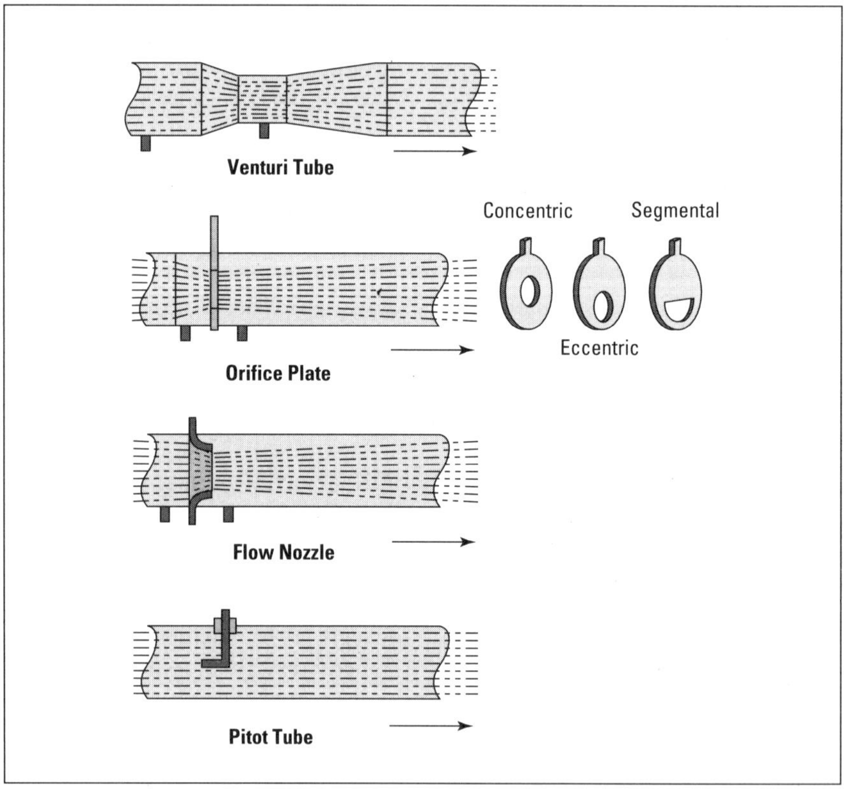

Figure 2. Four types of primary elements used for differential pressure measurement.

sure the average velocity in the pipe. A variation is the *averaging pitot tube*, which uses several ports to compensate for changes in the velocity profile.

The advantages of pitot and averaging pitot tubes are the very low permanent pressure loss that occurs and relatively low price. This makes these devices a good choice in larger-pipe-diameter clean-gas applications.

Other primary element designs exist, but are not discussed here because of their limited use.

Another style of flow meter based on pressure head principles, but different in design from differential-pressure-style units is the *target meter*. It measures the force (commonly referred to as drag) exerted by a flowing stream against an obstruction, normally a circular disk, placed in the flow stream. For turbulent flows, for which the target meter is best suited, the major component of drag is the pressure drag that results from a difference between the upstream and downstream sides of the disk.

On the upstream side of the disk, the fluid slows as it changes direction to pass through the annulus formed by the disk and the pipe wall, increasing the pressure. As the fluid passes through this annulus, its velocity increases, thus causing a decrease in pressure. The dynamics of the fluid passing the disk creates an area of turbulence downstream of the target, resulting in a lower pressure region directly behind the disk. The resulting force caused by this pressure drag is proportional to the square of the velocity of the fluid. This force can be measured by a variety of techniques including force balance, resistive strain gauge, and mechanical force gauge *(1)*.

Target meters are relatively low accuracy meters compared to other meter types. Accuracy may range from 0.5% to 5.0% of full flow, which makes target meters more suitable for monitoring than control applications. They are, however, suitable for use with dirty fluids, higher viscosity ones, and slurries that otherwise would be difficult to measure with standard differential-pressure meters because of plugging of pressure-sensing ports or internal mechanical parts.

■ ***Figure 3. Cutaway view of a turbine meter.***

Velocity meters

The volumetric flow rate of a fluid can be determined from its velocity, via:

$$Q = v A \qquad (2)$$

where Q is the volumetric flow rate, v is the velocity, and A is the cross-sectional area. Several types of velocity meters, including turbine meters, electromagnetic meters, vortex meters, and ultrasonic meters, are used in the CPI.

A *turbine meter* (see Figure 3) consists of a shaft-mounted, bladed turbine rotor that spins with a rotational speed proportional to the velocity of the fluid. The rotor speed is measured either by mechanical connection or by electronic pulse generation of each rotor blade.

Turbine meters are one of the most accurate types of meters available. While stated accuracy standards vary, such meters typically reach an accuracy of 0.15% of reading for liquid flows, and 1% for gas flows. Actual accuracy standards depend upon the physical properties of the fluid being measured and installation conditions. The meter must be installed with a proper length of straight pipe both upstream and downstream. Many times, flow straighteners are required to condition the fluid for proper operation. It also may be necessary to install strainers before the meter to prevent clogging or damage to the rotor bearings.

Turbine meters come in a large range of sizes, from fractional-inch models that can measure as little as 0.001 gal/min for liquids and 0.001 actual ft^3/min for gases to 30-in. models that measure over 50,000 gal/min for liquids and 25,000 actual ft^3/min for gases. Turbines also can operate in extreme pressures, subject to flange ratings, and temperatures ranging from cryogenic to 1,500°F.

Turbines, because of their internal bearings, are best applied to clean liquids and gases. With applications in clean, low-viscosity, lubricating liquids, turbine meters frequently operate for several years before maintenance is required.

Electromagnetic flow meters, commonly called magmeters, can measure electrically conductive liquids. Magmeters are based on Faraday's Law of Electromagnetic Induction, which states that a voltage is induced by a conductor moving through a magnetic field. The magnitude of the induced voltage, E, is directly proportional to the velocity of the conductor, V, the conductor width, D, and the strength of the magnetic field, B. In a magmeter, the three important parameters — electrode plane, direction of flow, and magnetic field — are at right angles to one another, thus reducing Faraday's Law to:

$$E = k\,B\,D\,V \qquad (3)$$

where k is a proportionality constant.

A magmeter is composed of two parts. A flow tube (Figure 4) is installed in-line with the process piping, while coils located on opposite sides of the flow tube create the necessary magnetic field. Liquid moving through the magnetic field generates a voltage that is detected by two electrodes. Because the magnetic field strength is kept constant and the electrode spacing is fixed, the only variable is the velocity of the conductive fluid. Therefore, E is directly proportional to V, resulting in a linear output.

Figure 4. How an electromagnetic meter works.

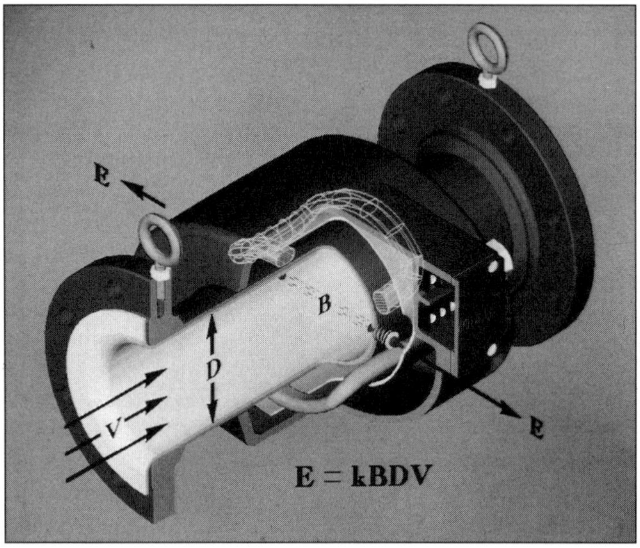

The second part of the magmeter system, the transmitter, drives the coils to generate the magnetic field, and also electronically conditions the voltage detected by the electrodes. The transmitter then produces both analog and frequency signals proportional to the liquid's velocity.

Magmeters are less sensitive to liquid velocity profiles than most other types of flow meters. To ensure accuracy over widely varying process conditions, however, flow tubes should be installed with five pipe diameters of straight pipe length upstream and two pipe diameters downstream.

For a magmeter to work, the fluid must have a minimum conductivity of 5 microSiemens/cm (5 micromhos/cm) and be nonmagnetic. Most water-based fluids and many chemicals can be measured by magmeters, but petroleum-based fluids generally cannot be.

Magmeters are capable of measuring a large range of flow rates. Standard sizes range from a fraction of an inch to 36 in. Some manufacturers build meters up to 96 in. on special order. Measured flow rates range from 0.01 gal/min to 500,000 gal/min. Different models provide accuracy from 0.25% of rate to 1.0% of full scale. Turndown for magmeters also is good, ranging from 10:1 to 30:1, depending on manufacturer and model.

Magmeters offer several major advantages. First, because they operate entirely outside the pipeline, magmeters create no additional pressure drop. Their nonintrusive design makes them a good choice for slurry, high-viscosity, and entrained-solids flows. Second, they can be used with a wide variety of corrosive chemicals. In order to operate, a magmeter must use a flow tube lined with a nonmetallic material; these liners are available in a choice of corrosion- and erosion-resistant materials including fluorocarbons, synthetic rubber, and ceramics. The electrodes also are available in a wide assortment of materials such as Type 316 stainless steel, high-nickel alloys, titanium, tantalum, nickel, zirconium, and platinum. Furthermore, because the velocity measurement is not influenced by fluid density and viscosity, a magmeter may be used for multiple fluid-flow measurements without recalibration.

The combined advantages of magmeters make them good choices for many difficult applications. They especially have found favor for use in fibrous-pulp and mining slurries. Reference 4 provides more details on magmeters.

Vortex meters rely on a phenomenon of vortex shedding known as the van Karman effect. As fluid passes a bluff body inserted into a flow stream, it separates and generates small eddies or vortices that are shed alternately along and behind each side of the bluff body (see Figure 5). These vortices cause areas of fluctuating pressure that can be detected by a sensor. The frequency of vortex generation is directly proportional to fluid velocity.

A minimum Reynolds number must be maintained for a vortex meter to operate properly. If the Reynolds number is too low, the vortices may not be shed at a consistent rate, impairing the accuracy of the meter. For most models of vortex meters, the minimum Reynolds number must be between 10,000 and 20,000. Reynolds number is strongly related to the viscosity of the fluid; this means that

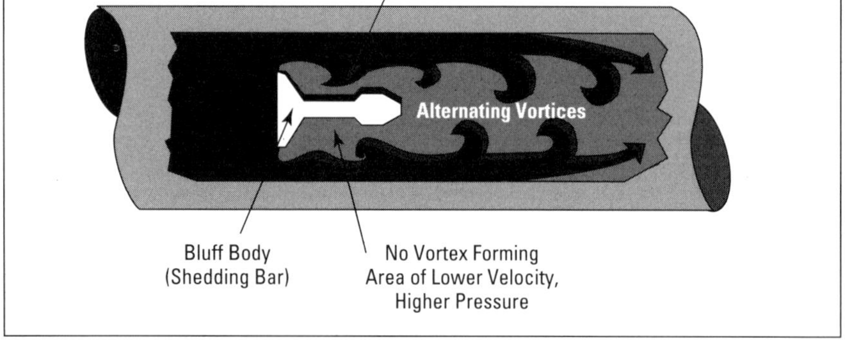

Figure 5. In a vortex meter, a bluff body in the flow stream creates alternating vortices, whose frequency is proportional to the fluid velocity. (Courtesy of Rosemount Measurement.)

meter accuracy will start to degrade at viscosity levels above 10 cp. Operation becomes questionable above 20 cp.

The maximum Reynolds-number limit is approximately 10,000,000. In practice, this limit concerns low density gases with pipe velocities approaching 250 ft/s.

Vortex meters are well suited for measuring clean liquids, gases, and steam. Normally, small amounts of solids do not affect the operation of the meter, but some designs are susceptible to plugging of their small sensing ports.

Expected accuracy standards are 0.75% of reading for liquids, and 1.5% of reading for gases. Meter turndown varies with manufacturer but generally should be in the range of 10:1.

Ultrasonic meters employ sound waves to determine the flow rate of fluids. The two types in general use are the Doppler and the transit time. These meters complement each other, as the Doppler suits fluids that have entrained gases or solids while the transit time handles clean liquids and gases.

In a *Doppler meter*, a transducer element transmits an ultrasonic sound wave of constant frequency into the fluid. Entrained solids or gas bubbles within the fluid reflect the sound back to a receiver element. Because the reflecting objects are moving, the frequency of the returned signal differs from that of the original signal by an amount proportional to the flow velocity.

A *transit time meter* introduces pulses diagonally across the fluid flow in the pipe in both the upstream and downstream directions. The difference in time for pulses to travel from a transmitter to a receiver is used to calculate the velocity of the fluid. (See Figure 6.)

Like magmeters, ultrasonic meters have the advantage of not intruding into the fluid flow stream. This makes them a viable option for flow measurement of slurries or highly corrosive liquids. Some models can be clamped to the exterior of the pipe — enabling flow measurement in existing lines without the need to penetrate the pipe.

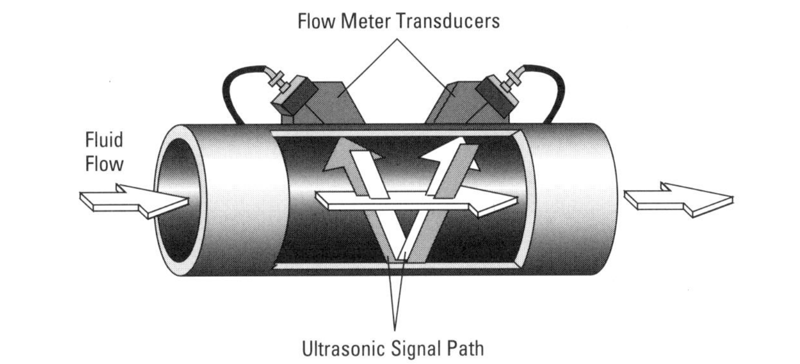

■ ***Figure 6. Ultrasonic meter using transit time measurement technique. (Courtesy of Panametrics.)***

Disappointing results were realized with ultrasonic flow meters in the past. The complexity of the measurement, and the effects of the pipe, flow profile, and environmental interferences were not fully understood when the meters first became commercially available. Improvements in flow meter designs and a better understanding of the critical importance of correct installation have made ultrasonic flow technology a more viable option *(5)*.

Doppler-style ultrasonic flow meters are less accurate than other flow meters, with accuracy typically 2–5% of full scale. These accuracy levels are more suitable for monitoring applications than control purposes. Accuracy of transit time meters is better, ranging from 1% of rate to 5% of fullscale reading, depending upon the manufacturer and the effort spent in calibration. Some manufacturers claim improved accuracy with models that use multiple acoustic paths vs. a single path. Always closely examine whether this higher accuracy really is required, as such models cost far more than standard meters.

Mass meters

These meters measure the mass flow rate of the fluid directly without the need for temperature or pressure compensation. They are divided into two types: thermal mass, and Coriolis.

The majority of *thermal mass meters* are used for low gas flow measurements. There are two designs — the heated element, and the temperature rise meter.

The *heated element* design measures the rate of heat loss to a flow stream. A heated element introduced into the flow stream gives up heat to gas flowing past it. The heat loss depends upon the temperature and size of the heating element, and the thermal conductivity, specific heat, density, and velocity of the gas stream. When the gas properties are known, the rate of heat loss is proportional to the square root of the velocity.

This heat loss can be measured in two ways. In the constant current method, the probe temperature varies with the flow rate. In the constant temperature method, the current necessary to keep the element at a constant temperature is measured and related to the gas flow rate. A common use for the heated-element design is air velocity measurement, and it may be referred to as an "anemometer" when used in this application.

Just as with a pitot tube, this style of flow meter is a point measurement device. So, if the sensor is not placed at the location of the average velocity of the flow profile, an erroneous measurement occurs.

In a *temperature rise meter*, a precision power supply directs heat to the midpoint of a sensor tube. Equidistant upstream and downstream of the heat input are temperature measuring elements. With no flow in the sensor tube, the heat reaching each temperature element is equal. With increasing flow, the stream carries heat away from the upstream element and toward the downstream element. The temperature difference between the two elements is measured, and this difference is proportional to the gas mass flow rate.

If the heater and the temperature sensors protrude into the pipe, they may be damaged by erosion or corrosion, and thus create sources for process fluid leakage. So, to obviate such problems, the heater and temperature sensors can be mounted on the outside of the pipe.

Two styles of temperature-rise flow meters exist. The *flow through* style allows the entire flow stream to pass by the heater and sensors. This design is used for liquids and slurries. A more common style, used exclusively for gases, is the *bypass* design. As shown in Figure 7, an orifice in the main line of the flow meter forces a constant proportion of the flow through a small capillary line that contains the heating element and the temperature sensors. The small size of the line minimizes the energy necessary to heat the fluid, and improves the meter response time.

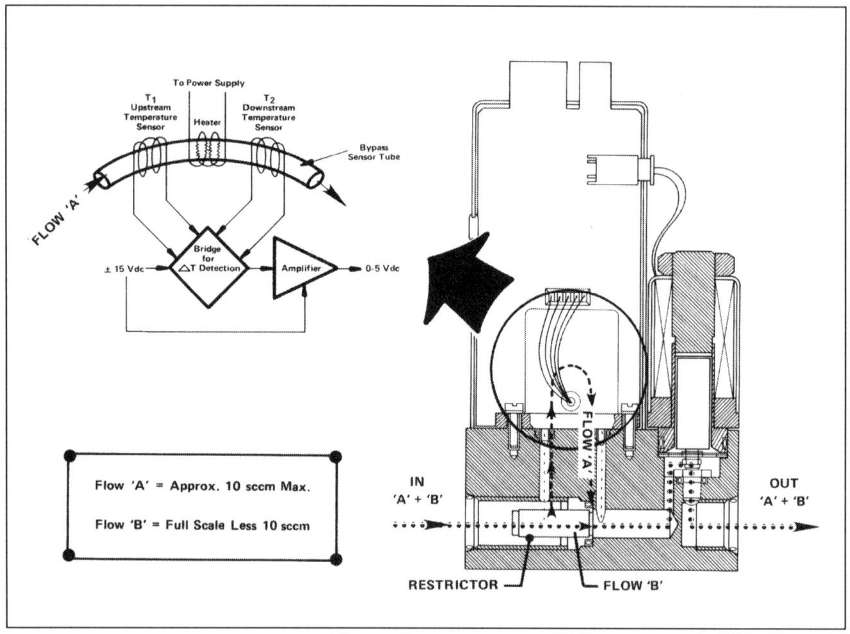

Figure 7. How a bypass temperature-rise type of thermal mass flow meter operates.

Thermal mass meters are used extensively in semiconductor manufacturing to accurately measure low gas flow rates. They also are employed in such applications as component leak testing, small gas purge flows, and gas flow measurement in pilot plants.

Accuracy standards for all types are 1% of full scale, with a turndown of 10:1. Accuracy will be influenced by changes in the composition and the moisture content of the gas being measured.

Coriolis meters are the most popular type of mass flow meter, and have become widely accepted in the CPI.

A Coriolis meter operates somewhat like a tuning fork — vibrating a sensing tube at its natural frequency. Mass flow through the tube causes a natural twisting action known as the Coriolis effect. Magnetic position sensors measure the amplitude of the deflection angle from this twisting action. (See Figure 8.) This measurement is converted to a voltage that is linearly proportional to the mass flow rate of the fluid passing through the tube.

An advantage of the Coriolis meter is that, by measuring the natural frequency of the tube, the density of the fluid in the tube can be determined at the same time as its mass flow rate. This makes it possible to infer the flow rate of one component in a two-component stream.

The meters are relatively easy to specify and install. Because they are unaffected by fluid density and viscosity, they are capable of measuring several different fluids without recalibration. And, when calibration is necessary, it can be performed with water instead of the process fluid without degrading accuracy. In addition, these meters are unaffected by the flow profile, and do not require minimum straight-pipe lengths on either side of the meter.

The major disadvantage of Coriolis meters is the high pressure drop that can occur. This may limit their use in higher viscosity applications, although such meters successfully have handled fluids with viscosity levels in the thousands of centipoise. Pressure drop varies greatly, however, among the different flow meter designs. It is a misconception that the pressure drop is directly related to the number of bends in the measuring tube. Rather, the tube diameter is more the determining factor. Indeed, straight tube designs typically have high pressure drops because of their use of small diameter tubes. Needless to say, manufacturers are developing lower-pressure-drop designs.

Coriolis meters are manufactured in a variety of shapes, sizes, and materials of construction *(6,7)*. These factors influence a meter's sensitivity; therefore, the performance characteristics of the many different Coriolis meters vary significantly. Meters come in sizes from 1/16 in. to six in., with flow rates from 0.01 lb/min to 25,000 lb/min. Accuracy standards range from better than 0.2% of rate to 0.5% of rate, and turndowns from 25:1 to 100:1.

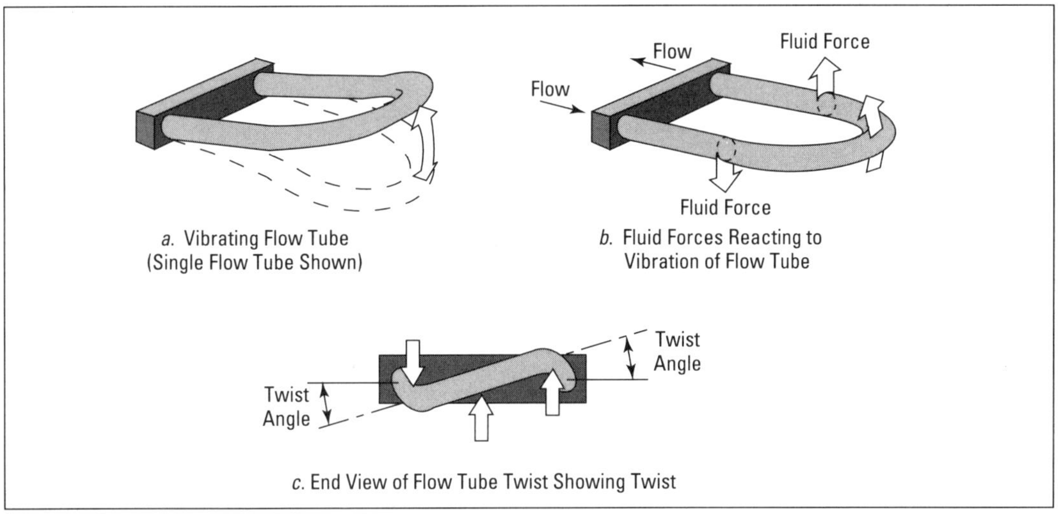

■ ***Figure 8. In a Coriolis meter, the twist angle developed relates to mass flow rate, while the frequency of vibration indicates the density of the fluid. (Courtesy of Micro Motion.)***

The nonintrusive design of such meters, along with the choice of multiple construction materials allow Coriolis meters to be used in a variety of applications. Their high accuracy has made them a prominent choice for critical applications such as reactant measurement and custody transfer.

Although Coriolis meters mainly are used in liquid applications, they also are applicable to gas flows. Where in the past these gas measurements were limited to pressures greater than 100 psig, advancements in technology have allowed at least one model to work at pressures considerably lower. Because the scope of applications continues to expand, manufacturers should be contacted for specific gas applications.

Volumetric meters

Positive displacement meters directly measure the volume of fluid flow. There are a number of designs, but all operate on the same principle: the flow is divided into segments of known volume, and the number of segments that pass through the meter are counted. Multiplying the volume of one segment by the total number of segments gives the volumetric flow rate.

Positive displacement meters either can be connected mechanically to local readouts or can transmit pulses representing a known volume of fluid to a centralized location. Common types of such meters are: bi-rotor (Figure 9), oval gear, sliding vane, disc, and piston.

These meters require the maintenance of a seal between two hard surfaces; therefore, they are better suited for viscous liquids than low viscosity ones. Thin fluids are more susceptible to slippage that decreases the meter's accuracy. The pressure drop developed by the meter increases, however, as the fluid viscosity rises.

Positive displacement meters are ideal for certain batching, blending, and custody transfer applications — especially for viscous petroleum-based fluids of less than 5 micromhos/cm that cannot be measured by magmeters. Positive displacement meters are widely used in oil-transmission pipelines.

Accuracy standards can be very good — 0.5% of rate. Flow rates vary from a fraction of a gallon per minute to thousands of gallons per minute. Turndown typically is 10:1.

Because they are mechanical devices with close tolerances and moving parts, positive displacement meters should not be used for dirty liquids and gritty slurries. Also, if heavy fluids coat the inner chambers of the meter, a reduced volume will be delivered by

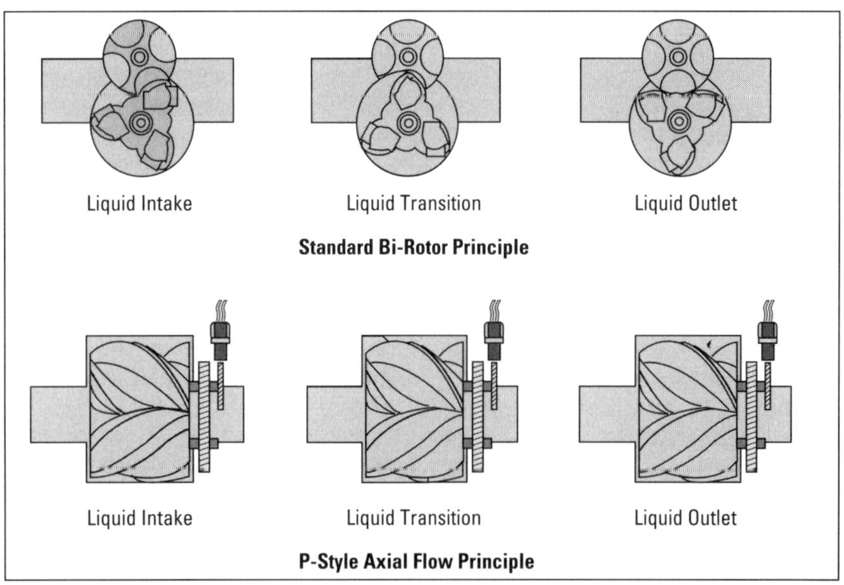

■ ***Figure 9. Operating principle of a standard bi-rotor meter. Courtesy of Brooks Instrument.***

each segment, thus causing measurement error.

Other meters

Two other designs — variable-area, and insertion meters — round out our discussion.

A *variable area meter*, also known as a rotameter, consists of a float inside a vertically mounted, tapered tube (see Figure 10). The float moves up and down in proportion to the fluid flow rate and the annular area between the float and the tube wall. With liquids, the float is raised by a combination of the buoyancy of the liquid and the velocity head of the fluid. With gases, buoyancy is negligible, and the float responds to the velocity head alone. The float reaches a stable position in the tube when the upward force equals the downward gravitational force exerted by the weight of the float. A change in flow rate upsets the force balance. The float then moves up or down, changing the annular area, until it again reaches a position where the forces are in equilibrium. Its height is directly proportional to the flow rate.

■ ***Figure 10. Typical variable area meter.***

The variable area meter has a linear scale, a relatively long measurement range, and low pressure drop. It is simple to install and maintain, and can be designed to accommodate a wide range of pressures and temperatures. While used mainly for local indication, it is adaptable to designs that transmit data to a centralized control room or alarm system.

Tubes usually are made of glass, but stainless steel and other metals are available. When using a metal tube, the flow rate is automatically read out by magnetic sensing or other means.

Variable area meters come in sizes from ⅛ in. to three in., with flow rates ranging from hundredths of a gallon per minute to hundreds of gallons per minute. Most variable-area meters have turndowns of approximately 10:1. Accuracy depends on tube design and length — typically for tubes over 20 in. long, it is 0.5% of rate.

A final type of flow meter is the *insertion meter*. It differs from an in-line meter in that it only measures a sample of the flow in the line, not the entire flow. An insertion meter consists of a point-velocity measuring device that is inserted into a flowing fluid. A variety of technologies, including differential pressure, vortex, electromagnetic, and turbine, are available. Pitot tube and heated-element thermal mass meters, which we already have discussed, are types of insertion meters. Insertion meters, because of their lower cost, normally are used in larger pipe sizes in noncritical applications.

The location of the sensing element in the fluid is very important to the accuracy of the meter. Many manufacturers recommend placing the meter at the centerline because that is the point of flattest velocity profile. With centerline positioning, however, velocity compensation is required to provide better accuracy. A second position that may be recommended is the critical position (which sees the average velocity in the pipeline). While this eliminates the need for compensation, it may be difficult to determine the exact location of this position. Also, the critical position varies with the Reynolds number of the fluid.

Making the right choice

Many different types of flow meters are available. So, a new engineer may feel overwhelmed by all the choices. Those with experience, however, find flow meter selection a relatively easy process. By following the guidelines in this article, checking reference materials such as those listed above, and observing the performance of flow meter technologies in actual service, you quickly should be able to achieve a good understanding of flow principles — and select appropriate meters. CEP

Literature Cited

1 **Spitzer, D. W.,** "Flow Measurement, Practical Guides for Measurement and Control," Instrument Soc. of America, Research Triangle Park, NC (1991).

2. **Miller, R. W.,** "Flow Measurement Engineering Handbook," 2nd ed., McGraw-Hill, New York (1989).

3. **Khandelwal, P. K., and V. Gupta,** "Make the Most of Orifice Meters," *Chem. Eng. Progress*, **89** (5), pp. 32–37 (May 1993).

4. **Rose, C., and G. E. Vass,** "Consider the Attractions of Magnetic Flowmeters," *Chem. Eng. Progress*, **89** (12), pp. 61–65 (Dec. 1993).

5. **Liptak, B. G.,** "Instrument Engineers Handbook," 3rd ed., Chilton, Radnor, PA (1995).

6. **Baker, R. C.,** "Coriolis Flowmeters: Industrial Practice and Published Information," *Flow Meas. Instru.*, **5** (4), pp. 229–246 (1994).

7. **Carpenter, B.,** "Choose the Right Material for Coriolis Mass Flowmeters," *Chem. Eng. Progress*, **86** (10), pp. 55–60 (Oct. 1990).

Further Reading

Labs, W., "Flowmeters: Accuracies are Better, Prices Lower," *I&CS*, pp. 29–33 (Feb. 1990).

Pomroy, J., "Selecting Flowmeters," *I&CS*, pp. 61–68 (March 1994).

J. W. DOLENC is chemical industry manager for Fisher-Rosemount, Eden Prairie, MN (612/828–3466; Fax: 612/828–7723). He has been with the firm for over six years. Previously, he worked for 12 years in process engineering and plant engineering groups at Colgate-Palmolive. He received a BS in chemical engineering from the University of Minnesota, and is a registered Professional Engineer in the state of New Jersey.

Consider the Attractions of Magnetic Flowmeters

Posing no obstructions to flow and offering high accuracy, these meters are a good choice for a variety of services. Here's a guide to when to select them.

Charlie Rose and Gabor E. Vass,
Fischer & Porter Co.

Magnetic flowmeters — or magmeters, as they are commonly called today — essentially are sections of pipe in the line to be metered, and offer no obstruction to flow. Units were introduced in the 1950s, with early designs developed primarily to meet the stringent metering needs of the pulp-and-paper industry, where process liquids could be highly corrosive or erosive and any obstruction in the line caused serious measurement problems. The magmeter met these tough metering requirements quite successfully and, with many design improvements and variations, is capturing an ever-growing portion of the flowmeter market — especially in the chemical process industries (CPI).

The basic elements of the magmeter are a length of straight, lined, nonmagnetic stainless steel pipe, generally of the same diameter as the line to be monitored; a set of electromagnetic coils external to the pipe that produce a magnetic field perpendicular to the flow of liquid through the pipe; a pair of electrodes located at right angles to both the magnetic field and the flow; and a signal converter. Electrodes usually are in direct contact with the fluid, but are nearly flush with the wall of the pipe and so do not obstruct fluid flow. If contamination or coating by process fluids is a concern, electrodes that can be removed for cleaning can be provided. And when removable electrodes are not suitable or contact with the fluid is not allowed, the meter can be fitted with capacitance-type electrodes, positioned behind the pipe liner. A typical flanged magmeter with a polytetrafluoroethylene liner is shown in Figure 1; note the liner protector, which safeguards the liner during installation and provides grounding to the process liquids.

■ ***Figure 1. Typical magmeter with flanged ends. Polytetrafluoroethylene liner electrically insulates the metal meter body from the liquid to be measured, and isolates the electrodes from each other.***

Magmeters have no moving parts, and offer accuracies to ±0.25% of flow rate. So, today, they are used widely throughout the CPI for many applications, including:

- chemicals, where units can handle highly corrosive liquids such as acids and bases, and difficult fluids such as polymers, latex, and emulsions, which tend to set up and plug other types of meters;
- pharmaceuticals, where a sanitary design and proper materials of construction prevent contamination of such metered liquids as coatings sprayed on tablets, flavoring added to cough syrup, and products flowing to bottling machines;
- foods and beverages, where sanitary designs that meet 3-A standards (developed by the International Association of Milk and Food Environmental

Sanitarians) and U.S. Food and Drug Administration regulations for food contact service are used for critical applications;

• mining, where the key application is measuring slurries — such as of iron ore, taconite, magnetite, pyrite, bauxite, phosphate, copper ore, and alumina — with abrasive solids in suspension;

• water and wastewater treatment, where both municipal and industrial plants employ units to measure water, wastewater, raw sewage, recirculated sludge, digester supernatant, settled sludge, and primary sludge; and

• pulp and paper, where applications have continued to grow, with increased emphasis on accuracy of measurement, as well as minimum maintenance. Mills today are measuring white liquor, black liquor, brown stock, bleached stock, bleaching chemicals, and paper-machine additives.

In addition, the magmeter's high accuracy has spurred its adoption for monitoring custody transfer, such as of taxable process fluids.

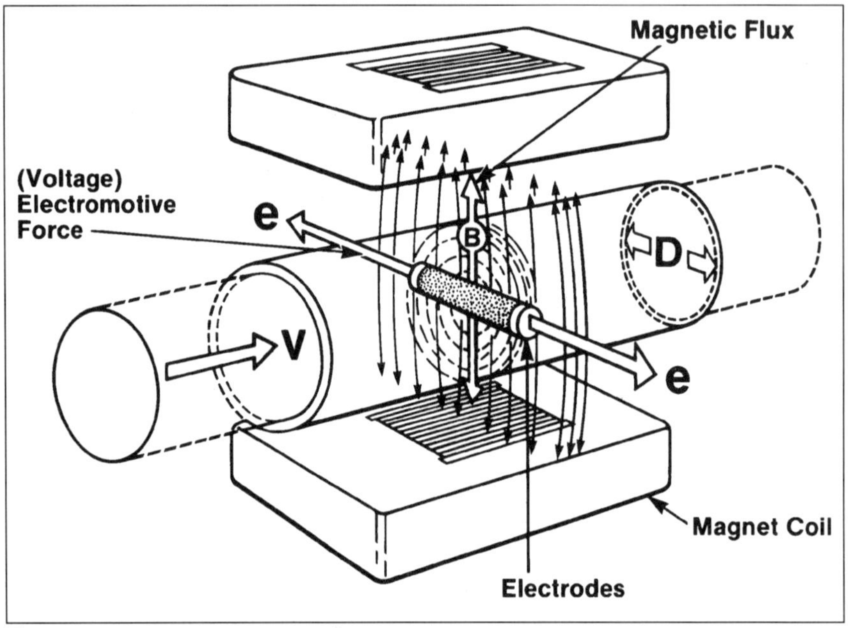

■ *Figure 2. How a magmeter works.*

Operating principle

Magmeters measure volumetric flow rate by utilizing the properties of a conductive liquid to generate an induced voltage as the liquid flows through a magnetic field. The amplitude of the voltage thus produced is directly proportional to the flow rate of the liquid. Some designs develop a characterized, nonuniform magnetic field that significantly reduces sensitivity to the effects of flow profile *(1)*.

The mathematical relationships involved in operation of a magmeter are based upon Faraday's Law of Electromagnetic Induction, as illustrated in Figure 2. A magnetic field, with a flux density B, is generated by a pair of magnet coils that are energized by an external source of electric power. The coils are arranged so as to cause the magnetic field to cross the flowing liquid at right angles. The liquid is flowing at velocity V through a section of pipe of inside diameter D. The pipe is part of the meter body and must be electrically insulated (usually by a liner) to prevent the flow signal from shorting out to the body or pipe in which the meter is installed. The voltage e, induced by the flowing liquid, is sensed by a pair of electrodes, located on a horizontal plane on opposite sides of the pipe. Faraday's Law for this arrangement may be stated:

$$e = k \times B \times D \times V$$

where k is a constant. In operating the magmeter, B and D are held constant; therefore, e is directly proportional to V, and this voltage (actually a small millivoltage) is transmitted to a receiver. A variety of receivers are available to convert the flow signal into suitable engineering units of flow rate, such as gallons or liters per hour. These devices, also known as signal converters or transmitters, are actually precise voltmeters and, today, can be microprocessor-based — enabling them to perform a wide range of measurement and control, as well as diagnostic and communication functions.

Note that the magmeter theoretically can measure the rate of the liquid flowing in either direction through the pipe. Such bidirectional flow measurement is actually a feature of certain models, and is programmable from a transmitter or a host computer.

Advantages and limitations

Compared to other flow measuring methods, the modern magmeter offers the following advantages:

1. no moving parts;

2. no pressure drop (providing the meter size equals the line size);

3. no measurement variation due to viscosity, temperature, and density of the fluid;

4. the ability to measure highly corrosive and solids-bearing liquids;

5. shorter runs of straight-run piping required upstream — 3 pipe diameters, as compared to the 10–20 needed for vortex and turbine meters, and at least 8 diameters for orifice meters;

6. higher turndown (or rangeability) — that is, the ratio of maximum to minimum operating capacity, which can be as high as 100:1, compared to 10:1 to 20:1 for other metering methods; and

7. a linear output signal, which requires no characterization.

In addition, magmeters are available in designs: that have interchangeable electronics regardless of meter size — thus reducing on-site stocking requirements; that can be used with clean-in-place procedures; and that can handle dirty liquids (by using ultrasonic cleaning of electrodes or electrodeless options).

Magmeters do have several limitations, though. Standard designs cannot meter liquids with very low electrical conductivities or those that are nonconducting — such as most petroleum and hydrocarbon products. Special designs, however, have been developed to measure liquids with conductivities as low as 0.05 microsiemens per centimeter (μs/cm) — permitting flow measurements, for example, of alcohols.

To give a reliable measurement, a magmeter's pipe section must be full of liquid, and the electrode axis always should be horizontal. The magmeter should not be positioned close to equipment that generates strong electromagnetic fields, such as some induction equipment; signal cable should not be run parallel to high power cable. In addition, because of the possibility of vacuum, magmeters with nonbonded liners should not be installed on the inlet side of a pump.

We have summarized some broad parameters that bracket the extremes of what can be obtained from this versatile flowmeter. (You should, of course, check the actual specifications of the particular magmeters that you are considering.)

Accuracy: From ±0.25% to ±1.0% of flow rate, with factory-calibration standard.

Design pressures: Up to 740 psig (5.0 MPa) at 100°F (35°C) standard, with higher ratings available.

Design temperature: From -40°F (-40°C) to 356°F (180°C) standard, with higher ratings optional.

Sizes: 0.04–120 in. dia. (1–3,050 mm dia.).

Flow range: 0.01–1,175,000 gal/min (37 ml/min. to 266,800 m^3/h).

End connections: Flanged, wafer, and sanitary, with custom ends in proprietary designs available on request in larger sizes.

Key factors in magmeter selection

A number of basic parameters should be considered when evaluating whether to install a magmeter.

Conductivity. A liquid's electrical conductivity is always the place to start, unless it is well known as being acceptable, along with fluid velocity. Most standard magmeters today can handle liquids with conductivities of at least 5 μs/cm. In smaller sizes (4 in. dia. or less), some models can handle conductivities down to 0.05 μs/cm; such applications, however, should be checked out with the vendor's applications specialists.

Materials. Parts of the meter that will be exposed to the flowing fluid — such "wetted parts" include the electrode (unless the meter is a special electrodeless type) and the liner on the inside of the meter pipe section — usually should be selected for their resistance to temperature, corrosion, and abrasion. Polytetrafluoroethylene is an excellent liner material overall for temperature and corrosion resistance, but does not fare very well in abrasive service. Rubber (such as neoprene or natural rubber) is better for handling abrasives, and does fairly well in corrosive service at moderate temperatures. Ceramics can provide high resistance to abrasion and also can offer high resistance to corrosion and temperature.

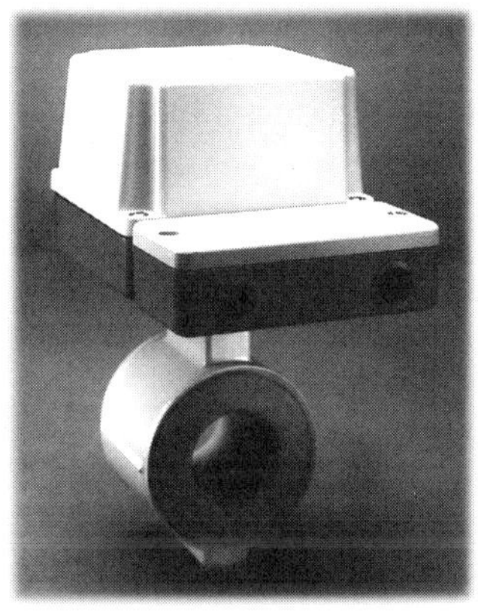

■ ***Figure 3. Typical wafer-style or flangeless magmeter. Note the four bolt holes in the body, used to mount it between a pair of pipe flanges.***

Sanitary construction. Some processes require a food-grade liner and electrode material that will not contaminate the metered process liquid. One design for such sanitary service has a cast ceramic meter tube and platinum electrodes; it has USDA approval and meets 3-A standards for milk and liquid milk products. Polytetrafluoroethylene and some rubbers also can meet the food-grade requirement.

Electrodes. Electrode material selection is less complex than liner selection. Corrosion-resistance is the primary consideration. One common model has zirconium as standard for the wetted part, with tantalum, type-316 stainless steel, and high-nickel alloys as options. As noted above, platinum can be used where any contamination of product is a concern.

Line pressure. Most standard magmeters are designed to withstand process pressures rated at ANSI Class 150 (275 psig at 100°F) or ANSI Class 300 (740 psig at 100°F).

Full-body vs. insertion design. Insertion magmeters now offer a fast method of obtaining flow data at lower installation costs than full-body meters. One such insertion design is said to be nonfouling and suitable for slurries and pulp stock, as well as potable and cooling water; it installs through 1-in. taps on pipes from 4–60 in. in diameter. Insertion meters, however, provide single-point measurement for constant flows. If the flow profile changes (for instance, from laminar to turbulent flow), errors will occur. Thus, most applications are better handled by full-body designs.

Flanged vs. wafer bodies. As shown in Figures 1 and 3, magmeter bodies commonly come in flanged and wafer (flangeless) styles. In the former, the flange with bolt holes is an integral part of the meter body; this was the original design of all magmeters and is still widely used — particularly in larger sizes for service at higher pressures. The wafer body offers real economies in meter manufacture and weight. It mounts between either ANSI Class 150 or 300 flanges, but for reasons of safety with higher pressures it normally is limited to sizes of 4 in. dia. or less. (The flange bolts are some 6–8 in. long and subject to potentially dangerous elongation in the case of fire.)

Pulsed-DC vs. AC power. Using pulsed direct current instead of alternating current to excite the magnetic coil has advantages in certain applications. The newer pulsed-DC systems have a stable zero feature and are not as susceptible to coating effects of the electrode, due to their high input impedance (greater than 10^{12} ohms); thus, they are being used for the majority of installations. A good feature of AC systems is that they have high signal-to-noise ratios, so they are better for liquids with air bubbles or entrained air, for slurries with high solids content, or, because measurements are taken 120 times/s, for fast batching applications.

Signal converter. Also known as a transmitter, this electronics module takes the extremely-low-level voltage signal from the meter's electrodes and amplifies, as well as conditions, it — providing a more useful signal, usually in the form of industry-standard 4–20 milliamp DC. For some applications where higher resolution is desired, the transmitter alternatively can supply a frequency output, such as 0–1,000 Hz or 0–10,000 Hz.

With either the current or frequency signal, because the output signal from the magmeter is linear, the lowest end of scale represents no flow and the highest end, maximum flow rate. In some newer transmitters, the user can program output to be either analog (such as 4–20 mA DC) or frequency, and operation at 120 V or 220 V, 50/60 Hz, or 24V DC.

In certain models, location of the signal converter can be integral with the meter or panel-, wall-, or pipe-mounted near the meter, or in a NEMA 4X box at some remote location up to 300 ft away.

Communications with central control rooms. This ability is becoming more and more the norm. Some newer transmitters are particularly user-friendly and have the capability, upon demand from a suitable remote control device, to provide access to all the information in their database and to change it, if necessary, via RS 485 interface or HART protocol. All transmitter functions, including self-diagnostics, can be performed from the remote terminal. In addition, flow data, as well as parameters and performance of the magmeter, can be collected and stored at the remote computer or control device for future evaluation.

Where to use and not use magmeters

The problem of meter selection has never been simple and continues to grow in complexity — as new types of flowmeters and major innovations in older methods add to the already great variety of choices. Then, too, plants are demanding even greater accuracies of measurement as the basis for closer control, especially where this can mean substantial savings in the metered fluid, such as during custody transfer. Maintenance, including diagnostics, as well as expected accuracy and rangeability, also should be factored in deciding between one metering method and another. Table 1 lists general criteria that should be considered.

So where do magmeters fit? We have spelled out the major advantages, limitations, and range of applicability of the magmeter to put its use in some perspective. These should provide basic "yes" and "no" guidelines for many applications. For example, the magmeter is confined to the measurement of liquids that must have a conductivity of at least 0.05 µs/cm. Therefore, to measure the flow of gases, vapors, or steam, as well as most hydrocarbons or other nonconducting liquids, use of a magmeter clearly is inappropriate.

When the standard magmeter's design temperature limit of 356°F (180°F) and pressure limit of 740 psig (5.0 mPa) are not adequate, units with higher temperature and pressure ratings can be specified, or an alternative is to use a vortex meter. Some newer vortex meters can measure liquids, gases, and steam at temperatures up to 580°F (300°F) and pressures up to 3,750 psig (255 bar). Such meters, however, are limited to line sizes from 1½–8 in., with the preferred flow range being from 50% to full-scale for the chosen meter size.

For services in which both magmeters and vortex meters can be considered, carefully weigh the tradeoffs. For example, while vortex flowmeters offer good price-performance ratios, they do require much longer straight runs of piping upstream and downstream, and introduce considerable pressure drop, compared to magmeters.

Table 1. Evaluation criteria for flowmeter selection.

Hardware Factors	
• Accuracy	• Inherent reliability
• Repeatability	• Initial meter cost
• Range and turndown	• Sizes available
Application Factors	
• Line size	• Criticality of the data
• Pressure/temperature of fluid	• Special material required
• Straight pipe or straighteners	• Viscosity and density
• Clean/dirty fluid	• Reversible flow
• Other fluid properties	• Ambient temperature
• Type of data required (mass or volume)	
Installation and Maintenance Factors	
• Shutdown possible	• Field vs. factory calibration
• Complexity (time to install)	• Preventative maintenance required
• Spare parts required	• Primary failure mode
Human Factors	
• Desired readout	• Cost after ownership
• Computer interface	• Plant traditions
• Capability of in-house maintenance staff	• Industry standards/norms
• Vendor credibility	

Source: Ref. 4.

Further, while magmeters are independent of Reynold's number, this number should be above 10,000 for a vortex meter. Finally, while the magmeter's accuracy is independent of viscosity, high viscosity with the vortex can present problems, especially at low Reynold's numbers.

Another common method of flow measurement that can handle nonconducting liquids, as well as gases and vapors at higher temperatures and pressures, is the tried-and-true differential pressure (DP) or head meter; it uses an orifice plate or other flow restriction, such as a flow nozzle or venturi. Its temperature and pressure limits are determined by the pipe materials and installation, but 2,000 psig usually is no problem.

Where a magmeter can be used, however, it offers several significant advantages over a head meter. For one, it offers a much wider rangeability — 100:1, compared to 4:1 for a given DP-transmitter span setting. Also, with no obstruction to flow, the magmeter has practically no pressure loss, while the DP meter causes a significant permanent pressure drop — thereby increasing pumping costs and possibly even requiring a larger pump on the installation. Against this, though, the DP meter generally is much lower in initial cost.

The DP meter measures flow rate via a square root relationship. So, while a DP unit can be used to measure low flows, it is difficult to get good readings below about 25% of scale. Further, this type of meter requires much longer runs of straight piping than the magmeter — 10–25 pipe diameters upstream and 5–10 downstream, in order to obtain accuracies within ±1% of maximum flow rate.

Generally, the DP meter can be a good choice for metering clean liquids, vapors and gases, low-velocity flows, and for cryogenic applications. The advent of "smart" transmitters, including those using the HART protocol for remote electronic communications, has given DP meters a new lease on life.

Another alternative method of metering that can handle gas flows (including steam), as well as liquids — provided the fluid is clean — is the turbine meter. With an accuracy of ±0.25% of rate for liquids, this is one of the most accurate liquid meters and is even used as a calibrating device for magmeters — installed temporarily in series with the magmeter and then removed, to be kept in storage for the next calibration.

There are, of course, many other types of flowmeters, such as Coriolis effect or thermal mass flowmeters, variable-area (rotameter), positive displacement, and ultrasonic units. Each type has features that may make it the best choice for a given application. A discussion of these, however, is beyond the scope of this article. Consult, for instance, Ref. 2 for some selection guidance.

Cost of meter ownership

The actual bottom line in meter selection is the total cost of operating the system over at least a year or two *(3)*. The initial cost of the equipment to get a certain accuracy of measurement may be relatively high. On the other hand, the accuracy of an initially cheaper method may pass muster at the start, but may deteriorate rapidly in service — due, say, to erosion of the measuring element such as an orifice. Also, the lower initial cost can be quickly lost in added maintenance costs.

Installation costs, including costs of power lines and signal cables, must be factored in. The cost of accessories like strainers and straightening vanes, if required, also must be added. Then, projected maintenance costs need to be estimated. Meters with moving parts are sensitive to wear and clogging from entrained solids, and thus may need more upkeep.

Complex electronic components in transmitters may demand special skills in plant maintenance personnel, requiring perhaps their attendance at training schools, or supplier's or in-plant training courses. Alternatively, the plant can have service contracts with outside organizations. These costs, while easily justified by improved metering, can add significantly to total expenses.

Then, too, consider the power required to operate a meter; this can be significant in certain cases. Older style magmeters in larger sizes, for example, had high electrical power requirements. DP meters can incur high pressure drops, resulting in considerable energy expenditures to overcome these losses.

Thus, figuring the actual cost of meter ownership is not necessarily a simple exercise — but having the selection knowledge in-plant, or knowing where to get it, can pay off handsomely.

More and more CPI companies are looking at overall costs when selecting flowmeters. And, undoubtedly, this has contributed to the increasing acceptance of magmeters. CEP

Literature Cited

1. "Magnetic Flowmeters," Fischer & Porter Co., Warminster, PA (1993).
2. "Practical Guides for Measurement and Control," D. W. Spitzer, ed., Instrument Soc. of America, Research Triangle Park, NC, pp. 540–572 (1991).
3. **Dimm, T.**, "Cost of Ownership Drives Flowmeter Evolution," *I&CS*, p. 19 (Jan. 1993).
4. **Rusnak, J.**, "The Fundamentals of Flowmeter Selection," *INTECH*, pp. 49–51 (Apr. 1989).

C. ROSE is senior product manager for magnetic flowmeters at Fischer & Porter Co., Warminster, PA (215/674-6362; Fax: 215/674-7184). He has worked for the company for almost seven years in a variety of marketing positions, specializing in flow, pressure, and level measurements. Previously, he worked for Honeywell in a market application position. He holds a BS in chemical engineering from Drexel Univ., and is working on an MBA at St. Joseph's Univ.

G. E. VASS is applications manager for magnetic flowmeters for Fischer & Porter Co. (215/674-6350; Fax: 215/674-7184). Previously, he worked for more than 12 years for Honeywell, Inc., in design engineering and market application positions, specializing in smart field instrumentation and systems integration. He studied electrical engineering at Temple Univ.

Use Restriction Orifices to Maintain Required Flow Rates

These devices can protect equipment, ensure safety, and boost process performance.

Pankaj K. Khandelwal
Uhde India Ltd.

A restriction orifice (RO) is another name for an orifice plate, that is, a plate with a concentric hole in it, inserted in between two flanges in a pipe. The reduced cross-sectional area of the RO lowers the downstream flow rate as the pressure drops across the opening. The required downstream flow rate and pressure drop mainly set the size of the orifice. The author has presented equations for determining flow and pressure drop for liquids and gases in an RO (*1*).

In the chemical process industries (CPI), ROs are used to reduce fluid pressures, often to keep processes either stable or safe during low flows. We will discuss some of the more common applications of ROS in this article.

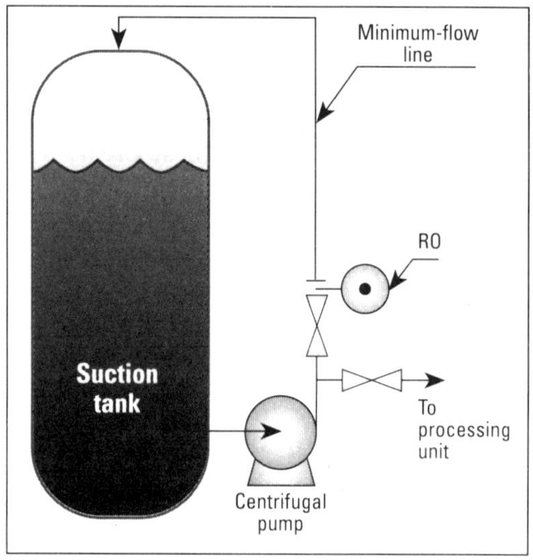

■ *Figure 1. An RO protects a centrifugal pump from cavitation.*

Maintaining minimum flow in pumps

Centrifugal pumps cannot run dry, and must have a minimum fluid flow to operate properly — to avoid cavitation, and subsequent mechanical damage to the pump.

To protect the pump, a minimum-flow line normally is created leading from the pump's main discharge line back to the suction tank. Sometimes, this line also functions as a recirculation line (see Figure 1). Usually, minimum flow is considered to be 10–15% of the rated capacity of the pump. This capacity meets the discharge pressure required to satisfy process requirements.

The RO creates a pressure drop that sets the minimum flow line for the pump. This orifice is always located near the pump discharge. The upstream pressure on the RO is fixed by the discharge pressure of the pump; the downstream pressure by that needed to attain the suction tank pressure. The RO is sized, based on this pressure drop across the RO, and the minimum specified fluid flow rate of the pump.

Safely operating critical processes

Certain critical processes require minimum fluid flow for safe operation during the emergency shutdown of a plant. For example, consider two-phase processes in which toxic gases, such as chlorine, are present. During normal operation, the chlorine is removed safely from the processing unit with the flowing liquids. But during an emergency shutdown, the chlorine collects in the unit and its pipeline headers. Hence, a

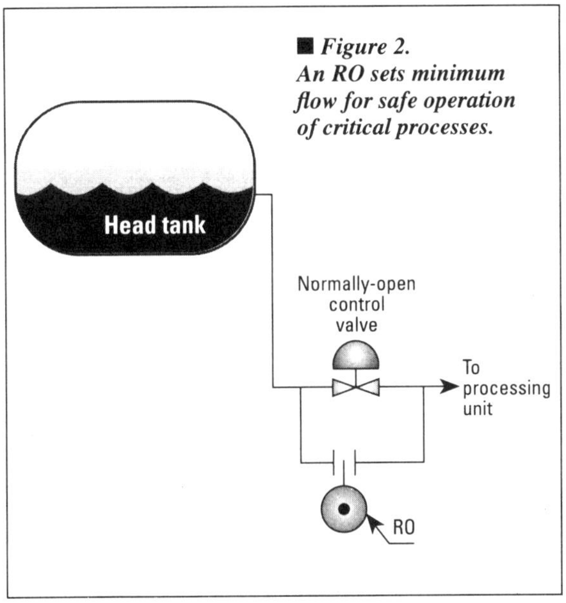

■ *Figure 2. An RO sets minimum flow for safe operation of critical processes.*

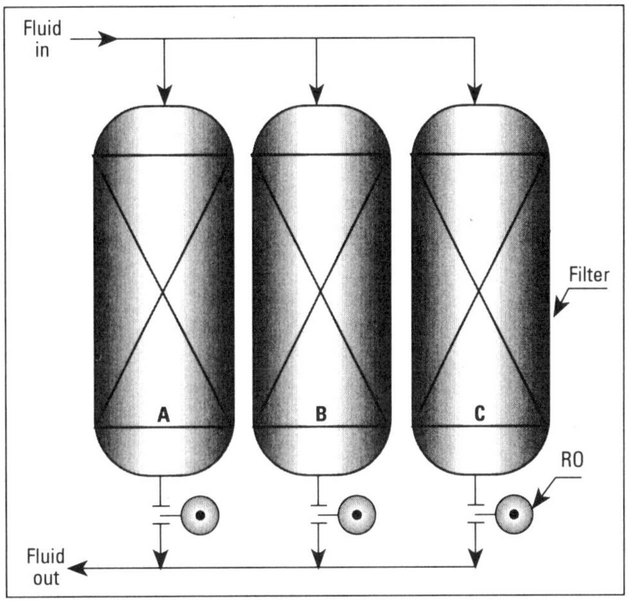

■ *Figure 3. In parallel filtration, ROs provide even flow rates.*

minimum fluid flow is recommended to flush away the collected gases.

One such application is shown in Figure 2, where the reactant fluid is fed to the processing unit from a feed tank. A minimum safe fluid flow rate is always maintained by using an RO. For normal operation of the processing unit, flow is controlled by a fail-to-close (normally open) control valve.

The upstream pressure on the RO is equal to the liquid head from the tank. The downstream pressure is set to meet the pressure requirement of the processing unit. Again, the RO is sized, based on its differential pressure and the minimum safe fluid flow rate.

Assuring proper filtration

During filtration, a layer of particles forms on the surface of the filter medium, creating a pressure drop. In every filtration system, there is a maximum specified pressure drop across the filter. Once that value is reached, the filter is backwashed.

For continuous filtration, two or more (say, *N*) filter units are used in parallel, depending on the flow rate and solids concentration. One unit is used as a standby. The operating cycle is set such that (*N* - 1) units are always on-line. Periodically, one dirty filter unit is taken out of service, and backwashed. A previously cleaned filter is put in operation.

Because pressure drop across the clean filter is less than that across the clogged filter, unequal flow distribution results. To avoid this, ROs are used in the downstream lines of each filter (Figure 3).

To show how the use of ROs evens out flow rates, calculations for Figure 3 appear in Table 1, which includes pressure drops and flow rates for two cases, one with and another without ROs.

Case 1 (without ROs) shows that whenever a dirty filter is replaced by the clean one, the fluid flow rate through the clean filter becomes high, compared to the dirty filter. In fact, the fluid flow rate exceeds the design value of 50 m^3/h. When this happens, the filter will malfunction and the desired performance will not be achieved. Case 2 (with ROs) minimizes this effect. Although introducing ROs in the pipelines leads to higher pressure drops, overall operation improves, because design flow rates are never exceeded.

Maintaining pump flows for multiple applications

In some CPI plants, a single pump sometimes is used for multiple applications. For example, the same pump may feed raw material to two different processing units (A and B), working independently at different times (see Figure 4).

What if the flow and pressure head requirements of both units are different? In

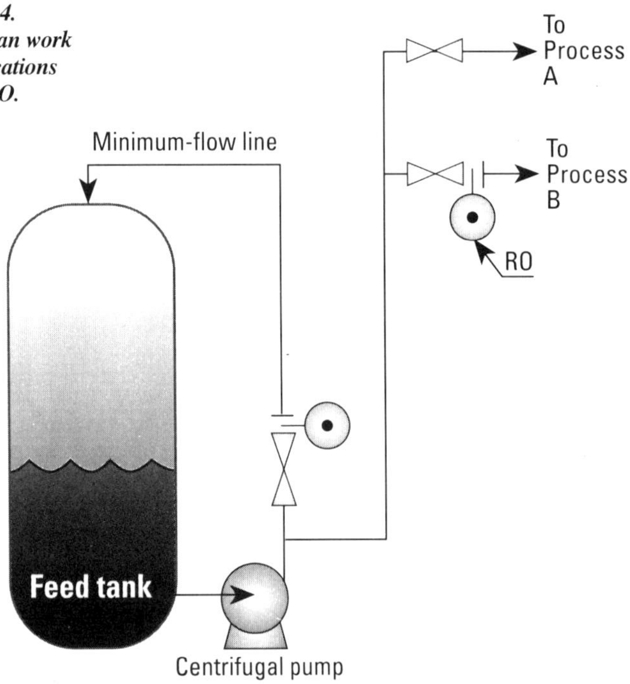

■ *Figure 4. A pump can work two applications with an RO.*

Table 1. Calculations show that ROs provide even flow rates in parallel filtration.

	End of last cycle			Start of next cycle		
	Filter			Filter		
	A	B	C	A	B	C
Case 1: without RO						
Pressure drop, bar	0.2	0.2	0.1	0.2	0.2	0.1
Flow rate, m^3/h	50	50	0	0*	33.3	66.7
Deviation from normal flow, %	0	0	—	—	-33.4	+33.4
Case 2: with RO						
Pressure drop, bar†	0.7	0.7	0.6	0.7	0.7	0.6
Flow rate, m^3/h	50	50	0	0*	46.1	53.9
Deviation from normal flow, %	0	0	—	—	-7.8	+7.8

Basis:
Total fluid flow rate = 100 m^3/h
Pressure drop across filter = 0.1 bar
Maximum allowable pressure drop across filter = 0.2 bar

* The clogged filter is backwashed.
† Total pressure drop across filter and RO (RO drop = 0.5 bar).

Table 2. An RO in the line to Process B meets pressure requirements.

	Process requirements		Pump supplies	
	Flow rate, m^3/h	Differential pressure, bar	Flow rate, m^3/h	Differential pressure, bar
Process A	5.0	2.0	5.0	2.0
Process B	3.0	1.8	3.0	2.2

certain cases, such requirements are met by the proper selection of pump characteristics. In others, this is not possible, and additional pressure drop is required. Here, an RO is used.

Table 2 shows how this situation might look like for Processes A and B. The pump suits the requirements of Process A, but requires an additional pressure drop of 0.4 bar for Process B. An RO placed in the discharge line feeding Process B will meet the pressure requirement.

Providing adequate agitation-air distribution in tanks

In the CPI, agitation air is used for applications including maintaining a homogeneous slurry in tanks, removing gases from the liquids, and purifying wastewater. Normally, compressed air requirements at different units are met by a centrally-located compressor and utility stations at different units. The compressed air typically will be at a pressure of 2 bar. While this pressure may be suitable for some plant applications, it is too high for agitation.

The required agitation pressure depends on the liquid head in the tank. Again, a reduced pressure drop is created in the agitation air line by inserting an RO before the tank. If a tank is large and requires several spargers, multiple air lines are drawn off of the main header. Each line will have an RO in it to reduce the pressure, and will uniformly distribute the air. CEP

Literature Cited

1. **Khandelwal, P. K., and V. Gupta,** "Make The Most Of Orifice Meters," *Chem. Eng. Progress,* **89** (5), p. 32 (May 1993).

P. K. KHANDELWAL is a senior process engineer for Uhde India Ltd., Bombay (011–9122–578–301; Fax: 011–9122–578–4327), where he does basic and detailed engineering. Previously, he worked for Birla Research Institute of Applied Sciences as a senior research engineer where he was involved with process development, pilot-plant design, and development of engineering software. The author of a number of technical papers, he holds a bachelor's degree in engineering (chemical) from the Univ. of Roorkee (India), and a master's in technology (chemical) from the Indian Institute of Technology, Kanpur.

Watch Out for Instrument Errors

Instrument faults are to blame for many control problems and plant upsets, but the actual culprits often are hard to identify or are completely overlooked.

Fred F. Sanders,
Monsanto Chemical Co.

Most companies have advanced control strategies to optimize plant performance and regulate key product parameters. In many cases, supervisory control, using laboratory data, is directly linked to process regulatory controls. Often, in addition, an intermediate control level provides an on-line estimate of some key product parameter. The strategy of these multitiered control systems frequently is quite complex, and the primary emphasis usually is on controller coding/configuration and system integration. This complexity as well as control interaction often make problem diagnosis difficult.

There are many examples of advanced controls that were adopted and then failed to meet expectations. Most frequently, control strategy or tuning is blamed; the advanced controls are scrapped, and the plant is returned to simpler, more manual operation.

Yet, in a chemical process, the accuracy and reliability of process and analytical measurements often are critical factors in control-system performance and, ultimately, in yields and product uniformity. Indeed, when we conducted historical studies on one of our key chemical processes to determine the primary causes of upsets (defined as excursions that were severe enough to result in the restriction or downgrading of product), we found that nearly two-thirds of these upsets could be traced to instrument faults. Table 1 shows the causes of typical sensor errors that we have observed.

Nevertheless, measurement integrity often is overlooked during troubleshooting. So, in this article, we will highlight its impact and explore some of the subtle ways in which instrument problems affect processes.

Table 1. Typical causes of sensor errors.

- Wrong sensor application
- Sensor/transmitter impulse line plugged
- Signal wiring not shielded or grounded
- Faulty electronics at controller input
- Induced current in signal wiring
- Electronics not properly grounded
- Sensor not correctly mounted
- Transmitter not properly calibrated
- Loose or faulty signal connectors
- Electronics faulty in isolator
- Faulty signal wiring
- Ambient-temperature effects

The downside of upgrades

To apply advanced controls, plants often upgrade parts of their control systems. Most common is the replacement of analog controls with digital controls and the addition of a distributed control system (DCS). To minimize cost, however, usually most field instruments are retained, existing cable trays are used or extended, and signal and power wiring are mixed within cable trays. Results sometimes can be disappointing — frequently because engineers forget that the overall reliability of control is only as good as the sensor inputs to the system.

Unfortunately, most sensor errors may go unrecognized, particularly if those signals are used in fast-feedback control loops, such as those for flow or pressure. In cascade-control systems, errors normal-

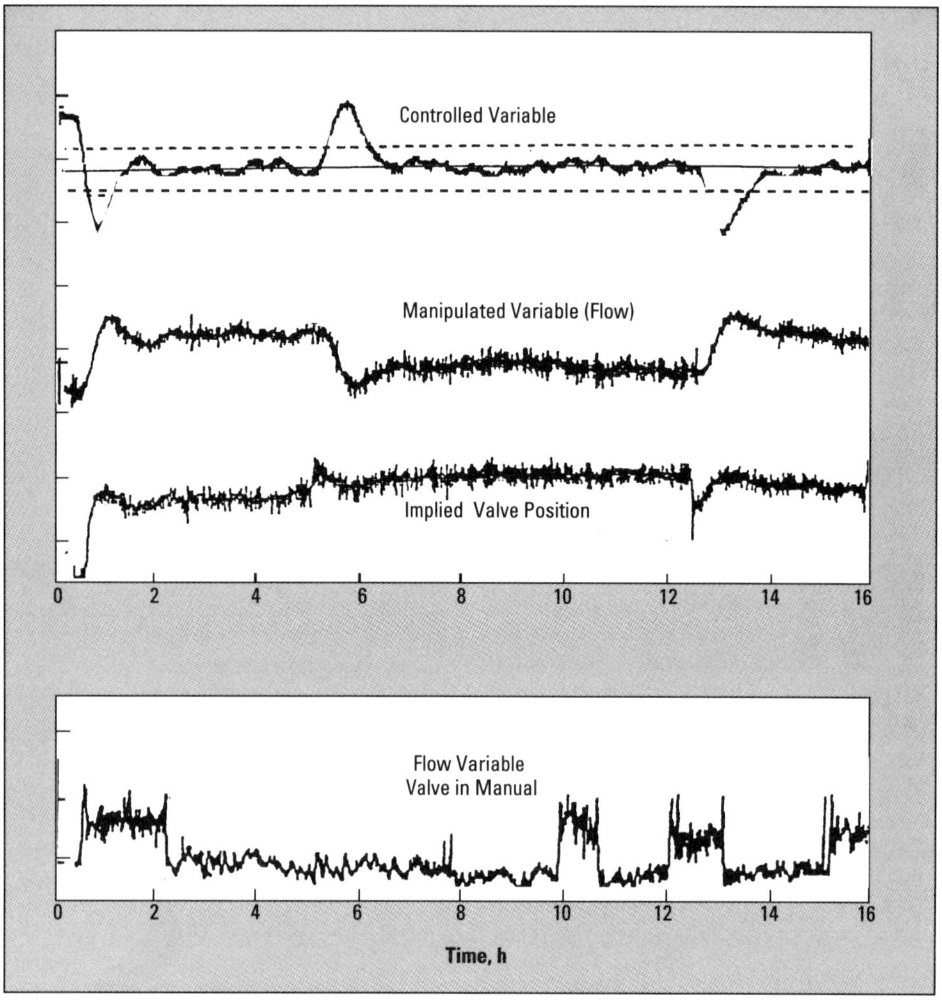

■ ***Figure 1. Implied valve position warns of flow measurement error.***

ly are passed on as disturbances to the next higher control level. In single-loop regulatory controls, overall process and product variability usually is increased. Often, secondary process upsets, caused by instrumentation problems, are observed long before the primary cause can be diagnosed properly. This sometimes results in a large amount of off-grade material being produced.

Some sensor errors can be identified easily. For instance, most control systems recognize extreme sensor faults, such as open circuits, short circuits, and the like. In addition, sensor rates of change, and mid-select strategies (which use the middle value of multiple inputs) provide intelligence at the controller level. Many process variables, such as levels and temperatures, may exhibit damped behavior because of the inertia of a process; therefore, significant changes beyond a variable's normal inertia often can be identified as suspect.

Spotting other instrument problems can be tough, though.

Total sensor redundancy is, perhaps, the most obvious and least desirable method of providing sensor validation. Process modeling and neural-network applications can provide information that often can allow sensor validation. In addition, some intelligence at the transmitter level is beginning to find its way into the market place.

Nevertheless, if your plant is like most, you likely will have to largely rely on your own insights, ingenuity, and intuition to pinpoint many instrument problems. After all, the impact of instrument errors can be insidious, as the following examples illustrate.

Wrestling with wiring

One of the most frustrating categories for troubleshooting is signal wiring. For instance, in one case, gas flow was being precisely controlled to several vessels. Unknowingly, electrical wiring for resistance heaters had been added to the existing signal cable tray. Flow signal errors of 1.5% were created in each of the vessels whenever the heaters fired. Not until product variability increased and common higher-level disturbances were examined was the problem finally identified. Many common-sense design practices, such as proper shielding and grounding of signal cable, can eliminate similar electrical-noise problems.

One of the best methods for detecting such problems is on-line comparison of flow readings to predictive flow models based on valve characteristics. The use of on-line expert systems in this area is expanding — facilitated by the availability of data currently provided by most DCSs and process data management systems.

Consider Figure 1, which plots the response of a poorly performing cascade-control system. The controlled variable (primary loop) is a critical process parameter. The output of its controller provides the setpoint for the manipulated variable (flow) in the secondary loop. When just the controlled and manipulated variables are observed, the source of the excursions is not clear. However, when the secondary-loop implied valve position, which usually is not checked, is compared to the manipulated variable, flow measurement errors are suspect, because of the gross disagreement between the two. Further checks with the flow valve in manual clearly show the measurement disturbance. This disturbance was traced to a faulty isolator.

Flow rates of high viscosity fluids often are deduced via gear-pump speeds. In DC-drive systems, armature voltage frequently is used inferentially as an indication of speed and, therefore, flow. In many cases, this type of indication is adequate. However, in critical applications where precise level control is important, this measurement (typically used for feedforward adjustments) is inadequate because of inherent drift errors, usually in the range of 1–2%. A much better choice is speed indication based on magnetic pickup sensors, despite the need for sensor alignment, shielding, and grounding.

Sensing process pressure is critical in many applications. Figure 2 provides an example of pressure indication from faulty wiring from the sensor to transmitter. The signal had been used as part of an advanced control strategy *for 4 years* — with poor results. The 700 ft of wire, pulled as part of the control-system upgrade, was never checked for integrity. Because of the intermittent nature of this signal error and the inherently high normal variability in the measurement, the error was not totally recognized. Sensors and transmitters, however, were changed out during equipment overhauls, connecting plugs were replaced, alterations to the existing control strategy (which didn't work) were suggested, and, at times, electricians were even called upon to attempt online recalibration.

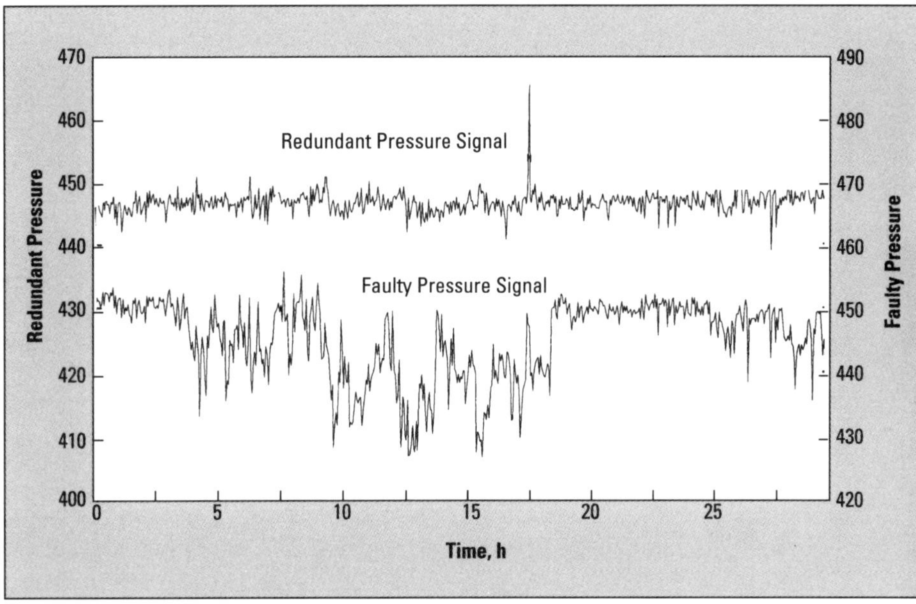

■ *Figure 2. Pressure error from faulty signal wiring.*

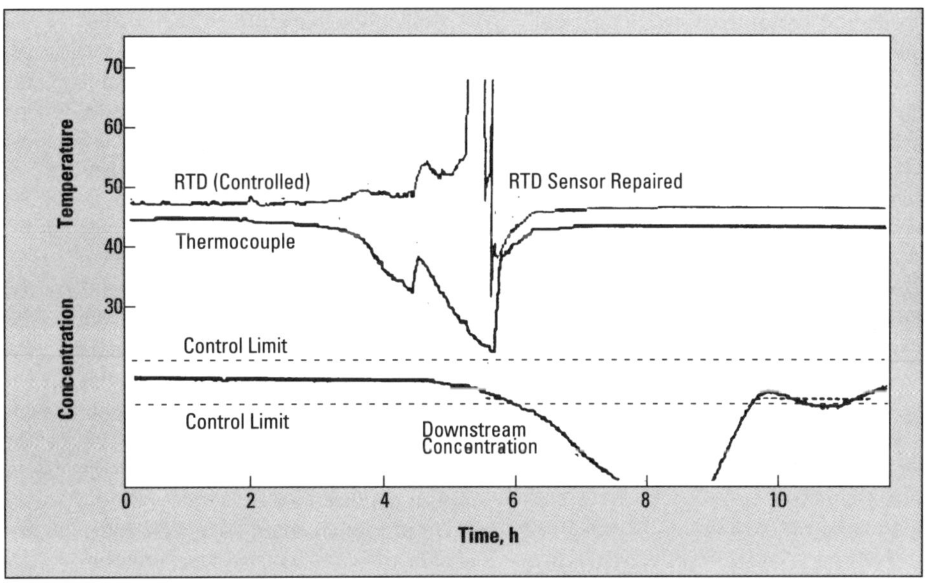

■ *Figure 3. Temperature sensor error from faulty signal connector.*

After local redundant measurement was added, the signal error became more apparent. When wiring resistance was checked, high resistance was found in one of the wires that supplied the excitation voltage to a Wheatstone-bridge strain gauge. Before replacing the six-conductor wiring, however, the connector was modified — replacing the faulty conductor with one of the wires used for calibration. This convinced skeptical manufacturing personnel of the correctness of the diagnosis, and permitted full use of the advanced control strategy — with major improvements in the control.

The dangers of drift

Some of the most common sensor errors that we have observed include drifts in pressure related to "slow fluid fill loss," and ambient temperature effects from expansion and contraction of the fluid-filled capillaries, typically mercury, sodium/potassium (NaK), or silicone oil. Although some developments, such as resonant-silicon sensor technology, offer higher resolution and accuracy, the "smart sensors" available today only provide for self diagnostics and autoranging, not actual calibration and drift detection.

Some of the new sensors, which we are now testing, however, do promise potential benefits over standard fluid-filled systems, especially with respect

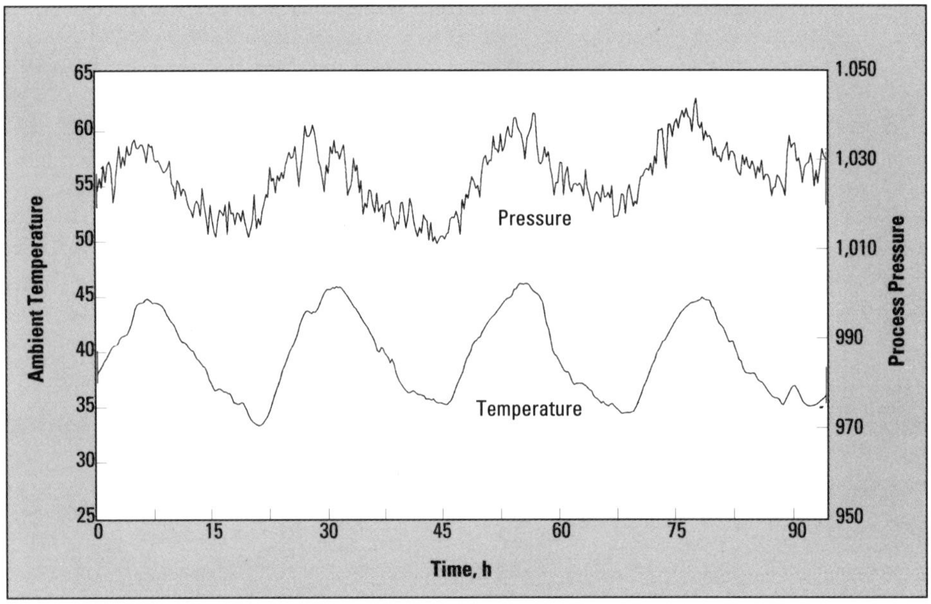

■ *Figure 4. Pressure error from ambient-temperature effects.*

to drifts from ambient-temperature effects. These sensors include:

- optical fibers, a light-emitting-diode light source, and an optical aperture system to measure diaphragm displacement;
- silicon-on-sapphire technology using piezoresistive silicon gauges and a resistance temperature detector (RTD) for temperature compensation, both mounted at the tip; and
- higher accuracy fluid-filled systems with built-in temperature compensation.

Of course, one of the best, but most expensive, ways to provide protection from many sensor drifts in critical applications is the use of redundancy. We have had success in using three sensors in intelligent on-line control strategies such as mid-select. This approach, however, can become very expensive, and thus only makes sense for the most critical applications.

To precisely control outlet concentration in an evaporator, an advanced control strategy was developed. Its success hinged on accurately measuring steam flow. Because this, in itself, often can be a challenge, some problems should have been anticipated — but weren't. The control strategy worked, at least most of the time. The times when upsets or excursions occurred, no one suspected the steam flow, because it was controlled in an inner cascade loop and always was close to its setpoint. Because of space limitations, the annubar used for measuring the steam flow had not been installed as recommended by the vendor. Unfortunately, even a misinstalled annubar will provide accurate readings sometimes — the problem again was intermittent and difficult to recognize.

Temperature control is important in many processes. RTDs, because of their accuracy and reliability, usually are chosen over thermocouples. Faulty signal connectors can lead to problems, however. In one case, condensate from a steam leak had accumulated on the connector, yielding a false temperature measurement. This condition, as in the other examples, initially was difficult to recognize. A redundant measurement from a thermocouple was available, but was not monitored routinely by the operator. The fault, which had caused variations and disturbances in downstream concentration for two weeks, only became apparent when the severity of the error prevented temperature control, as shown in Figure 3.

Assessing ambient conditions

Checking ambient conditions only on a daily or 24-h basis sometimes can mask the true source of problems. Day-to-night temperature changes, especially at those times of the year when the greatest differences exist, can prompt instrument errors. Figure 4 shows monitored pressure measurements from a sensor linked via a fluid-filled system to a strain gauge. The sensor/strain gauge, which did not have ambient-temperature compensation, was exposed to the changes in temperature as depicted on the graph. The pressure "zero shift" shown is well documented in the specifications supplied by the vendor. We have found, however, that these specifications often tend to underestimate the effects of ambient temperature, because they only consider fluid expansion characteristics in the transducer housing and not the flexible stem. In our example, ambient temperature effects were overlooked, because the sensors were indoors in an area that

Further Reading

Ender, D. B., "Process Control Performance: Not as Good as You Think," *Control Engineering,* pp. 180–190 (Sept. 1993).

Gans, M., and B. Palmer, "Take Charge of Your Plant Laboratory," *Chem. Eng. Progress,* **88** (9), pp. 26–33 (Sept. 1993).

Rowan, D. A., "On-Line Expert Systems in Process Industries," *AI Expert,* pp. 30–38 (Aug. 1989).

Sanders, F. F., "Continuous Control Gets On-Line Advisor, *INTECH,* pp. 32–35 (Aug. 1992).

Sanders, F. F., "Handling Process Disturbances and Measurement Errors," *Proceedings — Advances in Instrumentation and Control,* Anaheim, CA, Instrument Society of America, Research Triangle Park, NC (1994).

Southman, M. E., "Using Statistics and Process Knowledge Together for Improvement," presented at Process Control User Forum, Miami Beach, FL, Chemical Manufacturers Association, Washington, DC (Mar. 26, 1990).

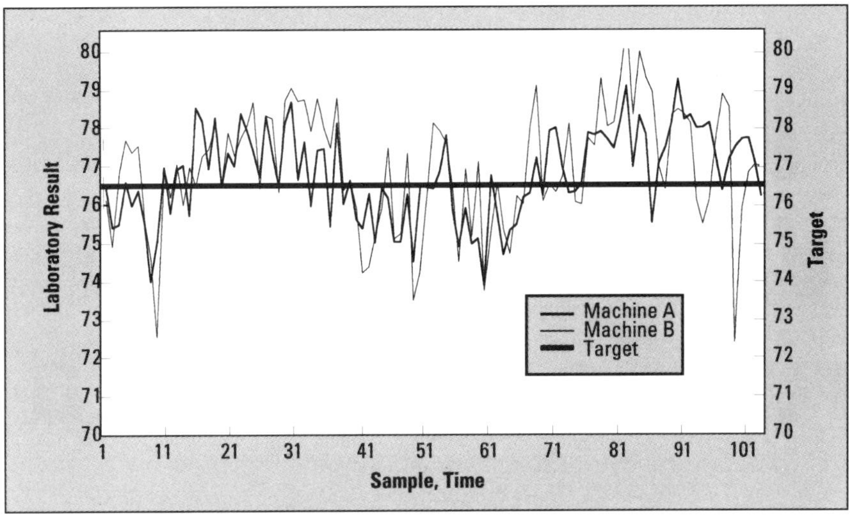

Figure 5. Laboratory analysis varies with outside temperature.

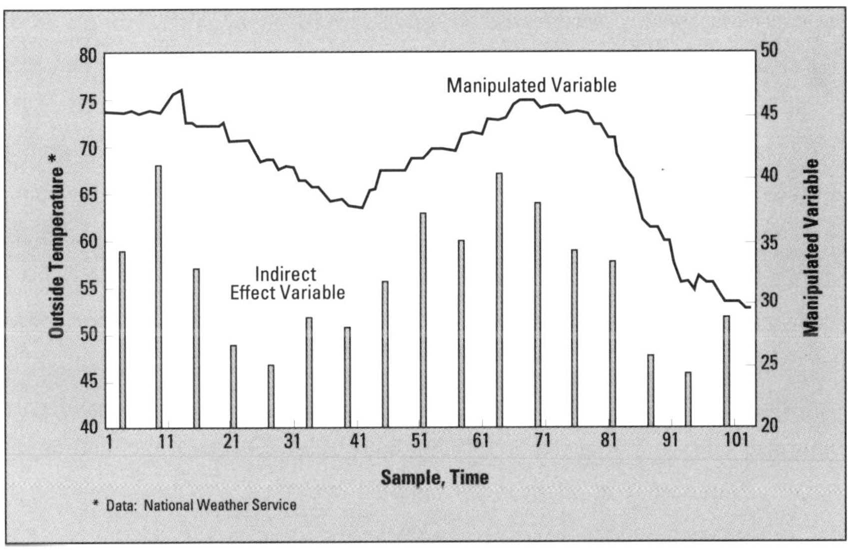

Figure 6. Process adjustments for controlling product parameter.

was thought to have a relatively constant temperature.

In another example, process temperatures were found to cycle when sensors that monitor the pressure of heat-transfer-fluid vapors were exposed to ambient temperatures.

Most companies use laboratory data as a basis for making adjustments to their process. In most cases, an appropriate set of standards insure that the results are accurate. However, just as process sensors should be validated for accuracy and reliability, the integrity of laboratory instrumentation and procedures should be subjected to the same scrutiny. Figure 5 shows a laboratory analysis of a particular product parameter for two separate machines, sampled approximately six times per day. Although there is high variability between samples, there also is a common data trend. Figure 6 illustrates the average adjustments made to the process in an effort to control the parameter in Figure 5 to target. The indirect effect variable shown is the average daily outside temperature. Note the apparent lagged correlation between the variables in Figure 6. Here, the measured laboratory results were found in error, being biased by uncontrolled humidity conditions in the laboratory. The outside temperature simply correlates with the humidity. In this case, the adjustments to the process actually were adding variability to the product. An important first step in achieving effective control was to establish a reference standard for the product parameter (which didn't exist), and determine its relationship to ambient conditions. Note, also, that a lower laboratory sample frequency would, most likely, have revealed the pseudo-cyclic data trend in Figure 5 as simple random laboratory error. This would have made identification of the underlying cause much more difficult.

Summing up

Significant reduction in product variability can be made through advanced controls. It is important to understand, however, the limitations of instrumentation, and the potentials for measurement error. When assessing the performance of multitiered systems, start at the lowest control level, and validate the controlled variables and closed-loop performance — only then proceed to the next higher level. Use adequate redundancy for critical measurements. Laboratory analysis is critical, as it often is used as the final quality check before releasing a product to the customer. So, make sure that adequate standards are available to insure the reliability of results. CEP

F. F. SANDERS is an engineering specialist in the Monsanto Chemical Co.'s nylon manufacturing plant in Pensacola, FL (904/968-8898; Fax: 904/968-8543), focusing on the development and application of advanced process controls. His experience includes process data management, optimization, and the development and implementation of expert systems for chemical processes. Holder of a BS in chemical engineering from the University of Louisville, he is a member of AIChE, and a senior member of the Instrument Society of America.

Succeed at On-Line Validation and Reconstruction Of Data

Many process measurements are contaminated by noise. A two-stage, on-line technique allows these measurements to be made sufficiently accurate for monitoring, control, and identification.

Ming T. Tham,
University of Newcastle upon Tyne
Andrew Parr,
Conoco Ltd.

Plant data obtained for process monitoring, control, and identification often are contaminated by random effects. Generally classified as noise, these disturbances emanate from various sources. Some stem from short-term process inconsistencies inherent to plant operation. Others are features of the instrumentation used or the measurement techniques employed — for instance, electromagnetic interferences, quantization errors, inconsistent laboratory analyses, and so on. Depending on their severity, these disturbances may render information contained in the measurements ambiguous. Consequently, running the plant based upon these measurements will lead to poor process performance. Stochastic systems theory may be used to account for random effects in designing control- and system-identification algorithms. Most control theoretic methodologies, however, are not sufficiently robust in dealing with large measurement errors. Clearly, for accurate process control and system identification, the underlying noise-free signal should be used. The question is how do we extract this signal from noisy measurements?

Kalman-filters, observers, and similar procedures have been proposed to enable the recovery of the underlying signals from noisy measurements *(1,2,3,4)*. These algorithms usually are developed assuming that a model of the process is available, the signal is stationary, there is no correlation between signal and noise, and the noise statistics are known. Such conditions, however, are seldom encountered in process plants. Feedback loops, time-varying characteristics, process interactions, and nonlinear behavior often restrict the applicability of published algorithms. Adaptive algorithms also are of limited use, because the adaptation mechanism, frequently based on the least-squares principle, is itself susceptible to errors in data. It actually was our desire to improve the robustness of adaptive algorithms under practical conditions that motivated this investigation. Moreover, there are numerous process units in a typical chemical plant, and hundreds of variables may have to be monitored. It therefore would be impractical to implement a Kalman-filter, say, to each measurement channel. The cost of model development would be phenomenal, while the computational load would hobble most process computers.

The limitations of current knowledge-based systems

Real-time knowledge-based systems (RTKBSs) also have attracted much interest for process fault detection, plant supervision, and control. They make use of a combination of *quantitative* (measured or derived) and *qualitative* process knowledge to implement or provide suggestions for improved plant operation. Quantitative methods may be precise but not necessarily accurate. Thus decisions based on their results may be grossly in error when confronted by gray problem areas. Qualitative approaches cope well with gray issues, by drawing upon expert knowledge about the process. The obvious penalty, though, is that the configured system will be process specific. Additionally, the knowledge base may be unwieldy and difficult to maintain. Even when these factors, together with associated costs are discounted, questions about the validity of the encoded knowledge and the reliability of reasoning procedures have yet to be resolved.

Thus, it often is asked whether the sophistication of some of these procedures is justified. Indeed, the most widely used signal-processing algorithm in the process industries is the low-pass filter: typically unit gain, first- or second-order lag functions (see, for instance, Ref. 5). These filters improve data quality by attenuating noise from measured signals. As will be shown later, however, low-pass filters are inefficient in dealing with large errors, such as from data spikes.

This article describes a simple, heuristically based, two-stage technique that can be used to improve the integrity of data collected from process plant. It is designed to deal with small random effects as well as data spikes. Depending on the behavior of the sampled signal, a variety of elementary statistical tests are made to assess the quality of each new data point. If the data point fails this *validation* stage — that is, there is a strong possibility that it is a spike and hence a bad data point — then a series of corrective actions (the *reconstruction* stage) are employed to obtain a better value. The technique is not process specific and, hence, is applicable generically to all process signals. It may be implemented as a stand-alone real-time procedure or as an integral part of a RTKBS system.

Low-pass filters

Consider an underlying stationary signal corrupted by random disturbances with uniform magnitudes and zero means, for example, quantization error. In this case, filtering via simple averaging is effective. The recursive form of the averaging filter has a structure equivalent to the first-order low-pass filter, which, in turn, can be shown *(6)* to be optimal for a first-order random walk process. The following recursive formulation is especially useful for on-line data processing:

$$y_k^f = (1 - \alpha)\, y_{k-1}^f + \alpha y_k \qquad 0 < \alpha < 1 \qquad (1)$$

$$y_k = y_k^0 + \xi_k \qquad (2)$$

where α is the filter constant, y_k is the measurement of y_k^0,which is corrupted by ξ_k, and y_k^f if the filtered signal. This algorithm commonly is used to reduce noise in process signals. Note that this filter does not eliminate the effects of data errors. Rather, errors are attenuated in an exponential manner at the rate determined by the constant α. This may be illustrated by considering two extreme cases. The error ξ_k will not feature in y_k^f if $\alpha = 0$, but neither will the underlying signal y_k^0. On the other hand, if $\alpha = 1$, then y_k^f will contain the full effect of ξ_k, that is, no filtering occurs. Thus, the closer that α is to zero, the heavier the filtering action.

By choosing α between 0 and 1, ξ_k will be removed within $1/\alpha$ samples.

While low-pass filters are quite effective in dealing with small magnitude random effects, they cannot cope efficiently with large signal spikes or *outliers*. Consider the case when $\xi_k = d$ at the time instant k, and $= 0$ otherwise. y_k^f then is calculated as a fraction of the measurement y_k, plus a fraction of the previous measurement, y_{k-1}^f. Thus, the error introduced by ξ_k is effectively reduced to $(1-\alpha).d$. At the time instant $k+1$, $\xi_{k+1} = 0$. However, y_{k+1}^f is still not equal to y_{k+1}^0, because it is a function of y_k^f. Nevertheless, the spike is further reduced to $(1-\alpha).\alpha.d$. Further attenuation of d takes place in the same manner as time proceeds. Although the filtered signal will be smooth, if the magnitude of d is large, it may be a while before it finally is removed.

When the noise-to-signal ratio is large, the requirement for heavy filtering could cause significant time lags in the signal. Thus, in choosing the parameters of low-pass filters, a compromise has to be made between the desired degree of noise attenuation and the additional time lag introduced. To recover the true nature of the measured signal, a *data validation* methodology that can distinguish between good and bad data is required. When data points are deemed bad, then this technique also should provide *signal reconstruction*, that is, an estimate of the true value. Other desirable properties of the procedure are that it must not be process specific; its mechanism must be sufficiently easy to understand to encourage application; and its algorithm should be simple to facilitate implementation without excessive coding on process computers.

Data validation and signal reconstruction

Several statistical techniques have been proposed for the detection of outliers; see, for instance, Refs. 7, 8, and 9. The statistics of the data, such as their distribution characteristics, often are assumed to be known, however, or, if unknown, can be estimated. Process environments usually defy accurate mathematical or statistical descriptions. A more severe limitation is that almost all the proposed methods are batch or off-line procedures. In contrast, data are collected and used in a sequential manner in process plants — for instance, for on-line feedback control, and process monitoring. Thus, it is not unusual to find that standard automatic control algorithms are complemented by coded information about perceived process behavior, such as maximum possible changes of various process variables, physical-chemical limitations, and so on. As with RTKBS approaches, the utility of such a solution, however, is compromised by the specificity of the technique.

The following procedure has been found to be quite effective in detecting outlying data points. We have found, through extensive simulations, that the efficiency of sequential outlier detection is dependent on the behavior of the signal. Therefore, the first task is to apply pertinent tests to extract the features of the signal. There are three typical behavioral patterns that can be observed in a signal, namely:

- a distinct trend;
- a trend that, due to noise, is not immediately discernible; and
- no trend.

Test A — testing for a distinct trend. Distinct trends in signals occur when the process is being moved from one operating level to another. This feature is detected easily by considering the signs of the changes between successive data points. Given three consecutive data points x_k, x_{k-1}, and x_{k-2}, then:

$$\text{If}\left\{sgn\,(x_k - x_{k-1}) = sgn\,(x_{k-1} - x_{k-2})\right\} \Rightarrow \text{trend in signal} \tag{3}$$

where $sgn\,(z<0) = -1$, $sgn\,(z>0) = +1$, and $sgn\,(z=0) = 0$. Because the on-line data validation and reconstruction procedure is applied to situations where measurements are noisy, it will be rare to observe $sgn\ (x_k - x_{k-1}) = sgn\ (x_{k-1} - x_{k-2}) = 0$.

Test B — testing for a trend that is not apparent. Sometimes, a process signal may exhibit slow drifts. The trend, however, may be masked by noise. In this case, detecting the trend is less straightforward. Suppose that we have collected a sequence of data points $\{x_k, k=1,\ldots n\}$, where n is even. If there is a trend, then we would expect the mean of half of the original sequence, $X_1 = \{x_1, x_2, x_3, \ldots x_{n/2}\}$ to be different from that of the other half, $X_2 = \{x_{n/2+1}, x_{n/2+2}, x_{n/2+3}, \ldots x_n\}$. We can employ the following test to determine whether $\bar{X}_1$, the mean of X_1, is equal to $\bar{X}_2$, the mean of X_2:

$$\text{If}\left(\left|\frac{(\bar{X}_1 - \bar{X}_2)\sqrt{n}}{2s}\right| > t_{\alpha,n-2}\right) \Rightarrow \bar{X}_1 \neq \bar{X}_2 \text{ at the } \alpha \text{ level of significance}$$

$$\Rightarrow \text{trend in signal} \tag{4}$$

where

$$\bar{X}_1 = \frac{2}{n}\sum_{k=1}^{n/2} x_k \qquad \bar{X}_2 = \frac{2}{n}\sum_{k=1+n/2}^{n} x_k$$

$$s^2 = \frac{s_1^2 + s_2^2}{2} \qquad s_1^2 = \sum_{k=1}^{n/2} \frac{\left(x_k - \bar{X}_1\right)^2}{n/2 - 1}$$

$$s_2^2 = \sum_{k-1+n/2}^{n} \frac{\left(x_k - \bar{X}_2\right)^2}{n/2 - 1}$$

For $\alpha = 0.005$ and $n = 10$, $t_{\alpha,n-2} = 3.355$. Those who are statistically inclined will recognize that we are applying Students t-test (see, for instance, Ref. 10) to test the null hypothesis that $\bar{X}_1 = \bar{X}_2$, when the standard deviations of the respective subsequences are unknown. Note that Test A is performed first, followed by Test B. If the conditions in Tests A and B are not satisfied, then it is obvious that there is no trend in the data sequence.

The features of the signal have to be extracted because the checks applied to determine whether the new data point is an outlier depends on how the signal is behaving. The manner by which the signal is reconstructed when a data point is found to be erroneous also depends on the tests that detected it.

Dealing with outliers when there is a distinct trend in the signal. To check whether a data point is an outlier when there is a distinct trend in the signal, the following test is performed. The average change in magnitude $\bar{\Delta}_k$ between each successive data points over a data window of n points is calculated, via:

$$\bar{\Delta}_k = \frac{1}{n}\sum_{i=1}^{n}\left|x_{k-i} - x_{k-i-1}\right| \tag{5}$$

The size of the latest change, $\Delta_k = |x_k - x_{k-1}|$, is then compared to a multiple of $\bar{\Delta}_k$:

$$\text{If}\left(\Delta_k > \beta\bar{\Delta}_k,\ \beta > 0\right) \Rightarrow x_k \text{ is an outlier} \tag{6}$$

If this condition occurs, then an estimate for x_k is made according to:

$$\hat{x}_k = x_{k-1} + sgn\left(x_k - x_{k-1}\right).\beta.\bar{\Delta}_k \tag{7}$$

The rationale behind this test is that if a trend has been established, then the magnitude of the change between two consecutive data points will be dictated by the nature of the process. The size of a typical change in signal thus can be extracted from historical data. Therefore, the test compares the size of the latest change in signal with the average of changes over a specified data window of n previous samples.

Detecting outliers when there is no trend. At the other extreme is the detection of bad data points when trends have not been indicated; this is perhaps the simplest operation. Here, $\bar{x}_k$, the mean, and, s, the standard deviation of the signal sequence over the last n samples are calculated:

$$\bar{x}_k = \frac{1}{n}\sum_{i=1}^{n} x_{k-i} \qquad s = \sqrt{\sum_{i-1}^{n}\frac{\left(x_{k-i} - \bar{x}_k\right)^2}{n\quad 1}} \tag{8}$$

and used in the test:

$$\text{If}\left(\left|x_k - \bar{x}_k\right| > \gamma.s,\ \gamma > 0\right) \Rightarrow x_k \text{ is an outlier} \tag{9}$$

If the condition indicates that x_k is a bad data point, then it is approximated by:

$$\hat{x}_k = \bar{x}_k + sgn\left(x_k - \bar{x}_k\right).\gamma.s \quad (10)$$

Equation 10 is similar to applying a statistical process control chart to monitor the behavior of the signal. The test essentially is an expression of the statistical confidence of whether the latest data point has the same properties as previous data points. Thus, the larger γ is, the less stringent is the test.

Dealing with outliers when the trend is not apparent. When an indistinct trend has been indicated by Test B, the following test is made to assess whether x_k is an outlier:

$$\text{If}\left(\left|x_k - x_k^m\right| > \delta.\bar{\Delta}_k^m,\ \delta > 0\right) \Rightarrow x_k \text{ is an outlier} \quad (11)$$

where x_k^m is the median of the signal sequence $\{x_{k-1}, i = 1,2,\ldots n\}$ and $\bar{\Delta}_k^m = \frac{1}{n}\sum_{j=1}^{n}\left|x_{k-j} - x_k^m\right|$ is the mean absolute deviation of x_k about the median. If x_k is detected to be an outlier, then it is approximated by:

$$\hat{x}_k = x_k^m + sgn\,(x_k - x_k^m).\delta.\bar{x}_k^m \quad (12)$$

Even when a trend is detected using Test B, we still are unsure of the general behavior of the signal. Thus, the test for an outlier is made by comparing the current data point with the median of the n previous points. In statistical analysis, the median often is used in place of the mean as a measure of "central tendency" when there are uncertainties associated with the data. Equation 11 therefore may be loosely regarded as the robust version of Eq. 9.

Equations 7, 9, and 12 are merely initial corrective actions taken to deal with the respective classes of faults in the data point. The procedure should not be terminated after these corrections, because there is still the possibility that a wrong decision has been made. To accommodate this possibility if any of the corrective actions has been invoked, the data sequence

Table 1. Smoothing algorithm.

Initialization:

$$x_{k-n}^{(1)} = z_{k-n+1} \quad x_{k-n}^{(2)} = z_{k-n+1}$$
$$x_{k-n}^{(3)} = z_{k-n+1} \quad x_{k-n}^{(4)} = z_{k-n+1}$$

Procedure: $(0 \le \alpha \le 1)$

for $j = k - n + 1$ to k

$$x_j^{(1)} = (1-\alpha)x_{j-1}^{(1)} + \alpha z_j$$
$$x_j^{(2)} = (1-\alpha)x_{j-1}^{(2)} + \alpha x_j^{(1)}$$
$$\bar{x}_j^{(1)} = 2x_{j-1}^{(1)} - x_j^{(2)}$$
$$x_j^{(3)} = (1-\alpha)x_{j-1}^{(3)} + \alpha\bar{x}_j^{(1)}$$
$$x_j^{(4)} = (1-\alpha)x_{j-1}^{(4)} + \alpha x_j^{(4)}$$
$$\bar{x}_j^{(3)} = 2x_{j-1}^{(3)} - x_j^{(4)}$$

next j

$$x_k^f = \bar{x}_k^{(3)}$$

$\{z_k|\hat{x}_k, x_{k-i}, i = 1, 2,\ldots, n-1\}$ is formed. Notice that the current (outlying) data point x_k has been replaced by its first approximation, $\hat{x}_k$. This then is processed by the smoothing algorithm given in Table 1.

The table presents the smoothing algorithm in a form that clarifies its mechanism. The procedure, however, can be condensed and made more efficient. The following three operations are particularly significant:

$$x_j^{(1)} = (1-\alpha)x_{j-1}^{(1)} + \alpha z_j \quad (13a)$$

$$x_j^{(2)} = (1-\alpha)x_{j-1}^{(2)} + \alpha x_j^{(1)} \quad (13b)$$

$$\bar{x}_j^{(1)} = 2x_{j-1}^{(1)} + x_j^{(2)} \quad (13c)$$

Equation 13a is the first-order low-pass filter already presented as Eq. 1. The output of this filter $x_j^{(1)}$ is processed again by another first-order low-pass filter, Eq. 13b, to give $x_j^{(2)}$. Because both filters have the same filter constant α, the lag l, between z_j and $x_j^{(1)}$, and $x_j^{(1)}$ and $x_j^{(2)}$ should be identical. In other words,

$$l_j = x_j^{(1)} - z_j \tag{14a}$$

$$l_j = x_j^{(2)} - x_j^{(1)} \tag{14b}$$

Subtracting Eq. 14b from Eq. 14a, and rearranging yields:

$$z_j = 2x_j^{(1)} - x_j^{(2)} \tag{15}$$

The reconstruction of z_j is exact only if the data sequence describes a perfect line *(11)*. Nevertheless, Eq. 15 can be used to provide an estimate $\bar{x}_j^{(1)}$ of z_j as in Eq. 13c. To provide an improved estimate of x_k when it is deemed to be an outlier, this operation is performed twice in succession on the data sequence to finally give the improved estimate of x_k as x_k^f.

Equations 13a–c describe a double exponential smoothing algorithm that suffers less from phase lag effects than Eq. 1. By including information about the trend, Brown's linear exponential smoothing algorithm results *(12)*. This also is used as the final stage of the validation and reconstruction procedure, to further improve noise rejection. It is applied to the reconstructed data point if an outlier has been detected, or the raw data point otherwise. The result then can serve for process monitoring, control, or identification. The data reconciliation procedure is illustrated in the simplified flow chart shown in Figure 1.

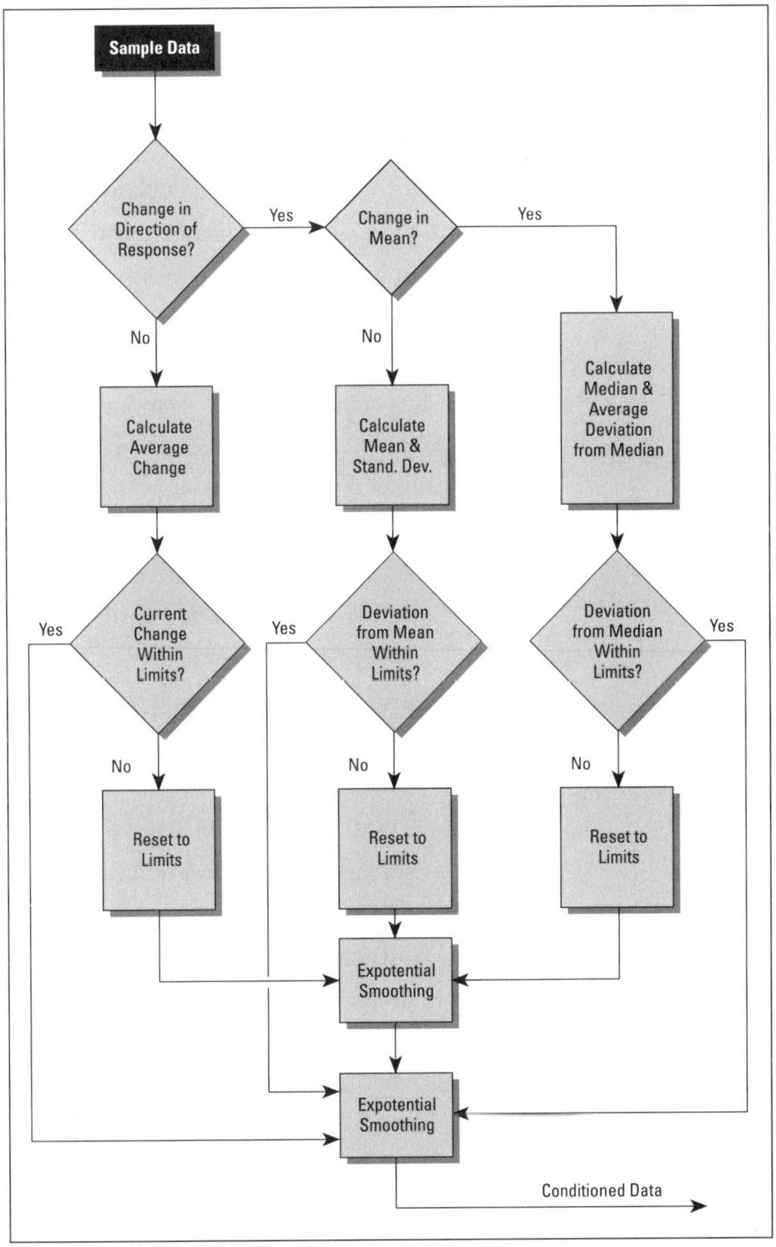

Figure 1. Simplified flowchart of data validation and reconstruction technique.

The value of the technique

It was not possible to evaluate the data validation and reconstruction technique on real measurements, because the underlying signals generally are unknown. Nevertheless, it has been tested extensively via simulation studies. The following sections illustrate the performances that could be achieved under open-loop and closed-loop conditions.

Open-loop performance. Consider Figure 2, which illustrates a signal sequence corrupted by large and small magnitude random effects. This sequence has been designed to reflect both large and small transients, as well as differing rates of change in signal evolution. The gray lines are measurements of the true underlying signal, which is the solid black line. In applying the validation and reconstruction technique, a

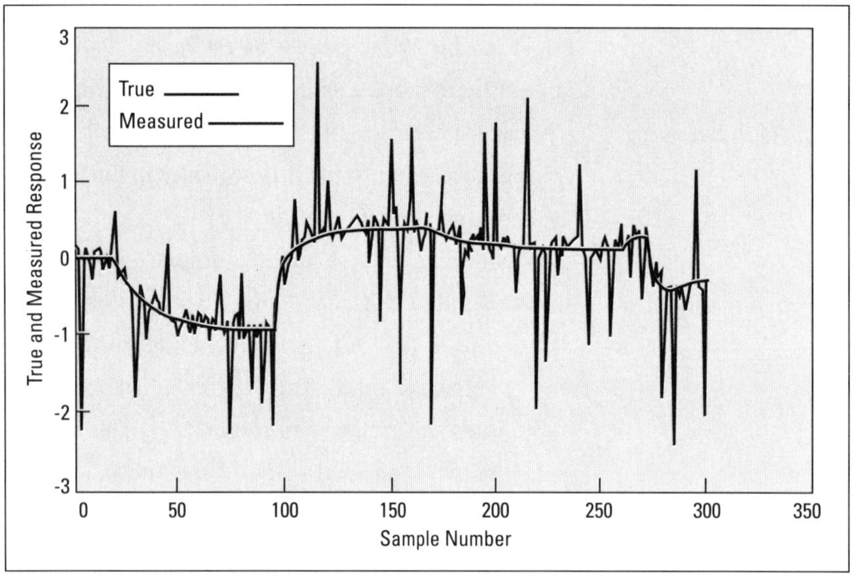

Figure 2. True underlying signal and its measurement.

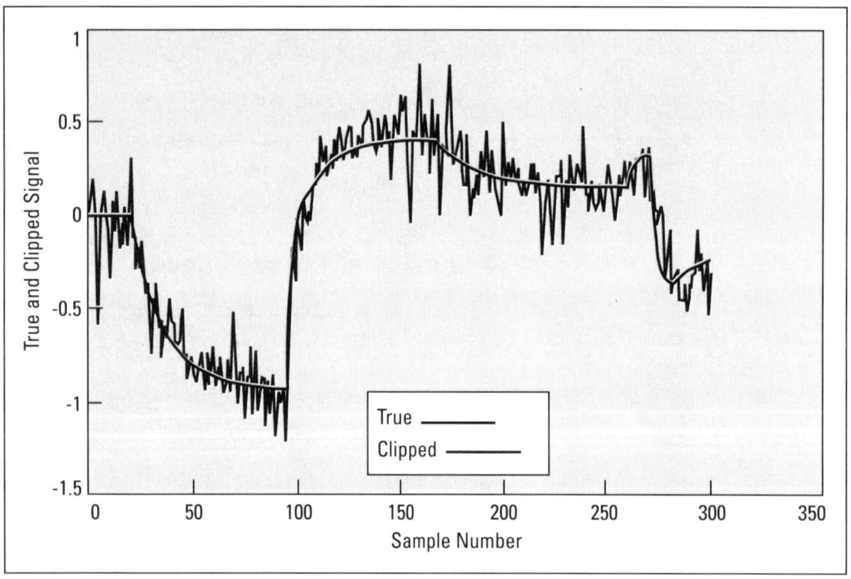

Figure 3. Effect of clipped measurement.

data window of the previous 10 samples was used to calculate the relevant statistics for feature extraction and validation checks.

Figure 3 shows the result of simply clipping points that were considered outliers by the algorithm — that is, applying Eqs. 7, 9, or 12. Notice that this initial stage of the reconstruction has resulted in significant signal recovery. This is enhanced further by the smoothing operations, as shown in Figure 4. These two figures show that the first stage of the reconstruction exercise removes large errors in the signal, while the second, smoothing stage (with $\alpha = 0.1$) attenuates the smaller magnitude high-frequency components of the measurement. The penalty of incorporating smoothing in the procedure is the introduction of a lag in the processed signal.

As shown in Figure 5, however, this lag is insignificant compared to the one incurred through the use of a first-order low-pass filter (Eq. 1) with filter constant $\alpha = 0.9$. Note that this value of the filter constant was fine-tuned to yield the smallest sum of squared error (SSE) between the filtered measurement and the true signal.

We can reduce the lag in the processed measurement by using a lead-lag filter, such as that described by Eqs. 13a–c. The result is shown in Figure 6. Although the lag between the filtered measurement and true signal is reduced, huge deviations from the true response still are very much in evidence. This is because the lead terms in the lead-lag filter tend to amplify spikes in the measurement, illustrating the compromise that has to be made in all signal processing applications.

The results also emphasize that signal processing techniques must be able to deal with large and small disturbances in an independent manner. In other words, candidate procedures should have the necessary degrees of

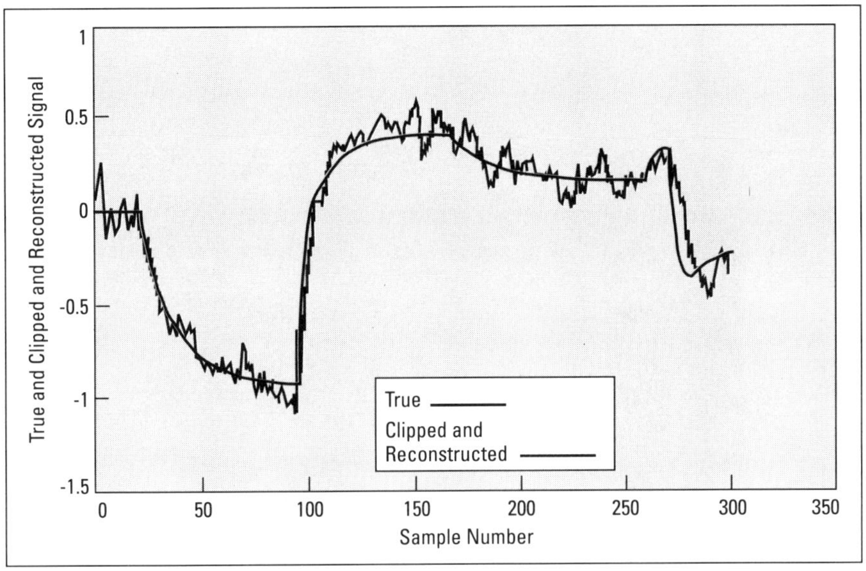

Figure 4. Use of clipped measurement and reconstruction.

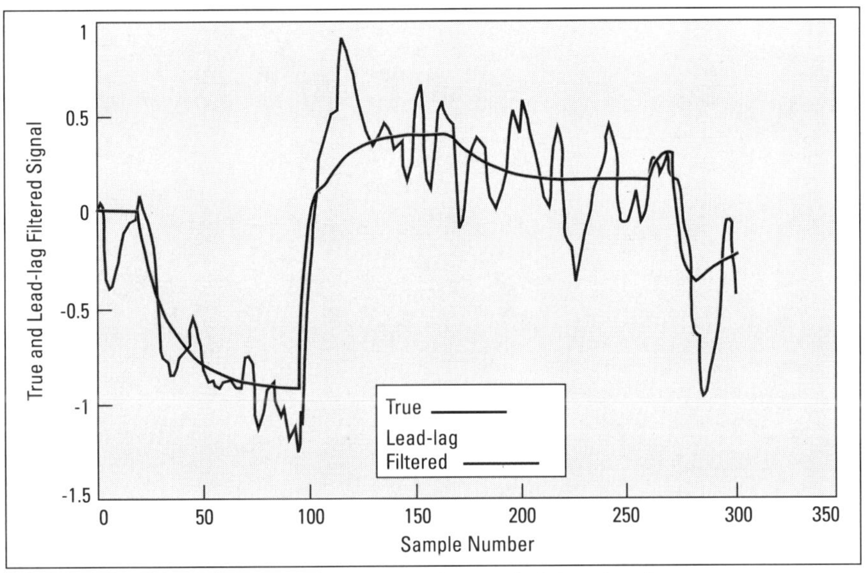

Figure 5. Lead-lag filtered measurement vs. true underlying signal.

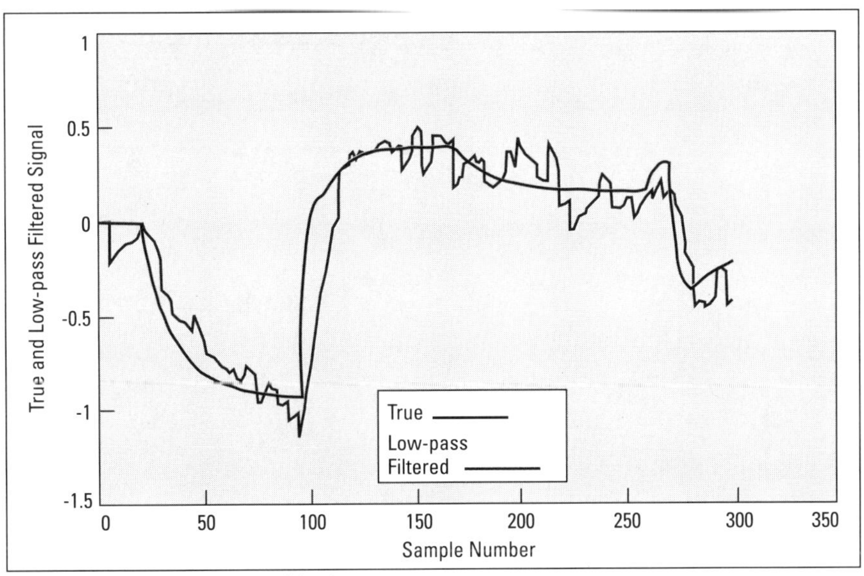

Figure 6. Effect of low-pass filtering.

freedom to deal with different disturbances that affect the measured response. For example, the second smoothing stage of the proposed methodology, used to deal with smaller disturbances, was found to be relatively insensitive to the choice of smoothing constant. Values of α between 0.1 and 0.4 resulted in similar performances. In contrast, changing the low-pass filter constant to 0.8 from 0.9 led to severe degradation in signal recovery. Figure 7 illustrates the information loss, as quantified by the SSE, that occurs using the described signal-processing techniques. This bar chart clearly shows that application of the validation and reconstruction procedure resulted in the least information loss; it is approximately three times better at signal recovery than the commonly used low-pass filter.

Closed-loop performances. The differences become more distinct when the two techniques are used with automatic controllers. Consider the block diagram shown in Figure 8; the process to be controlled is a simple first-order system with unit gain, a time constant of 10 time units, and a time delay of 2 time units. A proportional-plus-integral (PI) controller was used to track changes in set point and to reject a slowly increasing feed-forward load disturbance. The load disturbance was generated by passing a noisy sinusoid through a first-order process with unit gain, and time constant of 20 time units, as shown in the block diagram. In common with the previous examples, the output of this process also was contaminated by both small and large magnitude random disturbances, ξ and η, respectively.

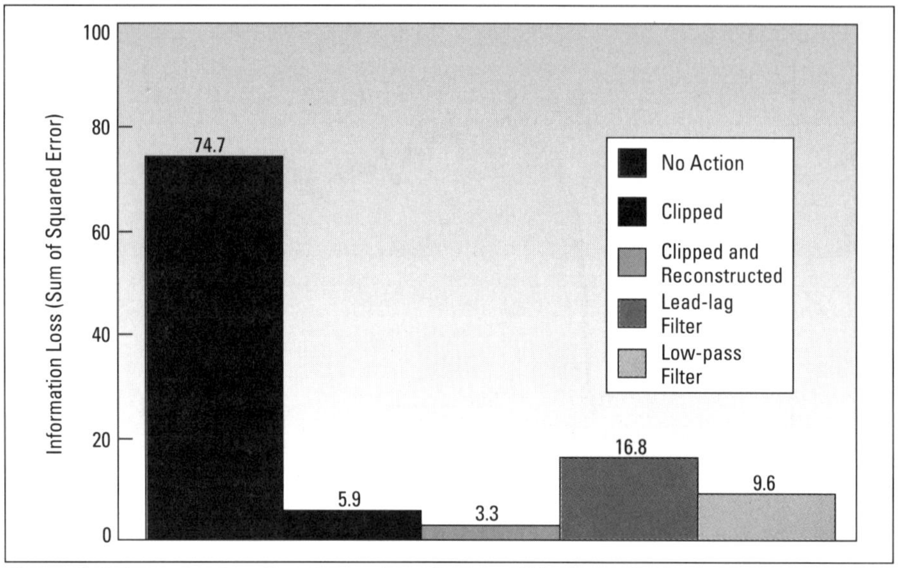

Figure 7. Information loss by using various data-conditioning techniques.

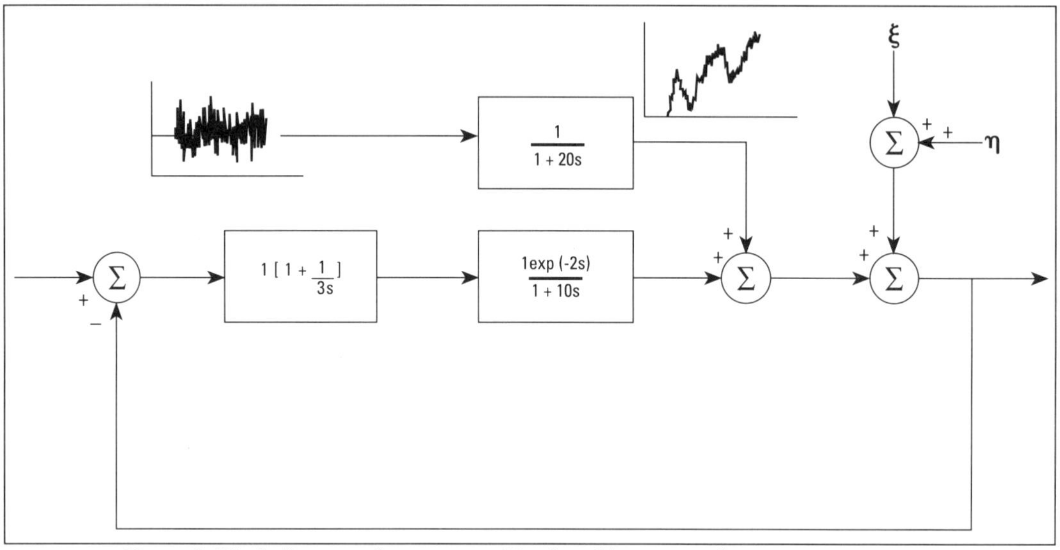

Figure 8. Block diagram of systems used in closed-loop examples.

The controlled responses and the evolution of the manipulated input, without feedback signal conditioning, are shown in Figures 9 and 10. In Figure 9, the black line is the underlying controlled signal, the lighter line is the measurement, and the dotted line is the set point (a convention also used in the subsequent figures). When the feedback signal was processed by a low-pass filter, the corresponding controlled output and manipulated input are shown in Figures 11 and 12. Comparing Figure 9 with Figure 11, we can see that use of the filter generally has improved the control of true process behavior. The inclusion of the filter

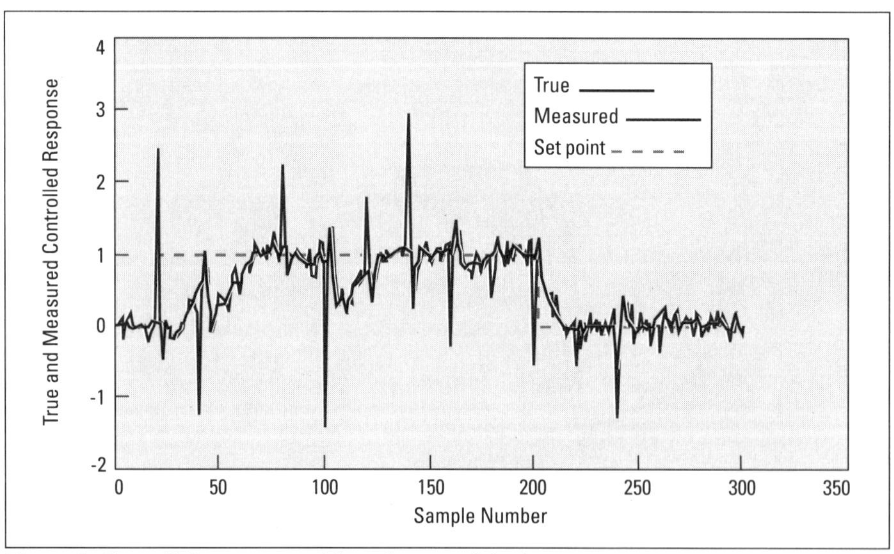

Figure 9. Controlled responses without data conditioning.

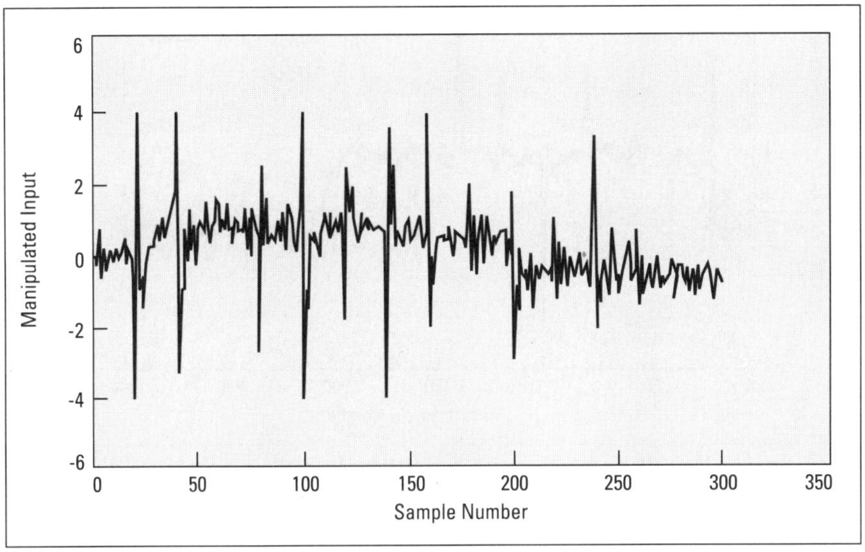

Figure 10. Manipulated input without data conditioning.

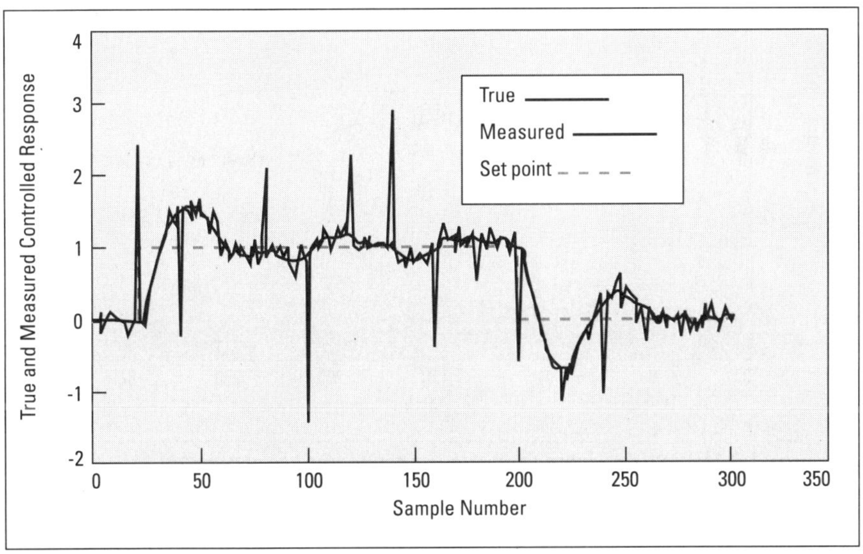

Figure 11. Controlled responses when low-pass filtering is used.

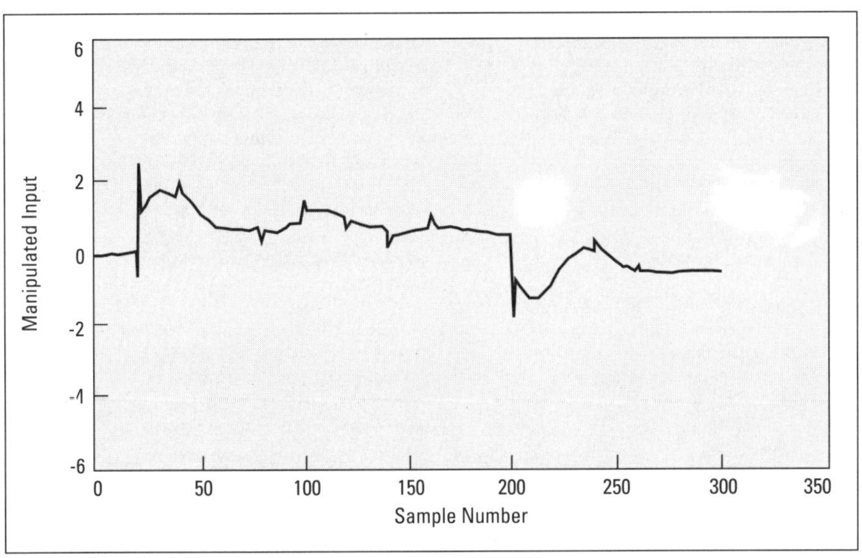

Figure 12. Manipulated input when low-pass filtering is used.

in the loop, however, increased the phase shift of the system, resulting in overshoots when the set point was changed. Nevertheless, an added benefit was that the manipulated input is smoother. This translates to less control activity — hence reducing wear and tear of the final control element.

The responses obtained when the low-pass filter was replaced by the data validation and reconstruction technique are shown in Figures 13 and 14. The enhancement in control performances over the use of the low-pass filter is immediately apparent. There was no overshoot to the first set point demand. On the second set point change, the overshoot was relatively small. The manipulated input was noisier, however, but this may be regarded as insignificant compared to the nature of the measurements. Note that the PI parameters were kept the same in the above comparisons. While fine-tuning may improve the performances of the various cases, the closed-loop study illustrates the power of the signal conditioning technique.

The payoff

System integrity can be significantly improved by avoiding problems caused by corrupted measurements. In designing procedures to recover the underlying signal from measurements contaminated by noise, however, it is important to differentiate between the effects of large and small magnitude disturbances. A two-stage technique has shown significant performance improvements over the commonly used first-order low-pass filter in simulated open- and closed-loop situations. The approach is not process specific, and is conceptually and computationally simple. It also should be easy to implement on most modern process computers, because they support the necessary mathematical operations.

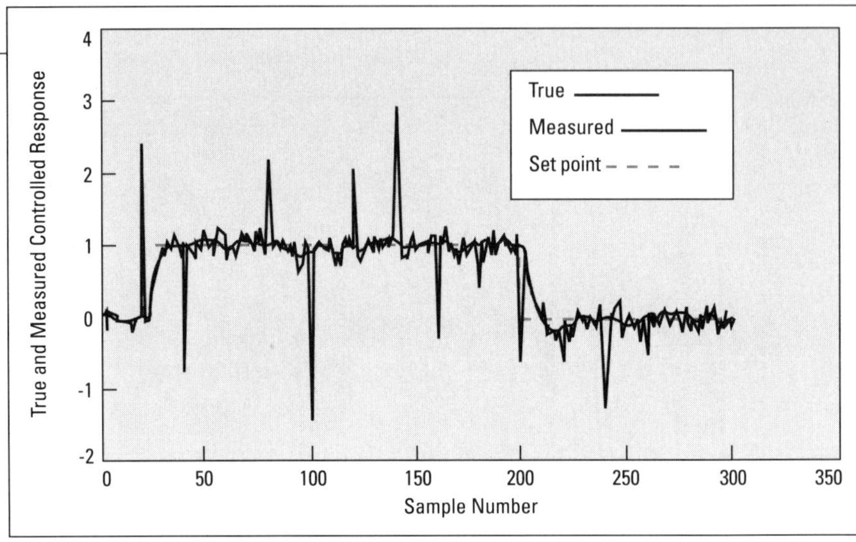

Figure 13. Controlled responses with data validation and reconstruction.

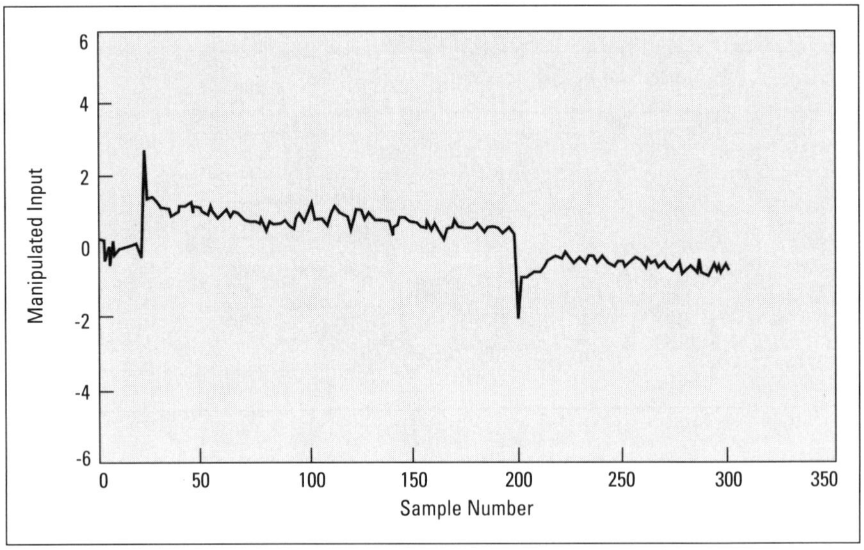

Figure 14. Manipulated input with data validation and reconstruction.

Acknowledgments

The authors acknowledge the support of the Department of Chemical and Process Engineering, University of Newcastle upon Tyne in carrying out this work. The encouragement of our industrial colleagues also is gratefully appreciated.

Literature Cited

1. **Cadzow, J. A.,** "Foundations of Digital Signal Processing and Data Analysis," Macmillan, New York (1987).
2. **Lynn, P. A.,** "An Introduction to the Analysis and Processing of Signals," 2nd ed., Macmillan, New York (1985).
3. **Mendel, J. M.,** "Lessons in Digital Estimation Theory," Prentice Hall, Englewood Cliffs, NJ (1987).
4. **Oppenheim, A. V., A. S. Willsky, and I. T. Young,** "Signals and Systems," Prentice Hall, Englewood Cliffs, NJ (1983).
5. **Fertik, H. A.,** "Tuning Controllers for Noisy Processes," *ISA Trans.*, 14 (4), p. 292 (1975).
6. **Young, P. C.,** "Recursive Estimation and Time Series Analysis," Springer-Verlag, New York (1984).
7. **Abraham, B., and G. E. P. Box,** "Bayesian Analysis of Some Outlier Problems in Time Series," *Biometrika*, 66, p. 229 (1979).
8. **Chernik, M. R., D. J. Downing, and D. H. Pike,** "Detecting Outliers in Time Series Data," *J. Amer. Stat. Assn.*, 77, (380), p. 743 (1982).
9. **Devlin, S. J., R. Gnanadesikan, and J. R. Kettenring,** "Robust Estimation and Outlier Detection with Correlation Coefficients," *Biometrika*, 62 (3), p. 531 (1975).
10. **Chatfield, C.,** "Statistics for Technology," 3rd ed., Chapman and Hall, London (1983).
11. **Gilchrist, W.,** "Statistical Forecasting," John Wiley, New York (1976).
12. **Gardner, E. S. Jr.,** "Exponential Smoothing: The State of the Art," *J. of Forecasting*, 4 (1), p. 1 (1985).

M. T. THAM is a Lecturer at the University of Newcastle upon Tyne, U.K. (0114491/222–6000; Fax: 0114491/261–1182). He is closely involved in the development of model-based estimation and control techniques for application to industrial processes — and has published more than 100 technical papers in these areas. His current interests include nonlinear process control, inferential estimation and control, object-oriented dynamic systems modeling, and statistical process control. He earned a BS (Honors) degree in chemical engineering, and a PhD in process control, both from the University of Newcastle upon Tyne. A Certified Engineer, he is a member of the Institution of Chemical Engineers.

A. PARR is a divisional process engineer at the Humberside Refinery of Conoco Ltd. in Grimsby, U. K. (0114446/957-1571; Fax: 0114446/955-5674). He received an ME (Honors) in chemical and process engineering from the University of Newcastle upon Tyne and is a graduate member of the Institution of Chemical Engineers.

Select the Right pH Measurement System

To achieve desired accuracy, reliability, and economy, the sensor and analyzer components in such a system must be carefully chosen for the specific application.

James W. Hayes,
Consultant

While widely used throughout the chemical process industries (CPI), pH measurement systems frequently are not fully understood. So, unfortunately, achieving accurate and reliable measurements remains a common problem *(1, 2)*. This article reviews factors to consider when selecting a pH system, including the impact of several key variables on measurement accuracy and sensor performance, and the availability of various component options that can help ensure an effective pH-measurement system.

Most chemical engineers know that pH measures the level of acidity, neutrality, or alkalinity of a solution. Defined as the negative logarithm of the hydrogen ion (H^+) concentration in a solution, pH is a dimensionless number:

$$pH = -\log(H^+)$$

Hydrogen-ion-concentration levels in solutions can vary over an extremely broad range from 1×10^0 to 1×10^{-14} moles per liter. The 0–14 pH scale provides a convenient way to express hydrogen ion concentrations of any magnitude. Because pH is a logarithmic function, an increase or decrease in pH value by only 1.0 unit represents a tenfold shift in concentration of H^+ ions.

To put this extremely broad range into perspective — and thus highlight the unique demands on instrument sensitivity and accuracy — let's compare the job of a pH measurement system with that of an atomic clock. An atomic clock retains accuracy to within 1 second over a span of 100,000 years. If the clock had the range of a pH system, however, it would need to cover a period of 3,168,800 years to be equivalent.

Table 1. Typical solution pH values.

pH	Solution
0	Hydrochloric acid (37%)
1	Hydrochloric acid (1.4%)
1–3	Human gastric system
2	Lemon juice
3	Beer
5	Cheese
7	Pure water
7.3–7.5	Human blood
8.4	Sodium bicarbonate
10	Milk of magnesia
11	Ammonia (1.7%)
12	Trisodium phosphate
14	Caustic soda (4%)

Considered neutral, pure water has a 7.0 pH value, reflecting a concentration of 1×10^{-7} moles of hydrogen ions in one liter of solution. Acidic solutions have higher concentrations of hydrogen ions, thus lower pH values. Solutions having pH values greater than 7 have hydrogen ion concentrations less than water, and are referred to as bases or caustic solutions. Typical pH values of some common solutions are listed in Table 1.

The basics of selection

The first consideration must be the measurement sensor itself. Fundamentally, the sensor's accuracy determines the overall accuracy and reliability of the control system. After all, the best analyzers or computer

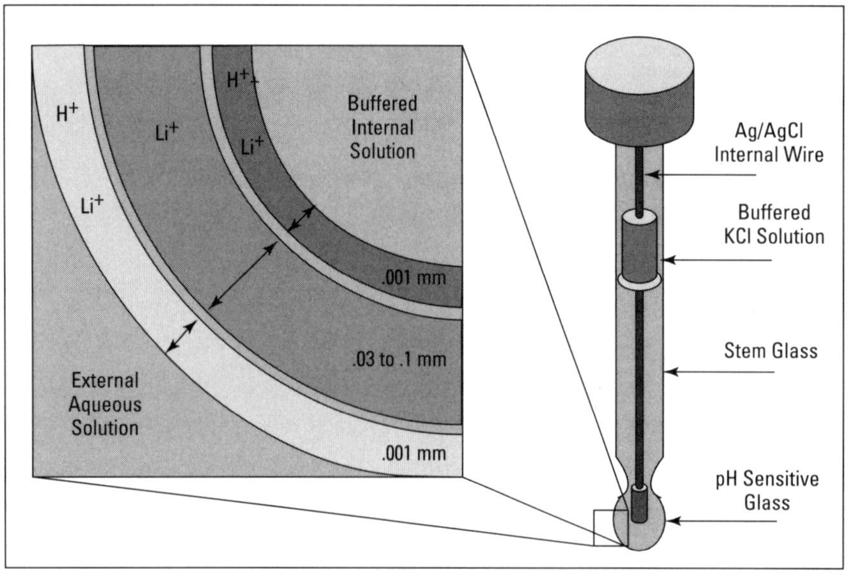

■ ***Figure 1. Details of a glass pH electrode.***

system fully equipped with predictive diagnostics and multiple outputs cannot correct for faulty measurement.

The sensor selected should provide optimum measurement accuracy, response time, and the best possible durability for the application. Sensor durability and reconditionability can have a major impact on overall system operating and maintenance costs. Process variables affecting selection of the sensor include:

- temperature;
- pressure;
- flow rate;
- viscosity;
- solids content;
- chemical composition; and
- chemical compatibility of the process and wetted sensor parts.

The role of the pH-sensor output signal must be the next consideration. Will the signal be used for local indication, by an analyzer that will handle control, or by a control computer? Must the system meet any safety certification requirements? Analyzer selection should be based on compatibility with the sensor output signal, the distance between the sensor and the analyzer, whether readout of pH values only or readout and control is needed, as well the degree of system self-diagnostics desired.

Then, the interconnecting cable and mounting hardware must be selected to suit the sensor-to-analyzer distance, and environmental requirements. The interconnect cable must transmit the low-level output signal from the sensor to the analyzer with the least-possible signal degradation due to impedance or externally caused electrical noise. Depending on the installed distance between the sensor and analyzer, it also may be necessary to provide signal pre-amplification close to the sensor. In some situations, the mounting or sampling systems become major operating and maintenance issues that also must be considered with the system.

SENSORS

A pH sensor consists of a galvanic cell with two electrodes — a special pH glass electrode that is sensitive to hydrogen ion activity, and a reference electrode that provides the comparative base for the voltage measurement. The reference electrolyte is a 7.0 pH solution.

With both electrodes immersed, an electrical circuit is completed through the electrolyte solution and reference junction. This produces a dc milliVolt (mV) potential across the hydrated glass membrane of the sensor. The magnitude of this potential is a function of the hydrogen ion activity in the solution.

Glass pH measuring electrode

Special, pH-sensitive glass electrodes have been commonly used for pH measurement since the 1930s. Such an electrode consists of a metallic wire conductor suspended inside a specially constructed glass tube that is filled with an electrolyte solution. A typical configuration (see Figure 1) uses a silver wire coated with silver chloride immersed in a buffered potassium-chloride solution.

The sensing area in the tip is most often sphere-shaped to provide the largest surface area for liquid contact and rapid response. This sensor tip is actually a membrane of specially-formulated, pH sensitive materials such as oxides of silica, lithium, calcium, and other elements. Water penetrates the glass, which seems to facilitate the movement of ions, forming a hydrated layer on both the inner and outer glass surfaces of the tip. The milliVolt potential, produced by the ion activity of the solution, is created across the interface between the pH glass and the external aqueous solution.

Reference electrode

The reference electrode is constructed of a nonconducting glass envelope, typically containing a silver/silver-chloride reference element. The glass envelope is filled with potassium chloride electrolyte or other internal buffer fluid.

Electrical continuity from the reference electrolyte to the process solution is through a porous junction, or frit, located at or near the bottom of the electrode. Junctions are made from a variety of compatible materials including ceramic, wood, conductive polymer, and sintered glass. Obviously, the frit material must be nonreactive to the process.

A design tradeoff exists between the porosity and surface area of the junction. On one hand, the junction structure should be sufficiently open to en-

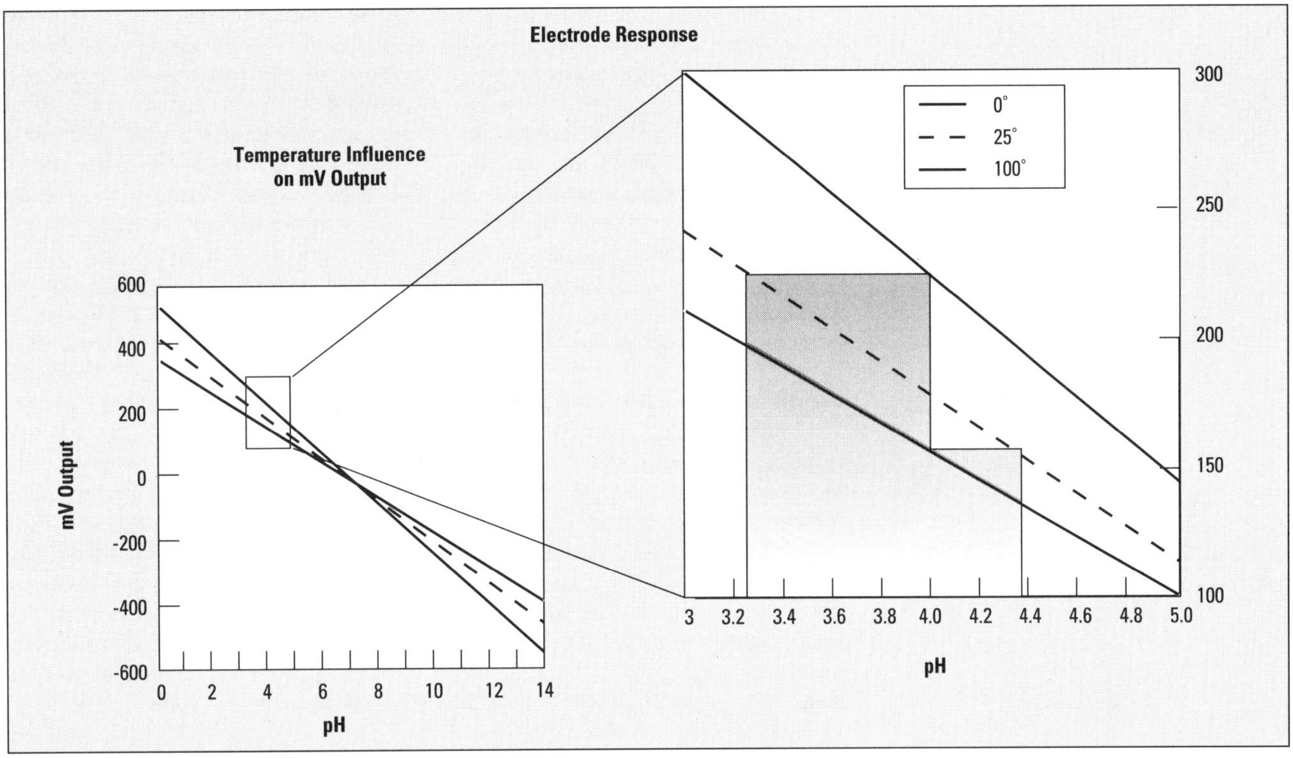

■ *Figure 2. Temperature effect on pH-electrode error.*

hance electron migration from the process solution through the junction. Increased junction porosity, however, also provides less resistance for process solution flow into the reference electrode, resulting in contamination of its internal electrolyte. Contaminated reference solution causes dramatic changes in indicated pH. Reference contamination is an age-old issue for long service life of pH probes. For sanitary applications, junction porosity must be limited to reduce the possibility of microbial contamination.

Outward leakage of internal electrolyte also is promoted with a more porous junction. The junction can foul if the electrolyte reacts with the process solution to form precipitates on its outer surface. In applications where the reference electrode experiences wide temperature variations, the reference solution actually may be expelled due to pumping action caused by process heating and cooling.

To combat the tendency for contamination, nonflowing, double-junction reference electrodes are available that provide an electrically conducting chamber between the reference electrode and the process solution being measured. The outer chamber can be filled with either the same solution as the reference electrolyte, or a different solution that is more compatible with the process. This design approach reduces electrolyte contamination, but also may increase the potential across the junction and decrease the speed of response for pH measurement.

Another approach for reducing the rate of contamination is to use a viscous gel or paste electrolyte in the reference electrode. Gel fills provide a longer life, but have a less stable junction potential than fluid junctions.

There are some process situations where any type of junction may become fouled and require frequent cleaning or replacement. In such cases, a flowing junction may be used: a reservoir of electrolyte is connected by tubing to the reference electrode and a pressure, slightly above process pressure, is applied to cause outward flow of the electrolyte solution through the junction. This flushes the junction and maintains its measurement accuracy.

An alternative that is frequently most economical is a reconditionable probe. Such a probe allows the contaminated reference solution to be flushed and replaced. At the same time, the reference junction is in a housing that also can be replaced for minimal cost. The cost of these materials is small compared to the cost of a disposable probe; so, a reconditionable probe may provide a far lower cost of ownership over an extended period. Plus, use of a reconditionable probe enables the selection of various, alternative reference solutions in highly specialized applications, such as for high operating temperatures or nonaqueous solutions.

The effect of temperature

In measuring pH, temperature is a factor. pH electrodes respond electrically in accordance with the Nernst equation:

$$E = E_0 + k\,T \log [H^+]$$

where E = total measured electrical potential; E_0 = electrode base potential; k

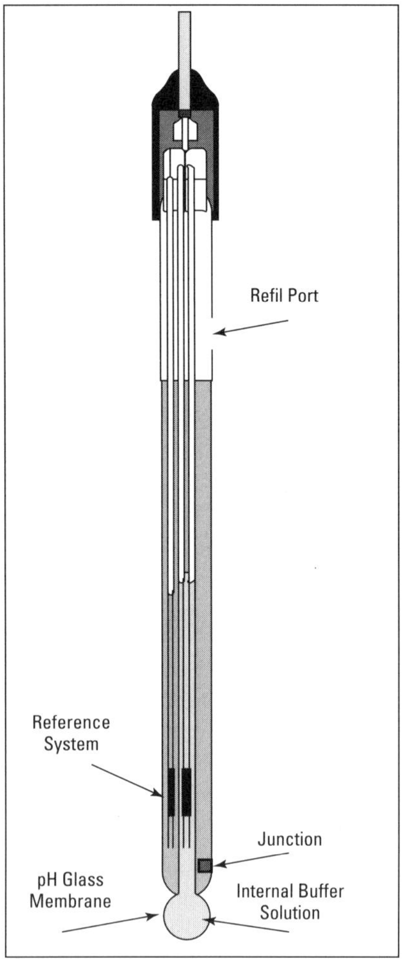

■ *Figure 3. Combination electrode packages both electrodes together.*

= a constant; T = absolute temperature of process solution; and H^+ = hydrogen ion activity. As this equation indicates, the measured electrical potential is a product of both the process solution temperature and hydrogen ion concentration. For a given H^+ concentration (solution pH), the measured potential will increase as the temperature rises, and decrease as the temperature falls. Thus, simple cooling of a sample can cause significant variations in the measured pH value. This relationship is linear and, therefore, can be automatically compensated for by the pH analyzer.

Note, though, that pH temperature compensation is a correction only for the electrode and does not relate to, or change, any temperature influence on the chemistry of the process solution being measured. In process troubleshooting, chemical changes frequently are misinterpreted as measurement errors.

The mV vs. pH vs. temperature graph (Figure 2) shows that the three temperature slopes intersect at an isopotential point of 0 mV and 25°C. The temperature influence on electrode potential increases dramatically when working at pH levels significantly away from the isopotential point.

Combination electrodes

For ease of use, the individual measuring and reference electrodes can be joined in a single, compact probe, as shown in Figure 3. These commonly used combination electrodes include an outer glass reference tube that concentrically surrounds the measuring electrode, except for its pH-sensitive glass membrane. The space between the measuring electrode and the outer glass tube contains the reference system and reference electrolyte. The reference junction is located near the bottom of the outer glass tube, and is constructed similarly to junctions used on individual reference electrodes.

Most manufacturers offer an optional, integral temperature compensator in their combination electrodes. Depending on the type of temperature device, it may be more economical to mount it separately. An important point about separate mounting is that calibration also must be temperature compensated. Several systems permit manual entry of a temperature value into the analyzer to overcome this limitation.

One note of caution relates to special application electrodes. The typical pH electrode has an ideal isopotential (zero) point at 7.0 pH. For example, an electrode with a zero point at 7 pH immersed in 7 buffer produces a 0 mV signal at 25°C. Special purpose electrodes are available with the zero point at other pH levels. Examples include the antimony electrode with its zero point at 4.6 pH, and Pfaudler's Glasteel electrode which has its zero point between 1 and 3 pH. These electrodes require the use of a pH analyzer configurable to their zero point.

Measurement-electrode end shapes

The nature of the solution to be measured affects the selection of the shape of the electrode. As shown in Figure 4, five typical end shapes are available for glass pH measurement

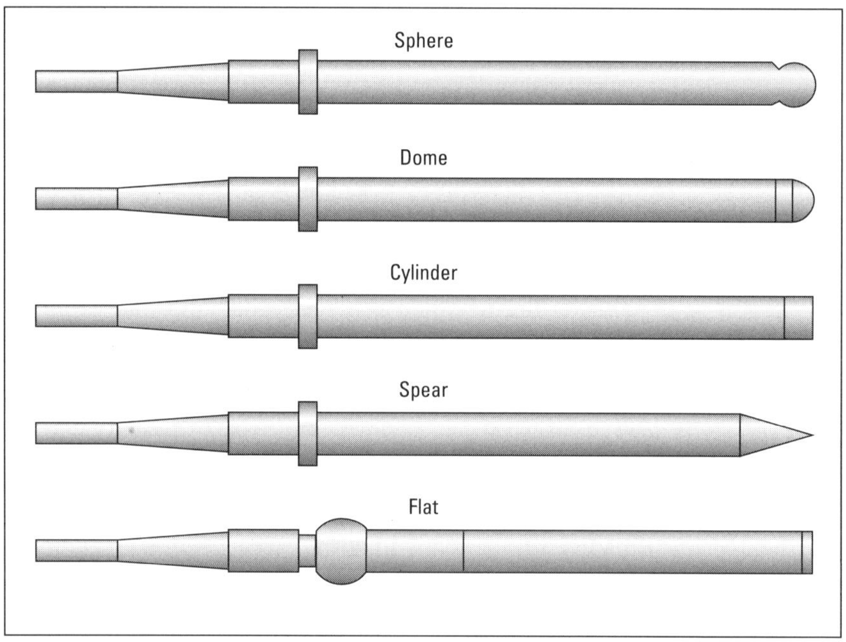

■ *Figure 4. Electrodes are available in a variety end shapes.*

electrodes: spherical, dome, cylindrical, spear, and flat. The sphere's large surface area and uniform wall thickness generally make it preferred in most applications. The large surface area reduces the possibility of coating or blinding, and minimizes response time. The sphere can utilize the wide variety of special glass types available.

In solutions with a high solids content or fibers, the dome shape is advisable. The dome is easier to clean and, therefore, also is suited to stations with automatic cleaning devices. This mechanically more durable electrode may be used in abrasive systems. The major limitation to domed probes is the reduced measuring area and slower response to process changes. Lower resistance glasses usually are used to maintain responsiveness — but this usually sacrifices chemical resistance and durability.

Seldom seen in process applications other than fermenters, the cylindrical electrode has medium electrical resistance, and thicker glass with slower response time.

The spear-shaped electrode has high electrical resistance and is used for stabbing into semi-solid media. This very thick, durable glass electrode shape is mainly used in the food industry.

Flat electrodes have enjoyed various surges of popularity. Their flat surface parallels flow, allowing the moving stream to effectively wipe the measuring surface. This comes at the expense of reduced measuring area, generally thicker glass, and limited glass-formulation alternatives. Also, flat electrodes have more stress at the edges and, usually, slower response time due to the additional resistance of the thicker glass and smaller area.

Several manufacturers promote "ruggedized" glass measuring electrodes. These electrodes are made somewhat more durable by using thicker glass. This thick glass, however, is significantly less responsive to pH change.

Be wary of glass electrodes said to be suitable for pH from 0 to 14. For reasonable performance, a range of 2 to 12 is more realistic. When typical conditions tend to either end of the range, glass selection becomes more of an issue. In concentrated acids, the glass may take on a charge that affects responsiveness.

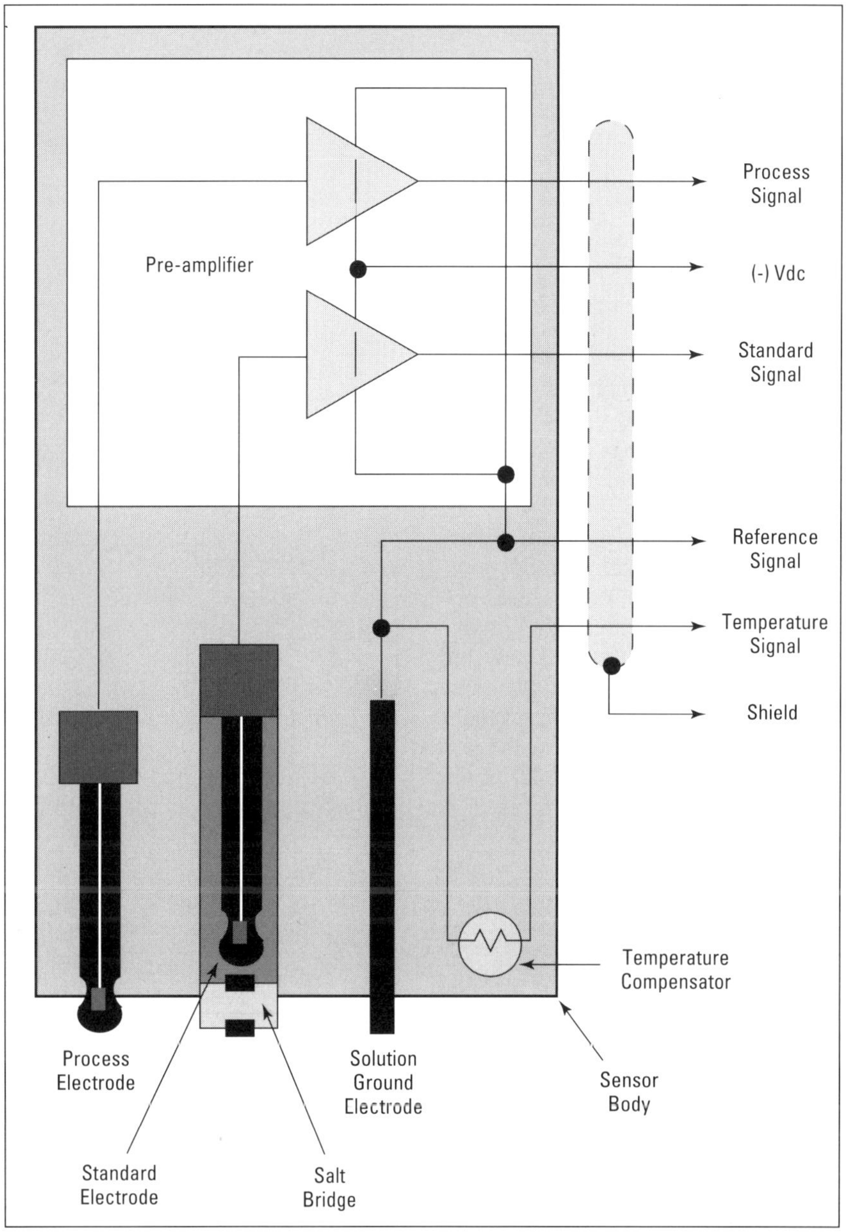

■ *Figure 5. Differential probe.*

Other options

The differential probe combines two glass pH electrodes, a solution ground electrode, a temperature compensator, and a preamplifier into a single assembly (see Figure 5). One glass electrode is mounted to be active with the tip exposed to the process fluid. The second electrode functions as the reference, isolated behind a replaceable double junction salt bridge and surrounded with buffered reference solution. This differential sensor provides extended probe operating life. The replaceable double junction slows contamination of the reference solution, which is buffered to reduce contamination effects, and can be easily reconditioned. The structure of the probe body also serves to protect the delicate measurement components in rugged process applications.

The differential technique compares the active and reference measurements to the solution ground rod, eliminating several operating problems such as ground loops. Additionally, the integral preamplifier reduces signal noise in the line. The tradeoffs are a larger mounting connection requirement and substantially greater initial investment. These systems currently are available from only a limited number of suppliers.

The Ion Selective Field Effect Transistor (ISFET) pH electrode uses a special transistor rather than pH glass to measure hydrogen ion activity. Theoretically, response time is faster because hydrogen ion activity is measured by field effect. Unfortunately, ISFET probes still incorporate the traditional reference system with its inherent weaknesses. Other drawbacks for this type of electrode include very small measuring area, delicate wiring connections, and problems with reference junctions. Also, it requires the use of a particular supplier's probes and analyzer, which are not readily compatible with any other supplier's components.

Iridium oxide sensors are intended for applications where glass is not compatible or suitable for the process. Iridium has very good repeatability and linearity, and can withstand temperatures as high as 200°C. Oxygen Reduction Potential (ORP) changes, however, cause problems for this type of electrode, because an ORP change is interpreted as a pH change. (For more on ORP, see Ref. 3.)

The impact of the stream

Obviously, when specifying pH measuring equipment, the wetted materials need to be compatible with the operating conditions. While an electrode may perform in severe applications, a mounting or probe body of epoxy or PVC may degrade quickly.

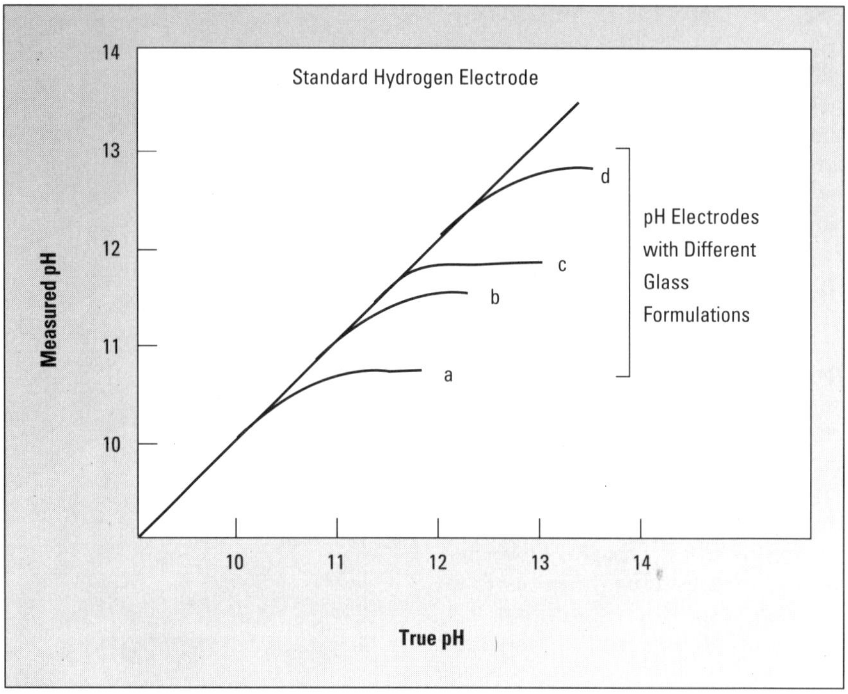

■ ***Figure 6. Sodium ions lead to errors in pH measurement.***

Other issues related to stream composition also are important *(4)*. Among the key factors to consider are:

Sodium ion error. The largest pH measurement errors occur at high pH levels, 12 and higher. Although the glass pH electrode responds very selectively to H^+ ions, in concentrated alkaline solutions there are fewer hydrogen ions and other cations like Li^+, Na^+, and K^+ may affect the ion exchange. This causes lower hydrogen ion activity and correspondingly lower pH indication than actual. The amount of interference decreases with increasing ion size. Lithium is not frequently found in solutions, and potassium causes very little interference due to its relatively large size. So, the most significant interference is from sodium ions. Most manufacturers have formulated pH glasses to reduce or minimize this sodium ion error.

Above 12 pH, sodium ion concentration often is high due to the presence of sodium hydroxide. Depending upon the exact glass formulation, sodium ion interference may take effect at a higher or lower pH value. Figure 6 shows how measured pH differs by glass formulation.

No pH glass formulation currently available has zero sodium ion error. Because some error will be present, it is important that the error be constant and reproducible. With many glass formulations, this is not possible because the electrode becomes sensitized to the environment it was in prior to the high pH. For example, the exact start point of an electrode's sodium ion error might be 11.5 pH after immersion in tap water, but 12.5 after being in an alkaline solution. Additional considerations include temperature and electrode age. Some sensors give a sodium ion error that is constant throughout their lifetime.

Silver concentration. Chemical reactions between the reference electrolyte and the process solution can form insoluble precipitates that can bridge the junction and increase electrical resistance. Hg^{++}, Cu^+, Ag^+, and Pb^{++} will form such precipitates by reacting with KCl electrolyte. This can be avoided by separating the reference electrolyte from the process solution with an intermediate electrolyte that does not react with these ions.

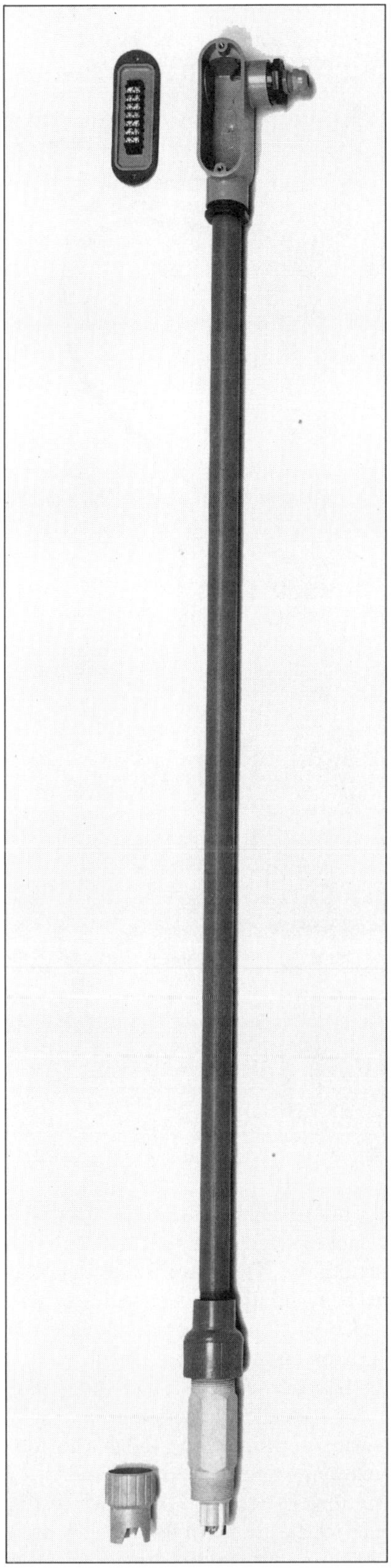

■ *Figure 7. Submersion sensor assembly.*

The formation of insoluble compounds on the Ag/AgCl reference electrode presents a more difficult problem due to the increasing solubility of silver chloride with increasing concentration of potassium chloride. Bromides, cyanides, iodides, and sulfides also react with silver chloride. Silver sulfides deposited on reference junctions have been shown to degrade electrode response time, greatly increase resistance, and cause erratic diffusion potentials that result in faulty pH measurements. Here, again, the remedy is to use an intermediate electrolyte containing no silver. This intermediate electrolyte periodically must be replaced due to diffusion of silver ions from the reference electrolyte into the intermediate solution.

This precipitate-formation problem can be reduced in those electrodes that have a replaceable salt-bridge junction. The replacement salt bridge must contain an electrolyte having no chloride content.

Organics. The presence of organic chemicals, such as tars, oils, and even alcohols, in the solution to be measured can cause rapid deterioration of pH measurement accuracy — due to formation of an organic coating on either, or both, the pH glass membrane of the measuring electrode and the junction of the reference electrode. The coating increases the impedance of the electrode, and blinds off that area of the membrane for hydrogen ion exchange. If left unattended, the galvanic cell formed by the pH measurement and reference electrodes will become an open circuit. Regular electrode cleaning may be required, and this may increase the frequency of electrode replacement due to wear, abrasion, or breakage of the sensitive ion-exchange surfaces.

Hydrofluoric acid. HF molecules severely attack all silicate glasses and dissolve them rapidly in acidic solutions. Therefore, glass pH electrodes have a short service life in such solutions. When a fluoride-containing solution is below 6 pH, an alternative to glass must be selected. The antimony electrode provides longer service life in these specific HF environments, but has relatively poor selectivity for hydrogen ions as well as a nonlinear response. Measurements by this method are not as reliable as those made with glass electrodes. Antimony electrodes should never be used in food processing applications due to the toxic nature of the metal. In these situations, iridium oxide probes can be considered. Glass electrodes always are preferable for use in fluoride-containing solutions in applications where pH levels remain above 6.0.

Solid particles or fibers. The presence of solid particles or fibers in the process solution can cause rapid deterioration of electrodes in a variety of ways. If the particles are hard and abrasive, impingement of the process stream on the sensor can scratch, wear, or break the pH glass membrane. This problem can be minimized by slowing the flow velocity in the location of the sensor. Very fine particles also may penetrate the porous surfaces of the reference junction and electrodes, eventually blinding them. Frequent washing of the pH membrane and reference will help to slow deterioration.

Stringy fibers in a flowing solution can plug sensors by wrapping around the electrode, or any protrusion into the liquid. Selection of dome- or flat-end electrodes without sleeves may reduce the potential for fiber entrapment and be mechanically more durable. The effective pH-measuring surface on these electrodes, however, is significantly less than on spherical-end shapes and, therefore, can be more quickly fouled if there is any tendency for electrode coating.

Viscosity. Highly viscous solutions can be difficult for accurate pH measurement if solution flow is low or the solution clings to the surface. Increasing the velocity past the sensor may help keep the surfaces clear. Excessive velocity, however, can result in entrapment of air, or cavitation at the downstream side of the electrodes, causing erratic sensing.

The effect of operating conditions

The pressure and flow rate of the process solution significantly impact the performance of sensors.

Pressure. With proper installation and sealing of sensors in a pipe or pressure vessel, pH measurements can be done with widely available hardware at operating pressures up to 100 psig. Higher pressures require specially designed hardware. Repeated pressure fluctuations over time, however, may cause migration of the process solution through the porous reference junction, contaminating the reference electrolyte and degrading measurement accuracy.

Flow velocity. Process solution flow around the pH sensor can be good or bad, depending on the specific application. For control applications, flow is necessary for transport of fresh solution past the sensor to detect pH level changes. Excessive flow rates, however, accelerate wear of the pH glass membrane, particularly if the solution contains abrasive particles. Forces from high flow velocities also can cause glass breakage. A maximum flow rate of 10 ft/s should not be exceeded.

At times it is advantageous to increase flow past an electrode, particularly one with a flat end oriented parallel to the flow stream. The wiping action of the stream promotes self-cleaning of the tip.

Calibration issues

Buffer solutions are used to calibrate the electrode/meter system and are very stable with regard to changes of pH level. These solutions are commercially available in a range of pH values. Most common are 4, 7, and 10 pH buffers. For greatest system accuracy, electrodes should be calibrated with buffer solutions at the electrode isopotential point, and at a point close to the anticipated pH of the process solution to be measured. A word of caution about the instability of some 10 pH buffers: at this pH, these buffers react readily with CO_2 in air, which causes the pH to shift downward. Also, 10 pH buffers supplied in plastic containers may degrade even though the container is not opened because of CO_2 permeating the plastic. An alternative to consider is the 9 pH buffer.

Buffer solutions are inexpensive and should be discarded after one use. Even slight contamination in the standard reference affects the pH system's performance. Re-using buffer may save a few pennies, but cause an entire batch of product to be out-of-tolerance.

Electrode mounting

Many mounting configurations for installation of pH sensors into tanks, piping, or sumps are commercially available. Very frequently, reliable performance depends on how and where the pH probe is mounted. A universal consideration in probe placement is that the sensor be continuously maintained in solution. If left exposed to air, the membrane can dry out and become unresponsive. (Soaking dehydrated probes in warm buffers often can restore their usefulness.)

Submersion sensor assemblies are used to immerse the probe through an opening at the top of a tank, sump, or stilling well. The arrangement consists of a sensor mounted onto a suitable extension pipe with an electrical junction box at the other end. A guard sleeve fitted to the end of the sensor can protect the electrodes during insertion and from solids in an agitated vessel. Figure 7 shows such an assembly.

For processes that normally cannot be shut down for routine maintenance, insertion probes are employed. These sensor assemblies (see Figure 8) allow servicing of the probes from pressurized process lines or at the bottom of vessels without interrupting operation. They also are suitable for installation near the bottom of a tank to enable sensor replacement without the need to empty its contents. Various sensor types and construction materials are offered, with selection based on compatibility with the process. Units are available for either manual or power-assisted operation.

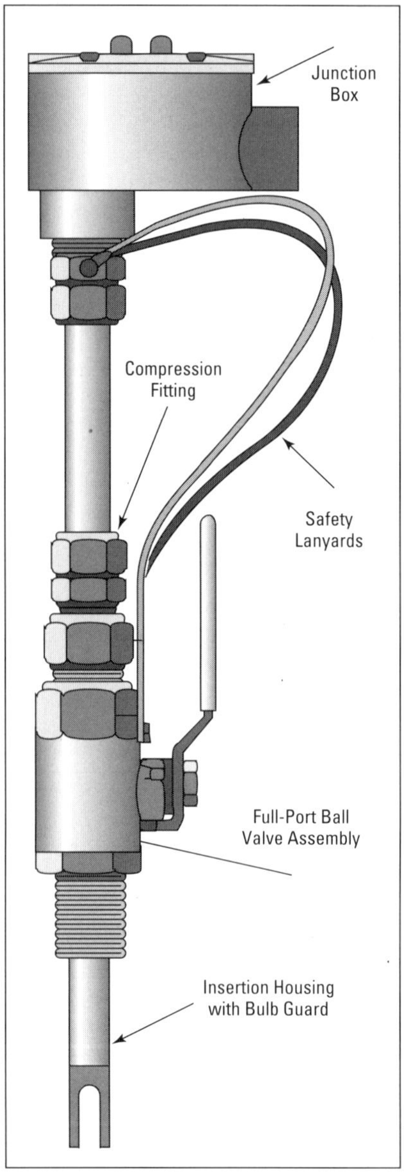

■ ***Figure 8. Insertion sensor assembly.***

In-line sensor assemblies (Figure 9) adapt to standard pipe-connection configurations. The sensor either can be attached through a tee with standard taper pipe threads, or held in place with a union and seal-type retainer.

Sanitary probes, including units whose electrodes periodically can be steam sterilized, are available to meet relevant regulatory requirements. A sterilizable assembly is shown in Figure 10; electrodes of this type are available for both atmospheric and pressurized applications.

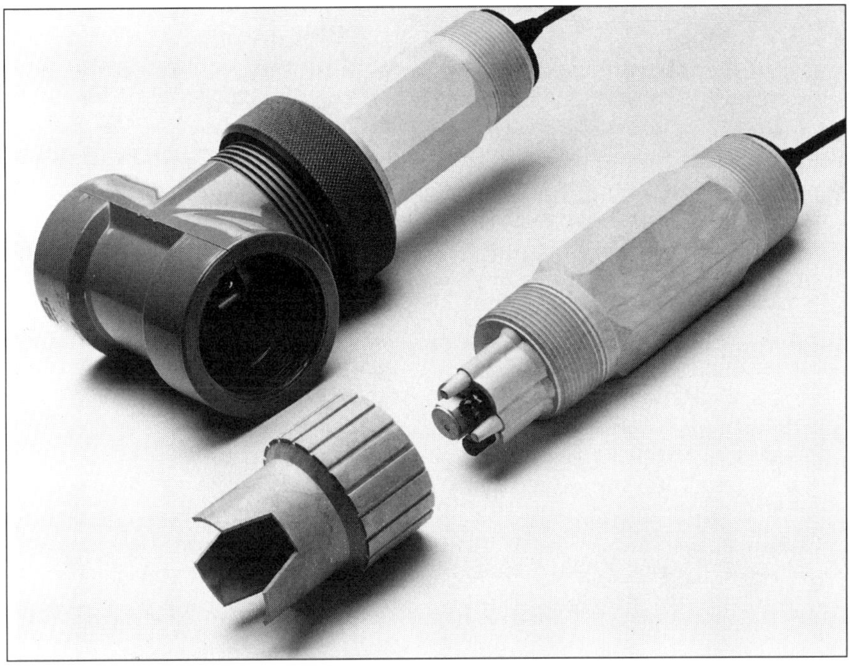

■ *Figure 9. In-line sensor assembly.*

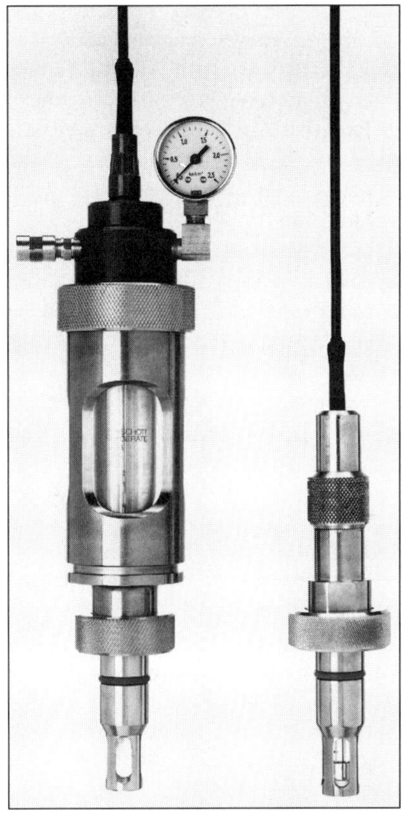

■ *Figure 10. Steam-sterilizable electrode assembly.*

ANALYZERS

An analyzer converts the pH-sensor output into a meaningful signal or visual indication. Common practice is to locate the analyzer near the probe for local indication and easy calibration, and then send the output to a central control area or computer. Obviously, the instrument's enclosure needs to keep the weather and any hazardous conditions out of the analyzer. System safety ratings and requirements affect the selection and cost of the analyzer.

Analog analyzer units with dial-type pointers for local indication may suffice if all that is needed is to keep the operator apprised of conditions. Advances in digital electronics, however, have made microprocessor-based units affordable. These units offer features not available with analog equipment.

Microprocessor-based analyzers come in a range of capabilities. Relatively simple, portable models have limited function and accuracy, but are sufficient for quick, ball-park checks — serving as an electronic version of a piece of litmus paper. More complex mounted analyzers add features for multiple inputs of pH, ORP, and process-fluid temperature measurement. They also can include probe diagnostics, multiple outputs for process controls or alarms, dc power supply for external components, computer interface capability, timer-controlled contacts for automatic sensor cleaning devices, and controller contacts that can be set either as minimum/maximum limit contacts or as controls. Provisions for analyzer- and system self-diagnostic capabilities, automatic two-point calibration, and automatic data-and-event logging functions also are being incorporated into some of the more comprehensive analyzer units.

Always keep the intended end-use of the measurements in mind. Buying unnecessary capabilities and options not only adds to cost, but actually can complicate operations. Remember, a capability is useful only if it solves an issue in the measurement and control application.

Also, pay attention to the interconnection between the sensor and analyzer. Depending upon the situation, the analyzer may be mounted a considerable distance from the probe location. A standard guideline is to keep sensor-cable length to 10 ft or less, because cable resistances and interferences can destroy the reliability of the system. By using sensors with pre-amplifiers or transmitters, however, the separation between the sensor and analyzer can be as much as 3,000 ft without signal degradation, while using relatively inexpensive cable.

Control requirements

Modern analyzers can produce measurement repeatability better than 0.01 pH. This degree of accuracy is necessary to meet or exceed the more stringent control requirements for some process fluids and waste streams. Whole books are available that discuss details of pH control *(5, 6, 7)*. Here, we will only highlight some considerations.

Adjustable high- and low-limit relays can provide automatic warning and control of out-of-specification process conditions. User-established limits can call for preventive actions (such as opening valves for reagent addition) or activate failure alarms. Limit control often is the best choice for pH control in large vessels; the long lag times effectively reduce the variability of the output product while remaining in specification.

The second easiest and most widely applied control method is a proportional electrical signal. The standard signal in the U.S. is a 4–20 mA current. The other proportional control method is by pulse, using either pulse duration or frequency.

Three-mode proportional-integral-derivative (PID) controllers best suit continuous flow applications with tighter limits. Singular or bidirectional PID controllers can be included as part of the analyzer hardware. These controllers also have user-defined contacts for either current output or pulse control. In actual applications, process variability makes tuning PID controllers difficult, sometimes even counterproductive.

Computer interface

The hot topic in instrumentation today is how best to get the process measurement information back to a central control system. The 4–20 mA current loop generally is employed but has its limitations. Communications protocols that allow interaction with the analyzers from remote locations continue to evolve. The HART protocol has broad backing, and a growing installed base. The Instrument Society of America and industry groups continue to develop promising fieldbus communications standards. Early planning for incorporating a communications system can prevent system incompatibilities later.

When installed in the plant, the data display screen can be especially important during daily operation. Most microprocessor-based analyzers use highly-visible liquid-crystal-display characters. Backlighting of the characters improves visibility in the darker corners of the plant. The displays often can be user-defined to select which functions will be continuously visible to the operator. For example, the main display can show current status for either pH, electrode potential (mV), ORP (mV), temperature (°C), or time of day. Sub-displays also can be configured to include such information as date, current input and output (%), controller output (%), calibration timing, as well as other data.

Diagnostics

Some of the latest generation of analyzers include self-diagnostic functions to expedite troubleshooting of the probe, analyzer, and the entire pH measuring system. On-screen displays can provide advance warnings and failure alarms to prevent manufacture of bad product or discharge of fluids beyond limits. By spending time during system configuration, the process engineer can better model acceptable performance ranges and determine at which point to start corrective actions. When an alarm is activated, pertinent diagnostic information and explanations can be displayed to focus the corrective action.

A feature incorporated in some analyzers is system-related self-diagnostics, which include continuous monitoring of the measuring and reference electrodes for breakage, aging, contamination, and other electrode conditions *(8)*. By using electrode statistics functions — which compare the current electrode operating conditions with the data from the first calibration — the life expectancy of the electrodes can be forecast, allowing the operator to change probes when they have aged but before they fail and, importantly, before they cause harm in the process.

Transmitters

Up to this point, we have concentrated on line-powered analyzer systems. These stand-alone units comprise a vast market segment. The continuing implementation of computer control systems, however, is increasing interest in analyzers that work directly with the computer. Such transmitters generally are powered by the process control computer. These loop-powered transmitters provide accurate measurement, but the central computer does the control and record keeping.

Some basic guidance

The measurement of pH is deceptively simple. Yet, pH systems often are troublesome. Careful consideration of the factors mentioned in this article will help you select the most appropriate system, and achieve accurate and reliable measurements. CEP

Literature Cited

1. **Horwitz, B. A.,** "pHrustrations of a Process Engineer," *Chem. Eng. Progress*, **89** (3), pp. 123–125 (Mar. 1993).
2. **McMillan, G. K.,** "Understand Some Basic Truths of pH Measurement," *Chem. Eng. Progress*, **87** (10), pp. 30–37 (Oct. 1991).
3. **McPherson, L.,** "Properly Measure Ions in Solution," *Chem. Eng. Progress*, **91** (2), pp. 54–57 (Feb. 1995).
4. **King, K., and J. Hayes,** "pH Measurement Without the Crystal Ball," presented at Annual Meeting, Chicago, Instrument Soc. of America, Research Triangle Park, NC (Nov. 1993).
5. **Bates, R. G.,** "Determination of pH — Theory and Practice," John Wiley, New York (1973).
6. **McMillan, G. K.,** "pH Measurement and Control," 2nd ed., Instrument Soc. of America, Research Triangle Park, NC (1994).
7. **Wescott, C. C.,** "pH Measurements," Academic Press, New York (1978).
8. **Nichols, M. L.,** "Improve the Performance of pH Measuring Loops," *Chem. Eng. Progress*, **90** (12), pp. 64–67 (Dec. 1994).

J. W. HAYES is a marketing consultant in Waukesha, WI (414/549-9451; Fax: 414/549-9431; e-mail: jwhayes@execpc.com). He has over 20 years of experience in the CPI, and specializes in strategic marketing programs that address end-user issues with product, training, or both. He received a BS in chemical engineering from Rose-Hulman Inst. of Technology, and an MBA from the Univ. of Missouri, St. Louis.

Take Charge of Your Plant Laboratory

You may be missing an important opportunity if you ignore your analytical laboratory until plant operating problems thrust it to the forefront.

Manfred Gans
Technology Evaluation & Development Associates, Inc.

Burton Palmer
Consultant, Industrial Analyst

The analytical laboratory of a process plant frequently is the most neglected resource at the facility. When the plant is running smoothly, managers often are interested mainly in analyses for product quality and specifications; to a lesser degree, they might keep track of the quality of the raw materials. Process stream analyses, however, are reduced to a minimum — because few managers know how to exploit these analyses to the fullest. And very few realize that certain situations scream out for special analysis, even at times when all seems well.

In contrast, when a plant gets into trouble, the laboratory suddenly becomes critically important. When we can't make capacity, yield, or specification product — or, for some mysterious reason, we use an excess of utilities, catalyst, or chemicals — or we face conditions in which our plant becomes unsafe, requires excessive operator attention, or suddenly fails to meet environmental restraints, "all hell breaks loose." In no time, the plant operating engineers and managers overload the laboratory with work. Then, the laboratory service becomes really bad — reinforcing the notion undoubtedly already held by many that the laboratory is unreliable and uncooperative.

Maybe, though, if we examine the laboratory service more thoroughly, we'll be able to use it better. After all, we don't wait for a major breakdown of mechanical equipment to organize our maintenance department, nor do we wait for a plant disruption to thoroughly acquaint ourselves with the maintenance department's functions, and its weaknesses, and strengths.

So, it behooves us to closely look into the capabilities and limitations of our plant laboratories. Amazingly, there seems to be almost no literature on the recommended interaction between operating engineers of chemical plants and their analytical laboratories. Perhaps this article will generate some new ideas.

Who is in charge?

The first questions that we, the plant operating engineers and managers, have to resolve in our minds and those of our superiors are:

- Who is in charge of laboratory services?
- Who sets the schedule for its work?
- Who judges the quality of the services that we receive?

The answer is obvious. We have to take charge. If we want a proficient plant operation, we can't allow the laboratory to function independently, with a program of work that does not meet the running needs of the plant and that produces results whose accuracy is so poor that we can't base our decisions on them. But being in charge forces us to thoroughly understand the scope, operation, and limitations of the laboratory.

Potential and problems

The plant analytical laboratory can be an important source of help during normal as well as troublesome operations. The laboratory can discover off-specification products or raw materials, measure key compounds throughout the process, and identify undesirable byproducts. Additionally, the laboratory can detect corrosion products and locate their source. The laboratory also can provide facilities and trained staff to operate bench-scale experiments.

Unfortunately, at many plants the laboratory instead inflicts additional problems. Often the required analyses are not issued in a timely fashion, nor are the data sufficiently accurate or precise enough for the needs of the plant. The analytical methods for additional tests needed during upset conditions are not readily available — or even identified. And requests to the laboratory for additional support frequently are countered with pleas for more instruments, staff, or overtime. (The laboratory really may not have all the required facilities.)

Consequently, when a plant is in trouble, it often has the added burden of coping with analytical data problems as well as operational ones. Yet, there is little opportunity to correct long-standing laboratory problems when the plant is in hectic troubleshooting mode. Thus, it is essential that we know the problems and the limits of our analytical services long before the plant gets into trouble.

So, let's examine the services that we are supposed to get from our analytical laboratory and set out a plan to critically evaluate the data we get from the laboratory and, if necessary, correct them.

Analyze your analyses

We can get information on the composition of our streams in three different ways, via:

1. laboratory analyzers;

2. on-line process analyzers; or

3. inference, that is, from plant material balances and physical properties, as well as knowledge of the process.

Inferred analyses can be a most useful tool for checking analytical data. Consider, for instance, the following dramatic example at a plant for the chlorination of methane. In that plant, the product stream from a high-temperature chlorination reaction was passed through a series of scrubbers and condensers where first the hydrogen chloride and then the various chlorinated hydrocarbons were removed from the unconverted methane. The scrubbed and cleaned unconverted methane then was recycled to the reactor. This recycle was supposed to contain approximately 3% methyl chloride but, soon after startup, the laboratory reported that our recycle stream contained 21% methyl chloride.

We, the operating engineers, categorically rejected that analysis. After all, we knew the pressures and temperatures of each scrubbing and condensation stage and, because no unforeseen component was ever reported, we could calculate from vapor pressure data the composition of each scrubbing and condensation stage. They *all* conformed to the process flowsheet, which had foreseen a 3% concentration of methyl chloride in the recycled methane.

For three weeks, the laboratory warned us each morning that our recycle was loaded with methyl chloride, and for three weeks we cavalierly rejected its finding, marking its reports with such choice comments as "nuts," and then displaying them in the control room.

Experts were sent in to find out what had gone wrong. Unfortunately, they trusted the laboratory results more than our vapor pressure equilibrium calculations. And they put forward a lot of complicated theories, but could not suggest a solution to the problem.

Then, a chemist, who had been sick, returned to the laboratory and pointed out that a certain instrument had been calibrated wrongly. When she recalibrated that instrument, the methyl chloride content of our recycle stream dramatically dropped to 3%.

We, the operating engineers, had made a major mistake too, though. We should have made up a

Basic Concepts

To engage in a sensible discussion with analytical chemists, we have to understand the terms that they use:

Precision reflects data scatter, and is measured by the estimated standard deviation. Given enough data, we ourselves can determine by statistical analysis the precision of the analysis for any one stream. This is not as formidable a task as it sounds. If we suspect excess scatter in any one stream analysis, we usually can send more samples from that stream and ask the laboratory to concentrate on that specific analysis. Thus, we get more data points for our statistics.

Accuracy indicates the difference between the *average* value of the data and the actual, true stream composition. The chemist, drawing on the literature, knowledge of the laboratory instruments, and experience, should be able to give us an idea of the accuracy that the laboratory can achieve for the analysis of a stream.

Bias denotes a nonrandom error. It can be caused by the presence of certain components of a mixture; it can be a "personal bias" — from the same chemist making the same error all the time; or it can be caused by an analytical instrument or the sampling method.

sample of known composition and sent it to the lab — or we should have suggested to the lab that it analyze samples of known composition. Either would have saved the company the expense of sending in all those experts; maybe we could have collected a bigger Christmas bonus that year!

Unfortunately, inference cannot be successfully applied in all cases, because it usually requires a wealth of physical-chemical data. So, we must rely on our laboratory staff and the data that they provide.

Carefully check the data

Scrutiny of the analytical data as received is important. Too often, such data are entered into log books or databases without further regard. Simple but effective comparisons can be made to evaluate the data when received. First, lists of the normal composition of each stream analysis should be available for quick comparison so that out-of-line values quickly can be noted. A second means of confirming the validity of the analytical data is the routine calculation of statistical standard deviations, which will show increasing data scatter when laboratory methods are getting sloppy. "Confidence levels" can reveal that data averages still have a wide error while comparisons of data sets may disclose significant differences. Third, simple process calculations, such as totaling stream components or checking balances across a unit, can be very helpful. Finally, comparing the laboratory data with inferred analytical information can be extremely useful, as we've already shown. In many cases, comparison with inferred information will provide incontrovertible evidence about the reliability of the analytical data.

As another important check, request the *full readout* — as opposed to the readout for the major components only — of all analyses done by chromatograph or mass spectroscopy. By requiring the chemists to routinely display full spectral readouts, including trace and unknown stream components, and possibly to explain all peaks, you may detect "unwelcome strangers," that is, interfering compounds, such as as the methane shown in Figure 1. Many subtle changes in a process or an individual unit operation often are revealed by changes in such full readouts of the individual streams. Monitoring the unmeasured and often unidentified minor components requires little additional effort and offers large potential rewards, as we will illustrate. To do this, you must take benchmark full-chart chromatograms and spectrograms as well as spectrograms of each stream when the plant is stable and operating within normal limits. Then, take subsequent full-chart chromatograms and spectrograms routinely to observe process changes.

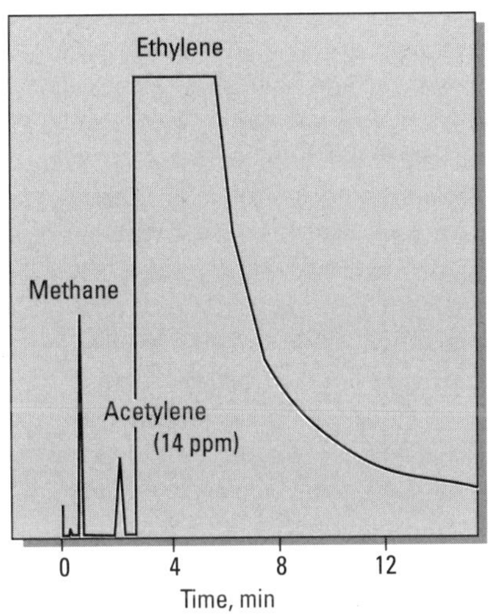

■ ***Figure 1.** Full readout chromatogram of trace acetylene in ethylene reveals "incidental" methane peak. Courtesy of Supelco Inc.*

Our term "unwelcome stranger" is somewhat of a misnomer. Even gross process changes do not usually involve new chemistry, only the relative concentrations will change — albeit sometimes very drastically. The real stranger usually comes from a source extraneous to the process, such as a leak into the stream or contamination from somewhere in the sampling or analysis system. So, a word of warning: confirm that the stranger is really in the process stream and not just in the sample or the "method" before spending large amounts of valuable time on its identification and elimination.

Also, look at the *metal content* of your stream. A reaction between the process stream and the materials of construction — commonly referred to as corrosion — might generate undesired byproducts. We can't deal with this vast subject here; suffice it to say that most chemists and operating engineers rarely look at the relatively easy to perform analyses for metal compounds. Once a plant has been run in and is operating steadily, concentrations of corrosion products in a bottoms or waste stream should remain constant. An increase in a corrosion product signals upset conditions. Metal balances should then be taken across specific, suspect equipment.

Errors and degree of accuracy

We have to realize that some error is present in all analytical data, irrespective of how they are generated. This error can be caused by the analytical method, by the sampling procedure, or by the stream itself. (For instance, if the sample is nonhomogeneous without the laboratory chemist being aware of the degree of nonhomogeneity, or if there is a chemical component, our "unwelcome stranger," in our stream that surreptitiously interferes with the chosen analytical methods, then we assign the error to the sample or stream itself.)

Realizing this limit on the degree of accuracy, we have to ask ourselves, "What is the maximum error that can be tolerated for normal operation?" Then, we have to ask whether this acceptable error is adequate for solving our plant operating problems and whether the laboratory can provide data of that accuracy. If the laboratory cannot meet the requirement, then either we have to change our troubleshooting strategy or else the laboratory has to use different analytical methods or different instruments. Perhaps, we will have to make use of external, specialized analytical services.

There are three types of errors: gross, random, and systematic. Gross errors are so large and obvious that the data are usually rejected out of hand. Even gross errors should be recorded, however, because a pattern of gross errors can signal laboratory or sampling problems. Random error, also called data scatter and measured as method precision, is present in every measurement and forms a normal distribution around a true theoretical value; it should be defined precisely by statistical methods. Systematic error or bias is caused by one or more variables acting in some nonrandom way. Both bias and random error coexist in all analytical data; only their respective magnitudes vary. We will discuss methods to deal with bias.

Symptoms of poor quality

To understand what needs to be done in the laboratory, we first must judge the quality of the analytical data given to us. Here are some symptoms of poor quality:

- poor precision, that is, excessive scatter of the data;
- data inconsistent with basic process principles (as in our chlorocarbons example) — in other words, the analytical data does not agree with inferred chemical composition data;
- a prior history of poor-quality data from a specific analytical method;
- disagreement in results between analytical methods; and
- data bias caused by a faulty analytical method.

Frequently, however, there are no obvious symptoms. Even if there are, though, we still face the task of convincing ourselves, the chemists in the lab, or management that an improvement in analytical service is called for.

If we must contend with poor precision, that is, excessive scatter, additional analyses of the *same* sample *will* give an average value with improved accuracy. This assumes that the poor precision was caused by an inadequate analytical method. If the scatter, however, is due to poor sampling procedure or the sample itself (nonhomogeneity, "unwelcome stranger," or the like), additional analyses of the same sample will not improve the accuracy.

Overcoming bias

Most analytical methods are theoretically capable of achieving the required accuracy and precision. (See sidebar.) Subtle changes in a method, however, as it used by different people over a period of time can increase errors. Method bias can become a vexing problem. Do the analytical data point to a real process problem or are the data wrong? Unfortunately, that is a very common question asked during critical plant periods. Remember, it is far easier to keep the analytical method free from significant bias than it is to find and retroactively correct analytical bias in plant samples. Therefore, insist that control samples of known compositions be measured and reported frequently, with their frequency determined by the criticality of the analyses. Second, whenever possible, routinely compare the analytical-method results with either a second method or with inferred plant data. Discrepancies in the control sample data or between methods must be investigated aggressively by the analytical group and the discrepancies must be eliminated.

Instituting strict documentation of methods and routine analyses of standard samples will prevent degradation of most analytical methods. Nevertheless, evaluating and using suspect analytical data remains one of the most common and annoying problems facing the plant operating engineer.

Analytical-method bias can be determined relatively easily by replicate analysis of known standard mixtures. Having repeatedly used the same analytical method for process streams and known standard mixtures, we now are able to calculate and agree on the difference between the true value and the normally measured value. We then can use this difference to correct all the stream analyses measured by the same method. This correction may have to be used cautiously if we are dealing with different streams that *might* contain different impurities (different unwelcome strangers). In other words, we have to be guided by our knowledge of the process. Furthermore, bias caused by the sampling method or by chemical component interferences cannot be corrected in this manner.

Evaluating laboratory performance

How can we quantitatively evaluate the analytical data generated by our laboratory? First, for a continuous process, we must make sure that the plant is running steadily. We then obtain simultaneous analytical data from as many streams in a troublesome section as we can, making use of both on-line and laboratory analyses.

We have to carefully select the streams in such a way that we can perform material balances for the components that we can analyze. Conceivably at this early stage if, for instance, the troublesome system is a distillation column, we might want to look only at key components.

This process of sampling and analyzing then is repeated at timed intervals.

Now, every time we obtain a complete set of analytical data, we perform a material or heat balance on that set of raw data. We make no attempt to correct or reduce (average) the data. We can, though, use reliable physical-chemical data to arrive at the material or heat balance.

Next, we plot the value of each component in each stream and the deviation for each component from material or heat balance against time (as the independent variable). Then, we examine these plots for:

• *Trends* (as opposed to scatter). If the data for any one component show a trend (value changes with time), then it is fairly safe to assume that an analyzer is going off calibration. Alternatively but less likely, the process is getting deeper into trouble.

• *Relation to plant upsets.* If our periodic data have a relation to plant upsets, then the analytical service is most probably adequate.

Having collected data at periodic intervals over a meaningful time period, we now select only those data that were taken during steady-state plant conditions. And, at this time, we should discard data that are completely out of line. Then, for each component of each stream, we calculate the average estimated standard deviation and the 95% confidence level of the average value. This will allow us to judge whether the data scatter is acceptable; if it is not acceptable, we have to try to locate the probable cause of error (as we will discuss).

Statistical tests are only approximate when applied to the relatively small number of data points that we usually get from plant samples. In other words, to use statistical tests with confidence, we need a lot of data. Nevertheless, the information from the tests can be very useful and should be routinely used when evaluating plant analytical data.

If the troublesome system is a batch process, we still go ahead and take periodic samples and do our calculation for components, as above. Results, of course, will change with time, but still should fit a precalculated or preestimated smooth curve.

These data will either convince us that the analytical methods are valid or they will give us a very powerful tool to convince the chemists and management that the analytical data are inadequate.

Locating the probable cause of error

If everyone is convinced that the analyses are faulty, we then have the task of locating the probable cause of error. It is absolutely possible — in fact, it is quite likely — that only one stream analysis or one analytical method is faulty. Using the plots that we have generated, we now identify for each stream and each component the analytical method used. We try to find the differences in precision among the different streams, the different components, and the analytical methods used to identify streams or components. Hopefully, we thus can pinpoint one stream, component, or analytical method that is obviously wrong. This suspicion can then be confirmed by using more rigorous methods of analysis or sending samples to outside specialized laboratories. In any case, our analytical effort should now be directed to critical streams only.

Plant-sampling systems are often ignored. Sampling commonly is delegated to the least experienced plant operators and analysts.

Good sampling is critical

The entire plant-laboratory analysis program rests on the assumption that the analyses are performed on samples that are truly representative of the process streams. Analytical data from nonrepresentative samples are completely worthless and misleading. No useful data can be retrieved from a nonrepresentative sample. Yet plant-sampling systems are often ignored. Sampling commonly is delegated to the least experienced plant operators and analysts. Frequently, written sampling instructions or organized training of the operators are lacking — as are periodic inspections of the plant sample points and containers. So, always consider a possible problem in the sampling system any time there is a question about the laboratory analyses.

Sampling errors cannot be corrected retrospectively, but usually at least can be detected by using material balances. Also, such errors probably will show up in our plot of analytical results vs. time. If we suspect or know that there are sampling errors, we have to closely examine the way that the samples are taken, transported, and handled in the laboratory. Look at the entire sampling system, including:

1. the plant-stream sampling point and the equipment used to draw the sample;

2. the sample container;

3. the sample transport;

4. the sample discharge system in the laboratory;

5. the laboratory sample storage;

6. the secondary laboratory sampling system — that is, the way the laboratory takes samples from the plant-stream sample; and

7. sample-container cleaning.

(Clearly, given this number of possible troublespots in a sampling system, a schedule for periodic inspection of sample containers and sample systems is essential.)

Next, we have to investigate the training that the plant operators received in the different sample-collection procedures. Conceivably, new techniques may improve the sample. Here, safety is a most important consideration — sample collection has led to too many accidents harming operators and equipment.

Finally, in our investigation, we have to ask ourselves:

- Are there any deadlegs in the sampling system?
- Is the sampling point correctly located to receive representative samples, or are we in a deadleg?
- If there is a deadleg (and almost all sampling systems have deadlegs), are the operators using the correct purge time?
- Are the materials of construction of the sampling point correct?
- Are there leaks in or near the sampling systems?
- Has the system been heat traced correctly? Too little heat tracing will lead to cold spots; too much may cause decomposition.
- Do we know the exact temperature and pressure at which the sample is taken, and are we making the correct allowance for these conditions?

There are some types of plant samples where the analyses are always difficult and potential errors abound. One example is the analysis of low-conversion, large-recycle operations. Also, each analytical method has an inherent error, usually about 2–4% relative to the concentration being measured. This points to an obvious problem when a process with a 15% conversion is measured with a method error of 3%. Usually, one must resort to measuring reaction products rather than the reactant and coping with the complication of reaction products in the recycle. Assays of high-purity products present the same problem, and most high-purity products are evaluated by indirect specification tests rather than a direct product assay.

The nonhomogeneous or multiphase sample also creates great difficulties. Techniques requiring great skill and care — and offering numerous opportunities for error — must be used throughout the sampling and the sample analysis procedure. The question always remains: Is the sample truly representative of the process stream?

Preventing analytical service problems

The services of an analytical laboratory really are analogous to those that we get from the maintenance department. There is no doubt that the concept, technology, and practice of preventative maintenance has greatly improved the on-stream time and performance of process plants in the last 20 years. So, in a parallel fashion, can we think of concepts, technologies, and practices for the analytical laboratory that will prevent errors in the analytical data and thus improve plant performance before we ever get into trouble? We believe the following steps will greatly contribute to such prevention of errors:

- Periodically, analyze known mixtures using the analytical methods regularly practiced and then correct all methods for excessive bias.
- Routinely use analytical-method control samples; for instance, send the laboratory two samples derived by halving an existing sample.
- Urge the laboratory at times when it is not too busy to check for interfering compounds. Because plant laboratories rarely contain the sophisticated analytical tools available in, for example, research establishments, these analyses may have to be done by an outside service.
- Carefully review and inspect each sample point and the entire sample collection, transport, and storage system.
- Check each sample stream for nonhomogeneity brought about by the presence of particles of different densities or several fluid phases.
- Finally, critically examine the routine and emergency analytical schedules to determine which analyses might be better performed by an outside laboratory equipped with better instruments than the ones the plant can afford. Given today's fast communications' capabilities, the use of such laboratories needn't incur undue delays.

■ ***Figure 2** Gas chromatograph coupled with a thermal desorption unit that is used as a sample preparation device. Courtesy of Supelco Inc.*

What to do when troubles have developed

During a troubleshooting period, first and foremost, the plant engineering staff must fully accept the task of defining the overall analytical laboratory objectives. As a first step, the operation engineering staff should review the entire sample schedule in detail and eliminate all those sample analyses not urgently required for plant operation. The purpose is to free analytical staff and instrumentation for the other analyses that will be necessary during the troubleshooting period.

Next, the engineering staff should decide on the specific analyses needed for the troubleshooting program. This means not only the streams to be analyzed, but the necessary components, the precision and accuracy required, as well as the approximate number of analyses and the acceptable analysis response time.

Now, the proposed analytical program can be presented to the analytical supervisor for joint review. Based on the laboratory's staffing and instrumentation, as well as methods and services available from other laboratories, a tentative analytical program can be agreed upon and implemented.

The program suggested above is an ongoing one, not a onetime task, and changes will be inevitable. Also, it will require effort from operating engineering staff already hard pressed with other duties. Significant advantages will ensue, however. The lines of responsibility are made clear. The senior operations engineering staff sets the objectives for the analytical laboratory. The laboratory carries out those objectives as best it can. Each communicates with the other as the program continues. Certainly, the laboratory always will be most efficient when it can work with a well-planned sample schedule.

When troubles have started, it is essential that all the steps concerning scrutiny, assessment, overcoming errors, and quantitative evaluation of analytical services be undertaken if they were not completed before. Moreover, these same steps must be applied to any extra analytical services called for as a result of troubleshooting.

Coping with analytical problems

Defining the laboratory agenda is a good first start, but other analytical problems can also surface. New analyses that the laboratory agreed to develop may not be available when needed. Or the laboratory may not be able to adhere to the agreed upon sample schedule and may fall behind in response time or in sample output. While these problems involve laboratory operations, they also affect plant operations — so, it is appropriate to consider steps that can be taken to alleviate them.

If new analytical methods are not available on-site for a troubleshooting program, consider sending samples to a consulting laboratory. Arrangements for quick response often can be made without undue costs or difficulties. In fact, the cost for outside services may be lower than developing plant analyses that will be used only for a short time. And, if your plant has access to a central corporate research and development laboratory, try to tap it as a source of help.

A shotgun approach to solving problems in a troubled plant is not a good one when the laboratory is involved.

Several strategies can be used to improve analytical response time. The sample collection, delivery, and work-up system can be modified for critical samples. Many analytical methods can be speeded up for the analyses of key components, which then can be reported as soon as analyzed. Decreasing the analysis time can also increase the potential number of analyses reported. Short-term rental of additional laboratory equipment should also be considered.

Dealing with poor data

Inevitably, some analytical data that cannot be duplicated will be produced during atypical plant conditions. Such data should be "retrieved" if at all possible. If the errors stem from poor precision or bias in the analytical method, then the data can be retrieved by retrospective analyses if the original samples have been saved and stored properly. Poor precision or a bias in the laboratory method can be improved by correcting the method problems and reanalyzing the samples. Sometimes, a bias found after the samples have been destroyed can be corrected by measuring the bias and correcting the data. Unfortunately, error caused by a problem in the sampling system cannot be corrected.

On-line analyzer tradeoffs

When we have successfully demonstrated that inefficient plant performance stems from the inadequacy or complication of one or more laboratory analyses, we'll soon face a recommendation that we switch to on-line analyses for one or more streams. Therefore, let's list the advantages and disadvantages of on-line analyzers, assuming that there is a suitable unit for our critical stream.

On the plus side:

• On-line analyzers have fast response times (as compared with laboratory analysis); thus, the frequency of sample taking and analysis usually can be increased greatly.

• Sample transport problems and operator errors are avoided.

• There may be some savings in operator and analytical labor costs.

• The results of on-line analyses can be integrated easily into the plant data system, if the analyzers have proven their reliability, we may even be able to use them in the plant automatic control and safety systems.

Against this, though:

• On-line analyzers generally need special instrument housing and climate control.

• They may require complex and expensive sampling systems whose design *cannot* be left to the instrument manufacturer. (In fact, those sampling systems and the analyzer are miniature chemical plants and, thus, they have to be designed with utmost care.)

• The analyzers also may continuously use expensive special gases and chemicals.

• Last, but by no means least, on-line analyzers require special maintenance skills. Are the plant's chemists and instrument technicians sufficiently experienced for that task?

An example

Consider this case history: Without apparent cause, a final product from a large chemical plant went off-specification for trace aldehyde. A predictable series of events followed. Additional confirming specification tests were made on more samples. The laboratory supervisor reviewed the entire test method and found no method error. Retained product samples from earlier operations were retested and found within specification, as earlier reported. The engineering staff, now justifiably alarmed, searched frantically for clues in the plant operating data. Engineering staff meetings were held frequently and each theory was explored and acted upon. Consequently, samples poured in from every section of the plant for rush analyses. Many samples required method modifications or development. Normal laboratory operations soon broke down. Laboratory personnel were now working 12–14 hour days, 7 days a week, with the inevitable consequences — analytical mistakes multiplied, and the laboratory became discredited.

Several thousand miles away, we received a summons with the following instructions: "Get over there and straighten out that plant laboratory!!!! It has screwed up the entire plant operation!" No mention was made of an aldehyde specification problem.

The resolution of the problem was anticlimactic. Once the real plant problem — that is, the product off-specification — was revealed and the process flowsheets were studied, it was apparent that the acetaldehyde had to get into the product as a heavy — that is, something less volatile than acetaldehyde itself. In other words, the distillation sequence and the distillation performance of the plant clearly indicated that acetaldehyde itself was too light a compound to stay with the product. So, to stay with the product, it had to be bound to some other component.

No benchmark full-readout chromatograms were ever taken in this plant, but chromatograms of the samples remained on the recorder chart rolls, which fortunately were not discarded. Tedious scrutiny of hundreds of chromatograms of samples taken throughout the plant before and after the off-specification acetaldehyde occurrence finally did reveal a relatively large increase in a minor, unidentified component in the column designed to separate acetaldehyde and some other lights from the product and heavies. This component showed up in the full readout, but it was too insignificant to be identified and reported. Reference to a basic organic text showed that paraldehyde, a trimer of acetaldehyde, is easily formed from acetaldehyde under acidic conditions. No paraldehyde was available at the plant site to confirm the unknown peak as paraldehyde. Paraldehyde is easily hydrolyzed, however, and shaking the sample with dilute acid for a few minutes completely removed the peak from the sample — a strong indication that paraldehyde was present in the original sample.

The engineering staff immediately acted on the assumption of the presence of paraldehyde in the process. They cleaned the distillation tower, which, as was then discovered, could have been contaminated earlier with sulfuric acid. Paraldehyde now dropped to earlier levels, operating conditions were optimized to control the level of paraldehyde, and an analytical method was quickly developed for routine analyses of the culprit. The plant quickly was returned to full operation making on-specification product.

What can be learned from this relatively minor incident?

1. A complete set of accessible benchmark chromatographic charts would have alerted the plant staff to the offending compound much earlier.

2. A thorough understanding of the plant's processing scheme, particularly its distillation section, together with some basic organic chemistry, would have pinpointed via inferred analyses the problem to the key section in the plant, saving much time and wasted effort.

3. A shotgun approach to solving problems in a troubled plant is not a good one when the laboratory is involved. The deluge of extraneous samples quickly overtaxed the laboratory and rendered it almost useless.

Take charge

If plant managers and plant operating engineers want to get full benefits from their laboratories, they have to take charge. They have to set schedules. They have to consider contract laboratory services. They have to assess the accuracy and the limitation of the data they get, as well as be prepared to initiate remedial measures if the service is inadequate for efficient plant operation. CEP

M. GANS is president of Technology Evaluation & Development Associates, Inc., Hoboken, NJ (201/792–3814; Fax: 201/792–5677). He has been a consultant to the United Nations since 1978, primarily involved with creating applied research and development institutes that can work with and troubleshoot for industry in developing nations. Until 1986, he was senior vice president of technology for Scientific Design Co.

B. PALMER is a consultant for industrial analytical operations in Tucson, AZ (602/749–9154; Fax: 602/886–3162). His clients have included the United Nations and Texas Eastern. Until 1986, he was manger of the analytical dept. of the Halcon-SD research group.

Use Graphical Techniques To Improve Process Analysis

Triangular diagrams can provide valuable insights and design assistance for a variety of unit operations — and can be effectively developed with simulator and data analysis software.

Lee R. Partin,
Eastman Chemical Co.

Process simulators are used extensively for the analysis of chemical plant flowsheets. The engineer synthesizes a flowsheet, sets the unit operation parameters, and inputs the information to the software. The simulator then performs a rating calculation of the specified process under the design conditions. It is often difficult, however, to find the proper operating conditions to meet specifications such as product purities or the constraints of existing equipment. Today's simulators provide options for converging on constraints and optimizing variables, but the solutions can be hard to obtain. The engineer is not guaranteed that a satisfactory solution exists — for instance, reaction-equilibria limitations may require a different reactor style, azeotropes can alter the desired distillation tower operation, byproducts may not have a way to be purged, the calculations may enter an infeasible operating domain, and so forth.

Graphical techniques can provide the answers. For example, the composite curve and grand composite curve of pinch technology for heat exchange network analysis *(1,2,3)* provide the engineer with knowledge of feasibility and a way to set realistic targets for energy usage and heat exchanger design. Such techniques also apply to other unit operations. In this article, we will detail graphical tools that provide a similar level of insight and design assistance for the synthesis and simulation of other unit operations. We will provide examples of their use, and we also will discuss the software requirements of the process simulator and data analysis tool.

We will focus on the triangular diagram. It is a powerful tool used to present the compositions of three-component mixtures. It shows the process capabilities graphically over the entire composition range of the three components. The engineer can quickly verify the accuracy of the data and inspect them to learn the important characteristics. It is also useful as a design tool. The engineer locates the desired product composition and visualizes the process to reach it *(4)*. This is an appealing extension to the process simulator which performs design by multiple ratings of the process with adjusted process parameters. User preference determines the choice between the right triangular and equilateral triangular format. The right triangular format is the same as a standard x-y plot with only a triangular portion active. This format may be more readily available in your data analysis software. It also has an advantage in that an axis can be expanded to give more accuracy where needed. The equilateral format is traditionally used. It is also an x-y plot but the y axis has been tilted to 60°.

Which plots are valuable?

Let's first look at the types of data that can be plotted for various unit operations. Consider the potential usefulness of each in your work. For example, a basic function of plotting the data is to verify that they are correct. Does the product decant properly? Are the azeotropes correct? Are the thermodynamic models okay? Plots provide an easy but powerful way to save you from fatal flaws in your simulations. Besides such validity checking, triangular diagrams also allow you to gain an understanding of the phenomenon controlling each selected unit operation. What are its capabilities? How can you exploit them? What are the limitations?

The plots provide a means to calculate some preliminary design information. You

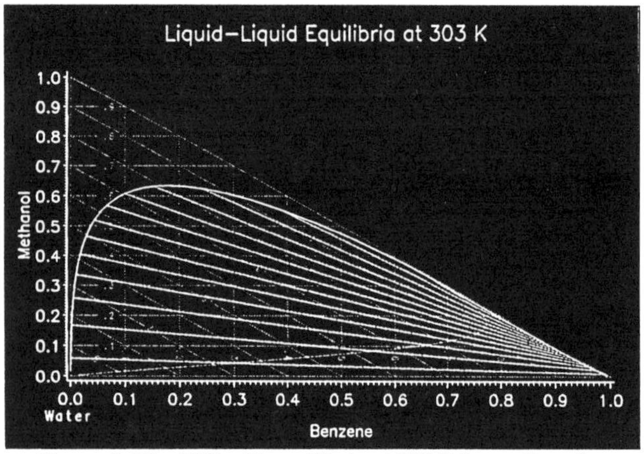

Figure 1. Liquid/liquid equilibria at 303 K for methanol, water, and benzene.

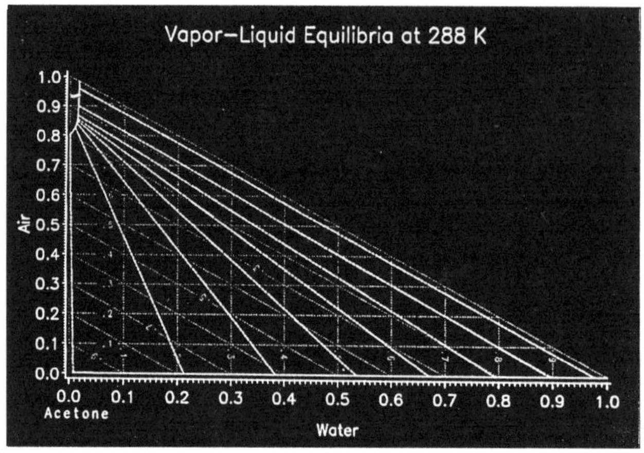

Figure 2. Vapor/liquid equilibria for air, acetone, and water at 288 K and 101 kPa.

locate the desired product and synthesize a flowsheet to produce it. At this point, you have a feasible flowsheet that is ready for detailed simulation and you have a preliminary design that simplifies the simulation task. You also have the tools for reviewing the process with your customers and peers. "What if" questions like variations in feed stock composition can be addressed visually. Keep these issues in mind as we continue.

Liquid/liquid equilibrium data. The liquid/liquid equilibrium plot is an excellent tool for understanding the capabilities of extractors and decanters *(4)*. It shows feasible ranges for overall feed compositions and allowable extract and raffinate compositions. Also, minimum flow requirements can be calculated. Within the feasible ranges, the engineer can design for the desired products and perform design calculations. The plot includes the two binodal curves of phase compositions up to the Plait point, the conjugate curve describing all tie lines between phases, and selected tie lines for a specified temperature and pressure. (The conjugate curve is the plot of the y value from one phase of the binodal curve vs. the x value from the other binodal curve. It is best viewed with the right triangular format.) Figure 1 presents data for methanol, water, and benzene as estimated by the Unifac group contribution method *(5)*. The plot closely matches experimental data from Dechema *(6)*.

The liquid/liquid equilibria calculations are difficult when the composition approaches the Plait point. An extrapolation technique is recommended for calculating the Plait point from data close to it *(7)*.

Vapor/liquid equilibrium data. The pertinent data for absorbers and scrubbers are the vapor/liquid equilibria data at a fixed temperature and pressure. Their usage and display are quite similar to the data for extractors. An example is presented in Figure 2 for scrubbing acetone from air with cold water.

The engineer using a process simulator for distillation calculations does not have the assurance that the desired products are feasible.

Solubility data. Another type of data is solubility data for a fixed temperature and pressure. Examples are *p*-hydroquinone solubility in mixtures of methanol and water, or the competing solubilities of sodium chloride and potassium chloride in water.

Distillation lines. These show the feasible distillate and underflow products for single-feed distillation columns over a wide range of feed compositions. A distillation line is calculated by stepping off stages for a tower at total reflux. A liquid composition is chosen as a starting condition. A bubble-point flash calculation is performed to determine the vapor in equilibrium with the liquid at a set pressure. The vapor composition then becomes the liquid composition for the next stage. The steps are repeated until the vapor composition does not change, signifying a pure component or an azeotrope.

I was introduced to the theory for distillation lines by Prof. Doherty and his research team at the Univ. of Massachusetts *(8,9,10)*. The lines are normally calculated by numerical integration of simple distillation equations and are also referred to as residue maps. A recent article *(11)* described the alternative style of multiple flashes which was chosen since it is quite easy to implement within a steady-state simulator. We will present an example that shows the importance of distillation lines.

Reaction lines. Triangular diagrams have also been promoted for showing reactor performance *(12,13)* by the plot of feed compositions to each of a series of continuous stirred-tank reactors (CSTRs). The user specifies temperature, pressure, residence time, and number of CSTRs in series. (Small residence times can be selected to approximate plug flow reactors.) Then, the calculations are performed over a wide range of initial feed compositions. The data are plotted on a triangular diagram, but there are usually more than three compounds in the reactor. For a given plot, three compounds are selected and the

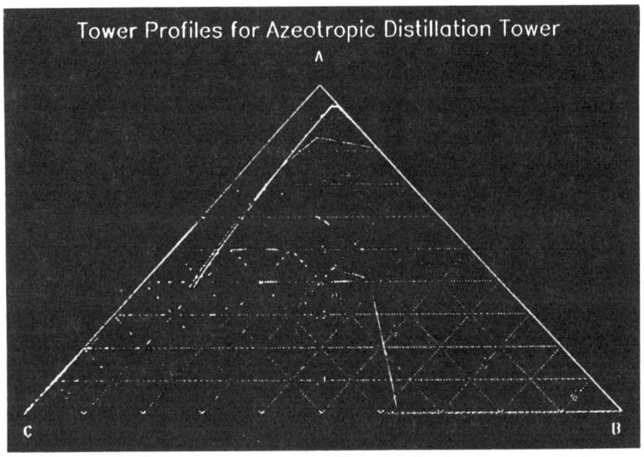

■ *Figure 3. Column section profiles via design approach for azeotropic tower.*

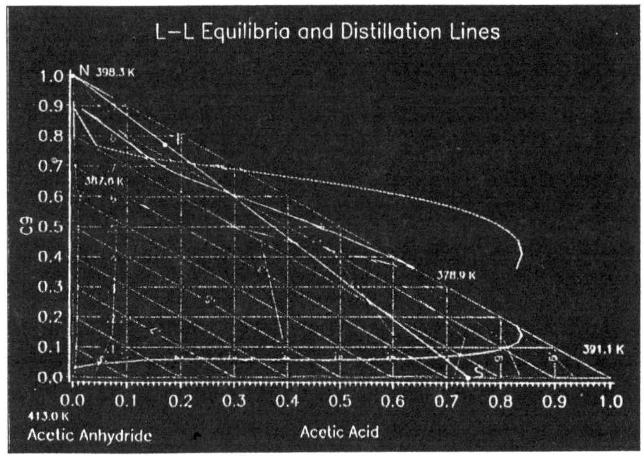

■ *Figure 4. Liquid/liquid equilibria (303 K) and distillation lines (101 kPa).*

composition is normalized. The selection of compounds is based upon their importance. Hopefully, a single plot can show the major feeds and products. If not, then you can consider ways to limit complexity such as fixing the ratio of certain feeds. It may be desirable to present the data in two triangular diagrams and have color-coded curves to link the data. The polyester production example that we will discuss does this. The feeds and major product are contained on one plot and a second plot details the product composition. The goal is to capture the most significant information. Each case will have its own special requirements.

The plots are both design and synthesis tools for establishing the desired feed ratio of reactants, number and type of reactors, and operating parameters.

Lines for unit operations in series. The calculation style for reactors can also be applied to other unit operations that can be placed in series. One example is crystallizers. Multiple crystallizers can be operated in series with differing operating conditions to optimize the crystal size distribution and purity of the product.

Another common example is distillation involving multiple flashes in series. There is an exciting transformation when you view distillation towers as sections of multiple flashes compared to the typical rating approach of a process simulator which requires a full input of the tower geometry and operating conditions. With a given product purity and energy usage, you can calculate a profile for a section of a tower as a set of flashes in series. The number of stages is set high to give the full range of data. Figure 3 shows an example of rectifying and stripping section profiles for a difficult-to-simulate azeotropic tower. The desired bottoms product is nearly pure component C. Several stripping section profiles are presented with different distributions of A and B as impurities in the C product. The distribution of impurities makes a dramatic impact upon the shape of the profile. To have a feasible single-feed tower, the rectifying and stripping sections must cross. Examination of Figure 3 reveals that only some of the rectifying section profiles give feasible solutions. Traditionally, the engineer using a process simulator for distillation calculations does not have the assurance that the desired products are feasible — and this can lead to numerous hours being spent in desperation trying to obtain the desired product purities.

> ***There is an exciting transformation when you view distillation towers as sections of multiple flashes compared to the typical rating approach . . .***

Figure 3 was calculated using a network of mixers, splitters, and flashes to model distillation trays in series. There are a great deal more design calculations that can be performed, as shown by the work at the Univ. of Massachusetts *(14,15,16)*. The design approach provides the engineer with knowledge of feasibility along with minimum reflux requirements. Then, given the reflux ratio and purity specifications, the design proceeds to calculate the number of stages, feed stage, and tower profiles. Hopefully, simulation software firms will start implementing the new techniques.

Multiple data sets. Showing a number of data sets on the same plot can provide insights and enhance decision making. For instance, multiple temperatures for liquid/liquid phase envelopes are useful for determining the proper operating temperature. Likewise, multiple pressures for distillation lines provide input into the desired operating pressure, especially when azeotropes and distillation boundaries exist. Reaction line plots at varying temperature and pressure conditions are a good way to visualize when and how much to change operating conditions for cases with multiple reactors in series (for instance, polyester processes). Also, the synthesis of combined unit operations is facilitated by plotting multiple data on a single graph as shown in the following example.

Using the plots — an example

A recent project took advantage of the diagrams for process synthesis,

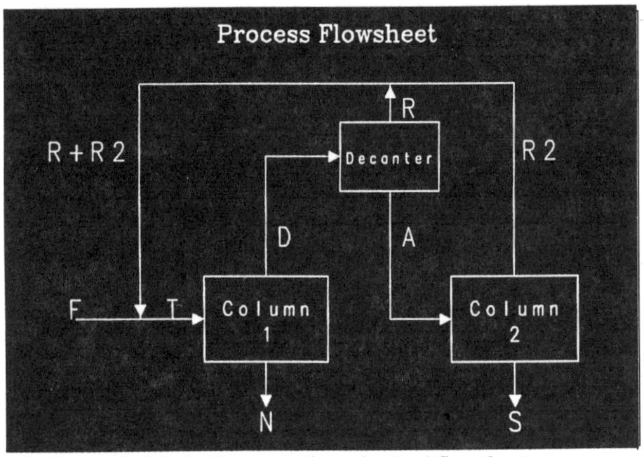

■ *Figure 5. Block diagram of synthesized flowsheet.*

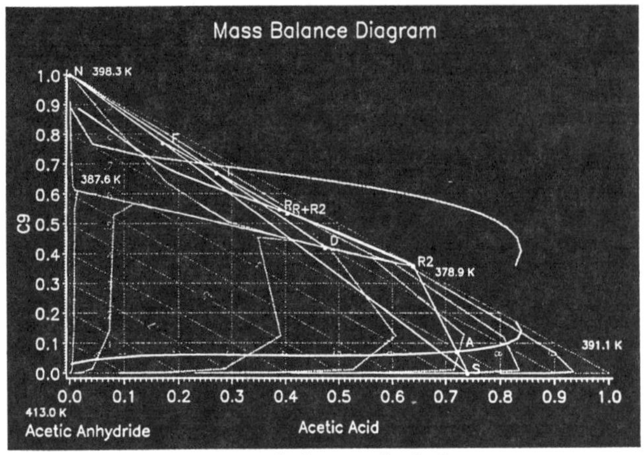

■ *Figure 6. Process mass-balance calculations.*

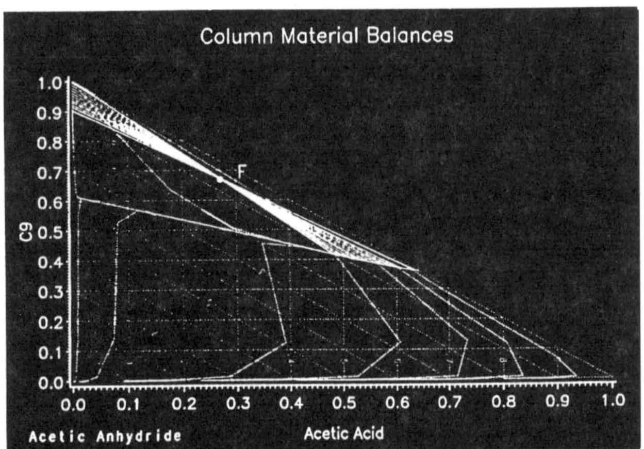

■ *Figure 7. Wide range of potential column mass-balance lines.*

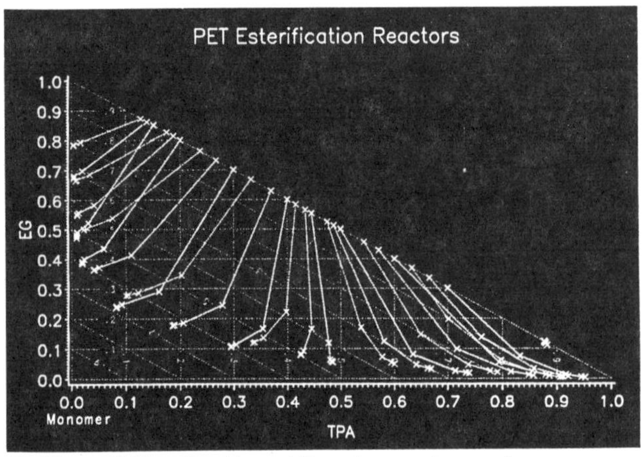

■ *Figure 8. Reaction lines at three temperature and pressure conditions for esterification in polyethylene terephthalate production.*

and the design of the process material balance. The project included a step for the purification of a branched, nine-carbon alkane (C9) with a boiling point of 398 K (125°C) from a feed stream that also contained acetic acid and acetic anhydride. Figure 4 shows the triangular diagram with the distillation lines at 101 kPa (1 atm) and the liquid/liquid equilibria data at 298 K (25°C). The overall material balance line for the step is also shown: point F equals feed stream, point N equals C9 product and point S equals acetic acid-acetic anhydride product. The distillation lines show that there is a distillation boundary from the acetic anhydride-C9 azeotrope to the acetic acid-C9 azeotrope which breaks the diagram into two regions. Distillation columns operating in one region cannot produce a product that significantly crosses into the other region. The material-balance line crosses the distillation boundary. Therefore, a single column will not perform the separation.

The liquid/liquid equilibria crosses the distillation boundary. This observation was used to synthesize a flowsheet of two columns and a decanter as a feasible process. (See the block diagram in Figure 5 and the mass balance diagram in Figure 6.) The triangular diagram was then used to design the process mass balance. The design value of S, the second column bottoms stream, was set by drawing the overall material balance line. The distillation lines in the area of point S show that the acetic acid-C9 azeotrope is a potential overhead stream from the second column. Its material balance line is drawn from S to near the binary azeotrope for point R2. The line crosses the binodal curve at the column feed point A. Point R from the decanter is related to point A through the tie line as given by the conjugate curve. The decanter feed point D lies along the line from A to R. A point was selected near the intersection of the tie line with the distillation boundary. This gets the distillate well within the decanting region. The line ratio principle is applied to set the quantities of each stream and to complete the addition of streams for R + R2 and T.

The design features of the triangular diagram were beneficial for completing the engineering study. It allowed for the quick calculation of the above. Next, the diagram was a graphical tool for discussing the process with the customer. "What if" issues were addressed by visual inspection. Questions arose about the variability of the feed flow rate and acetic anhydride concentration. The process fails at low acetic anhydride concentrations due to a loss of phase

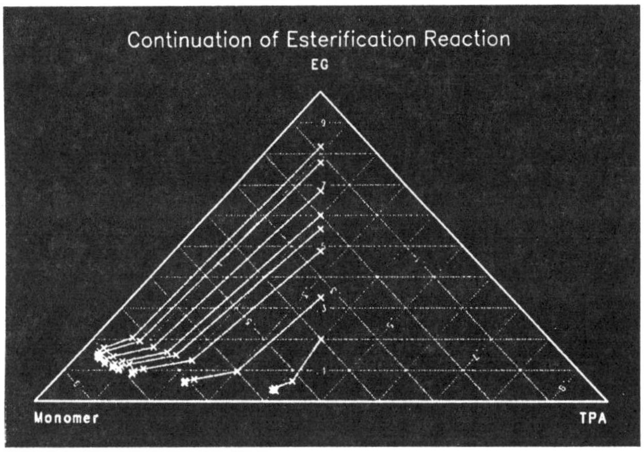

■ *Figure 9. Reaction lines for downstream reactor conditions.*

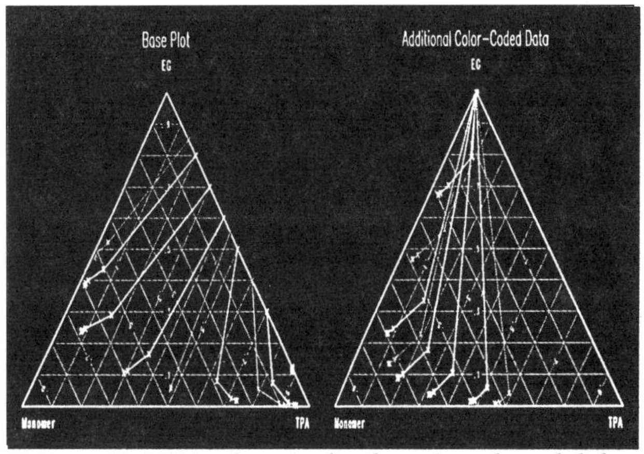

■ *Figure 10. Additional reaction line data using color-coded plots.*

split within the decanter. This resulted in a new proposal in which fluctuating C9 sources with low impurity levels would be treated in a different section of the process. The feed point F then moves to within the two-phase region and goes directly to the decanter.

An added benefit resulted from the diagram approach. When detailed distillation calculations were performed, convergence failures were encountered. Figure 7 presents the mass balances for a wide range of distillate rates with the feed at point F. By inspection, it was found that convergence failure occurs when the distillate rate approaches a turning point of the material balance line. For example, at high distillate rates, the bottoms product is essentially pure C9 and the distillate falls on the straight line projection from pure C9 and F. As the distillate rate is dropped, the distillate composition extends toward the distillation boundary. Convergence problems occur as the composition tries to pass through the boundary. Convergence was again possible when the distillate rate was dropped significantly below the break point. Solutions for all cases were achieved by performing a series of solutions which slowly approached the break point from both sides.

Another example

As a second example, consider the application of plots to the synthesis of reactor networks. The production of polyethylene terephthalate (PET) is performed in a series of CSTRs for reacting terephthalic acid (TPA) and ethylene glycol (EG). It is a liquid-phase reaction with a vapor purge of the water byproduct. The reaction line concept of plotting the feeds to the series of reactors with a set temperature, pressure, and residence time fits well for the application. Figure 8 shows an example set of calculations. Of the several components present in the reactor, TPA, EG, and the monomer product were selected for the plot as the most important for the initial esterification reactors. The initial reactor feed of TPA and EG lie along the diagonal white line of zero monomer mole fraction. The impact of the feed ratio of EG to TPA is quite evident from the wide range of products obtained for the different curves.

L. R. PARTIN is a research associate at Eastman Chemical Co., Kingsport, TN (615/229–5716; Fax: 615/229–3017). He has been active in the process simulation field for the past 14 years, starting with the programming of an in-house simulator and membership on the Industrial Advisory Committee of the Aspen Project.
His assignments have included the management of process-modeling tools and the full range of technology development, from establishing basic concepts, and operating pilot plants, to engineering new facilities. He is currently developing a continuing education seminar on applied chemical process simulation for presentation through the Dept. of Professional Development at the Univ. of Wisconsin in April 1993. A registered professional engineer in Tennessee, he earned a BSChE from the Univ. of Kentucky and an MSChE from Purdue Univ.

It is also evident that the first reactor in a series has a large conversion and that the third and higher reactors add little conversion. A reaction equilibrium constraint results in a limiting conversion as displayed by the end points of the curves.

The engineer must select operating conditions for the reactors. Additional temperature and pressure specifications have been plotted on Figure 8 as yellow and blue curves for comparison to the original white curves. This type of data is beneficial in selecting the operating conditions.

Since additional reactors at the same operating conditions provide a limited conversion, additional reactors should shift the operating conditions. Figure 9 shows a series of plots with a significant change in the temperature and pressure.

Three components are normalized for a given plot but multiple color-coded plots can display more composition information. Figure 10 contains the original data from Figure 8 along with a second plot containing the components that are the majority of the polymer: monomer end groups, carboxylic acid end groups, and the internal repeat unit. The curves are matched to the first plot via the color code.

Reactor performance can be quantified through calculated parameters. Some important results for the polyester reaction are the conversion of TPA, the degree of polymerization, the inherent viscosity, and the rate of a side reaction that forms diethylene glycol (DEG). Contour maps are an

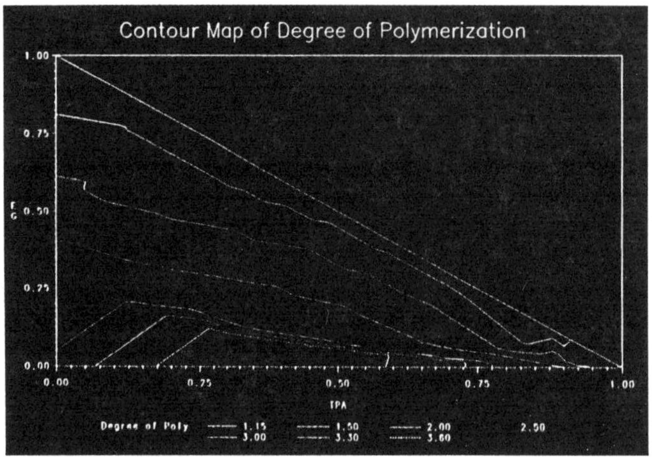

■ *Figure 11. Contour map of degree of polymerization.*

excellent way to plot the results on the same triangular coordinates. The degree of polymerization is presented in Figure 11 for the same data set as shown in white in Figure 8. Therefore, compositions from a data point on Figure 8 can be translated into its degree of polymerization via Figure 11. Another good use of a contour map is to identify operability limitations. For example, a plot of DEG formation rates would display regions of high formation rates that would be unacceptable. A problem was encountered with the plotting of the contour map which can be seen in Figure 11. The analysis software forced the plotting of results throughout the variable domain. This led to erroneous data extrapolation past the reaction equilibrium boundary.

Software specification and development

There are several capabilities that are needed from the process simulator:

- thermodynamic models;
- data banks;
- physical property estimation;
- data regression;
- phase equilibria estimation;
- phase calculations, both vapor/liquid, and vapor/liquid/liquid;
- automated calculation of liquid/liquid phase envelopes, and distillation lines; and
- CSTR calculations.

The main requirements center on the calculation of physical properties and phase equilibria. The simulator must also provide a way to program the calculations and print the information in a desired format. The examples with multiple unit operations in series are performed with standard flowsheet models. We used Aspen Plus.

A menu-driven application was programmed to facilitate the analysis and plotting of the data (based on SAS software). The plotting functions were available in a high level language along with a tool for quick programming of a menu-driven system. The application interactively reads the data files. The user selects the amount and types of data to place on a given plot. Therefore, the user can mix data types such as liquid/liquid equilibria and distillation lines on a graph. Four basic plotting styles were implemented. Triangular diagrams are plotted in both right triangular and equilateral formats. When appropriate, data are plotted in three-dimensional or contour map format. The 3D plots are acceptable for showing general trends but contour maps are better when values are to be read from the plot. The selected data analysis tool needs these capabilities and it must be able to handle the plotting of a large number of curves on a single graph as evident from the example plots.

To sum up

Triangular diagrams give the process engineer insights into the process capabilities in an easily interpreted graphical form. They are tools for the synthesis and design of chemical process flowsheets. A software system was programmed to perform the calculations via additions to a process simulator and an application within a data analysis package. The software has aided engineering studies and the understanding unit operation capabilities. CEP

LITERATURE CITED

1. **Linnhoff, B., and D. R. Vredeveld,** "Pinch Technology Has Come of Age," *Chem. Eng. Progress,* **80** (7), p. 33 (July 1984).
2. **Linnhoff, B., G. T. Polley, and V. Sahdev,** "General Process Improvement Through Pinch Technology," *Chem. Eng. Progress,* **84** (6), p. 51 (June 1988).
3. **Smith, R., and B. Linnhoff,** "The Design of Separators in the Context of Overall Process," *Chem. Eng. Res. Des.,* **66**, p. 195 (May 1988).
4. **Skelland, A. H. P.,** "Diffusional Mass Transfer," Wiley, New York (1974).
5. **Magnussen, T., P. Rasmussen, and A. Fredenslund,** "UNIFAC Parameter Table for Prediction of Liquid-Liquid Equilibria," *Ind. Eng. Process Dev.,* **20**, p. 331 (1981).
6. **Sorensen, J. M., and W. Arlt,** "Liquid/liquid Equilibrium Data Collection 2, Ternary Systems," DECHEMA, Frankfurt, Germany (1980).
7. **Treybal, R. E., L. D. Weber, and J. F. Daley,** "The System Acetone-Water-1,1,2 Trichloroethane, Ternary Liquid and Binary Vapor Equilibria," *Ind. Eng. Chem.,* **38**, p. 817, (1946).
8. **Doherty, M. F., and J. D. Perkins,** "The Behavior of Multicomponent Azeotropic Distillation Processes," *IChemE Symp. Ser.,* **56**, 4.2/21 (1979).
9. **Doherty, M. F., and J. D. Perkins,** "On the Dynamics of Distillation Processes – 3, The Topological Structure of Ternary Residue Curve Maps," *Chem. Eng. Sci.,* **34**, p. 1,401 (1979).
10. **Doherty, M. F.,** "The Presynthesis Problem for Homogeneous Azeotropic Distillation has a Unique Explicit Solution," *Chem. Eng. Sci.,* **40** (10), p. 1,885 (1985).
11. **Stichlmair, J., J. R. Fair, and J. L. Bravo,** "Separation of Azeotropic Mixtures via Enhanced Distillation," *Chem. Eng. Progress,* **85** (1), p. 63 (Jan. 1989).
12. **Klusacek, K., R. R. Hudgins, and P. L. Silveston,** "Triangular Diagrams Teach Steady and Dynamic Behavior of Catalytic Reactions," *Chem. Eng. Educ.,* 176 (Summer 1989).
13. **Wei, J., and C. D. Prater,** "Advances in Catalysis," **13**, Academic Press, New York (1965).
14. **Julka, V.,** "A Geometric Theory of Multicomponent Distillation," Ph.D. Thesis, Univ. of Massachusetts, Amherst (1992).
15. **Julka, V., and M. F. Doherty,** "Geometric Behavior and Minimum Flows for Nonideal Multicomponent Distillation," *Chem. Eng. Sci.,* **45**, p. 1,801 (1990).
16. **Fidkowski, Z. T., M. F. Malone, and M. F. Doherty,** "Nonideal Multicomponent Distillation: Use of Bifurcation Theory for Design," *AIChE J.,* **37** (12), p. 1,761 (Dec. 1991).

Use Process Simulation to Improve Plant Operations

A steady-state model can provide quantitative information about operations, as well as insights for revamps, but it needs to be developed with operations specifically in mind.

Alexandre Dimian,
University of Amsterdam

Today, companies in the chemical process industries (CPI) rely heavily on process simulation to maintain and improve their competitive positions. By integrating several types of software, such as steady-state flowsheeting, dynamic simulation, energy integration, computational fluid dynamics, and expert systems *(1)*, process simulation can indeed serve as a potent tool.

Simulation has been defined *(2)* as "a process of designing an operational model of a system and conducting experiments with this model for the purpose either of understanding the behavior of the system or of evaluating alternative strategies for the development or operation of the system. It has to be able to reproduce selected aspects of the behavior of the system modeled to an accepted degree of accuracy." This definition is fully consistent with the flowsheeting (systems) approach in process engineering *(3,4)*.

The use of flowsheeting in design, however, differs from its use in operations — and this, as we will explain, should be reflected in the development of the simulation procedure.

In design, attention focuses on the main elements of material and heat balances, on equipment investment, and, more generally, on process economics. The systems approach permits the evaluation of feasibility and global plant integration — always for a "predicted" behavior of the reaction system. While a deeper systems analysis of the plant would be worthwhile, considering that the basic design could be responsible for more than 80% of the cost of investment and operation, a detailed simulation is constrained, however, by the project schedule and lack of data.

In operations, in contrast, attention centers mainly on product flow rate and specifications, but also on plant troubleshooting, controllability, and maintenance. The performance of reactors and separation systems impose the rules of the game. They are interdependent and time variable to some extent. Only a detailed plant simulation enables an understanding of these interdependencies and their quantitative evaluation. Thus, the exact knowledge of a *detailed* material and energy balance is by far more important in operations than in design. Even the flow rates of trace impurities are relevant, because they may impact equipment maintenance and environment protection. The material and energy balance as well as the operational characteristics of a plant are highly interconnected, and well suited for a systems analysis. Using commercial flowsheeting software, it is possible to produce a computerized tool that will permit us to learn or even "mirror" the plant behavior under different operating conditions or with different raw materials and product specifications. We call this tool the steady-state plant-simulation model.

Developing such a model is a preliminary and necessary stage in achieving real-time plant optimization *(5,6)*— which involves treating data reconciliation and rigorous simulation simultaneously by means of optimization techniques, whose objective is to maximize process profitability.

In this article, however, we focus on the steady-state model, which is simpler to build, and has a wide variety of applications in its own right; it can be used directly in revamping and a wide variety of other engineering projects.

The use of flowsheeting in design, however, differs from its use in operations — and this should be reflected in the development of the simulation procedure.

Plant-simulation-model objectives

The plant simulation model should mirror the behavior of a complex plant (that is, its material and energy stream network, and equipment performance), subject to constraints in feedstocks, products, equipment capacities, operational parameters, and utilities consumption.

Its objectives include to:

1. provide a comprehensive report of material and energy streams;

2. determine the correlation between the chemical reactors and the separation system;

3. study the formation and separation of byproducts and impurities;

4. support preventive maintenance by tracking performance of key equipment over time, and its relation to the buildup of impurities;

5. improve the robustness of plant operation;

6. assess how to eliminate wastes and prevent environment pollution;

7. evaluate plant flexibility to changes in feedstocks or products;

8. validate process instrumentation and enhance process control;

9. update process documentation and prepare future investments; and

10. optimize the economic performance of the plant.

To achieve these goals, the rigorous simulation model must be tuned. This demands reliable plant data: stream compositions, temperatures, pressures, and so on. These are obtained from test runs during steady-state operation.

Now, let's briefly examine some of the technical elements encountered in building a steady-state plant-simulation model.

Flowsheet recycle structure

To develop a plant-simulation model, we must first develop a Simulation Flow Diagram. It is close to but not identical with the Process Flow Diagram (PFD), because of modeling limitations or flowsheeting convergence constraints. This means that the simulator should be provided with capabilities assuring flexibility both in modeling as well as in flowsheeting procedures — for instance:

- transfer of information between units (such as heat exchanger duty, and mechanical work), or between streams;
- full access to stream and unit variables, both in retrieving as well as in modifying mode;
- powerful user programming capabilities; and
- flowsheet split/merge features.

Advanced commercial simulation systems also come with intelligent graphical user interfaces, which speed the development of error-free simulation problems and provide some help with thermodynamics and modeling.

Nevertheless, the simulation of a large plant produces a complicated flowsheet structure. This can become even more complex, if flowsheet controllers (as discussed below) are introduced.

Unfortunately, the technical literature provides little information about structural analysis of a large plant for simulation purposes. So, here, we will identify a workable approach.

First, the recycle structure of a large flowsheet may be analyzed by means of some simple structures, as sketched in Figure 1. These typically are structured loops, crossed loops, and interconnected loops.

We call *structured loops* those recycle structures where the information flow is independent of other loops. There is no input of information to an internal unit from an external loop, but output is possible. These structures are robust in operation.

Crossed loops generally are for process/process heat exchangers. The convergence treatment is made easier by using transmission of information techniques, in this case, the heat exchange duty.

Interconnected loops exchange both input as output information flows, so they are highly coupled.

The intensive use of *controllers* is another important feature in the simulation of plant operation. Controllers are special modeling units used to manipulate flowsheet variables to achieve a desired result on a process unit or stream. Typically, they serve to regulate unit-operation parameters or adjust input-stream flow rates to meet unit or stream specifications. The action of a controller or design specification unit is illustrated in Figure 2.

In *feedback mode*, the controlled variable is regulated by an upstream manipulated variable. Thus, the controller creates a new recycle loop between (1) the unit varying the manipulated variable, (2) the unit or stream whose property is being measured, and (3) the controller itself. The controller can be placed either outside or inside the recycle loop. If the controller is placed within a recycle loop, however, care should be taken to avoid conflicts in the convergence calculation.

In *feedforward mode*, the downstream variable is set to a specified

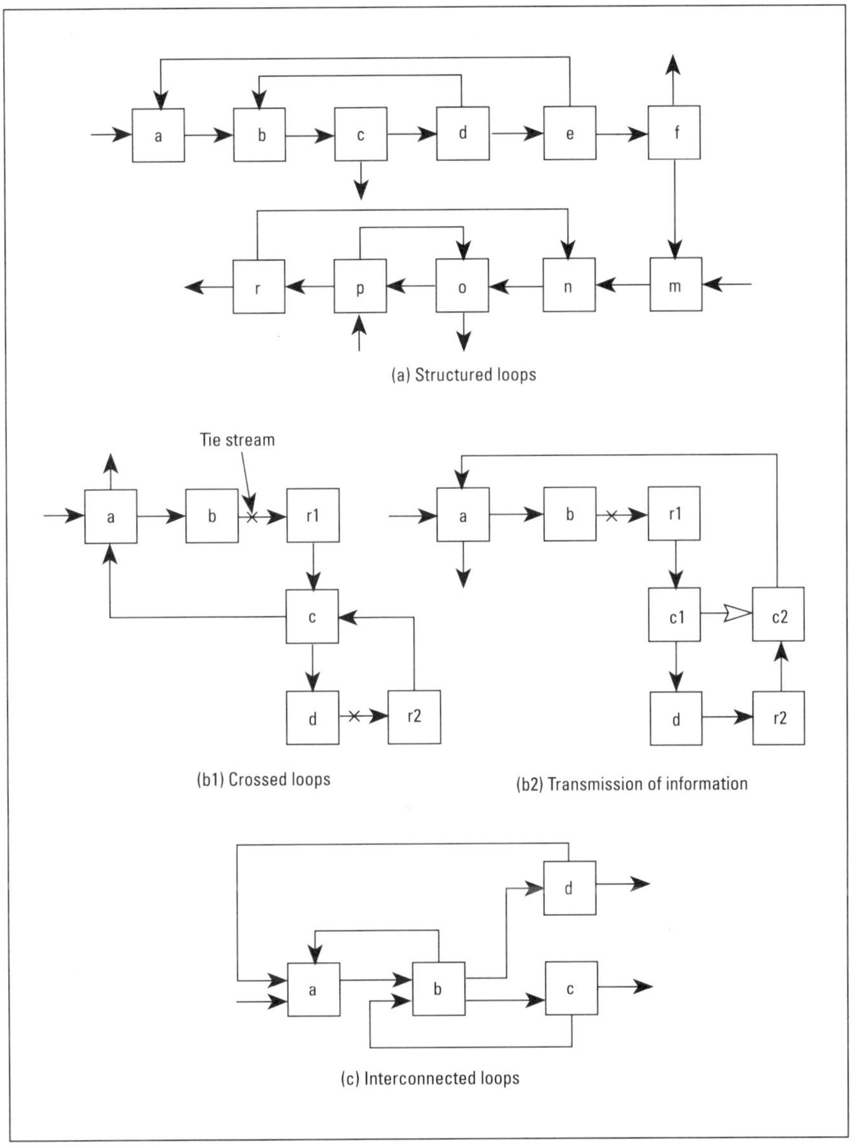

■ ***Figure 1. Simple flowsheet structures.***

value based on an upstream measured variable. This feature also can be used to pass information and eliminate recycle loops.

A particularly powerful flowsheeting feature in simulating the operation of a plant is expressed in ASPEN *(7)* by means of *design specifications*. They represent an imposed value on the flowsheet (that is, a set stream property or unit parameter) or a function of some flowsheet variables. For each design specification, the user must select a manipulated variable, which can be either a unit input variable or an input stream. The main differences from a normal multivariable feedback controller are: (1) the ability to access any desired simulation variable, (2) full programming capabilities, both in formulating specifications as well as in manipulating variables, and (3) the possibility of simultaneously treating tear streams and design specifications in the same convergence blocks. This feature enables the linking of the control features of the simulation model with the main plant-control objectives of the operating plant.

Flowsheet decomposition

Decomposition involves selecting appropriate tear streams. The basic concept in tearing has been reviewed recently *(4)*. Most commercial simulators provide automatic procedures for tear-stream identification and for generation of the calculation sequence. The user also may choose the tear streams, and then let the simulator generate the calculation sequence. Manual sequence definition may be necessary, however, if the controller or design-specification loop crosses the corresponding recycle loop.

The user must evaluate the reliability of the computation sequence, because the automatic techniques minimize the number of recycle streams but do not account for the feasibility of the individual units. It generally is best if recycle computations start before key units, like distillation columns and chemical reactors. Good starting values are necessary in cases where the flowsheet is sensitive to the accumulation of some components. Convergence problems, however, seldom are linked to a lack of extreme initialization accuracy of the tear streams.

A *transmission of information* technique may be used to reduce the number of the tear streams, and to increase robustness in computations. The most typical situations concern the transfer of duty between the tube and shell sides of heat exchangers, and the transfer of work between expanders and compressors.

Some simple rules should be kept in mind when decomposing the flowsheet of a complex plant:

- convert cross loops in linear sequences by means of the transmission of information method;
- avoid large interacting recycle loops in preliminary simulation, or use stream manipulators;
- partition the flowsheet into large blocks, whose functions are relatively independent and which are linked by a minimum number of streams;
- use only the minimum number of design specifications necessary before

obtaining a first converged solution; and

- analyze the relationship between the purge streams and the behavior of the separation system.

Structured loops are easy to converge if the physical plant behavior is correctly managed. Interconnected loops are more difficult; sometimes the numerical sensitivity of convergence procedures can mask the unstable behavior of physically interdependent systems. Simultaneous convergence is recommended if the loops are close on the flowsheet. Stream manipulation can be of help in the case of very large recycles.

The convergence method and its software implementation clearly are important. In my view, however, it is more critical to correctly structure the computation sequence and assure robust convergence inside all individual loops, surveying the behavior of each unit. If the model still shows an extreme sensitivity, this fact could reflect plant operation difficulties. For example, the trend to continuously accumulate some components might be explained by some bottlenecks in the separation system and the need to periodically purge some streams.

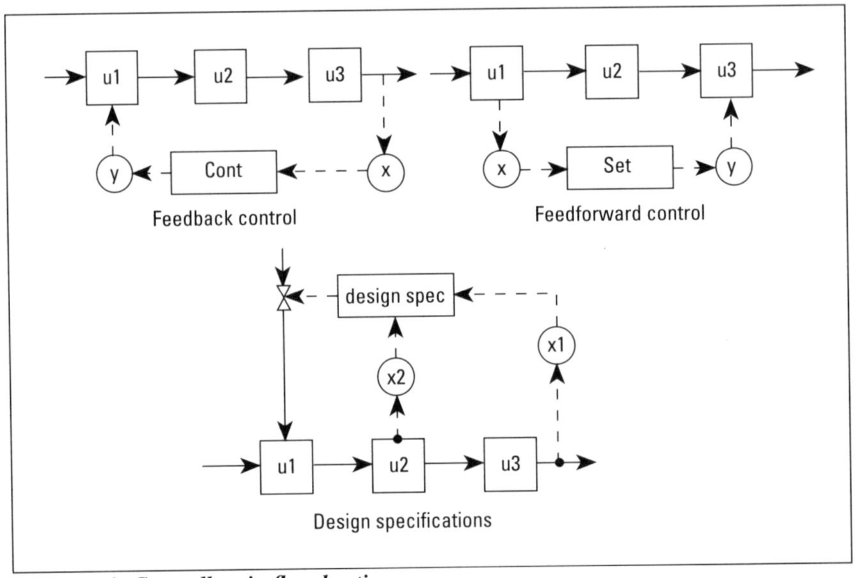

■ *Figure 2. Controllers in flowsheeting.*

Figure 3 illustrates the simplified, generic structure of a complex plant flowsheet, such as is often encountered in the CPI. Typically, the flowsheet might contain more than 50 items, 75–150 streams with 10–20 components, chemically defined or undefined, or gathered in fractions. Visual inspection shows that the plant can be analyzed by means of the simple structures that we have already described. Several large interrelated recycle loops may exist, each of them

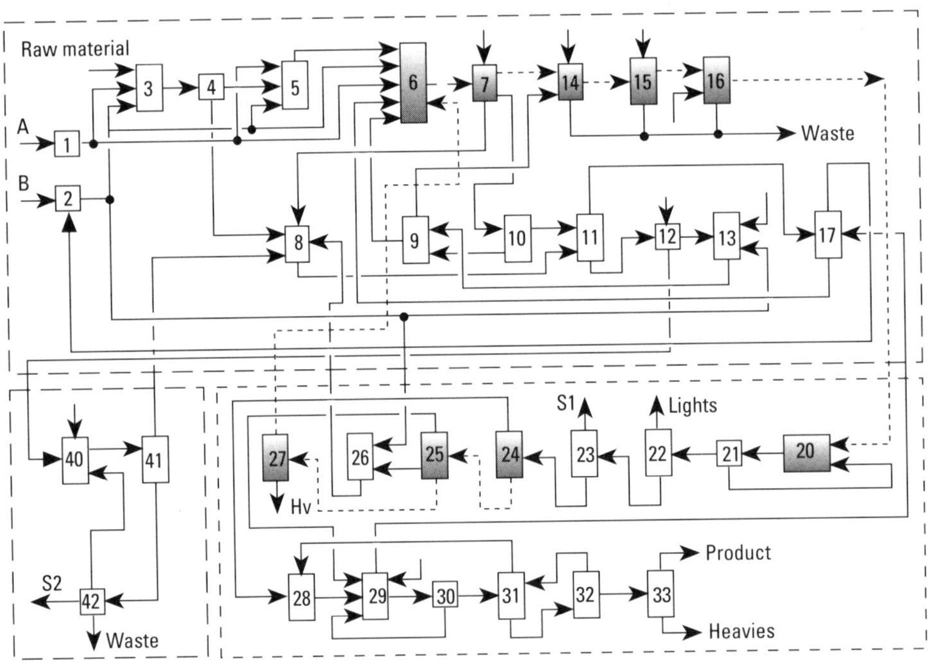

■ *Figure 3. Simulation of a large plant.*

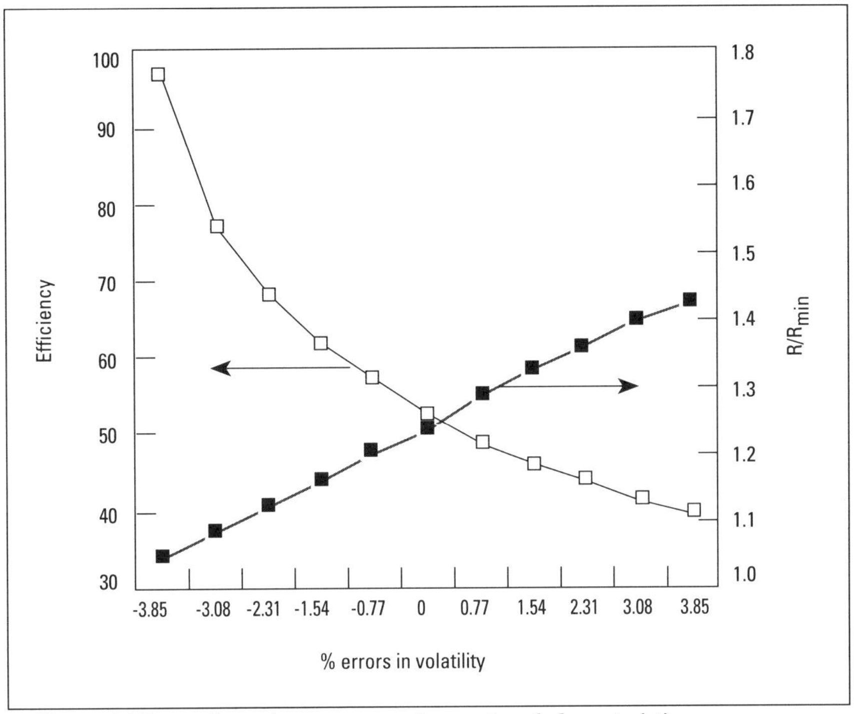

■ *Figure 4. Thermodynamic accuracy and operational characteristics.*

containing other process recycle or controller loops. To facilitate the analysis, we have split the flowsheet in three parts, A, B, and C. The criteria used in this first decomposition are (1) relative functional independence, and (2) minimum number of tear streams.

First, the complete simulation models of the three parts are analyzed, converged, and tuned separately. Later, they are merged into a global simulation model. Its robustness is tested during the sensitivity study after the final tuning.

Thermodynamic study

A thermodynamic study is essential for assuring the reliability of the plant simulation. Indeed, the uncertainty generated by errors in vapor/liquid equilibrium (VLE) data might be more important than the available precision in estimating separation efficiency or operating characteristics *(1,8)*. For instance, consider Figure 4, which gives the results obtained with the Smoker method based on the following real data: feed of a 50/50 A/B mixture; 99% A distillate recovery; a reflux/distillate ratio of 8:1; 66 actual plates; and a relative volatility 1.3 with an absolute error of ± 0.05. The figure points up the significant influence of VLE accuracy in evaluating the separation characteristics — if the relative volatility is underestimated by 2%, the global efficiency goes from 50% to 68%, while for an overestimation of 2%, it drops to 45%. If the minimum reflux conditions are reached, the efficiency determination has no meaning.

Consequently, the thermodynamic parameters have to be accurate and consistent in the working domain. I recommend performing the identification procedure directly on original equilibrium data rather than retrieving them from the literature. The accuracy in enthalpy computation also has to be checked when specifications based on thermal balance are applied, inasmuch as that the improvement in the calculation of *K* values doesn't necessarily produce accurate *H* values.

Because a complete thermodynamic study can take considerable staff resources, its extension has to be tailored to simulation needs. Missing data might be replaced by industrial measurements of phase equilibrium or by organizing a simple experimental setup.

Tuning the material balance

The final document that has to be developed is the *comprehensive stream report*. The accurate description of plant material streams is necessary to calculate both the global economic indexes, as well as the contribution of individual process stages or equipment items. Moreover, the stream report permits an understanding of the main problems in operation and maintenance and, when systematically performed, real-time plant behavior. It also offers a quantitative basis for communication among different members of the plant team involved in operations, technical services, and plant management.

Developing the report requires tuning of the material balance, including updating of the accuracy and consistency of analytical control procedures. This tuning also has diagnostic value. For instance, inconsistency or unbalance in some components may actually originate from the poor performance of another piece of equipment. Thus, the propagation of a default and its consequences on the whole plant can be followed quantitatively. This is represented in Figure 3 by the shaded equipment. There, a badly designed reboiler of Column 24 produces an impurity, which by fault propagation will affect the selectivity of Reactor 20.

After tuning a base-case study, the plant simulation model can be used to investigate the flexibility and robustness of plant behavior to different process variables. Such a sensitivity study enables an understanding of the main difficulties in operations, the interdependence among the recycle loops and among different equipment.

Chemical reactors. Flowsheeting experience reveals that the chemical reaction system is the key element in any material balance. Nevertheless, reactor modeling in flowsheeting software is not as developed as that for distillation. The difficulty in formulating reliable industrial models and the lack of kinetic data are sometimes advanced as an explanation. Thus, in my view, a reactor simulation with detailed kinetics and a realistic flow model may be executed better with specialized software. In fact, in flowsheeting, we only need an accurate description of the transformation linking the input and the output of the reaction system. This simplification doesn't diminish the role of the reaction system in flowsheeting; it just emphasizes the idea that attention has to be focused on its behavior in the system.

This again highlights the differences between design and operations. In the design mode, the modeling of chemical reactors focuses on the main product rates, neglecting quantitatively unimportant reactions. In fact, the time behavior of the reaction system, like catalyst activity, or impurities generation, is difficult to estimate. Indeed, the complex behavior of an industrial chemical reactor generally is not predictable in the design phase. In contrast, in operations, a detailed knowledge of the reaction system is available, but its interaction with other equipment items is still unknown, unless a comprehensive simulation is performed.

The ability to describe the output composition of a complex reaction system for given reactor operating conditions as function of variable input stream is the key feature that we need in modeling the chemical reactor in flowsheeting. Three options are available in commercial simulators:

1. stoichiometric description;

2. yield description; and

3. regression analysis.

In the first case, a set of stoichiometric, independent chemical reactions has to be formulated, and the molar extent of reaction used as the reaction coordinate. Accordingly, the material balance is closed, both on the mass as well as on the atomic constraints; the material balance will be consistent over the whole flowsheet.

The second method is useful when a stoichiometric description of a complex reaction system in not possible, but a mass product distribution is available. It can handle chemically undefined components, such as lights, heavies, or grouped components (as determined by chromatographic or other analysis), and often found in many industrial processes (cracking, oxidation, and so on). The stream report should handle these components the same way that plant documents do. If the reactor output and input aren't linked by a stoichiometric relationship, a discontinuity in material balance will be introduced. The mass balance is respected, but not the atomic balance. Because the product distribution is independent of the reactor input composition, the effect of the recycle on the reactor behavior will be cancelled. The user should carefully examine the stream report and finely adjust the yield coefficients until the errors in a component's material balance are minimized.

Nonlinear regression models are necessary when the relationship is more complex than just a stoichiometric description. Different effects, such as components concentration, temperature, catalyst influence, and operating parameters, also may be considered. Their implementation within a simulator requires advanced features in accessing flowsheet variables and programming.

Separation units. At a preliminary stage, component splitters and flash units may be used to obtain an approximate material balance. Even at this level, however, some convergence problems may be encountered — if there are strong loop interactions. Such approximations may provide some insights, but ultimately we must incorporate more and more rigorous models and accurate thermodynamics.

In this context, one of the most sensitive problems is how to introduce the separation efficiency, to match specifications in distillation operation units. There are two possibilities:

1. working in theoretical trays and global tray efficiency; and

2. using Murphree efficiency.

The first mode is preferred by some practitioners *(8)*, although Murphree efficiency also may be useful in tuning and monitoring plant performance. Mass-transfer rate-based models introduced recently in commercial simulation *(7,9)* also may help in describing composition effects on separation efficiency, especially in the case of high-purity distillation.

The treatment of impurities

The handling of impurities, often neglected or unknown in the design phase, is one of the most important aspects of plant operations. Some impurities have their origin in the raw materials; others are produced by thermal decomposition or other actions in the process. They cause drops in performance, equipment fouling, and product losses, and may lead to environment pollution. Only a global plant simulation allows the quantitative identification of their sources, buildup mechanism, and consequences on equipment performance. Thus, such a simulation can be particularly useful for retrofit projects, where it may provide insights that lead to a relative reduction in impurity levels.

Consider a process that can be described by the following simplified, general schema (which, by the way, is relatively common in the CPI).

1. Intermediate products:

Raw materials $\rightarrow X + S_1 + Imp_1$

2. Main product:

$X \rightarrow P + S_2 + Imp_2$

3. Secondary-product recovery:

$S_2 \rightarrow X + S_3 + Imp_3$

where P is the main product, X is an intermediate product, and *Imp* designates impurities — each of which can include a variety of components such as I_1, I_2, and I_3 — and S secondary products.

The impurities can consist of low, intermediate, and high volatility components. The intermediate volatility ones are more difficult to han-

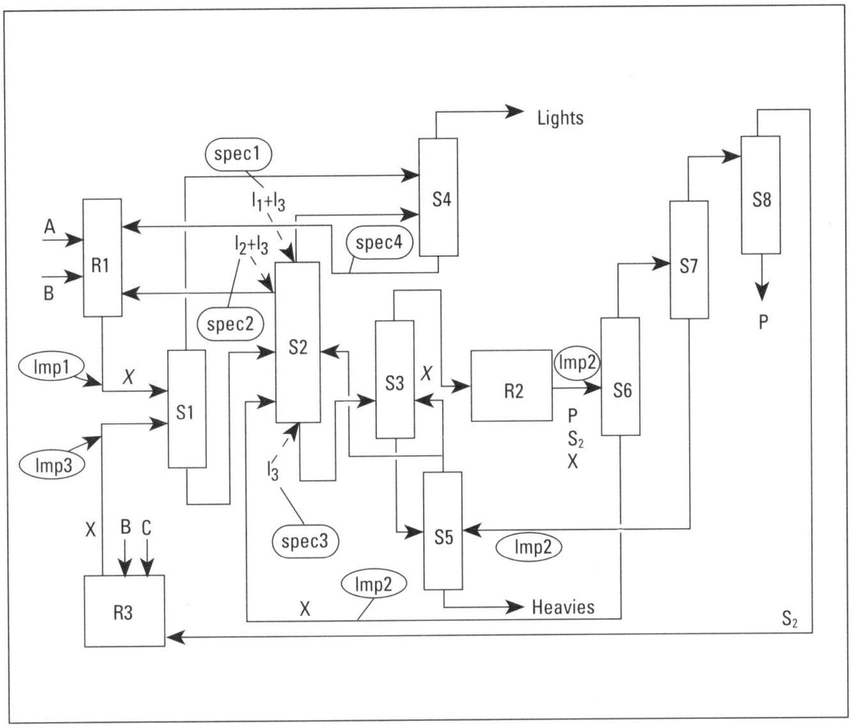

■ ***Figure 5. Treatment of impurities by simulation.***

dle, as will be shown in the following example.

There may need to be limits on impurity concentration, for example, to prevent polymerization or technological risks. On the other hand, some impurities may be useful for their catalytic or inhibitor action. Therefore, an important number of variables have to be "controlled" as explained before, creating new convergence loops. Impurities can be eliminated physically (via purge streams), or chemically.

The treatment of impurities in a real plant following the above reaction schema is shown in Figure 5. The three reaction steps take place in reactors R1, R2, and R3. Accordingly, three main recycle loops are created, which are interconnected by means of streams issued from the associated separation systems. These loops cross in the separation unit S2, which actually is a large distillation column.

Visual analysis suggests that S2 may be responsible for impurities management over the whole flowsheet, because most of the lights and all of the heavies are collected here. (They then are eliminated in S4 and S5, respectively.) In fact, the operation of S2 is subject to several constraints on impurities treatment, noted here with specie, whose significance is as follows:

spec1 = maximum concentration of light I_1 polymerizable impurity, which can be removed in the Lights;

spec2 = maximized concentration of intermediate impurity I_2, which has to be destroyed chemically in R1; and

spec3 = optimum concentration of a light impurity I_3 with catalytic action in the R2 reaction step.

Distillation, sidestream, and reflux flow rates are manipulated to meet these specifications. Even if the simulation of the isolated S2 unit doesn't raise any particular problem, its integration in the flowsheet is very difficult.

Intensive simulations, however, make clear the major role of S4 in making the material balance converge. The operation of this unit, actually a small distillation column, is constrained by a specification, *spec4*, concerning the minimum recovery of valuable components in the bottom product. This unit controls not only the output of light impurities, but also the concentration level of intermediate impurities I_2 and I_3, and, consequently, the specifications of the unit S2. Moreover, the separation performance of S4 is affected by the polymerizable impurity I_1. The chemical destruction of the light I_1 polymerizable impurity prevents fouling of S4, as well as improves the operational robustness of the whole flowsheet. This operation will create new heavy impurities. The appropriate flowsheet modifications can be examined using the actual plant simulation model.

The plant simulation model

Developing a simulation model of a large plant requires both a significant initial investment, as well as continuous improvement, updating, and maintenance of the simulation tool. Life-cycle models are powerful and practical techniques that can be used to manage complex systems like those involved in computer-based systems engineering *(2)*.

Three life-cycle forms may be considered: (1) the V-model; (2) the waterfall model; and (3) the spiral model. Based on my experience, the V-model (see Figure 6) is the most convenient for managing the development of a plant simulator.

In the V-model, the basic ideas are to decompose the work into a number of tasks, and then separate them into "specification" (design), and "production" (coding) tasks. The left side of the V represents tasks for the refinement of the specifications, while the right side represents production and assembly tasks.

In this approach, management and quality assurance are carried out together. The correctness of each step is "verified" before proceeding to the next one. Each production task is "validated" against the corresponding specification task by following the V-cycle, as depicted in the figure. Here, verification means that the product meets quality characteristics, such as consistency, completeness, and correctness, while validation means that the product satisfies the specification requirements. Unforeseen project problems may be managed by properly cycling through specification-implementation-testing procedures.

A generic framework of a V-model is presented in Figure 6 *(2)*. It shows the seven key tasks in developing the simulation model:

1. *Requirements specification.* Here, the so called "specifications," such as systems capabilities, operational constraints, the software environment, and so on, are defined.

2. *Architectural design.* This answers the question of how the system has to be designed to fulfill the above specifications. The output gives a preliminary design framework, like architecture, data structures, components, and interfaces.

3. *Detailed design.* This looks at each data structure, component, and interface. The output is a complete definition and algorithm for each designed component.

4. *Implementation and unit test.* The data structures and component algorithms are implemented, using the programming language and operation system defined in the specifications. Tests of individual components or on a limited scale also are performed.

5. *Integration testing.* This task executes the progressive assemblage of components and data structures.

6. *System testing.* Here, the system is validated with respect to its functions and against the specifications. It's common to call this an alpha test.

7. *Acceptance testing.* The conformity and robustness of the product is checked against the requirements formulated in the first task. This often is called a beta test — until final acceptance.

Tasks 1–3 may be included in the "specification" phase, while Tasks 4–7 handle the "production." It is worth noting that the "implementation and unit test" task, which corresponds to a coding (programming) activity in a software project, is an important task, but not the most important one from investment point of view. Typically for software, initial programming accounts for less than 10% of overall costs, while maintenance generally accounts for more than 50%, testing and integration 20%, and specification and design 20%. Also, programming usually is responsible for only about one-third of all errors, against two-thirds for conceptual mistakes.

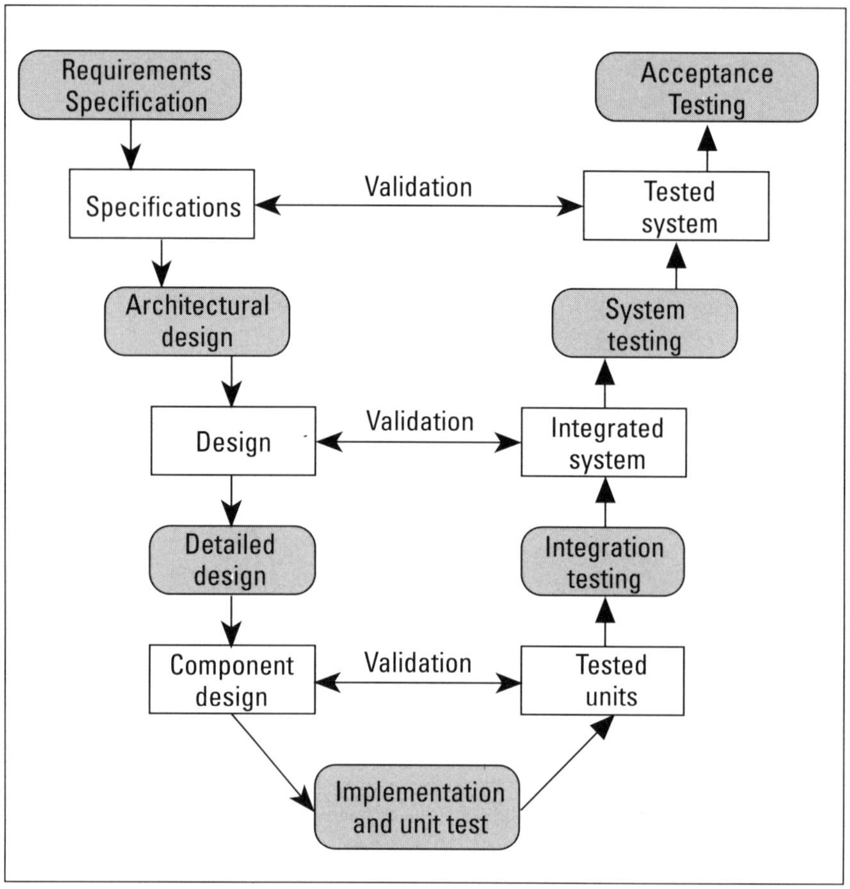

■*Figure 6. V-model life cycle of a plant simulation model.*

A. DIMIAN is an associate professor in the department of chemical engineering at the University of Amsterdam, Amsterdam, the Netherlands (0113120/525–6034; Fax: 0113120/525–5604), where he teaches process design and integration, and specializes in process simulation. Before joining academia, he spent 10 years as a consultant in computer applications for petroleum and chemical engineering, working for major French companies and with leading process simulation systems. He has over 26 years of experience in industry and teaching, and has published about 30 technical articles, and four books. He received an MS and PhD from the Polytechnic Institute of sBucharest, both in chemical engineering. He is a member of AIChE.

Applying the approach

The general framework presented above generally is valid in the development of any complex computer-based applications in process engineering, such as static/dynamic plant simulation models, or model-based training systems. The V life-cycle concept may lead to a reliable and maintainable software tool. When adapting this approach for a plant-simulation model, however, some specific elements should be considered.

Requirements specification involves developing a clear definition of the objectives of the plant simulation model — for instance, the material balance, energy balance, stream physical properties, utilities balance, catalyst consumption, impurities accumulation, heat exchanger fouling, and the like. This task requires a comprehensive analysis of process technology and process data. A collection of benchmarks from typical test runs, and comprehensive technical reports have to be provided to perform the final acceptance simulations. This step also may handle computer science aspects, like hardware and software requirements, network usage, the graphical user interface, and connection to a database-management system.

Architectural design should analyze the input/output structure of the model, and the definition of design specifications or other control loops. The structured PFD decomposition has to be carefully performed, and the convergence properties studied. A prototype model using black boxes and only the material-balance mode could be tried at this level. Complete definitions of chemical components and general thermodynamic methods also have to be given.

Detailed design consists of a full description of the individual components — in this case, the unit operation blocks: design characteristics, simulation model, unit specifications, thermodynamic model, thermodynamic parameters, and performance.

Implementation and unit test means the coding of each unit operation, using a keyword language or a graphical input mode. The calibration of performance of some units, like stage efficiency in distillation, have to be studied at this level. This activity requires intensive usage of thermodynamic tools and true conformance with plant measurements. It generally takes more time than just "programming" in software technology.

Integration testing mainly checks the behavior of the plant model in flowsheeting mode: the robustness of the convergence of the specification and recycle loops, the presentation of reports and graphics, the link with database systems, and so on.

System testing means calibration of the plant model against given benchmarks and the validation of the global behavior. Thus, the differences between simulated results and actual measurements have to be minimized by adjusting some parameters. Sensitivity studies also may be performed at this level, to explore the organic link between different system parts. Inconsistencies in the specification or in the system design may be discovered here. Thus, the V-model enables a straightforward identification and treatment of these problems.

To sum up

Process simulation can provide critical leverage to companies in the CPI. So, not surprisingly, flowsheeting is becoming much more common in process operations. Indeed, advances in both computer hardware and software are enabling the use of powerful simulation tools directly at plant sites.

The systems approach is the only way to understand the subtle interrelations among raw materials, product specifications, and equipment operation in a complex plant. With the help of a powerful commercial all-purpose simulator, it is possible to build a reliable plant simulation model that can be used both in operations and preventive maintenance, as well as in retrofitting studies.

A proper PFD decomposition, the structured treatment of the convergence loops, and the accuracy of thermodynamic models are the most important conditions for a reliable simulation model.

For a large plant, the development of a comprehensive simulation model requires a considerable investment. The project may be optimally managed by adopting a V-model-based life-cycle methodology. CEP

Literature Cited

1. **Dimian, A.,** "State of the Art in Process Simulation," Informatique Petroliere Scientifique seminar documents, IPS, Paris (1988–1993).
2. "Systems Engineering — Principles and Practice of Computer-Based Systems Engineering," B. Thomé, ed., Wiley, New York (1993).
3. **Douglas, J. M.,** "Conceptual Design of Chemical Processes," McGraw-Hill, New York (1988).
4. **Mah, R. S. H.,** "Chemical Process Structures and Information Flows," Butterworths, Sevenoaks, U.K. (1990).
5. **Canfield, F. B., and P. K. Nair,** "The Key of Computed Integrated Processing," presented at ESCAPE-1, Elsinore, Denmark (May 1992).
6. **Lauks, U. E., R. J. Vasbinder, P. J. Valkenburg, and C. van Leeuwen,** "On-line Optimization of an Ethylene Plant," presented at ESCAPE-1, Elsinore, Denmark (May 1992).
7. "ASPEN User Manual," release 8, Aspen Technology, Cambridge, MA (1988).
8. **Kister, H. Z.,** "Distillation Operation," McGraw-Hill, New York (1992).
9. **Ryan, J. M., C.-L. Hsieh, and M. S. Sivasubramanian,** "Predict Misting and Bubbling in Towers," *Chem. Eng. Progress,* **90** (8), p. 83 (Aug. 1994).

Make the Most of Your Replacement Distributed Control System

The issues involved in installing one generation of distributed digital control for another significantly differ from those involved in replacing an analog system.

Gerald R. Brown,
Bechtel Corp.

■ *Typical alarm system wiring. Your worst nightmare!*

It is hard to imagine a major process-related project being implemented today without a distributed control system (DCS). Such systems evolved from being the expensive experiments of the mid-1970s to the *de facto* standard of the 1990s. Some of the early systems, now approaching two decades old, still are in productive service.

While admittedly functional, many of these older systems have problems. Among the most discernible are the obvious afflictions of age common to complex electronic systems — including circuitry and component failures (often due to a combination of neglect and heat), mechanical fatigue of circuit boards and connectors (frequently due to improper maintenance procedures), and the wearing out of the ever-visible operator keyboards and cathode ray tube (CRT) displays (due to constant use). Less apparent are the nagging concerns of many operations and maintenance departments, often the prime justifications for replacement:

1. availability of replacement parts;

2. scarcity of qualified engineers and technicians; and

3. inability to implement newer techniques, applications, or technology.

Problems with aging systems

These concerns, in essence, deal with the distinct issues of reliability and flexibility. The core question to be asked regarding reliability is: "How long can I expect to continue operating my plant/process before operations (profitability) is impacted by my inability to secure replacement parts or technical assistance?" As an illustration, consider the plight of plants struggling to obtain specialized parts for the computer systems that often accompanied a DCS. These computer-based systems were never manufactured in significant quantities and the manufacturers may no longer even provide replacement parts.

The flexibility issue is less objective in asking the question: "How much is my operation (again, profitability) negatively impacted by not being able to do something now available with newer systems?" For example, the common practice of integrating a programmable logic controller (PLC) to the DCS was a technological leap into the unknown 15 years ago. Other issues might include redundancy, remote input/output (I/O), and information integration *(1)*.

Integration issues, particularly real-time cost and quality information, also are often-cited reasons for replacing older systems. For many production departments, the DCS is the information backbone for both quality and cost control *(2)*. In this situation, the justification process can be made much more objective and quantitative as actual production data are used in the analysis leading to a system replacement decision *(3,4)*.

Another issue often considered is the investment loss if a system is scrapped. This cannot be quantified in an accounting sense since electronic systems are fully depreciated long before they are considered "aged," and ongoing maintenance and training are usually expensed against operating budgets in the year of occurrence. The built-up experience, knowledge, training, and comfort level with the installed DCS, however, are difficult to assess. They seem much easier to evaluate, though, if the experiences are negative; frustration, even fear, can easily outweigh familiarity.

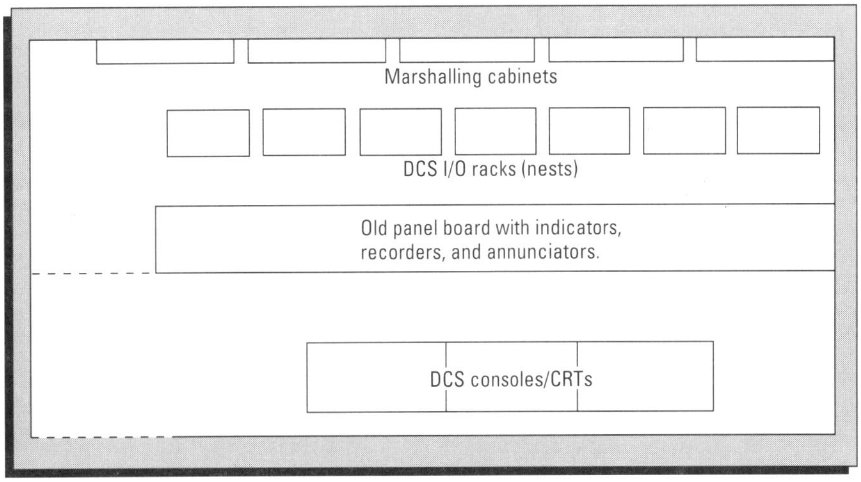

■ ***Figure 1. Existing system layout.***

The transition to a new DCS could be as simple as one-for-one replacement, but that is unlikely.

Planning the transition

After you have made the decision to replace your aging DCS with this year's model (the fun part), you must come to grips with the consequences of that decision. You now have to plan and execute the replacement project (the work part). You will face many technological obstacles. Before focusing on those, however, consider the psychological impediments that may be thrown in front of the project — these may pose very real roadblocks to effectiveness, even success. You must overcome the natural, prevalent, and frustrating resistance to change. Most likely, people are comfortable, even happy, with the current situation. You will need to move beyond viewing the project as merely "replacing in kind." Indeed, it is unlikely that you can buy a system as limited as the one you are replacing.

The incessant march of technological progress in the DCS marketplace (marketing of features) presupposes that you are buying a quantum leap in capability and flexibility in your new control system. The first step in planning the transition should be to throw away the old paradigm and see the new system for its capabilities, not the capabilities of the one that you are replacing. Paramount in this analysis is how the new system will interact with people — that is, how well the system meets the future needs of the operators and managers *(5, 6, 7, 8)*.

This vantage allows you to see the new system installed and doing new, more, and better things. The decision to install a DCS in 1978 wasn't predicated on doing the "same old thing." The stage was set for innovation and you have the opportunity to resume that trend.

Once you can envision the new system doing its new things, plan its installation with that end in mind. Don't start with planning engineering (configuration) or design (installation) until you have a firm and detailed cutover plan developed with operations in mind. From that plan, back into a logical construction sequence. And, finally, from the construction planning needs, develop the engineering/design plan. Of course, training of people to accept and use the new system must accompany the whole process. In logical order, the planning steps are:

1. start with the end in mind;

2. plan the cutover from the operation's point of view;

3. back into the construction plan;

4. develop the engineering plan to support the construction needs; and

5. educate and train the people to take advantage of the new system's capabilities.

Figure 1 illustrates how, even though the existing field wiring, marshalling cabinets, and marshalling wiring will be reused, the existing DCS I/O racks (cabinets) must be removed one-by-one and replaced with a new DCS cabinet, while the remaining cabinets are still functional. In this example, the designers do not have the luxury of removing all the cabinets at the same time because that would require a plant shutdown.

A planned, orderly cutover to the new system lends itself well to a comfortable transition for the operators, assuming they have had the appropriate training in advance. Many times the engineers involved in this type of on-line transition assume the operations department is as familiar with the control equipment as the engineering department. You must evaluate the training needs of all people involved, including engineering, operations, maintenance, and management. This critical evaluation early in the project will help avoid the occurrence of "first-time users of sophisticated control systems who [experience] a disappointing difference between expected and realized improvements in process control" *(9)*. The people who will use the new system need more than just the mechanics of button pushing, they need to see and understand your vision for the new system's operation.

Software engineering

Once you begin the engineering portion of the project, it will become quickly obvious that the heart of the control system's software effort is intangible in classic engineering practice. The algorithms, function blocks, or sequences necessary to configure a DCS are rarely documented in a manner allowing them to be seen, heard, or felt. Instead, they are represented as "bubbles" on a process and instrumentation diagram (P&ID), or blocks on a loop drawing, or not shown at all.

To make the transition from an existing system's software to a newer system, you must pick a consistent methodology to depict the intent and scope of the software functions. One successful method is to use a combination of P&ID symbology (ISA S5.1), complex loop diagrams with descriptions for those loops too complex or cumbersome to be shown on a P&ID, and a relational database index.

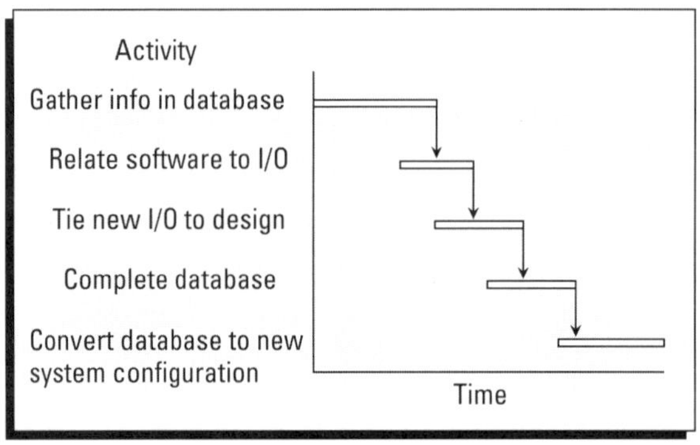

■ *Figure 2. Engineering activities.*

The index becomes the depository for all data pertaining to the other two "master" documents. All DCS functions and subsequently, all tags, are shown on either a P&ID or a complex loop diagram. All tags are listed in the index. The index then becomes the document from which you manage the engineering, design, construction, and cutover activities. Note the significance of the database effort as an integral part of the engineering activities depicted as a generic Gantt chart in Figure 2.

The starting point for your index is the existing system's database — potentially one database for the software functions and a second one for the I/O. The I/O, after all, are where the real world (transmitters, valves, and so on) meet the software world. The I/O database (including its relation to the software database) forms the core of the physical design. This physical design drives the entire engineering process, because the scope typically involves lifting the field wiring from the older system and re-terminating it on the new system. The database merely serves as the depository of the wiring information. Once the existing database is complete, by either survey or faith in your plant's documentation, the transition to the new system can begin.

The transition to a new DCS could be as simple as one-for-one replacement, but that is unlikely. Regardless, one tenet must remain throughout the project: all I/O and software tags in the existing database must be accounted for, in some manner, in the new system's configuration. This relating of the existing I/O and software functions in the index provides a common point of reference for the engineering of the new DCS software and the physical design.

Figure 3 is a typical database printout of a common temperature-to-flow cascade control loop. Note how the old DCS had three tags, while the new DCS requires four. Yet, only two are common to both systems. This is a common situation, because many of the old DCSs had

NEW TAG	OLD TAG	SERVICE	FUNCTN	PNT TYP	EU LOW	EU HI	EU DESC	LOOP	MOD	PROC	PNT
2FC156	FIC-156	Chilled Water Flow	Softwr	PID	0.0	2,500.0	GPM	01	05	—	—
—	FY-156A	Square Root — Old	—	—	—	—	—	—	—	—	—
2FI156	—	Chilled Water Flow	E-AI	AI-SORT	0.0	2,500.0	GPM	01	05	03	06
2FY156	—	Chilled Water Flow	E-AO	AO-LIN	0.0	100.0	% OUT	01	05	12	02
2TI159	TI-159	Cooler Temp	E-AI	AI-LIN	0.0	100.0	DEGF	04	02	01	01

■ *Figure 3. Example datbase (new functions and addresses shown).*

hardware I/O and software "slots" that were linked.

In the newer systems, the separation of control functions and I/O no longer require this formal relationship. Instead, the I/O now require a separate tag as an independent software function (in this case, the I/O tags take the same name as the field device). As further example, the new system has square root extraction as a function of the analog input point configuration, whereas the old system had this function as a separate algorithm (actually, a separate electronic module).

Also note the "function" and "point type" fields allow identification of a point's intrinsic function and differentiation of its extrinsic type (between real I/O and software tags). This manner of classification provides assurance that:

1. all existing I/O and software tags are identified;

2. all new software (engineering) functions are identified;

3. engineering can relate old tags/functions to new tags/functions; and

4. design and construction can relate old I/O to new I/O.

Physical design

The functional design stage is the real heart of the replacement project. It is the overlap between the definition stage of the project, where your feasibility analysis matured into a specification, and the implementation phase, where startup, commissioning, and maintenance are the focus *(10)*. The transition planning and software engineering efforts, shown on the Gantt chart in Figure 4, are precursors to actually installing the system. The physical design is ultimately manifested in the construction package.

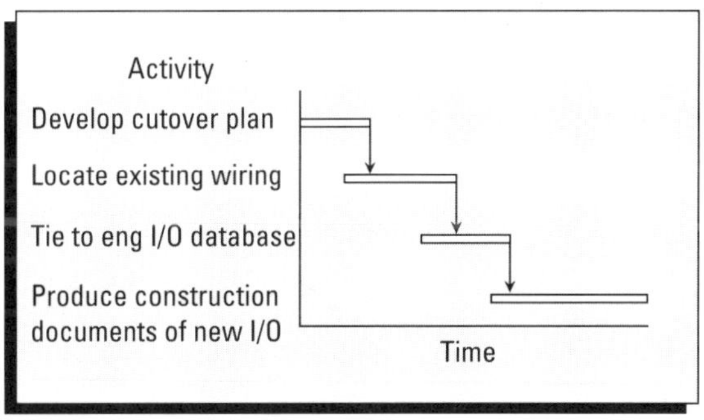

■ *Figure 4. Design activities.*

Construction in an existing, running plant necessitates a different focus than a grassroots project. Operation of the plant usually determines the schedule dates as well as the cutover sequence. The design must follow the construction plan, which follows the cutover sequence.

Recall the situation depicted in Figure 1: The design includes the physical aspects of removing the existing cabinets one-by-one, as well as the operational aspects of maintaining an operating plant while part of the plant is running on the old DCS and part on the new DCS. Obviously, assigning I/O to wiring without considering operations and cutover will lead to a lot of unpleasant surprises in the cutover phase, possibly even unplanned shutdowns.

The design activities focus on the physical aspects of installing the hardware and connecting the I/O to the existing wiring. Unlike a grassroots project, replacing an installed DCS requires considerable attention to the available space and the length and routing of the existing cables. Space issues often necessitate removing pieces of the old system in order to install pieces of the new system. This piecemeal installation approach requires careful planning from both a process operations and physical installation viewpoint; both systems must be functional at each subsequent stage of removal and replacement. Use of the engineering index, on a tag-by-tag basis, is the preferred method for maintaining integrity of both systems. Because existing field cabling is reused, the length and location of those cables will determine the location and content of the new termination (or marshalling) cabinets. From these requirements, location drawings, loop drawings, termination lists, and installation details are prepared.

Cutover

In developing the cutover sequence, you must address issues of process operations, convenience, and safety. Process operations relates to the ability to continue production during a cutover. For example, a control loop on the old system may be put in manual while the transmitter input is moved from the old to the new system. Convenience involves whether you are able or willing to operate from two different systems simultaneously. In this example, the operator would have the transmitter on the new system and the valve on the old system for a period of time. Safety considers the viability of maintaining a safe operation during the transition. How well can the operator move between the two systems, and how fast can that person react to a process upset shown on the new system when the valve is still controlled by the old system? The response may be acceptable for pH control of a cooling tower basin, but not for cooling water flow to a reactor.

A "hot" cutover, one where loops are switched between the old system to the new system one-at-a-time and the process continues to run, is common for DCS upgrades, especially in noncritical processes. In the hot cutover case, the goal may be to leave the old system running as long as possible as a backup. As loops are moved over to the new system, reliance on the new DCS grows. For this reason, the transition to the new system should be based on the operator's ability to comfortably operate the plant, not the capability of construction to swap over the loops. Note how the attention paid to training early in the project will be rewarded during this critical phase *(11)*.

In applications where a mistake can create a dangerous situation or can cause a shutdown that is difficult to recover from, a planned shutdown is required. This involves a "cold" cutover, one where the process is shut down, and is typical where the DCS is involved in safety shutdown functions. In these applications, because the potential for error is greatly increased, it is advisable to leave as much of the old system intact as possible until after the plant has restarted. This backup provides a margin of safety for protecting production.

Seize the opportunity!

Existing physical conditions, such as space and wiring, along with the limitations of maintaining process operations make the task of engineering and designing a DCS replacement much different than a grassroots installation. While there is a distinct loss of flexibility in assigning I/O independently from process and physical constraints, the flexibility gained with new features, functions, and capabilities make the new DCS much more valuable and usable than the one being replaced. Though the initial goal may be to replace the DCS "before it is too late," the engineers replacing the aged system have the opportunity to turn a replacement project into an enhancement opportunity by continuing the tradition of innovation. ♦ CEP

Literature Cited

1. **Hoffman, R., and B. Wolske,** "Cement Plant Enters Second Generation of Automation," *Control*, p. 24 (Sept. 1990).
2. **Bartos, F.,** "Quality Control Seeks a Plantwide Image in Process Industries," *Control Eng.*, p. 50 (June 1991).
3. **Bansal, S.,** "How to Approach a Plantwide Control Project," *Control*, p. 42 (Jan. 1991).
4. **Beard, M.,** "Getting the Most From Your DCS," *Control*, p. 26 (Oct. 1990).
5. **Babb, M.,** "What DCS Operators Want," *Control Eng.*, p. 53 (Nov. 1991).
6. **Wallace, T.,** "The Future of DCS Operator Interfaces," *Control*, p. 36 (Oct. 1990).
7. **Rice, G.,** "The Working Environment of the Operator Now and in the Future," *TAPPI J.*, p. 120 (Nov. 1988).
8. **Collins, D.,** "Control Room Instrumentation," *Measure. & Control*, p. 193 (Sept. 1989).
9. **Price, V.,** "Control Training: Who Needs It?," *InTech*, p. 170 (Sept. 1990).
10. **Harvey, L., and S. Bansal,** "Select the Right Systems Integrator," *Control*, p. 62 (Feb. 1991).
11. **Dennis, H.,** "Designing a DCS System," *InTech*, p. 34 (Sept. 1991).

G. R. BROWN is a project engineer for Bechtel Corp., Houston, TX (713/235–4059; Fax; 713/235–5019). He has worked for 10 years at Bechtel and for four years for a DCS manufacturer as an application engineer. He has 16 years of experience in engineering, design, construction, and startup of chemical, refining, and pulp and paper projects, including dozens of DCS and PLC installations. He has a AS in mechanical engineering from the University of Louisville and a BS and an MBA from the University of Phoenix. He is a registered professional engineer in Texas and Mississippi, and a registered professional control systems engineer in California.

Rethink the Role of Control Valves

Some units now have intelligence—and this can markedly improve process operations.

Barry A. White,
Valtek StarSystems

Control valves have been used to regulate the flow of fluids in pipes as long as most of us can remember. They have been the third, and most overlooked, part of the control loop. Different chemical manufacturers, however, have improved their processes by using intelligent control valves to measure and control the flow of their fluids. They have seen dramatic improvements in the ability to measure their process conditions and to protect the process fluids in the pipes, which have helped their profitability in these times.

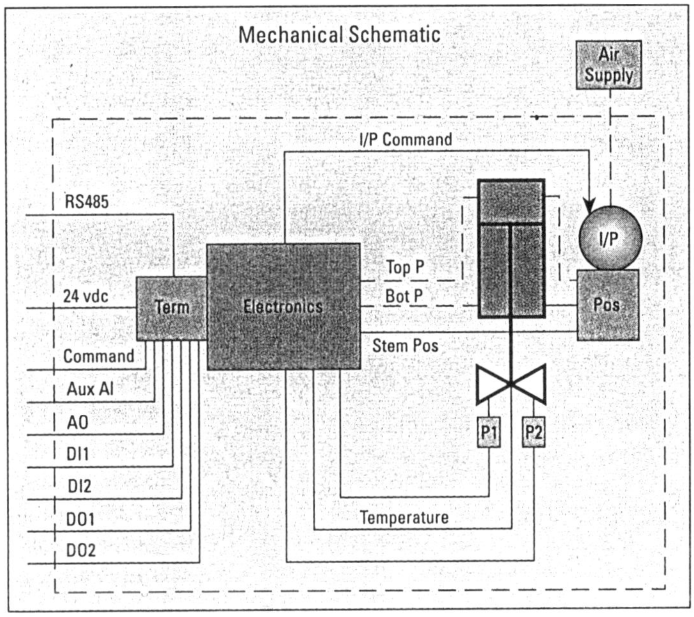

■ *Figure 1. Intelligent Control Valve System Schematic Diagram.*

As you know, a control valve is a variable restriction in the line that traditionally has received its position command from a separate controller, either a single-loop controller, or part of a distributed control system. It has performed its assigned task willingly without much complaint, as long as the designers specified the internals properly. And this was about as much interest the designers had in control valves—just making sure they did not wear out too fast. But what the designers overlooked was that the control valve is the only part of the control loop that actually affected the process. The transmitters and controllers just measured and judged how good the process was doing; they did not have any actual effect on the process.

Several control valve manufacturers looked at their industry and how their customers were using their products and decided to change the situation. Specifically, they saw that the defense and automotive industries, for example, had made major advances in using rugged sensors and microprocessors in their products to meet demands caused by external sources. So these control valve manufacturers applied these technologies to their products and developed intelligent control valves.

Intelligent control valve systems

First of all, what does intelligence in a control valve system mean? There has been some confusion in the use of the terms smart and intelligent. Smart generally means that the device has a digital microprocessor that can talk to a host over a digital communications protocol and can handle simple tasks such as diagnostics and calibration. We have defined intelligent to include smart devices, but also to cover devices that can perform all three actions of a control loop: measurement, control, and process regulation. These valves also should be able to monitor the process and bring it to a safe condition if there should be a failure or upset condition. We generally have found that intelligence offers significant benefits in three areas: performance, maintenance, and safety. As we continue, I will point out those components that differentiate a truly intelligent control valve system from a merely smart device.

At the minimum, an intelligent control valve must have sensors that measure the process temperature, upstream and downstream pressure, fluid flow rate, and all the valve conditions (stem position, cylinder pressures, positioner characteristics). It needs a powerful microprocessor to analyze these conditions and to handle communications with a remote computer or distributed control system (DCS) console/controller. The program in the microprocessor must be capable of acting as a single-loop controller, interlock device, diagnostician, calibrator, and backup device. Figure 1 shows a block diagram of such a system. Of course, this assumes that all these components are put onto a first-class control device, since they do add some cost that might not be justified on a general purpose on-off valve.

What are applications for intelligent control valve systems?

The original application for these control valve systems was those processes that had high labor costs, expensive process materials, and very sophisticated control needs. Examples of these include offshore oil production platforms, where the cost of stationing personnel is very high; pharmaceutical plants, where the chemicals are very valuable and may be hazardous; and aerospace applications, where process response times are measured in milliseconds. As we have learned in actual applications, these industries are very interested in intelligent control valve systems (ICVS), but the chemical process industries and textile industries have control problems that are just as daunting. Specific problems that the ICVS has solved include measuring the flow of vent material from nylon autoclaves, controlling the flow of dimethyl terephthalate (DMT) bottoms from a still, and preventing backflow in polymerization reactors.

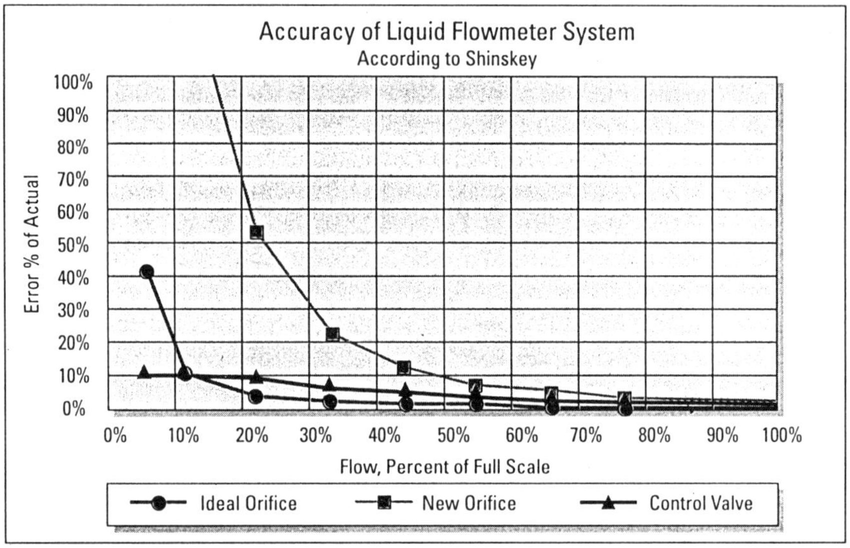

■ *Figure 2. Accuracy of head meters.*

In addition to better control and lower installation costs, intelligent valve systems with process sensors can be used to diagnose process problems. For example, by looking at the relationship between upstream pressure and flow, an intelligent control valve can detect pluggage or pump wear as the system degrades. It can also detect seat wear in the valve and alert the maintenance staff to replace the trim.

Finally, the intelligent control valve can be programmed to monitor the process and bring it to a safe condition in the case of a process upset or other problem. For example, if communications with the DCS are lost during a batch, this system could be programmed to hold the batch at a safe point and not take a chance on ruining the material or causing a condition that could endanger operating personnel.

Flowmeters

Since these applications are based on using the control valve to measure flow, one ought to briefly review how a control valve can be used to measure flow. Historically, the control valve industry has used Bernoulli's fluid flow equations to determine what size valve the customer should use to control his process. The customer would give the supplier two or three sets of conditions—upstream pressure, allowable pressure drop, fluid density, and flow—and the supplier would solve the equations to get what is called the C_v or flow coefficient, which is a measure of how much fluid will pass through a valve at a given pressure drop.

If these parameters can be measured in a control valve, then it could be used as a flowmeter. Upstream and downstream pressure sensors accurately measure pressure drop and, when coupled to the temperature measurement, give the density of the process fluid (either liquid or gas). The flow coefficient vs. travel relationship is measured for each trim set and stored in microprocessor memory.

An intelligent control valve system has two advantages as a flowmeter. The first is that the system can measure flow accurately over the controllable range of the valve, normally 30:1 for a globe valve. The second is that because the flow path through a control valve is repeatable, the traditional restrictions on straight runs of pipe upstream and downstream of the control valve were reduced. Traditional head-type flowmeters have a flow turndown limited by the turndown of the differential pressure transmitter and the square root flow/pressure relationship to about 3:1. The varying restriction of a control

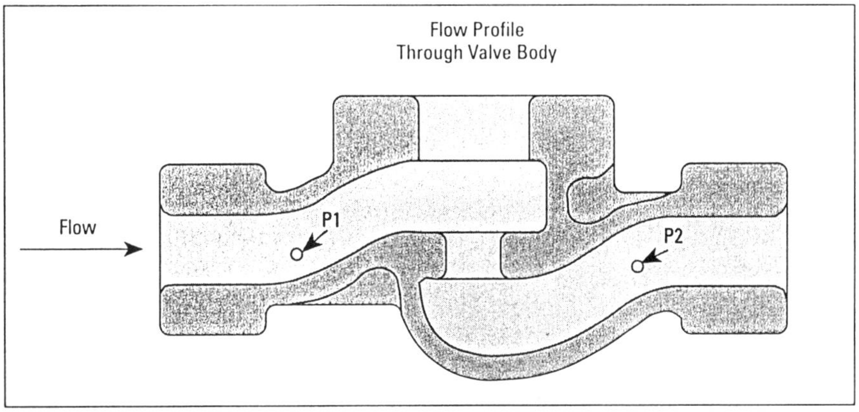

■ *Figure 3. Flow profile.*

valve means that the pressure drop across the plug will stay in the measurable range for the full stroke of the valve. This is shown in Figure 2, where the accuracy of the flowmeter is plotted against normalized flow for orifice plates and an intelligent control valve system. As you can see, the orifice error increases outside of the specification band around 30% of full-scale flow, while the intelligent valve error flattens out at low flows.

The flow profile through a globe valve is shown in Figure 3, along with the location of the pressure measuring points. The pressure points are inside the valve where it starts to turn the flow through the throttling zone. Since the flow profile is now set by the valve body, and the coefficients and pressure corrections are measured for each valve, elbows and restrictions upstream of the valve body do not affect the flow measurement like they would an orifice plate or vortex meter.

Control schemes based on energy could give us much more sensitive control.

Examples

Nylon autoclave vent valve. One of the last steps in making nylon flake is to complete the diamine reaction and cross-link the polymer while driving off water. One customer does this step in batch autoclaves, as shown in Figure 4. It has learned a good temperature and pressure profile for the batch to keep a high enough diamine concentration in the autoclave to make quality product. However, the firm likes to have information about the flow rate of material through the pressure control valve during the batch so it can determine the best batch profile. As you can imagine, the piping on top of an autoclave is quite congested and it does allow room for a reasonably priced flow meter in 6 in. pipe. The company modified the pressure control valve on the vent to be an intelligent control valve and has been able to measure the temperature, pressure, and flow through the vent valve during the batch.

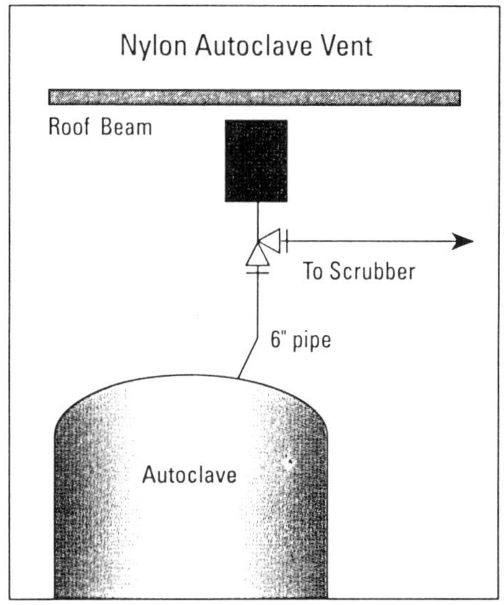

■ *Figure 4. Autoclave Schematic.*

Figure 5 is a profile showing the end of a boiloff that we ran during the startup period; these are not actual process data. In this case, the profile is stored in the intelligent valve's memory during the batch and is downloaded to a PC after the batch is complete. This means that there is no additional load on the DCS during

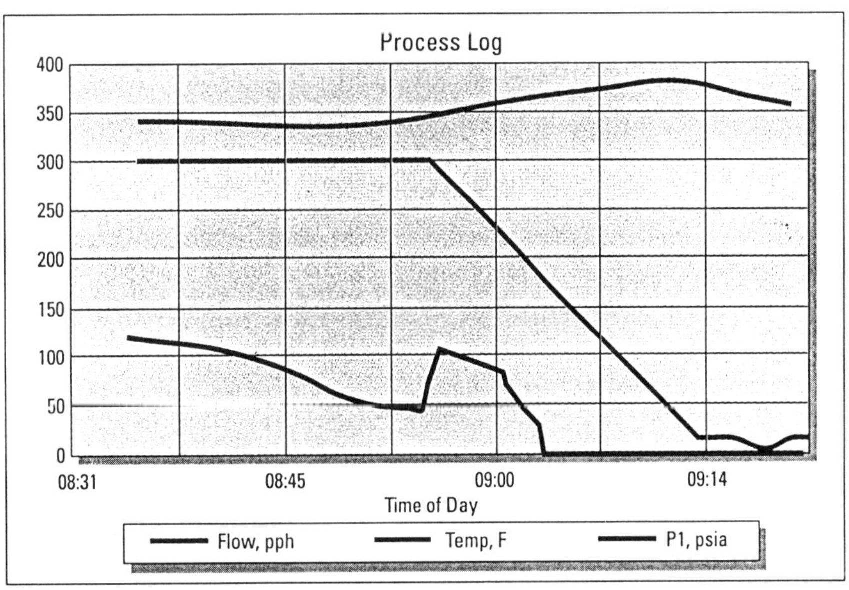

■ *Figure 5. Process Profile.*

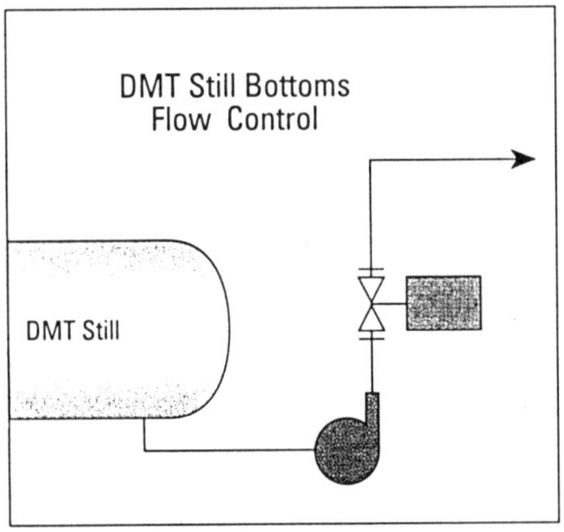

■ *Figure 6. Schematic of DMT Still.*

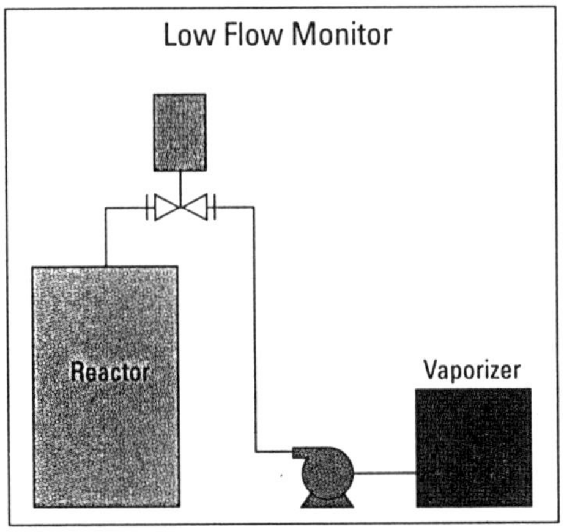

■ *Figure 7. Low Flow Cutoff.*

the batch, which was also important to the user. As a result of gathering process data, the user was able to shorten batch times. The temperature response was slow, since the temperature probe measures the valve body, and the user augmented the temperature measurement with one inside the thermocouple.

DMT bottoms. Another customer was distilling dimethyl terephthalate as part of making polyester intermediates. That firm also had a very tight piping geometry at the bottom of the still, so did not have room for a conventional flowmeter (Figure 6). It also could not find a conventional flow meter that could accurately measure the flow rate. In this case, the fluid is very viscous and could plug standard pressure sensors. Flush-mounted pressure sensors were installed and the system has been measuring flow for more than 18 months with no downtime. This has been the only meter the customer has found that will stand up to high temperatures, viscous fluids, and the close quarters underneath the still.

Polymerization

Reactor low flow cutoff occurs by bubbling gaseous ammonia into the polymer melt. The ammonia is stored as a liquid and must be vaporized before pumping it into the reactor (Figure 7). Whenever the flow drops below a certain rate, due to a problem in the pump or vaporizer, the pressure in the reactor pushes the polymer back up into the gas feed line, where it plugged the feed line. This had caused quite a bit of downtime since the feed tube had to be replaced after each pluggage. The intelligent valve system was installed, programmed to close on low ammonia flow. Since it has been installed, there have been no feed tube pluggages. The secret of the program was to adjust the valve opening so the pressure drop across it was constant. As the flow decreases during the batch, the valve will close down as if the flow dropped below the trip point. Other alternatives that had been tried relied on measuring the pressure in the feed tube or ammonia supply system, but they were slow to prevent the polymer backup.

But look at options besides intellegent systems

These four examples were successful because they had two important elements: a control valve and the need to measure flow. In some cases one might be tempted to use an intellegent control valve because it seems like an elegant solution. But we always have to justify the choice. The hardware price of an intellegent system is about the same as the price of the flow meter, controller, and control valve that it replaces. The installation cost is usually lower, since the number of components and connections is less than the traditional system. However, the process engineer should ask if these measurements are needed, or if the logic of the control could be handled by something less expensive, such as a low-end programmable controller. After you have evaluated the alternatives completely, you will find that your plant control strategy will probably include a range of solutions from traditional single-loop controllers to supervisory control systems using intellegent control valves.

The future

These applications have been quite successful for their users. As far as what could be done with future generations of intelligent control valves, we see, first of all, low-cost truly distributed control systems where the three parts of the control loop—measurement, control and final control element—are all located in the field, and just sending status information to a low-cost computer screen rather than using banks of expensive controllers and input-output (I/O) cards in an environmentally controlled room. Then, with the increased response of these integrated loops, they can either control the process tighter or handle other functions, such as calculating energy balances around the process. Where we now control by measuring mass flows through a plant, control schemes based on energy could give us much more sensitive control. CEP

B. A. WHITE is the sales and marketing manager for the Star Systems division of Valtek International, Provo, UT (801/489-2563; Fax: 801/489-2599). His career spans 25 years as outside salesman for Rust Instrumentation, Salt Lake City, senior applications engineering specialist, Fisher Controls, Marshalltown, IA, process control engineer, St. Joe Minerals, Monaca, PA, and research engineer, DuPont, Wilmington, DE and LaPorte, TX. He received a BS and MS in chemical engineering from Oklahoma State University. He is registered as a professional engineer in Delaware and is member of the American Chemical Society and Society of Mining Engineers.

Improve Control Valve Response

Improperly estimating the impact of control-valve dead band and stick/slip can lead to incorrect valve selection and tuning, and, thus, degraded loop performance. A new method can help you avoid such mistakes.

Gregory K. McMillan,
Monsanto Co.

Without a smart valve positioner that gives actual valve position (*AVP*) feedback, a user really doesn't know how a control valve behaves. When there is a loop problem, valve position usually is the one piece of information that is missing. The stroke indicated on the operator console or control panel gives the implied valve position (*IVP*), which likely is not the *AVP*, given the realities of a mechanical device exposed to the process and plant environment.

The situation is even stranger when you consider that the operator knows the status of the transmitter and controller from indication of the measurement and setpoint, but has no ability to see what directly affects the process — the AVP and the associated change in valve flow coefficient.

Why does this matter? A lack of understanding of the consequences of valve responses, coupled with a desire to reduce hardware cost, may lead to incorrect decisions, such as the use in industrial processes of on/off, laboratory, or commercial-grade valves and dampers characterized by high friction or slop in linkages for throttling. To make the situation worse, positioners may be omitted to reduce the initial and spare-parts cost *(1)*. Unfortunately, the true cost of the final element in terms of lower yield, quality, and capacity, due to greater loop variability from poor responses, usually is not recognized.

Actual valve position

AVP is the position of the valve trim (that is, the plug, disk, ball, or the like), not necessarily the position of the actuator stem. This is a distinction that is particularly important for rotary valves where twist, flex, and slop in linkages occur. Furthermore, due to the use of actuator-position feedback, smart positioners and digital controllers may not accurately show or tightly control *AVP*.

AVP may not match the implied valve position (*IVP*) at any point due to nonlinearities such as:

1. dead band;
2. stick/slip;
3. prestroke dead time; and
4. stroking time.

We will describe each of these types of behavior. A mismatch also can be caused by incorrect bench and calibration settings, but these are beyond the scope of this article.

Deadly dead band

Let's define dead band as the change in signal required to change the direction of stroke under actual installation conditions that include the effect of accessories (for instance, I/P transducers and boosters), and typical maintenance practices. Figure 1 shows dead band as occurring at zero and full travel as part of a hysteresis cycle, but it actually occurs at any position whenever the valve changes between opening and closing. Dead band, however, usually is largest at the shutoff because of the increased friction of the seating surfaces. It also can happen when the valve stops moving, due to a constant controller output, although this may not strictly be part of the theoretical definition.

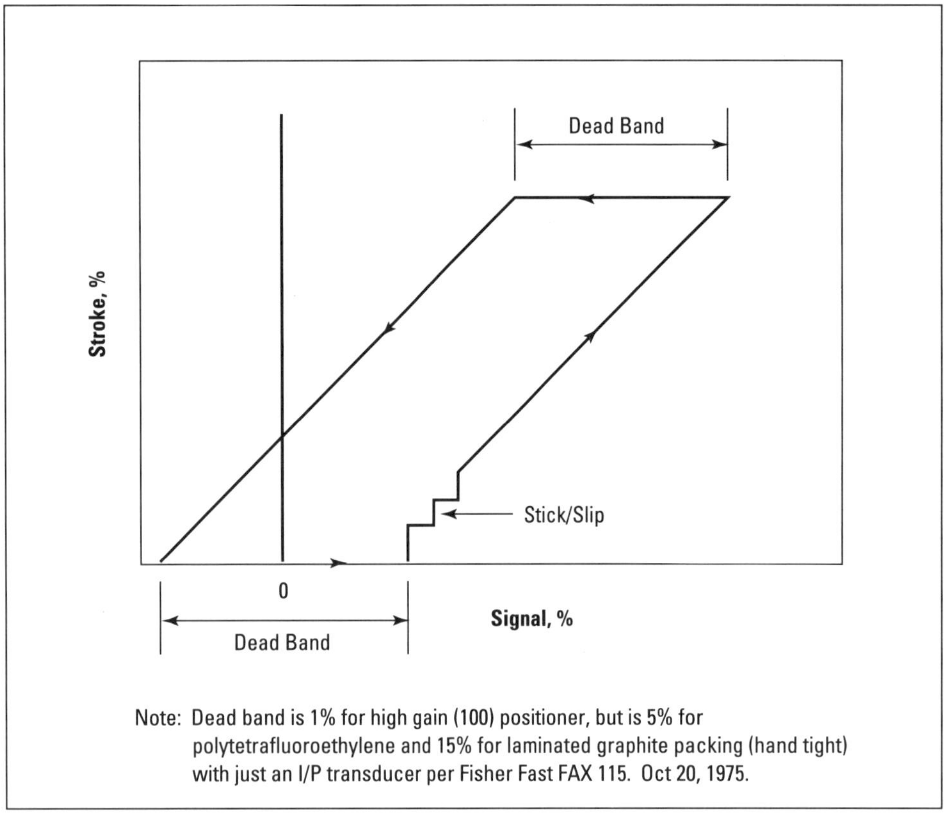

■ *Figure 1. Valve dead band and stick/slip.*

Dead band is important because the lag until the controller output has changed enough to exceed the dead band creates loop dead time.

Dead time, as defined here, is the time delay between a disturbance and the ability of the controller to see and correct for that upset. If the dead time were zero, an infinite controller gain would be permissible and perfect control would be possible. While this, of course, is not practical, any additional dead time makes the loop more sensitive to nonlinearities and disturbances *(2)*. Table 1 shows that the control valve is the biggest source of dead time for flow, level, and pressure loops, and is, in general, a major source for inline loops. These ratings presuppose an analog transmitter with minimal filtering and a controller gain detuned to be about one-half of that used to achieve a quarter amplitude damping (that is, $K_x = 0.5$).

The amount of dead time created is equal to the dead band divided by the rate of change of the controller output. Thus, the dead time increases as the rate of change of the controller output decreases — and this rate decreases with the controller gain. If you consider that the allowable controller gain decreases when the dead time increases, there is a confusing cycling effect. If you also take into account that the rate of change of the controller output depends upon the rate of change of a load upset or setpoint, estimating dead time from dead band may seem hopeless. This may help explain why dead-band effects largely are ignored in tuning methods and control theory.

The key to developing an equation to estimate valve dead time from the observed open-loop dead time is to realize that the change of controller output in the dead band follows the initial response, which resembles a ramp for all processes, whether self-regulating, integrating, or runaway. This pseudo-integrator approximation, where the ramp rate (that is, integrator gain) is the process gain divided by the loop time constant, facilitates the use of a simple estimate of the controller gain as the inverse of the product of the integrator gain and loop dead time. The rate of change of the process variable, which also is the integrator gain, multiplied by the controller gain is the rate of change of the IVP. Fortunately, the integrator gains cancel out; so, the rate of change of output depends upon the total loop dead time.

Table 1. Relative sizes of loop dead times.

Application Type	Contribution (1 = largest, 4 = smallest) of			
	Distributed Control System	Valve	Process	Sensor
Liquid flow	2	1	4	3
Liquid level	2	1	4	3
Gas pressure	2	1	4	3
Reactor pH	4	2	3	1
Neutralizer pH	4	3	1	2
Inline pH	2	2	4	1
Reactor temperature	4	3	1	2
Inline temperature	3	2	4	1
Column temperature	4	3	1	2

Basis: analog transmitter with minimum filter, and a controller gain detuned to give smooth response ($K_x = 0.5$).

The development of a suitable equation (Eq. 6 below) involves the substitution of estimates of *AVP* and rate of change of *IVP*, and rearrangement to solve for the valve dead time (TD_v) as a function of dead band (*DB*), the required total change in actual valve position (ΔAVP), and the observed open-loop dead time (TD_o):

$$TD_v = \frac{DB}{\dfrac{\Delta IVP}{\Delta t}} \tag{1}$$

$$\begin{aligned}\frac{\Delta IVP}{\Delta t} &= K_c \times \frac{\Delta PV}{\Delta t} \\ &= \frac{K_x}{K_o \times TD} \times K_o \times \Delta AVP \\ &= \frac{\Delta IVP - \dfrac{DB}{2}}{TD} \times K_x \end{aligned} \tag{2}$$

$$TD = TD_o + TD_v \tag{3}$$

$$TD_v = \frac{DB}{\dfrac{K_x \times \left(\Delta IVP - \dfrac{DB}{2}\right)}{TD_o + TD_v}} \tag{4}$$

$$\begin{aligned} TD_v \times \left(\frac{K_x \times \left(\Delta IVP - \dfrac{DB}{2}\right) - DB}{K_x \times \left(\Delta IVP - \dfrac{DB}{2}\right)} \right) \\ = \frac{DB}{K_x \times \left(\Delta IVP - \dfrac{DB}{2}\right)} \times TD_o \end{aligned} \tag{5}$$

$$\begin{aligned} TD_v &= \frac{DB}{K_x \times \left(\Delta IVP - \dfrac{DB}{2}\right) - DB} \times TD_o \\ &= \frac{DB}{K_x \times \Delta AVP - DB} \times TD_o \end{aligned} \tag{6}$$

where

$$K_c = \frac{K_x}{K_o \times TD} \tag{7}$$

$$\frac{\Delta PV}{\Delta t} = K_o \times \Delta AVP \tag{8}$$

$$\Delta AVP = \Delta IVP - \frac{DB}{2} \tag{9}$$

Three major assumptions are made in the development of Eq. 6. The first is that the upset is fast. Slower upsets will reduce the excursion rate and increase the dead time from dead band. The second assumption is that the integral- and derivative-mode contributions are small compared to the proportional-mode contribution to the controller output while in the dead band. If not, the controller gain should be multiplied by an expression of the combined response of the three modes. For the proportional plus integral mode, the expression would be 1 + $(TD/t_i)/2$. The third assumption is that the controller is tuned aggressively for nearly a quarter amplitude response. If the controller is tuned with a lower gain to provide a slower, but smoother recovery, the controller gain should be multiplied by a factor that approximates the degree of detuning (for instance, $K_x = 0.5$).

The validity and implications of Eq. 6 were explored via trend recordings generated via an Advanced Continuous Simulation Language (ACSL) program to model the simple generic loop depicted in Figure 2. Figures 3a–b show that, for a fast load upset, the equation correctly predicts how the dead time for a proportional-integral-derivative (PID) controller increases from 0.5 min to 2.0 min as the dead band is increased from 0 to 5% for a 10% required ΔAVP. In this case, the contribution from the integral mode partially offsets the effect of the reduction in gain used to provide a smooth, stable

The increase in loop dead time can be extremely large, and is drastically altered by the size of the required ΔAVP.

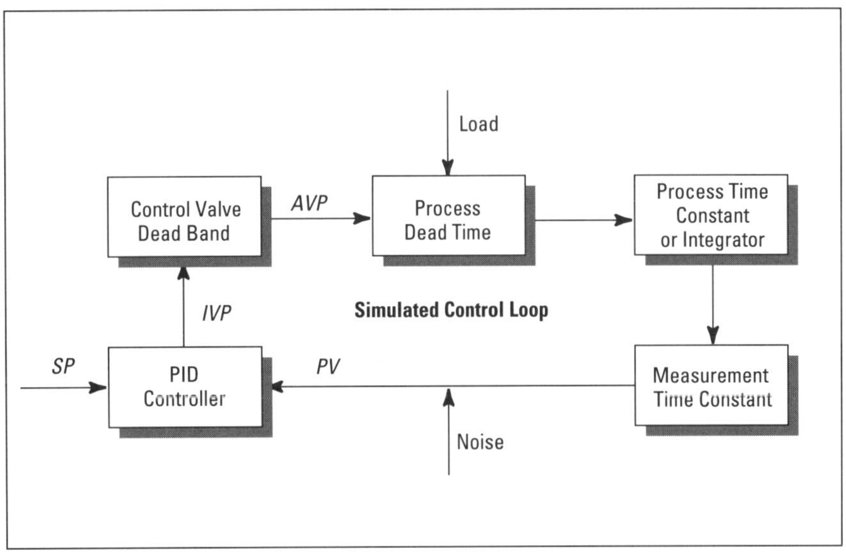

■ ***Figure 2. Model of a simple, generic loop.***

response. Figures 4a–b show that, for a slow load, the dead time from dead band is larger. A slow setpoint change (a self-inflicted upset) also would likely increase the dead time.

Tentative tips

Equation 6 shows that the increase in loop dead time can be extremely large, and is drastically altered by the size of the required ΔAVP. This increase, plus the difference between AVP and IVP, has many implications. Some of the problems and solutions are as follows:

1. Large amplitude *slow* noise will create dead time from dead band due to reversals of stroke. Figures 5a for fast noise and Figure 5b for slow noise illustrate how fast reversals cancel out. The scan frequency should be as fast as the noise frequency, and a filter should be used to keep the changes of controller output within the dead band of the control valve.

2. The window of allowable gains for non-self-regulating processes gets smaller as the valve dead band increases. Figures 6a–d show how the response gets more oscillatory for level loops tuned with a low controller gain, despite minimal reset action. A high performance (high gain) positioner should be used for control valves in level, total dissolved solids, batch, pH, fermenters, and exothermic reactor loops.

3. The use of a filter to slow setpoint changes, to reduce overshoot, creates more dead time from dead band. A head start should be used to initialize the control valve position at the value needed to achieve setpoint.

4. The period of oscillation of the closed-loop method for controller tuning, where the integral and derivative modes are turned off, will depend upon the distance of the actual gain from the ultimate gain and, thus, the degree of dampening of the oscillation. Attenuation greater than quarter amplitude should be used for estimates of loop period only if conservative settings of the loop-reset setting are permitted.

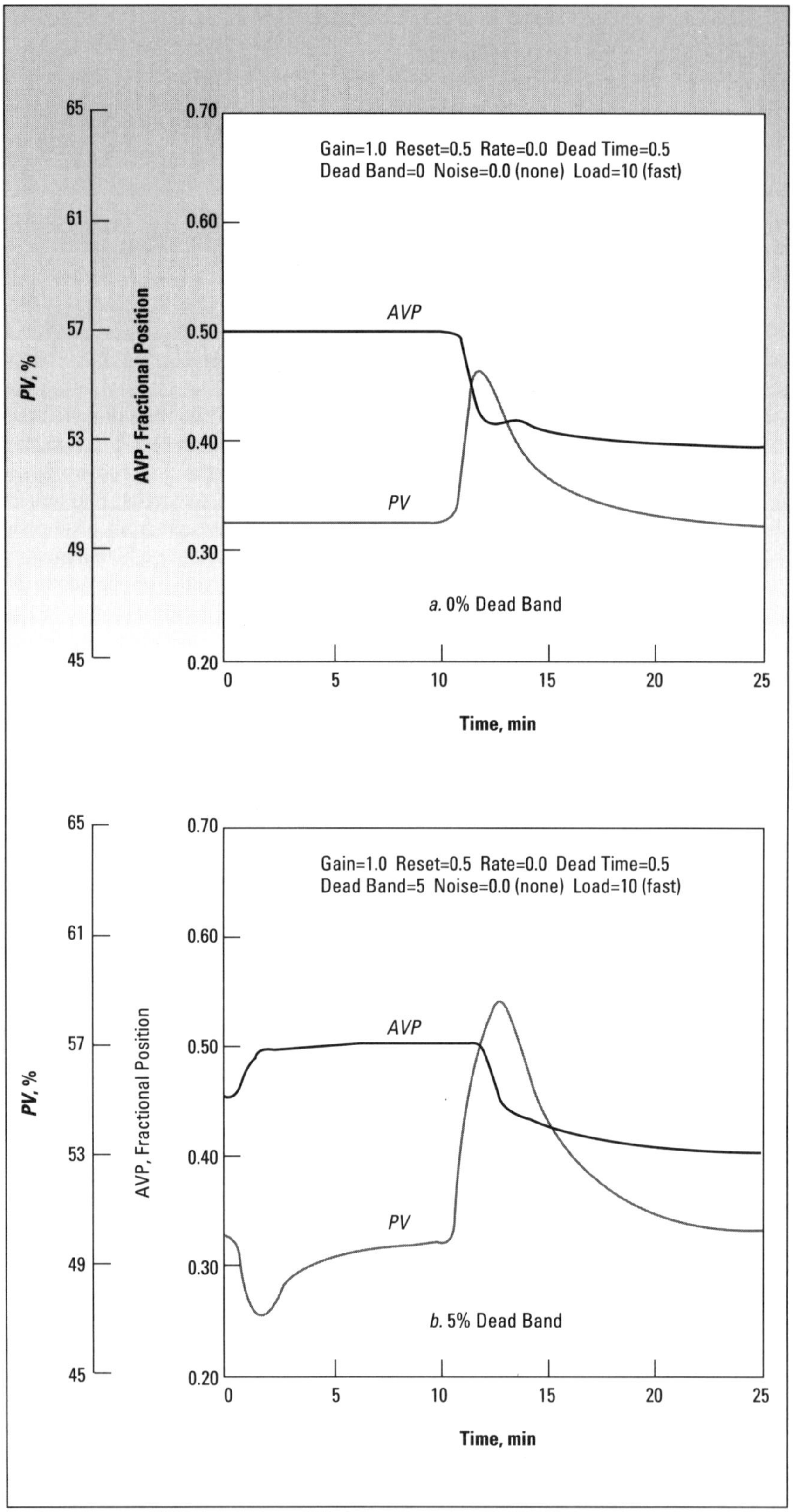

■ ***Figure 3. Effect of dead band for a fast load upset at time = 10 min.***

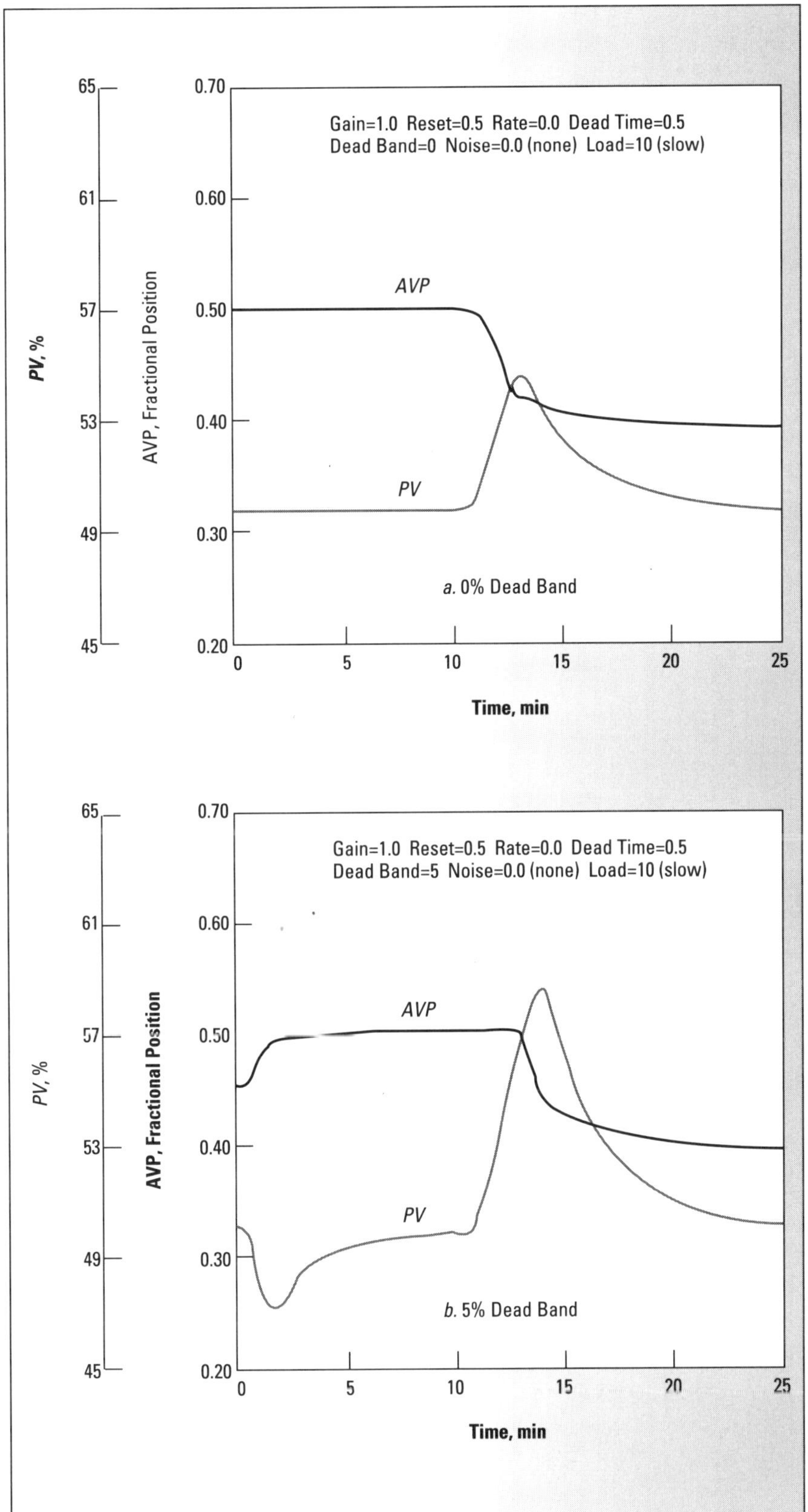

■ *Figure 4. Effect of dead band for a slow load upset at time = 10 min.*

5. The computed loop dead time and gain for both the open-loop and relay methods of controller tuning can be significantly underestimated, due to the omission of valve dead time and ΔAVP. The dead band should be measured and either Eq. 6 or, instead of a step, a ramp based on the pseudo-integrator approximation should be used. Also, ΔAVP should be estimated as ΔIVP - $DB/2$.

6. Large valves or steep installed-flow characteristics will increase the valve dead time from dead band due to a decrease in the required change of the valve position. If the required stroke approaches the dead band, the dead time from dead band is greater than the observed open-loop dead time and the valve becomes the largest source of dead time. Use linear trim for nearly constant pressure drops, and equal percentage trim for highly variable pressure drops. Also, to avoid oversizing of the control valve, the maximum flow requirement should be near the point at which the installed flow characteristic flattens.

7. Keeping packing more than hand tight can cause excessive valve dead time due to high friction. Stems should be polished, and packing should be replaced to stop leaks. Do not tighten packing with wrenches or "Channel Lock" style pliers.

8. Packing used for high temperature applications or to reduce environmental emissions can cause excessive valve dead time due to high packing friction. A digital controller with an adjustable positioner gain and diagnostics might help you deal with the consequences of poor valve response, but it is better to get rid of the source of the problem. Consider extension bonnets and bellows seals to eliminate the need for the high friction packing.

9. Intermediate links between the actuator and trim can cause a grossly inaccurate estimate of valve dead band from stem position *(3)*. The dead band for rotary valves should be based upon flow or pressure measurements. The output of the controller should be increased in steps of 0.5% until a change in the process variable is noted *(4)*. This works best for a flow or pressure measurement because the process dead time is less than a second. Note that load disturbances, noise, and loop dead-time measurements can cause inaccurate results. Alternatively, to infer the percent dead band, take the percent difference in a process measurement between equal initial and final controller outputs divided by the open-loop gain.

10. Positioners with low gains will cause a larger dead time due to a larger dead band and offset. Use high performance (high gain) positioners.

11. If a booster with a dead band greater than 1% is used instead of a high performance positioner, the valve dead time will be larger even though the valve stroking time may be faster. Install a positioner and a booster bypass with an adjustable needle valve. The needle valve should be opened just enough to stop the oscillations caused by the positioner output being dead ended into a small booster volume instead of a large actuator volume. (See Figure 7.) The use of boosters without positioners on fast loops has resulted in butterfly disk wander due to positive feedback *(2,5)* from the booster's high outlet-port sensitivity and erratic positioning from diaphragm actuators *(6)*. Response tests of control valves for overpressure protection and surge prevention should use a ramp rather than a step in the I/P signal to show the amount of dead time from dead band *(5)*.

12. If the observed loop dead time is much less than a second for a loop where the variability must be reduced, the use of a typical control valve and scan time seriously limit any possible improvement. Consider a variable speed drive (VSD) and a fast scan rate (for instance, 0.05 s), or analog controller. Test results for liquid flow and pressure loops show that the controller gain can be significantly increased and, subsequently, the control errors can be greatly reduced by a VSD and analog controller *(2,7)*.

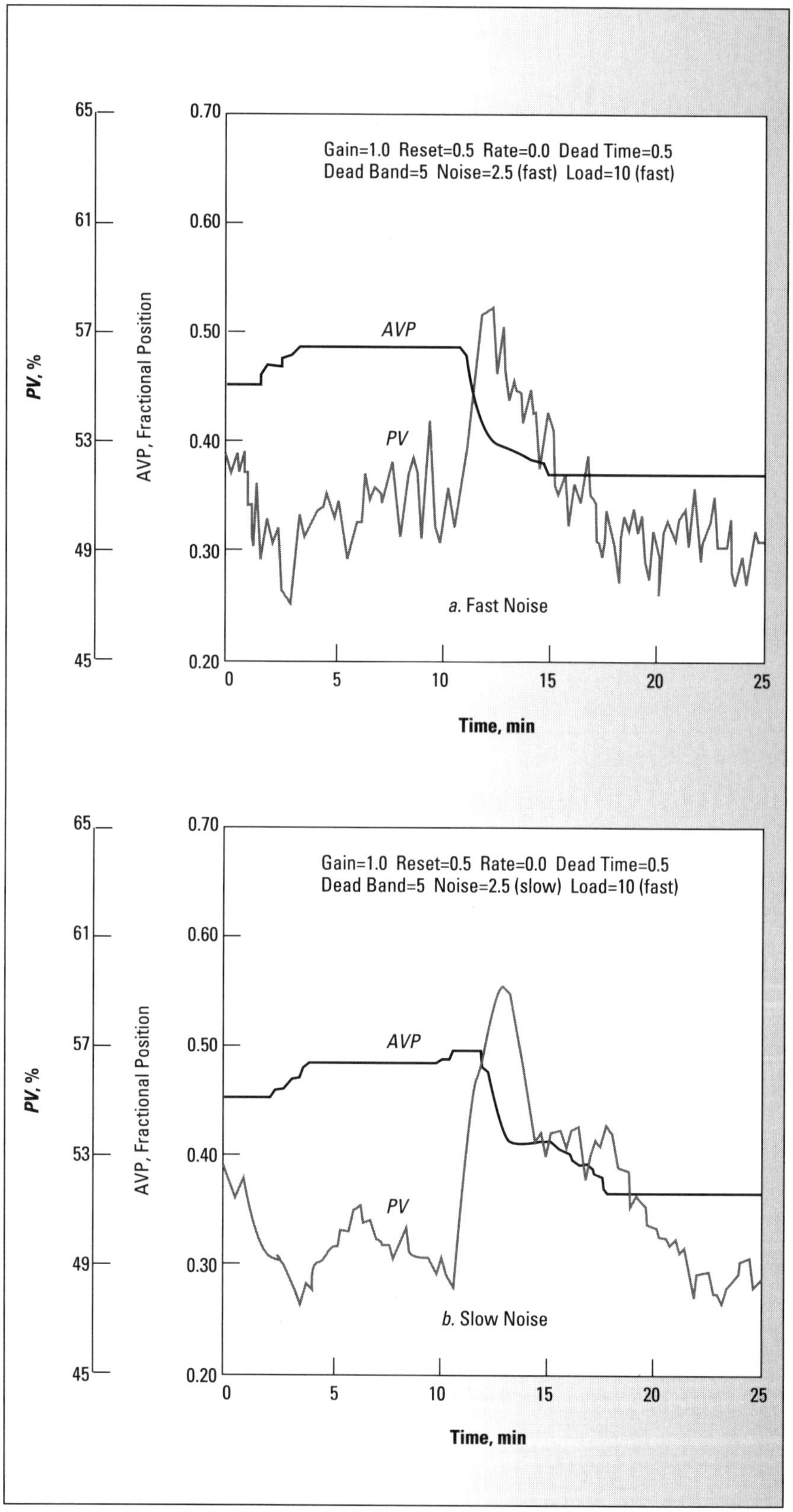

Figure 5. Effect of dead band for fast and slow noise for upset at time = 10 min.

Stick/slip

Stick/slip normally is not measurable for control valves in good condition in a position above breakaway from the seat. It shows up as staircasing in a feedback measurement of flow or valve position, as indicated in Figure 2. The consequences can be worse than dead band. Not only does the sticking action cause dead time, but the slipping action means that the valve may never be able to get near the setpoint due to excessive loss of final-element resolution from large steps in the stroke. Also, there is a much greater propensity to develop rapidly growing fluctuations and instability from slip. In contrast, dead band by itself tends to reduce oscillation frequency and amplitude in self-regulating processes. The degree of slip is difficult to detect from changes in process measurements such as temperature and pH, because nonlinearities make the magnitude of the change a function of operating point.

Valveless actuators

If you ignore dead band, the stroke response can be approximated as a prestroke dead time plus a velocity-limited exponential stroking time, as illustrated in Figure 8. The prestroke dead time and stroking time shown are determined by the amount of time it takes for a pressure change of sufficient magnitude to occur in the actuator to cause stem movement — they, therefore, depend on the actuator volume, and the flow coefficients of pilots and restrictions in the pneumatic signal based on information in the Fisher laboratory report for problem 1,283 *(2)*. The values cited by manufacturers are for actuators not connected to a control valve. There are not many of these installations. For this and other reasons, prestroke dead time and

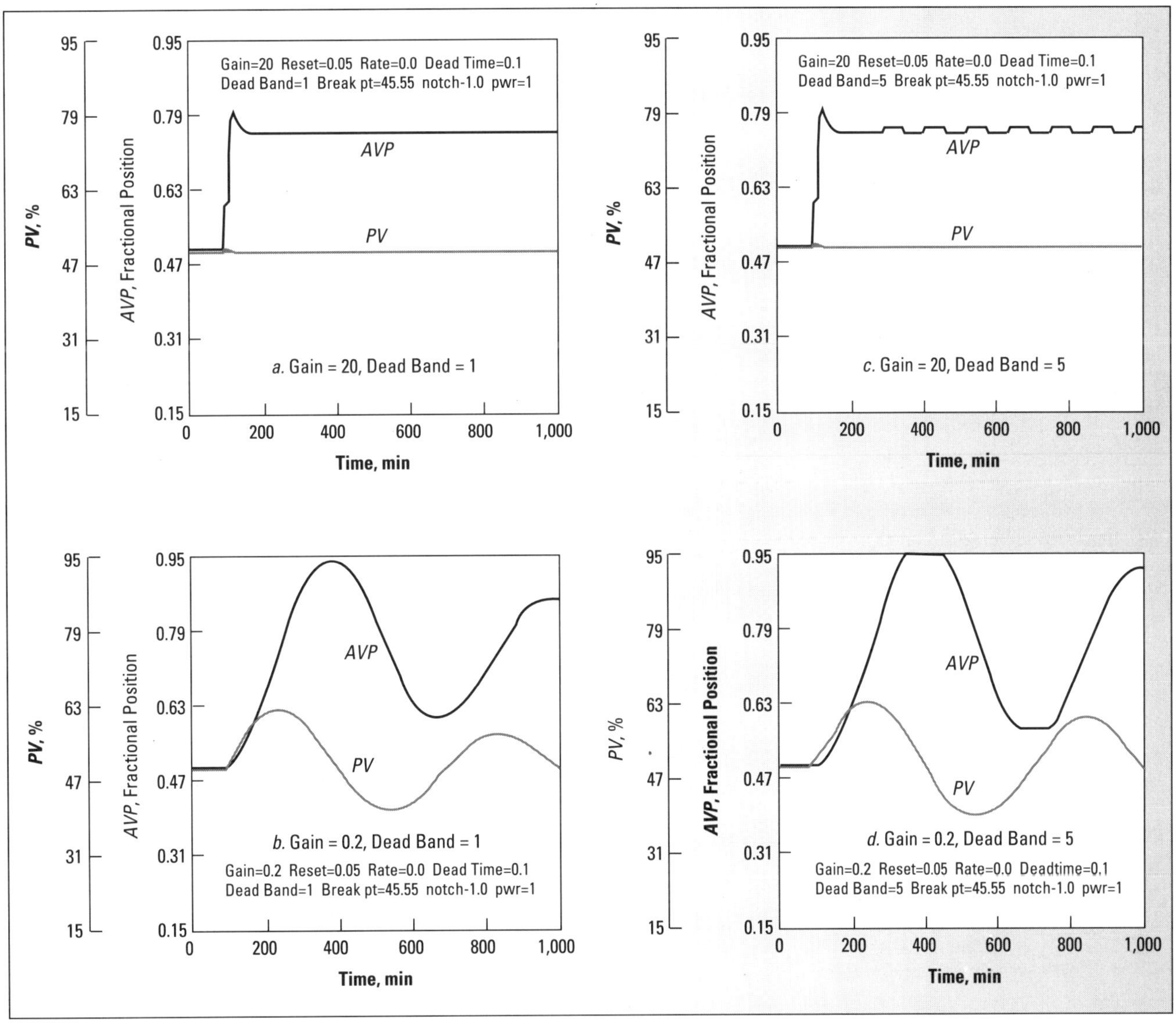

■ ***Figure 6. Effect of dead band on gain window for level loop.***

■ *Figure 7. Installation of booster, adjustable bypass, and positioner.*

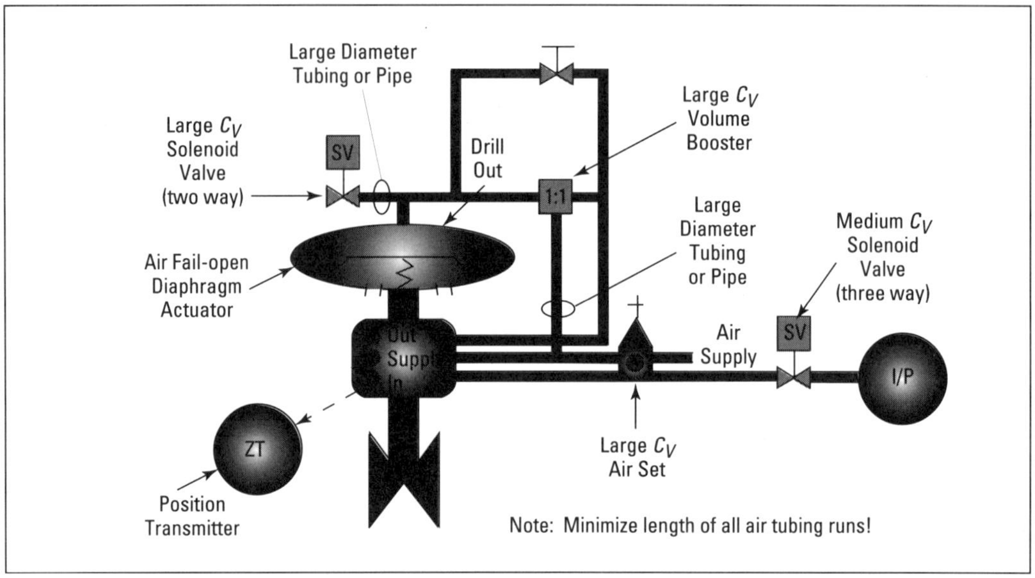

stroking time often are not important. The exceptions are large actuators and control valves for overpressure protection and surge prevention. For most applications, the rate of change of controller output is slower than the slewing rate of the control valve; so, stroking time does not create a time lag larger than a small inherent time constant. (The product of the control-valve stroking time and the ratio of the inherent time constant to this stroking time usually is less than one second.) Thus, electric actuators and VSDs with a slower slewing rate can improve a fast loop's performance due to less dead time from dead band.

Most VSDs have no measurable dead time *(5)*. If some drive slip exists, tachometer feedback can be added to create a secondary speed loop. This speed loop should use high gain, proportional only action. Its tuning and benefit is analogous to that of the positioner.

A cursory analysis might suggest that low-cost VSDs might mean the demise of control valves. However, the needs for providing low flows to equipment at high static pressures, positive shutoff, and fail-safe action, as well as the necessary environmental protection for the VSD, means that wholesale replacement of control valves by VSDs is unlikely in the near future.

A software solution

If you accurately know the dead band, a rather simple kicker algorithm can be added to the configuration for a distributed control system or programmable logic controller that will increment and decrement the signal to the control valve by one-half of the dead band for the transition to an increasing and decreasing signal, respectively. Any noise, however, will kick the output. So, a somewhat more complicated algorithm for a noise band must be added. Whereas noise bands are relatively easy to implement on the measurement signal, they conceptually are difficult to implement on the output signal due to the lack of a reference point such as the setpoint. A reference point in the controller output can be established by capturing the output for the transition to the positive direction. The noise band should be set from observation of the fluctua-

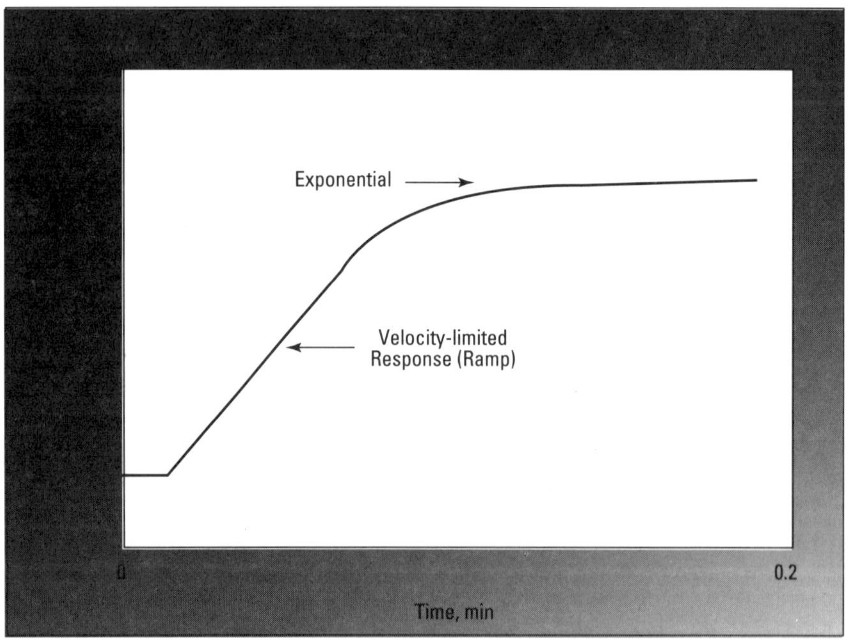

■ ***Figure 8. Prestroke dead time and stroking time constant.***

tions of the controller output after the controller's filter, gain, and rate settings have been tuned. Obviously, the noise band must be much less than the dead band for this algorithm to provide much benefit. Also, too large of a kick would cause something similar to slip that is worse than dead band on a equal magnitude basis.

The short-term fix

Using a valve positioner seems an obvious short-term fix — that is, until you ponder why some plants do not use positioners. They do this deliberately, not out of lack of understanding, perhaps because plant staff in the past have had to bypass positioners. Such bypassing is not that uncommon in my experience.

A simple, if preposterous, example provides a reasonable explanation as to why some loops are worse with positioners. Let's take a process loop, and add an inner (secondary or slave) flow controller to create a cascade loop. Furthermore, let's assume that the flow transmitter has a long, complicated, and confusing calibration procedure, that the sensor is attached to the transmitter by an external mechanical linkage that is subject to torsional error and linkage slop and is liable to physical and environmental damage, and that the flow controller has a fixed gain that causes instability for some loops. You certainly might expect that the original loop would be better off in some cases without the cascade-to-flow control. One such cascade-loop failure may outweigh numerous successes, especially when maintenance may involve hours exposed to the elements to adjust an antiquated mechanism. While this situation seems absurd for flow loops, it does exist for valve-position controllers (valve positioners). Thus, while valve positioners are recommended, they should be accompanied by training, maintenance of linkages, and attention to the tentative tips mentioned before. Also, new smart positioners or digital valve controllers must address these issues and provide fast calibration methods, and protected, reliable feedback connections that give actual valve position.

Back to the future

The control valve actuator basically has not changed much in last 40 years. When you consider all the advances made in transmitters and controllers over the last four decades, the present performance of control valves is inconsistent with the accuracy and precision of modern instrumentation *(8)*. Fieldbus opens up some possibilities to determine which manufacturers have the best valves. It also enables predictive maintenance from control valve signatures and stroke history, and the use of a flow controller at the valve to further enhance valve performance and diagnostics. A step change in valve performance might involve digital actuators or electromagnetic fields. CEP

Nomenclature

AVP	=	actual valve position
ΔAVP	=	change in actual valve position, %
DB	=	dead band from valve hysteresis, %
IVP	=	implied valve position
ΔIVP	=	change in implied valve position, %
K_c	=	controller gain, dimensionless
K_o	=	pseudo-integrator open-loop gain, 1/min
K_x	=	fraction detuned for total loop dead time
ΔPV	=	change in process variable, %
Δt	=	change in time, min
t_i	=	integral time, min/repeat
TD	=	total loop dead time, min
TD_o	=	dead time seen in open-loop test, min
TD_v	=	dead time from control-valve dead band, min

Literature Cited

1. **McMillan, G. K., S. Weiner, et al.,** "How to Become an Instrument Engineer," part 1.523, Instrument Soc. of America, Research Triangle Park, NC (1994).
2. **McMillan, G. K.,** "Tuning and Control Loop Performance," 3rd ed., Instrument Soc. of America, Research Triangle Park, NC (1994).
3. **Coughran, M. T.,** "Valves: Testing for Peak Performance," *InTech*, pp. 58–61 (Oct. 1994).
4. EnTech Control Valve Dynamic Specification, version 2.1, EnTech Control Inc., Toronto, Ont. (March 1994).
5. **McMillan, G. K.,** "Centrifugal and Axial Compressor Control," Instrument Soc. of America, Research Triangle Park, NC (1983).
6. **Arant, J.,** "Control Valve Notes and Positioner Myths," Part I, *Control*, pp. 34–39 (Nov. 1993), and Part II, pp. 24–30 (Dec. 1993).
7. **Shinskey, G.,** "Flow and Pressure Control Using Variable Speed Motors," *Proceedings*, 2nd Control Engineering Conf., Chicago (May 1982).
8. **Miller, F.,** "Control Valves — The State of the Art," *Valve Magazine*, pp. 12–16 (Fall 1993).

Acknowledgment

This article was developed from a presentation at the Instrumentation Symposium for the Process Industries (Jan. 1995), sponsored by the chemical engineering dept. of Texas A&M University, College Station, TX.

G. K. McMILLAN is a Fellow in the Monsanto Co.'s Integrated Manufacturing Control Group in St.Louis, MO (314/694–6039; Fax: 314/694–4503; e-mail: gkmcmi@ccmail.monsanto.com), where he specializes in process control improvement. He earned a BS in engineering physics from the Univ. of Kansas, and an MS in electrical engineering from the Univ. of Missouri at Rolla. He is a Fellow in the Instrument Soc. of America, and received its Kermit Fischer Award for outstanding environmental service through pH control in 1991. He was selected Control Engineer of the Year for the Chemical Industry in 1994 by *Control* magazine. He is author of ten books including "pH Measurement and Control," "Tuning and Control Loop Performance," and "Advanced Temperature Measurement and Control."

Improve the Sequence Descriptions for Batch Control Projects

By pushing for — and using — the standardized control reference model proposed here, you will bolster communication among process and control engineers

John L. Hedrick,
Automation & Control Technologies, Inc.

In batch process design projects, one of the many responsibilities of the process engineer is to develop control sequence descriptions. Control engineers regard these descriptions as requirements' specifications for their part of the project. Unfortunately, the typical sequence description is inadequate on projects using modern batch operation and control systems. As a result, control engineers are forced to rework them into a more usable form. This often leads to design errors in meeting user requirements. It also causes problems at project commissioning, because it is difficult to reconcile the system implementation with the original sequence descriptions.

This situation could be avoided if sequence descriptions were structured for use by both process and control engineers. This is especially true in complex applications that require the automation of flexible formulations, concurrent process operations, common resource arbitration, and fail action and recovery. A properly structured functional requirement specification (FRS) for sequence control would originate in conceptual design and persist throughout the design effort, growing in detail throughout the project life cycle. With today's emphasis on continuous improvement, as-operating design data should be available throughout the plant life cycle.

An object-oriented technique that we will describe can facilitate sequence control design and implementation from concept to operations within a computer-assisted engineering data management (EDM) environment. It has been effective in promoting shared responsibilities for concurrent engineering among process and control engineers.

Process and control engineering

In the chemical process industries, a capital improvement project generally is executed by a project team, which aims to develop the optimum solution based on objectives, priorities, and constraints set by management. The team usually consists of staff from a variety of disciplines and effective communications, particularly between process and control engineers, can be a problem *(1)*. In recent years, the pressure for shorter design cycles and increased flexibility in plant operation have further complicated the task.

A key participant in the process design is the process engineer — who works with research personnel and others to select the right components, streams, unit operations, and process topology. The process engineer also is responsible for estimating equipment costs for the economic analysis. The whole team participates in creating the ultimate solution. This requires team sharing and analysis, performed in a cooperative environment from concept development through commissioning and into operation.

An important function throughout a project is design capture — which means depicting the design details in a format that is precise and accurate. Process flow diagrams (PFDs) and piping and instrument diagrams (P&IDs) are examples of design capture media.

The key to design capture is version control to insure that team members reference the correct design. The major classifications during the life cycle of a facility are as-designed, as-built, and as-operating.

Between these versions there are many "red-line" revisions during a typical project.

The goal of EDM is to provide computer assistance for making use of as-operating versions the norm rather than the exception. Most plant retrofit and expansion projects begin with a massive effort to bring the facility design data up to as-operating versions. This could be avoided, or at least minimized, with an efficient and easy to use EDM system for design capture and version control that is used for both design and maintenance.

Changes are inevitable in the design process and may occur at any time. When changes occur, it is important that all affected parties are notified and that they have adequate response time to adjust accordingly. The activity of resolving conflicts when changes occur is important to the success of any project. Once conflicts are resolved, there should be a formal mechanism to designate the designs and changes as the authorized versions.

It is my experience that most of these activities are done at startup of the process control system. Well-intentioned project teams have experienced a lot of rework during startup to resolve conflicts that should have been corrected months earlier. Startup is the worst possible time to resolve conflicts and capture designs on large projects like those found in modern process control systems. Thus, the resulting control system is usually of poorer quality and more expensive than necessary. In control system design, if the process engineer and the control engineer would exchange precise requirements and strategies early and continuously throughout a project, much of the process control system rework at startup could be avoided *(1)*.

Control engineers would like the requirements' specification from process design to closely match the structure and operation of the actual system. Process engineers would like to avoid computer jargon. If a method were available to fulfill both wishes, engineers would be able to express the design more precisely and accurately among one another and, more importantly, they would be able to verify the design at commissioning more easily. An EDM environment to fulfill this goal is possible.

The process control reference model

Improving the flow of engineering design data between the process engineer and the control engineer requires a common frame of reference. It must support the operation and control design activity with terminology, structure, functionality, and integration in a form that all project disciplines understand. The traditional reference model depicting process control as a set of individual loops and relay logic interlocks is adequate for simple and low-level control applications, but fails when used with modern computer integrated manufacturing (CIM). These systems demand high-level automation and data management like that found in supervisory or batch cycle control applications.

The need for an improved reference model is especially acute in a process that requires sequence descriptions written by the process engineer. Most sequence descriptions are based on a linear, stepwise control model. This model is inadequate for applications that require the automation of concurrent and parallel unit operations, batch scheduling, fail action and recovery, and shared resource arbitration. If the process engineer would provide a functional requirements' specification that conforms to actual application requirements and matches a modern process-control reference model, the interdisciplinary communication problem could be improved significantly.

What does modern control look like in terms of a reference model? In 1984, the TC-4 committee of the Purdue workshop produced the first attempt at establishing a reference model for advanced batch control. This work has continued through the efforts of industry specialists working in the SP88 committee of the Instrument Society of America (ISA). On a parallel path, the European technical community has addressed the issue through the NAMUR working party. The current version of the reference model is documented in the SP88 review draft *(2)*. We can use the SP88 batch reference model to describe control in continuous and sequential processing applications, as well. After all, control of continuous, sequential, and batch processes is quite similar, especially in unit sequencing and at the lower control levels connected to the measurement and final control elements.

To be effective, the reference model must be used by both process and control engineers. It must be system independent to avoid vendor bias. It is also critical that the model's viability will last. For example, the basic dynamics of proportional-integral-

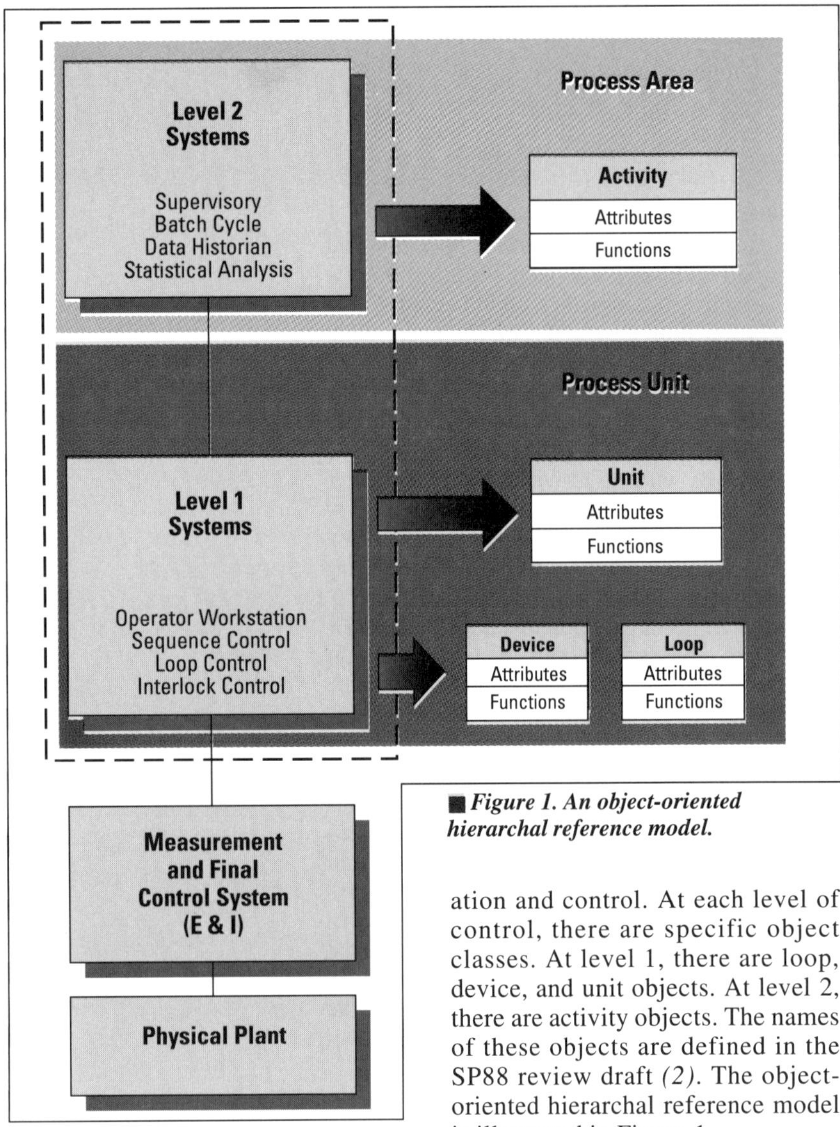

■ ***Figure 1. An object-oriented hierarchal reference model.***

derivative (PID) loop control have persisted for over 50 years in spite of numerous technology changes (for instance, from pneumatic flappers and nozzles, to electronic operational amplifiers, to digital microprocessors). PID remains the primary reference model for single-loop regulatory control. Likewise, ladder diagram logic is still the best representation of interlock control.

Recognizing the importance of CIM in modern process plants, the overall model structure is based on the Purdue CIM reference model *(3)*. It classifies control systems into hierarchal levels that include the operation and control of the entire facility. Real-time control is associated with levels 1 and 2 of the Purdue model. Borrowing from the SP88 work, we can extend the Purdue model to include object orientation *(4)*. This is a very powerful concept in representing system operation and control. At each level of control, there are specific object classes. At level 1, there are loop, device, and unit objects. At level 2, there are activity objects. The names of these objects are defined in the SP88 review draft *(2)*. The object-oriented hierarchal reference model is illustrated in Figure 1.

An object class is composed of a set of attributes (for instance, measured variable, setpoint, output, alarms, and mode) and functions (such as performance monitoring, operation support, and control enforcement). For regulatory and discrete control, a loop object regulates a controlled variable to a setpoint and a device object sets process elements to prescribed states. For process unit control, a unit object provides sequence control. For process area control, the activity object provides supervisory and batch cycle control. Object identification (or instantiation) ranges from simple to complex types, depending on the degree of performance monitoring, operation support, and control enforcement required in a particular application. An application may not require the full model. In the typical continuous process, level 1 loops and devices are adequate unless startup and shutdown, or supervisory control is required.

Control flow diagrams

The foundation of the model is the discrete and regulatory control that is required in all processes. When designing the control strategies for discrete and regulatory control, the fundamental questions in the mind of the control engineer are:

- What are the controlled and indicated loops and devices?
- What are the control associations among them?
- What are the interlock cause/action associations?
- How do these relate to the process flowsheet?
- What is operated from the field and what is operated from the control room?

As mentioned earlier, an important activity in a design project is design capture and version control. For discrete and regulatory control, the established design standard is the ISA symbology *(5)*. This standard format includes a reference to the process (for example, a PFD).

Using the ISA standard symbols combined with PFD or P&ID symbology, design capture and version control for loop and device objects is in the form of what we can call a control flow diagram (CFD). A CFD depicts the instantiated loop and device objects and the associations among objects such as interlocks, cascades, ratio, override, and stream routing.

Figure 2 shows a CFD for a waste-treatment unit. Loop objects are required for level indication, feedstream flow, and pH control. As for device control, a simple device object is required on the main tank to indicate the run status of a field-operated agitator. Around the additive tank, device objects are required for metering the additive into the tank, and for discharging the additive tank to the main tank.

The CFD is considered a conceptual design media. More detailed functional requirements specifications — P&IDs, loop diagrams, electrical schematics, loop descriptions, and so forth — must be developed before the control system can be generated and deployed.

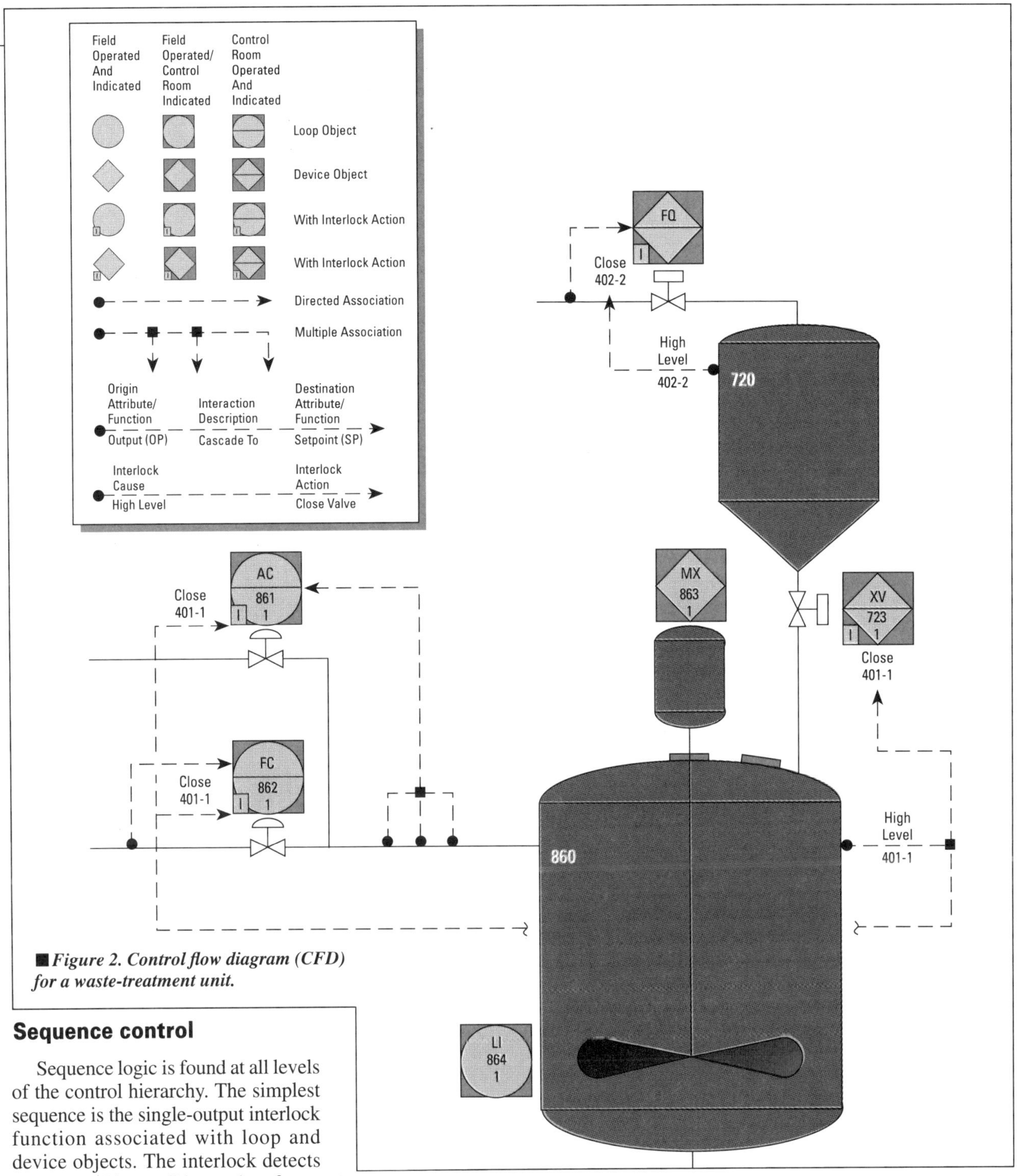

■ *Figure 2. Control flow diagram (CFD) for a waste-treatment unit.*

Sequence control

Sequence logic is found at all levels of the control hierarchy. The simplest sequence is the single-output interlock function associated with loop and device objects. The interlock detects an undesirable condition and performs a prescribed action on the object to prevent an unwanted process event. After the condition clears, the consequences of the action, such as valve closure, must be reset to continue routine operation. The sequence executes in a cycle, driving process equipment through a cycle of states. This is the essence of sequence control.

When discussing sequence control for an application, it is usually in the context of driving a set of process equipment in a coordinated cycle. In the case of the unit object at level 1, sequence control manipulates device and loop setpoints and parameters to accomplish the prescribed control objective. Unit objects usually are associated with a single process unit such as a batch reactor, mixer, or fermenter, but may be used to sequence multiple pieces of equipment such as those in a conveying system.

At the activity level, sequencing is associated with batch cycle control. A batch activity manipulates the unit objects to control the movement of batches through one or more process units based on a recipe.

Figure 3 illustrates sequence control using the terminology of SP88 *(2)*. In the figure, the process is depicted as a state diagram and the sequence control as a simplified sequential function chart diagram *(6)*. The initial state of a process unit or area is idle (that is, not making product). When the unit begins operation, it is considered active (that is,

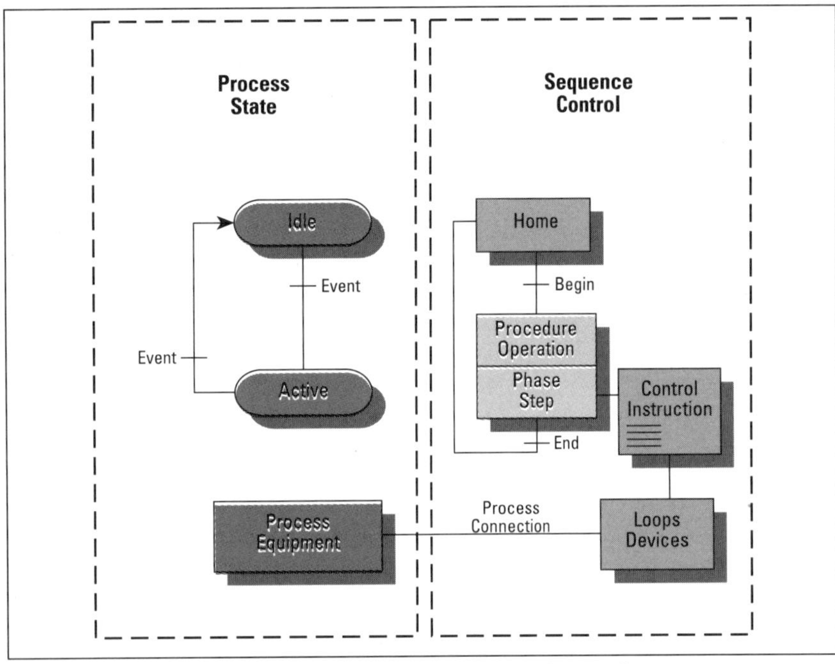

■ *Figure 3. Process state and how it relates to sequential control.*

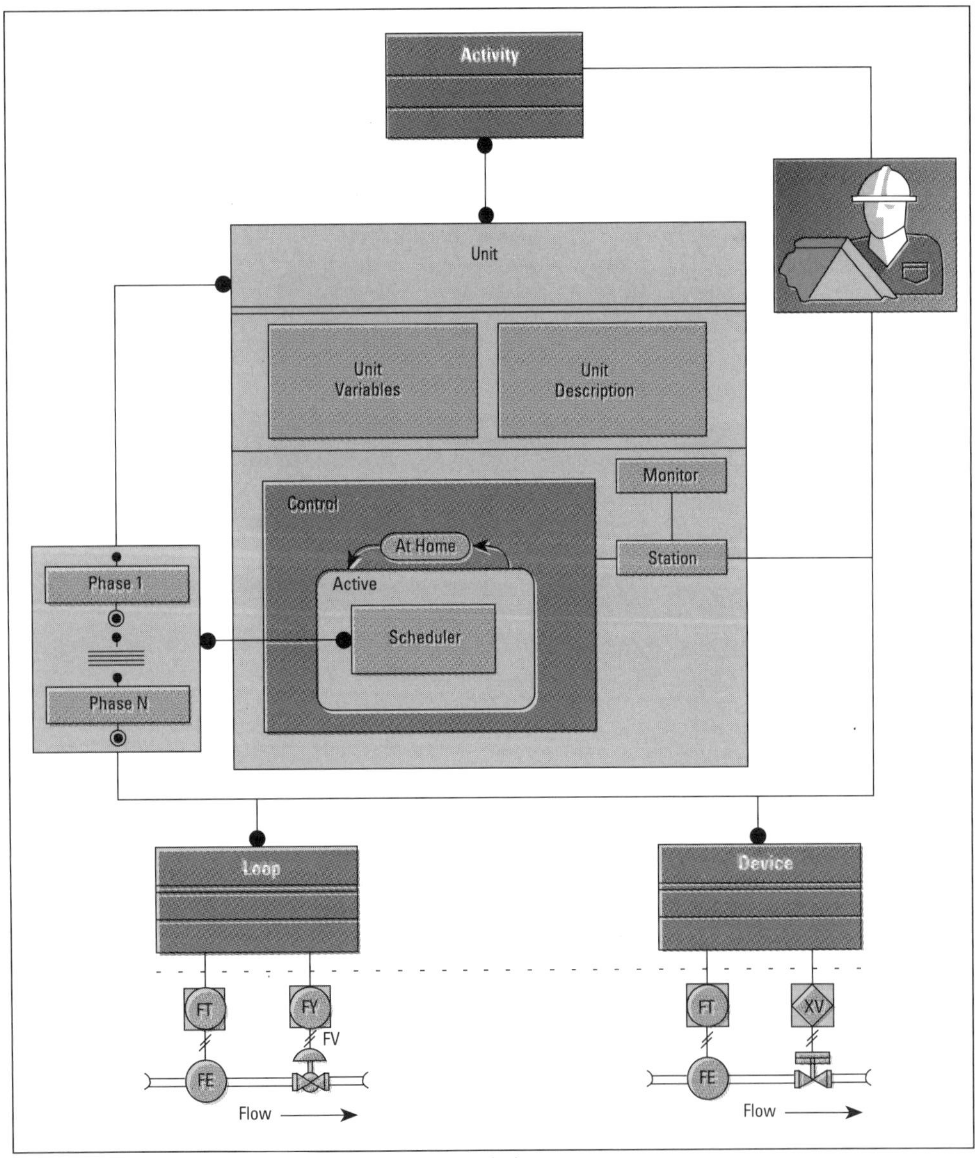

■ *Figure 4. Unit object class and associated phases.*

making product). In a typical application, there are many intermediate and subordinate states within the active state of the process.

In the figure, sequence control is depicted as a cycle with an initial position called home. A manual or automatic event causes the sequencer to move from the home position through its sequential logic. At the endpoint, the sequencer returns home and awaits the next start command. The process state is a function of the sequence control actions.

In activity objects, the sequence control is performed by procedures composed of subfunctions called operations. Unit objects are composed of phases made up of subfunctions called steps. Steps, in turn, consist of control instructions that manipulate loop and device objects, coordinate operator intervention, perform computations, and monitor for abnormal process conditions.

In a complex batch application, several activity objects may be used to coordinate the batches through multiple unit objects. Control system technology is available to provide fully automated advanced batch control. The challenge for the designer is to understand user requirements in enough detail to implement the right functions for the application. Sequential control of this type cannot be described by one long, linear stepwise string of control instructions.

The unit object model

In the interest of space, we will discuss only the unit object, and its phases and steps. By presenting a glimpse of the overall model, however, we will see why the use of linear sequence descriptions to express sequence control requirements should be abandoned. We propose an alternative that will improve the technical exchange between the process engineer and the control engineer.

Definition of a unit object. A

unit object class provides operation support, performance monitoring, and control enforcement for a group of process equipment. Operation support is the human/machine interface that includes the look and feel window to the process state. It includes the functions that warn of abnormal conditions and that allow the operator to command the unit and its phase control. Control enforcement consists of the actual sequential control functions. Performance monitoring is the alarm generation, process unit calculations (such as rate, inventory, and key measures), and historical data-management functions. Figure 4 depicts the unit object class and associated phases. It also shows the relationship with the operator, loop and device objects, and the activity.

Device vs. unit. The control objective of a device is very similar to that of a unit. Devices perform their task at the lowest level of access by the operator. If the overall control of a set of process units has a single point of entry, then it could be considered a device. This grouping of control in a monolithic form is undesirable in times of process upset, during checkout, or routine maintenance. If for some reason the full automatic sequencing cannot be performed, operating personnel must be allowed to use the loop and device objects to complete a sequence or to bring the process down gracefully. The proper allocation of device, loop, and unit control is a critical consideration in design for operability.

Phase. Each unit object has one or more sequential control schemes associated with it. The individual schemes are called phases. A phase is activated by an operator or by an activity object. The unit phase scheduler initiates the phase execution. The step's control instructions manipulate the process unit equipment through the loop and device objects. The phase is depicted in Figure 5 as a state function diagram.

A typical phase has three distinct parts: the scheduler, the routine step logic, and the fail action and recovery logic. When activated, the phase changes from the inactive to the executing state. While in the executing state, the phase may be in one of several substates. When routine step logic is being performed, the phase is in the normal state. The normal state has a substate called hold that is activated from external commands — from the operator or activity — and should only be valid at the beginning of a phase. If a failure occurs or the operator stops the unit object, the phase will branch automatically to the fail state. When the phase ends, it returns to the inactive state. At this point, the unit object may activate another phase or return to the at-home state.

The typical phase is composed of multiple steps. Usually, the first step (or set of steps) performs actions, such as setting the initial loop and device setpoints, and modes of control, that are required to prepare the process unit for the rest of the phase cycle. More-complex setups include operator actions such as selecting raw material tanks, pumps, and meter runs. The operator may be requested to enter a laboratory analysis, set manual valves, and start auxiliary equipment around the unit. Setup includes checking all permissives that must be satisfied before proceeding.

After setup, the phase performs the sequence-control actions to operate the process unit. This is the operating part of the phase, and includes control instructions such as metering, weighing, and profiling. Some of these control actions may continue through multiple steps as so-called concurrent tasks. The length of a phase varies from a few minutes to many days.

The last part of a phase is the set-down step(s) to prepare the process equipment for the next cycle. Typical control tasks include unit clean out, pump stops, loop and device deactivation, and vessel drain-out or purge. As in the case of setup, this can be quite elaborate and involve operator action responses.

Simple vs. complex phase logic. A simple phase is a fixed cycle with end conditions that are serial and perform a single process-control task. Serial logic means the step end conditions occur essentially at the same time or one at a time. Complex phase logic involves concurrent multiple control tasks (that is, more than one unit control task executes at the same time) where the end conditions may not happen in unison or one at a time. The charge of multiple reactants into a reactor without knowing which reactant will end first is an example of a complex sequence-control task.

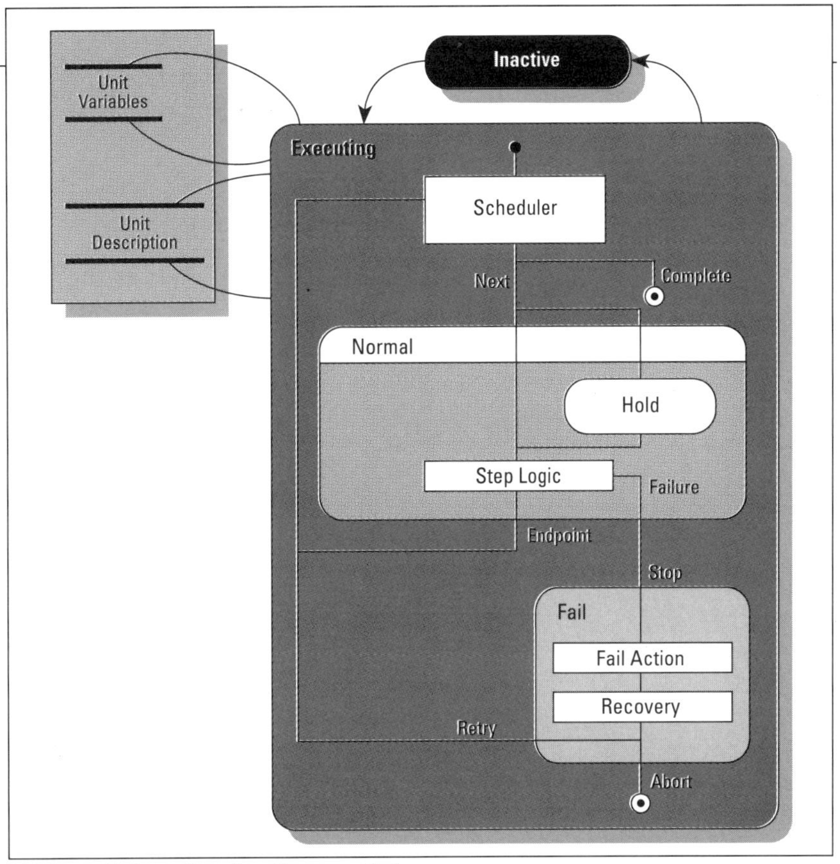

■ ***Figure 5. State function diagram of phases.***

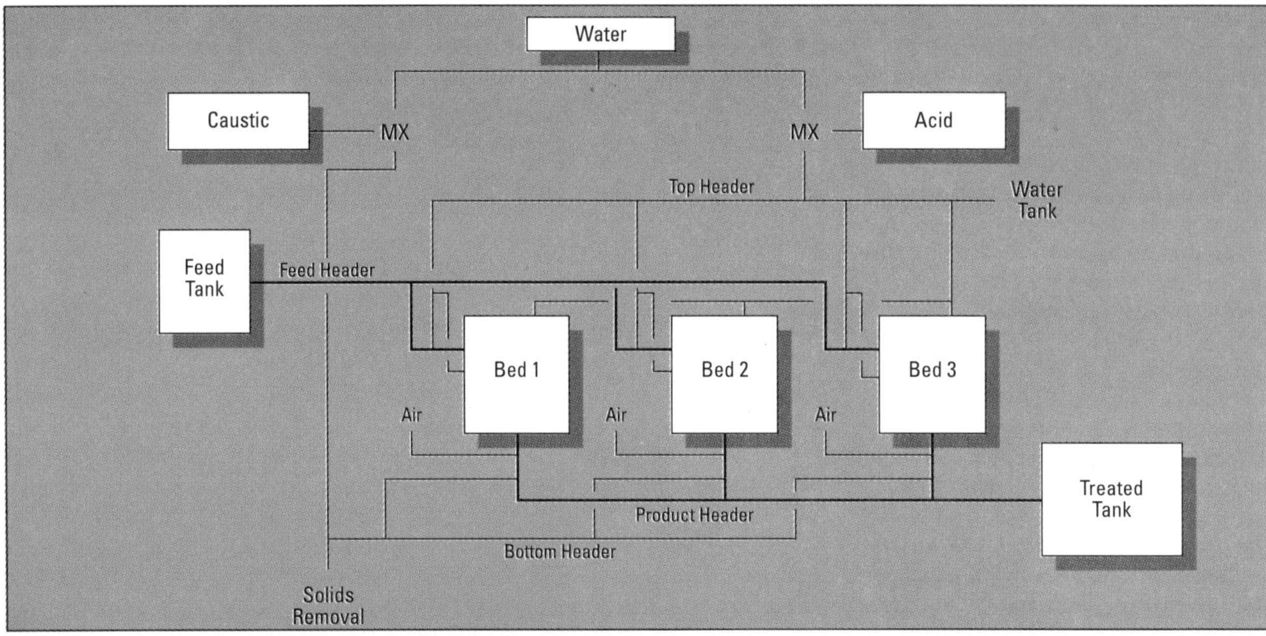

■ ***Figure 6. Feed-stream cleanup via a multiple bed system.***

Phase fail action and recovery. Phases should include fail action and recovery functions because things can go wrong in routine plant operations. During plant operation, batches can be saved if the proper fail action and recovery is implemented. This, however, can change a simple sequence into a very complex one.

Fail actions are triggered by fail conditions that are monitored by the unit object if the condition is unit wide, or by specific phases if the condition is only of concern during them. Once a fail action is triggered, appropriate loop and device objects are placed in a safe position. It is then the operator's job to troubleshoot the problem. A critical aspect of good unit object design is to provide accurate information for use in the diagnosis. Once the cause is identified, the recovery logic should allow safe restart or a proper abort path.

Aborting the phase returns control to the unit where the cycle may be terminated, necessitating a manual operation to finish the cycle. A restart recovery action returns the phase to the scheduler — which moves the phase backward, forward, or to the last step, resuming normal automatic control (that is, recovery).

Area operation design

The capabilities of control systems have increased dramatically in recent years. The challenges of just a few short years ago seem insignificant today. For instance, simple cascade control was once difficult to implement and operate; now it is straightforward. Likewise, self-tuning dynamic matrix control and batch recipe control now are commonplace. Are we taking advantage of modern control systems? Sadly, the answer in many plants is no. This capability is seldom considered in the routine cycle of the engineering design.

To take full advantage of modern control systems, a project team must plan and seriously consider automating the higher levels of the reference model from the start of a project. We can call this the design of automated area operations. It is a systematic approach to create the required control solution at the unit and activity control level.

What are area operations? Let's define them as the activities associated with the control of entire process areas and whole process units to produce the desired product. Specifically, they are high-level functions that were once the exclusive responsibility of the operator as described in an operating procedure manual. Today, they can be automated as part of the control system.

The fundamental questions that concern a control engineer about the design of automated area operations are:

1. What functions are required to operate the process?

2. How much flexibility is required?

3. Which are routine, special, or emergency?

This leads to some typical topics of discussion in reviewing the requirements with process engineering:

- process areas included in the application scope;
- field vs. centralized area operations;
- startup, shutdown, production, and turndown operations;
- routine, special, and emergency operations;
- concurrent operations;
- operator jurisdiction;
- type of control enforcement;
- type of performance monitoring;
- data collection and reporting;
- data exchange; and
- control object associations.

Area operation design is an iterative activity that includes the following major steps:

Developing a process state description. Describe process states and reference them to a physical plant view, usually in the form of a process flowsheet or simplified P&ID. The design capture and version control is composed of process state abstracts and, in the more-complex applications, a state diagram.

Performing an area operation design. Using the operation and control reference model, identify activity and unit objects, their attributes and functions, and the associations among objects. The design capture and version control for this step is in the form of a state function diagram, function abstracts and, in complex applications, an operation flow diagram.

Iterating on step 2. Consider flexibility, variability, special functions, emergency functions, setup functions, and what is manual vs. automated.

A multi-unit application

The PFD in Figure 6 represents a process that requires multiple unit objects to fully automate the area operation. The process removes solid waste from a continuous feed stream. The feed stream is passed through beds that contain material that collects undesirable components. Over time, a bed accumulates enough undesirable material that it must be taken off-line and regenerated — removing the waste material. The time to do this may be detected by differential pressure across the bed, by an in-stream measurement, or by laboratory analysis. In any case, operating personnel must switch out the spent bed and replace it with a fresh bed. After the spent bed is taken off-line, the bed material is regenerated through a sequence of water, acid, and caustic washes. After regeneration, the bed is ready for the next switchover to replace another spent bed.

The process is designed so that only one bed is operating on the feed stream at a time. Also, only one bed may be regenerated at a time, because the three beds share common water, acid, and caustic supply headers. The two operations, however, may be performed in parallel as concurrently executing control tasks. Even though this is a continuous process, it has a significant amount of sequence control for automated switchover and regeneration.

Samples of area operation design capture for this application are depicted in Figures 7–9. The process state diagram of Figure 7 illustrates the major process states and reflects the constraints of one bed at a time in the on-line and regenerating states. Figure 8 is an object diagram that defines the unit object instantiations (identifications) for this process, showing the number of units and their associations. Figure 9 is a sequential function chart that depicts the regenerating phases.

Unit object detail design

One of the results of area operation design is the separation of manual operations from the ones to be auto-

■ *Figure 7. State diagram for the cleanup process.*

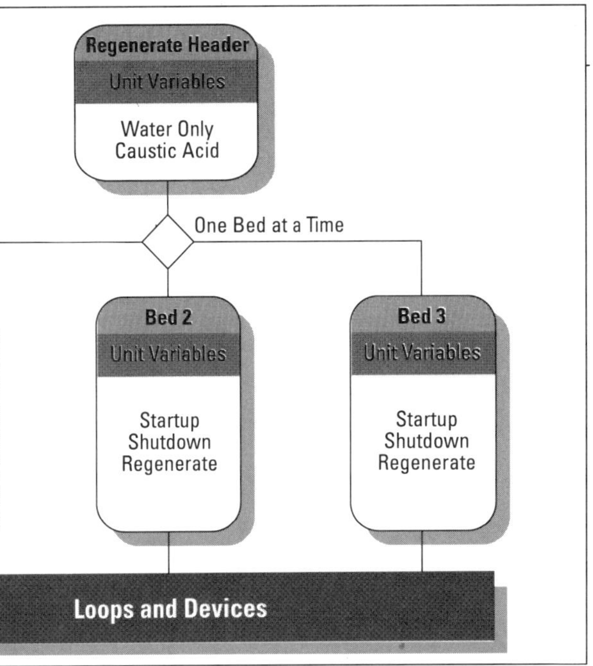

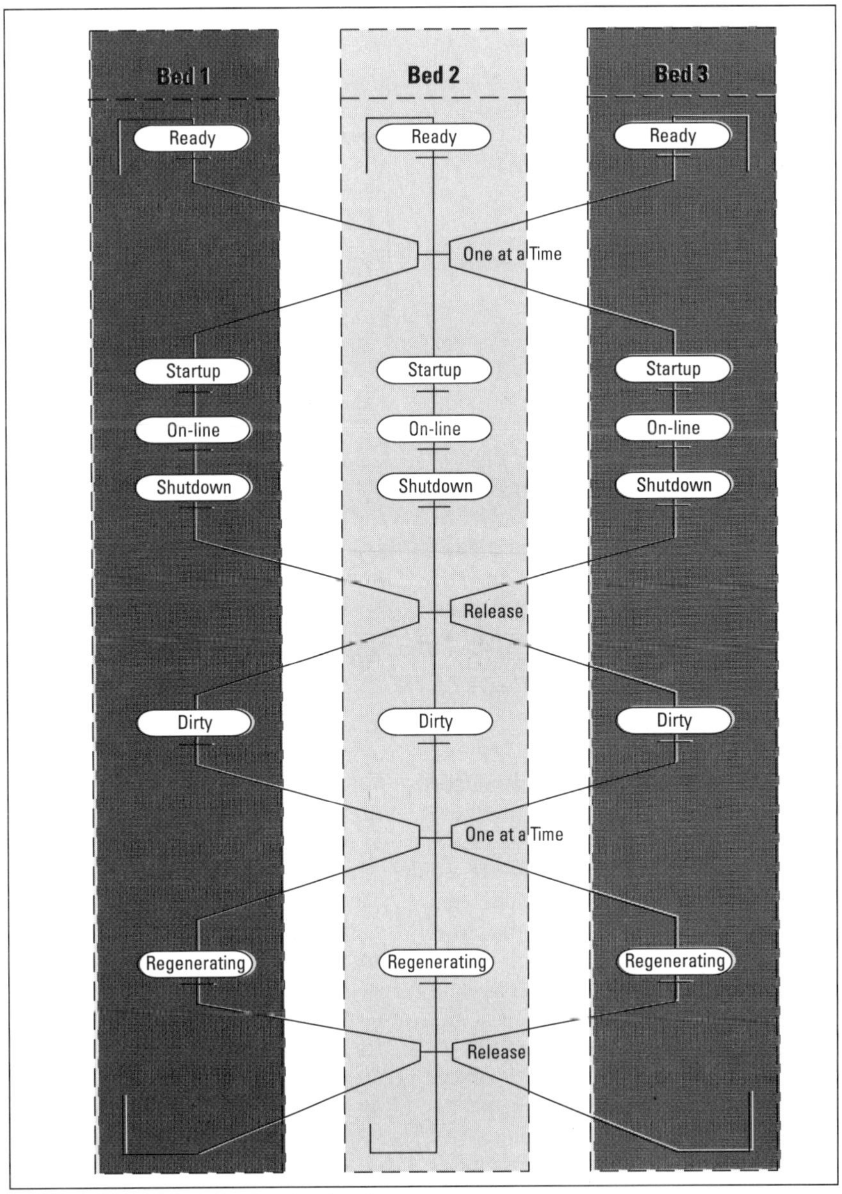

■ *Figure 8. Object diagram for the cleanup process.*

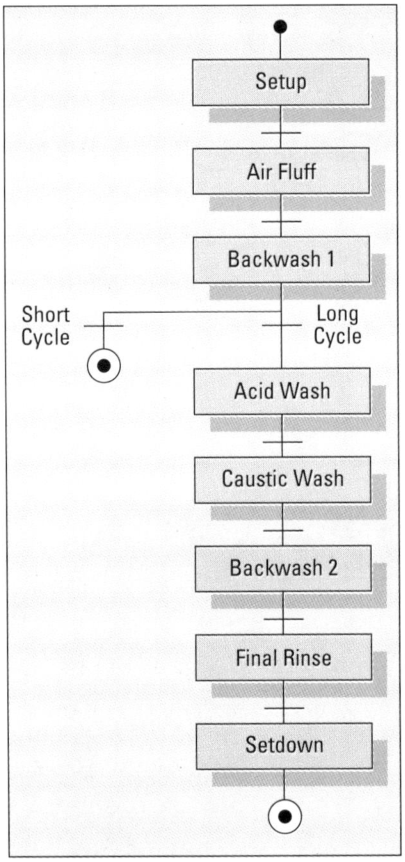

■ ***Figure 9. Sequential function chart for bed regeneration.***

mated. The automated operations are carried forward into the detail activity and unit design phase of the project. The detail design format follows the structure and dynamics of the reference model, and is composed of individual documents and diagrams for each activity and unit that was identified in the area operation design. The flow of data from area operation design to detail design is cumulative. That is, data from the area operation design effort is expanded and further refined in the detail design.

In our example, the unit FRS is composed of unit descriptions for each bed object and one for the regenerate header object. The detail FRS contains a unit association and shared resource description that includes the loop and device list in the unit description and a definition of the methods to arbitrate bed startup and shutdown, as well as the use of the regenerate headers. Since the beds are identical, a single set of phase descriptions using pseudo-names from the unit variables and unit descriptions are required. The phases of the regeneration unit must also be defined. The detail FRS is developed for the next step in the project — the system selection, application generation, and deployment activities.

Application generation

If the preceding design methodology is used, the FRS is in a form that is more efficient for the control system engineer to use than the linear, stepwise format of the past. It is easy to see that a modular FRS will facilitate a modular implementation that can be referenced back to the requirements' definition developed in the area operation design. To the control system engineer, modularity means a more robust and maintainable system for the life cycle of the facility. By structuring the full design cycle around an industry-standard reference model, the process engineer will be able to more easily verify that the implemented control system meets the application requirements.

With the availability of modern simulation packages for early system test, it is possible to verify the control implementation and the design well before startup. The combination of a structured design and simulated system test has proven itself on numerous projects. It has minimized or eliminated rework and reprogramming — meaning lower engineering and programming costs, as well as shorter time to the optimal operation and control solution. This should be the objective of all team members.

The payoff

The point of this article is simple: push for an industry-standard or at least company-standard control reference model — and use it. Eventually, the interaction among all team members will improve and so, too, will the overall quality of the solution. **CEP**

LITERATURE CITED

1. **Sheffield, R. E.,** "Integrate Process and Control System Design," *Chem. Eng. Progress,* **88** (10), p. 30 (Oct. 1992).
2. "Batch Control Systems, Models & Terminology, SP88 Standard," Draft 2, Instrument Society of America, Research Triangle Park, NC (Aug. 25, 1991).
3. "A Reference Model For Computer Integrated Manufacturing," prepared by the CIM Reference Model Committee of the Purdue International Workshop on Industrial Computer Systems, Instrument Society of America, Research Triangle Park, NC (1989).
4. **Rumbaugh, J., M. Blaha, and W. Lorensen,** "Object-Oriented Modeling And Design," Prentice Hall, Englewood Cliffs, NJ (1991).
5. "ANSI/ISA-S5.1-1984 And ISA-S5.3-1983 Standard," Instrument Society of America, Research Triangle Park, NC (1989).
6. "IEC 848, 1988 Standard," International Electrotechnical Commission, Geneva, Switz. (1988).

FURTHER READING

Fisher, T. G., "Batch Control Systems," *ISA Trans.*, **28** (3), (1989).

Hughes, T. A., "Programmable Controllers," Resources for Measurement and Control Series, Instrument Society of America, Research Triangle Park, NC (1989).

Hughes, T. A., "Programmable Controllers," *ISA Trans.*, **29** (2) (1990).

Hedrick, J. L., and G. Severns, "Planning Control Methods for Batch Processes," *Chem. Eng.*, p. 69 (Apr. 18, 1983).

Hedrick, J. L., and J. A. Ruder, "Practical Considerations For Robust Process Control Systems," Proceedings of the 15th Annual Advanced Control Conference, Purdue Univ., West Lafayette, IN (Sept. 11–13, 1989).

Hedrick, J. L., "Estimating Software Engineering for a Quality Project," presented at Instrument Society of America Symposium, St. Louis, MO (May 15–17, 1990).

Hedrick, J. L., "Batch Control Engineering for Distributed Control Systems," Proceedings of the 16th Annual Advanced Control Conference, Purdue University, West Lafayette, IN (Sept. 24–26, 1990).

J. L. HEDRICK is president of Automation & Control Technologies, Inc., St. Louis, MO (314/993–4080; Fax: 314/872–8686), a process control and automation consulting firm. He has over 24 years of experience in the field, including, when he worked for Fisher Controls, the functional design of Provox batch control products. He has developed training courses, given seminars, published articles, and now is developing an independent learning module on batch control for the Instrument Society of America (ISA) —in which he is a senior member. He received a BS in physics from the Univ. of Missouri (Rolla).

Achieve Total Quality Control of Continuous Processes

Use this systematic approach to combine plant process control with statistical process control.

Pradeep B. Deshpande,
University of Louisville

Russell E. Hannula,
Rohm & Haas Kentucky, Inc.

Mohan A. Bhalodia,
Exxon Research and Engineering Co.

Clay W. Hansen,
Exxon Chemical Co.

Concepts of statistical process control (SPC) now are widely used in the continuous process industries to measure product-quality variability. On-line SPC also helps to identify when a process is out of statistical control, prompting personnel to investigate and identify the causes and take corrective actions. Strict adherence to SPC often results in decreased variability in product quality over the long run.

Stochastic and engineering process control also have seen a number of significant developments in recent years. It is now possible to develop minimum-variance control laws from stochastic control principles which, under ideal conditions, lead to closed-loop data that may be used directly for statistical monitoring *(1)*. In some cases, this approach has led to rich dividends *(2)*. In others, though, minimum-variance control laws cannot be employed because they give rise to excessive manipulated variable movements. It is possible to damp out the manipulated variable movements by adjusting the controller parameters — but autocorrelation in the output data increases, and, therefore, traditional SPC rules can no longer be directly applied. In recent years, however, techniques have appeared that enable the designer to analyze and massage the autocorrelated data to make them amenable to SPC analysis.

Where a process is adequately represented by first order with dead time (FODT) type process models, feedback control laws now are available whose tuning constants may be adjusted, in terms of manipulated variable movements, to improve quality or to reduce costs, consistent with product specifications. With these control laws, for appropriate choice of tuning constants, it is possible to obtain minimum variance performance (in stochastic control sense), or to trade this off in favor of smoother manipulated variable movements.

Alternatively, the popular proportional-integral-derivative (PID) approach with feedforward control and deadtime compensation is available for controlling a wide variety of processes, including those that are potentially open-loop unstable, integrating types, and so on. This approach, while being attractive for several types of processes, also results in output data that are autocorrelated. As previously mentioned, however, we now know how to massage the closed-loop data and make them amenable to SPC analysis.

All in all, as we will explain in this article, technology now is available that permits us to combine the essential features of engineering and statistical process control for total quality control of continuous processes. This involves a two-part procedure. First, a suitable control law is designed to maintain the quality variable within specifications in the presence of disturbances. Then, the output time-series data are analyzed and suitably massaged as necessary so that a quality control chart (such as Shewhart or cumulative sum, CUSUM) can be prepared. The usual SPC rules then can be applied to maintain process performance at the optimum level.

Let's look at the regulatory control of a single-loop system. For illustration, we'll consider three types of control strategies — standard PID with feedforward control and deadtime compensation; the recent Q-PID control law; and a stochastic design.

Standard PID

This type of controller continues to be the most popular in industry. To be effective such a controller, however, must be tuned properly. This can be complicated because the tuning constants present in the control law interact with one another. Furthermore, the system performance degrades as the process dead time increases.

There are several approaches to tuning a PID controller. (See, for instance, Refs. 4–9.) Some involve open-loop testing, while others are based on closed-loop experimentation. The settings that result are meant to satisfy certain specified optimization criteria, such as minimum integral of the squared error (ISE) and quarter decay amplitude ratio.

Because the performance of the system deteriorates as the deadtime in the loop increases, compensation is applied to remove the deadtime from the characteristic equation of the system and improve performance. There are two ways to achieve deadtime compensation. One, due to O.J.M. Smith *(10)*, is called the Smith Predictor, while the other, due to C. F. Moore *(11)*, is called the Analytical Predictor.

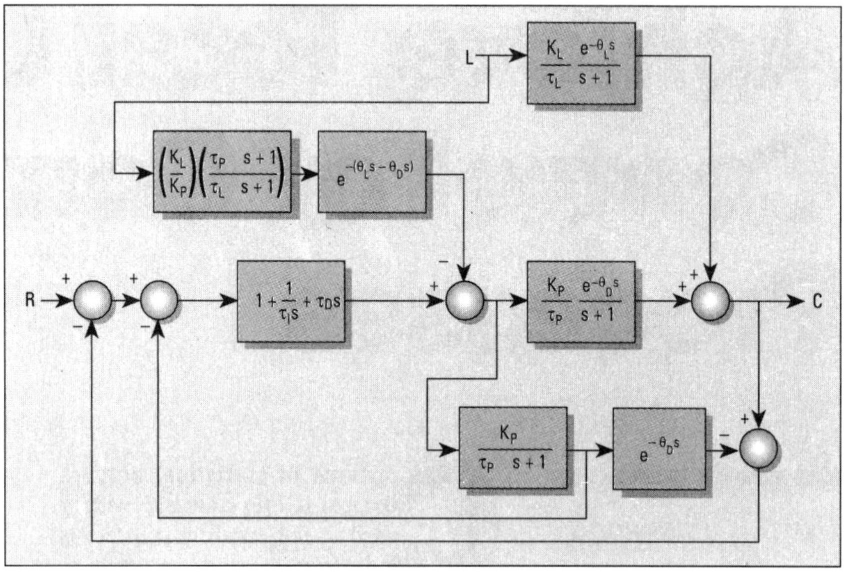

■ *Figure 1. Conventional PID control.*

In real-life applications, disturbances invariably are present and the controller must compensate for the negative effect of these disturbances on the process output. In those applications where disturbances can be measured, feedforward control can be employed. Figure 1 depicts the block diagram of the closed-loop system showing PID control with deadtime compensation and feedforward control, as commonly used in industry. The Smith Predictor approach for deadtime compensation is shown but, if desired, Moore's analytical Predictor may be employed instead. Blocks to implement PID control, lead lag, and deadtime compensa-

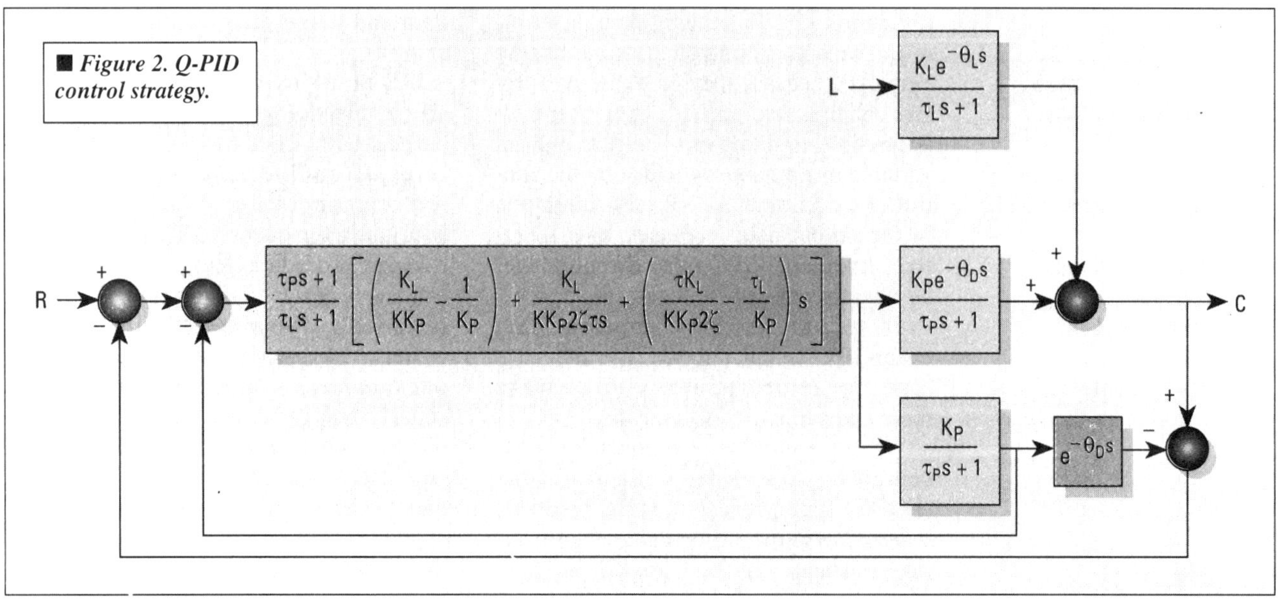

■ *Figure 2. Q-PID control strategy.*

tion come standard with modern distributed control systems.

The Q-PID approach

In the context of total quality control, certain properties are desirable for a control law:

- the capability of maintaining the quality variable within the required specifications in the presence of load disturbances and modeling errors.
- a transparent effect of the tuning constants on dynamic performance and the ability to adjust the tuning constants to improve quality or decrease manipulated variable movements, or vice versa, consistent with the product specification requirements and customer needs.
- PID terms, if possible, because of the familiarity and continuing popularity of this type of control.

The Q-PID controller has these properties *(3)*. Following the introduction of a disturbance, the process output is specified to return to the set point via a second-order trajectory (which is implicit in the Ziegler-Nichols formulas). Perfect control is theoretically feasible. If the process and load dynamics are assumed to be FODT models, the Q-PID control law becomes

$$G_c(s) = \frac{\tau_P s + 1}{\tau_L s + 1}\left[\left(\frac{K_L}{K K_P} - \frac{1}{K_P}\right) + \frac{K_L}{K K_P 2\zeta\tau s} + \left(\frac{\tau K_L}{K K_P 2\zeta} - \frac{\tau_L}{K_P}\right)s\right] e^{\theta_d s} \tag{1}$$

It contains PID-type terms, a lead lag network and deadtime compensation. The terms ζ and τ are tuning constants — with ζ a measure of damping, and τ a function of the settling time. In general, specifying the performance in the direction of perfect control (minimum variance in the context of stochastic control) increases the cost in terms of manipulated variable movements. The proper choice of ζ and τ depends on the quality specifications; there is no point in specifying tighter control than what the specifications call for because of the higher costs that would follow. A block diagram of the Q-PID based closed-loop control system is shown in Figure 2.

Both the conventional and the Q-PID approaches involve a PID controller, a lead lag network, and deadtime compensation, but differ in their result — according to an assessment of their performance on a simulated FODT process. The PID controller in the standard approach was tuned by the Cohen-Coon method. In the Q-PID approach, the parameters ζ and τ were selected to deliver about the same output transients as those under standard PID control. Figures 3a and 3b show the results with both approaches — on the basis of perfect modeling and a +30% error in plant gain. Q-PID delivers about the same type of output response but the manipulated variable movements are stronger. As shown in Figure 3c, these movements can be reduced by adjusting the parameters of the lead lag network.

When there is a factor of three error in the deadtime, both systems yield instability. Due to the transparent relationship between tuning constants and dynamic performance in the Q-PID approach, the system could be readily brought back to stability. Tuning with the standard approach may be somewhat harder, although this comment is based on limited experience.

In general, the standard approach may be able to withstand larger modeling errors than the Q-PID, but the loop-shaping capabilities of Q-PID will be helpful in quality control applications where it is desired to hold the quality variable within a certain band in the presence of disturbances. Also, the Q-PID approach requires one less controller for implementation, a fact that may have significance in dedicated single-loop applications. A self-tuning version of Q-PID now has been developed.

Stochastic control systems design

It is known that for identical model structures (G_P and G_L) and identical closed-loop performance specifications (for instance, minimum ISE/variance), there really is no difference between the design of feedback controllers for deterministic or stochastic disturbances *(1)*. Nonetheless, it is important to be familiar with how controllers are designed for stochastic disturbances because with this approach the closed-loop data can lend themselves directly to statistical monitoring. For more background, see, in particular, the classic texts on stochastic control of Box and Jenkins *(12)* and Astrom *(13)*, and the article by MacGregor *(1)*.

Consider the block diagram of a single-loop linear system shown in Figure 4. The process is perturbed by stochastic disturbance, called *noise*, which is obtained by passing white noise, a_t, (having zero mean and a constant variance σ_a^2) through a suitable model structure, such as a first-order lag, an integrating-type load such as a ramp, and so on. The disturbance model structure is selected such that it is representative of the real-life situation. In fact, plant testing with pseudo-random binary sequence signals followed by time-series analysis can help identify the models that would be needed for designing this type of controller. The purpose of the exercise then is to design a digital control law that will satisfy a certain optimization criterion. For reasons that will be clear shortly, we select *minimization of variance* as that criterion. From Figure 4 the output, C_t, is related to the manipulated variable, M_t, and the noise, N_t, according to

$$C_t = \frac{\omega(z^{-1})}{\delta(z^{-1})} M_{t-F-1} + N_t \tag{2}$$

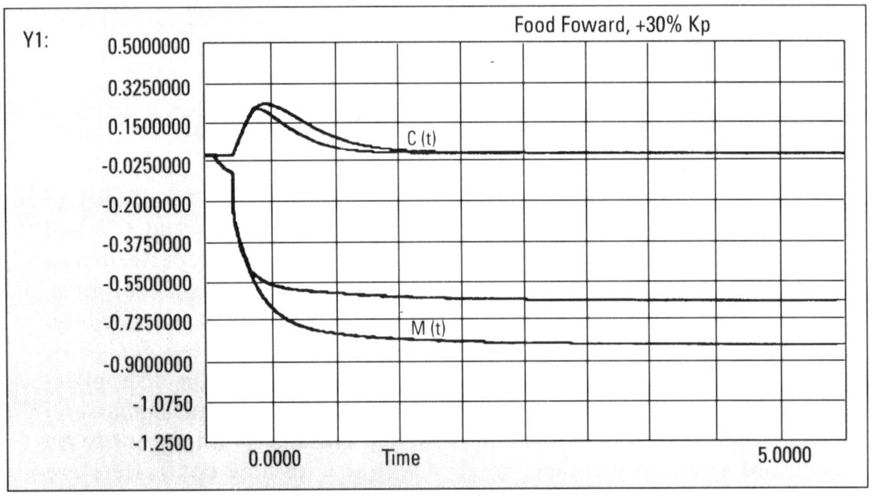

(a) Standard Approach
(Perfect modeling results shown for comparison)

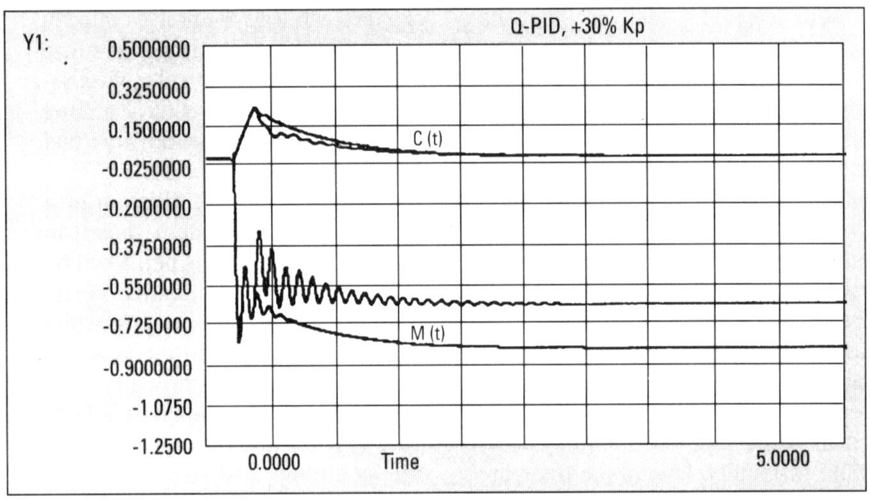

(b) Q-PID Approach

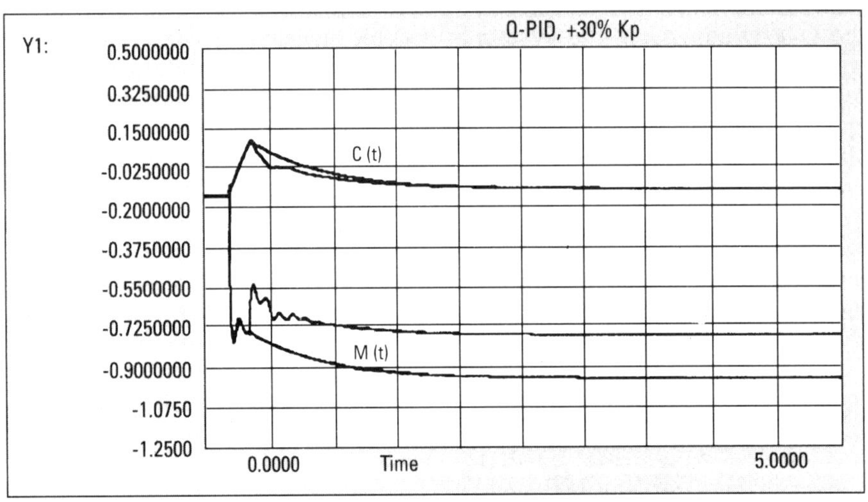

(c) Filtering used to reduce high-frequency oscillations

■ *Figure 3. Regulatory responses for 30% modeling error in plant gain.*

where F = time delay in terms of number of sampling periods. Equation 2 can be equivalently written as

$$C_{t+F+1} = \frac{\omega(z^{-1})}{\delta(z^{-1})} M_t + N_{t+F+1} \quad (3)$$

Minimum variance control requires that C_{t+F+1} be set to zero. So,

$$M_t = -\frac{\delta(z^{-1})}{\omega(z^{-1})} N_{t+F+1} \quad (4)$$

The control law given in Eq. 4 cannot be implemented, however, because it requires knowledge of N_t at F+1 sampling periods in the future. So, instead, we must use an estimated value, $\hat{N}_t$.

Now, let us assume that the noise model is adequately described by

$$N_t = \frac{\theta(z^{-1})}{\phi(z^{-1})\nabla^d} a_t \quad (5)$$

where

$$\nabla^d = (1 - z^{-1})^d \quad (6)$$

Usually, the parameter d = 0, 1, or 2. It permits the designer to describe nonstationary type of disturbances.

We then can derive

$$N_{t+F+1} = \frac{\theta(z^{-1})}{\phi(z^{-1})\nabla^d} a_{t+F+1}$$

$$= \psi_1(z^{-1})\, a_{t+F+1} + \frac{L(z^{-1})}{\phi(z^{-1})\nabla^d} a_t \quad (7)$$

where $\psi_1(z^{-1})$ is a polynomial of order F, and the elements of L may be obtained by solving the equality

$$\theta(z^{-1}) = \psi_1(z^{-1})\,\phi(z^{-1})\,\nabla^d + L(z^{-1})\,z^{-(F+1)} \quad (8)$$

This equality will be satisfied only if the order of $L(z^{-1})$ is max $\{q - (F+1), (p + d + 1)\}$, where p is the order of $\phi(z^{-1})$ and q is the order of $\theta(z^{-1})$. The coefficients of

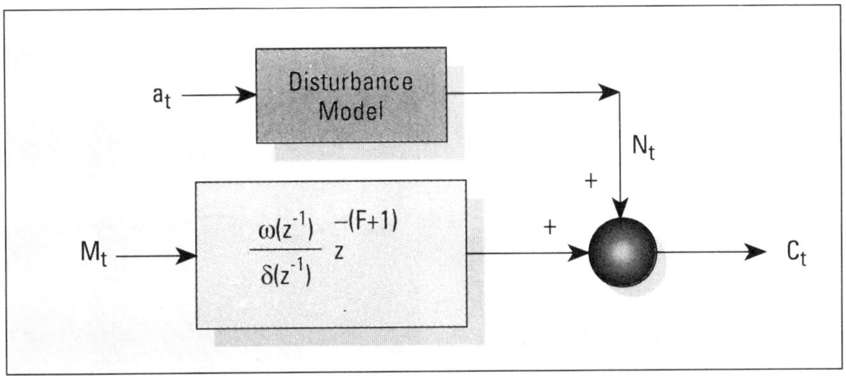

■ *Figure 4. Stochastic representation of a single-loop system.*

$L(z^{-1})$ may be obtained by equating the coefficients of z^{-k}. These manipulations will be clear from the example that follows.

The forecasted disturbance now can be written as

$$\hat{N}_{t+F+1} = \psi_{II}(z^{-1})\, a_t = \frac{L(z^{-1})}{\phi(z^{-1})\nabla^d}\, a_t \tag{9}$$

This ultimately leads to

$$\nabla^d M_t = -\frac{\delta(z^{-1})}{\omega(z^{-1})}\frac{L(z^{-1})}{\phi(z^{-1})}\frac{1}{\psi_I(z^{-1})}\, e_t \tag{10}$$

where

$$e_t = C_t - C_t^{set} \tag{11}$$

Equation 10 is the minimum-variance stochastic control law for single-loop systems.

An example

Now, let's consider the batch polymerization reactor problem reported by Vander Wiel and coworkers *(2)*. The process produces a polymer resin of specified intrinsic viscosity measured at the completion of each batch. The amount of catalyst in gallons added to each batch is the manipulated variable. The analysis of experimental data suggests that the input-output model is adequately represented as

$$C_t = \beta M_{t-1} + \frac{1-\theta(z^{-1})}{1-\rho(z^{-1})}\, a_t \tag{12}$$

where the second term on the right hand side represents the noise model, N_t. Following the suggested procedure for designing controllers for stochastic disturbances, we set C_{t+1} in Eq. 12 equal to zero, giving

$$M_t = -\frac{1}{\beta}\left[\frac{1-\theta(z^{-1})}{1-\rho(z^{-1})}\right] a_{t+1} = -\frac{1}{\beta} N_{t+1} \tag{13}$$

By substitution, we ultimately can derive

$$M_t = -\frac{1}{\beta}\,\frac{\rho-\theta}{1-\rho(z^{-1})}\, e_t \tag{14}$$

or equivalently

$$M_t = \rho M_{t-1} - \frac{1}{\beta}(\rho-\theta)\, e_t \tag{15}$$

It can be shown that substitution of Eq. 15 back into Eq. 12 gives

$$C_t = a_t \tag{16}$$

That is, the closed-loop output is distributed according to a normal distribution having zero mean and a constant variance. Thus, the output time series already is in a form suitable for control charting. Note, however, that for processes with deadtime under minimum-variance control, the data points every F sampling periods need to be used for control charting because, in this instance, the autocorrelation reduces to zero at lag F.

In general, the choice of minimum variance results in excessive movements of the manipulated variables. In this case, however, M_t simply refers to the gallons of catalyst that must be added based on viscosity measurements and, therefore, there are no identifiable penalties such as valve wear and tear, excessive load on the manipulated variable circuits, or the like. Indeed, Ref. 2 reports rich dividends from this strategy.

Nomenclature

a	= normally distributed random variable
C	= controlled variable
$C_{xx}(k)$	= estimated autocovariance at lag k
$C_{xx}(0)$	= variance
e_t	= forecast error, $A - P(t-1)$
E	= error (set point – measured value)
F	= delay expressed as number of integer sampling periods
G	= transfer function
k	= kth sampling instant
K	= gain
L	= load
M	= manipulated variable
N	= noise
$r_{xx}(k)$	= estimated autocorrelation function at lag k
s	= Laplace transform operator
X	= observed data
z	= z-transform operator

Subscripts

c	= pertaining to controller
d	= pertaining to process
i	= pertaining to integral mode
L	= pertaining to load
P	= pertaining to process
t	= pertaining to time

Greek letters

$\delta(z^{-1})$	= polynomial in z^{-1}
$\omega(z^{-1})$	= polynomial in z^{-1}
ε_t	= prediction error, $A - P$
ϕ_p	= autoregressive polynomial of order p, $(1+\phi_1 z^{-1} + \phi_2 z^{-2} + \ldots + \phi_p z^{-p})$
θ_q	= moving average polynomial of order q, $(1+\theta_1 z^{-1} + \theta_2 z^{-2} + \ldots + \theta_q z^{-q})$
σ	= standard deviation
θ	= deadtime
ρ	= autocorrelation function
τ	= characteristic time constant

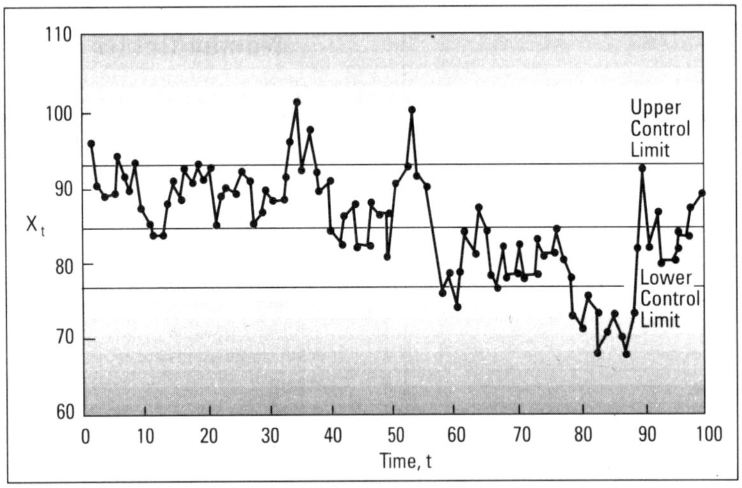

■ *Figure 5. Shewhart control chart for a chemical process.*

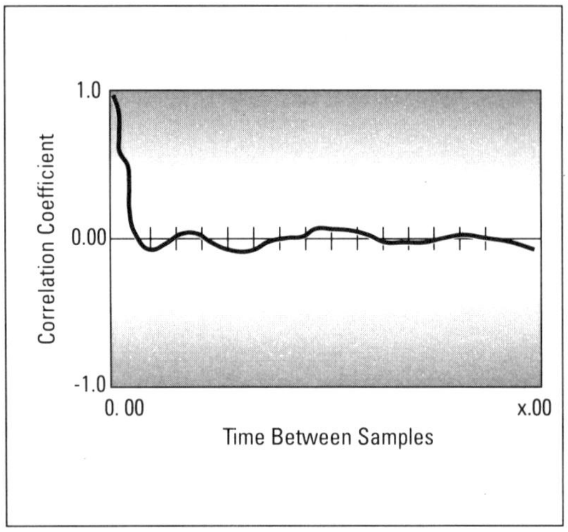

■ *Figure 6. Typical autocorrelogram.*

Statistical monitoring

Where minimum-variance control cannot be implemented and the output data are autocorrelated, they still can be massaged so that SPC rules can be applied.

First, let's assume that the feedback controller has been designed properly and correct tuning constants have been selected. Thus, the process gives a product of acceptable quality in the presence of load disturbances and modeling errors. We want to maintain the continuous process under statistical control.

As we have seen in the section on stochastic control, under ideal conditions (no time delays, no modeling errors and no constraints on the manipulated variable movements), the closed-loop process operating under the command of a minimum-variance control law naturally leads to output data that are normally distributed having a zero mean and a constant variance. The output data thus can be used directly in preparing control charts (Shewhart, CUSUM, or the like). In continuous processes, however, the choice of minimum-variance control often cannot be specified, because it would entail excessive manipulated variable movements that would invariably pose operating problems for the plant. Furthermore, the quality requirements in specific situations may not call for such tight control, in which case it would be wiser to select tuning constants that will dampen the oscillations. These comments notwithstanding, Ref. 2 presents an ingenious application involving a batch polymerization process where minimum-variance control yielded excellent results.

Greater-than-minimum variance

Now, let's examine a case in which the variance is greater than the minimum. Under this condition, the output data become autocorrelated. Thus, the traditional SPC rules cannot be applied directly. As an example, consider Figure 5, a Shewhart control chart that has been prepared from a set of 100 observations of a critical parameter *(14)*. These observations are taken every few minutes using an on-line sensor, and the process standard deviation is estimated from a moving-range control chart. The lag1 sample autocorrelation coefficient for these observations is $r_1 = 0.86$. Note in Figure 5 that there are many out-of-control signals. Ref. 14 points out that, due to the highly autocorrelated nature of the data, many false alarms have resulted. That is, no assignable causes could be found. Intuitively, what occurs in situations such as these is that the standard deviation calculated from the moving-range chart underestimates the variation in the process data. This results in control limits that are too close together, leading to frequent false signals of an assignable cause of variation.

The problems arising due to autocorrelation may be overcome in one of two ways.

In the first approach, an autocorrelogram is prepared showing how the autocorrelation coefficient varies with the sampling interval; see Figure 6 *(15)*. From such a plot, a sampling interval can be selected for which the autocorrelation coefficient is sufficiently small. Then, a control chart can be prepared using the selected sampling interval — for which SPC rules may be applied as usual — to detect the presence of assignable causes and to maintain the process under statistical control. A potential drawback of this approach is that the selected sampling interval may be too large, meaning that the process could go out of control before the next data point becomes available. For additional details on how to test for correlation using the autocorrelation function and to compensate for it in the design of control charts, consult Ref. 16.

In the second approach, the thrust is to fit an appropriate time series model to past observations and then apply the control charts to the stream of residuals from the model; see, for instance, Ref. 14. Thus, if A represents the observation and P the predicted value from an appropriate model fitted to historical data, then, the residuals $\varepsilon_t = A - P$, representing the prediction error, will behave as independent and identically distributed random variables *(14)*. Several time-series models have been suggested for this purpose. One is an autoregressive integrated moving average (ARIMA) model that is of the form

$$\phi_p(z^{-1})\,\nabla^d A = \theta_q(z^{-1})\,a_t \quad (17)$$

Another is the exponentially weighted moving average (EWMA) statistic. In this instance, the sequence of one-step-ahead forecast errors

$$e_t = A - P\,(t-1) \quad (18)$$

are deemed to be independently and identically distributed and may be used to prepare control charts to which SPC may be applied as usual. Here, $P\,(t-1)$ is the forecast of P at time instant t, made at time instant $(t - 1)$.

The EWMA approach is said to have computational advantages over the exact ARIMA approach and will be adequate provided the observations are positively autocorrelated and the process mean does not drift too quickly. For additional details, see Ref. 14 and the discussion following that article.

As an illustration, consider the data shown in Figure 5. They suggest the underlying model to be ARIMA (1, 0, 0) or AR (1) and may be written as

$$A = \xi + \eta A_{t-1} + a_t \quad (19)$$

The fitted model utilizing all 100 observations is

$$P = 13.04 + 0.847\,A_{t-1} \quad (20)$$

Analysis of the residuals from this model indicates that the process is in control.

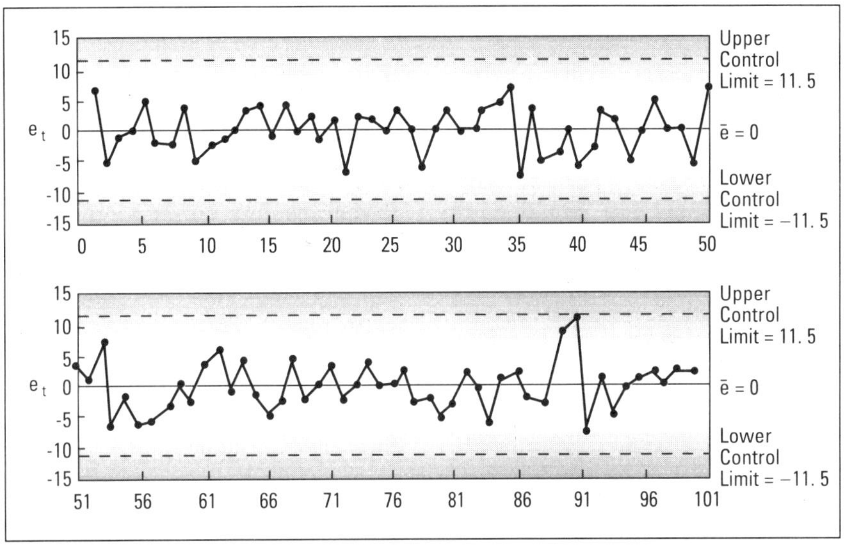

■ ***Figure 7. One-step-ahead prediction errors from the EWMA and the Shewhart control limits.***

Figure 7 shows for the same set of 100 observations a control chart that is based on the one-step-ahead forecast errors and the EWMA statistic. Again, the process is seen to be in control for the most part, with only one data point (No. 91) out of the control limits.

As we have previously mentioned, the specification of minimum variance may cause excessive movements of the manipulated variable. In such cases, one of two approaches may be followed. In the first, a suitable filter is employed in conjunction with the feedback controller. In the second, an optimization problem may be formulated involving the variances of the input *and* output. The solution of the problem then gives constrained inputs (that is, manipulated variable moves) that satisfy the specified criteria. This type of approach has been used in the stochastic control literature. Dynamic matrix control is an example of the strategy from the deterministic control literature.

The specification of minimum variance may cause excessive movements of the manipulated variable.

Dealing with faults

A variety of assignable causes can lead to out-of-control points on the control chart. For some, the remedy is in the domain of instrumentation and control. For others, the remedy may lie elsewhere. Some commonly encountered assignable causes in the domain of instrumentation and control are:

- a malfunctioning control valve or sensor;
- changes in dynamic process parameters, such as gain, time constants, and deadtime, due to equipment fouling, catalyst decay, and so forth; and
- an increase in system nonlinearities.

Technologies such as on-line identification, self-tuning and adaptive control, and expert systems may provide a framework for eliminating the detrimental effects of these assignable causes.

In some cases, the assignable causes may be known but suitable measurements are not possible. For example, in certain polymerization reactors, low-level impurities in the monomer feed can cause wide swings in the reactor outputs, but the impurities are not measurable. In such cases, the only remedy may be the absolute elimination of the cause — such as completely removing such impurities from the raw materials.

While in this article we have focused on single-loop systems, the ideas can be extended to multivariable systems as well. This requires a suitable (multivariable) controller, however — one whose tuning constants could be adjusted to maintain each quality variable of the multivariable system within specified limits. Such a study has been completed, and the results will be communicated in due course. CEP

P. B. DESHPANDE is professor and former chair of the chemical engineering department at the University of Louisville, Louisville, KY (502/588–0885; Fax: 502/588–6355). He has over 22 years of academic and industrial experience, consults for various companies, and gives continuing education courses around the world. He is the author, coauthor, or editor of four books on process control, has published 65 papers, and also has developed a number of control algorithms. He has received several awards for his work, including the 1990 Donald P. Eckman Award in process control education from the Instrument Society of America. He is a member of AIChE, ACS and is a senior member of the Instrument Society of America.

R. E. HANNULA is manager of the total quality systems dept. at Rohm and Haas Kentucky, Inc., Louisville, KY. He has over 20 years of experience with that company, General Electric, and Aluminum Co. of America. He specializes in quality management systems and the application of statistics for process and product quality improvement. He holds BS and MA degrees in mathematics from the University of Michigan and MS and PhD degrees in statistics from Carnegie-Mellon University.

M. A. BHALODIA is a senior staff engineer with Exxon Research and Engineering Co., Florham Park, NJ. He has over 20 years of experience in development and implementation of process control and plant computing technology, real-time knowledge-based systems for process monitoring and diagnosis, process alarm management, and prevention of major upsets. He received a BS from Maharaja Sayajirao University of Baroda, India, an MS from Michigan State University, and a PhD from the State University of New York, Buffalo, all in chemical engineering.

C. W. HANSEN is an applications engineer at Exxon Chemical Co., Baton Rouge, LA. He obtained a ME in chemical engineering from the University of Louisville in 1992.

Acknowledgments

The first author is indebted to Daniel L. Ash, retired president of Rohm and Haas Kentucky, Inc., for prodding him to find a way to address the quality control issues in the appropriate chemical engineering courses. The assistance of Dr. S. S. Tambe of the National Chemical Laboratory, Pune, India, in the preparation of the manuscript also is appreciated.

Literature Cited

1. **MacGregor, J. F.,** "On-Line Statistical Process Control," *Chem. Eng. Progress*, **84** (10), p. 21 (1988).
2. **Vander Weil, S. A., W. T. Tucker, F. W. Faltin, and N. Doganaksoy,** "Algorithmic Statistical Process Control: Concepts and an Application," *Technometrics*, **34** (3), p. 286 (1992).
3. **Deshpande, P. B.,** "Improve Quality Control On-line with PID Controllers," *Chem. Eng. Progress*, **88** (5), p. 71 (1992).
4. **Cohen, G. H., and G. A. Coon,** "Theoretical Considerations of Retarded Control," *Trans ASME*, **75**, p. 827 (1953).
5. **Ziegler, Z. G., and N. B. Nichols,** "Optimum Settings for Automatic Controllers," *Trans. ASME*, **64**, p. 759 (1942).
6. **Smith, C. L.,** "Digital Computer Process Control," Intext Publishers, Scranton, PA (1972).
7. **Fertik, H. A.,** "Tuning Controllers for Noisy Processes," *ISA Trans.*, **14** (4), p. 292 (1975).
8. **Garcia, C. E., and M. Morari,** "Internal Model Control 1 — A Unifying Review and Some New Results," *Ind. Eng. Chem. Proc. Des & Dev.*, **21**, p. 308 (1982).
9. **Chien, I. L., and P. S. Fruehauf,** "Consider IMC Tuning to Improve Controller Performance," *Chem. Eng. Progress.*, **86** (10), p. 33 (1990).
10. **Smith, O. J. M.,** "Close Control of Loops with Dead Time," *Chem. Eng. Progress*, **53** (5), p. 217 (1957).
11. **Moore, C. F.,** "Selected Problems in Design and Implementation of Direct Digital Control," PhD Thesis, Louisiana State University, Baton Rouge, LA (1969).
12. **Box, G. E. P., and G. M. Jenkins,** "Time Series Analysis: Forecasting, and Control," Holden Day Publishers, Oakland, CA (1976).
13. **Astrom, K. J.,** "Introduction to Stochastic Control Theory," Academic Press, New York (1970).
14. **Montgomery, D. C., and C. M. Mastrangelo,** "Some Statistical Process Control Methods for Autocorrelated Data," *J. Quality Tech.*, **23** (3), p. 179 (1991).
15. **Yeager, R. L., and T. R. Davis,** "Reduce Process Variation with Real-time SPC," *Hydrocarbon Proc.*, **71** (3), p. 89 (1992).
16. **Allison, P., and N. Dumont,** "Effects of Correlation on Statistical Process Control Charts," *Pulp and Papermaker*, **5**, p. 38 (1990).
17. **Box, G., and T. Kramer,** "Statistical Process Monitoring and Feedback Adjustment," *Technometrics*, **34** (3), p. 251 (1992).
18. **Harris, T. J.,** "Optimal Controllers for Nonsymmetric and Nonquadratic Loss Functions," *Technometrics*, **34** (3), p. 298 (1992).

Should You Use Constrained Model Predictive Control?

CMPC already has provided significant paybacks at a number of plants. Here's when to consider it.

Pradeep B. Deshpande, Jeffrey A. Caldwell, and Srinivas S. Yerrapragada,
University of Louisville

Mohan A. Bhalodia,
Exxon Research & Engineering Co.

Constrained Model Predictive Control (CMPC) is the most significant control technology to arrive on the market since proportional-integral-derivative (PID) control was invented over 50 years ago. CMPC has been applied successfully in the chemical process industries (CPI) — in a number of instances yielding hundreds of thousands of dollars per year in additional yield or throughput, or in energy savings. Payback periods of a year or less are not uncommon.

A number of CMPC variants are available today, including Dynamic Matrix Control (DMC), Identification and Command (IdCom), and Goal Maximizing Controller (GMAXC), among others. Reference 1 and the *CEP Software Directory* list commercial CMPC software packages.

While CMPC is an attractive choice in many applications, it may not be appropriate in others. In this article, we review the essentials of CMPC and illustrate through several examples where the technology may be applied with best results.

CMPC in a nutshell

This type of strategy is ideally suited for multivariable control applications. The essential features of the approach are:

1. The strategy combines optimization with feedback/feedforward control and provides constraint-handling capabilities. Nonsquare systems can be handled.

2. For a process with more outputs than inputs, the user may hold the outputs within the specified upper and lower bounds. Tighter control of certain outputs relative to others may be specified as necessary. Optimization with respect to specific inputs or outputs may be carried out.

3. For a process with more inputs than outputs, the manipulated variables may be allocated on the basis of an economic optimization index.

4. For a multivariable process with equal number of inputs and outputs, the strategy leads to excellent regulatory and servo control.

5. Last but not the least, CMPC, in conjunction with total quality control concepts *(2)*, can considerably enhance communication with management related to improving quality, boosting throughput, and minimizing energy consumption.

The underlying theory

CMPC is a model-based multivariable control strategy that combines feedback/feedforward control with optimization *(3, 4, 5, 6, 7)*. This combination gives CMPC strategies their constraint-handling capabilities. Because economic plant operation often is located in the neighborhood of operating constraints, CMPC has become increasingly popular in the CPI in recent years.

A number of manifestations of model predictive control are possible. Here, let's look at a simpler one to illustrate the concepts. Consider a linear multivariable system schematically shown in Figure 1. The process outputs Y are related to the manipulated variables M, measured disturbances L, and unmeasured disturbances D according to:

$$Y = K_P M + K_L L + D \qquad (1)$$

where Y, M, L, and D are in deviation form

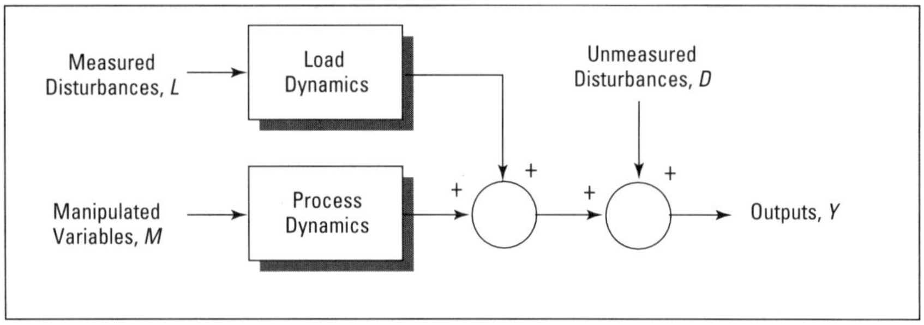

Figure 1. Schematic of open-loop process.

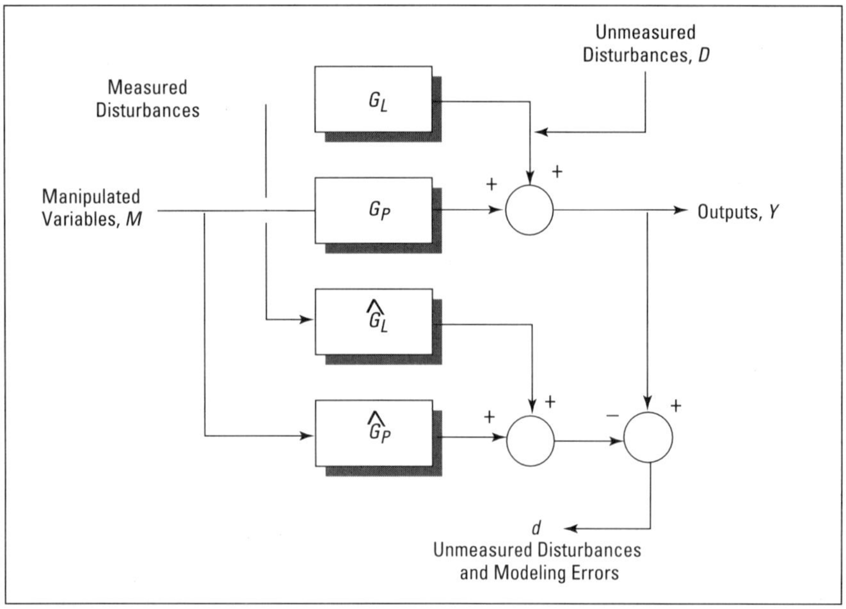

Figure 2. Recovering feedback signal for model predictive control.

and K represents a matrix of steady-state gains. Note that Eq. 1 is free of dynamic terms. For an illustrative 2-output, 3-input, 1-measured-load variable problem, Eq. 1 may be expanded to give:

$$y_1 = b_1 + K_{11}m_1 + K_{12}m_2 + K_{13}m_3 + K_{L1}l + D^1 \quad (2a)$$

$$y_2 = b_2 + K_{21}m_1 + K_{22}m_2 + K_{23}m_3 + K_{L2}l + D^2 \quad (2b)$$

with:

$$b_1 = y_{1s} - K_{11}m_{1s} - K_{12}m_{2s} - K_{13}m_{3s} - K_{L1}l_s \quad (3a)$$

$$b_2 = y_{2s} - K_{21}m_{1s} - K_{22}m_{2s} - K_{23}m_{3s} - K_{L2}l_s \quad (3b)$$

where y, m, l, and D are in absolute form and the subscript s denotes steady state. Formulating an optimization problem on this basis, using linear programming (LP), then the optimization index may be written as:

$$Min\ J = [MEGA\{C_1(V_1^U + V_1^L) + C_2(V_2^U + V_2^L)\} + C_3m_1 + C_4m_2 + C_5m_3] \quad (4)$$

subject to:

$$K_{11}m_1 + K_{12}m_2 + K_{13}m_3 - y_1 = -b_1 - d^1 - K_{L1}l \quad (5a)$$

$$K_{21}m_1 + K_{22}m_2 + K_{23}m_3 - y_2 = -b_2 - d^2 - K_{L2}l \quad (5b)$$

$$y_1 + S_1^U - V_1^U = y_1^U \quad (5c)$$

$$y_1 - S_1^L + V_1^L = y_1^L \quad (5d)$$

$$y_2 + S_2^U - V_2^U = y_2^U \quad (5e)$$

$$y_2 - S_2^L + V_2^L = y_2^L \quad (5f)$$

$$m_1 \le m_1^U \quad (5g)$$

$$m_1 \ge m_1^L \quad (5h)$$

$$m_2 \le m_2^U \quad (5i)$$

$$m_2 \ge m_2^L \quad (5j)$$

$$m_3 \le m_3^U \quad (5k)$$

$$m_3 \ge m_3^L \quad (5l)$$

In these equations, S refers to slack/surplus variables, V to violation variables, and C to cost penalties. The term *MEGA* represents a large number, say 10^6, and d^i is the effect of the unmeasured disturbances and modeling errors on the output i. The superscripts U and L refer to an upper bound and a lower bound, respectively. Owing to the presence of violation variables, the bounds on the controlled variables are treated as soft constraints that may be violated; the bounds on the manipulated variables, however, are hard constraints and are never violated.

An examination of Eqs. 5a–l reveals that the terms on the right-hand side involve upper and lower bounds on the controlled and manipulated variables, the biases b_i, and the terms d^i. The bounds are user specifications and, because the mathematical model of the process is known, the biases can be calculated. The terms d^i are feedback signals as shown in Figure 2, and can be calculated at each control interval by sampling the process outputs and subtracting from them the effect of the manipulated variables and the measured load variables on the controlled variables calculated from a mathematical model of the process. Note that the feedback signals d^i account for the unmeasured disturbances D^i as well as modeling errors. A linear program then may be invoked to solve Eqs. 5a–l, giving the manipulated variables to apply to the process.

Some comments on how Eq. 4 carries out the optimization are in order:

- As written, LP will minimize the objective function; maximization of the objective function may be achieved by making it negative.
- The magnitude of *MEGA* is very large. Thus, the principal aim of the optimizer will be to minimize the violation variables. In other words, the optimizer will try its best to maintain the outputs within their respective upper and lower bounds. If the optimizer is unable to find a solution where the violation variables are zero, then it will minimize the violation variables.
- By assigning different costs to the outputs, the user can strive to achieve tighter control of some outputs relative to others.

• If a solution can be found where all the violation variables are zero (that is, all the outputs are within their respective upper and lower bounds), then the term *MEGA* drops out of the picture and the optimal solution can be driven by the costs on the manipulated variables.

• The user may minimize or maximize an unmeasured output provided that its steady-state gains with respect to the manipulated variables are known or can be estimated. In this instance, these gains become the costs on the manipulated variables.

The foregoing approach can be made more attractive by predicting the process outputs for a number of sampling instants into the future, correcting the predicted outputs for the presence of disturbances, and requiring the corrected future outputs to obey the specified upper and lower bounds. Such predictions are made with the help of step- or impulse-response coefficients that are determined from process reaction curves. The use of step- or impulse-response coefficients is what gives CMPC the power to anticipate the detrimental effect of problematic situations and take corrective actions. The simpler procedure outlined above, however, adequately illustrates the salient features of CMPC.

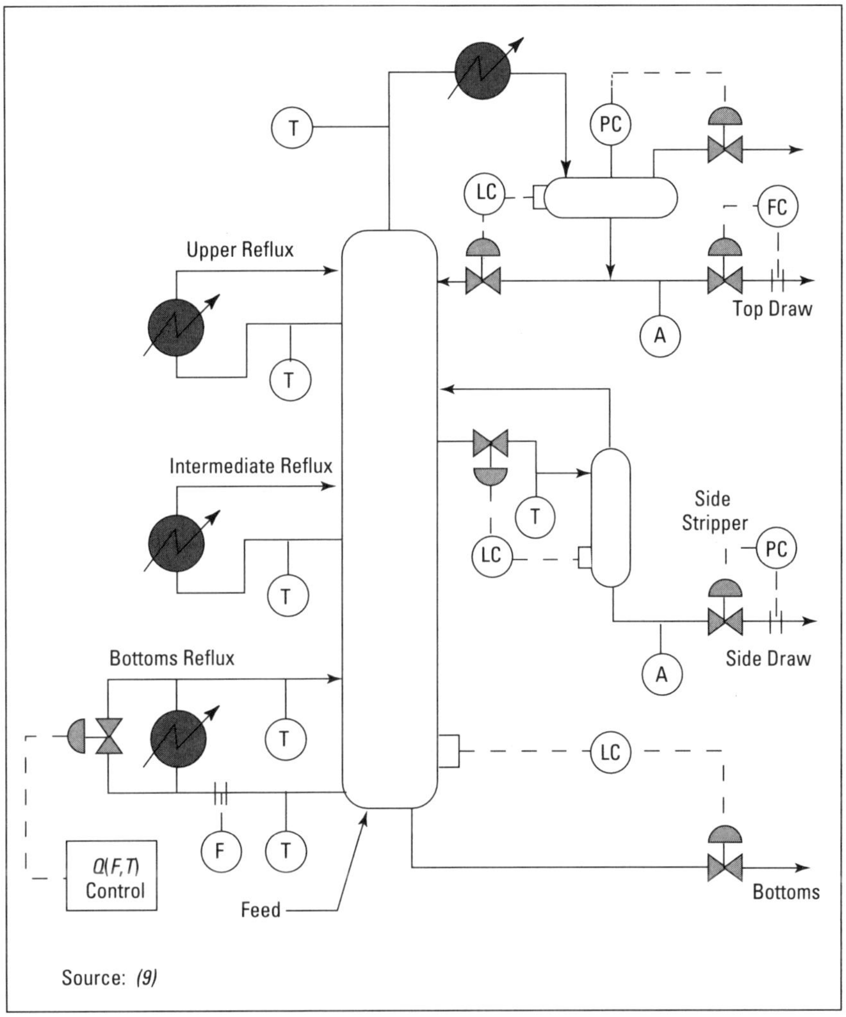

■ *Figure 3. Heavy oil fractionator.*

The many manifestations of CMPC

The arrangement in Figure 2 essentially depicts an Internal Model Control (IMC) structure *(3)*. What is missing, of course, is the controller. In traditional IMC, the controller represents the invertible part of the process model, and a filter is added to the feedback path to achieve robustness. In CMPC, the deterministic controller is replaced by an optimization program. The many CMPC products on the market differ in the way the optimization is carried out. There are several possible scenarios:

1. Perhaps the simplest approach is to take the manipulated variables from the solution described in the previous section and apply them to the process. This will ensure steady-state decoupling but, because it does not look into the future, it may allow constraint violations to occur. We call this approach LPIMC.

2. LPIMC can be made more powerful by predicting the process outputs a number of control intervals into the future, and requiring that all the predicted outputs (corrected for the presence of disturbances) be contained within their respective upper and lower bounds. A linear program can then be invoked to calculate a set of future moves such that a suitable optimization index is minimized (or maximized). The first move may be applied to the process and the entire procedure repeated at the next control interval to achieve CMPC. We call this approach Predictive LP (see also Refs. 4 and 5).

3. When the optimization index is quadratic in nature, using a procedure similar to that for option 2, a quadratic program may be employed for optimization *(6)*. This approach is known as Q-DMC.

4. A variant of the approach in option 2 is IdCom *(7)*. Here, outputs are predicted as in option 2 but, rather than calculating a set of successive moves, several intermediate moves are blocked (that is, held constant), and a gradient-searching optimization procedure is employed.

5. A newer approach to CMPC involves the use of a neural network as the model, and a nonlinear optimization routine to calculate the optimal MPC controller outputs that satisfy nonlinear constraints. Only a few applications of this approach exist so far (for instance, see Ref. 8), but it could have potential for complex nonlinear systems.

Several of these approaches have been applied successfully in industrial applications.

To bring out the salient features of CMPC, let's look at a few examples.

Control of a fractionator

First, let's consider a multivariable process with more outputs than inputs — a heavy oil fractionator *(9)* schematically shown in Figure 3. The unit includes two product draws and a bottoms product; two temperature points; three pumparounds with reflux temperatures; top and side draw flow rates that can be manipulated; and a bottom pumparound that can be manipulated for steam generation. The dynamic model relating the controlled variables to the manipulated variables and load variables consists of first-order with dead-time transfer functions shown in Table 1.

The control objectives are to:

1. maintain the top and side draw product end points at 0.0 ± 0.005 at steady state; and

2. maximize steam generation, given a number of constraints:

- all draws must be within 0.0 ± 0.5;
- the bottoms reflux duty must be held within 0.0 ± 0.5;
- the move size must be constrained to within ± 0.05/min;
- the bottom reflux draw temperature has a minimum value of -0.5;
- the top end point must be maintained within 0.0 ± 0.5;
- the upper and intermediate reflux duties must stay in the range of 0.0 ± 0.5; and
- the duties of the top two pumparounds are subject to disturbances from other columns.

The input data for the CMPC program used to solve this are shown in Table 2. The regulatory performance of the control strategy was evaluated by

Table 1. Process transfer functions $(Ke^{-\theta s})/(\tau s + 1)$.

	Top Draw	Side Draw	Bottoms Reflux Duty	Intermediate Reflux Duty	Upper Reflux Duty
Top end point					
K	4.05	1.77	5.88	1.20	1.44
τ, min	50	60	50	45	40
θ, min	27	28	27	27	27
Side end point					
K	5.39	5.72	6.90	1.52	1.83
τ, min	50	60	40	25	20
θ, min	18	14	15	15	15
Top temperature					
K	3.66	1.65	5.53	1.16	1.27
τ, min	9	30	40	11	6
θ, min	2	20	2	0	0
Upper reflux temperature					
K	5.92	2.54	8.10	1.73	1.79
τ, min	12	27	20	5	19
θ, min	11	12	2	0	0
Side draw temperature					
K	4.13	2.38	6.23	1.31	1.26
τ, min	8	19	10	2	22
θ, min	5	7	2	0	0
Intermediate reflux temperature					
K	4.06	4.18	6.53	1.19	1.17
τ, min	13	33	9	19	24
θ, min	8	4	1	0	0
Bottoms reflux temperature					
K	4.38	4.42	7.20	1.14	1.26
τ, min	33	44	19	27	32
θ, min	20	22	0	0	0

Table 2. Input data for the heavy oil fractionator.

Number/Variable	Steady-State Value	Weight	High Limit	Low Limit
Controlled variables:				
1 Top end point	0.0	10.0	0.005	-0.005
2 Side end point	0.0	10.0	0.005	-0.005
3 Top temperature	0.0	1.00	0.000	0.000
4 Upper reflux temperature	0.0	1.00	1.000	-1.000
5 Side draw temperature	0.0	1.00	0.000	0.000
6 Intermediate reflux temperature	0.0	1.00	1.000	-1.000
7 Bottoms reflux temperature	0.0	1.00	1.000	-0.500
Manipulated variables:				
1 Top draw	0.0	1.00	0.500	-0.500
2 Side draw	0.0	1.00	0.500	-0.500
3 Bottoms reflux duty	0.0	1.00	0.500	-0.500
Measured load (used for testing purposes):				
1 Intermediate reflux duty	0.0	—	0.500	0.500
2 Upper reflux duty	0.0	—	0.500	0.500
Other pertinent information:				
Control interval	2 min			
Number of future moves calculated	8			
Move penalty	10 (for all manipulated variables)			

subjecting the process to a step change of 0.5 in each of the two load variables. The resulting closed-loop responses are plotted in Figures 4a and b. Figure 4a shows that the process outputs are within the desired specifications at the new steady state, while Figure 4b indicates that the bottom reflux duty (manipulated variable no. 3) has been reduced, implying that steam production has been maximized (see Figure 3).

A deceptively simple system

Now, let's consider a 2×2 problem *(10)* whose dynamics are described by the transfer function matrix:

$$\begin{bmatrix} Y_1 \\ Y_2 \end{bmatrix} = \begin{bmatrix} \dfrac{1}{7s+1} & \dfrac{a}{7s+1} \\ \dfrac{a}{7s+1} & \dfrac{1}{7s+1} \end{bmatrix} \begin{bmatrix} M_1 \\ M_2 \end{bmatrix}$$

First, let $a = 0.2$. This gives a condition number of 1.5. (The condition number of a matrix is determined from singular value decomposition, a concept from linear algebra. It represents the ratio of the largest to the smallest singular value of a system. The larger the condition number, the more inaccurate the matrix inverse. Control strategies such as MPC that implicitly or explicitly utilize the inverse of the process gain matrix in controller calculations are apt to not work as well for systems with high condition numbers.) Then, let $a = 0.99$, giving a condition number of 199.

In both instances, the process outputs initially are at rest and we wish to drive the first output to 1.0. The set point of the second output remains at 0. The input data file for this problem is shown in Table 3. The closed-loop responses for the two case studies are depicted in Figures 5a and b. Note that in the second example very large changes in the manipulated variables are required. In practice, such large changes are not possible. If the manipulated variables are constrained, say, to within a range of ±10, control will be lost. This example illustrates that MPC may not be suitable for ill-conditioned systems such as high-purity distillation towers, which can have very high condition numbers.

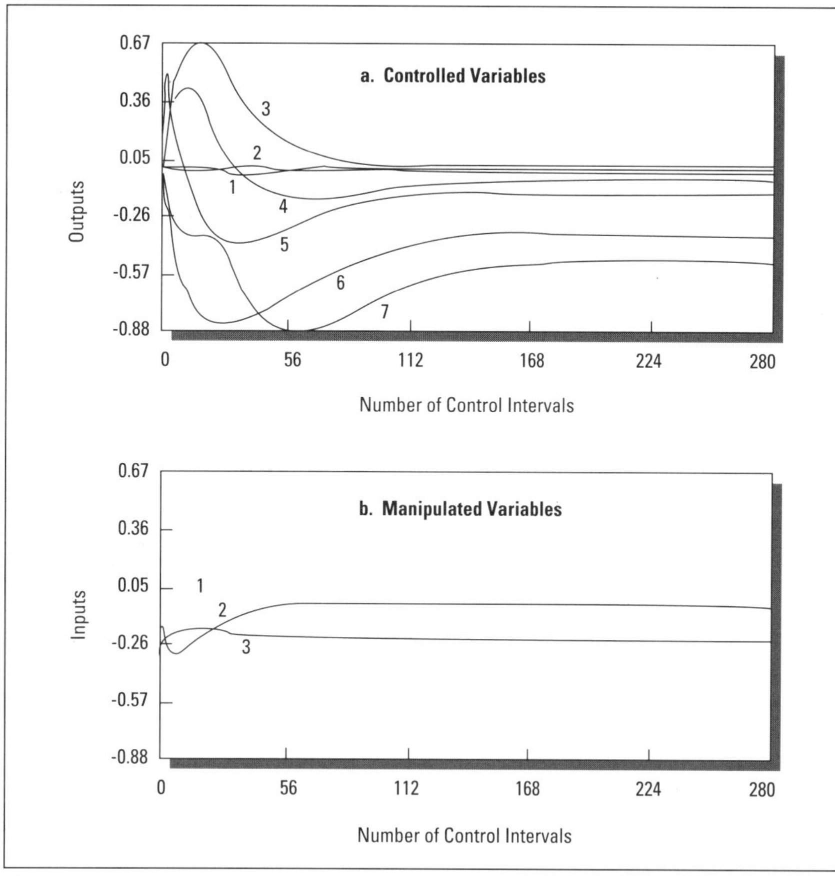

■ *Figure 4. Regulatory control of heavy-oil fractionator.*

Table 3. Input data for the 2x2 problem.

Controlled variables:				
Number	Steady-State Value	Weight	High Limit	Low Limit
1	0.0	1.0	1.0	1.0
2	0.0	1.0	0.0	0.0
Manipulated variables:				
1	0.0	1.00	100.0	-100.0
2	0.0	1.00	100.0	-100.0
Other pertinent information:				
Control interval		1 min		
Number of future moves calculated		5		
Move penalty		1 (for both manipulated variables)		

A two-output three-input problem

Now, let's consider a problem *(4)* where there are more manipulated variables than controlled variables: an infinite combination of three inputs can produce the same values of two outputs. Here, economic optimization involves allocating the manipulated variables (MVs) such that the objective function is optimized. As a start, let's specify that the first MV is to be maximized. Assume that the process is initially at rest and that we wish to drive the first output to 1 while specifying that the second output remain at

0. The input data files for this problem are given in Table 4. The costs of the three MVs are -1, 0, and 0, respectively, implying that we wish to maximize the first MV and do not care what happens to the second and the third MVs as long as they remain within their upper and lower bounds. The closed-loop responses in Figure 6a show that the controller is driving the first MV toward its upper constraint of 1. Figure 6b illustrates the results if the costs on the MVs are changed to 1, 0, and 0, respectively. In this instance, as desired, the controller has driven the first MV towards its minimum constraint.

Practical considerations

A dynamic mathematical model is the backbone of all CMPC strategies on the market. As is evident from Figure 2, without the model block it would be impossible to reconstruct the feedback signals d^i for further processing by the CMPC controller. Thus, the relevant issue is what kind of a model to employ.

Most processes of any commercial significance are nonlinear; some are mildly nonlinear, and others, such as pH systems, are extremely nonlinear. There are basically two approaches to modeling — either theoretical, or based on experimental testing.

In the theoretical approach, a detailed mechanistic mathematical model is developed utilizing unsteady-state mass, energy, and momentum balances. For complex systems, a large set of nonlinear differential/algebraic equations is likely to result. This set of equations could serve as the model block in Figure 2. In many cases, such a set may not reflect reality sufficiently well — because fundamental models do not readily account for the delay effects of mixing. Indeed, there is evidence that the open-loop response of fundamental models often is much faster than the equivalent experimental system (see, for instance, Ref. 11). Frequently, though, fundamental models can provide valuable insights. Such a model could serve as the model block in Figure 2 but, in this instance, a nonlinear optimization routine would be appropriate. The choice of nonlinear models in conjunction with nonlinear optimization procedures in a CMPC strategy is a subject of intense research, but industrial applications to date are based, for the most part, on linear models.

The experimental approach ranges from simple step or pulse testing to a sophisticated testing procedure in which all the inputs are perturbed over a long period of time (even a week or more), and the resulting input/output data analyzed to obtain a dynamic model. This sometimes is referred to as the "black box" approach. Independent of what is done with the resulting model, it has been observed that the test itself provides valuable insights about the plant to operators and engineers — because such tests force the personnel to "think" multivariable.

On the commercial side, there are two types of experimental testing procedures described in the published literature. One adopted by DMC Corporation, called [DMI] (for Dynamic Matrix Identification), leads to a step-response model of the multivariable plant *(12)*. The other is based on time-series analysis, which proceeds further and calculates a transfer function model of the multivariable plant *(13)*. Some investigators feel that the step-response representation more accurately reflects reality, given the observation of some unusual step response shapes whose characterization by transfer functions may require very high order models. This conclusion must be taken as tentative, though; more will be able to be said as experience mounts.

Some exciting developments are taking place in the field of neural networks. Neural networks can model linear and nonlinear systems with equal ease. They may serve as the model block in Figure 2 (see, for example, Ref. 8).

Nomenclature

- b = bias value
- C = cost penalties used in optimization
- d = effect of unmeasured disturbances and modeling errors on output
- D = effect of unmeasured disturbances on output
- J = objective function used in optimization
- K = matrix of steady-state gains
- L = measured process disturbance (deviation form)
- l = measured process disturbance (absolute value)
- m = process input (absolute value)
- M = manipulated variable vector (deviation form)
- S = slack/surplus variables in optimization
- V = violation variable
- y = process output (absolute value)
- Y = process output vector (deviation form)

Greek letters:

- τ = time constant
- θ = dead time

Table 4. Input data for the 2-output 3-input problem.

Number	Steady-State Value	Weight	High Limit	Low Limit
Controlled variables:				
1	0.0	1.0	1.0	0.0
2	0.0	1.0	0.0	0.0
Manipulated variables:				
1	0.0	-1.00	1.00	-1.00
2	0.0	0.00	1.00	-1.00
3	0.0	0.00	1.00	-1.00

Other pertinent information:	
Control interval	1 min
Number of future moves calculated	5
Move penalty	10 (for all manipulated variables)

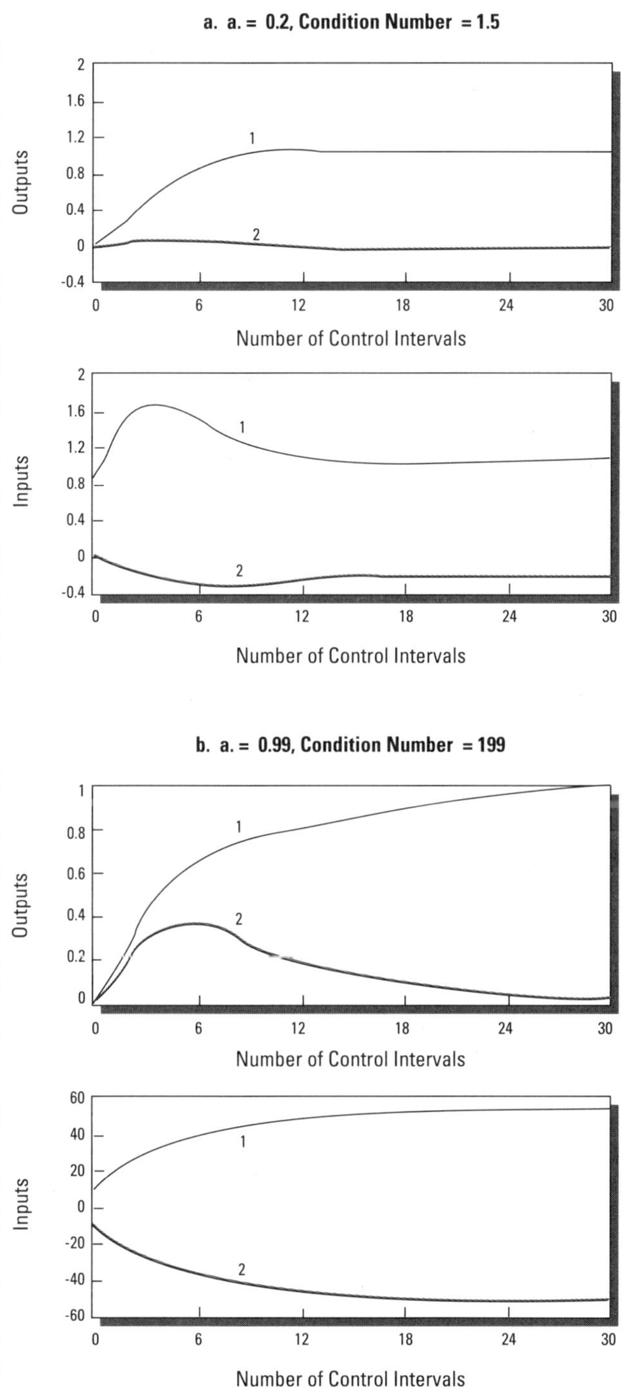

Figure 5. Control of a 2x2 system.

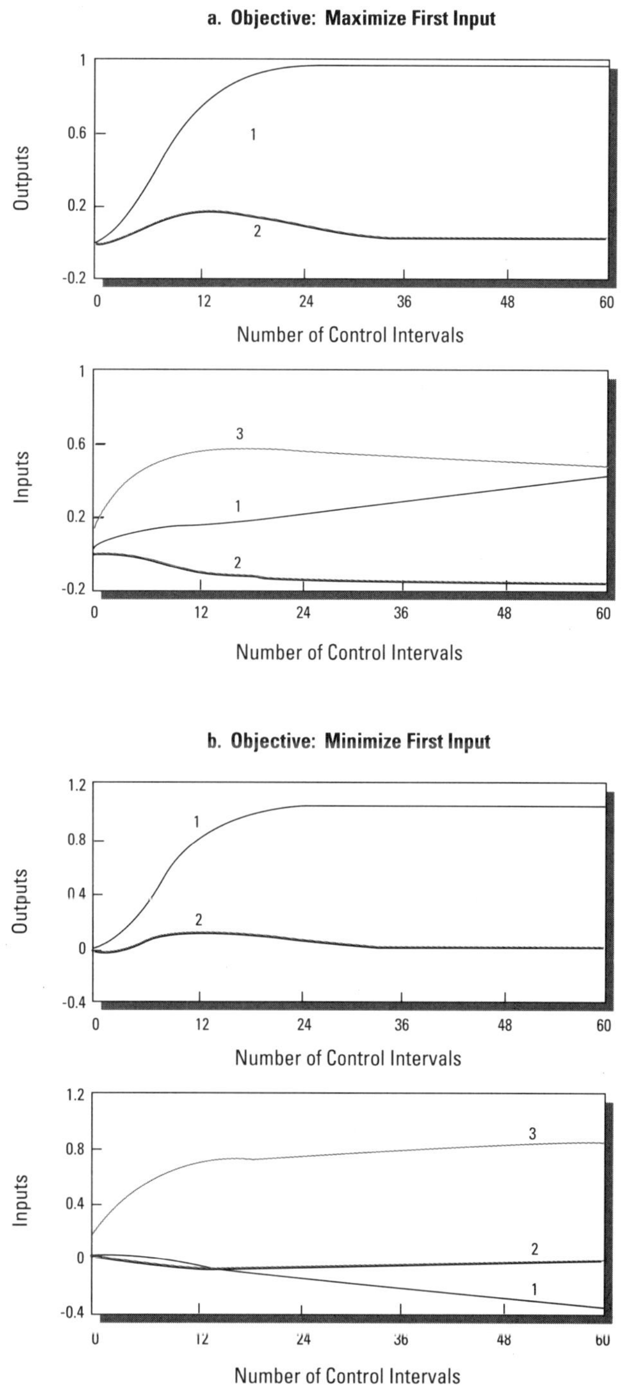

Figure 6. Control of the 2-output, 3-input problem.

Literature Cited

1. **Ayral, T. E.,** "Criteria for Comparing MPC Software Products," *In Control*, **6** (3), p. 1 (May-June 1993).
2. **Deshpande, P. B., R. E. Hannula, M. A. Bhalodia, and C. W. Hansen,** "Achieve Total Quality Control of Continuous Processes," *Chem. Eng. Progress*, **89** (7), p. 59 (July 1993).
3. **Garcia, C. E., and M. Morari,** "Internal Model Control 1. A Unifying Review and Some New Results," *Ind. Eng. Chem. Process Des. Dev.*, **21**, p. 308 (1982).
4. **Morshedi, A. M., C. R. Cutler, and T. A. Skrovanek,** "Optimal Solution of Dynamic Matrix Control with Linear Programming Techniques (LDMC)," *Proceedings*, American Control Conference, Boston, MA, pp. 199–208 (June 1985).
5. **Cutler, C. R., and B. L. Ramaker,** "Dynamic Matrix Control — A Computer Control Algorithm," presented at AIChE National Meeting, Houston, TX (Apr. 1979).
6. **Garcia, C. E., and A. M. Morshedi,** "A Quadratic Programming Solution of Dynamic Matrix Control (QDMC)," *Chem. Eng. Commun.*, **46**, p. 73 (1986).
7. **Richalet, J., A. Rault, J. L. Testud, and J. Papon,** "Model Predictive Heuristic Control: Applications to Industrial Processes," *Automatica*, **14**, p. 413 (1978).
8. **Graettinger, T. J., N. V. Bhat, K. Heckendorn, and J. S. Buck,** "Model Predictive Control Using Neural Networks," presented at AIChE National Meeting, Atlanta, GA (Apr. 1994).
9. **Prett, D. M., and M. Morari,** "The Shell Process Control Workshop," Butterworths, Stoneham, MA (1987).
10. **Skogestad, S., P. Lundstrom, and M. Hovd,** "Control of Identical Parallel Processes," presented at AIChE Annual Meeting, San Francisco, CA (Nov. 1989).
11. **Wong, Y. H., P. R. Krishnaswamy, W. K. Teo, B. D. Kulkarni, and P. B. Deshpande,** "Experimental Application of Robust Nonlinear Control Law to pH Control," *Chem. Eng. Sci.*, **49**, p. 199 (1994).
12. **Cutler, C. R., and F. H. Yocum,** "Experience with DMC Inverse for Identification," *Proceedings*, International Process Conference IV, South Padre Island, TX, p. 297 (1991).
13. **Box, G. E. P., and G. M. Jenkins,** "Time Series Analysis," Holden Day Publishers, Oakland, CA (1976).

Initial CMPC applications have tended to be at large CPI facilities already equipped with modern distributed control systems (DCSs) like Honeywell TDC-2000s and TDC-3000s. Typically, the processes under scrutiny were operating under traditional advanced controls that include such concepts as deadtime compensation, feedforward and cascade control, some decoupling, override instrumentation, and so on. The CMPC implementations have tended to be on DEC VAX computers. This requires interface equipment and software to communicate between the workstation and the DCS. Multitasking is also a must. Some facility must be provided to easily switch between traditional advanced controls and CMPC.

The ever increasing power and versatility of the personal computer (PC) have allowed CMPC tools to migrate to PCs in recent years. Indeed, a number of manufacturers provide PC-based data acquisition and control systems. So, now, it is possible to consider CMPC for smaller process units. Here, too, multitasking environments provided by operating systems such as Windows and Windows NT from Microsoft are essential.

To achieve the opportunities for enhancing throughput, minimizing energy consumption, and improving quality control that CMPC offers will require adequate attention to personnel issues. Each control room will have to have on hand an engineer, probably with an advanced degree, who is intimately familiar with CMPC concepts. Also, operators will have to undergo appropriate training. Finally, continuing education will have to become a routine feature as engineers and operators transfer in and out of assignments.

P. B. DESHPANDE is Alumni Research Professor and former chair of the dept. of chemical engineering at the Univ. of Louisville (502/852–0885; Fax: 502/852.–6355), where he heads the Center for Desalination: Modelling, Optimization & Control. He is also president of Simulation and Advanced Controls, Inc., a Louisville-based company specializing in advanced control services. He has over 23 years of experience in academia and industry. Author or coauthor of four books on process control and over 75 papers, he has developed a number of control algorithms. He has received several awards for his work, including the 1990 Donald P. Eckman Award in process control instrumentation, given by the Instrument Society of America (ISA). He is a member of AIChE, ACS, and a fellow of ISA.

M. A. BHALODIA is a senior staff engineer for Exxon Research & Engineering Co., Florham Park, NJ. He has over 20 years of experience in developing and implementing process control and plant-computing technology, real-time knowledge-based systems for process monitoring and diagnosis, process alarm management, and prevention of major upsets. He received a BS from Maharaja Sayajirao Univ. (Baroda, India), an MS from Michigan State Univ., and a PhD from State Univ. of New York at Buffalo, all in chemical engineering.

S. S. YERRAPRAGADA is currently a post-doctoral research associate in the chemical engineering dept. at the Univ. of Louisville. He has three years of industrial and four years of research experience, and now specializes in the areas of modeling, real-time simulation, control, and optimization. He obtained a bachelors degree from the Regional Engineering College (Warangal, India), and an MS and PhD from the Indian Institute of Technology (Delhi).

J. A. CALDWELL is a PhD candidate in the dept. of chemical engineering of the Univ. of Louisville, where he received his BS and ME degrees in chemical engineering. He currently is working on a research project involving model predictive control of an industrial process. He has had industrial experience with Exxon Chemical Co., and Louisville Water Co. He is a member of AIChE, and ISA.

Practical Guidelines for Heat Transfer in Mixing Vessels

James Y. Oldshue
Oldshue Technologies International, Fairport, NY 14450

Heat transfer is a common application in many mixer systems. Basic data is available in published literature, but there are always many questions that come up that relate to actual installed heat transfer surfaces in mixing tanks, and the effect of various mixing impellers and flow patterns.

NEW FLUIDFOIL IMPELLERS

There are three types of flow patterns available in vertical cylindrical mixing vessels. The radial flow flat blade turbine (Rushton turbine) has been used for 50 years or more in mixing vessels. Its use in a mixing tank with helical coils is shown in Figure 1 Reliable data on this configuration is available in the literature, and Figure 2 gives a typical correlation.

Data has been taken on this impeller in the mixing vessel using thermo-couples placed around various orientations of the coils, and thermo-couples being placed carefully at the mid-point in the bevel thickness of the tube, and data taken on the heat transfer coefficient on a local basis based on many thermocouple readings throughout the helical coil.

Thirty to forty years ago, the axial flow turbine became more common in mixing vessels. It gave increased flow at the same diameter and power of the flat blade turbine. A related impeller, the marine-style constant pitch propeller, was not used extensively industrially, but good pilot plant data was obtained as shown in the curves in Figure 2. Both the flat blade turbine and the marine propeller gave an angled discharge in a radial direction from the impeller, rather than a straight axial flow as might be surmised from the term, axial flow, and these impellers were often correlated (along with the flat blade turbine) with a dimensional N_{HU} number plus other groups versus Reynolds number.

At first glance at Figure 2, one might think that the flat blade turbine was a better heat transfer device until the fact that at same Reynolds number, the flat blade turbine draws 30 times for power than the marine propeller. When the marine propeller is sped up to draw the same power, and the corresponding Reynolds number, and the heat transfer relationships are approximately the same as a flat blade turbine. So within most industrial accuracy, it can be considered that at the same power and same diameter, with the axial flow impellers requiring a higher speed, the overall heat transfer coefficient on the mixer side is approximately the same.

In order to increase flow at the expense of shear rate at a given power level, fluidfoil impellers have been introduced at which the general appearance is shown in Figure 5 along with the other impellers referred to previously in Figure 3 and 4. The fluidfoil impeller typically has a camber to blade shape, a carefully constructed pitch increase or blade angle increase from the tip blade angle to the hub with the goal of getting a very axial flow pattern coming down the impeller as shown by the laser velocity measurement in Figure 6. A comparable laser velocity profile for the axial flow turbine, Figure 3, is shown in Figure 7.

Extensive data are not available on the fluidfoil impeller to the degree shown in Figure 2, but what has been observed is the following:

1. With helical coils, the axial flow pattern turns 180° at the bottom of the tank, flows upward along the coil, and the lower turns on the coils tends to blank off effective flow from the other turns in the helical coil bank so that the heat transfer performance is, at best, about the same as with the axial flow impellers, and on many occasions can be 5 to 10% less on the mixer size coefficient when compared to the same power and diameter.

2. There are many very important reasons for using the fluidfoil impellers, since many processes are optimized by maximizing the flow from the impeller and minimizing the shear rate. Heat transfer to helical coils is not one of these reasons.

3. In a jacketed tank, the increased flow from the fluidfoil flow impeller at a given impeller diameter does increase the velocity across the heat transfer surface to some degree and heat transfer results are somewhat improved by the use of the fluidfoil impeller compared to the axial flow turbine.

GEOMETRY FOR HELICAL COIL INSTALLATIONS

Heat transfer coefficients are promoted by the turbulence generated by the heat transfer coefficient generated by the heat transfer surface itself. Correlations shown in studies similar to that in Figure 2, indicate that the heat transfer coefficient is inversely proportional to the tube diameter to the square root. This means that the heat transfer coefficient on a 2" O.D. tube in approximately 40% higher than it is on a 4" O.D. tube. This phenomena should be considered in evaluating the effect of tube diameter in the helical coil. It can be used to evaluate the heat transfer including both the area of the tubes as actually installed and the coefficient as a function of the actual tube diameter itself. In general, the tube diameter should be as small as possible commensurate with reasonable pressure through the helical coil for the inside flow of cooling or heating fluid.

It is also normally desirable to put at least one-half of the baffle width inside the helical coil bank toward the impeller. For example, if the tank were 10' in diameter, there would typically be four 10" wide baffles, and at least 5" of baffle projection should be placed inside the coil toward the impeller, Figure 1. It is typical to use four baffles with radial and axial flow turbines, and only three baffles with the new fluidfoil impellers.

The largest gain for heat transfer with impeller power is in going from natural convection up to forced convection through all the tubes in the system as shown in Figure 8. All the correlations in the literature are similar to those in Figure 2, are done in the forced convection region and the exponent on power is about 0.22. Since this only applies to the mixer side coefficient, which may be only one-third of the total heat transfer resistance, the improvement of total heat transfer is not economically obtained by increasing the power of the mixer alone. This means that if the power required for other mixing requirements is below that which is required for forced convection through the heat transfer tubes, increasing the power to the mixer at that point is economically justified. If the power required in the rest of the process requirements is well above the power level for forced convection of the heat transfer, then that power is, of course, used to calculate the heat transfer coefficient obtained and take whatever heat transfer, then that power is, of course, used to calculate the heat transfer coefficient obtained and take whatever heat transfer is obtained in that particular power level.

Putting in second, third, fourth or even five banks of coils in a reactor can be done. In order to insure forced convection and adequate temperature and uniformity through all the banks of the tubes, the flow and normally the power as well must be increased to make this occur. Under that circumstance the heat transfer between the various banks of helical coils will be similar.

It is quite typical to use one tube diameter spacing between adjacent tubes in a coil bank. If the coils are placed closer, such as one-half of a tube diameter. Heat transfer will be increased if both conditions are in the forced convection range, but it will take more flow and therefore more power to obtain forced convection through the more tightly placed tubes. This normally should not be continued beyond 0.5 tube diameter space, since we then start interfering with blending and flow patterns in other parts of the tank which can effect the overall process result.

OTHER HEAT TRANSFER SURFACE GEOMETRIES

Jacketed tanks do not normally develop as high a heat transfer film coefficient as helical coils since the turbulence generation is due to boundary layer flow and not a tube placed in a flowing stream. The axial film coefficient itself is higher with an unbaffled tank than with a baffled tank, but practical limitations of swirling and vortexing of unbaffled tanks limits these phenomena. In a baffled tank situation, the heat transfer coefficient on a jacketed tank is between .6 and .8 for the coefficient on the helical coils. Lots of variations are possible on the fluid flow pattern inside the jacket, and also with the use of jackets composed of semi-circular tubes wrapped around the tank.

The use of hollow wall baffles to circulate fluids to allow heat transfer as possible, but normally there is not enough surface area available to make this a major heat transfer boost. In the pilot plant or the laboratory, this can sometimes be done because the heat transfer requirements per unit volume are lower and can be affected by baffles that are internally cooled.

The use of vertical tubes (Figure 9) to baffle the tank and allow heat transfer can also be done. Figure 10 shows that situation. The vertical tubes work fine, but the amount of area that can be installed in a tank this way compared to helical coils is a relatively small fraction.

In modification of the hollow baffles is the use of hollow, embossed flow conduits and plates, sometimes referred to as plate coils or thermal panels, or other proprietary names and is shown in Figure 10, many of these can be placed in a plant size mixing vessel, from 20 - 40 in number.

The heat transfer coefficient on the surface of the embossed conduit and some of that on helical coil, and the metal webbing between the fluid conduits also is of value in the heat transfer process. Proprietary data is usually needed to evaluate the effect of these heat transfer surfaces.

EFFECT OF DISPERSED GASES & SUSPENDED SOLIDS

Gases dispersed in mixing vessels have a wide ranged effects on heat transfer. Part of the consideration is whether the tank is essentially in a bubble column mode in which the gas is predominantly flowing through the vessel relatively undispersed by the impeller and the gas flow pattern creates the fluid flow pattern. This can be a bubble column that does not employ a mixer, or can be a flooded situation in a tank equipped with a mixer. A major calculation concern is that under these conditions, what is the effective density and viscosity of the fluid to use in the heat transfer calculation. That consideration is left to others, and we will discuss here the situation where the gas is "well dispersed" by the impeller. This obviously has a range of definitions all the way from where the impeller power is roughly equal to the expansion power of the gas all the way to where the impeller may have 10 to 100 times more power than the expanding gas stream. My general observation is that when gas is added to a mixing vessel, the power often decreases at a given speed, but with the newer impellers and other sparge rings, power can either be the same or increase. Regardless, if you use the power that is actually being applied during the gassed condition, and calculate the speed of the impeller that is required to put that amount of power in with the same impeller configuration, then that speed can be used as a Reynolds number in the correlation like that seen in Figure 2 to obtain the heat transfer coefficient. I have not seen evidence that the gas bubbles markedly blanket the film surface of the heat transfer device to lower by insulating the heat transfer performance. The former consideration that considers coalescing systems, in which the gas does not make up a major fraction of the fluid mixture and is less than roughly 10 to 15% of volume. As you get into higher viscosities and other surface tensions, then the gas bubbles become part of the rheology of the system as well as the thermal conductivity and a proper approach to the average viscosities and other thermal properties has to be made.

Suspended solids also have quite a range of process operations. In aqueous systems where the solids are particulate and/or less than 30% by volume, the solid particles actually penetrate the heat transfer fluid film contaminates their inertia energy to disrupting that film, and in many cases, can actually improve heat transfer at a given power level.

If the solid particles are fibrous or biological, then they start to impart a viscosity effect and also an effect on thermal properties. Now we have a situation where these average properties must be used to estimate the heat transfer effect. For example fermentation in biological organisms, a typical case being penicillin, as the concentration of solids increases, the heat transfer coefficients decrease largely because a change in the apparent viscosity of total fluid medium. The same thing holds true in the minerals industry of very high concentrations of solids, particularly where there is a appreciable amount of -250 mesh material which can effect the apparent viscosity of the fluid mixture.

EFFECT OF HELICAL COILS ON CIRCULATION AND BLENDING

Helical coils do tend to lengthen the blending to the outer most part of the tank out to the coil. There is little definitive data on how much poorer blend times are behind the coils than exists in the center portion of the tank. However, the main characteristics of mixing vessels is that the blend time is increased dramatically on scale up, and this becomes a major problem in comparing pilot scale results with full scale. If one then adds to that phenomena the fact that pilot tanks do often require heat transfer and coils are now put into the full scale reactor, that compounds the problem of blend time increase.

Particularly in biological systems, the effect of blending of ingredients added to the reactor can effect dramatically the biological processes occurring and blend time concerns are a major factor in considering what role blending may have on the overall scale up situation.

It normally is not worth the expense to put in dummy helical coils in the pilot plant since the blend time is going to increase anyway, but that certainly could be done if flow pattern results are desired.

SCALE UP & SCALE DOWN - EFFECT OF NON-GEOMETRIC SIMILARITY

If the heat transfer correlation in Figure 2 applied to small and large tanks with the impeller and coils being geometrically scaled, it turns out that the power would have to increase per unit volume by ratio of the tank scale by one-half exponent to have the same heat transfer coefficient. That is a purely academic exercise since that does not at all guarantee the same total heat transfer per unit of volume of vessel nor is it necessary to scale up the heat transfer tubes diameters themselves geometrically. In general, the tube diameter on the full scale tank should be set at whatever size is compatible with pressure drops from the coils. The design of the mixer determines for all the process requirements as well as heat transfer, then the heat transfer coefficient calculated from data like Figure 2. Within the required surface area, is estimated to see whether that lies within the practical range of one to five banks on helical coils, or within the range of jacket parameters involved, or perhaps of helical coils and jackets.

Several other publications have discuss the subject of geometric and non-geometric similarity on the scale up of other mixing requirements, and heat transfer is just one of the many aspects of that overall scale up situation.

OVERALL HEAT TRANSFER & GUIDELINES

There are different methods of calculating the effect of heating or cooling on the heat transfer coefficient. Some investigators have used an exponent based on the ratio of viscosity between the viscosity at the bulk temperature and the viscosity at the average temperature of the film. others use the average viscosity at the surface temperature of the heat transfer device. One must be careful in making sure of the definition of the temperature at which the viscosity was evaluated in the film. Figure 11 is a reasonable approximation of the heat transfer for organic materials at various viscosities for heating and cooling. Obviously, without specifying many other parameters, this chart can only be an estimation for a good starting point.

To evaluate the effect of the entire heat transfer mechanism on the overall coefficient, usually an estimate must be made on the different film coefficients, a temperature profile through the heat transfer process made and then definement of the heat transfer coefficients obtained on the individual surfaces.

Sometimes for temperature sensitive materials, the prediction of surface temperatures are important and reliance must be then made on how certain temperatures will be effected in the impeller region when overall fluid velocity is higher than it is in other portions of the heat transfer surface where the flow patterns are much less vigorous.

Other factors not treated in this particular paper are; heat transfer in viscous materials would make for helical impellers, which is normally a convection mechanism through the impeller and the tank wall, the case of heat transfer to open surfaces which is important in large tanks, and also the heat transfer situation with surface aeration devices in waste treating which are operating to provide gas-liquid mass transfer, as well as have an effect on the heat transfer coefficient to large bodies of fluid in open reservoirs. Another case is where the reactor is essentially full of vertical tubes, almost becoming a skill and tube exchanger, with the mixer in the center providing whatever mixing action is needed throughout a multiplicity of very closely packed vertical tubes.

Figure 1 - Typical helical coil installation with flat blade turbine.
Figure 2- Correlation of the Nusselt number versus Reynolds number for helical coils.
Figure 3 - Typical axial flow turbine.
Figure 4 - Flat blade turbine.
Figure 5 - A310 fluidfoil impeller.
Figure 6 - Typical laser scan for the A310 Impeller.
Figure 7 - Typical laser scan for the A200 Impeller.
Figure 8 - Typical jacketed tank.
Figure 9 - Typical vertical tube installation.
Figure 10 - Typical plate coil installation.
Figure 11 - Coefficients for organic fluids for estimating heating and cooling coefficients.

D	Impeller diameter
T	Tank diameter
N	Impeller speed
h	Heat transfer coefficient
μ	Viscosity

ρ	Fluid density
k	Thermal conductivity
d	Tube diameter
P	Power

References:

Oldshue, J.Y., "Fluid Mixing Technology", McGraw Hill, 1983.

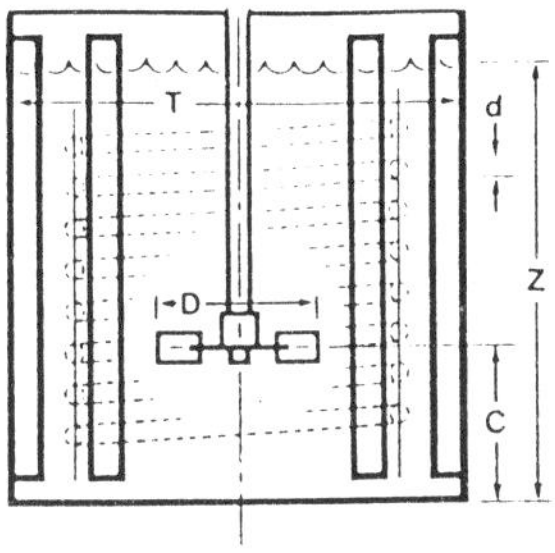

Figure 1

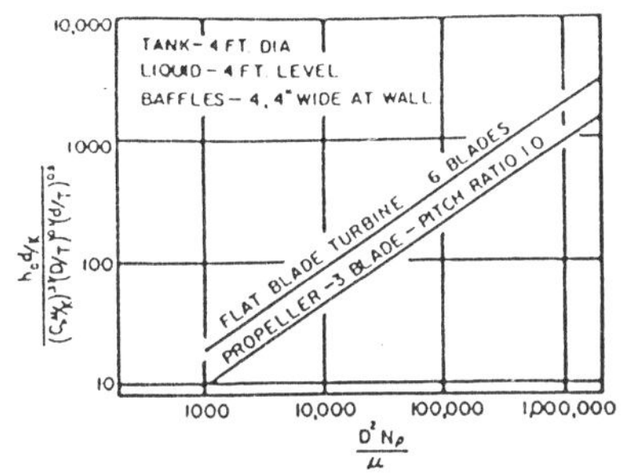

Figure 2

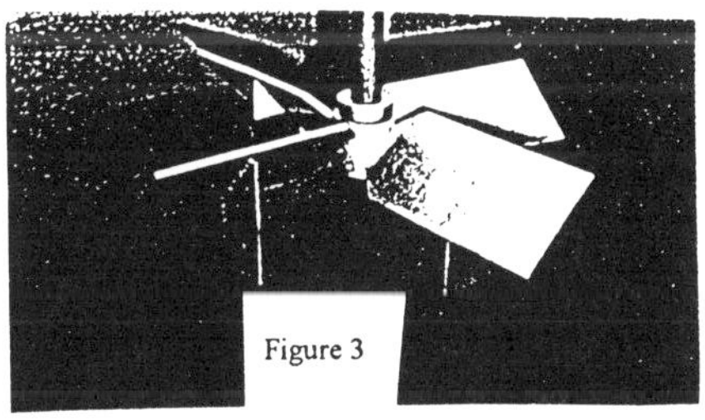

Figure 3

Figure 4

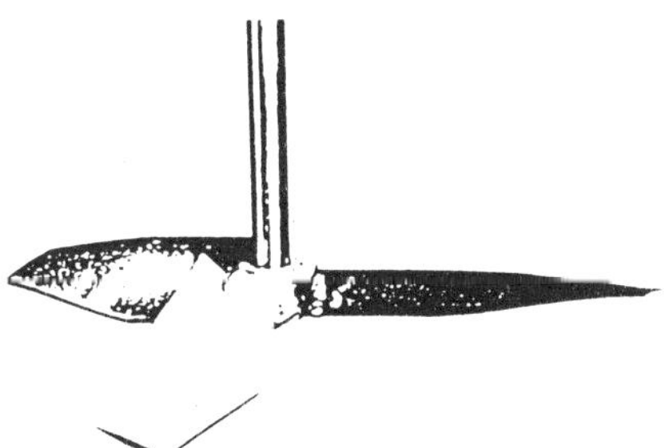

Figure 5

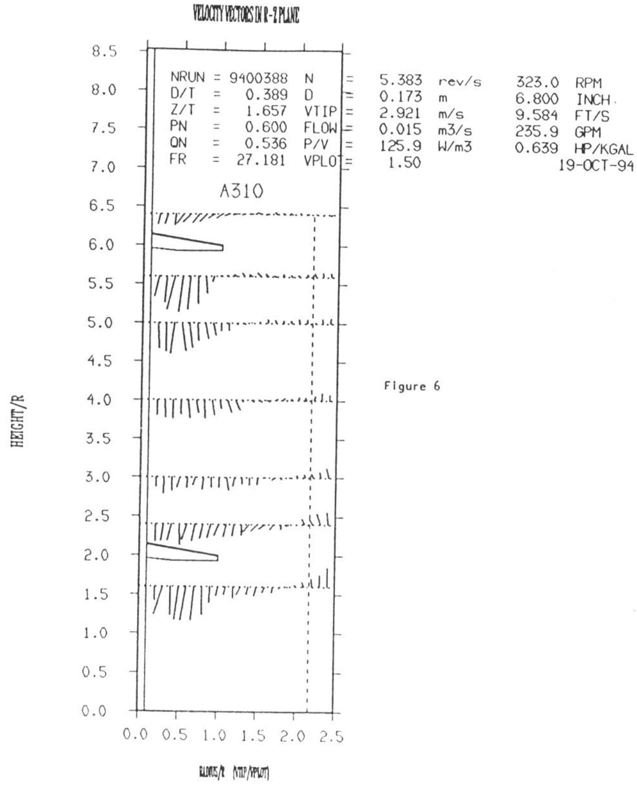

Figure 6

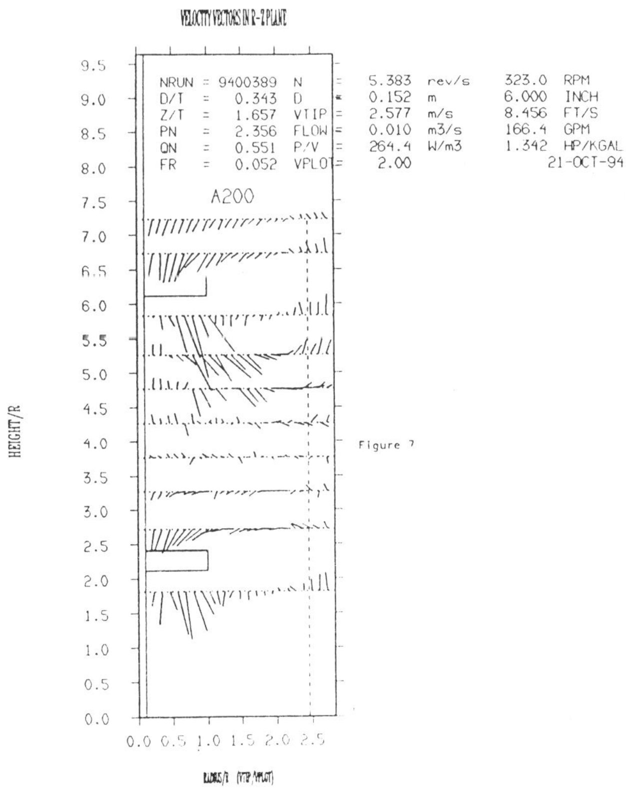

Figure 7

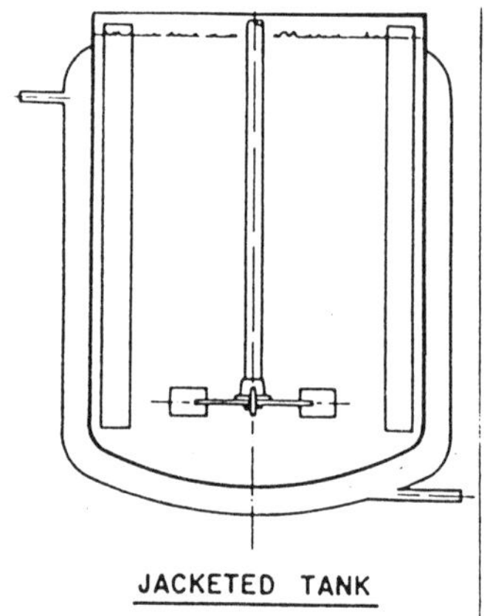

Figure 8

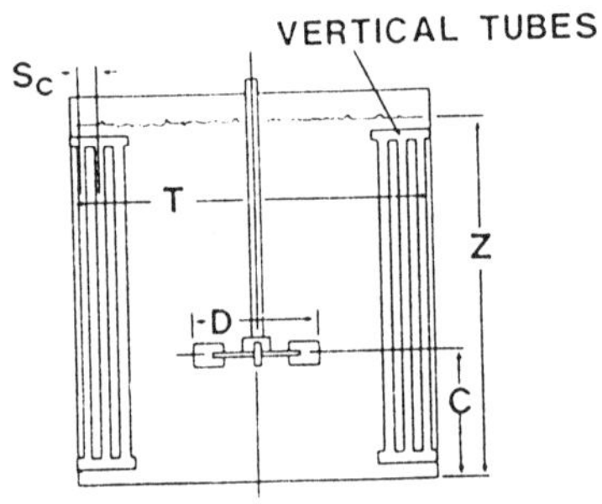

Figure 9

Figure 10

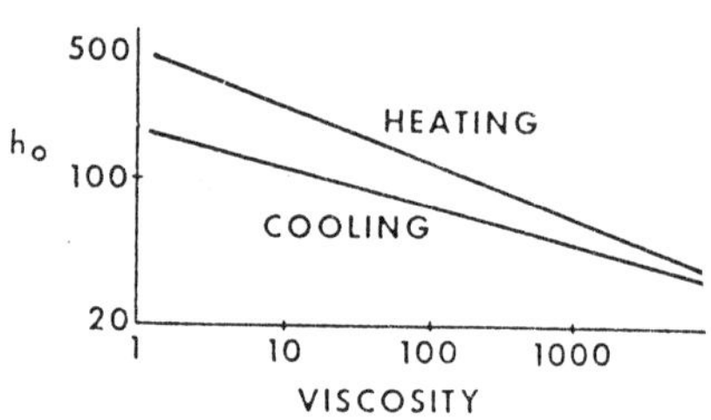

Figure 11

Consider Magnetically-Driven Pilot Mixers

New rare-earth magnets have made these mixers competitive with mechanically-driven ones for laboratories and pilot plants.

Michael T. Walsh and Vincent M. McCalley,
Pressure Products Industries, Inc.

For more than 30 years, the choice between mechanically-driven and magnetically-driven mixers for laboratory and pilot-plant use was relatively simple. For the majority of applications, mechanical mixers were selected due to their low capital cost, dependable performance, and high level of available power. Magnetic mixers could be cost-justified only for applications involving either high pressures — above 500 psi (3.5 MPa) — or hazardous or toxic materials.

Over the past several years, however, the introduction of a third generation of magnetic mixers has made the choice less obvious. The major force in bringing about this change has been advances in magnetic technology.

The first magnetically-driven mixers, in use during the 1960s and early 1970s, used ceramic magnets. The idea was new, but the power was low and the price was high. The second generation, in use from the mid 1970s to the present, was based on early rare-earth magnets — such as low-energy samarium-cobalt and cesium-cobalt. These mixers raised the power level by a factor of ten, but effectively raised, rather than lowered, the cost of magnetic mixing.

The latest generation, which began to have an impact on the market in 1987, takes advantage of the most recent advances in magnetic materials and design. These newest mixers, which are based on rare-earth magnets, provide higher levels of power in smaller mixers, more cost-effectively than the earlier generations. Such mixers are suitable for use on vessels ranging in size from 50 mL to 4,000 L.

As a comparison, consider a first-generation magnet that produces a torque of 3 in.-lb. The same size of third-generation magnet produces a torque of 60 in.-lb.

Now, both mechanical mixers and magnetic mixers are suitable for many research and process development applications. This means the choice must be weighed differently than it had been in the past, considering not only mixer characteristics, but also the changing dynamics in today's labs and pilot plants. Several factors affect the choice of mixers:

- Tightening safety and environmental restrictions;
- The use of non-Newtonian and high-viscosity materials, which demand high power output;
- Interfacing the mixer with microprocessor control systems; and
- Increasing concern over maintenance costs and costs of retrofitting.

Safety and environmental concerns

Environmental and safety regulations have been tightening. Also, corporate standards for worker safety are often even stricter. The ultimate goal, of course, is elimination of all potential leaks and workplace contamination.

One step toward this goal is selecting appropriate mixers, so that mixing of even the most viscous or difficult-to-agitate compounds is possible without exposing laboratory personnel to leaks.

Mechanically-driven mixers have an inherent potential for allowing fugitive leaks to the atmosphere. These mixers transmit power by means of gears between the outer housing and the mixer shaft, or by direct drive or belt drive from an electric or air-driven motor. This means that the shaft must be sealed to prevent process leaks. Such seals are located on rotating parts of the mixer, thereby creating a pathway for leakage (see Figure 1).

Most mixers use either mechanical packing or mechanical seals. Mechanical packing operates on the principle of controlled leakage. Through the force of a gland nut or gland plate, packing within a stuffing box is forced into contact with the mixer shaft. Lubricant within the packing eases the contact load; when lubricant is lost, the packing must be replaced. Such seals can minimize leakage at lower pressures.

Mechanical seals provide more positive sealing and minimize maintenance. They prevent leakage by means of two seal surfaces, one stationary and one rotating in close contact with it. The two surfaces are lapped to a flatness of 0.00001 of an inch. There is a close running clearance between the two surfaces and a lubricant film in between. This liquid is kept continuously cool and clean with a flush lubrication system.

Mechanical seals can be used reliably in applications where pressure may reach as high as 4,000 psi (28 MPa), provided a secondary seal, such as packing, or a double mechanical seal is used. However, mechanical seals are subject to catastrophic failure and fugitive leakage, and the use of lubricants may cause contamination of the process.

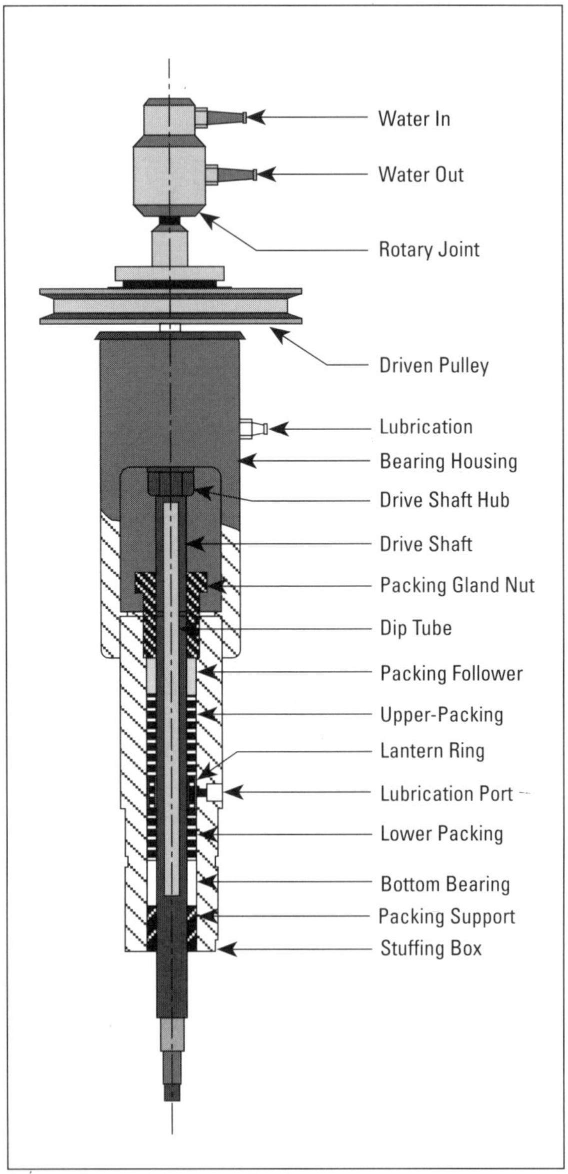

■ *Figure 1. Mechanical mixers have seals that can leak.*

While the purchase price of mechanical mixers is less than for magnetic mixers, the price increases as the type of seal or seals required becomes more complex. When comparing price, it is advisable to review all the requirements of the application, particularly the tolerance for leakage or contamination, as the magnetic mixer may be worth the premium in price.

Magnetic mixers are designed without mechanical seals or packing, eliminating the potential for leakage. The typical design is based on the use of two magnet sets (see Figure 2). The driver magnets are located in the outer housing, which is not subject to contact with process materials. Through direct coupling or via a V-belt, an electric or air-driven motor rotates the outer housing and driver magnets, which in turn rotate the driven magnets and the mixer shaft. The driven magnets are attached to the mixer shaft and are completely encapsulated in a seal-welded sheath to avoid contact with process fluids. The shaft assembly is enclosed in a pressure housing that retains process fluids. When the mixer is operating within its specified operating parameters, failure of the pressure housing is not a concern.

The pressure housing typically provides pressure ratings from vacuum to 6,000 psi (41 MPa) and temperature ratings to 650°F (343°C), depending on the materials of construction. Special designs have been used to temperatures as high as 1,100°F (595°C). Additionally, magnetic mixers do not require lubrication, thus providing no potential for contamination of the process.

The question of power

In research and process-development applications, high power in mixing is required when: the materials have a high viscosity; are non-Newtonian (requiring high-turbulence mixing to avoid stratification); are dilatant (that is, they instantaneously set to a solid during agitation); and, most importantly, have unknown power requirements. In the last case, high power means added assurance.

To provide the high torque required in such situations, gear-driven mechanical mixers have an advantage. In a 7.5 hp size, such a mixer can deliver up to 700 in.-lb of starting torque. The advantages of higher torque capability and lower capital cost have made mechanical mixers a common choice for applications such as mixing slurries and making up polymers.

Mechanical mixers do have drawbacks, however, when factors other than power are considered as well. Dilatancy can cause serious damage to mechanical mixers. And potential for leakage and contamination is a major concern when hazardous materials are involved.

Now, in these applications magnetic mixers present a cost-effective alternative. The typical second-generation magnetic mixers, with early rare earth magnet technology, were not cost-effective in developing high torque. In fact, either the cost per unit of torque became prohibitive, or the size of the mixer required was out of scale with the rest of the system.

The new generation mixers, with the latest magnet technology, can be purchased for a price of as little as \$50 per in.-lb of torque. The new mixers are also available in sizes to match the scale of any system. A mixer capable of delivering up to 60 in.-lbs of static torque at 2,500 rpm would typically be less than 15 in. high (excluding shaft), less than 5 in. in diameter, would weigh about 20 lb, and could be purchased for about \$3,000.

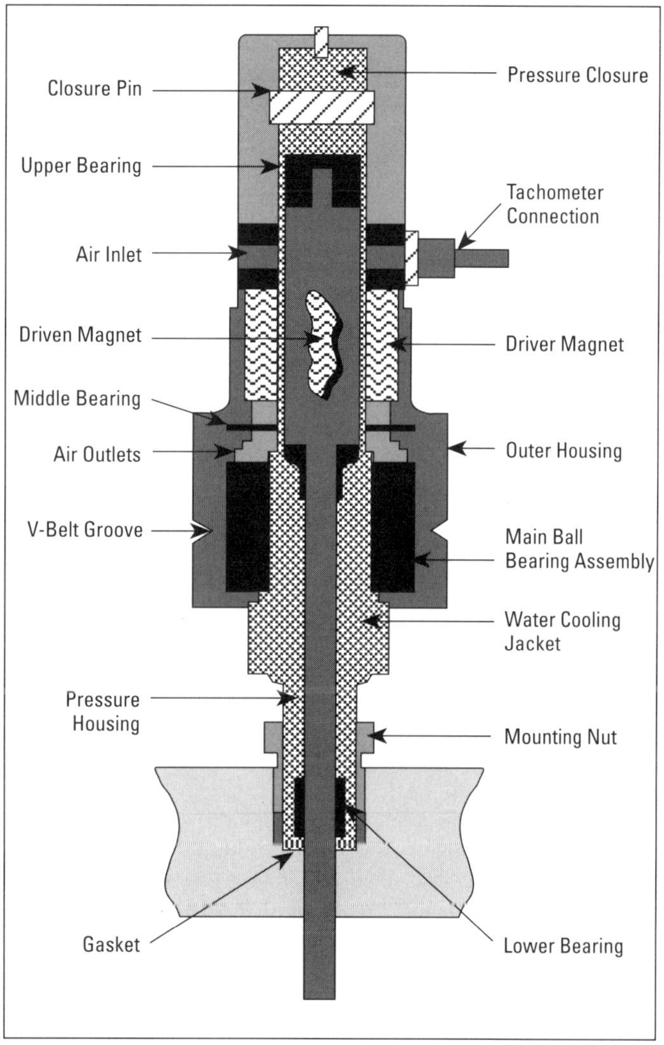

■ ***Figure 2. Magnetic mixers do not have mechanical seals or packing, eliminating the potential for leaks.***

Overload protection

All mixers should be protected by fuse for overload conditions, but this may not be sufficient for dilatant materials. This instantaneous increase in viscosity — setting to a solid — as a result of pressure or agitation, is a hazard in research applications.

A fuse or breaker on a mechanical mixer may not react quickly enough to prevent damage to the mixer shaft or drive system. Other forms of overload protection can be layered on to a mechanical mixer, but at added expense.

In a magnetic mixer, however, protection is inherent. The magnetic coupling of driver to driven magnets acts as a clutch, providing time for a shaft stoppage alarm and shutdown-command program to protect the mixer from damage. Such programs are standard functions of the tachometer typically supplied with with such mixers.

When a magnetic mixer is damaged, the choice to repair or replace is made by determining the parts that have been damaged. If the bearings and gaskets have been damaged, it is more practical and cost-effective to replace them. If, however, the damage is to the magnets, it may be more cost-effective simply to replace the mixer.

Accuracy for microprocessor control

Both mechanical and magnetic mixers are available with motor speed controls and tachometers. Such systems allow visual monitoring and adjustment in the lab. Additionally, alarms and shutdown commands can be configured for magnetic mixers, while mechanical mixers are typically protected against severe overload by a fuse.

In most applications, mixers must also be able to interface with digital control systems, based on mainframe, mini- or microcomputers. Both mechanical and magnetic mixers are used with such systems. In either case, tachometers and motor controls can be equipped with RS-232 ports or other connections for interface with control systems.

For the most accurate control, it is advisable to use a mixer where the actual operating speed of the mixer shaft is measured. In mechanical mixers and older-design magnetic mixers, motor speed was measured and a correlation to shaft speed was made.

This method has an inherent problem, however, since the shaft can actually halt while the motor continues in operation. In the latest generation of magnetic mixers, this problem has been solved. Shaft speed is read with a Hall-effect sensor from the rotating driven magnets. The Hall effect is the development of a varying magnetic field causing a current in a conductor.

As budget restrictions for research and development continue, it is not simply the initial cost of equipment that must be reviewed, but maintenance costs must be considered as well.

In mechanical mixers, this same method can be used, as a magnet can be attached to the shaft and the measurement made with the same type of sensor. The cost difference up-front for this accuracy of control is merely the cost of the sensing device. The long-term cost of not using such a device may be the replacement cost of the mixer if it is destroyed due to an undetected shaft stoppage.

Maintenance costs

As budget restrictions for research and development continue, it is not simply the initial cost of equipment that must be reviewed, but maintenance costs must be considered as well.

Where mixers are used for hundreds of hours at a time, maintenance costs for mechanical mixers are set. Approximately four times a year, each mixer must be disassembled, its seals repacked and the unit reassembled. This requires an average of 32 hours of a machinist's time per mixer. Using \$50 per hour, the total cost of the procedure is \$1,600, resulting in a typical yearly maintenance cost of \$6,400. Additionally, for mixers that require a flush lubricant system, the external lubrication system must also be maintained.

By eliminating both these requirements through their nonlubricated, packless, sealless design, magnetic mixers can provide a return on investment in less than a year in many cases. The maintenance required by magnetic mixers is minimal. The replacement of worn bearings, typically necessary after 3,000 to 6,000 hours of operation, generally entails less than an hour of a machinist's time and a cost for parts of less than \$200 — the approximate price of a prepackaged parts replacement kit from the manufacturer.

Because the containment housing is exposed to process fluids, magnetic mixers may require periodic disassembly and cleaning. With the latest generation of mixers, this disassembly has been simplified so that it can typically be accomplished in 5–10 minutes.

Also, the inner surface finishes are smooth with gradual transitions to avoid product buildup and allow easy cleaning. When sterile conditions are necessary, magnetic mixers can be autoclaved up to 300°F (158°C). Cleaning and autoclaving costs are minimal, as disassembly time is short; when an autoclave is available at the user's facility, this procedure is quick and inexpensive. This is so even if a source for the service must be found off-site.

To reap such maintenance savings, little additional expenditure is necessary. Magnetic mixers can be retrofitted to older reactors and vessels simply. Adaptor mounts are typically available with the mixer; thus, no tooling of the vessel cover is required. Power for the mixer may come from an in-line motor, either air-driven or electric, or from a new or existing motor. In the case of an existing motor, the only change necessary would be a new pulley and V-belt to transmit the power.

Expanding applications

As magnetic mixers and mechanical mixers have become more competitive for general purpose use, the range of magnetic mixer applications is expanding. Safety and environmental restrictions are driving the development of larger, economical magnetic mixers to exploit their inherent leak-tight operation. Conversely, smaller-scale mixers are needed for laboratory use as sample sizes and, therefore, reactor sizes are commonly decreasing.

Designs such as side- and bottom-entry magnetic mixers are available or under development to provide flexibility. In the use of side-entry mixers, however, special consideration must be given to the overhung load of the impeller and shaft. If necessary, additional bearing support must be added.

For bottom-entry mixers, the cleanliness of the process must be given consideration, since process fluids can enter the containment housing. Lip seals can be added to the mixer if this is considered a problem. CEP

M. T. WALSH is product manager for Pressure Products Industries, Inc., Warminster, Pa., (215/ 675-1600; Fax 215/ 443-8341). Walsh has worked on the application of high-pressure equipment for chemical research and processing for more than 13 years. His experience includes consultation with both corporate and academic research personnel on mixing topics. Walsh holds a BSME degree from Purdue University.

V. M. McCALLEY is chief engineer for Pressure Products Industries, Inc. He has 13 years of experience in the engineering and design of stirred reactors and high-pressure fluid-handling equipment for laboratory and pilot plant applications. McCalley holds a BSME degree from Temple University, and is a member of the American Society of Mechanical Engineers.

Properly Choose Mechanical Agitators for Viscous Liquids

Turbine and helical-ribbon mixers are the most commonly used mixers. Calculations, modeling, and economics are all factors in selection.

André Bakker and Lewis E. Gates,
Chemineer, Inc.

High-viscosity mixing applications occur in most chemical process industries (CPI) plants. For instance, the polymer production industry must blend high-viscosity reaction masses to thermal and chemical uniformity. This industry must also blend small amounts of low-viscosity antioxidants and colorants into polymer streams. The personal-care products industries encounter many high-viscosity mixing applications in the preparation of creams, lotions, pastes, and drugs. Other high-viscosity applications occur in the production of food, paint, drilling mud, and greases, to name a few. Viscosities can be in the range from about 1 Pa-s all the way up to 25,000 Pa-s in some extreme cases. The materials to be mixed in these applications can be very important economically.

Mixing can occur in pipeline systems with motionless mixers, or in vessels using mechanical agitators, depending on the application and the process requirements. A wide variety of both motionless mixers and mechanical agitators is available to handle specific mixing problems and fluid types. One or more agitators can be used to blend a tank. Each agitator can have one or more shafts, with one or more impellers that do not necessarily have to be of the same type. Impellers can be top-entering, side-entering, or bottom-entering.

This article gives an overview of designing the most commonly used agitator for blending applications: a top-entering agitator with a single shaft. The agitator can be equipped with multiple turbine-style impellers of different design, or with helical-ribbon or anchor-style impellers to optimize the agitator for the specific application and blending problem on hand.

■ ***Industrial-scale helical-ribbon mixer with supporting cross-bars.***

Various overviews of liquid blending can be found in the literature, see for example, Fasano and Penney *(1),* Bouwmans *(2),* and Fasano *et al. (3).* These articles focus mainly on blending of liquids in the turbulent regime: impeller Reynolds numbers of around 10,000 and higher.

Although turbulent blending will be briefly discussed here also, this article will focus on blending in the laminar and transitional regimes. Also, we will discuss the special requirements for blending non-Newtonian fluids, with and without yield stress. We will first discuss the flow patterns and applicability of different impeller types and will then present some design guidelines.

Turbine agitators

A wide variety of turbine-style impellers is available. Figure 1 lists the most-common ones: the pitched-blade turbine, high-efficiency impeller, disc turbine, and simple straight-blade turbine. Well-designed turbine impeller systems can be used up to viscosities of about 50 Pa-s, depending on the scale, application, and process requirements.

Each impeller has its own characteristics and its own area of application. Straight-blade impellers and disc turbines tend to be used to create zones with high shear rates, such as in dispersion. High-efficiency and pitched-blade impellers, and pitched-blade turbines tend to be used at Reynolds numbers larger than about 100, where a good overall circulation flow is important. The impeller Reynolds number is defined as:

$$N_{Re} = \frac{\rho N D^2}{\mu_{eff}} \quad (1)$$

For Newtonian fluids, the effective viscosity μ_{eff} is simply the dynamic viscosity. For non-Newtonian fluids, the viscosity will vary with shear rate throughout the volume of the tank. To calculate the average viscosity experienced by the impeller, an effective shear rate S_{eff} is calculated by Metzner and Otto *(4)*:

$$S_{eff} = KN \quad (2)$$

Here, N is the impeller rotational speed in s^{-1} and K is an impeller constant. Table 1 lists K-values for several different impeller types. Substituting the effective shear rate in the viscosity equation for the particular fluid then gives the effective viscosity, which can be used to calculate the impeller Reynolds number. Several other methods of calculating the effective shear rate have been reported in the literature. Most of these are more complicated than the method used here, while not always more accurate.

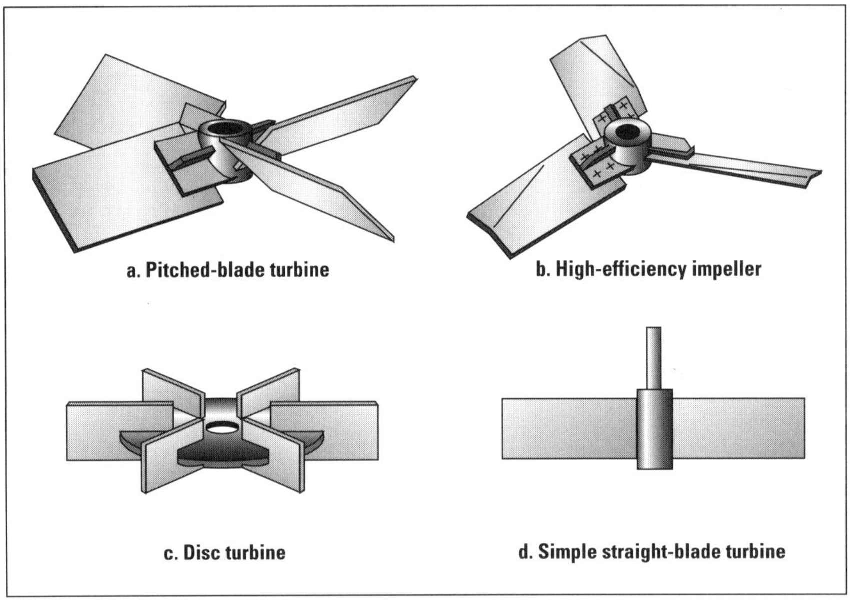

■ ***Figure 1. A wide variety of turbine agitators is available to handle viscosities to about 50 Pa-s.***

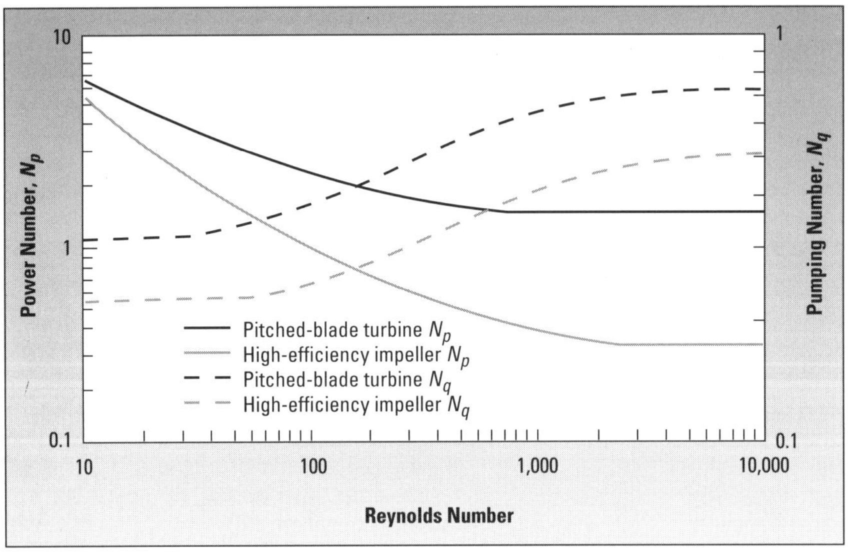

■ ***Figure 2. Power number and pumping number as functions of Reynolds number for a pitched-blade turbine and high-efficiency impeller.***

Two other dimensionless numbers that are often used to characterize an impeller are the power number and the pumping number. The power number N_p is implicitly defined by the following equation:

$$P = N_p \rho N^3 D^5 \quad (3)$$

When the power number of an impeller is known, the power draw of the impeller can be calculated using this equation, given liquid density, impeller speed, and diameter.

Table 1. *K*-values for effective shear-rate model (Eq. 2).

Impeller Type	*K*
High efficiency	10
Pitched blade	11
Straight blade	11
Disc turbine	11.5
Anchor (D/T = 0.98)	24.7
Helical ribbon (D/T = 0.96, P/D = 1)	29.4

The values for the anchor and helical ribbon impellers are for the specified standard geometries only. For other geometries, Eqs. 12 and 17 should be used.

The pumping number N_q is implicitly defined by:

$$Q_l = N_q N D^3 \tag{4}$$

When the pumping number of an impeller is known, the pumping rate of the impeller Q_l can be calculated for a given diameter and speed.

Both the impeller power number and pumping number depend on a variety of factors, such as the ratio of impeller to tank diameter D/T, the impeller off-bottom clearance ratio C/T, and the impeller Reynolds number. Figure 2 shows both power and pumping number correlations for both the high-efficiency impeller and pitched-blade turbine, using a standard geometry: $D/T = C/T = 1/3$.

For Reynolds numbers above 10,000, both N_p and N_q show little variation with Reynolds number. In this regime, the flow is considered to be fully turbulent. When the Reynolds number decreases, the pumping number decreases, while the power number goes up. This regime is usually called the transitional regime.

At very low Reynolds numbers ($N_{Re} < 10$), the power number is inversely proportional to the Reynolds number. In this Reynolds number regime, the flow is considered laminar. The exact Reynolds numbers at which the transitions occur are a function of impeller type, size, and number. Grenville *et al.* *(5)* correlated the Reynolds number at the boundary between the turbulent and transitional regime as follows

$$N_{Re,\,t} = \frac{6{,}370}{N_p^{1/3}} \tag{5}$$

Flow patterns

The rate of mixing in a tank is related to the overall flow pattern. Flow patterns can be effectively studied using modern techniques, such as computational fluid mixing (CFM), digital particle image velocimetry (DPIV), and laser Doppler velocimetry (LDV).

As an example, we will discuss the flow patterns created by the pitched-blade turbine. These patterns will be illustrated with results of CFM simulations. In these simulations, the impellers were modeled using LDV data obtained by Wang *et al.* *(6)*. The results of the simulations are shown in Figure 3. Color denotes the local velocity magnitude. The fluid in the red regions moves fast, while that in the blue regions moves slow.

When the flow is turbulent, the pitched-blade turbine creates an axial jet (Figure 3a). Although most axial-flow impellers can be used pumping upwards also, they are mainly used pumping downwards as shown in the figure. The impeller creates a typical axial flow pattern, with one main circulation loop. A smaller circulation loop with weak flow forms near the shaft, under the impeller. The highest velocities are found in the outflow of the impeller, near the blade tip. Lower velocities are found far away from the impeller, for example, near the liquid's surface.

When the Reynolds number is decreased, for example, by lowering the

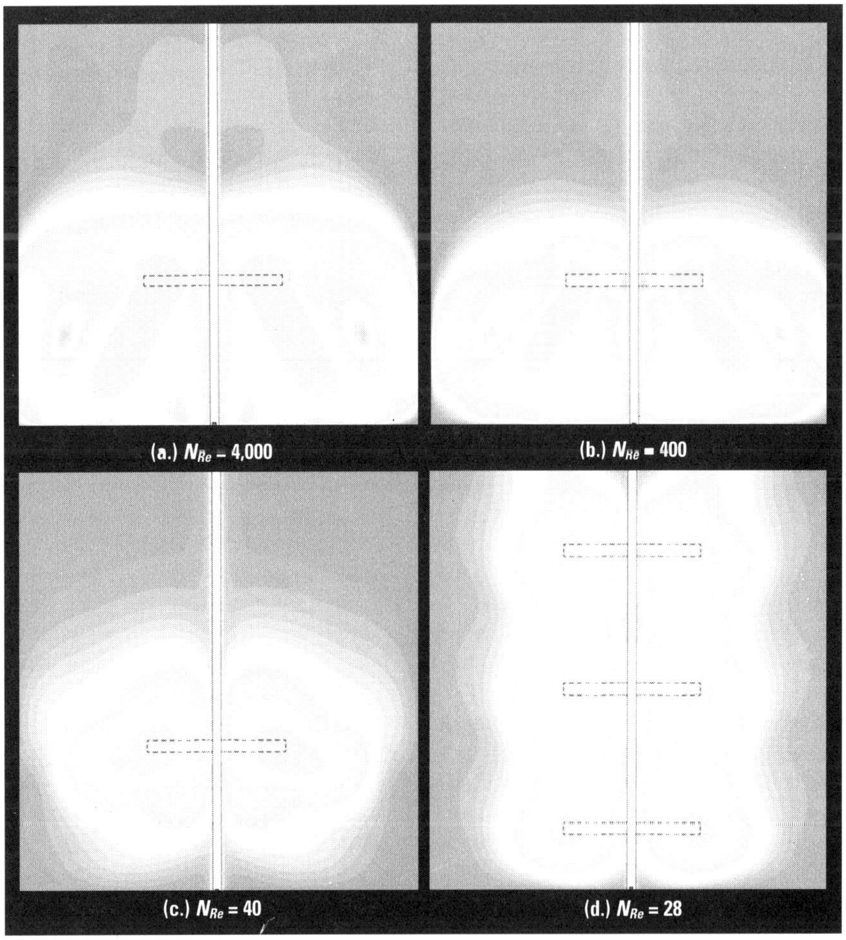

■ ***Figure 3. Simulation of the flow patterns created by a single pitched-blade turbine in (a.), (b.), (c.), and by a three-impeller system in (d.).***

speed or by increasing the liquid viscosity, the outflow from the impeller changes. The flow pattern gradually becomes more radial and, as a result of the decreasing pumping capacity, the average velocity in the tank decreases also. Figures 3b and 3c show the flow pattern at Reynolds numbers of 400 and 40.

At the lowest Reynolds number shown here, the impeller pumps mainly in the radial direction. The liquid velocities are very weak farther away from the impeller. Obviously, such a flow pattern is not desirable to achieve good overall mixing.

To increase the rate of mixing in the tank under these circumstances, a different impeller system must be used. The overall mixing in the tank can be improved by using more impellers. Also, the diameter of the impeller should be large enough to create significant liquid movement at the wall.

Figure 3d shows a redesigned impeller system for this tank. The number of impellers has been increased to three. The impeller speed has been decreased such that power draw remains the same. It can be seen that now a more-uniform velocity profile is being found compared to Figure 3c. The velocities near the wall are still low, however.

This example stresses the importance in high-viscosity systems of having an impeller system that physically agitates the whole volume of the tank. In a tank with a low-viscosity fluid where, under fully turbulent conditions, only one impeller would be sufficient, multiple impellers may be needed for high-viscosity systems.

When agitating viscous fluids in the transitional or laminar regime, it may not be necessary to use baffles, as should be done when the flow is turbulent. On an industrial scale, full baffling is usually recommended for fluids with a viscosity lower than about 5 Pa-s, to convert swirling into axial and radial circulation.

Full baffling means the use of four baffles with a width of *T*/12 and offset from the tank wall a distance of *T*/72. For more-viscous fluids, a smaller number of narrower baffles may be used, or no baffles may be used at all. This, however, also depends on the process requirements. The viscosity of the fluid may increase significantly during the process. For optimum performance at the start of the process, full baffling may be needed, while the performance may be better without baffles when the process nears completion.

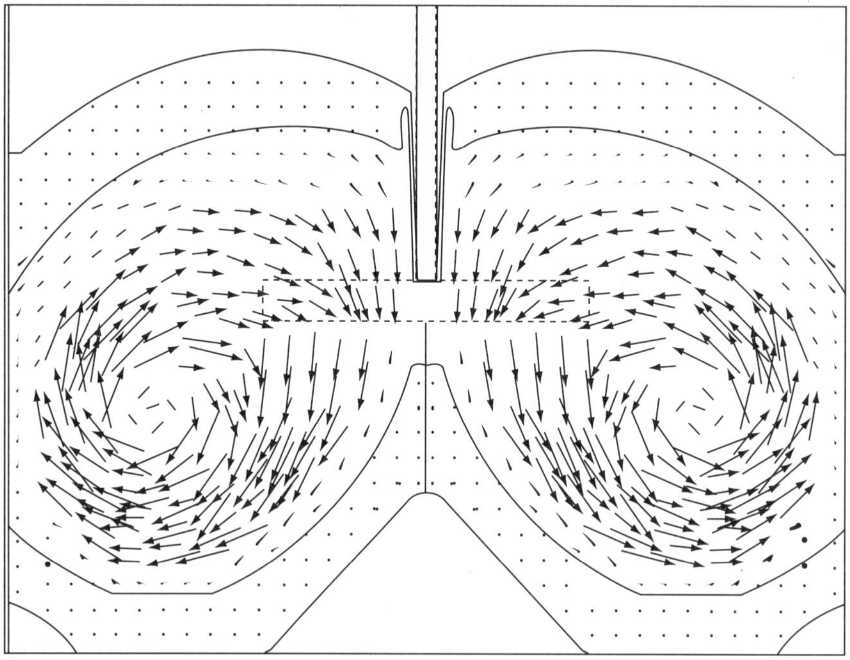

■ ***Figure 4. Simulation of the flow pattern in a tank filled with paper pulp, a yield-stress fluid.***

An optimum baffle arrangement for the process as a whole thus needs to be determined. An alternative might be to mount the impeller off-center or to use a rectangular tank to create an effect similar to baffling, without creating the slow-moving zones in the wake of the baffles. In such cases, special testing may be required and exact guidelines cannot be given.

Effect of rheology

Changing the rheology of the fluid will obviously affect the flow pattern created by the impeller. Figure 4 shows a similar tank and agitator as in the previous figures, but now filled with paper pulp, which has a yield stress (see Bakker and Fasano *(7)*). In the blue region around the impeller, the fluid is turbulent. The flow is laminar in the blue-green region. In the areas that are colored yellow, the shear stresses are below the yield stress and the fluid is stagnant.

The impeller creates a cavern in which the fluid is moving relatively fast and where the flow is turbulent. However, in the bulk of the fluid, where the shear stresses are below the yield stress, the fluid is not moving at all. Increasing the diameter of the impeller or the impeller speed will increase the size of the cavern.

With yield-stress fluids, the engineer must make sure that the shear stresses in the whole volume of the fluid are larger than the yield stress. A calculation procedure for sizing impellers for such applications is beyond the scope of this article. However, there is sufficient information available in the literature to calculate cavern sizes for various impeller systems (Solomon *et al. (8),* Elson *et al. (9),* and Etchells *et al. (10)*).

In low-Reynolds number ($N_{Re} \leq 400$) laminar flow, it is not always sufficient to create movement throughout the whole volume. It is important to avoid the creation of self-circulating or segregated zones, in which the fluid only slowly mixes with the rest of the tank. The existence of such segregated regions in viscous systems has been found for various impeller systems, as mentioned by Nagata *(11)*.

If segregated zones form in an industrial reactor, the product quality and consistency may deteriorate. Muzzio and Lamberto *(12)* suggest varying the impeller speed to create time-dependent flow patterns. This prevents the formation of steady self-circulating flow loops.

In general, at low Reynolds numbers, when the use of helical impellers is not yet necessary or not possible, it is best to use multiple, large-diameter axial flow impellers. Multiple feed-pipes adding directly into the impellers are preferred. This gives the best guarantee that the fluid is evenly mixed throughout the whole liquid volume. Both laboratory tests and CFM modeling can be used to study the flow patterns in detail.

The above description of mixing mainly applies to steady-state flows, or when the overall fluid flow is a lot slower relative to the time-scale at which mixing occurs. Although this description is fairly accurate under laminar flow conditions, the situation is more complicated when the flow is turbulent.

In turbulent flow, eddies are present with a wide variety of time-scales that are shorter than the blade-passage frequency. Also, large-scale chaotic variations in the mean flow can occur with a period that is much longer than the impeller rotational frequency, as shown by Myers *et al. (13)*. The high-frequency turbulent eddies provide mixing action termed turbulent diffusion. In contrast, long time-scale fluctuations in the overall mean-flow field provide additional, quasi-random stretching and folding action in the fluid that can be very effective in decreasing large-scale concentration gradients.

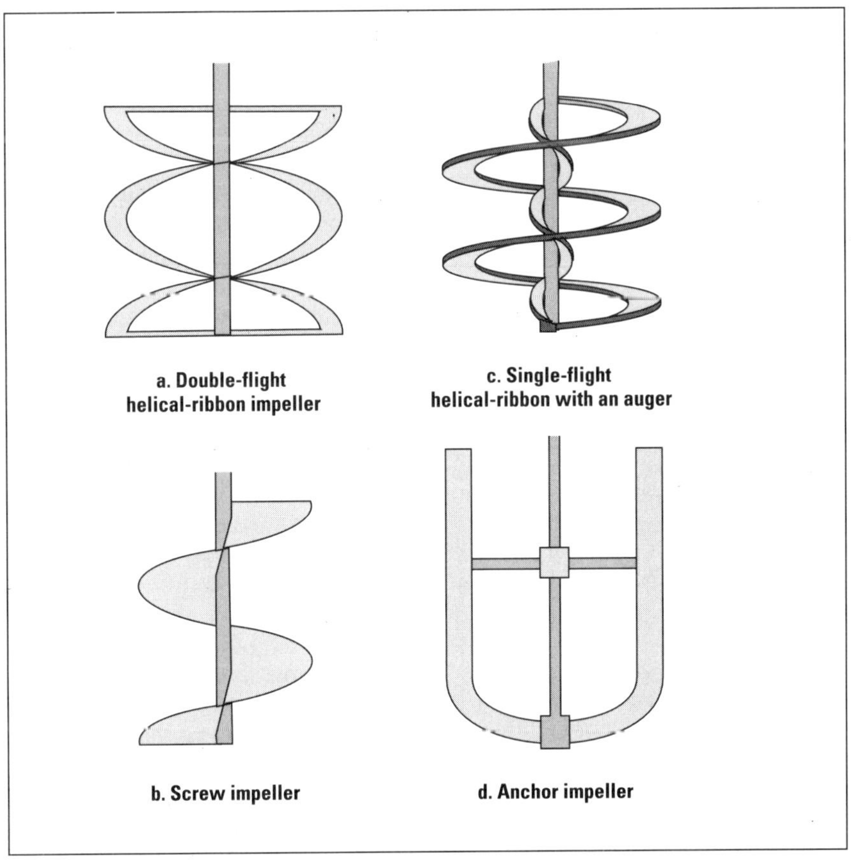

■ ***Figure 5. Helical-ribbon and anchor impellers provide an alternative to turbine impellers.***

Helical-ribbon and anchor impellers

An alternative to selecting a turbine impeller system is the use of a helical-ribbon impeller or an anchor impeller (Figures 5a and 5d). These impellers sweep the whole wall surface of the tank and physically agitate most of the fluid batch. As a result, they can be used at much lower Reynolds numbers ($N_{Re} \leq 400$) than the open-style turbines.

Helical ribbons are primarily used when very viscous materials are to be mixed, starting at viscosities of about 20 Pa-s and higher, depending on the scale. Helical ribbons have been successfully used for viscosities up to 25,000 Pa-s. These impellers often have a diameter that is very close to the inside diameter of the tank, and, therefore, are sometimes called close-clearance agitators. They guarantee liquid motion all the way to the wall of the tank, even for these very viscous materials.

Helical impellers can also be manufactured in a different configuration, with the helical impeller blade wrapped closely around the shaft (Figure 5b). Such impellers are also called helical-screw impellers, and are sometimes used in combination with a draft tube.

When these screw impellers are combined with a close-clearance helical ribbon as shown in Figure 5c, the screw is sometimes called an auger. The power draw from an impeller is proportional to the diameter to the 5th power, as given in Eq. 3. Since the diameter of the auger is much smaller than the diameter of the outside helical ribbon, the power draw of an auger is often less than 1% of that of a helical ribbon. When correctly designed, an auger can be used to improve the mixing near the shaft, without significantly increasing the power draw and torque requirement for the agitator. If the auger is incorrectly designed, and the pumping capacity of the auger does not match the flow created by the helical ribbon impeller, the auger may actually block the liquid flow near the shaft and hamper the mixing process.

The cover photo shows a flow pattern created by a helical-ribbon impeller at $N_{Re} = 10$, as calculated using CFM. Again, the velocity vectors point in the direction of the local liquid velocity. Both the length of the vectors and the color show the local velocity magnitude. Red denotes a large velocity and blue, a small one. Only the axial and radial velocity components are shown.

(a.) 1.0 min (b.) 1.5 min (c.) 3.5 min (d.) 5.0 min

■ ***Figure 6. Results of a blend time experiment with a laboratory-scale double-flight helix.***

The helical ribbon is rotating clockwise (seen from the top of the tank) and is pumping downward at the wall, although these devices can be built to pump upward at the wall also. Circulation loops form around the helical blade. Part of the liquid that is pumped downwards by the top helix reaches the bottom helix. The rest of the fluid is circulated back through the center of the tank.

Even at this low Reynolds number, the fluid in the tank is moving at sufficiently high velocities to guarantee good liquid movement and top-to-bottom mixing. The flow pattern of the helical ribbon depends on many geometrical parameters, such as pitch, blade width, and helix diameter. For example, if the gap between the helix and the tank wall is too large, liquid may circulate back upwards at the tank wall, thus decreasing the velocities in the center of the tank. CFM models can be used to accurately predict such effects and to optimize the mixer design for the application on-hand.

Figure 6 shows the result of a laboratory-scale blending test, using a double-flight helical ribbon. A tracer fluid is added at the liquid surface and slowly mixed in. The photographs were taken after 1.0, 1.5, 3.5, and 5.0 min. The helical ribbon is rotating at 6 rpm.

Fluid is pulled down at the wall and circulated back up through the center. The complicated flow pattern generated by the helix is clearly visible from the intricate shape of the striations and surfaces seen in the figure.

The helix is usually mounted with several supporting cross-bars (see photo on p. 25). These cross-bars also physically sweep the liquid, adding to the mixing action. It is recommended to extend the helix all the way to the liquid surface and to make the clearance between the bottom cross-bar and the tank bottom as small as possible.

Another option is to add anchors at the top and bottom. This prevents the formation of slow-moving or stagnant

Design example: Helical-ribbon or turbine?

A chemical reaction vessel requires an agitator that will blend the process liquids to 95% uniformity in less than 5 min. Both a helical-ribbon impeller and a turbine impeller system will be designed to meet this criterion, and will be compared. The reactor information is as follows:

Tank dia.	$T = 2.5$ m
Liquid level	$Z = 2.5$ m
Liquid viscosity	$\mu = 25$ Pa-s
Liquid density	$\rho = 1{,}200$ kg/m^3

A standard helical ribbon impeller with $D/T = 0.96$ and $W/D = 0.1$ will be designed. The required impeller rotational speed can be calculated from the required blend time and Eq. 13:

$$N = \frac{75}{t_{b,\,95\%}} = \frac{75}{300} = 0.25\ \text{s}^{-1}$$

The closest impeller speed that is commercially available is $N = 0.275$ s^{-1} = 16.5 rpm. This is the speed that will have to be used in the reactor, and the blend time is then:

$$t_{b,\,95\%} = \frac{75}{0.275} = 4\ \text{min}\ ,\ 33\ \text{s}$$

To verify that the Reynolds number is in the range of validity of the blend-time correlation, the Reynolds number is calculated as:

$$N_{Re} = \frac{1{,}200 \times 0.275 \times 2.4^2}{25} = 76 < 100$$

The power number is then calculated from Eq. 9 to be:

$$N_p = \frac{350}{76} = 4.61$$

And the power draw is then:

$$P = 4.61 \times 1{,}200 \times 0.275^3 \times 2.4^5 = 9{,}160\ \text{W}$$

The required torque is calculated as:

$$M = \frac{P}{2\pi N} = \frac{9{,}160}{2\pi\left(0.275\right)} = 5{,}301\ \text{N-m}$$

Designing a turbine impeller system that meets the same blend-time criterion at this low Reynolds number requires making several iterations through the equations presented in this article, and using the blend-time correlations presented by Fasano and Penney *(1)*, and Fasano *et al.* *(3)*. A suitable system was found to consist of three pitched-blade turbines with a dia. of 1.45 m, operating at 0.933 s^{-1} = 56 rpm.

This system has a calculated blend time of 4 min, 44 s. The Reynolds number is 94, which is pushing the lower limit at which these impellers can still be used. The power draw of this impeller system is $P = 41{,}230$ W, and the torque requirement is $M = 7{,}033$ N-m.

Comparing the two systems reveals that both have approximately the same blend time. Because of the larger power draw and torque requirement of the turbine agitator it will require a more expensive motor and gear box. A helical ribbon impeller itself will be more expensive than three pitched-blade turbines. The total capital cost of the whole agitator with the helical ribbon (including shaft, motor, and gear box) is about 10% higher than when pitched-blade impellers are used. However, the power draw is less than a fourth of the pitched-blade impeller system, resulting in a significant reduction in electricity and operating expenses.

The final decision on which impeller system is to be used may also depend on future plans. If it is anticipated that, at some point, liquids with a viscosity higher than the design viscosity of 25 Pa-s will be used, then the helical-ribbon impeller will be the best choice. Conversely, if it is anticipated that future processes may use lower-viscosity materials, the pitched-blade impeller system may be used.

zones near the liquid surface and the vessel bottom.

When it is important to have high velocities at the tank wall, for example, in heat-transfer applications, it is recommended to make the clearance between the helix and the tank wall as small as possible. Further, the helix can be equipped with scrapers that physically remove the fluid from the tank wall.

Agitator design

The previous sections gave qualitative discussions about the flow patterns generated by the different impeller styles and the effects on mixing in the tank. This section gives more quantitative design guidelines, both for designing turbine agitators and close-clearance agitators such as helical ribbons and anchor impellers. The above sidebar presents a worked-out example in sizing both types of mixers.

The required agitator size is often determined, based on previous process experience, by means of laboratory tests, or by means of computer simulations. An often-used design aid is the scale of agitation S_A defined by Hicks *et al.* *(14)*:

$$S_A = 32.8\,\frac{N_q N D^3}{\frac{\pi}{4} T_{eff}^2} \qquad (6)$$

where:

$$T_{eff} = \left(\frac{4}{\pi} V_l\right)^{1/3} \qquad (7)$$

Here, T_{eff} is the effective tank diameter and V_l is the total volume of liquid. S_A provides a 1–10 range of agitation intensity with 1 being mildly mixed, and 10 being very violently mixed. Agitation levels of 1 and 2 are characteristic of applications that only require low fluid-velocities to achieve the process result. Levels of 3–6 are most common in the CPI. Values of 7–10 are characteristic of applications requiring high fluid velocities, such as in critical reactors. Gates *et al.* *(15)* present guidelines on how to relate S_A to specific process applications.

The scale of agitation is not suited for use with fluids with a large yield-stress or which are viscoelastic. Viscoelastic fluids tend to develop flow normal to the direction of shear. This can lead to "counterintuitive" effects in the flow field, such as fluids climbing up a rotating shaft, and fields in which the direction is reversed when compared to Newtonian fluids. Such fluids require special testing and design considerations when being agitated, and are not discussed in detail in this article.

When laboratory tests are performed, the scaled-down system is usually geometrically similar to the large-scale industrial vessel. To achieve the same mixing intensity in both the laboratory and industrial vessel, S_A is usually kept constant. To achieve the same qualitative flow pattern, the Reynolds number, as defined in Eq. 1, should be kept the same on both scales. If possible, both S_A and the Reynolds number should be kept constant by using a model fluid with a lower viscosity on the laboratory scale.

When studying vortex motion at the liquid surface, it is necessary to keep both the Reynolds number and the Froude number constant at both scales. The Froude number is defined as:

$$N_{Fr} = \frac{N^2 D}{g} \quad (8)$$

The impeller speed at the laboratory scale can be calculated from the impeller speed at the large scale and the Froude number. A laboratory model-fluid has to be chosen such that the Reynolds number is the same on both scales. Other commonly used scaling criteria are constant power-per-unit-volume, which is often used in mass-transfer and dispersion applications, and constant tip-speed, which is similar to a constant S_A, provided that the Reynolds number is kept constant also.

The power draw of the pitched-blade turbine and high-efficiency impeller is easily calculated using Figure 2. In multiple-impeller systems, the power draw of each impeller is usually calculated as if the impeller were the only one in the tank. Then the individual power draws are added to obtain the overall draw. In situations where the impellers are mounted closely together and pump in the same direction, the overall power draw may actually be less, but adding the power of each impeller gives a usable, somewhat conservative estimate of the power-draw and torque requirements.

In power-draw calculations for non-Newtonian fluids, the effective viscosity μ_{eff} is substituted for the dynamic viscosity. The effective viscosity is calculated from the particular viscosity vs. shear-rate curve and the effective shear rate S_{eff} from Eq. 2.

When testing non-Newtonian fluids, care should be taken that the viscosity measurements are done in the appropriate shear-rate range. For example, the popular power-law model is often only valid over a range of about two decades of shear rate. This should be taken into account when doing lab testing at a small scale and a relatively large shear rate, and building the final plant on a large scale, at much lower shear rate. To avoid costly errors, viscosity curves should always be measured in the shear-rate range and temperature range in which they will be used in the plant.

Nomenclature

C = impeller to bottom clearance, m
D = impeller dia., m
f_{geo} = geometry correction factor in power draw equations, dimensionless
g = gravitational acceleration, m/s^2
H = impeller or helix height, m
K = impeller constant, dimensionless
M = torque, N-m
n_f = number of flights for a helix
N = impeller rotational speed, 1/s
N_{Fr} = Froude number, dimensionless
N_p = impeller power number, dimensionless
N_q = impeller pumping number, dimensionless
N_{Re} = Reynolds number, dimensionless
$N_{Re,t}$ = Reynolds number at boundary between transitional and turbulent flow, dimensionless
P = impeller power draw, W
P_i = pitch of a helical ribbon impeller, m
Q_l = impeller pumping rate, m^3/s
S_{eff} = effective shear rate experienced by the impeller, 1/s
S_A = scale of agitation, dimensionless
W = blade width, m
$t_{b,\,95\%}$ = blend time to 95% uniformity, s
T = tank dia., m
T_{eff} = effective tank dia., m
V_l = liquid volume, m^3
Z = liquid height in tank, m

Greek letters

ρ = density, kg/m^3
μ = dynamic viscosity, Pa-s
μ_{eff} = effective dynamic viscosity in power-draw calculations, Pa-s

Blend time

Blend time is another important quantity, often used in the design of chemical reactors and mixing tanks. An extensive discussion of blend time calculations for axial flow impellers such as the high-efficiency impeller and the pitched-blade turbine is presented by Fasano *et al.* *(3)*. The algorithms developed are limited to impeller Reynolds numbers larger than 100. For lower Reynolds numbers, the blend time increases so fast with increasing viscosity that accurate correlations are not available. This is also the regime where the use of helical-ribbon impellers should be strongly considered. If it is nonetheless necessary to estimate the blend time for such a highly viscous system equipped with turbine agitators, it is recommended to perform a laboratory test. The blend time at the large scale can then be calculated, assuming that the product of impeller speed and blend time is the same at both scales.

The power draw of a *helical-ribbon impeller* is similarly calculated from the power number, as follows.

For $N_{Re} \leq 100$:

$$N_p = \frac{350}{N_{Re}} f_{geo} \quad (9)$$

For $N_{Re} > 100$:

$$N_p = 1.12 \exp[6.15 - 1.37 \ln N_{Re} + 0.0613(\ln N_{Re})^2] f_{geo} \quad (10)$$

where:

$$f_{geo} = \left(\frac{D}{P_i}\right)^{1/2} \left(\frac{H}{D}\right) \left(\frac{W/D}{0.1}\right)^{0.16} \left(\frac{D/24}{T-D}\right)^{1/2} \left(\frac{n_f}{2}\right)^{1/2} \quad (11)$$

In Eq. 11, pitch P_i is the vertical climb of a single flight of the helix in 360 deg. rotation. The N_p curve is shown in Figure 7 for a standard design with $f_{geo} = 1$.

The Reynolds number is calculated using the viscosity at the effective shear rate. For helical ribbon impellers the effective shear rate is given by:

$$S_{eff} = 25\left(\frac{D}{T}\right)^{1/2}\left[\frac{P_i}{\left(\pi^2 D^2 + P_i^2\right)^{1/2}}\right]^{-0.152} N \qquad (12)$$

Accurate blend-time correlations are scarce, and there is a significant amount of scatter in the data presented in the literature, especially when it comes to the effect of changes in the geometry. For a standard double-flight helical-ribbon impeller ($P/D = 1$, $W/D = 0.1$, and $D/T = 0.96$) the blend time for 95% uniformity can be estimated from the equation below.

For $N_{Re} \leq 100$:

$$t_{b,\,95\%} = \frac{75}{N} \qquad (13)$$

The power number for *anchor impellers* is given by the equation below.

For $N_{Re} \leq 10$:

$$N_p = \frac{400}{N_{Re}} f_{geo} \qquad (14)$$

For $10 < N_{Re} < 10{,}000$:

$$N_p = 1.05\exp[5.64 - 0.783\ln N_{Re} - 0.0523(\ln N_{Re})^2 + 0.00674(\ln N_{Re})^3] f_{geo} \qquad (15)$$

where:

$$f_{geo} = \left(\frac{H}{D}\right)\left(\frac{W/D}{0.1}\right)^{0.16}\left(\frac{D/49}{T-D}\right)^{1/2} \qquad (16)$$

The N_p curve is shown in Figure 7, for a standard design with $f_{geo} = 1$. For the helical ribbon, $D/T = 0.96$ and $P_i/D = 1$. For the anchor impeller, $D/T = 0.98$. The effective shear rate for anchor impellers is given by:

$$S_{eff} = 25\left(\frac{D}{T}\right)^{1/2} N \qquad (17)$$

The blend time for 95% uniformity for anchor impellers of standard geometry ($W/D = 0.1$, $D/T = 0.98$, and $Z/T = 1$) can be estimated by Eq. 18.

For $100 < N_{Re} < 10{,}000$:

$$t_{b,\,95\%} = \exp(12.9\,N_{Re}^{-0.135}) \qquad (18)$$

Comparing Eq. 13 with Eq. 18 shows that at a Reynolds number of 100, it will take an anchor impeller more than 13 times as long to achieve 95% uniformity as a helical-ribbon impeller operating at the same speed.

Both with helical-ribbon impellers and with anchor impellers, the power input can be so high that cooling is required to remove the excess heat. Heat-transfer calculations are beyond the scope of this article. A good source of heat-transfer correlations is Nagata *(11)*.

Conclusions

This article gives an overview of the major considerations when mixing viscous fluids in industrial situations. Much information is available in the lit-

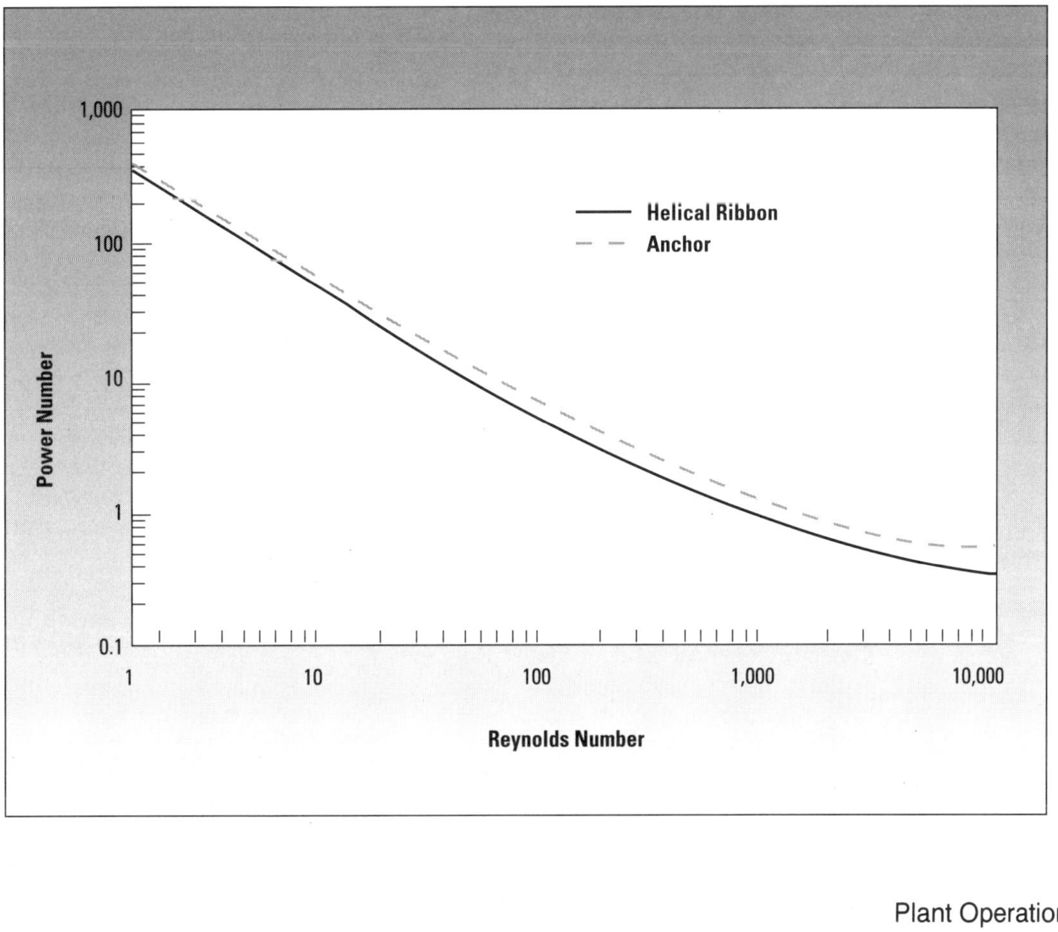

Figure 7. Power number as a function of Reynolds number for a helical-ribbon impeller and an anchor impeller, both of standard geometry.

erature on mixing of low-viscosity fluids, but less is offered on mixing of higher-viscosity fluids. This is due to the smaller number of applications, the complexity of studying high-viscosity mixing, and the often-unusual rheological characteristics of viscous process fluids. While the number of applications is smaller, the dollar value of equipment for high-viscosity applications is higher because mixing equipment costs rise quickly with increasing viscosity.

The technology base for design of high-viscosity mixers is much more limited than for design of low-viscosity mixers. While this article has summarized important design information for high-viscosity mixing, the authors strongly encourage the use of computer simulations, and laboratory and pilot-plant studies to supplement and confirm the recommended approaches given here. Careful analysis of all relevant mechanical and process design parameters, and capital and maintenance cost is required in all cases where viscous materials are to be mixed on an industrial scale.

Historically, mixing used to be a mainly empirical field of study. Recently, more fundamental studies have been performed. Also, the advances in CFM mixing have made it possible to accurately predict flow fields, generating information that can be used in evaluating the mixing performance of an agitator to a level of detail that was unheard of a decade ago. CEP

Acknowledgments

The authors wish to acknowledge the contributions of James W. Nordmeyer, Kevin G. Walsh, Sue M. Anders, Steve D. Stocker, and Julian B. Fasano at Chemineer, Inc., and Kevin J. Myers at the University of Dayton. All computer simulations in this article were performed using Fluent from Fluent, Inc.

A. BAKKER is principal research engineer at Chemineer, Inc., Dayton, OH (513/454-3292; Fax: 513/454-3379). He has been with the company since 1991, after completing his doctoral dissertation on gas-liquid mixing. His specialization includes computational fluid dynamics in mixing processes. Bakker holds an engineering degree and a PhD in applied physics from Delft University of Technology, The Netherlands.

L. E. GATES is vice president of engineering and development at Chemineer, Inc., Dayton, OH (513/4454-3277; Fax: 513/454-3379). During his 27 years with Chemineer, he has held positions in engineering, sales, and marketing. Prior to joining the company, he was a research engineer at Du Pont. Gates holds BChE and MS degrees in chemical engineering from Ohio State University, and is a PE in Ohio.

Literature Cited

1. **Fasano, J. B., and W. R. Penney,** "Avoid Blending Mix-Ups," *Chem. Eng. Progress,* **87**(10), pp. 56–63 (Oct. 1991).
2. **Bouwmans, I.,** "The Blending of Liquids in Stirred Vessels," PhD dissertation, Delft University of Technology, Delft, The Netherlands (1992).
3. **Fasano, J. B., *et al.*,** "Advanced Impeller Geometry Boosts Liquid Agitation," *Chem. Eng.,* **101**(8), pp. 110–116 (Aug. 1994).
4. **Metzner, A. B., and R. E. Otto,** "Agitation of Non-Newtonian Fluids," *AIChE J.,* **3**(1), pp. 3–10 (1957).
5. **Grenville, R., *et al.*,** "Blending of Miscible Liquids in the Turbulent and Transitional Regimes," paper presented at Mixing XV, 15th Biennial North American Mixing Conference, Banff, AL, Canada, sponsored by North American Mixing Forum (NAMF)/AIChE (June 18–23, 1995).
6. **Wang, M. H., *et al.*,** "Effect of Reynolds No. on the Flow Generated by Pitched Blade and High Efficiency Turbines," paper presented at Mixing XV, 15th Biennial North American Mixing Conference, Banff, AL, Canada, sponsored by North American Mixing Forum (NAMF)/AIChE (June 18–23, 1995).
7. **Bakker, A., and J. B. Fasano,** "A Computational Study of the Flow Pattern in an Industrial Paper Pulp Chest with a Side Entering Impeller," "Process Mixing: Chemical and Biochemical Applications — Part II," *AIChE Symp. Series,* **89**(293), pp. 118–124 (1993).
8. **Solomon, J., *et al.*,** "Cavern Sizes in Agitated Fluids with a Yield Stress," *Chem. Eng. Communications,* **11**(1–3), pp. 143–164 (1981).
9. **Elson, T. P., *et al.*,** "X-Ray Studies of Cavern Size and Mixing Performance with Fluids Possessing a Yield Stress," *Chem. Eng. Sci.,* **41**(10), pp. 2555–2562 (1986).
10. **Etchells, A. W., *et al.*,** "Mixing Bingham Plastics on an Industrial Scale," paper presented at Fluid Mixing III, *Proceedings,* European Federation of Chemical Engineering Publication Series No. 63, Bradford, U.K., sponsored by IChemE, Rugby, U.K. (Sept. 8–10, 1987).
11. **Nagata, S.,** "Mixing — Principles and Applications," Chapter 4, "Mixing of Homogeneous Liquids," Halsted Press (John Wiley), New York (1975).
12. **Muzzio, F. J., and D. J. Lamberto,** "Using Time-Dependent Flows to Enhance Mixing in Stirred Tank Systems," paper presented at Mixing XV, 15th Biennial North American Mixing Conference, Banff, AL, Canada, sponsored by North American Mixing Forum (NAMF)/AIChE (June 18-23, 1995).
13. **Myers, K. J., *et al.*,** "A DPIV Investigation of Flow Pattern Instabilities of Axial-Flow Impellers," paper presented at Mixing XV, 15th Biennial North American Mixing Conference, Banff, AL, Canada, sponsored by North American Mixing Forum (NAMF)/AIChE (June 18–23, 1995).
14. **Hicks, R. W., *et al.*,** "How to Design Agitators for Desired Process Response," *Chem. Eng.,* **83,** pp. 102–110 (Apr. 26, 1976).
15. **Gates, L. E., *et al.*,** "Application Guidelines for Turbine Agitators," *Chem. Eng.,* **83**, pp. 165–170 (Dec. 6, 1976).

Consider Mechanical Design of Agitators

Accounting for hydraulic and mechanical forces will help eliminate problems of undue vibration and ensure a properly-designed support structure.

Julian B. Fasano, Jerry L. Miller, and Steve A. Pasley,
Chemineer, Inc.

Agitators correctly designed from a process standpoint must be just as accurately designed from a mechanical standpoint. Many of the mechanical factors involved in the design of an agitator were covered in the 1970s by Hill and Kime *(1)*, and Ramsey and Zoller *(2)*. Newer tools and information suggest a revisit of this area.

Most agitators consist of a shaft and impeller system, as shown in Figure 1. Engineering analysis may require that some agitators use a bearing at the bottom of the tank to additionally support the shaft. Shaft design must accommodate hydraulic and mechanical system loads and must avoid proximity to the natural frequency. Therefore, appropriate design requires a determination of the shaft and impeller natural frequency. Of prime concern is the lateral natural frequency.

Dynamic methods for accurately determining natural frequency are discussed, as well as forcing sources and their frequencies. Appropriate support for an agitator and the effects of structural stiffness are addressed, in addition to improvements in mechanical seals, fastener locking techniques, and materials of construction. Knowledge-based systems (KBS) and finite-element analysis (FEA) have permitted a more precise examination of designs and optimization. These and other areas of advancement have improved the quality, life, and operation of agitators.

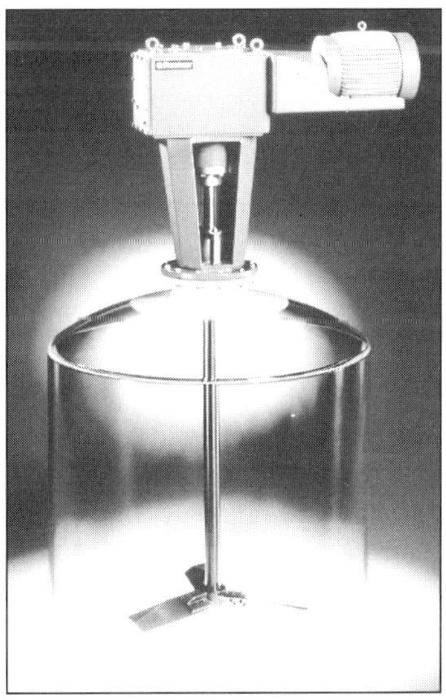

Figure 1. Most agitators consist of a shaft and impeller system.

Dynamic critical-speed calculation, deflection, and stress

Critical speed or natural frequency is the frequency of free vibration of a system. It is the frequency at which an undamped system with a single degree of freedom will oscillate upon momentary displacement from its rest position. The operating speed of the shaft and impeller system must be sufficiently far from the system's natural frequency to prevent deflections that would cause the shaft material to yield.

Ramsey and Zoller *(2)* have covered the elements of natural frequency for a shaft and impeller system using a lumped-mass, static technique for computing the critical speed of a shaft and impeller system.

Static models generally assume that there is an infinitely stiff structure to which the agitator is mounted. When the structure's stiffness is sufficiently low, the static model is compromised, and accurate predictions of natural frequency cannot be made. Because a shaft and impeller system cannot be considered to be rigid in the operating frequency range, its distributed properties need to be taken into account. For agitators, one must combine the properties of a distributed and lumped system. The transfer matrix method is an example of such a technique.

The transfer matrix method *(3)* is used to calculate the critical speed and static response of the shaft design. This matrix is composed of the mass and elastic characteristics of each span. This matrix is multiplied by the deflection, slope, bending moment, and shear at the station on one end of the span to calculate the deflection, slope, bending moment, and shear at the station on the other end of the span. This calculation for each span is shown below in matrix form. Each span i has station i - 1 on one end of the span and a station i on the other end:

$$\begin{vmatrix} -w_d \\ s \\ M \\ V \end{vmatrix}_i = \begin{vmatrix} U_{11} & U_{12} & U_{13} & U_{14} \\ U_{21} & U_{22} & U_{23} & U_{24} \\ U_{31} & U_{32} & U_{33} & U_{34} \\ U_{41} & U_{42} & U_{43} & U_{44} \end{vmatrix}_i \cdot \begin{vmatrix} -w_d \\ s \\ M \\ V \end{vmatrix}_{i-1} \quad (1)$$

The $\boldsymbol{U}$ matrix (which is the transfer matrix) is unique for each span and is composed of the shaft properties of the span as well as lumped masses or springs at station i for span i. An example of a lumped mass is an impeller, and an example of a spring is a bearing. This matrix is always a 4 × 4 square matrix, no matter how many spans are used because the matrix is used to define individual spans starting with span 1 and ending with span n (for n stations).

Additionally, since the first station is always station 0 (which is one of the two boundary conditions for the shaft), there is no span 0 and thus no transfer matrix for span 0. The method for calculating the critical speed is to first multiply each individual transfer matrix for each span by the previous matrix. In other words, if there are n stations, the transfer matrix for the shaft design would be as follows:

$$U = U_n U_{n-1} U_{n-2} \ldots U_1 \quad (2)$$

The following general relationship now exists:

$$\begin{vmatrix} -w_d \\ s \\ M \\ V \end{vmatrix}_n = \left| U \right| \cdot \begin{vmatrix} -w_d \\ s \\ M \\ V \end{vmatrix}_0 \quad (3)$$

The shaft design is in terms of the boundary conditions (deflection, slope, bending moment, and shear at stations 0 and n) and the overall transfer matrix. The advantage of using the transfer matrix method can be seen in Eq. 3 because this matrix equation applies to any shaft design, regardless of the number of spans, lumped masses, or springs. For computational purposes, the bending moment and shear are defined as zero at the boundaries. Rewriting Eq. 3 reflecting the known boundary conditions yields:

$$\begin{vmatrix} -w_d \\ s \\ 0 \\ 0 \end{vmatrix}_n = \left| U \right| \cdot \begin{vmatrix} -w_d \\ s \\ 0 \\ 0 \end{vmatrix}_0 \quad (4)$$

The transfer matrix $\boldsymbol{U}$ is a function of frequency. So, there are four simultaneous equations and five unknowns (the deflection and slope at stations n and 0, and the frequency). Examining Eq. 4 closely yields the following equations:

$$0 = U_{31} - w_{d0} + U_{32} s_0 \quad (5)$$

$$0 = U_{41} - w_{d0} + U_{42} s_0 \quad (6)$$

Assuming the deflection and slope at station 0 are non-zero for a non-trivial solution, the only way that Eqs. 5 and 6 can be solved is if the following condition exists:

$$0 = U_{31} U_{42} - U_{32} U_{41} \quad (7)$$

The only unknown variable associated with Eq. 7 is the frequency. The lowest frequency (greater than 0 rpm) which satisfies Eq. 7 is the first critical speed. Again, the advantage of using the transfer matrix method can be seen with Eqs. 5, 6, and 7, since these equations apply to any shaft design regardless of the number of spans and stations.

The method for determining the deflection is very similar to calculating the critical speed. However, the hydraulic forces on the shaft are now taken into account. Also, the frequency is a known value. Hydraulic forces are determined based on the impeller torque, so speed (or frequency) is already included in the magnitude. Consequently, a forced response at the shaft speed would not be appropriate because the results would reflect the effect of frequency twice.

Therefore, it is necessary to determine the forced response for the static condition (frequency equals zero). From the bending moment for each

Nomenclature

- a = amplitude of vibration, in.
- d = shaft dia., in.
- D = impeller dia., in. or m
- F_h = hydraulic force, lb_f or N
- F_{Hi} = hydraulic force at station i, lb_f or N
- F_m = material factor (= 1.00 for Type 316 stainless steel)
- i = station number
- k_b = gearbox bearing stiffness, lb_f/in. or N/m
- k_s = support stiffness, lb_f/in. or N/m
- L = shaft length: impeller to first bearing, in.
- M = bending moment, lb_f/in. or N/m
- n = number of stations
- N = impeller rotational speed, rpm or rev/s
- N_c = natural frequency, 1/min or 1/s
- N_{Fh} = hydraulic force number, dimensionless
- s = slope, radians
- S_b = bearing span, in.
- t = head thickness, in. or mm
- t_r = recommended head thickness, in. or mm
- $\boldsymbol{U}$ = transfer matrix
- v = vibrational velocity, in./s
- V = shear force, lb_f or N
- w_d = deflection, in. or m
- W_e = equivalent weight, lb
- W_p = projected height of impeller blade, in. or m

Greek letter

- τ = impeller torque, in.-lb_f or N-m

station along with the torque from the impeller(s), the tensile and shear stresses can be calculated for each station. The first sidebar presents an example of static vs. dynamic agitator shaft analysis.

Agitator support structures

This is too often an overlooked design consideration. Critical-speed computations often assume that an agitator will be mounted on a sufficiently stiff structure such that vibrational computations will not be seriously compromised. As support stiffness decreases, the natural frequency decreases and becomes dependent on the degree of stiffness. Several significant problems may occur with low-stiffness supports:

- It could be difficult to accurately predict the first natural lateral frequency of the shaft and impeller system. Since agitator designs are generally conducted such that the operating-to-natural-frequency ratio is less than one, as stiffness falls, the ratio becomes closer to unity;
- The natural frequency or frequencies of the support and its members may become excitable by the various forcing functions or forcing-function harmonics of the agitator; and
- The transmissibility or dynamic-to-static force ratio imparts substantially more load to supporting members than used for design.

The techniques for providing a sufficiently stiff structure will be discussed in a later section.

Forcing sources and frequencies

Agitator forcing sources are either mechanical or hydraulic. The frequency at which these forces are applied produces forced vibration. The primary mechanical sources are imbalance and shaft straightness. Other mechanical sources are due to parallel and axial misalignment, bearings, and gears, among others. As long as the shaft is straight to within 0.005 in./ft (0.417 mm/m) and the shaft and impeller system is balanced to within International Standards Organization, ISO 1940, G-40 (1975-05-01) balance grade or better, the controlling vibrational sources are typically all hydraulic. Table 1 identifies the major hydraulic sources and their characteristic frequencies.

Table 1. Dominant hydraulic vibration sources and frequencies.

Source	Disturbing Frequency
Imbalanced hydraulic load	N
Blade-passing frequency	(No. of blades)N
Vortex-shedding frequency	$(0.1\text{–}0.3)\pi ND/W_p$

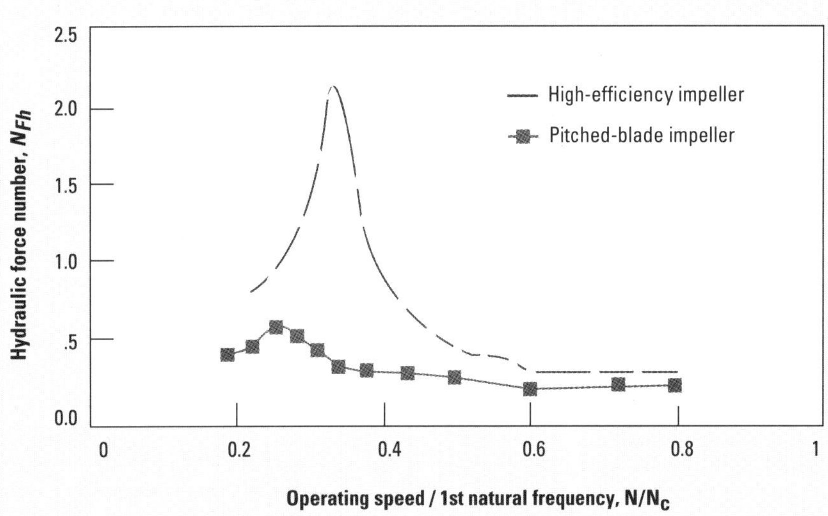

Figure 2. Effect of the blade-passing source on the hydraulic force number.

If the natural frequency of a spring system coincides with the forcing frequency or a higher-order harmonic of the forcing frequency, undamped movement in the system will increase to excessive amplitudes. This undesirable condition is known as resonance and should be avoided.

The imbalanced hydraulic load is a directionally, normally distributed force randomly applied radially to the impeller. This force is discussed by Hill and Kime *(1)*. The blade-passing source has been known for some time. However, no one has quantified its effects. The blade-passing source superimposes itself upon the imbalanced hydraulic load. Normally, this is not a significant problem. Nevertheless, when the disturbing frequency is near the natural frequency of the shaft, reinforcement occurs. A dimensionless method for evaluating the hydraulic forces to an impeller is by the dimensionless hydraulic force number:

$$N_{F_h} = \frac{F_h D}{\tau} \quad (8)$$

Figure 2 shows how the dimensionless hydraulic force behaves for both a pitched blade and HE-3 high-efficiency impellers. Clearly, one can see that it is not advantageous to operate a pitched-blade impeller in the vicinity of $N/N_c = 0.25$, nor an HE-3 impeller in the vicinity of $N/N_c = 0.33$. The notion that "the lower the N/N_c ratio the better" is not necessarily valid in this case. If operation in these ranges is unavoidable, design consideration must be given to the amplified hydraulic forces.

Vortex shedding has been recognized since da Vinci observed it in 1513. A great deal of study has been done on the flow patterns around immersed blunt bodies. At low Reynolds numbers, flow lines can follow the body shape and separate from the boundary. The separation, as predicted by Bernoulli's theorem, causes a zone of increased local velocity and a corresponding localized low-pressure zone. The result is a flow pattern which tends to curl back on itself and form a vortex which remains in a fixed position relative to the body. As the Reynolds number is increased, the vortex increases in size and intensity, separates from the body, and travels downstream. This phenomenon will occur first on one side of the body, then on the other in a regular repeating fashion.

On a rotational body, such as an impeller blade, radial and axial forces are strong enough to prevent side-to-side or top-to-bottom vortex reversal patterns. However, there is sufficient evidence to suggest that the local intensity does fluctuate in a repeating pattern. The intensity of the vortex system is highly dependent on the body shape. Fasano, *et al. (4)* depicted how much smaller this pattern is for a high-efficiency impeller than for a pitched-blade turbine.

As shedding frequencies are on the order of 2 to 12 times the shaft rotational frequency, resonance with the first lateral frequency is not a problem. The only concern is when the vortex shedding frequency is resonant with the support structure. This effect is such an infrequent problem that it is generally dealt with on an application-to-application basis. In addition, shedding frequencies are usually substantially less than blade natural frequencies.

The agitator and supporting structure from a mass-spring-damper system can be viewed as that shown in

Static vs. Dynamic Agitator Shaft Analysis

Agitator applications generally require the use of long, cantilevered shafts which, if not designed properly, can have inherent problems with deflection and stresses without any consideration of the impeller loads. Every shaft design has a critical speed, which depends upon the shaft material, diameter, length, and the impeller weight(s). The effect of operating at or too near the critical speed was presented by Ramsey and Zoller *(2)*.

Although the damping effect on the shaft may keep the displacement from becoming excessive, a suitable shaft design would have a critical speed which is sufficiently far from the operating speed to avoid mechanical failure.

Thus, the importance of accurately predicting the critical speed cannot be overemphasized. Ramsey and Zoller provide the following relatively simple static critical speed equation for the design of a shaft and impeller system.

$$S_{crit} = \frac{0.388 \times 10^6 d^2 F_m}{L\left(W_e\left[L + S_b\right]\right)^{1/2}}$$

This equation is based on static conditions and does not take into account the effect of the shaft speed, coupled with other influences such as shear deflection and rotary inertia. Additionally, this equation is based on a rigid structure supporting the shaft. The error associated with the equation depends upon the support stiffness of the agitator. The agitator gearbox (including the bearings used to support the shaft) is designed to withstand the loads associated with mixing applications and thus is sufficiently rigid. However, the gearbox is always attached to either mounting beams or a vessel head/nozzle. The overall rigidity of the system depends on the design of the support mounting. An example of a shaft design showing the support system can be seen in the figure.

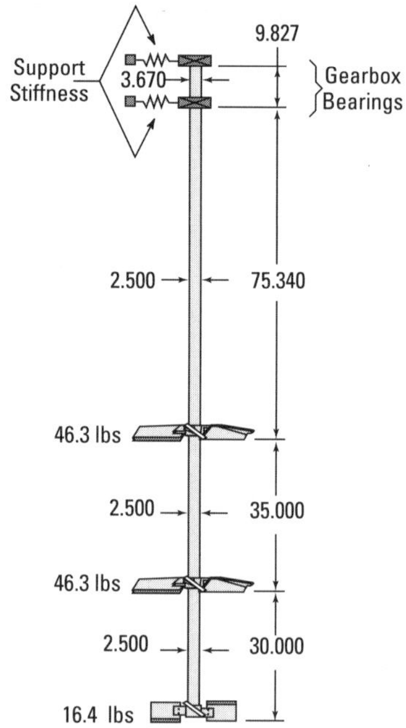

A shaft design showing the support system.

The static equation model assumes that the support spring stiffness is rigid. In reality, lateral stiffnesses of structures used to support agitators can generally vary from 2.5 x 10^4 to 1.0 x 10^6 lb_f/in. Dynamic shaft models can handle the various support stiffnesses encountered, and the effect of mounting stiffness on natural frequency is shown in the table for the shaft and impeller system shown above.

The static model cannot predict critical speed for a less than rigid support structure. Experiments have shown the dynamic model is much more accurate for predicting critical speeds, and consequently, to achieve a mechanically sound shaft design, the dynamic model for determining critical speed should always be used.

Effect of mounting stiffness on natural frequency.

	Critical Speed, rpm	
Stiffness, lb_f/in.	Static Model	Dynamic Model
50,000		113.2
100,000		126.8
200,000	140.0	135.7
500,000		141.9

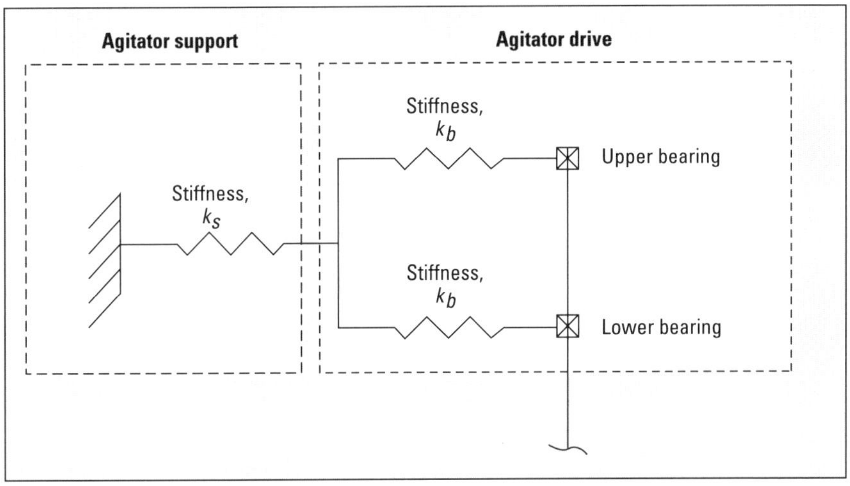

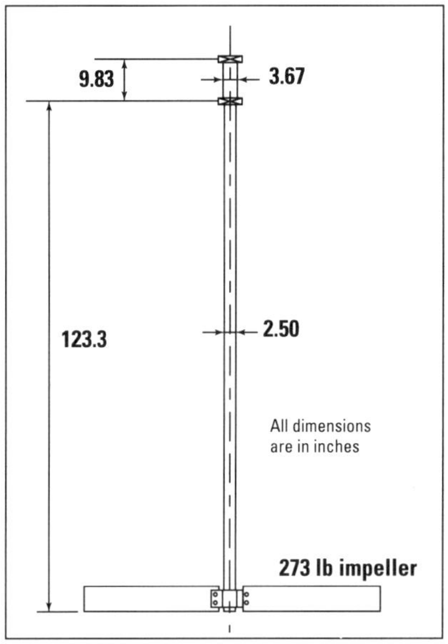

Figures 3 and 4. *Stiffness schematic, left, and agitator shaft and impeller assembly schematic.*

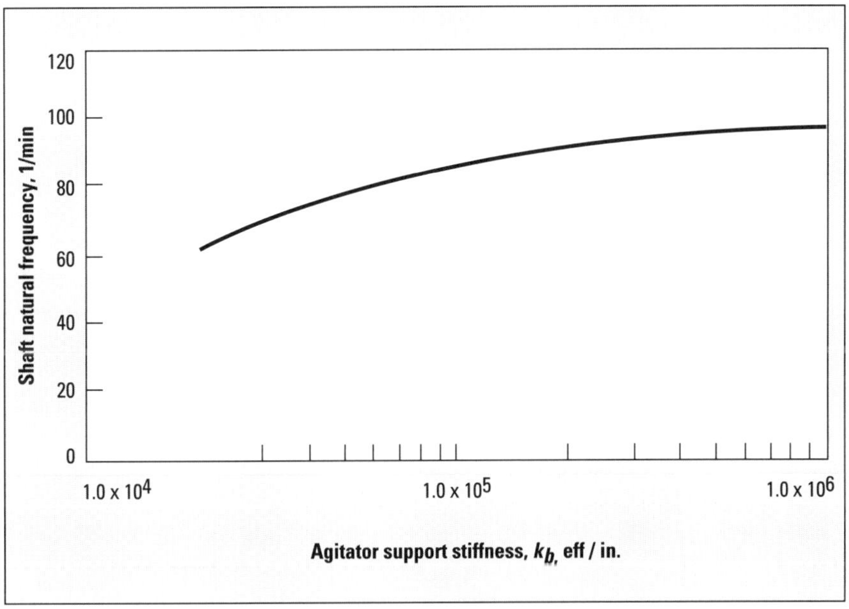

Figure 5. The effect of stiffness on shaft natural frequency.

Table 2. General machinery vibrational severity.

Description	Peak Velocity, in./s (mm/s)
Extremely smooth	< 0.0049 (0.124)
Very smooth	0.0049–0.0098 (0.124–0.249)
Smooth	0.0098–0.0196 (0.249–0.498)
Very good	0.0196–0.0392 (0.498–0.996)
Good	0.0392–0.0785 (0.996–1.99)
Fair	0.0785–0.157 (1.99–3.99)
Slightly rough	0.157–0.314 (3.99–7.98)
Rough	0.314–0.628 (7.98–16.0)
Very rough	> 0.628 (16.0)

Figure 3. The effective spring constant for the bearings becomes:

$$k_{b,\,\text{eff}} = \frac{1}{2\left(\dfrac{1}{k_s} + \dfrac{1}{2k_b}\right)} \tag{9}$$

Gearbox bearing support stiffness k_b is on the order of 1×10^6 lb_f/in. (175 MN/m), and support stiffnesses k_s can range from 2×10^5 to 1×10^6 lb_f/in. (35.0 to 175 MN/m). Thus, support stiffnesses k_s will most often dictate the "effective" bearing stiffnesses.

As an example, an agitator shaft and impeller assembly, represented schematically per Figure 4, has a natural frequency based on the overall stiffness $k_{b,\text{eff}}$, as shown in Figure 5. It is evident that once the effective bearing stiffness drops below 1×10^5 lb_f/in. (17.5 MN/m), the natural frequency begins to drop.

How stiff is stiff enough?

Stiffness has its price. Making a support structure stiffer will nearly always be reflected in higher costs. Knowing to what extent one must supply support will depend on acceptable levels of vibration. The dynamic loads of an agitator cause deflections and twisting of the support system. Depending on the vibrational frequency and location, these deflections and rotations can be translated into vibrational velocities. Numerous sources

Table 3. Recommended beam sizes.

Agitator Torque, in.•lb$_f$ (KN•m)	6 (1.83)	10 (3.05)	Tank Dia., ft (m) 15 (4.57)	20 (6.10)	30 (9.14)
			Beam Size, in. × lb/ft		
2,500 (0.280)	C 5 × 6.7	C 7 × 9.8	W 12 × 14	W 12 × 19	W 18 × 35
4,000 (0.450)	C 6 × 8.2	W 8 × 10	W 12 × 14	W 14 × 22	W 18 × 40
8,500 (0.960)	C 7 × 9.8	W 10 × 11.5	W 12 × 19	W 16 × 26	W 21 × 49
17,000 (1.90)	C 7 × 9.8	W 12 × 14	W 14 × 22	W 18 × 35	W 24 × 61
24,000 (2.70)	W 8 × 10	W 12 × 14	W 16 × 26	W 18 × 40	W 24 × 76
33,000 (3.70)	W 8 × 10	W 12 × 16.5	W 16 × 31	W 21 × 44	W 27 × 84
59,000 (6.70)	—	W 14 × 22	W 18 × 35	W 24 × 55	W 30 × 99
87,000 (9.80)	—	W 14 × 22	W 18 × 40	W 24 × 61	W 30 × 116
135,000 (15.3)	—	W 16 × 26	W 21 × 44	W 24 × 76	W 33 × 130
225,000 (25.4)	—	W 18 × 31	W 24 × 55	W 27 × 84	W 36 × 160
350,000 (39.6)	—	—	W 24 × 61	W 30 × 99	W 36 × 194
525,000 (59.3)	—	—	W 24 × 76	W 30 × 116	W 36 × 260

C = American Standard Channel; W = Wide-Flange Beam.

have provided guidelines on vibration severity. Suggestions from IRD Mechanalysis, Inc., "Audio-Visual Customer Training Manual," p. 10, IRD, Columbus, OH (1975), appear in Table 2.

Most designs for agitators fall within the 0.05–0.50 in./s (1.27–12.7 mm/s) range. Historically, the 0.20–0.30 in./s (5.08–7.62 mm/s) design range is a reasonable compromise between support structure cost and acceptable vibration levels. Each designer must apply these principles when designing an agitator support. Because torsional deflection of the support tends to control dynamic deflection, the minimum allowable lateral support stiffness tends to be a function of the agitator size. Smaller agitators require a less stiff support than do larger ones.

The prediction of the vibrational characteristics of structures can be quite complex. Nonetheless some of today's finite-element analysis (FEA) /structural software with modal-analysis capability can greatly simplify the design of agitator supports. The modal analysis of a head mounted agitator is shown in Figure 6. The mesh has been made large and the deflections exaggerated for visual effect. It is easily observed that the angular deflection source is at the point where the nozzle meets the vessel head. This points out the need for adequate head stiffness, especially in the vicinity of the nozzle.

There are basically only two types of support systems for tanks: a beam system support or a head mounted support. The most common support is the beam system which spans the top of the tank and is either supported from the vessel straight side or from a frame tied to ground. Table 3 provides guide-

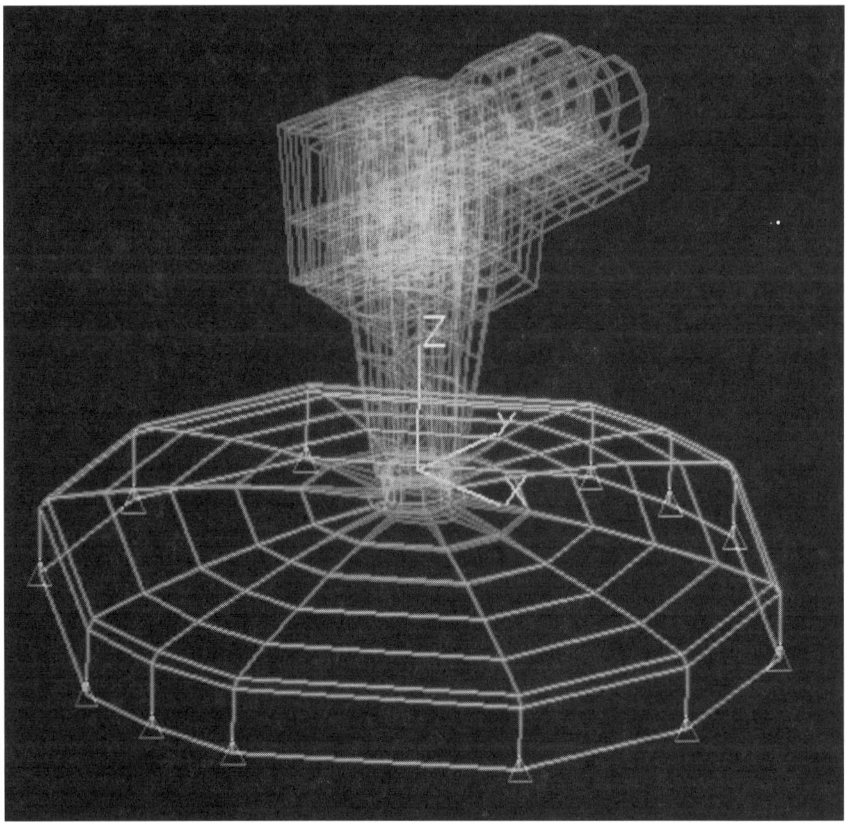

Figure 6. Modal analysis of a head mounted agitator.

lines for the overhead portion (Figure 7) of the beam support system for various size agitators and tanks. The vertical support system, which can vary enormously from application to application, should also be included in the design analysis.

Head-type supports can be subdivided into the nozzle type or the pad type. Tables 4 and 5 provide the recommended thicknesses t_r required to properly support an agitator; reinforcement of the nozzle/pad may be required. Table 4 provides the recommended course of action.

Sketches are provided in Figure 8 which describe the backup plates and gussets recommended for this type of support. Recommended head thicknesses for the nozzle support and pad support are provided in Tables 5 and 6, respectively. The agitator mounting nozzle/pad reinforcement dimensions can be found in Table 7.

These support recommendations are for agitators with typical bending loads for the invested torque. Under certain circumstances, bending loads can be considerably larger and may require a more rigid structure. It is up to the user to ensure that the agitator support structure meets the intended vibrational and strength criteria. The second sidebar presents an example of the effect of vessel-head thickness on vibration.

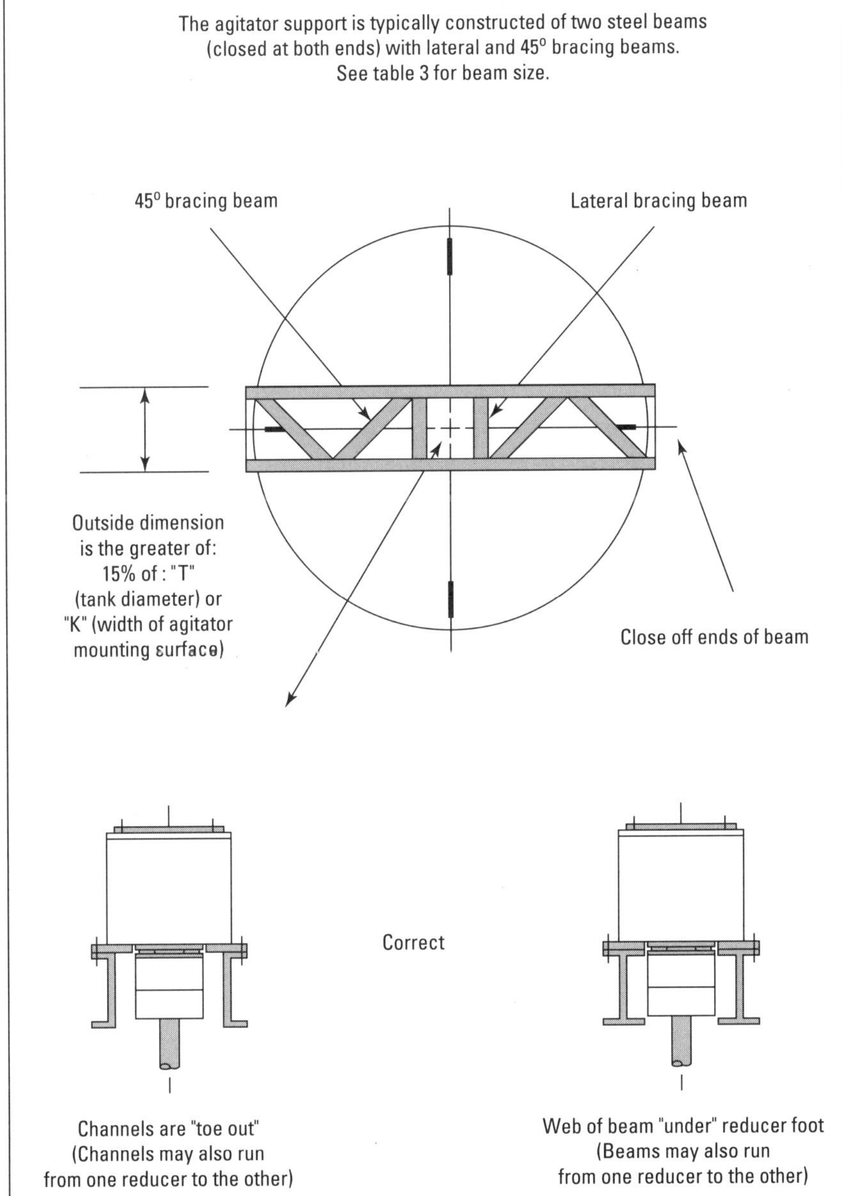

Figure 7. Correct overhead portion of the beam support system.

Mechanical seals

The requirements for sealing vessels have changed enormously over the last 20 years. The U.S. Environmental Protection Agency (EPA) is implementing the National Emissions Standard for Hazardous Air Pollutants under the authority of the 1990 Clean Air Act Amendments. EPA's goal is adequate containment of fugitive emissions. Mechanical seals are an effective way of eliminating or containing fugitive emissions *(5)*.

Mechanical seals can be classified as either pusher type or bellows type. Table 8 describes the major characteristics of these seals. Bellows seals are inherently balanced, and designs are available up to 300 psi (2.07 MPa) differential pressure. Because the secondary seal is not required to function dynamically with shaft deflection, bellows seals are generally quite good in agitator service.

However, pusher seals, which tend to be less costly, will most often perform the same function. And, as a consequence, pusher seals are by far the most common mechanical seal used in the chemical process industries (CPI).

The life span of a mechanical seal is dependent upon face pressure, face velocity, run-out, and contaminants. The greater any of these factors, the shorter the life of the seal. For most top-entering agitators, contamination can be safely eliminated, since the seal generally sees only the overhead vapor. If the seal materials are selected properly, the seal will preferentially wear out before corroding. Thus, if shaft speed is increased, either the face pressure or the run-out will have to be decreased to maintain the same seal life.

The most significant advancement in agitator mechanical seals in the last 20 years is the development of the dry-running seal. Dry-running seals are customarily good for 100–200 psi (0.689–1.38 MPa) up to 200-ft/min (1.02-m/s) face velocity, and run-outs of up to ⅛ in. (3.18 mm).

When the process can tolerate a minimal amount of overhead vapor leakage, these seals are often a very good alternative. Usually, either the rotating or stationary element must be carbon. When leakage of the overhead vapor cannot be tolerated, pressurized dual seals should be used. Dual, pressurized seals inject a barrier liquid into the seal chamber. Trace amounts of barrier fluid are then allowed to leak both into the vessel and past the outside seal. Barrier liquids are selected to be both environmentally safe and compatible with the process fluid(s).

Ramsey and Zoller *(2)* discuss the advantages and characteristics of a double, cartridge-type pressurized seal for agitators utilizing a bearing for run-out control. This type of cartridge design has been used to seal off pressures in excess of 1,000 psi (6.89 MPa).

Today, many off-the-shelf cartridge designs are available. These cartridges must be able to tolerate large seal deflections if they are to be used for agitator service. Most of these off-the-shelf designs, however, were designed for pump service, and neither contain a bearing, nor have run-out clearances large enough for agitator use. Therefore, caution should be exercised in using or specifying off-the-shelf cartridge seals for agitator service.

An example of the deflection which a mechanical seal without a closely mounted run-out control bearing must undergo is represented in Figure 9. Notice that sufficient clearance must be available between either the shaft or the shaft sleeve and the seat to accommodate this run-out. It should also be observed that the secondary seal must axially displace it-

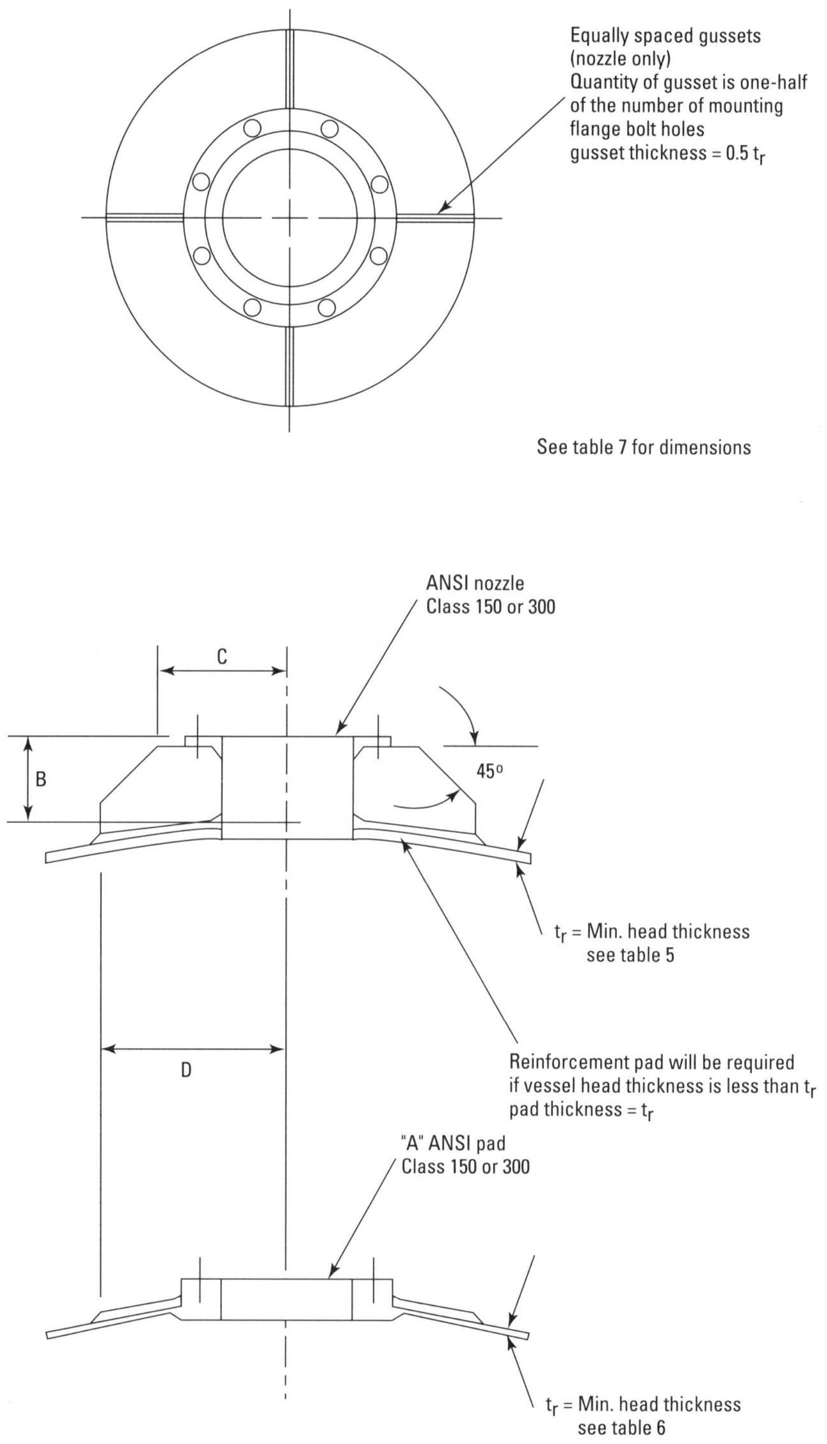

Figure 8. Recommended backup plates and gussets.

Table 4. Head-mounting recommendations.

If	Nozzle	Pad Reinforcement
$t \geq t_r$	No reinforcement required	No reinforcement required
$t_r/2 \leq t < t_r$	Add backup plate and gussets	Add backup plate
$t < t_r/2$	Increase head thickness to $t_r/2$ min., then add backup plate and gussets	
	Increase head thickness to $t_r/2$ min., then add backup plate	

Table 5. Recommended head thickness for nozzle-mounted agitators.

Agitator Torque, in.•lb$_f$ (KN•m)	ANSI Nozzle Size, in.	Vessel Dia., ft (m) 5 (1.52)	7 (2.13)	9 (2.74)	12 (3.66)	15 (4.57)
		Head Thickness, in. (mm)				
2,500 (0.280)	8	0.187 (4.76)	0.187 (4.76)	0.250 (6.35)	0.312 (7.92)	0.312 (7.92)
4,000 (0.450)	8	0.250 (6.35)	0.312 (7.92)	0.375 (9.53)	0.375 (9.53)	0.437"(11.1)
8,500 (0.960)	8	0.312 (7.92)	0.437 (9.53)	0.500 (12.7)	0.500 (12.7)	0.562 (14.3)
17,000 (1.90)	12	0.250 (6.35)	0.312 (7.92)	0.437 (11.1)	0.500 (12.7)	0.562 (14.3)
24,000 (2.70)	12	0.312 (7.92)	0.437 (11.1)	0.500 (12.7)	0.625 (15.9)	0.750 (19.1)
33,000 (3.70)	12	0.437 (11.1)	0.562 (14.3)	0.625 (15.9)	0.75 (19.1)	0.875 (22.2)
59,000 (6.70)	16	—	0.437 (11.1)	0.562 (14.3)	0.625 (15.9)	0.750 (19.1)
87,000 (9.80)	16	—	0.562 (14.3)	0.625 (15.9)	0.750 (19.1)	0.875 (22.2)
135,000 (15.3)	20	—	0.437 (11.1)	0.562 (14.3)	0.687 (17.5)	0.875 (22.2)
225,000 (25.4)	24	—	—	0.562 (14.3)	0.750 (19.1)	0.875 (22.2)
350,000 (39.6)	24	—	—	0.562 (14.3)	0.750 (19.1)	0.875 (22.2)
525,000 (59.3)	30	—	—	—	0.750 (19.1)	0.875 (22.2)
825,000 (93.2)	30	—	—	—	1.00 (25.4)	1.25 (31.7)

Table 6. Recommended head thickness for pad-mounted agitators.

Agitator Torque, in.•lb$_f$ (KN•m)	ANSI Pad Size, 150 lb.	Vessel Dia., ft (m) 5 (1.52)	7 (2.13)	9 (2.74)	12 (3.66)	15 (4.57)
		Head Thickness, in. (mm)				
2,500 (0.280)	8	0.125 (3.18)	0.125 (3.18)	0.125 (3.18)	0.125 (3.18)	0.187 (4.76)
4,000 (0.450)	8	0.125 (3.18)	0.187 (4.76)	0.187 (4.76)	0.250 (6.35)	0.312 (7.92)
8,500 (0.960)	8	0.187 (4.76)	0.250 (6.35)	0.250 (6.35)	0.312 (7.92)	0.375 (9.53)
17,000 (1.90)	12	0.187 (4.76)	0.187 (4.76)	0.250 (6.35)	0.250 (6.35)	0.375 (9.53)
24,000 (2.70)	12	0.250 (6.35)	0.250 (6.35)	0.312 (7.92)	0.375 (9.53)	0.437 (11.1)
33,000 (3.70)	12	0.250 (6.35)	0.312 (7.92)	0.375 (9.53)	0.437 (11.1)	0.500 (12.7)
59,000 (6.70)	16	—	0.250 (6.35)	0.312 (7.92)	0.375 (9.53)	0.437 (11.1)
87,000 (9.80)	16	—	0.312 (7.92)	0.375 (9.53)	0.437 (11.1)	0.500 (12.7)
135,000 (15.3)	20	—	0.312 (7.92)	0.375 (9.53)	0.437 (11.1)	0.500 (12.7)
225,000 (25.4)	24	—	—	0.437 (11.1)	0.437 (11.1)	0.562 (14.3)
350,000 (39.6)	24	—	—	0.375 (9.53)	0.437 (11.1)	0.562 (14.3)
525,000 (59.3)	30	—	—	—	0.500 (12.7)	0.625 (15.9)
825,000 (93.2)	30	—	—	—	0.625 (15.9)	0.750 (19.1)

*Vessel head thickness *t* may be reduced if the pad pressure class is greater than 150 lb.

self in the plane of bending to adjust to shaft bending.

The introduction of perfluoroelastomers, such as Kalrez, permits the use of O-ring secondary seals in corrosive or high-temperature (500°F, 260°C) applications. The O-ring can roll up and down on the shaft or shaft sleeve to accommodate shaft deflection. This has practically eliminated the gouging out of the shaft or shaft sleeve which occurs with polytetrafluoroethylene seal rings, and has significantly increased seal life in highly corrosive and high-temperature applications.

Table 7. Agitator mounting nozzle/pad.

ANSI Nozzle/ Pad Size, in.	B	C	D
	Reinforcement Dimensions, in. (mm)		
8	6 (152)	8 (203)	12 (305)
12	8 (203)	11.5 (292)	17 (432)
16	8 (203)	13.5 (343)	19 (483)
20	12 (305)	17.5 (445)	26 (660)
24	12 (305)	19.5 (495)	28 (711)
30	12 (305)	24 (610)	31 (787)

Table 8. Basic mechanical seal classifications.

Pusher Type

Types of Pressure Compensation	Methods of Loading	Types of Secondary Seal
Unbalanced	Single spring	O-ring
Balanced	Multiple spring	V-ring,
		Wedge ring

Bellows Type

Bellows Material Types	Types of Secondary Seal
Metal	O-ring
Elastomer	V-ring
Polytetrafluoroethylene (PTFE)	Wedge ring,
	Packing

Locking techniques for bolts and nuts

The very nature of the environment in a CPI vessel increases the difficulty of fasteners doing their job. Locking techniques have been used as insurance that fasteners will perform as intended, and, although locking techniques have not changed significantly in the last 20 years, our knowledge about the effectiveness of locking techniques has improved substantially. Many of the locking techniques used more than 20 years ago have been shown to be less effective than they were once thought to be. The most dramatic improvement has been in free-spinning-type locking fasteners.

Agitators impose a vibrational load on fasteners which can cause fasteners to loosen. The physics of vibrational loosening of fasteners is well covered by Junker *(6)*. Vibrational loosening will occur when there is relative motion between the thread flanks and the contact surfaces of clamping and clamped parts. This relative motion will not exist unless the transverse load can overcome the clamped friction force. Thus the basic rule for avoiding fastener loosening is to ensure that there is sufficient clamping load.

Predicting the effects of processing fluids on fasteners is beyond an exact calculation for most chemical processes. In these situations, use is recommended of fasteners with locking elements that tend to resist the internal off-torque which occurs when inherent

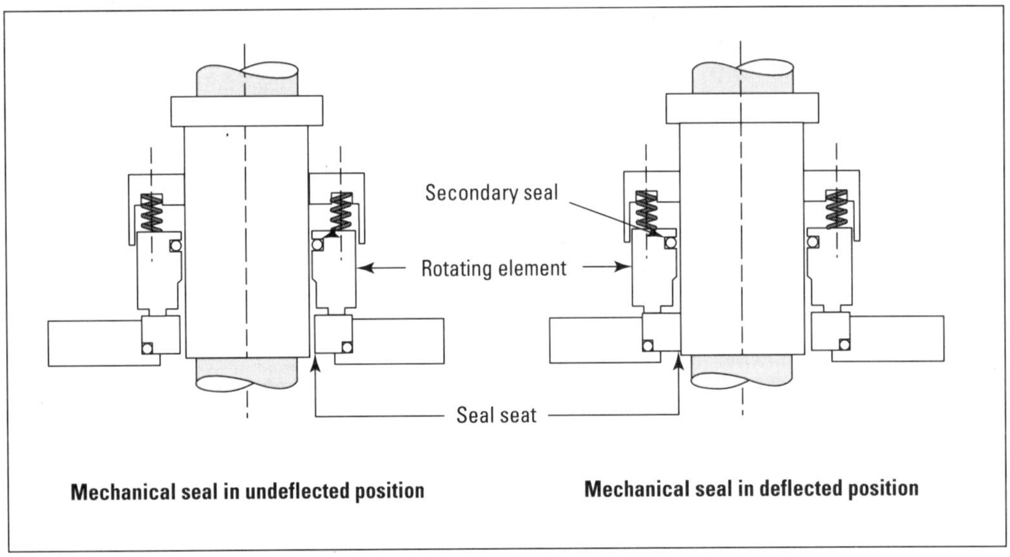

Figure 9. Deflection of a mechanical seal without a closely mounted run-out control bearing.

An Example of the Effect of Vessel Head Thickness on Vibration

When an agitator nozzle on the head of a tank is subjected to a static torque, bending moment, agitator weight, and internal pressure, there can be calculated a static stress at the nozzle-to-head junction. Under dynamic conditions, such as an actual agitator under operation, the bending moment is variable with a characteristic frequency equal to the rotational speed of the agitator. This forcing function causes the mass of the agitator to rock about a point near the nozzle-to-head interface. The continuous acceleration/deceleration of the agitator mass induces an additional load on the support structure, and the dynamic stress can become much larger than the static stress. The ratio of dynamic/static stress is dependent upon the magnitude of rocking and the speed of the agitator, or the vibrational velocity: $v = 2\pi Na/60$

The table shows an example of head thickness effects, comparing the vibration velocities for different head thicknesses. The added stress at the nozzle/head joint for the ¼-in. thick head would likely cause premature failure of the joint. The ½-in. thick head provides a vibrational velocity in the range normally found to be acceptable. And while the ¾-in. thick head is vibrationally very acceptable, it would be more expensive than the ½-in. thick head. The combined criteria of engineering acceptability and minimized cost suggests that the best choice is the ½-in. thick head.

The effects of head thickness on vibration.

Agitator Power, hp	10
Impeller Rotational Speed, rpm	84
Tank Head Style/Dia., ft	Elliptical/12
Bending Moment, in.-lbf	7,500
Distance: Tank Head to Top of Agitator, in.	54
Gussets/Backup Plate	None

At Top of Agitator.

Head Thickness, in.	Vibrational Velocity, in./s	Vibrational Deflection, in.
¼	0.97	0.11
½	0.25	0.028
¾	0.10	0.012

self-locking is lost. Bickford *(7)* suggests that locking devices can be categorized into six groups:

1. Anaerobic adhesives;

2. Spring- or tooth-type lock washers;

3. Lock wires, lock plates, or pins;

4. Prevailing torque fasteners;

5. Free-spinning lock nuts or bolts; and

6. Welded fasteners.

Group 1-type locking cannot be used universally in chemical process tanks because of chemical incompatibility. Group 2 devices require about five times as many cycles as a plain nut-and-bolt connection to loosen, all else being equal. Although they are of limited effectiveness because their stiffness is so much less than that of most fasteners, Group 2 devices are widely used because of their relatively low cost. Group 3 devices can effectively prevent the total loss of the nut, which may be extremely important, but they are not very effective in preventing a substantial loss of preload in the fastener.

Group 4 devices create an interference fit between the threads of the nut and bolt. Metal-to-metal interference fits create a usability problem, and many are single-use assemblies. Beam-type nuts can be reused if the beams formed by the crown slotting are not damaged. Beam nuts, however, are difficult to find in most CPI alloys. Some Group 4 devices have plastic inserts which create the interference, but have a very limited number of applications in the CPI because of chemical incompatibility.

Group 5, free-spinning types, are the most effective reusable type for preventing loosening. This group includes double nutting, which is the most cost-effective means of locking a nut-and-bolt assembly on an agitator. When the first nut is intended to develop the preload, two full nuts are used. The recommended practice is to apply the full preload torque to the first nut. Then, holding the first nut, tighten the second nut to the same full preload torque.

Also included in Group 5 are the free-spinning forms. These are bolts which can be freely run down into tapped holes or nuts, and only create an interference when they are torqued down. An option for preloaded connections is the serrated nut and bolt which provides serrations that dig into the joint surface and resist rotation. Serrated nuts and bolts are difficult to find in many CPI alloys, though.

Group 6, the locking of fasteners by welding, can be accomplished by either welding the bolt head and nut to the clamped pieces, or welding the nut to the bolt. This is the most secure loosening-prevention technique available. Removal and replacement are the major drawbacks. Grinding out the welds can be labor-intensive and difficult without damaging the fastener components. Often, the fasteners must be replaced.

Materials of construction

New alloys are relatively slow to impact on agitator design and manufacture. When new alloys become available in the marketplace, the form availability is limited. It is generally difficult to find bars of sufficient diameter and length for agitator service. If, after several years, the alloy has proven its worth in the marketplace, this situation changes and the appropriate forms become available.

This should be taken into account when specifying the materials of construction of an agitator. The most difficult components to obtain are the fasteners. Alloys must have established themselves well in the CPI before fasteners are available. If fasteners in the desired alloy cannot be found, it is more economical to use an off-the-shelf fastener in a galvanically more-corrosion-resistant material.

All alloys have American Society for Testing and Materials (ASTM), or American Iron and Steel Institute (AISI) compositional ranges. Alloy manufacturers today can often hold the composition of an alloying ingredient close to the low end of the range. This provides for a much higher degree of consistency, but often does not provide the corrosion resistance of the identical material it may be replacing.

For example Type 317L stainless steel has an allowable molybdenum range of 3–4%. If you were to buy this material today, you may find the composition in 99 out of 100 pieces to lie in the 3–3.25% range. About 20 years ago, an agitator made out of 317L would, on the average, have had a molybdenum content closer to 3.5%. On the surface, this compositional difference may not seem significant, but molybdenum plays a very strong role in chloride corrosion resistance and a reduced amount of this element could dramatically shorten the life of the agitator. Many designers are aware of this already and are specifying more-corrosion-resistant alloys. All engineers should be aware of this marketplace reality.

The largest impact for materials has been in the area of dual-phase stainless alloys. The two most popular in agitator service are the duplex stainless steels Ferralium 255 and Alloy 2205. They are about 30% stronger than Type 316 stainless steel and approximately equivalent to 904L stainless steel in corrosion resistance. High chromium and molybdenum contents provide the duplexes with excellent pitting and crevice-corrosion resistance, particularly in chloride environments *(8,9,10)*. Duplex stainlesses are limited in ductility and toughness and require special care in welding.

Relatively new are the 6%-molybdenum superaustenitic stainlesses which possess strengths of the duplex stainlesses and corrosion resistances about midway between the duplexes and the high-nickel-based alloys. Welding is not a problem with molybdenum-enriched filler metals. Examples of these are AL6XN and 254 SMO. Sorell *(11)*, Kane *(12)*, and Whitcraft *(13)* present concise reviews of the various alloy groups.

Knowledge-based systems

The mechanical design issues discussed above as well as other mechanical criteria can be incorporated into a KBS. The advantages of a KBS in agitator design are discussed by Bakker, *et al. (14)*. A key point to remember is that an agitator design is an iterative combination of: (1) desired process results; (2) mechanical requirements; and (3) cost. There are very few individuals who have the capability of combining these three elements effectively, whereas, a KBS has the ability of doing this faster and more accurately than any individual. An engineer using a KBS can design a mechanically sound agitator which not only will provide the required process result, but will do so at the lowest cost. CEP

J. B. FASANO is director, parts and field service for Chemineer, Inc., Dayton, OH (513/454-3263; Fax: 513/454-3379). He has been with Chemineer for 24 years, with primary emphasis on R&D and custom equipment design. From 1984 to 1995, he served as technical director. He holds a BSc in chemical engineering from the University of Dayton, an MSc in chemical engineering from Lehigh University, an MBA from the University of Dayton, and is currently pursuing a doctorate in materials engineering, also from the University of Dayton. Fasano is a PE in Ohio. He is a member of AIChE.

J. L. MILLER is principal mechanical engineer for Chemineer, Inc., Dayton, OH. He has been performing mechanical engineering research and design work for Chemineer for over 10 years, currently in the Product Engineering Department. Previously, he was employed by Copeland Corp. for five years as a research engineer. Miller graduated from Purdue University with a BSME and MSME, and is a PE in Ohio.

S. A. PASLEY is manager, product engineering for Chemineer, Inc., Dayton, OH. He has been with Chemineer since 1977, and has held positions as corporate engineer, supervisor of design and drafting, senior product engineer, and senior applications engineer. From 1974 to 1977, he was a mechanical engineer for Du Pont. Pasley holds BSME and MSME degrees from Purdue University.

Literature Cited

1. **Hill, R. S., and D. L. Kime,** "How to Specify Drive Trains for Turbine Agitators," *Chem. Eng.*, pp. 89–94 (Aug. 2, 1976).
2. **Ramsey, W. D., and G. C. Zoller,** "How the Design of Shafts, Seals and Impellers Affects Agitator Performance," *Chem. Eng.*, pp. 101–108 (Aug. 30, 1976).
3. **Pestel, E. C., and F. A. Leckie,** "Matrix Methods in Elastomechanics," McGraw-Hill, New York (1963).
4. **Fasano, J. B., *et al.*,** "Advanced Impeller Geometry Boosts Liquid Agitation," *Chem. Eng.*, **101**(8), p. 110 (Aug. 1994).
5. **Adams, William V.,** "Stop Leaks Economically with Advanced Mechanical Seals," *Chem. Eng. Progress,* **90**(6), pp. 24–31 (June 1994).
6. **Junker, Gerhard H.,** "New Criteria for Self-Loosening of Fasteners Under Vibration," Society of Automotive Engineers, *SAE Trans. 1969,* No. 690055, SAE, Warrendale, PA.
7. **Bickford, John H.,** "An Introduction to the Design and Behavior of Bolted Joints," Marcel Dekker, New York (1981).
8. **Redmond, James D.,** "Part 1: Selecting Second-Generation Duplex Stainless Steels," *Chem. Eng.*, **93**(20), pp. 153–155 (Oct. 27, 1986).
9. **Redmond, James D.,** "Part 2: Selecting Second-Generation Duplex Stainless Steels," *Chem. Eng.*, **93**(22), pp. 103–105 (Nov. 24, 1986).
10. **Quik, J. M. A., and M. Geudeke,** "Switch to Duplex Stainless Steels," *Chem. Eng. Progress,* **90**(11), p. 49 (Nov. 1994).
11. **Sorell, G.,** "Choose the Best Alloy for Incinerator Heat Exchangers," *Chem. Eng. Progress,* **90**(3), pp. 49–60 (Mar. 1994).
12. **Kane, R. D.,** "Super Stainless Steels Resist Hostile Environments," *Advanced Materials and Processes,* **144**(1), pp. 16–20 (July 1993).
13. **Whitcraft, P. K.,** "Try Aerospace Specialty Alloys in the CPI Plant," *Chem. Eng. Progress,* **91**(5), p. 59 (May 1995).
14. **Bakker, A., *et al.*,** "Computerizing the Steps of Mixer Selection," *Chem. Eng.*, **101**(3), pp. 120–129 (Mar. 1994).

The Potential of Reaction Engineering

Emerging developments promise to make a major impact on chemical engineering. Among these are short-contact-time reactors that improve the cracking of feedstreams, and techniques that promise better ceramics and more-effective hazardous-emissions cleanup.

L.-S. Fan, J. Rathman, and A. Ghosh-Dastidar,
The Ohio State University

Alan W. Weimer,
Dow Chemical, USA

Shoichi Kimura,
Oregon State University

Particulates reaction engineering encompasses a wide variety of gas, liquid, and gas/liquid reactions with particulates. Reactions can be either catalytic or noncatalytic, and the particles can vary from nanometer to millimeter in size. New areas of application continually are being discovered.

In this article, we will explore the engineering challenges — involving complex fluid/particle and particle/particle interactions — that are typical of the particulate reaction field. We will concentrate on four key areas — ultrafast reactions, ceramics production using fluidized beds, aerosol reactions, and colloidal reactions — because they exemplify the nature of these challenges. Suffice it to say, though, that other areas are equally significant in terms of ongoing research and potential commercial applications.

Four key areas

The field of ultrafast particulates reaction engineering has developed rapidly over the past few years. A number of researchers are working on gas/solid, and gas/atomized liquid reactions occurring over a millisecond (ms) timescale, and with the particulates typically in the 1–100 μm size range. Applications stretch from pollution control to catalytic cracking. For example, in controlling SO_2 emissions from coal-fired boilers, a viable low-cost technology is furnace-sorbent-injection (FSI), in which calcium hydroxide or carbonate particles (<10 μm) are injected into the upper-furnace region of the boiler to capture SO_2. Even though the residence time of the injected sorbent in the favorable reaction temperature window (800–1,200°C) varies from 1–2 s, the reactions are much faster, and most of the SO_2 capture takes place in the initial 100–200 ms *(1)*. Therefore, to improve this sorbent injection technology, it is essential to understand the rapid and complex SO_2/sorbent interaction that involves three overlapping mechanisms, namely, calcination, sintering, and sulfation.

Some other important applications of ultrafast reaction engineering can be found in the petroleum refining and petrochemical industries. For instance, in fluidized catalytic cracking (FCC), appreciable selectivity differences in final products stem from small differences in the distribution of residence times, and residence time itself. Optimum reaction times for obtaining desirable products may lie well below 100 ms, while conventional fluidized beds have much longer residence times. Another potential application is ultrafast pyrolysis (ultrapyrolysis) of heavy oil — cheap carbonaceous fuels such as biomass, high volatile coals, petroleum residuals, and tar sand bitumen can be converted to high-value-added feedstocks, usable by conventional refineries to produce transportation fuels and petrochemicals. Desirable residence times for such high-temperature pyrolysis typically are between 70–500 ms. Short-contact-time fluidized reactors (SCTFRs), which are similar to the entrained-flow reactors (EFRs) used for SO_2/sorbent studies *(2)*, normally are used for ultrafast reactions.

Ceramic powder production is also an important application of particulates reaction engineering. Various methods have been developed to produce fine ceramic powders; these may be classified into three groups, namely, solid-phase, liquid-phase, and vapor-phase synthesis. A typical commercial example of solid-phase synthesis is the carbothermal reduction of metal oxides to produce carbide ceramics, such as silicon carbide, and boron carbide. Nitride ceramics, including silicon nitride, aluminum nitride, and boron nitride, are synthesized by introducing gaseous reactants containing nitrogen or ammonia in the system to nitride reduced metal. To produce submicrometer powders, solids usually must be pulverized after synthesis. Residence time in the reactor for solid-phase synthesis is fairly long, on the order of hours. Current commercial processes primarily are of the batch type.

Synthesis of monosize submicrometer ceramic powders also is carried out by various liquid-phase synthesis processes. Solution precipitation is a commercial method to produce silicon nitride from silicon tetrachloride dissolved in a solvent and ammonia. The intermediate product, silicon diimide, is thermally decomposed into silicon nitride. Alkoxide hydrolysis to produce high-quality metal oxides, such as alumina, silica, and titania, is another application of precipitation. An emulsion method, which employs aqueous solutions of water-soluble ceramic precursors emulsified with organic fluids, also is used to produce complex ceramic compounds. Titanates, oxides, and superconductors have been prepared experimentally by evaporating water from an emulsion, then drying, and calcining the slurry. In addition, the sol-gel method is used for liquid-phase synthesis of ceramic powders.

In "aerosol synthesis," product solids are precipitated from vapor-phase chemical reactions. Precipitates from gases may form films, whiskers, and powders. Various types of reactors, such as furnace reactors, flame reactors, laser reactors, or plasma reactors, are used; these are heated in different ways and also operated at different temperatures. Residence time differs significantly from one reactor type to another — typically ranging from <2 ms in a laser-beam reactor to tens of ms in a flame reactor to seconds for furnace reactors.

Another important area is colloidal reaction engineering, which includes the application of very fine powders, finely subdivided multiphase systems (for instance, micelles, microemulsions, soluble polymers, and aerosols), films, fibers, foams, pores, surface irregularities, and large molecules. Many of the factors that are important in these "microheterogeneous" systems are the same as those considered in conventional heterogeneous reaction systems. There are several other factors, however, that are unique to colloids, and that often are exploited in the reaction processes.

Now, let's look at each of these areas in more detail.

Ultrafast reactions

In the SCTFRs commonly employed for ultrafast particulates reactions, one of several important design considerations is the provision of a region for rapid mixing between particulate and gas phases, thereby reducing the particle heating time (to <5–10 ms) to the desired reaction temperature. Because typical reaction times in such reactors range from <100 ms to a few hundred ms, any delay in particle heat up will adversely affect the reaction selectivities. And, in kinetics experiments, any such delay will result in an underestimation of the reaction rate. To achieve the necessary rapid mixing, most reactors use jet impaction. In the mixing region (mixor), a number of jets of gaseous or particulate feedstock (liquid or solid) are impacted onto jets of a heated gaseous or particulate reactant (thermofor) at high velocity. Such impaction causes nearly instantaneous mixing, resulting in rapid heat and mass transfer at the points of jet impaction.

After intense mixing, the reacting gas/particulates enter a quasi-plug-flow, transported reactor (downer). The bulk of the reaction takes place in this reaction zone, with desired products formed in a few hundred ms. For optimum selectivity, strict control of the residence time is required. Reactor systems are designed so that it is easy to alter their residence times, so the units can be used for different products and selectivities. Residence-time change is achieved either by varying the reactant feeding rates or, more conveniently, by shortening or lengthening the reaction zone by moving a gas/solid separator up and down the reactor.

The last stage of an ultrafast reactor is the phase separator, in which the reaction is quenched. Extremely rapid quenching is essential to halt any further reaction, which, if not stopped, certainly would alter the selectivity, and thereby degrade product quality. In the case of kinetic studies, additional solid conversion in the sampler or separator would result in an overestimation of the reaction rate. Therefore, quenching time is short — typically varying between <5 ms and approximately 30 ms, depending on the quench mechanisms and process requirements.

Now, let's examine two examples of ultrafast reactors and their applications.

Ultrafast kinetics of SO_2/sorbent reaction. In FSI technology, calcium-based sorbents, such as $Ca(OH)_2$ and $CaCO_3$, are injected into the upper-furnace region of a boiler to capture SO_2 from flue gas. The injected sorbents (<10 μm) rapidly calcine at high temperatures to produce highly porous CaO, which, in the oxidizing environment of the furnace, reacts with SO_2 to form $CaSO_4$. Sintering occurs simultaneously, reducing the surface area and pore volume of the reactant CaO. Equilibrium considerations favor the formation of $CaSO_4$ in a temperature window of 850–1,200°C, in which the sorbents typically spend about 1–2 s. All the mechanisms,

Shipping Instructions for BOOKSRUN

Print this Packing Slip and enclose inside the front cover of the book.

Please ship this item no later than Tue Sep 22, 2026.

Ship to:

ALIBRIS APEX DC 76901962-35
APEX
800 AVONDALE AVE.
GRANDVIEW HEIGHTS, OH 43212-3473
UNITED STATES

PN #	Item ID	Alibris ID	Media Type	Title / Author	Seller List Price	Order Date
76901962-35	Plant Operation M-20	B084276229	BOOK	Plant Operation and Optimization (Practical Engineering Perspectives) Robertson, J. L. and Nalven, G **Very good 1996 hardcover no dust jacket as issued name in book Volume Q-2 top edge of pages has some age spotting text is clean and free of markings 371 pages M-20.**	$63.99	Sep, 10 2026

76901962-35

however, are extremely fast under furnace conditions, and the bulk of the reaction takes place within a few hundred ms of sorbent injection *(1)*.

A comprehensive theory capable of explaining these highly complex and ultrafast phenomena has yet to be developed, because of the experimental difficulties associated with studying such reactions. Obtaining short-time-scale kinetic data, however, is essential for complete understanding of this sorbent/SO_2 interaction, and thus for improving the FSI process. Realizing the importance of short-time kinetic data, a host of researchers have employed EFR systems to perform SO_2/sorbent reactions. One of the most sophisticated reactor systems (Figure 1), which has the capability of accurately quantifying the residence time of the reacting solid, has been developed by Raghunathan and coworkers *(3)*.

Their EFR consists of a reactor tube assembly within a high-temperature furnace, a powder feeder, water-cooled injection and collection probes with microcomputer-interfaced data acquisition systems, and a particle separator/classifier. Reaction temperature is maintained by a three-zone, 1,500°C vertical furnace; the temperature of each zone is controlled independently by temperature controllers. While traveling up the annular region, the reactant gas is heated to the reaction temperature and enters into the reaction zone through an entrance block. Sorbent powder is injected into the incoming hot gas stream just above the entrance block. The hot impinging gas stream effectively destroys the momentum of the gas/particle stream, causing severe turbulence — instantaneous mixing and particle heat up are achieved. The gas/solid mixture is quenched very rapidly with a stream of cold N_2 purge gas. It then passes through a cascade cyclone sampler that provides *in situ* classifications of the samples in various size ranges. An optical system provides on-line measurement of residence time, and avoids the error associated with gas-flow approximation for ms time scales. The residence time of the sorbent is determined on-line from a cross-correlation of two signals *(3,4)*. Residence times as short as 10 ms can be measured with great accuracy.

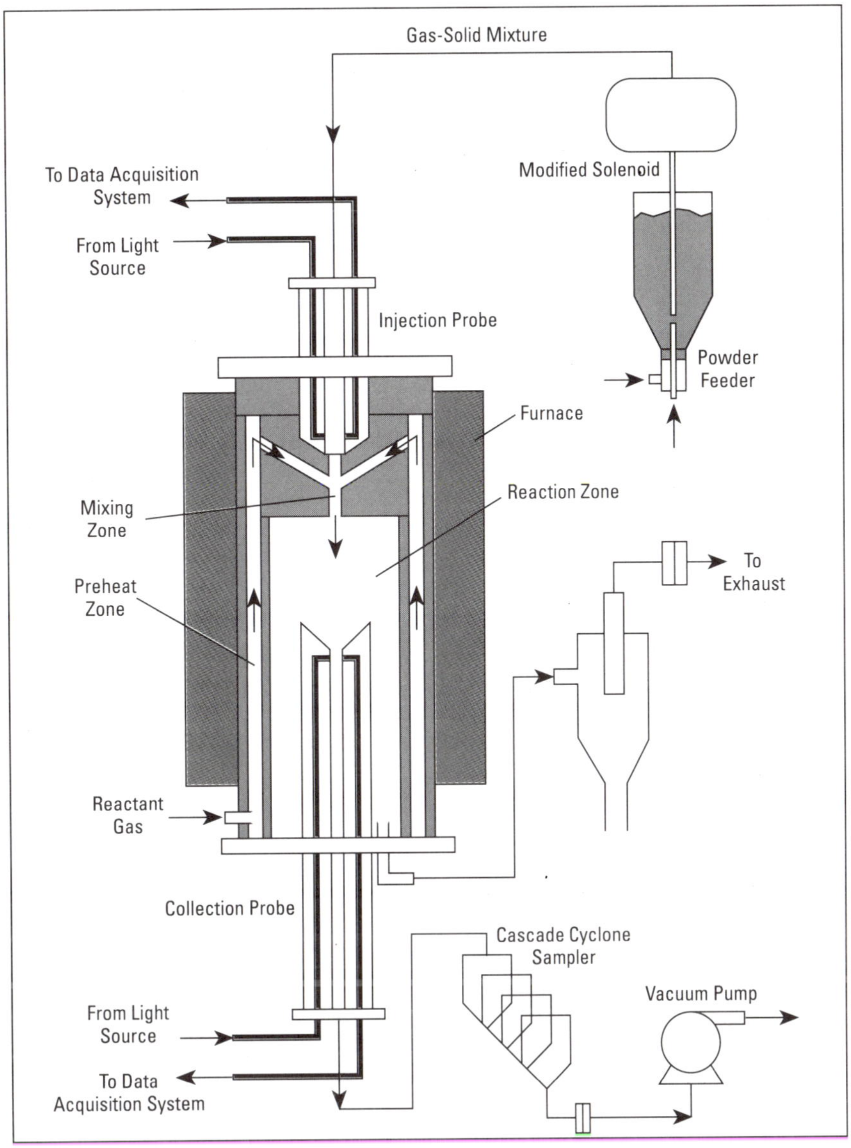

■ ***Figure 1. Schematic of high-temperature EFR (4).***

Using such a reactor system, short-time calcination and sulfation data for $Ca(OH)_2$ particles have been obtained by Ghosh-Dastidar and coworkers *(5)*. In Figure 2, $Ca(OH)_2$ conversion to $CaSO_4$ (SO_2 = 3,900 ppm, Ca/S = very low) is plotted against residence time ranging from <10 ms to about 300 ms. Multiple data points obtained for the <50 and 100 ms time scales for this reaction and also for $Ca(OH)_2$ calcination and CaO sulfation have led to insights into the complex, ultrafast SO_2/sorbent interaction phenomena. Short-time data, acquired for pore volume and surface area changes during calcination, also have shown the opposing effects of calcination and sintering on pore structure development.

Ultrapyrolysis of heavy feedstocks. A promising alternative to hydroliquefaction of heavy carbonaceous fuels is ultrapyrolysis of these feedstocks. Experimental results indicate that the process has tremendous potential for upgrading fuels such as biomass, high volatile coals, petroleum residuals, and tar sand bitumen, and obtaining favorable distribution of gaseous intermediates.

There are several advantages of ultrarapid pyrolysis over conventional methods. It is well known that the molecular structure of natural, heavy carbonaceous fuels is not strongly cross-linked. In fast pyrolysis, the feedstocks disintegrate into smaller reactive fragments by a thermal shock before the structure has time to become cross-linked and refractory, as they could with slow pyrolysis, thus leading to a lot of undesirable tar and coke. Other advantages of ultrapyrolysis are compactness of equipment due to fast kinetics, and the variety of feedstocks that can be handled.

Conventional fluidized beds are not well suited for high-temperature fast reactions, because the mixing times for particulates in these reactors are too long for reactions occurring over the ms scale. Residence times of the vapor products also are large, which causes thermal degradation of the products.

Vogiatzis and coworkers *(6)* have built a mini pilot plant on the general concept of a SCTFR to study the pyrolysis of Cold Lake bitumen feedstock over a range of temperatures (500–900°C) and residence times (70–500 ms). Hot nitrogen is used as the thermofor, which impacts on atomized liquid feedstock (preheated to 200°C) in the mixor. After the reactor section, the product stream is rapidly quenched with a cold N_2 stream to <300°C, a temperature at which no significant reaction takes place. Results suggest that a more-desirable product distribution, giving lower solid-residue yields and higher concentrations of olefins in the total gas yield, is obtained at short residence times (<150 ms) and high temperatures (>800°C). The product liquid can be recycled, thereby increasing the yield of olefinic compounds. A moderate removal of sulfur as hydrogen sulfide gas also is achieved within such a short pyrolysis time. If the reaction proceeds for longer residence times, however, the product gas quality deteriorates because of secondary cracking of the C_3–C_4 olefins.

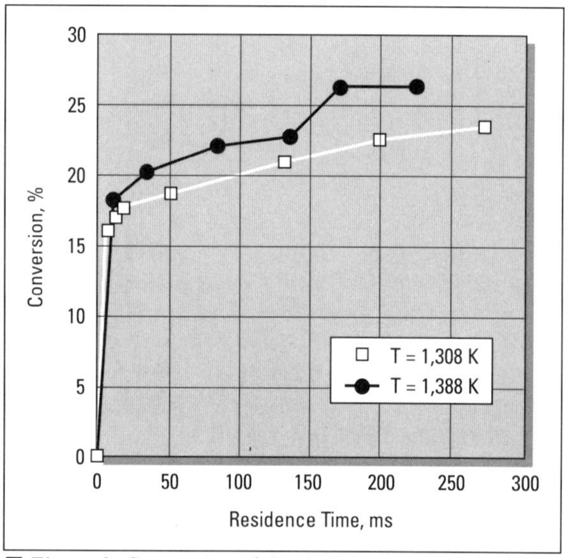

■ ***Figure 2. Conversion of $Ca(OH)_2$ vs. residence times during sulfation at different temperatures (5).***

Ceramic powder production

There has been growing interest recently in the production of nonoxide ceramics, such as nitrides and carbides. Traditional production methods rely on batch systems. Several new processes that apply fluidized-bed reactors (FBRs) appear to be ideally suited to continuously produce the powders.

In spite of the many advantages of FBRs for carrying out gas/solid reactions, there are some problems associated with such systems for the production of ceramic powders. The most difficult problem stems from the fine powder's cohesivity, which causes poor fluidizability. As a result, some fines tend to agglomerate and form clusters, leading to the collapse of the fluidized bed into a packed bed. The size reduction required in preparing fine ceramic powders from fluidizable particles still is a major element of total production cost. Therefore, for further cost reduction, it is essential to discover new fluidization methods that do not require size reduction.

As we will discuss, state-of-the-art technologies now are tackling such problems.

Silicon nitride. Si_3N_4 is expected to be the prime material for making components that can operate under severe mechanical and thermal stress conditions. It can be amorphous or crystalline, and crystals can exist in two forms, α and β. For high-temperature applications, however, the α-form is the most desirable raw material.

Direct nitridation in batches using fixed beds is one of the commercial methods for producing α-silicon nitride in large quantities. The exothermic nitridation reaction, which takes place at an optimum temperature in the range of 1,200–1,400°C, may be controlled so that the product is predominantly α-silicon nitride. (It takes at least 24 h to convert 99% of the silicon — higher reaction temperature shortens the reaction time, but the reaction becomes more difficult to control because of the high heat of the nitridation reaction.) The quality, however, is not uniform — often resulting in levels of the α-phase varying from batch to batch, and even from location to location in a particular batch. The batch systems also inherently suffer from difficulties in automating the handling of raw and product powders, labor intensiveness, and a long heating and cooling period — all of which lead to an increase in production cost.

A variety of methods, including the application of a vertical furnace, a pusher-type tunnel furnace, and a rotary kiln, have been proposed to solve these problems. None of these, however, can simultaneously solve the problems of production efficiency and quality of silicon nitride. Recently, a process applying fluidized beds has been introduced for continuous production of silicon nitride *(7)*.

In the fluidized-bed process, porous silicon particles of several hundred μm in size are prepared from partial sintering of μm-size silicon grains with the aid of a suitable binder such as polyvinyl alcohol. To prevent the agglomeration of these silicon particles in the FBR, silicon nitride particles of the same size are mixed at

10–30 weight % with porous silicon before introduction into the reactor.

The experimental data of Figure 3 for the FBR show how the nitridation reaction is controlled by programming the reaction temperature *(8)*. The reaction first is initiated at 1,300°C and, when the overall conversion of silicon reaches 50%, the temperature is raised to 1,390°C. Since the surface of silicon is coated with silicon nitride, particles are well fluidized even at a temperature close to the melting point of silicon (1,400°C), and nitridation is completed in a much shorter time. Because of the seed effect, the content of α-phase in the final product remains the same as that in silicon nitride produced at the constant temperature at which the reaction is initiated.

Silicon carbide. This is another material that is used for high temperature engineering applications. The crystal exists in two forms, namely, α and β. All current silicon carbide production relies on the Acheson electric furnace process. This is a batch process involving the charge-resistance heating of a mixture of silica sand and solid carbon to a temperature of around 2,000°C. It suffers the same problems, such as nonuniform product distribution from very large crystals to fine grains (even containing some amount of unreacted raw materials), as those in silicon nitride production. Very recently, a novel process that essentially is a continuous version of Acheson process utilizing a unique fluidizing technique has been introduced *(9)*.

Silica sand and petroleum coke are the raw materials. Free flowing granular solids are introduced continuously into a closed vessel. A nonreacting gas, such as nitrogen, is used to fluidize the mixture particles. An electric potential is applied between the vessel and a centrally positioned electrode to heat the contents of the fluidized bed to a temperature of around 1,900°C. (See Figure 4.) The relative feed rates of silica and coke, as well as the retention time of the mixture, are adjusted so that the content of silicon carbide in the product is about 95%, with the balance mainly free carbon.

The product silicon carbide, generally free flowing grains smaller than 3 mm in average size, is removed continuously from the fluidized-bed vessel. It undergoes various conventional post treatments for carbon removal, size reduction, and chemical purification.

Cobalt/tungsten-carbide nanophase composite. Nanophase materials, the size of whose building blocks is on the order of nanometers (1 nm = 10 Å), have gained attention from material scientists and engineers because the materials take on a rich variety of properties if the arrangement of the matter is controlled on a very small scale. Nanophase composites are made of different nanophases of materials or structures, and may be ceramic/ceramic, metal/ceramic, and ceramic/polymer. A number of nanophase composites, including a metal/ceramic nanocomposite used for making cutting tools and other wear-resistant devices, are commercially produced. One of the recent advances of particulate reaction engineering is producing nanophase composites using a fluidized-bed process.

The traditional method for making

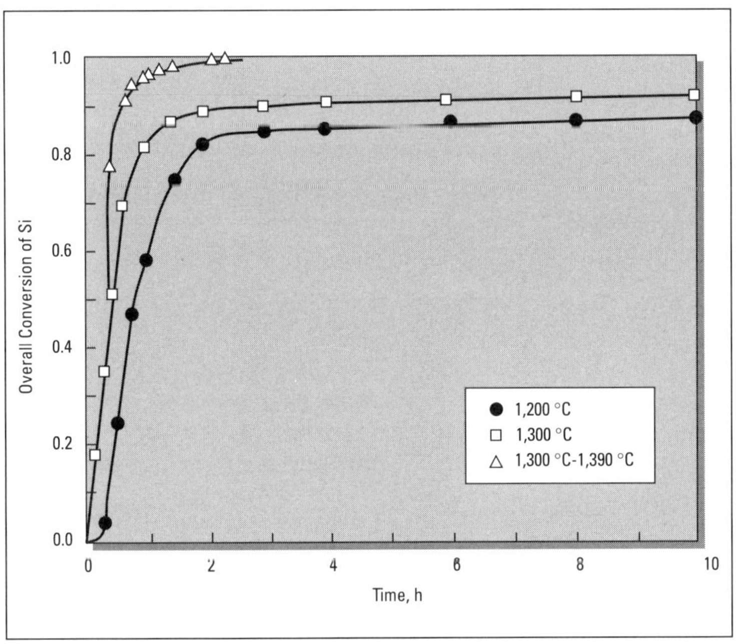

Figure 3. (top) Formation of silicon nitride vs. reaction time at different temperatures (7).

Figure 4. (right) Schematic of FBR used in the production of silicon carbide composite (9).

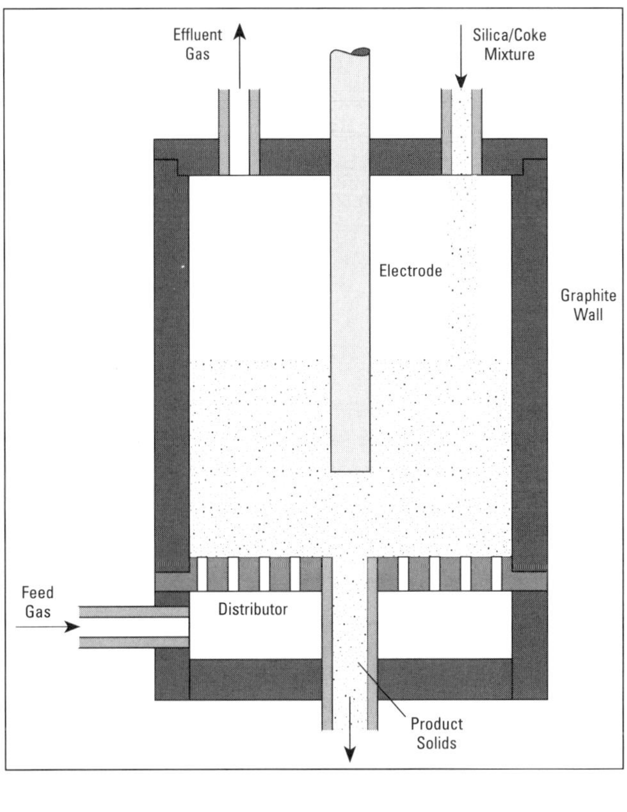

cobalt/tungsten-carbide nanocomposite has been crushing, grinding, blending, and consolidating tungsten carbide and cobalt powders. The minimum grain size of the final product is about 500 nm. In contrast, a new manufacturing method using a spray conversion process *(10)* allows the tungsten and cobalt grains to be smaller than 50 nm in diameter — leading to a stronger and tougher nanophase structure.

The new process, illustrated in Figure 5, starts by dissolving salts of tungsten and cobalt, such as ammonium metatungstate, and cobalt nitrate, in an aqueous solution. This solution is spray dried to produce a fine precursor powder of tungsten mixed at the molecular level. The powder, having sizes on the order of tens of μm or smaller, then is treated in an FBR operated batchwise to form the nanophase cobalt/tungsten-carbide composite.

The conversion process consists of several steps, with the reaction temperature going up in each step:

1. Decomposition. Water and ammonia first are removed from the powder while it is heated up to 150°C.

2. Calcination. The powder is dehydrated in the temperature range of 150–260°C with evolution of vapor, yielding nanodisperse mixed-metal oxides.

3. Reduction. The mixed-metal oxides are reduced with hydrogen, yielding nanodisperse alloy of tungsten and cobalt.

4. Carburization. The tungsten-cobalt alloy powder is carburized with a CO-CO_2 gas stream, forming nanodisperse cobalt/tungsten-carbide grains of several tens of nanometers or smaller in size, incorporated into larger spherical particles 5–50 μm in diameter. The formation of impurity tungsten oxide or mixed W-Co carbides is avoided by controlling the CO/CO_2 ratio. The large heat-transfer coefficient in the fluidized bed allows for good control of the reaction temperature, which is especially important in this step.

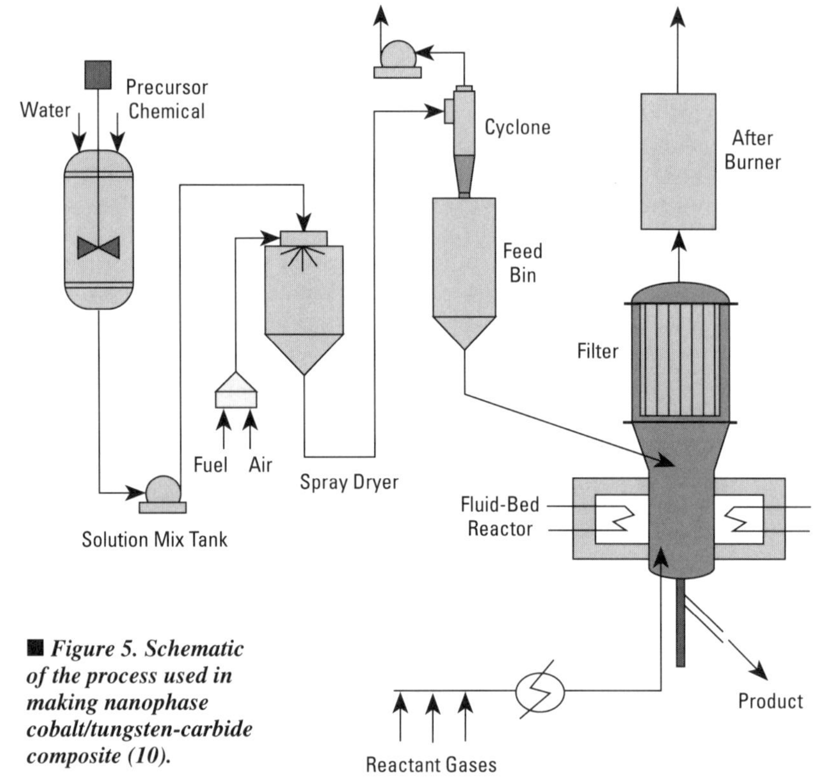

■ ***Figure 5. Schematic of the process used in making nanophase cobalt/tungsten-carbide composite (10).***

Aerosol processes

The synthesis of fine ceramic particles by aerosol processes *(11)* consists of rapid heating and reaction of gas/vapor (gas-to-particle conversion) or particle/droplet (particle-to-particle conversion) at high temperatures in dilute-phase flowing systems. Because the reaction time is short at high temperatures, it is possible to achieve fast reaction rates while minimizing particle growth.

In the gas-to-particle conversion route, particles are synthesized from individual molecules in the gas phase. The particle synthesis process is driven by the generation of molecules by chemical reaction from precursor gases or by the rapid cooling of a superheated vapor. In the particle-to-particle conversion process, particles or droplets containing particulate or molecular precursors are generated for subsequent reaction. The generation of precursor particles can be done independently of the reaction step. One commonly used method for synthesizing dried precursor particles directly is spray drying. In this process, precursors containing organic or inorganic reactants are prepared by the rapid evaporation of solvent from droplets generated by rotary, pressure, ultrasonic, or other types of atomizers. The synthesized precursor particles are typically of the 1–150 μm size range. Then, in an aerosol reaction step, these particles are dispersed and reacted in a flowing gas stream to form products. Other methods for preparing fine precursor powders include crystallization or standard drying methods, followed by a grinding step to reduce particle size.

Large-volume aerosol processes are being used commercially for the manufacture of carbon black and titania, and currently are being investigated for making various carbide, boride, and nitride nonoxide ceramic powders. These developmental processes promise either to synthesize intimately mixed ultrafine powder precursors, which can be reacted further by conventional methods to products, or to synthesize nonoxide powders directly by driving rapid reactions to completion.

To illustrate the diverse material-synthesis reactions that can be carried out in aerosol reactors, we will look at one gas-to-particle conversion route, and a particle-to-particle conversion.

SiC precursors. A high quality β-SiC powder can be obtained by heating a homogeneous C/SiO_2 powder precursor from an upstream aerosol reaction process *(12,13,14)*. The precursor is prepared by spraying a mixture of $SiCl_4$ and heavy oil into a fuel-rich air/propane flame. Fumed SiO_2 is produced by the reaction:

$$SiCl_4 + 2H_2O \Rightarrow SiO_2 + 4\ HCl$$

Fine carbon is generated simultaneously by pyrolysis of the heavy oil injected into the oxygen-containing flame. The aerosol reactor (shown in Figure 6) is a refractory-lined cylindrical tube, which is heated to a temperature of approximately 1,700°C. The fine aerosol mixture of SiO_2 and C is collected in a downstream bag filter.

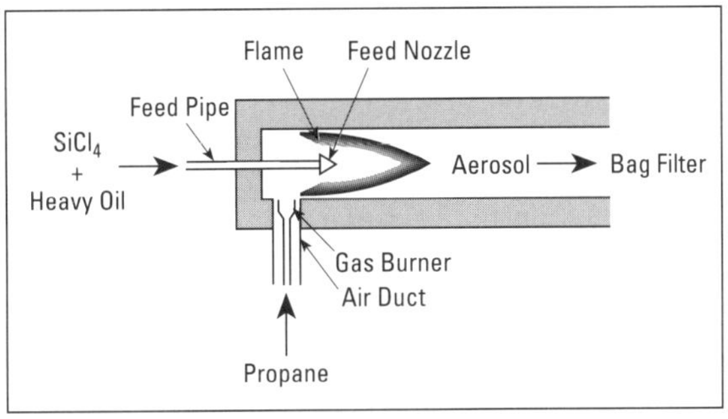

■ ***Figure 6. Schematic of aerosol flow reactor to manufacture C/SiO_2 chemically mixed precursor (13).***

This precursor powder exhibits a unique mixture morphology in terms of its surface area. C/SiO_2 possesses a surface area of only 45 m^2/g, whereas SiO_2 alone after carbon removal has a surface area of 490 m^2/g, and carbon following SiO_2 removal shows a surface area of 459 m^2/g.

The precursor powder is mixed with a binder, pelletized, and heated at 1,800°C for 30 min in a N_2 atmosphere to obtain β-SiC powder of < 0.1 μm average diameter (Figure 7). The powder surface area is typically in the range of 15–20 m^2/g, and can be pressureless sintered to near theoretical density.

Boron carbide. This material is conventionally manufactured by the carbothermal reduction of B_2O_3 in a batch electric-arc-furnace process, according to the reaction:

$$7C + B_2O_3 \Rightarrow B_4C + 6CO$$

Reaction and cool down take place over an extended period of time (days), because of the slow rate of heat conduction that controls the process. The large energy requirement for reaction (1,635 kJ/mol B_4C) further limits the overall heat-transfer/reaction process.

It recently has been shown *(15)* that the intrinsic reaction rate is extremely fast and that the reaction can be completed in seconds when heat-transfer resistances are minimized. In addition, rapid heating rates promote nucleation while minimizing particle growth, and allow submicron particles to be manufactured directly *(16)*.

Given these findings, a continuous "rapid carbothermal reduction" aerosol reactor process *(17,18)* has been developed to rapidly and uniformly heat fine precursor particles of C/B_2O_3 to reaction temperature and then minimize the reaction times at the high temperatures. The precursor is introduced with argon into a graphite reactor tube (Figure 8) to produce high-purity B_4C powder having a surface area of approximately 20 m^2/g. Rapid heating rates are achieved by thermal radiation and allow the solid reactants to completely convert to B_4C in seconds. For example, the reaction is completed in approximately 1 s at 1,900°C. The products can be densified into parts, the hardness of which exceeds that exhibited by parts fabricated from the best commercially available powder and nearly equals that from expensive laser synthesized powders.

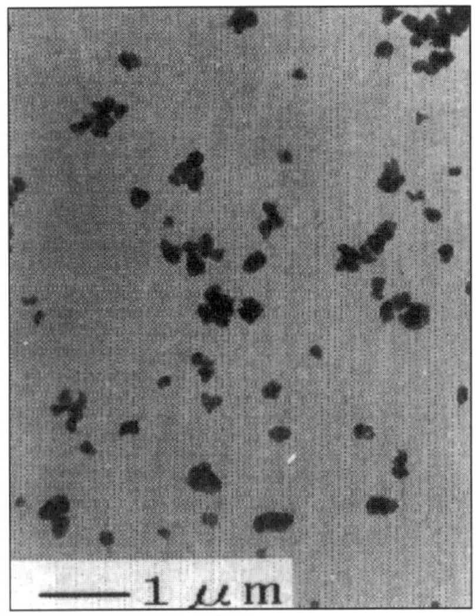

■ ***Figure 7. Photomicrograph of SiC powder prepared from chemically mixed C/SiO_2 precursor (13).***

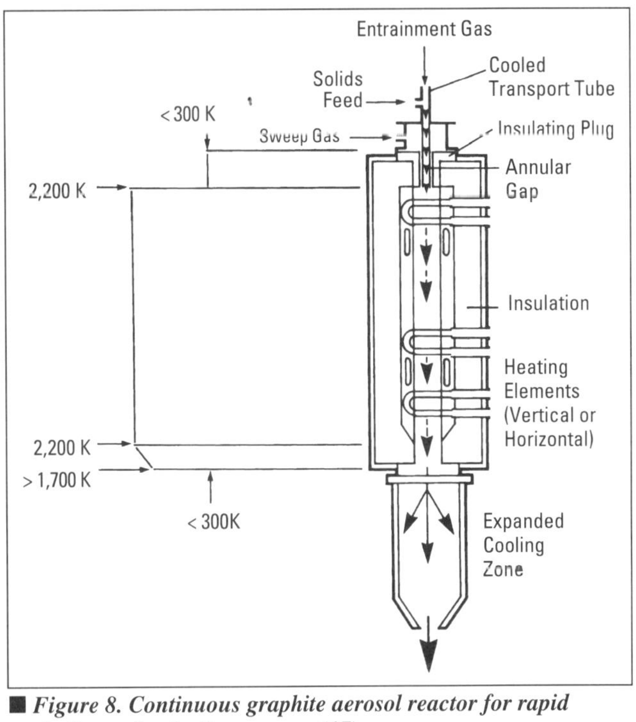

■ ***Figure 8. Continuous graphite aerosol reactor for rapid carbothermal reduction process (17).***

Colloidal systems

A colloid conventionally is defined as any particle that has at least one linear dimension between 1 nm and 1 μm. Classification of a particle as a colloid depends only on particle dimension — not chemical composition or physical state — and so colloid engineering includes the application of very fine powders, finely subdivided multiphase systems (for example, micelles, microemulsions, soluble polymers, and aerosols), films, fibers, foams, pores, surface irregularities, and large molecules. Many of the factors that are important in these "microheterogeneous" systems are important for conventional heterogeneous reaction systems. Other factors unique to colloids arise, however, and often can be exploited. Examples include "size quantization effects" that are observed in photocatalytic processes, as described next.

Semiconductor colloids as photocatalysts. Photo-induced reactions involving both organic and inorganic compounds can be accomplished using semiconductor colloids *(19)*. Reactants first are adsorbed onto the solid surface of a semiconductor, which then is excited with radiation of an appropriate intensity and wavelength. The resulting charge-transfer process between the semiconductor and adsorbed reactant (the "acceptor") can be employed to reduce or oxidize the reactant. This process is attractive from several standpoints, including its wide applicability and the fact that semiconductor materials are relatively inexpensive.

Use of colloidal semiconductor particles also offers potential advantages related to size quantization effects. The optical and electronic properties of semiconductors are essentially independent of size for particle diameters >10 nm. These properties for particles of the same composition but with diameters in the range 1–5 nm, however, often are much different — as particle size approaches the nanometer scale, the material no longer behaves as bulk semiconductor. These semiconductors are called "quantized" particles, to indicate that the optical and electronic properties are size-dependent, and are molecular, rather than solid-state, in nature.

The effect of particle size on optical and electronic properties can be remarkable. Henglein *(20)* notes the simple but illustrative example of cadmium phosphide, which is a black material in bulk form, but which can be made in various colors by preparing particles in the range 2–10 nm. Size quantization effects are very important from an applications standpoint, because quantized particles can photocatalyze reactions that bulk semiconductor cannot. In the extreme case, a reaction favored in one direction in the presence of bulk semiconductor is favored in the reverse direction in the presence of quantized particles of the same semiconductor material. Thus, markedly different types of reaction processes can be accomplished using the same semiconductor material as a photocatalyst and varying the particle size as needed.

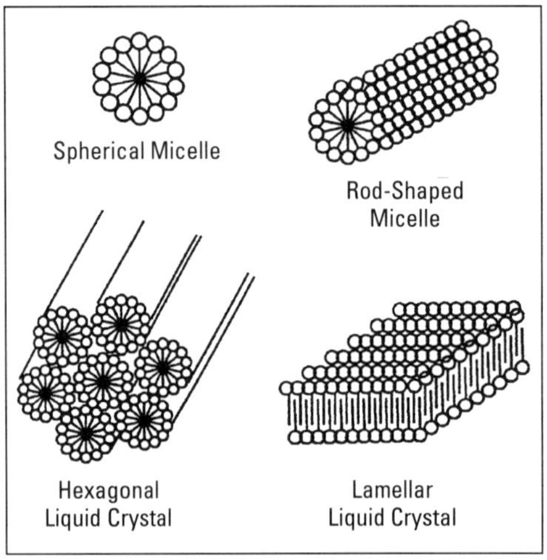

■ ***Figure 9. Different types of colloidal aggregates that may form in aqueous solutions of surfactants.***

In addition to semiconductor materials, colloidal solids have found great utility in many other areas. For example, colloidal metals commonly are employed in redox reactions that are initiated by either electromagnetic or ionizing radiation *(21)*.

Micellar catalysis. Surfactants are compounds containing both lipophilic and hydrophilic moieties in a single molecule. When dissolved in a solvent, surfactants concentrate at interfaces between phases, including vapor/liquid, liquid/liquid, and liquid/solid interfaces. At sufficiently high concentrations, surfactants also form varied aggregate structures within the bulk solvent phase; several possible aggregate structures for surfactants in aqueous solution are illustrated in Figure 9. The type of aggregate that forms in a particular solution depends on many factors, including surfactant structure and concentration, electrolyte concentration, and temperature. At least one of the physical dimensions of these aggregates is in the colloidal regime and, although solutions containing liquid crystals are anisotropic, aqueous solutions are macroscopically homogeneous in the concentration range employed for micellar catalysis.

It long has been known that aqueous solutions containing surfactant aggregates are capable of solubilizing significant amounts of compounds that normally are water insoluble *(22)*. Many reaction schemes previously required a combination of lipophilic and hydrophilic reactants. Micellar catalysis, however, offers a technique of performing such reactions under homogeneous conditions, with no need to add an organic solvent. The lipophilic reactant is solubilized into surfactant micelles in aqueous solution and its reaction with a hydrophilic reactant occurs in the vicinity of the micelle surface *(23)*. The hydrophilic reactant often is ionic and can be concentrated electrostatically near the micelle surface by selecting an appropriate (anionic or cationic) surfactant. The catalytic mechanism in micellar catalysis is believed to be due primarily to concentration effects *(24)*, that is, solubilization and electrostatic binding of reactants result in high localized concentrations of reactants, and therefore a considerable enhancement of the observed rate of reaction.

Much of the earlier work on micellar catalysis was aimed at using micelles as "simple" models for enzymes *(25)*. This analogy has proven to be somewhat useful, despite the fact that rate enhancements achieved by micellar catalysts generally are less than those observed for enzymes, and micelles are much less specific than enzymes. In any event, micellar catalysts remain as potential materials in their own right — as possible replacements for organic solvents in industrial reaction processes.

Admicelles: two-dimensional solvents. In dispersions of solid powders in aqueous solutions, surfactants adsorb at the liquid/solid interface. Depending on surfactant structure and concentration, an adsorbed surfactant bilayer may form in the presence of a hydrophilic solid. Bilayer structure is very similar to that of lamellar liquid crystals, illustrated in Figure 9, the difference being that lamellar liquid crystals exist in bulk solution whereas bilayers form at the interface between two phases. Adsorbed surfactant bilayers thus are similar to micellar aggregates, and commonly are called admicelles (for "adsorbed micelle"). For example, as is the case for micelles, compounds that are not water soluble can be effectively solubilized by admicelles.

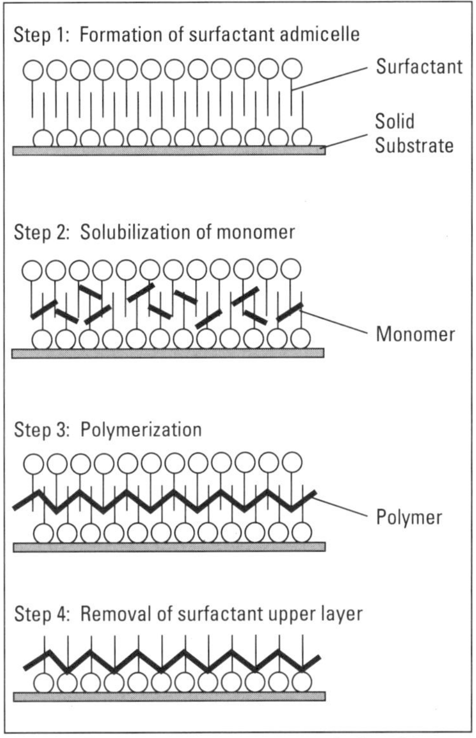

■ ***Figure 10. Formation of ultrathin polymer film on a solid substrate by polymerization in surfactant admicelles.***

Wu and coworkers *(26)* have developed a process, illustrated in Figure 10, for applying thin films of polymers to the surface of porous mineral oxides to make these materials more hydrophobic. The first step is adsorption of surfactant from aqueous solution onto the solid to create an admicelle. The mineral oxide (for instance, alumina powder) is initially hydrophilic — so an anionic surfactant (such as sodium dodecylsulfate, SDS) forms a bilayer with its hydrophilic head groups oriented toward the solid surface and toward the bulk aqueous solution. The extent of admicelle formation is maximized by controlling the pH and surfactant concentration. The second step is solubilization of monomers (like styrene) into the admicelle — the admicelle acts as a two-dimensional solvent for the reactant, concentrating it near the surface of the solid mineral oxide. The third step is the polymerization reaction inside the admicelle, to form a thin polymer film (for example, polystyrene). The mineral oxide with an ultrathin polymer/surfactant film then is removed from solution and dried. Depending on the desired surface modification, the powder can be used as-is, or washed with an appro-

L.-S. FAN is professor and chair-elect of the department of chemical engineering at the Ohio State University, Columbus, OH (614/292-7907; Fax: 614/292-1929), and chair of the Particle Technology Forum, an interdisciplinary group under the aegis of AIChE. His research interests are fluidization, powder technology, and particulates and multiphase reaction engineering. He is author or coauthor of two books, one monograph, and more than 160 articles. He is U.S. editor of Powder Technology, and a consulting editor of AIChE Journal. He has received several awards, including the 1993 Alexander von Humbolt Research Award for Senior U.S. Scientists. He holds a BS from National Taiwan University, and MS and PhD degrees from West Virginia University, all in chemical engineering, and an MS in statistics from Kansas State University. He is a member of AIChE.

J. RATHMAN is an assistant professor of chemical engineering at Ohio State University, Columbus, OH (614/292-3760; Fax: 614/292-3769; e-mail: rathman@kcg11, eng.ohio-state.edu). His research interests are in the area of colloids and surfaces, and include chemical reactions in colloidal media, and rheology of surfactant and dilute polymer solutions. Before joining academia, he worked for Conoco, Inc., and the Clorox Co. He holds a BS in chemistry from Montana State University, and MS and PhD degrees in chemical engineering from the University of Oklahoma. He is a member of AIChE.

A. GHOSH-DASTIDAR is a postdoctoral research associate in the department of chemical engineering at Ohio State University, Columbus, OH (614/292-6591; Fax: 614/292-1929). His research area is in high-temperature gas/solid reactions, and environmental reaction engineering. He currently is working on SO_2 and toxic metal removal from flue gas. He received a BTech from the Indian Institute of Technology (Kharagpur), an MS from the University of Akron, and a PhD from Ohio State University, all in chemical engineering.

A. W. WEIMER is a research associate with Dow Chemical, USA, Midland, MI (517/636-6183; Fax: 517/638-7092). Since joining the company in 1980, he has contributed to the development and scaleup of a broad spectrum of chemical processes involving hydrocarbons, inorganic chemicals, mixed alcohols, liquefied petroleum gas from syngas, methyl methacrylate monomer, polycarbonate resin, and nonoxide ceramic powders. His primary research interests are high-temperature kinetics, and multiphase reactor engineering. He is coinventor and developer of a number of processes for the manufacture of carbide, nitride, and boride nonoxide powders, and currently is responsible for the development of a low-cost carbothermal nitridation process for the production of high-quality, sinterable Si_3N_4 powder. He received a BS from the University of Cincinnati, and MS and PhD degrees from the University of Colorado, all in chemical engineering. A member of AIChE, he is a registered professional engineer in the state of Colorado.

S. KIMURA is associate professor of chemical engineering at Oregon State University, Corvallis, OR (503/737-4831; Fax: 503/737-4600). His research interests include chemical reaction engineering, and high-temperature materials. He is principal investigator for a project on fluidized-bed nitridation of silicon powder. Prior to joining OSU, he was on the faculty at Osaka University. He received BS and PhD degrees from Osaka University, and an MS from Oregon State University, all in chemical engineering. He is a member of AIChE, the American Ceramic Society, and the Society of Chemical Engineers of Japan.

Literature Cited

1. **Milne, C. R., and D. W. Pershing,** "Time Resolved Sulfation Rate Measurements for Sized Sorbents," *Proceedings*, 4th Annual International Pittsburgh Coal Conference, Pittsburgh, PA, p. 109 (1987).
2. **Bassi, A. S., C. L. Briens, and M. A. Bergougnou,** "Short Contact Time Fluidized Reactors (SCTFRs)," *Proceedings*, 4th International Conference on Circulating Fluidized Beds, Somerset, PA, p. 17 (1993).
3. **Raghunathan, K., A. Ghosh-Dastidar, and L.-S. Fan,** "High Temperature Reactor System for Study of Ultrafast Gas-Solid Reactions," *Rev. Sci. Instrum.*, **64** (7), p. 1,989 (1993).
4. **Raghunathan, K., A. Ghosh-Dastidar, and L.-S. Fan,** "A Technique for the Study of Ultrafast Gas-Solid Reactions for Residence Times less than 100 ms," *Rev. Sci. Instrum.*, **63** (11), p. 5,469 (1992).
5. **Ghosh-Dastidar, A., S. K. Mahuli, R. Agnihotri, and L.-S. Fan,** "Ultrafast Calcination, Sintering, and Sulfation of Hydrated Lime," *Proceedings*, 10th Annual International Pittsburgh Coal Conference, Pittsburgh, PA, p. 946 (1993).
6. **Vogiatzis, A. L., C. L. Briens, and M. A. Bergougnou,** "Selected Applications of Ultra-Rapid Fluidized (URF) Reactors: Ultrapyrolysis of Heavy Oils and Ultra-Rapid Catalytic Cracking," *AIChE Symp. Ser.*, **85** (270), p. 69 (1989).
7. **Shimizu, M., H. Fukuoka, and M. Fukuhira,** "Preparation of Silicon Nitride Powder," U.S. Patent 5,073,358 (1991).
8. **Jovanovic, Z., S. Kimura, and O. Levenspiel,** "Effects of Hydrogen and Temperature on the Kinetics of the Fluidized-Bed Nitridation of Silicon," *J. Amer. Ceram. Soc.*, **77** (1), p.186 (1994).
9. **Goldberger, W. M., A. K. Reed, and R. Morse,** "Synthesis and Characterization of HSC Silicon Carbide," *Cer. Trans.*, **2**, p. 93 (1989).
10. **Whyte, J. R., Jr., T. R. Parr, B. H. Kear, L. E. McCandlish, and B. K. Kim,** "Preparation of Nanophase WC-Co Composite Powders in a Fluidized Bed Reactor," *AIChE Symp. Ser.*, **88** (289), p. 116 (1992).
11. **Pratsinis, S. E., and T. T. Kodas,** "Manufacturing of Materials by Aerosol Processes," in "Aerosol Measurement — Principles, Techniques, and Applications," K. Willeke, and P. A. Baron, eds., Von Nostrand Reinhold, New York, p. 721 (1992).
12. **Murakawa, N., M. Nakajima, K. Muruyama, K. Isogaya, and K. Yoshida,** "Synthesis of an Easily Sinterable beta-SiC Powder," in "High Tech Ceramics," P. Vincenzini, ed., Elsevier, Amsterdam, p. 501 (1987).
13. **Murakawa, N., and M. Nakajima,** "Synthesis of SiC from Fine SiO_2-C Mixed Powders," *Nippon Serramikkusu Kyokai Gakujutsu Ronbunshi*, **99**, p. 704 (1991).
14. **Yoda, Y., M. Horita, K. Muruyama, N. Murakawa, K. Isogaya, F. Nakamura, and N. Kawamura,** "Process for Preparing Metal Carbides and Precursors Thereof," U.S. Patent 4,752,456 (1988).
15. **Weimer, A. W., W. G. Moore, R. P. Roach, J. E. Hitt, R. S. Dixit, and S. E. Prasnitis,** "Kinetics of Carbothermal Reduction Synthesis of Boron Carbide," *J. Amer. Ceram. Soc.*, **75** (9), p. 2,509 (1992).
16. **Rafaniello, W., and W. G. Moore,** "Producing Boron Carbide," U.S. Patent 4,804,525 (1989).
17. **Weimer A. W., R. P. Roach, C. N. Haney, W. G. Moore, and W. Rafaniello,** "Rapid Carbothermal Reduction of Boron Oxide in a Graphite Transport Reactor," *AIChE J.*, **37** (5), p. 759 (1991).
18. **Weimer, A. W., W. G. Moore, and R. P. Roach,** "Method for Producing Uniform, Fine Boron-Containing Ceramic Powders," U.S. Patent 5,194,234 (1993).
19. **Kamat, P.,** "Kinetics and Catalysis in Microheterogeneous Systems," M. Gratzel, and K. Kalyanasundaram, eds., Chapter 11, *Surfactant Sc. Ser.*, **38**, Marcel Dekker, New York, p. 375 (1991).
20. **Henglein, A.,** "Small-Particle Research: Physicochemical Properties of Extremely Small Colloidal Metal and Semiconductor Particles," *Chem. Rev.*, **89**, p. 1,861 (1989).
21. **Mulvaney, P., F. Grieser, and D. Meisel,** "Kinetics and Catalysis in Microheterogeneous Systems," M. Gratzel, and K. Kalyanasundaram, eds., Chapter 10, *Surfactant Sc. Ser.*, **38**, Marcel Dekker, New York, p. 303 (1991).
22. **Romsted, L.S.,** in "Micellization, Solubilization and Microemulsions," K. L. Mittal, ed., Plenum Press, New York, p. 509 (1977).
23. **Fendler, J., and E. Fendler,** "Catalysis in Micellar and Macromolecular Systems," Academic Press, New York, p. 43 (1975).
24. **Bunton, C.,** "Kinetics and Catalysis in Microheterogeneous Systems," M. Gratzel, and K. Kalyanasundaram, eds., Chapter 2, *Surfactant Sc. Ser.*, **38**, Marcel Dekker, New York, p. 13 (1991).
25. **Hiemenz, P.,** "Principles of Colloid and Surface Chemistry,"Chapter 8, Marcel Dekker, New York, p. 455 (1986).
26. **Wu, J., J. Harwell, and E. O'Rear,** "Two-Dimensional Solvents: Kinetics of Styrene Polymerization in Admicelles at or near Saturation," *J. Phys. Chem.*, **91**, p. 623 (1987).
27. **Wu, J., J. H. Harwell, E. A. O'Rear, and S. D. Christian,** "Application of Thin Films to Porous Mineral Oxides Using Two-Dimensional Solvents," *AIChE J.*, **34** (9), p. 1,511 (1988).

priate solvent to remove the upper layer of surfactant. If the powder is not washed, its surface remains hydrophilic, but it now has a net negative charge in aqueous solution, as opposed to the net positive charge of an untreated mineral oxide in the same aqueous solution. If the upper surfactant layer is removed, on the other hand, the solid surface becomes hydrophobic, and therefore easily wetted by nonaqueous liquids. The polymer film is not washed off during the surfactant removal step, because of extensive network formation during polymerization throughout the porous structure of the solid substrate.

This process represents a powerful application of a colloidal structure, the admicelle, in reaction engineering. An obvious requirement of this technique is that the surface area of the porous solid remain high. For the formation of polystyrene films on porous alumina powder in SDS admicelles, Wu and coworkers *(27)* produced alumina with a hydrophobic surface that retained the general pore structure with only a small decrease in surface area (58 m^2/g) relative to the untreated alumina powder (95 m^2/g).

Opportunities and challenges

We have highlighted a number of promising applications of particulate reaction engineering.

FSI technology allows ultrafast capture of sulfur dioxide from boiler flue gases. Ultrapyrolysis hold potential for converting heavy petroleum fractions, coal, and biomass. Production of nonoxide ceramics, including nanophase composites, by fluidized-bed processes promises higher quality products. Aerosol processes already are utilized for manufacture of high-purity particulates. Colloidal reaction engineering is opening up the prospects of semiconductor photocatalysts, micellar catalysts, and two-dimensional solvents.

All in all, we hope that this article has provided a sense of the exciting developments taking place — and the key role that chemical engineers are playing.

Optimize Performance of Fluidized-Bed Reactors

Achieving plug flow is often desirable. Here is an equation enabling a simplified solution for a structured-grid packing to help approach plug flow.

Gianni Papa,
Snamprogetti, SpA

Frederick A. Zenz,
PEMM-Corp.

In processes involving contact between dilute- and dense-phase streams, it is often desirable to minimize backmixing and approach countercurrent plug flow of both phases with an attendant concentration or temperature gradient. This should be done using the most economical and smallest size equipment.

Figure 1 illustrates such contacting arrangements, along with their limitations. Optimal performance means high throughput capacity, maximum use of vessel volume, and minimum cost of structural details. This performance is achieved at countercurrent or near plug flow in conventional gas/liquid packed towers (see Figure 1f). It seems that such flow should be equaled in gas/powdered-catalyst systems via filling arrangements shown in Figures 1b and 1g, with commercial ring, saddle, or other random-packing materials.

In general, this has not been true. This is because particulates can lodge in unaerated, stagnant zones within a packing's interstices *(1,2)*. However, the flow maldistribution inherent with such stagnation can be minimized using a structured packing with defined, uniformly distributed flow areas and by operation as near as possible to flooding. This is the practice in conventional packed absorption and distillation columns.

Such internal grid-type packings have been developed for specific applications through trial-and-error testing in so-called cold model apparatuses. This is done by determining their relative effectiveness in retarding the rate of top-to-bottom mixing in well-fluidized beds which would otherwise be nearly instantaneously homogeneous.

Such internal structures, as illustrated in Figure 2, are usually chevron or disc-and-donut grids in fluid-catalytic crackers or fluid-coke strippers. The structures can also be horizontally oriented tube bundles in vessels containing exo- or endothermic reactions *(3)*.

Design equations

A mechanistic engineering design of such internals is possible. This comes from the realization that the downflow of the dense phase through the grid restrictions is simply by gravity. Similarly, the upflow of the dilute phase (gas) as bubbles is correspondingly only the result of gravity-induced displacement *(4,5)*.

Equation 1 defines the minimum free area required for any specified flow of either phase, or conversely, the maximum or flooding flow rate able to pass through a specified area:

$$\frac{\text{lb / s}}{A} = \left(\frac{g}{\tan\theta}\right)^{1/2} \rho D^{1/2} \left(\frac{\Delta\rho}{\rho}\right)^{1/2} \qquad (1)$$

When applied to countercurrent streams of a dilute phase (gas) and a dense phase (liquid or bulk solids), Eq. 1 is rewritten for each phase in conventional rate units:

$$\frac{ACFM \text{ of gas}}{A_G} = 60\left(\frac{g}{\tan\theta}\right)^{1/2} \left(\frac{D_W - D_D}{2}\right)^{1/2} \left(\frac{\rho_D - \rho_G}{\rho_G}\right)^{1/2} \qquad (2)$$

$$\frac{\text{lb / s of catalyst}}{A_D} = \left(\frac{g}{\tan\theta}\right)^{1/2} D_D^{1/2} \left(\frac{\rho_D - \rho_G}{\rho_D}\right)^{1/2} \rho_D \qquad (3)$$

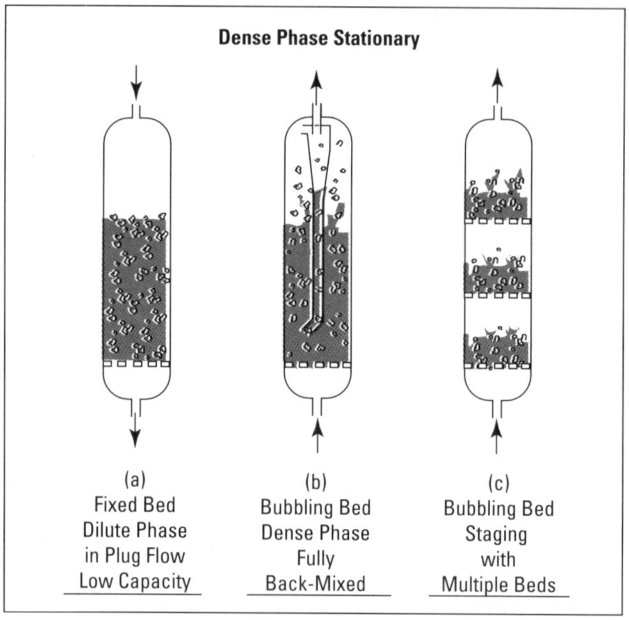

Figure 1. *Two-phase reactor arrangements.*

In Eq. 1, D represents the narrowest dimension of the flow area A required by the specified stream rate. D may be the diameter or annular width of a circular area, as in Figure 2b, or the narrowest width of a slot, as in Figures 2a and 2c.

The simultaneous change in D and A with change in the flow rates of the two phases results in the asymmetry of a plot of Eq. 2 vs. Eq. 3, despite equal intercepts when either of the flows approaches zero *(5)*.

This asymmetry, and its dependence on D, conveniently can be resolved for design purposes by expressing the flow areas in a slot-type grid as:

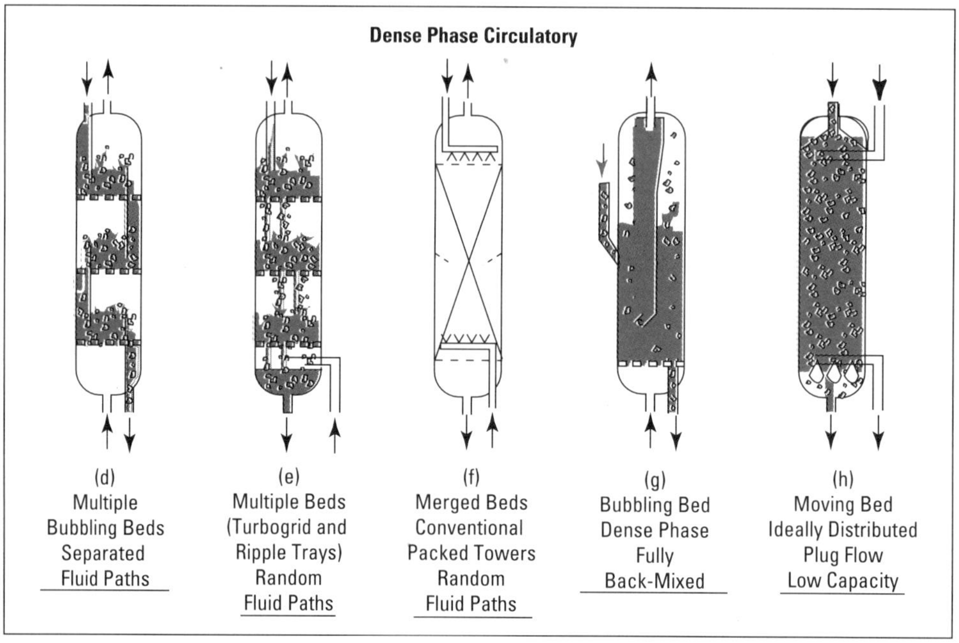

$$A_G = SN\left(D_W - D_D\right) \qquad (4)$$

and:

$$A_D = SND_D \qquad (5)$$

where, as illustrated in Figure 2d: N is the number of 1-ft-long slots in a square foot of grid area; S represents the total grid or reactor area; D_W is the total width of the slot; D_D is the width of the dense-phase flow stream; and $(D_W - D_D)$ is the sum of the widths of the dilute-phase flow streams on either side of the slot.

Figure 2 shows that the dilute-phase bubbles out from under the edges of the grid members forming the slot, in the same manner as gas bubbles out from under the skirt of a conventional bubble cap on a distillation tray. In an envisioned uniform array, operating with initially well-distributed flows, the dense phase must therefore pass down through the center portion of the slot, between the upflowing gas on either side.

Substituting Eq. 5 into Eq. 3 and solving for D_D:

$$D_D = \left[\frac{\text{lb / s}}{SN}\left(\frac{\tan\theta}{g}\right)^{1/2} \frac{1}{\rho_D}\left(\frac{\rho_D}{\rho_D - \rho_G}\right)^{1/2}\right]^{2/3} \qquad (6)$$

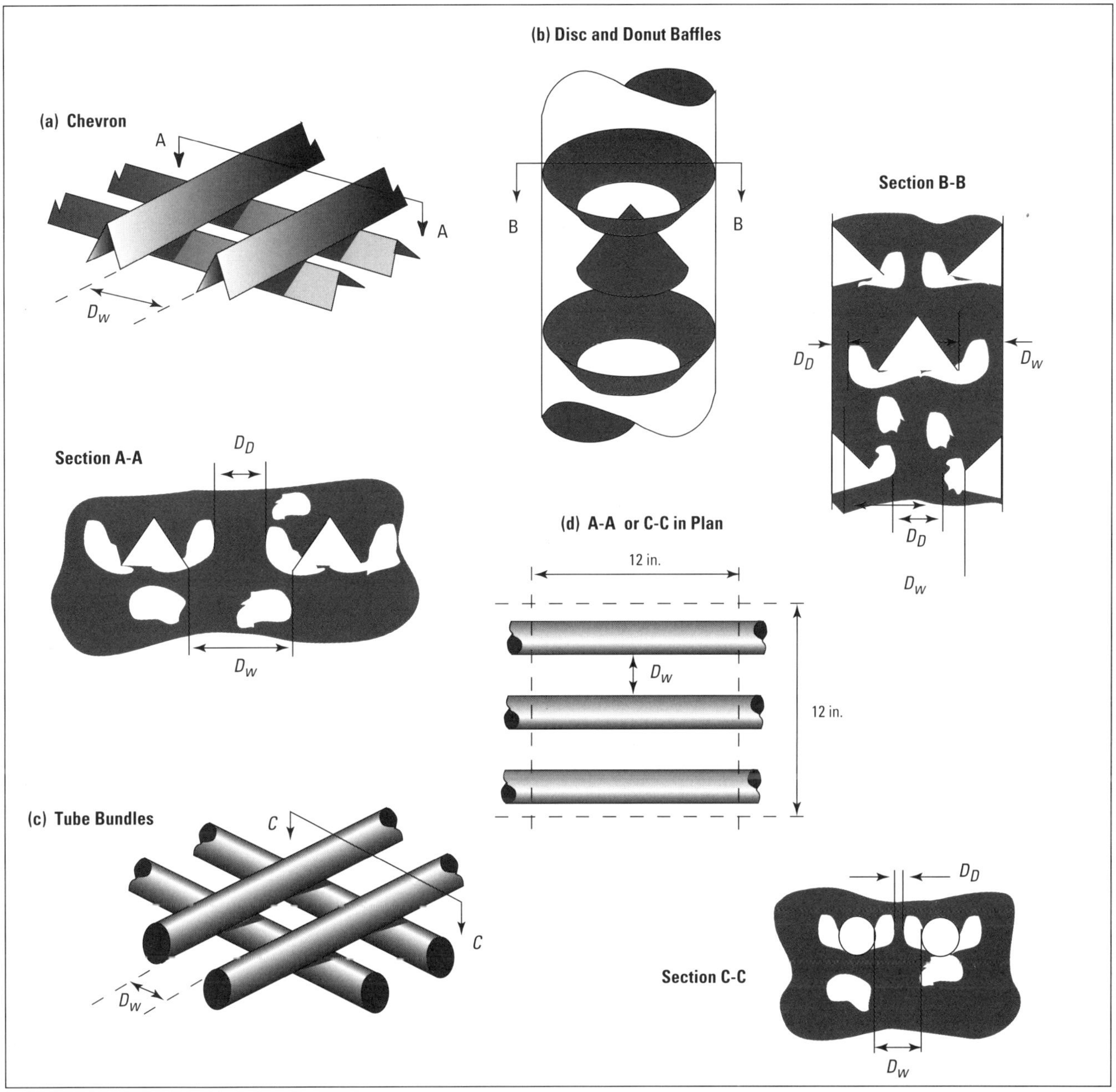

■ *Figure 2. Reactor internal grids retard top-to-bottom mixing.*

Nomenclature

A	=	flow area, ft^2
A_D	=	area occupied by downflowing dense phase, ft^2
A_G	=	area occupied by upflowing dilute gas phase, ft^2
$ACFM$	=	actual cubic feet per minute
D	=	narrowest dimension of flow area, ft
D_D	=	dense phase flow width, ft, defined in Figure 2
D_W	=	total flow width, ft, defined in Figure 2
f	=	fraction of grid area open to flow
g	=	gravitational acceleration, ft/s^2
N	=	number of 1-ft long slots/ft^2 of grid area, see Figure 2
S	=	total grid or reactor area, ft^2, see Figure 2

Greek letters

θ	=	angle of internal friction, deg.
ρ	=	density of flowing stream, lb/ft^3
ρ_D	=	dense phase bulk density, lb/ft^3
ρ_G	=	gas or dilute phase density, lb/ft^3

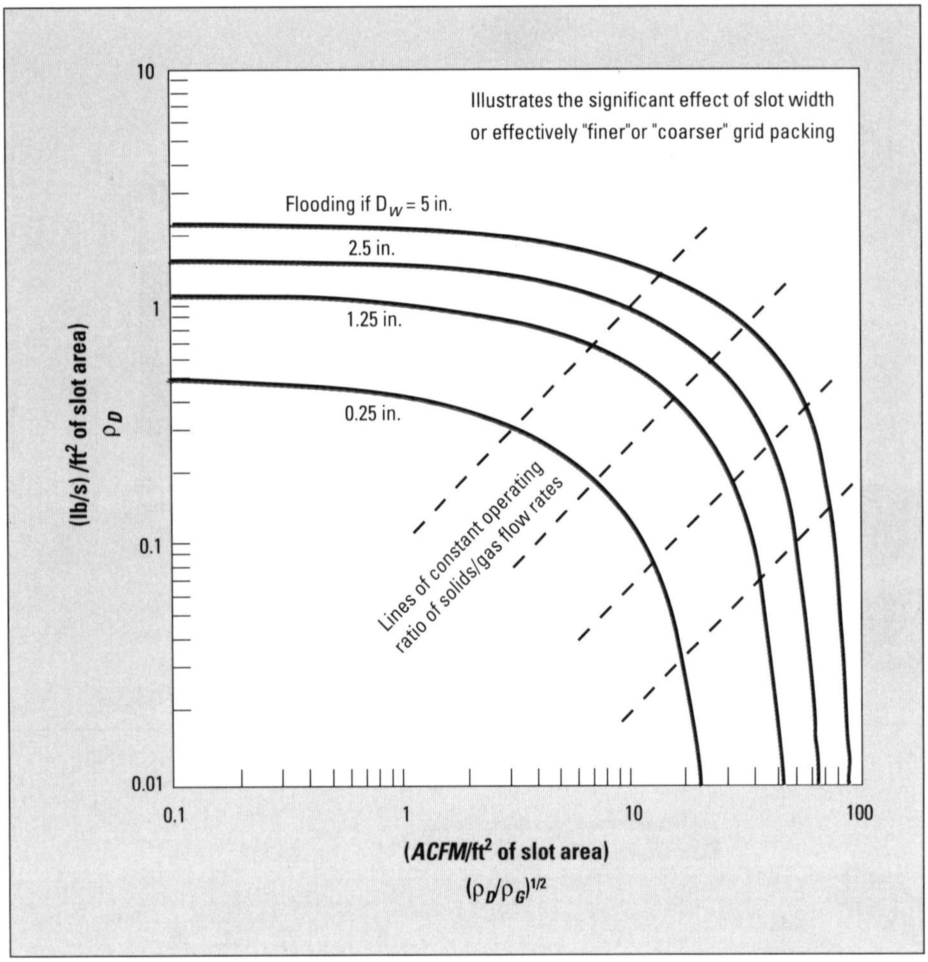

■ ***Figure 3. Representation of Equation 10.***

Substituting Eqs. 4 and 6 into Eq. 2, multiplying both sides by $S/\rho_D^{1/2}$ and simplifying:

$$\frac{ACFM}{\rho_D^{1/2}} = 60NS\left(\frac{g}{2\rho_G \tan\theta}\right)^{1/2}\left[D_W - \left(\frac{\text{lb}/\text{s}/\rho_D}{NS\left[\frac{g}{\tan\theta}\right]^{1/2}}\right)^{2/3}\right]^{3/2} \qquad (7)$$

Since $N = f/D_W$, Eq. 7 can be written as:

$$\frac{ACFM}{S\rho_D^{1/2}} = \frac{60f}{D_W}\left(\frac{g}{2\rho_G \tan\theta}\right)^{1/2}\left[D_W - \left(\frac{D_W(\text{lb}/\text{s})/S\rho_D}{f\left[\frac{g}{\tan\theta}\right]^{1/2}}\right)^{2/3}\right]^{3/2} \qquad (8)$$

or:

$$\frac{ACFM/Sf}{(\rho_D/\rho_G)^{1/2}} = \frac{60}{D_W}\left(\frac{g}{2\tan\theta}\right)^{1/2}\left[D_W - \left(\frac{D_W(\text{lb}/\text{s})/\rho_D Sf}{\left[\frac{g}{\tan\theta}\right]^{1/2}}\right)^{2/3}\right]^{3/2} \qquad (9)$$

which per square foot of slot area simplifies to:

$$\left(\frac{ACFM/\text{ft}^2}{(\rho_D/\rho_G)^{1/2}}\right)^{2/3} = 60^{2/3}\left(\frac{g}{2\tan\theta}\right)^{1/3} D_W^{1/3} - \left(\frac{60}{\sqrt{2}}\right)^{2/3}\left(\frac{\text{lb}/\text{s}\times\text{ft}^2}{\rho_D}\right)^{2/3} \qquad (10)$$

Let θ = 70 deg. This is the value for the solid that was tested, and applies to a wide range of solids and liquids. Also, Eq. 10 is fairly insensitive to the value of θ. Rewriting Eq. 10 yields:

$$\frac{ACFM/\text{ft}^2}{\left(D_W[\rho_D/\rho_G]\right)^{1/2}} = 27.63 - 12.16\left(\frac{\text{lb}/\text{s}\times\text{ft}^2}{\rho_D D_W^{1/2}}\right)^{2/3} \qquad (11)$$

Equations 10 and 11 are represented by Figures 3 and 4, respectively.

Equation 11, which is easily computerized, represents the flooding correlation *(5)* analytically, fully accounting for the asymmetry in its original representation. Figure 3 represents several solutions with D_W as a parameter. Along any 45-deg.

Figure 4. Representation of Equation 11.

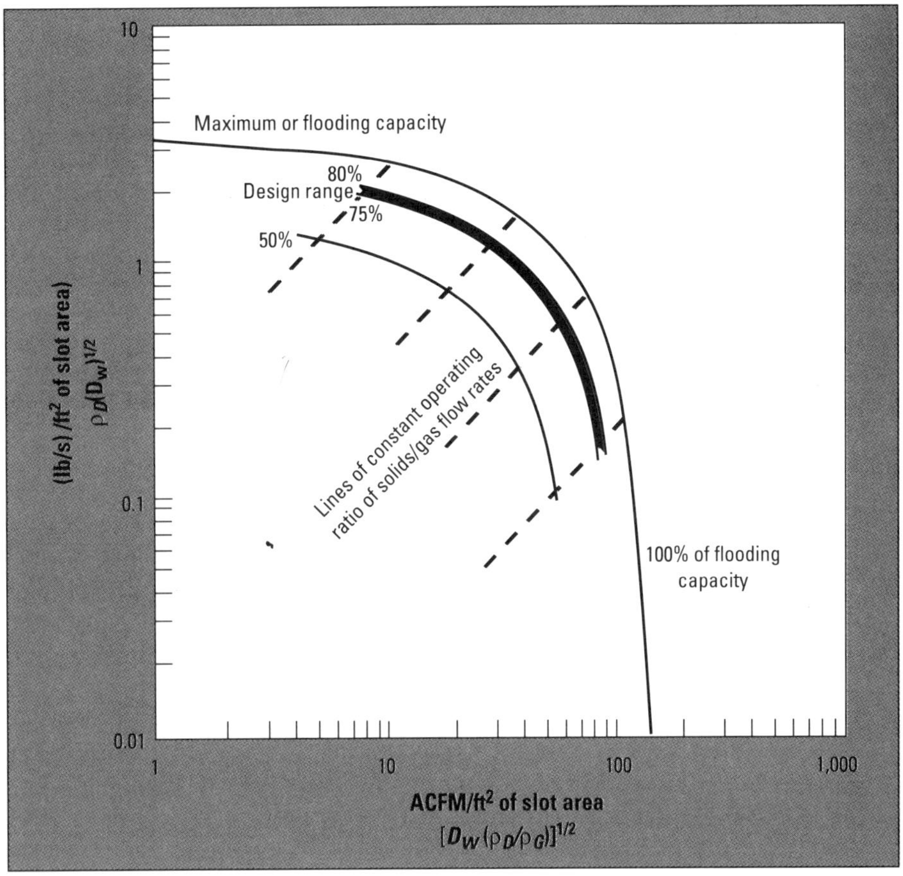

line, shown as dotted lines in Figures 3 and 4, is operation at a constant ratio of flow rates. Theoretically, at the intersection with the flooding curve, both phases approach plug flow through the vessel if the grids are stacked in close vertical array. This is true in the process whether the dense phase is in forced circulation through an external loop, as in Figure 1g, or internally mixed by natural circulation as in Figure 1b. For conventional packed-tower operation — Figure 1f — prudent design would dictate a vessel area or grid design such that the desired flows would be 75–80% of the grid's flooding limit.

Literature Cited

1. **Zenz, F. A., and D. F. Othmer,** "Fluidization and Fluid-Particle Systems," pp. 168-173, Reinhold, New York (1960).
2. **Harrison, D., and J. R. Grace,** "Fluidization," J. F. Davidson and D. Harrison, eds., Chapter 13, Academic Press, London (1971).
3. **Emelyanov, I. D.,** ***et al.,*** *Khim. Prom.,* **12**, p. 919 (1976).
4. **Zenz, F. A.,** "Flowing Solids Displace Bubble Voids," *Hydrocarb. Proc.,* **46** (4) pp. 171-175 (April 1967).
5. **York, J. L.,** ***et al.,*** "Solve All Column Flows with One Equation," *Chem Eng. Progress,* **88** (10), pp. 93-98 (October 1992).

Further Reading

Papa, G., and F. A. Zenz, "Correlating Throughput and Backmixing in Fluidized Beds," *Hydrocarb. Proc.,* **74** (1), p.81 (Jan. 1995).

G. PAPA is a process engineer at Snamprogetti SpA, Milan, Italy (011-3902-52022468; Fax: 011-3902-52038605). He joined Snamprogetti in 1985, the year he received his degree in industrial chemical engineering from Politecnico di Milano (Italy). Since 1986, he has been involved in fluidized-bed dehydrogenation technology.

F. A. ZENZ is a director of PEMM-Corp., Cold Spring, NY (914/424-3220; Fax: 914/424-8376), a non-profit industrial consortium. He is a professor emeritus at Manhattan College, a Fellow of AIChE, the 1985 recipient of the AIChE Award in Chemical Engineering Practice, and 1986 recipient of *Chemical Engineering's* Personal Achievement award. In the past 52 years, he has authored 90 papers, two books, 19 U.S. patents, and has been a consultant to more than 195 foreign and domestic corporations. He received his PhD in chemical engineering from the Polytechnic University, New York.

Successfully Scale Up Catalytic Gas-Fluidized Beds

Proper management of the scale-up process, in many instances, can essentially eliminate scale-up risks, while minimizing capital expenditure.

Behzad Jazayeri,
Fluor Daniel Inc.

Catalytic gas-fluidized-bed (CGFB) processes offer several advantages over fixed-bed reactor systems including: temperature uniformity; continuous catalyst regeneration; large-scale single-train operation; ability to operate within the flammability envelope, thus allowing more compact equipment; and less complex reactor heat-exchange systems. CGFBs already have proven themselves in a variety of applications, such as acrylonitrile, fluid catalytic cracking (FCC), maleic anhydride, phthalic anhydride, vinyl chloride, and synthetic oil. So, in process development, it clearly makes sense to investigate the suitability of a CGFB process.

Commercializing any new process can be a time-consuming effort, often requiring substantial capital expenditure and scale-up risk. Developing a CGFB process adds the complexity of fluidization and the challenge of catalyst development. In this article, I provide a generalized procedure to address these issues in an optimal, cost-effective manner. This procedure has been used to successfully bring several CGFB processes to commercial reality.

Getting started

Once an initial catalyst formulation with promising characteristics has been identified, a decision to continue the research program needs to be made. At this stage, only limited conversion/selectivity data are available. A preliminary process scheme using one (or two) possible reactor concepts should be developed for initial economic, safety, and environmental assessments. This task should be carried out by a small core team with authority to obtain support from all disciplines on an as-needed basis. The core team should include an experienced facilitator, the lead research engineer/chemist, and a lead process engineer as full-time members, and, as part-time members, a lead mechanical engineer, representatives of marketing and manufacturing, and other specialists as needed. This core team should report to a senior management "review panel" made up of representatives of research and development, engineering, marketing, and manufacturing with authority to fund projects.

The ability to conceptualize at this stage is very important. The lead engineers assigned must be seasoned with a broad-based background and exposure to a variety of processes. If they have no prior experience with fluidized-bed systems, the engineers should seek input from outside consultants or experienced engineering and construction companies.

The economics developed at this stage should be treated with a special perspective. Experience indicates that catalyst characteristics (yield and makeup rate) are likely to improve with time as new formulations become available. Therefore, using the initial data as a guide, assume the most optimistic characteristics (yield and makeup rate) for the catalyst — but realistic prices for feedstocks, byproducts, and utilities. Catalyst costs (if a major contributor to variable cost) should be obtained from catalyst vendors unless good in-house experience is available. The total installed cost (TIC) estimate should include an allowance factor, the magnitude of which depends on the complexity of the process and on the degree of innovation.

If the economics are poor, consider the following:

- If variable costs are high, can they be reduced realistically? How much does the reaction yield need to be improved? How much must catalyst cost or makeup rate be cut? What is the sensitivity to assumed raw materials and utility costs? Are process economics dependent on the export value of utilities (steam/electrical power), or byproducts?
- If the process is capital intensive, can it be simplified realistically? For instance, is it possible to use lower-cost materials of construction?

Then, presuming that a decision to commit to further research is made, the following parallel tasks should be performed:

Understand the chemistry. Bench-scale tests should be devised to answer key questions on the chemistry and kinetics of the process.

Develop a reactor model. Modeling the reactor at this stage can be a useful developmental tool *(1,2)*. Initially, a simple kinetic model that accounts for hydrodynamics and key reaction pathways is sufficient; more complex models can be used later on. Ideally, the simple model's structure should be designed for easy expansion. The reactor model can be used for sensitivity analysis of parameters such as temperature, pressure, space velocity, and so on. The model also can give guidance as to what type of reactor or what fluidization regime should be considered for the process.

Select the right reactor. Following these preliminary steps, you should be in a position to choose a type of reactor. Several options are available:

- conventional (sometimes referred to as fixed fluid bed, FFB), operating in the bubbling, transition, or turbulent regime;
- riser;
- downer (downflow riser);
- entrained bed; and
- quick contact.

The last four types usually are favored for short-contact-time reactions.

In a conventional reactor operating in the bubbling regime, passage of gas imparts a "boiling liquid like" behavior to the catalyst, the bulk of which remains within a generally well defined, dense-bed boundary *(3)*. Most of the gas travels through the dense bed as "bubbles" or "voids." Bubble eruption at the bed level results in catalyst entrainment. The entrained material must be captured and returned to the dense bed. One or more cyclones (internal or external) usually are used to capture and return the entrained solids. A "freeboard" space is provided between the dense-bed level and the inlet to the cyclone(s) to allow for maximum solid de-entrainment and to limit the solid loadings to the cyclone(s). This freeboard represents a low-catalyst-density zone where reactions can continue. Depending upon the reaction chemistry and the reactant(s) concentrations, the freeboard can be a lower-selectivity reaction zone. In the bubbling mode, solids are well mixed, and the gas behavior is between well mixed and plug flow.

As the operating velocity is increased, the transition, and, eventually, the turbulent regimes are reached. Gas essentially becomes plug flow, and the solid behavior begins to approach plug flow. The dense-bed level becomes less distinct, and more catalyst shifts to the freeboard zone, resulting in a reduction in the dense-bed density. Bubbles become less distinct and eventually disappear *(3)*. Gas/solid contact improves substantially. At some velocity limit, the entire reactor becomes the bed and only a small density difference exists between the top and bottom. This condition is the start of riser operation. Technologies that operate within the transitional or turbulent regimes include phthalic anhydride and acrylonitrile.

In the riser regime, all catalyst particles are fully entrained *(3)*. It, therefore, is possible to adjust the gas and solid residence times independently within a wide range. The action of gravity results in slip between the solids and gas. This slip is a function of the gas velocity and the solid flux *(4)*, and asymptotically approaches a low value at very high velocities. FCC units and the SASOL Synthol synthetic oil process are examples of riser technology *(3)*.

Presence of slip in risers has created interest in downer reactors *(5)*. In these, gas and solids are fed to the top of the reactor and are allowed to flow downward. Gas/solid separation still is required. Recent observations *(6)* indicate that negative (solid faster than gas), as well as positive slip can occur, depending upon the velocity. Downers operate within a limited range of bed densities.

The entrained bed is a riser operating with very low solid holdup.

Quick contact reactors are risers or downers with millisecond contact times. An example is the QC process *(7)*. Quick-contact reactors operate within a narrow bed-density range.

In assessing options, always keep in mind that a major challenge in design of commercial CGFBs is the prevention of gas bypassing due to formation of gas voids (bubbles).

Once a suitable reactor is selected, a simple dynamic model of the process can be used to simulate abnormal operations, such as startup, shutdown, and emergency conditions, to identify safety related issues.

Fine-tune the catalyst. Catalyst characteristics, such as average size, size distribution, sphericity, fines content, density, and flowability, are important parameters in scale-up. Nonspherical particles show poor fluidization characteristics and tend to achieve sphericity through attrition when used in fluidized beds. Because of this, spherical (or nearly spherical) particles are used exclusively in CGFBs.

Geldart *(8)* developed a method for classifying powders using average particle size and density (see Figure 1). This method provides a powerful tool for designing catalysts even though the original work is only applicable to ambient conditions. Geldart's Group A powders are best suited for use in CGFBs, as they generally give smaller voids. Group C powders tend to be cohesive and susceptible to defluidization. Group B powders form larger voids and can perform differently in small-scale units than in commercial units. The A/C Group (shown by the broad band between the A and C regions) has some of the characteristics of both these groups. The D Group should not be used in CGFBs.

For Group A powders, the transition regime starts at about 1 ft/s, and turbulence at about 2–3 ft/s. Riser operation can occur at about 7 ft/s. For Group B powders, transition occurs at about 4–6 ft/s, turbulence at around 7–10 ft/s, and riser operation at approximately 20 ft/s. With Group B powders, choking *(9)* limits the applicability of risers to lower-solid-loading systems. Use of downer, entrained-bed, and quick-contact reactors generally is restricted to systems where low catalyst holdup in the reactor is acceptable.

Usually, as the density of catalyst increases, its average size should be reduced so as to maintain Group A behavior as much as possible. Note, however, that cutting the average size usually raises catalyst production cost but lessens attrition. Group B powders generally require more complex reactor design for control of bubble growth. Therefore, the higher production cost for Group A catalyst must be balanced against the higher makeup rate and more complex design required for Group B catalyst.

Catalysts that can become sticky pose special problems. Sometimes, addition of a small quantity of an inert substance can reduce this problem to manageable levels. (This technique also may be helpful if a catalyst exhibits poor flowability characteristics.)

The fines fraction (that is, particles of 10–45μ dia.) can have a profound impact on the performance of CGFBs with Group A catalysts *(10)*. A fines content of 20–35% is recommended when the particle density is in the 1–2 g/cc range.

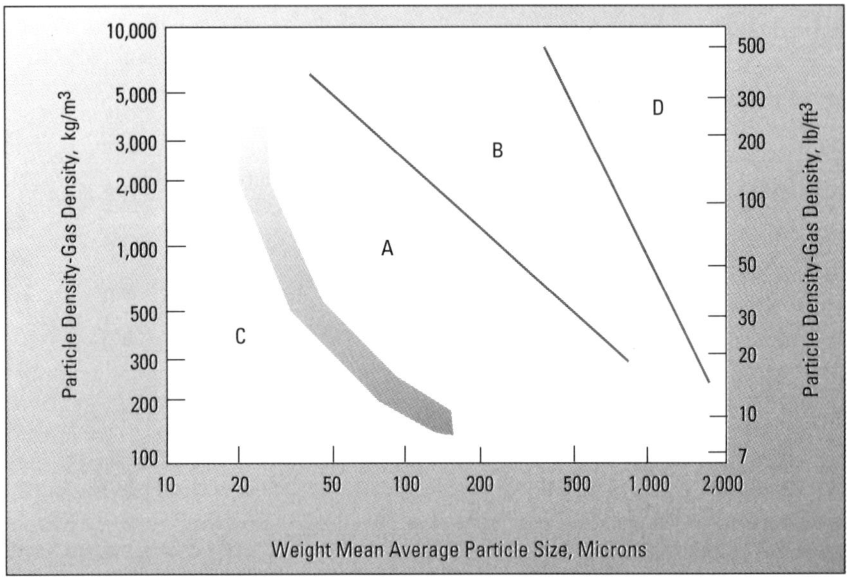

Figure 1. Classification of powders according to Geldart (8).

A fines concentration larger than 35% generally should be avoided — because either economics suffer despite the incremental performance improvement, or poorer fluidization occurs. A large fraction of fines below 10μ, and more generally 5μ, should be avoided unless the catalyst is extremely dense.

To keep startup costs down, there is a tendency not to de-fine (that is, dedust) the initial charge of the catalyst. The argument for this approach is that commercial cyclones have low recovery efficiencies for very small particles (below 10μ) and will therefore classify the catalyst during normal operation. This approach is very sound and applicable to many situations, but also has caused significant delays in the startup of some plants. In these plants, before classification could occur, the cohesive, very fine material either had coated heat-transfer surfaces or plugged cyclones. The lesson here is to verify all assumptions by proper testing before reaching commercial-stage operation.

There are tradeoffs that can be made between fines content and operating velocity. Increasing the operating velocity improves gas/solid contact due to turbulence *(10)*, which reduces the need for fines. This approach has been successfully used in regenerators of FCC units that operate with Group A catalyst at velocities of about 3–6 ft/s *(11)*. Risers and entrained beds that operate at even higher velocities and are considered "bubbleless" can be considered as providing the best contact mechanism.

An increase in operating velocity results in greater catalyst entrainment and therefore higher solid traffic through cyclones (or other gas/solid separation devices). This increased traffic usually means greater catalyst attrition and a higher makeup rate. Also, boosting the operating velocity usually cuts the catalyst holdup per unit volume of reactor. This reduction in holdup volume can diminish the reaction selectivity for oxidation reactions — where the relative magnitude of nonselective homogenous reactions to selective heterogeneous ones is a strong function of the available catalyst surface area per unit volume. Therefore, the option of trading off increased velocity for reduced fines content may have limited use for oxidation reactions or when the catalyst is expensive.

A size distribution of about 7–20:1 (based on the 2nd vs. 98th weight percentiles) is recommended for most applications. Avoid the use of narrower distributions as these result in poor fluidization characteristics. A distribution broader than 20:1 seldom can be justified — as it provides little or no performance enhancement.

Evaluate safety and environmental aspects. CGFBs pose general safety and environmental issues common to any process plant, and a comprehensive discussion of these is beyond the scope of this article.

CGFBs used in oxidation reactions, however, pose special safety issues due to their operation within the flammability envelope. Such situations demand careful study of the potential for runaway reactions, and after-burning due to oxidation reactions, either at the exit of the reactor *(12)* or, for recycled processes, in the recycle gas loop. Modeling may be employed for very simple systems but seldom is reliable for more complex ones. For example, attempts to model a system containing three combustibles and three inerts resulted in several hundred possible reaction pathways. In these situations, the use of test bombs and hot-flow reactors with synthesized streams is recommended. Note that flammability envelopes and autoignition temperatures generally shrink as pressure and temperatures are increased. "Cool flame" phenomena can result in strange behavior as process temperature is changed. This is due to formation of different species at different temperature intervals. After-burning is more likely to occur in larger equipment. In addition to normal operation, carefully evaluate startup, shutdown, and other abnormal operating cases for potential of after-burning and explosion, and, as needed, develop solutions.

There are several oxidation processes operating in the explosive envelope that have been commercialized successfully using CGFBs. Despite this, let me re-emphasize that a new process must be evaluated carefully. The available catalyst surface area per unit volume, and the corresponding reaction potential are the two key variables.

Entrained solids in the reactor exit can inhibit or control after-reactions *(12)*. In this situation, solid separation devices become the point where after-burning can initiate.

From an environmental standpoint, deactivated (spent) catalyst or catalyst fines ideally should be reprocessed or reclaimed.

Progressing to the next stage

If updated catalyst-performance and TIC data are promising, the next step is to evaluate the design of a laboratory-scale unit.

Within limitations, the laboratory (bench) unit should be a scaled-down design of the commercial system. At a minimum, it should operate over the same temperature and pressure ranges as the commercial process. Evaluate the impact of recycle streams on reactor performance. Avoid synthesized feeds; instead, use feedstocks with the same impurities and with comparable composition to that of the commercial process (unless testing shows the impurities to be inert to the process and catalyst). Check the impact of processing steps, such as regeneration/reduction, stripping with steam (or other gases), thermal shock, and the like, on the catalyst.

The lab unit should demonstrate the long-term catalyst-performance characteristics of the commercial unit. Therefore, control the rate of catalyst makeup so as to simulate the expected catalyst turnover rate of the commercial reactor. A common mistake is to use improperly designed cyclones in small lab units. This leads to excessive catalyst loss, which then is made up using fresh material. The catalyst tested in such units always will have an average life that is shorter than that of the commercial facility, masking possible longer-term deactivation problems.

At this point, if needed, tune the reactor model by using updated data.

Certain design aspects (such as of the cyclone *(13)*, baffles, or bubble-growth inhibitors *(14)*) are difficult — if not impossible — to test in small lab units. Here, a cold-model test unit can play a cost-effective role. Consider the use of outside facilities, such as those at the Institute of Gas Technology (Chicago), PEMM-Corp. (Cold Spring, NY), or Particulate Solid Research Inc. (Chicago). They often can perform experiments at much lower cost and in shorter time frames than in-house sources.

At the completion of this stage of development, carry out a final review of the process design, as well as economics, safety, and environmental issues. Critical equipment should be sized and engineered to a high level of detail. The reactor internals should be detailed out completely, because these items usually are fabrication intensive.

Remember that fluid beds are subjected to cyclic forces due to the motion of solids. These forces can become very large. The design of fluid-bed reactors and their supporting structure requires specialized know-how. Using a "total system" approach to design, along with advanced design tools, can eliminate items such as expansion joints, while minimizing the total cost of supporting structures (some of which can be several hundred feet tall and support millions of pounds of catalyst and equipment).

Going directly from laboratory to commercial scale?

Assuming favorable economics, the next step depends on a variety of factors.

The need to make a large quantity of product for performance testing in downstream processing steps is one reason for opting for a demonstration or process development unit (PDU). Design concepts that cannot be tested in small units or modeled by cold-test units are other reasons for use of PDUs. Buildup of trace compounds to levels as low as parts per billion in recycle streams can lead to poor catalyst performance, off-colored product, and other problems. Simulation models can be helpful, but really cannot replace a PDU for demonstrating the effect of recycle streams. A PDU therefore is strongly suggested when recycle streams are present. Always bear in mind, though, that the function of a PDU is to show the envisioned commercial operation and not to obtain fundamental data.

Some processes have been scaled up directly from small (a few in. dia.) units to large (several ft dia.) commercial ones with little or no loss in performance (that is, conversion and selectivity). In other cases, the performance of large (several ft dia.) PDUs have been found to be superior to that of much larger commercial units. These discrepancies are due partly to the chemistry of the process (fast vs. slow, competing reactions, and so on), and partly to unavoidable differences in the design of large- and small-scale units, of which the design of internals and gas distributors are the two most critical factors. Small PDUs utilize distributors with small holes at small spacing. In commercial plants, maintenance considerations limit nozzle spacing to much larger distances, resulting in larger nozzle sizes. The extent of nonselective reactions, which can occur at the boundaries of the issuing gas jets, therefore is likely to be higher in the commercial plant.

Literature Cited

1. **Dutta, S., and B. Jazayeri,** "Alternate Reactor Concepts for Oxidative Coupling of Methane," presented at 7th International Fluidization Conference, Brisbane, Australia, pp. 445–453 (May 1992).
2. **Dutta, S.,** "Fluidized and Moving Bed Desulfurization by Zinc Titanate," *AIChE Symp. Ser.,* **90** (301), pp. 157–169 (1994).
3. **Kunii, D., and O. Levenspiel,** "Fluidization Engineering," 2nd ed., chapters 2 and 3, Butterworth-Heineman, Stoneham, MA (1993).
4. **Jazayeri, B.,** "Estimate Pressure Drop and Solid Holdup in Risers Handling FCC-type Catalyst," *Hydroc. Proc.,* pp. 93–95 (May 1991).
5. **Dewitz, T. S.,** "Downflow Fluidized Catalytic Cracking System," U. S. Patent No. 4,797,262 (Jan. 10, 1989).
6. **Auebert, E., D. Barreteau, T. Gauthier, and R. Pontier,** "Pressure Profiles and Slip Velocities in a Co-Current Downflow Fluidized Reactor," presented at Circulating Fluid Bed 4 Conference, Hidden Valley, PA (Aug. 1994).
7. **Gartside, R. J.,** "QC — A New Reaction Systems," *Proceedings,* Fluidization VI Conference, Banff, Alta., pp. 25–32 (1989).
8. **Geldart, D.,** "Types of Gas Fluidization," *Powder Tech.,* pp. 284–292 (July 1973).
9. **Matsen, J. M.,** "The Rise and Fall of Recurrent Particles: Hydrodynamics of Circulation," presented at Circulating Fluidized Bed 2 Conference, Compiègne, France, p. 7 (Mar. 1988).
10. **Pell, M., and S. P. Jordan,** "Effects of Fines and Velocity on Fluid Bed Reactor Performance," *AIChE Symp. Ser.,* **84** (262), pp. 68–72 (1987).
11. **King, D.,** "Fluidized Catalytic Crackers: An Engineering Review," presented at 7th International Fluidization Conference, Brisbane, Australia, p. 18 (May 1992).
12. **Graham, J. J.,** "The Fluidized Bed Phthalic Anhydride Process" in "Loss Prevention," *CEP Tech. Man.,* (4), pp. 31–35 (1969).
13. **Lewnard, J. J., B. E. Herb, T. R. Tsoa, and J. A. Zenz,** "Effect of Design and Operating Parameters on Cyclone Performance for Circulating Fluidized Bed Boilers," presented at Circulating Fluidized Bed 4 Conference, Hidden Valley, PA, pp. 636–641 (Aug. 1994).
14. **Dutta, S., and G. D. Suciu,** "An Experimental Study of the Effectiveness of Baffles and Internals in Breaking Bubbles in Fluidized Beds," *J. Chem. Eng. Jap.,* **25** (3), pp. 365–348 (1992).
15. **Volk, W., C. A. Johnson, and H. H. Stotler,** "Effect of Reactor Internals on Quality of Fluidization," *Chem. Eng. Progress,* **58** (3), pp. 44–47 (Mar. 1962).
16. **Kunii, D., and O. Levenspiel,** "Fluidization Engineering," 2nd ed., Butterworth-Heineman, Stoneham, MA, p. 150 (1993).
17. **Dry, R. J., and M. P. Judd,** "Fluidized Beds of Fine, Dense Powders: Scale-up and Reactor Modelling," *Powder Tech.,* (43), pp. 41–53 (1985).

The need to test on a larger-scale PDU thus also depends on the expected deviation in performance between large- and small-scale units, which are design specific. In coming up with this estimate, experience is just as important as the prediction made by a good reactor model.

If a decision to build a PDU is made, test only what is absolutely needed and rely on available expertise and knowledge to predict equipment performance deviations for the commercial plant. Build the unit inside an existing production facility if possible. This strategy reduces the investment cost. It also allows the PDU to be used as a production unit to recoup a portion of the investment, and for operator training.

The PDU (and often the first commercial facility) requires a much higher level of instrumentation and diagnostics than commonly used in mature commercial facilities. This fact often is overlooked in preliminary cost estimating.

Designing large-scale CGFBs

A number of general factors should be considered when designing a commercial unit.

Is there a size limitation? The answer is no from a process point of view. A 60-ft-dia. commercial unit can be built to perform as well as, say, a 4-ft-dia. one. Often, however, the investment for several smaller reactors is less than that of a single, large unit due to such cost factors as field vs. shop fabrication, available over-the-road transportation limits, the need for thicker shells for larger vessels, and so on.

What should be done about backmixing? Backmixing frequently has been mentioned as the Achilles' heel of fluidized beds. The solution most often suggested is to use baffles to retard backmixing.

Backmixing only is important if it results in a loss of usable products, if it provides a product slate different from that desired, or if it prevents a closer approach to equilibrium. If the catalyst is very selective and the relative rate of heterogeneous reactions is fast compared to that of homogeneous ones, then backmixing may not be an issue. For instance, an experiment was conducted where the desired product of an exothermic reaction was deliberately spiked into the base of a highly backmixed fluid-bed test unit. The reaction chemistry involved several highly exothermic sequential and parallel reaction pathways, some of which resulted in the destruction of the desired product. The net yield of the desired product increased!

Use baffling only as an absolute last resort. If baffling retards solid backmixing (the mechanism by which gas is backmixed), then it also will promote size classification and create an axial temperature profile. (We already have discussed the importance of proper fines content.) Solids recycle loops to correct for unwanted classification are seldom cost-justified, and are problematic to operate. Avoid the use of baffles in oxidation reactors. Baffles create pockets of low-solid regions, which can undergo uncontrolled burning.

How should the hydraulic diameter concept be used? In 1962, a group at Hydrocarbon Research Inc. suggested this concept *(15)* — namely, that placement of internals reduces the hydraulic diameter of the reactor, and thus can make the performance of commercial-size reactors the same as that of smaller laboratory units. The hydraulic diameter was defined as "four times the free cross-sectional area of the bed divided by the wetted perimeter of all vertical surfaces exposed to the bed."

More recently, Kunni and Levenspiel *(16)* also have proposed the use of this concept.

Commercial experience clearly shows that internals improve the performance of large reactors. Unfortunately, for some systems, this improvement only is observed when a large number of internals are used (that is, small hydraulic diameter). For others, placement of only a few internals is sufficient to bring the performance of the commercial reactor in line with that of the small laboratory unit. Further confusing the issue, a wide range of hydraulic diameters can produce essentially the same reactor performance. Also, there is a critical hydraulic diameter below which performance deteriorates rapidly. This last observation is logical in that a fluid-bed reactor packed to its limit with internals will exhibit fixed-bed characteristics due to the hindrance of solid motion.

Personal communications with others practicing in this field has revealed two schools of thought on how internals affect reactor performance. The first school holds that internals result in a reduction in the global solid-circulation patterns (referred to as gulf streaming), which can create gas channels and gas bypassing. The second school suggests that internals break up or slow down voids. Either mechanism leads to increased gas/solid contact, and improved performance, as has been observed. Unfortunately, these observations do not apply in every case. For example, there are large (20 ft dia. and greater) units in operation that have very few internals placed on a spacing of several ft. The void size within these reactors rarely exceeds one inch. It is difficult to to see how either of the two schools of thought can apply to these systems.

What are the important design issues? Gas and solid distributions, and proper design of the internals are very important. This is one subject that is closely guarded by the industry and one area where access to prior experience outweighs access to any predictive tools — and it is the key issue to scale-up. It is sufficient to state that: design of the internals is critical, and beware of the hydraulic diameter concept *(17)*. CEP

B. JAZAYERI is a principal process engineer in the Technology Planning Group of Fluor Daniel Inc., Irvine, CA (714/975–7195; Fax: 714/975–3961). He specializes in the application, scale-up, and design of fluidized-bed-reactor systems, and has functioned as an engineering liaison/consultant on several projects involving such units — including DuPont's maleic anhydride transport-bed process. Prior to joining Fluor Daniel in 1988, he was a senior process engineer at Badger. He received a BS from University College, London, and an MS from Massachusetts Inst. of Technology, both in chemical engineering, and an MBA from Boston Univ.

Acknowledgment

I would like to thank A. Avidan, S. Dutta, W. Green, D. King, T. Knowlton, M. Pell, and A. Rajguru for their valuable insight and support.

Accurately Size Batch Reactor Relief Systems

During a chemical reaction, the relieving flow rate is not constant, but varies with factors such as reactor inventory, pressure, and temperature.

Frank Turner and Siegfried H. Richter,
John Brown Engineers and Constructors Inc.

Relief valves and rupture discs are common devices that protect process equipment from damage due to overpressure. Sizing a relief valve or rupture disc is not difficult for most nonreactive applications, once the engineer determines which cause of overpressure produces the largest relieving rate. However, when the vessel's contents undergo a chemical reaction, calculating and deciding upon the sizing basis for a relief device becomes far more complex.

When a partially filled batch reactor undergoes an uncontrolled exothermic reaction, several things occur simultaneously. The heat of reaction warms the contents and both temperature and pressure rise. The reaction changes the composition in the reactor, converting reactants into products and byproducts. As the temperature rises, the rate of chemical conversion accelerates. Eventually, the reactor pressure reaches the set pressure of the relieving device, which then opens. The relieving flow rate is not constant, but changes with the reactor inventory, composition, pressure, temperature, and the limitations of the relief system.

The problem is to find a method for sizing the relief system that models these factors. Ideally, the method used should provide suitable information for safe sizing of the relief device or rupture disc, related piping, and any downstream separation or recovery equipment. This article describes a procedure for generating computer programs that simulate runaway reactions. A recent reactor design project is used to illustrate the process, demonstrating the need for using an integrated approach.

Traditional relief device sizing

Vendor literature, the American Society of Mechanical Engineers (ASME) Boiler and Pressure Vessel Code, and American Petroleum Institute (API) standards provide a traditional approach to sizing relief valves and rupture discs. With this approach, the engineer estimates the required discharge rate and selects a valve model and size that has a capacity larger than the needed flow rate, given the relieving temperature, pressure, and composition. This method is acceptable for most nonreactive systems, but can grossly oversimplify the situation for a reactor. The accuracy of this method is even more uncertain when two-phase flow occurs during venting.

When using the traditional method to size a relief device for a reactive system, the engineer typically makes several implied assumptions. First, the peak relieving rate occurs immediately after opening of the relief valve or bursting of the rupture disc. Second, this peak venting rate is not affected by changing composition. Third, the relief device operates independent of any influence from the physical piping installation. Each of these assumptions can be incorrect, as will be discussed.

Required information

In preparation for design of the reactor relief system, a considerable amount of information must be assembled. This is true whether using a commercial relief-system program such as AIChE's SAFIRE or when developing a simulation program internally. Any assumptions made during the collection of information could have significant impact on the suitability of the final design and must be considered carefully. Some analytical testing may be required to generate the necessary information. The following section describes some of the design considerations and types of information which are required for simulating a batch reactor relief system.

Relief system design

Design of the relief system should be performed as early in the design phase of a project as possible. The large flows which may be produced in a relieving situation can result in large-diameter lines, vapor/liquid separators, and vent stacks. Allowing the plant layout to become finalized before designing the relief system makes redesign of nearby systems an unwelcome possibility. Large pressure drops in the discharge header and the methods needed to deal with the reactor discharge can result in relocation of the reactor. In some plant designs, the need to relieve a reactor safely has resulted in complete redesign of the reactor structure. Problems of this nature can be minimized by early design of the relief system. Minor process or layout changes can be resimulated as part of finalizing the detailed design.

In general practice, the relief valve or rupture disc is located at a minimum distance from the reactor, often mounted directly on a reactor nozzle. The selection of either a relief valve or rupture disc is made by the designer based on process considerations. Discharge piping should be kept as short as possible, leading to a discharge point suitable for the system. As the computer program uses a method that depends upon knowing the equivalent length of piping between the reactor and the relief device, this piping configuration must be fixed before the calculations can be begun.

Preliminary line sizing is adequate, as the impact of line size should be studied during the detailed design. Approximate equivalent length of a rupture disc holder with burst disc must be obtained from the rupture disc vendor. Equivalent length values for fully open relief valves are harder to obtain. DIERS uses an isentropic nozzle in its modeling of relief valves in the SAFIRE program. Discharge line routing downstream of the relief valve or rupture disc is also important, but not critical to the calculations. (This assumes that critical flow is forced by system design to occur at the discharge of the relief device, as will be discussed later.)

The volume of the reactor and ancillary equipment is required. If liquid entrainment is assumed, then the entire volume of the reactor system can be considered to be full of foam. This might extend into any overhead condenser, depending upon the location of the relief device.

The methods for disengagement of entrained liquid and discharge of vent gases must be given consideration. The computer simulation of the reactor produces information that will allow final sizing of the equipment and piping, but preliminary location of a knockout pot and suitable venting point must be made early in the plant design. These considerations are particularly important for flammable or toxic systems and when solids, high-melting-point material (polymers), or viscous material is discharged.

Chemical and physical properties

For each chemical in the reactor, certain basic information will be required. This list includes molecular weight, specific heat of vapor and liquid, latent heat, vapor and liquid viscosity, and liquid specific gravity. All pure-component properties should be available as functions of temperature, which may change significantly during the simulated runaway reaction.

Adiabatic vapor/liquid equilibrium calculations are a critical part of these simulations. At high pressure, vapor/liquid equilibrium will probably not follow ideal gas law *K*-values. Use of an equation of state to account for pressure effects on vapor/liquid equilibrium will need to be considered. Activity coefficient information may also be required to account for nonideal liquid-phase mixtures. The methods used should be able to account for any inert or nonvolatile components which may be present.

Reaction kinetics will need to be estimated over a wide range of conditions. The effects of changing temperature, pressure, and composition will have to be calculated during program execution. Any side reactions or competing reactions must be considered. For instance, significant degradation of a chemical at high temperature into volatile components could seriously affect the calculated relieving rates and operating conditions.

The venting characteristics of the system, if known, should be considered. Does the system foam, contain surfactants, operate as a slurry, or have significant viscosity? If so, liquid entrainment is likely during relieving. When in doubt, the design engineer should assume liquid entrainment for plant design.

Operation

A detailed understanding of the batch process is essential. Knowing the sequence of chemical addition, rate of addition, temperature, and pressure during each stage, and equipment used during each step allow a more accurate simulation of the runaway reaction.

For example, a reaction which begins as a final ingredient is slowly added should be considered to begin during the ingredient addition. Loss of electrical power, possibly stopping agitation and reactor cooling, would have slight impact if very little of the last reactant has been added. This same loss of power at the end of the reactant addition might result in a runaway reaction, but the predicted composition of the batch should take into account the partial reaction which took place during the addition of the final reactant. In the example project, the initial conditions were found to have a significant impact on the maximum venting rates.

The design engineer should also consider simulating upset conditions, such as over- or undercharging of any component. If manual addition is used in a batch, complete omission of an ingredient or a double charge might be conceivable. Imagine the effect of a double charge of catalyst on a reactor.

Simulation model

Having assembled the information described above, the design engineer must establish a credible scenario resulting in reactor relief. When necessary, several scenarios should be studied to ensure that the plant design covers the worst-case situation. In considering each scenario, the engineer should consider the entire situation, as some situations produce indirect effects which can worsen the relieving rate. Consider the following examples:

A reactor might experience a runaway reaction if electrical power to the plant is lost, stopping the agitator and cooling-jacket circulation pumps. If the power outage is assumed to occur at the point when the batch is at its highest level of reactants, one basis for calculation can be developed. If the power outage is assumed to be local, caused by damage due to a fire external to the reactor, then heat input from fire might also be considered. Fire was not considered significant in the example plant since the reactor had a heat-transfer-fluid filled jacket and heavy insulation, and was protected by a fire sprinkler system.

Consider a reactor designed with internal cooling coils or an external jacket. The engineer must consider if the scenario should assume heating of the vessel walls and jacket coolant. The example project had a reactor which used both an internal coil and external jacket for cooling. Loss of power to the agitator and buildup of thick slurry along the reactor wall and coil was thought to limit heat transfer to the reactor vessel wall during the several minutes needed for the reactor to empty during a runaway. The calculations conservatively ignored heat pickup by the vessel and jacket fluid.

A double charge of catalyst to a batch would probably not result in loss of agitation or jacket cooling. This scenario should then assume cooling occurs normally, building to the design cooling rate of the cooling system during a runaway. Assuming a double charge of catalyst and simultaneous failure of power, for instance, probably would be considered too unlikely a double jeopardy.

A final consideration is whether or not to assume liquid entrainment. Entrainment was known to occur in the design example and was modeled as a uniform froth. Visualize a reactor pressurized with shaving cream and allowed to vent; the froth is initially very rich, high in liquid phase. As the contents of the reactor empty, the liquid-phase portion of the foam decreases. At the end of the reaction, the proportion of liquid to vapor phase is at its lowest value. Because the vessel is assumed to stay full of foam, there is no time when only vapor is relieved.

Other estimates of liquid entrainment are given as options in the SAFIRE program from DIERS, but choosing among them should be based on some knowledge of the actual reactor system.

The design engineer must establish a credible scenario resulting in reactor relief.

Optimization of the programs

Because of the complexity of these programs and the huge number of repetitive calculations performed, the computer system used should have either a 486 or Pentium processor. Use of a compiled language package should also be regarded as mandatory. Choice of language is left to the engineer. The work done on the example project was done using a 486-33 computer and Borland C++ (using C source code). Execution times could run over an hour. The methods used were not optimized for speed, but were chosen to reliably converge on real solutions.

As described, the programs operated as a series of nested calculations. This arrangement is the coding algorithm most likely to converge, the easiest to code, and the simplest to logically visualize. The shortcoming of such an approach is long execution times.

In performing each adiabatic flash, simultaneous solution of several nonlinear equations would shorten the program execution. This option was not explored as a discontinuity in the heat balance formula exists at the pressure where the relief valve opens and closes, or the rupture disc bursts. Such a discontinuity might make convergence difficult. If the engineer has time to explore options, this might significantly reduce the execution time of the program.

In theory, all of the nested loop calculations could be solved simultaneously by nonlinear matrix methods. This approach would involve significantly more complex programming and debugging than the programs developed for the sample project, but should substantially reduce execution time. Since these programs are typically designed only for one plant or process, high program speed should be considered a lower priority than avoiding a lengthy programming and testing exercise. Commercial programs like SAFIRE may have been optimized along these lines.

All variables were defined as double precision. This allows use of very small convergence values for each iterative calculation. If the allowed convergence error values were too coarse for reactor pressure, adiabatic flash temperature at known pressure, and fraction of vapor formed in each vapor/liquid equilibrium, the programs might not adequately converge on an accurate solution in each time period. This could be detected by plotting flow vs. time. The changes in flow, once the rupture disc bursts or a relief valve opens, should result in a smooth curve. If such a plot has a cyclic or wavy appearance, the convergence criteria are too coarse.

Example: reactor operation and design

This example uses a two-stage batch process, in which the reactants are mixed with a heavy organic solvent and water. This multicomponent system produces a slurry containing both product and reaction byproducts. Heating initiates and accelerates the reaction in an agitated vessel, and the reaction is under temperature control by means of a tempered external cooling system.

The first reaction stage occurs at about 430°F and below 100 psig, during which more than 80% of the reactants are consumed. These operating conditions produce a controllable reaction rate and are well below the reactor design temperature and pressure. After an injection of water into the reactor, the second reaction stage takes place at 490°F and 250 psig. These operating conditions drive the reaction to near completion in a reasonable time of several hours.

The effects of a runaway reaction are highest during the first reaction stage, when most of the reactants have not been converted. As with most reactions, the kinetics slow as reactant concentrations fall. Consequently, the two-stage reaction operation was developed. Because the reaction is highly exothermic, a possible runaway situation might be initiated by a loss of reactor cooling. Loss of cooling could have several causes, including control failure, electrical failure, and cooling-water failure.

The reactor is a closed vertical vessel, with system pressure set by the vapor pressure of the vessel contents. The vessel is designed for 430 psig and 600°F with a two-speed agitator. Heat transfer is via a tempered, recirculated heat-transfer fluid, using an internal coil and an external jacket.

In the final plant design, two relief device systems protect the reactor. The first consists of two rupture discs, set in series, to provide high-pressure protection. Two rupture discs in series were chosen to provide a safeguard against loss of the batch due to corrosion or fatigue of the first disc.

The second system consists of one pressure safety valve, set for 125 psig, to protect the system during the low-pressure reaction stage. The reactor piping design includes a block valve between the reactor and the pressure safety valve, allowing isolation from the reactor during the second stage reaction. Use of a pressure safety valve minimizes the venting rate that a runaway reaction would create during the first reaction stage. Computer simulation of this system shows that relieving at 125 psig would have a peak mass-flow rate about 10% of the maximum rate expected through the two rupture discs at 350 psig.

Because the organic solvent has an open-cup flash-point temperature of 7°F, any release to atmosphere could result in a fire or explosion. To minimize this possibility and also consequent cleanup problems, the relief valve and rupture discs are connected to a knockout drum, from which a vent stack releases escaping vapor above the plant structure.

Reaction kinetics and physical considerations

The reaction occurs with the following kinetic formula:

$$dC_n/d\theta = 2k(C_1C_2)^{0.5}$$

where: $k = A \exp(-E/RT)$; A = frequency factor; C_n = byproduct concentration at time θ; C_1 = Reactant 1 concentration at time θ; C_2 = Reactant 2 concentration at time θ; E = energy of activation; k = reaction constant; R = gas constant; T = reactor temperature; and θ = reaction time.

The reaction kinetics are thus considered dependent on temperature and liquid-phase concentrations, but independent of system pressure. This formula is assumed valid to near completion and at all reactant concentrations.

Batch composition is significant only in that water in the liquid phase is present in a small concentration compared to the organic solvent. Water is the most volatile chemical present, and the organic solvent is lower in volatility. All reactants are intermediate in volatility and the reaction products are nonvolatile.

Wilson equation activity coefficients compensate for liquid-phase nonideality for water, the organic solvent, and one of the reactants. Because coefficients were not available for the remaining reactant, ideality was assumed. The resulting calculated vapor pressure at different temperatures closely matched data provided by the licensor. Consequently, vapor-phase ideality was assumed, eliminating the need to use an equation of state in the simulation.

General approach

A computer program was unavailable that would both rigorously model the reactor system and handle time-dependent vapor/liquid/solid multicomponent equilibria in the reactor and through the relief vent system. Significant liquid losses occurred during discharge from a low-pressure relief valve on a similar plant; this aggravated the problem. Any program used for these calculations must include an accurate method for two phase flow. Some programs exist for modeling multicomponent unsteady-state systems, but were unavailable for use or review. Consequently, John Brown developed the programs used for the reactor relief system design.

The computer programs simulate the batch during short time increments of 0.1 min. Different time intervals were tried on preliminary computer runs. Increments of 1.0 min resulted in only approximate solutions, due to the short relieving time. Time intervals of 0.1 min produced a reasonable number of data points, allowing for an accurate simulation of changing reactor conditions and mass-flow rates during venting.

Table 1. Equivalent length for rupture-disc assemblies.

Nominal Rupture Disc, in.	8		6	
Choke Tube Dia. = Throat Dia., in.	7.625		5.875	
L/D for One Blown Disc	75 max	30 min	75 max	30 min
Choke Tube Equivalent Length, without Acceleration Loss, ft	97	53	87	43
Inlet Acceleration Loss, ft	24	24	17	17
Total Equivalent Length, ft	121	77	104	60

Table 2. Area and pseudo diameter for safety valves.

Safety Valve Size	Orifice Size	Orifice Area, in.^{2}	Pseudo Orifice Dia., in.
4P6	P	6.38	2.85
6Q8	Q	11.05	3.75

Increments of 0.01 min resulted in reactor operating conditions and venting rates similar to the 0.1 min increment, but required much more time per calculation. Within each interval, the rate of reaction was calculated using the reactant concentrations and temperature from the previous interval. Preceding relieving, the heat of reaction results in a temperature and pressure rise. This temperature is calculated by trial and error within the program, such that the sensible heat used in the temperature rise must balance the heat released by the reaction. The chemical composition is revised to reflect the chemical conversion during the interval, and the saturation pressure is calculated.

When the calculated pressure exceeds the set point pressure of the relief valve or the burst pressure of the rupture disc, venting begins. Several relieving scenarios were modeled for both rupture discs and relief valves, assuming either vapor flow or a homogeneous two-phase froth.

For simulation purposes, the batch runaway reaction was assumed to occur totally adiabatically. Electrical failure in the plant would approximate this condition, as jacket cooling and batch agitation would be lost.These calculations ignore heat transfer to the reactor steel and jacket oil, since the presumed loss of agitation would limit heat transfer. The assumption of adiabatic operation is considered a conservative choice, since any actual heat absorbed by the reactor steel and jacket oil would lengthen the time to onset of venting and lessen the consequent venting flow rate.

Venting from a reactor could be modeled in several ways. One approach is to simulate a rupture disc or a relief valve as a simple orifice, but this approach does not recognize pressure losses in the piping upstream and downstream of the relieving device. In our plant design, the rupture disc assembly, consisting of two rupture discs with holders separated by an elbow, has 6 ft of vertical pipe between the reactor and the primary rupture disc. This assembly was modeled as a choke tube with an inside diameter equal to the interior diameter of the rupture disc holder. Choke tube equivalent length was calculated from the resistance coefficients of each pipe component following Crane's Technical Paper 410, "Flow of Fluids Through Valves, Fittings, and Pipes."

Using equivalent length ratios provided by a rupture disc vendor, the choke tubes modeled for the rupture disc assemblies resulted in the dimensions given in Table 1.

Critical flow exists at the exit of the choke tube when the built-up downstream backpressure is below the critical flow outlet pressure of the material traveling through the choke tube. This is guaranteed by proper design of the discharge piping and equipment downstream of the second rupture disc, using the peak flow rates predicted by the programs we created. In the project, 12-in. pipe was used for the discharge header to prevent overpressure of the low-pressure relief valve.

Conceptually, an open pressure-relief valve could be modeled similarly. Although conventional sizing equations treat a relief valve as an orifice, several coefficients factor into the equation to make the predicted flow conform to laboratory test results. This approach has proven accurate for single-phase vapor or liquid flow.

In any detailed simulation involving pressure relief valves, any fluid traveling through the valve must follow a tortuous path. A conventional relief valve contains three or more throat areas that are connected in series, each of fixed or variable size. These throats include the inlet nozzle below the disc, the area between the seating surface and the moving relief valve disc, the huddling chamber formed between an adjusting ring and the disc skirt, and the outlet nozzle. In addition, between these areas of restricted flow there are direction changes, contractions, and expansions through which any fluid must travel. Because of these factors, each type of relief valve has different flow characteristics. Disc lift limiters, ring adjustment, spring adjustment, and variable overpressure below the disc serve to further complicate accurate modeling of relief valves.

For these calculations, it was assumed that the relief valve remained fully open during the entire relieving scenario. This assumption is valid when the relieving pressure exceeds the design overpressure (in this case 10% above the relief valve set pressure per API for non-fire situations). We further assumed the governing throat area to be the same size as the nominal relief valve orifice area. The resulting data are in Table 2.

Assuming that a fully open relief valve can be modeled as a choke tube, the choke tube inside diameter was set equal to the diameter of the relief valve orifice, with a *L/D* of 175. To account for inlet piping and fluid acceleration losses, another *L/D* ratio of 80 was added. Thus, the total equivalent length of a relief valve was assumed to be: $L = 255D$. This is only a preliminary estimate and preferably should be confirmed by the vendor or by independent testing. In our case, extensive testing was judged not to be required because the reactor was ultimately protected against overpressure by the rupture disc assembly. Overpressure above 110% of the safety-valve set pressure due to underestimation of the equivalent length would not endanger reactor safety.

When reactor pressure reaches the set point of a relief valve, the valve opens and allows vapor to escape.

When a rupture disc or a relief valve is modeled as a choke tube, the reactor pressure is variable. Material leaves the reactor at a rate that is affected by the pressure drop in the discharge system. With a runaway reaction, the most likely limitation is reaching critical flow at some pinch point in the system, such as at the second rupture disc holder or the relief valve.

If modeling only vapor, compressible-gas-phase formulas are suitable for calculating reactor pressure from the downstream point where the gas velocity becomes sonic, knowing the equivalent length of the line between the reactor and the relief device. The downstream vapor line, separation system, and stack would therefore have to be designed not to produce backpressure high enough to affect the formation of sonic flow at this constriction.

When two-phase flow is assumed, the calculations are more involved. As pressure drops in passing through a pipe, the amount of vapor increases due to partial flashing of the liquid. This results in a variable pressure drop per foot of equivalent length.

Reference 1 describes a method for modeling such a system (see Reference 2 for background information). Simply described, for each time increment the procedure starts out by calculating the critical-flow outlet pressure at the pinch point for gas and liquid leaving the reactor at an assumed reactor pressure. The article includes a formula for calculating the equivalent length of pipe that produces a certain pressure drop, say 1 psi. Using this method with small pressure steps allows calculation of line pressure drop from the relief device backwards to the reactor, providing a check on the assumed reactor pressure. If the assumed and calculated reactor pressures do not agree, a new reactor pressure is assumed and the calculations repeated. The method also permits adjustment of both gas- and liquid-phase quantities, along with their physical properties, for the pressure found in each equivalent length increment. This is a significant departure from normal shortcut or two-phase pressure drop procedures.

This results in a trial-and-error calculation, searching for an assumed reactor pressure in each time increment that would produce the correct amount of vapor and liquid needed for critical flow in the exit of the choke tube.

There are several advantages to performing these calculations, which are not performed with any known shortcut method for sizing relief valves or rupture discs. The first is having a means to develop a rational basis for the system design, reflecting the most demanding scenario being considered. A second one is, for each time increment, knowing in detail the flow rate and composition of material to be handled by any downstream knockout drums or recovery systems. An additional advantage is knowing the reactor conditions when the reaction ends.

Relief valve simulations

Constant pressure simulations, single-phase venting — When reactor pressure reaches the set point of a relief valve, the valve opens and allows vapor to escape. This flow may be two-phase. When not undersized, the relief valve theoretically can maintain a constant vessel pressure by throttling the flow. If grossly oversized, the valve would vent more vapor than the batch is creating, causing reactor pressure to fall. As the pressure drops below the set point pressure, the relief valve would reseat and stop venting until pressure rises to the set point pressure again.

The program makes an adiabatic flash calculation at a specified reactor pressure. This calculation figures out the resulting vapor rate/composition and batch temperature, such that the sensible and latent heat changes equal the heat produced by the reaction. The program revises the chemical composition at the end of each time increment to reflect the chemical conversion and venting losses. Calculated reactor temperature and pressure, along with the quantity and composition of the vented material, are stored in a file for later design of downstream equipment. Program operation continues until a reactant is exhausted.

Initial simulations focused on operation of a 4P6 pressure relief valve and assumed only vapor venting. Figure 1 shows plots of reactor temperature against time for several pressure safety valve (PSV) set point values. For each set-point pressure case, the reactor temperature rose during venting and exceeded the design temperature of the vessel. In fact, this temperature rise exceeded the temperature limit possible for carbon steel, 650°F even when set pressures were reduced to pressures only slightly above the second stage reaction pressure. A larger size relief valve would not reach the maximum vessel temperature limit.

Clearly, finding results like this would not have been possible using any shortcut procedure. This ability is particularly important when the vessel walls are made of dissimilar materials, such as clad metal or glass-lined steel. The high-pressure relief valve design was replaced with the dual rupture disc arrangement described earlier.

Repeating these calculations for the low-pressure relief valve produced the temperature plot shown on Figure 2. The temperature does not exceed 650°F making use of a relief valve for low-pressure relief an acceptable design. Figure 3 shows the vapor rate leaving the low-pressure relief valve. Both temperature and venting rate increase as time passes, results which would invalidate any relief valve design based on initial venting estimates.

Initially, for first stage reaction at low pressure, the vent gas has a high proportion of water, with smaller amounts of reactants and solvent. Increasing temperature and flow rate are probably due to the rising concentration of heavy solvent in the reactor. The organic solvent has a lower vapor pressure and latent heat than water. As the water leaves the reactor, the temperature is forced upward and greater amounts of organic solvent boil off to remove the heat of reaction. This increase in temperature also accelerates the reaction rate and vapor formation. The maximum venting rate occurs as the reaction dies due to a lack of reactants.

Variable-pressure simulation with flashing two-phase venting — Because two-phase venting is known to occur in this system, flashing flow simulations were run, assuming a homogeneous mixture of vapor and liquid in the reactor. With this scenario, the worst case flows occur when the relief valve is assumed to be completely open during the bulk of the reactor venting. The limiting case for discharge rate then would be critical flow in a choke tube modeling the relief

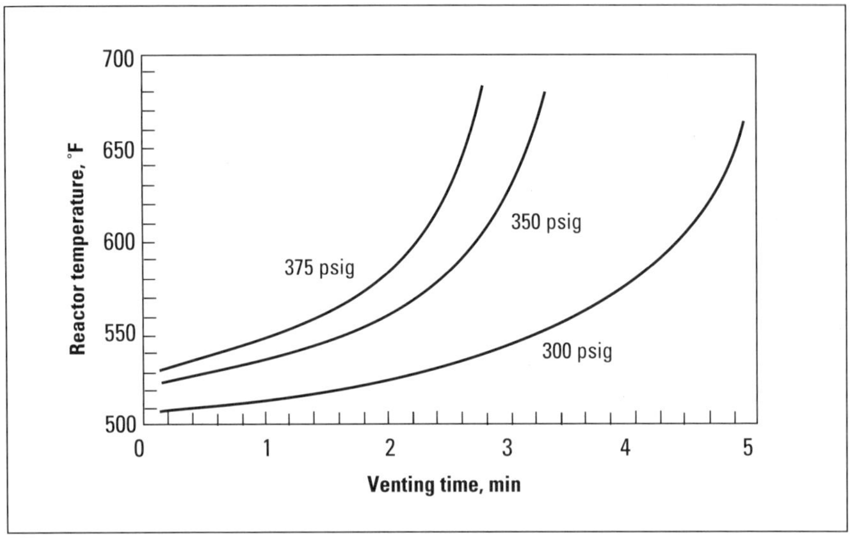

■ *Figure 1. For high-pressure process safety valve (PSV), temperatures exceed limit for carbon steel.*

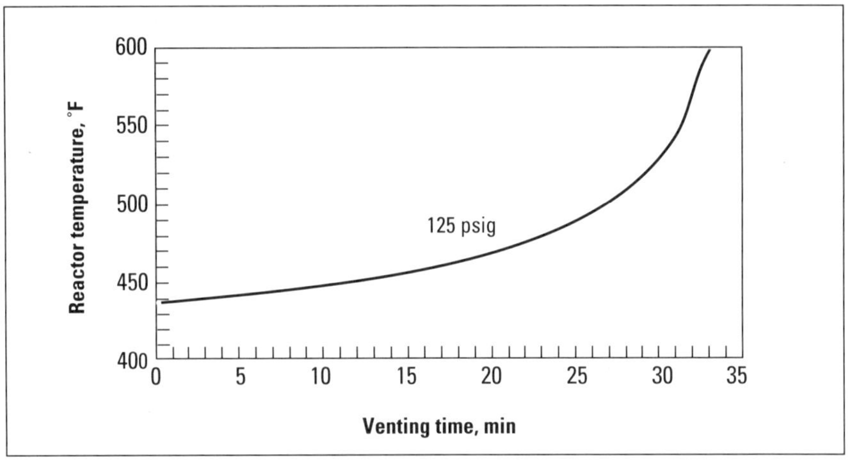

■ *Figure 2. Low-pressure relief valve yields acceptable temperature rise.*

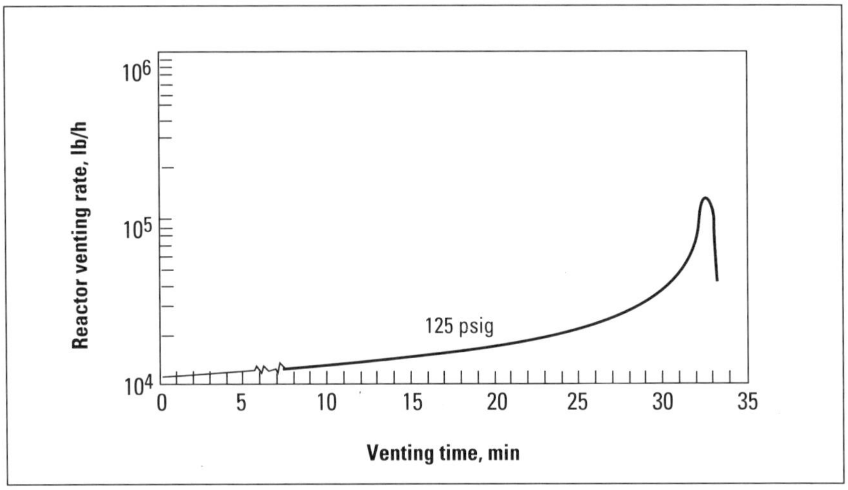

■ *Figure 3. Low-pressure vapor relief valve; maximum venting rate occurs near end of run.*

valve, as described above. Since reactor pressure during venting is not held constant, the design simulation is run with an assumed choke tube diameter.

Simulations were run using choke tubes with dia. of 2.85 and 3.75 in. for relief valves with 4- and 6-in. inlets, respectively. These sizes were selected after some preliminary trials. Since the reactor is designed for 430 psig, the low-pressure relief valve could be sized to allow reactor pressure to rise significantly above the initial set point value of 125 psig without exceeding the vessel design temperature. This is an important consideration, since a smaller relief valve would reduce the peak venting rate, decreasing the size and cost of downstream recovery and separation equipment.

Figures 4, 5, and 6 show the results of these simulations. Figure 4 shows that both relief-valve sizes keep the batch from exceeding the design temperature of the reactor. These values peak at temperatures below the maximum temperature. Loss of reactants vented with the two-phase flow shortened the venting time, reducing the water loss and resulting temperature rise.

Figure 5 shows the calculated reactor pressure during relief venting. As expected, Figures 4 and 5 both show higher temperature and pressure values for the smaller relief-valve size. The drop in temperature and pressure at the right side of each curve is due to the exhausting of one reactant. Venting beyond the peak reflects adiabatic flashing of the reactor contents, with no heat of reaction.

Figure 6 illustrates the venting rate from each relief-valve simulation. While the values are approximate due to uncertainties in the relief valve equivalent length, they do show a significantly larger venting rate than the vapor-only simulation. As expected, the smaller relief valve produces a lower venting rate and longer time for completion of the relieving event. These differences show that system design should include the effects of two-phase venting in any design. Traditional design methods do not allow such a

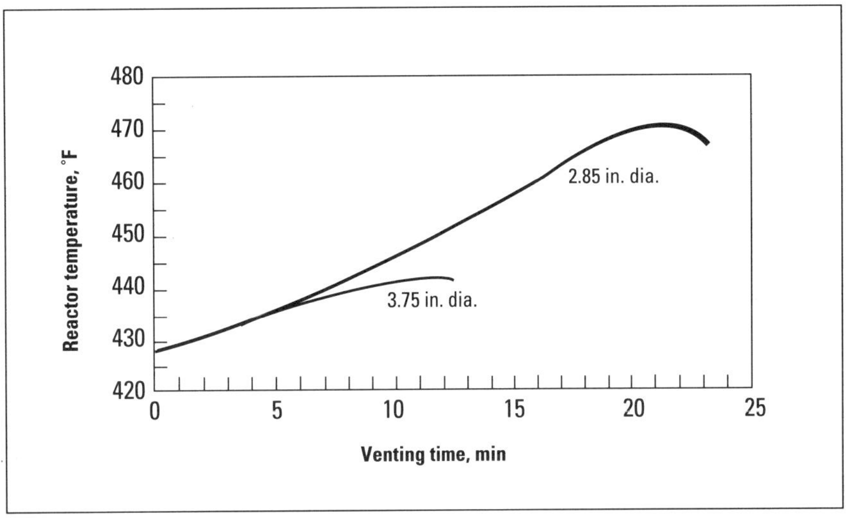

■ *Figure 4. Both relief valves keep the batch from exceeding the design temperature of the reactor.*

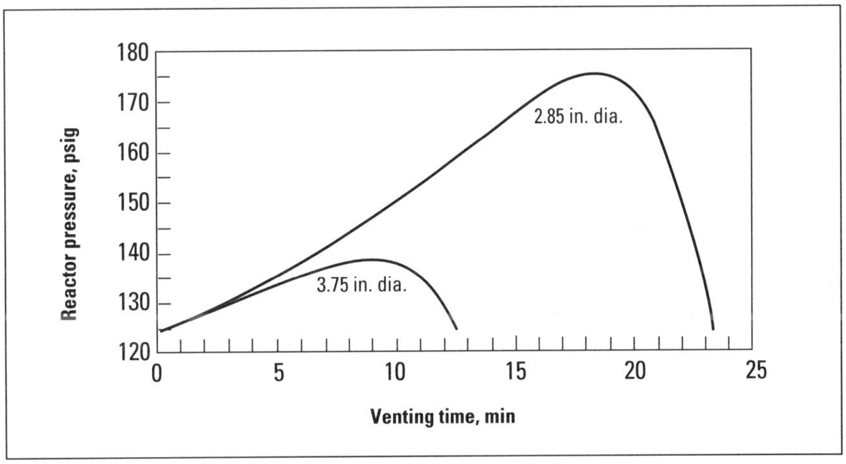

■ *Figure 5. Variable, two-phase low pressure; reactor pressure during relief venting reaches peak values.*

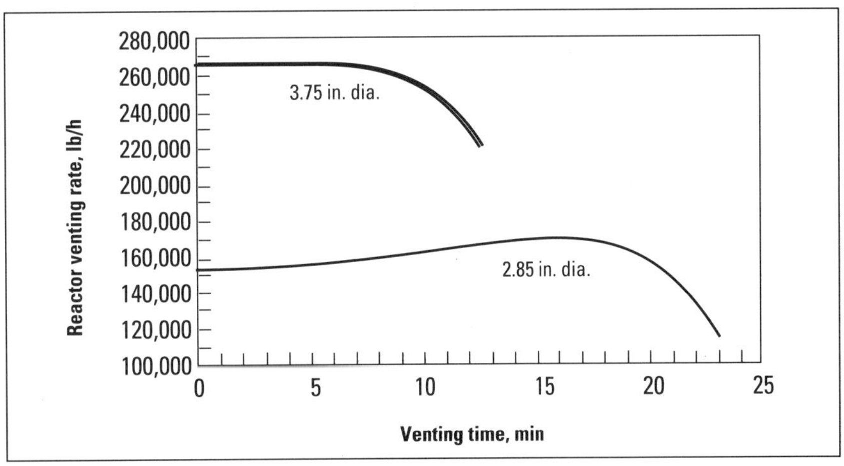

■ *Figure 6. Variable, two-phase low pressure; venting time varies with relief valve pseudo-diameter, back-calaculated from nominal orifice area.*

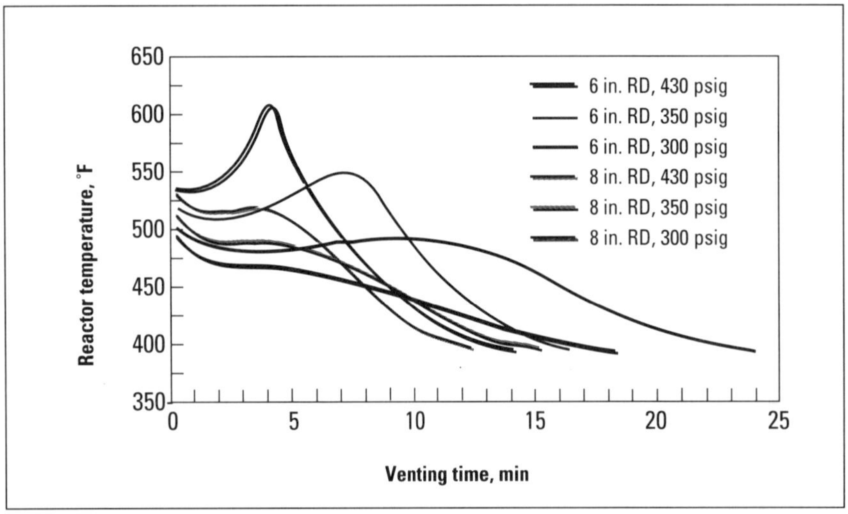

■ *Figure 7. Reactor temperature profiles for high-pressure venting, compressible gas.*

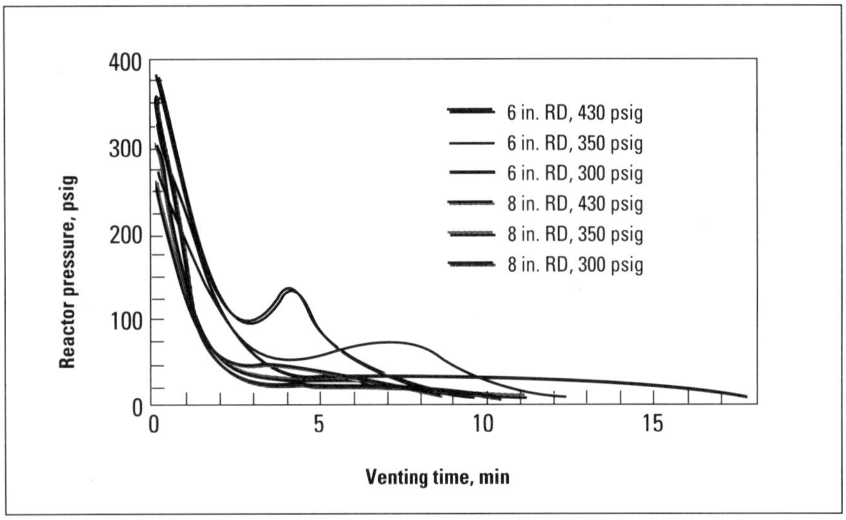

■ *Figure 8. Reactor pressure profiles for high-pressure venting, compressible gas.*

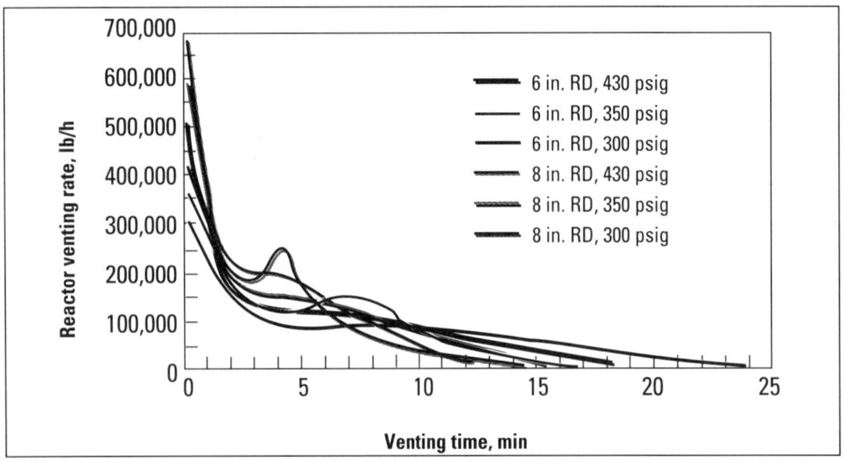

■ *Figure 9. Initial venting rates are different for each case, with 8-in. lines having significantly higher rates than 6-in. lines.*

calculation and would probably result in improperly sized separation equipment downstream of the relief valve.

Rupture disc simulation

When reactor pressure reaches the burst pressure of a rupture disc (RD), vapor escapes from the reactor. As with relief valves, this vapor may be accompanied by significant amounts of liquid. The limiting factor is critical flow at the rupture-disc support, which restricts flow from the reactor. Two sets of programs were developed, simulating a compressible gas-phase and two-phase homogeneous froth.

The programs perform an adiabatic flash calculation at assumed reactor pressures. This calculation determines the batch temperature, vapor flow rate, ratio of gas to liquid in the vent stream, and chemical composition, such that the sensible and latent heat changes equal the heat produced by the reaction. Chemical composition is revised at the end of each short time increment to reflect the chemical conversion and venting losses. Calculated reactor temperature and pressure, along with the quantity and composition of the vented material, are stored in a file for later design of downstream equipment. The program ends when a reactant is exhausted.

Compressible gas, single-phase venting

The program used for compressible gas-phase relieving is a simplification of the basic program used for two-phase relieving. The vapor discharge from the reactor, calculated at an assumed pressure, is used to calculate the sonic velocity at the rupture disc holder. Repeated calculation of the equivalent length in 1 psi increments allows a back calculation of the reactor pressure. If the back calculated pressure does not match the assumed reactor pressure, a new pressure is assumed and the calculations are repeated. These calculations were repeated for 6- and 8-in. rupture disc holders, assuming rupture disc set pressures of 300, 350, and 430 psig.

Figures 7, 8, and 9 show reactor temperature, pressure, and venting rate for

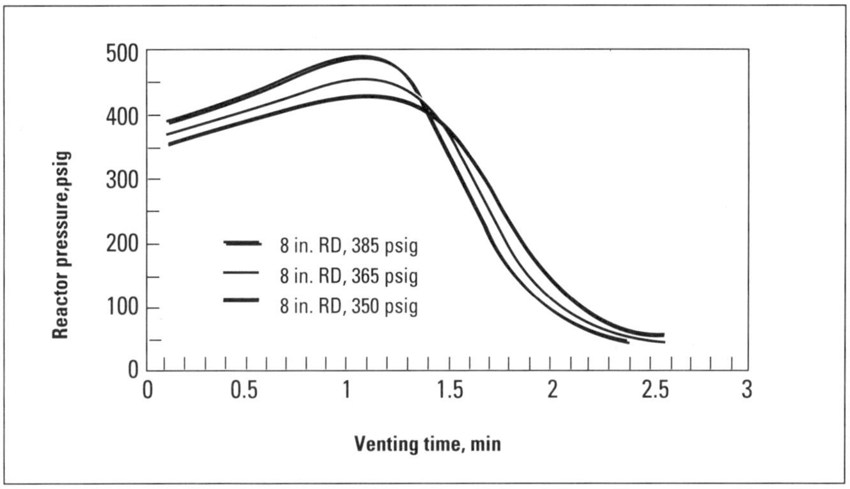

■ *Figure 10. Reactor pressure peaks roughly midway through the venting run.*

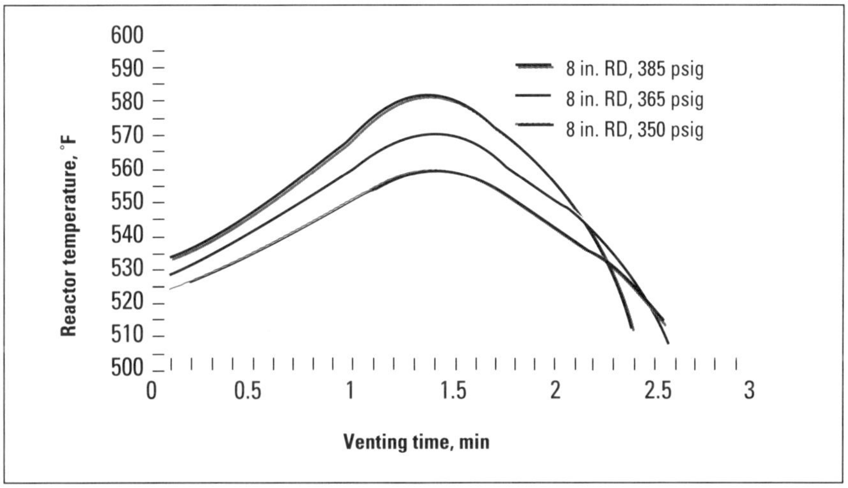

■ *Figure 11. Venting temperature for various rupture disc settings.*

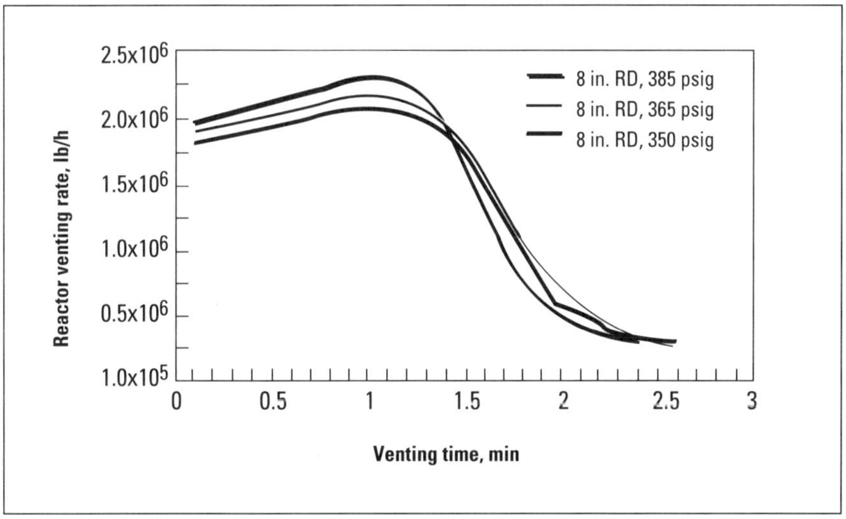

■ *Figure 12. Reactor venting rate for flashing flow through rupture disc at various bursting temperatures.*

each size and rupture disc burst pressure. While reactor pressure is no problem, Figure 7 shows that 6-in. rupture discs set at the reactor design pressure of 430 psig would result in the batch temperature briefly exceeding the coincident vessel design temperature of 600°F. Since reactor pressure is shown to have dropped to about 140 psig when this temperature is reached, this size could be considered acceptable if two-phase venting were not an issue. All other designs are more conservative.

The temperature and pressure peaks, calculated for the 6-in. rupture discs set at 350 and 430 psig, reflect changing kinetics and composition within the reactor. Initial vent gas has a relatively large proportion of water, allowing a rapid pressure drop. As the water concentration falls, the vapor rate set by the heat of reaction and adiabatic flashing due to pressure change results in an increase in organic solvent boilup. For the 350 and 430 psig rupture discs set in 6-in. lines, this increasing vapor rate causes the temperature and pressure to rise briefly. Depletion of reactants due to the ongoing reaction and loss with the relieving vapor soon slows the heat generated to a point where pressure and temperature again begin to fall.

Figure 9 shows the calculated relieving rates, based only on compressible vapor. Initial venting rates are different for each case, with 8-in. lines having significantly higher rates than 6-in. lines. Burst pressure was also significant in establishing peak venting rate. The higher the burst pressure, the larger the flow.

Compressible gas, with flashing two-phase venting

Relief through a rupture disc with two-phase flow was simulated using a program similar to the one employed for modeling low-pressure relief valves. The simulation was changed only to reflect the equivalent length of a choke tube based on the diameter of the rupture-disc holders and to continue calculation below the rupture-

Design of the equipment downstream of the rupture discs and relief valve is straightforward.

disc burst pressure. Rupture discs with a diameter of 8 in., the size of the reactor vent nozzle, were studied with set pressures of 350, 365, and 385 psig.

Two-phase flow greatly affects rupture disc relieving. Figures 10, 11, and 12 show reactor pressure, temperature, and relieving rate that increase after venting begins. A 385 psig rupture disc resulted in pressures of 490 psig and temperatures only slightly below the vessel design temperature. While an 8-in. rupture disc, set at 430 psig is acceptable for vapor relief in this system, flashing flow would result in exceeding both the design temperature and pressure of the vessel. This is a finding that would not be anticipated using traditional rupture-disc sizing methods.

Comparing Figures 9 and 12 shows that the relieving rate is considerably increased when flashing two-phase venting is considered. It should be noted that Figure 9 reflects a vapor stream, while Figure 12 shows a two-phase stream that is predominantly flashing liquid. Vapor rates leaving the reactor during two-phase venting are very low. Liquid chokes vapor flow by flashing. This phenomenon results in accumulated backpressure, producing the increase in reactor temperature and pressure. Discharge of much of the reactor batch as liquid, not chemical conversion, results in the eventual drop in reactor temperature and pressure.

Considering the uncertainties in these calculations, the final reactor design uses 8-in. rupture discs with a burst pressure of 350 psig.

Discharge piping, knockout tank, and discharge stack

Design of the equipment downstream of the rupture discs and relief valve is straightforward. The task is made easier in that the simulations have established a database of reactor discharge flows, temperatures, pressures, and compositions for each relieving condition. This information allows the engineer to calculate the percent flash at downstream system pressures. The resulting data permit stack sizing based on peak flash vapor rate and sizing of the relief system to prevent excessive backpressure. Vapor/liquid separators can be sized for the worst case of vapor and liquid flow. Thrust loading on the reactor relief discharge system can also be calculated.

Limitations

The only significant limitation is thought to be the lack of accurate choke-tube equivalent lengths for different rupture discs and relief valves. Given this limitation, simulation of rupture disc assemblies and relief valves as choke tubes is probably not unreasonable. Hydraulic flashing flow testing of relief valves by the manufacturer would be a considerable aid. CEP

Literature Cited

1. **Richter, S.H.,** "Sizing Relief Systems for Two-Phase Flow," *Hydrocarb. Proc.*, pp. 145-152 (July 1978).
2. **Richter, S. H.,** "For Flashing Fluids, Try Control Valve Sizing by Computer," *Petro/Chem Engineer,* pp. 23-26 (December 1968)

Acknowledgement

Special thanks are extended to Mel Pearson (John Brown, Houston) for his assistance during all phases of this project.

F. TURNER is a principal process engineer with John Brown in Bridgewater, NJ (908/429-9100; Fax: 908/429-9010). He has 14 years of experience in various projects including plants for polyethylene, polybutylene terephthalate, acrylonitrile-butadiene-styrene polymer, gas purification, Bisphenol A, and titanium dioxide. Recently he has worked as a Hazards and Operability Study (HAZOPS) leader. He is a graduate of Pennsylvania State University with a BS degree in chemical engineering.

S. H. RICHTER, now retired, was a principal process engineer with John Brown in Houston, TX, formerly in Lakeland, FL, and Cleveland, OH. He has worked on many projects, including plants for caprolactam, ethylene glycol, monomers, and various polymers. He was active on the API subcommittee on pressure relief systems and was chairman of the contractor's group of this subcommittee from 1983 to 1987. A graduate of Staatliche Ingenieur-Schule, Koethen, Germany, Richter holds an MS degree in mechanical engineering from Case Institute of Technology. He is a registered professional engineer in Ohio and a member of AIChE, as well as the author of various publications on flashing fluid flow.

Scale Up Heat Transfer Based on Reaction Calorimetry

Calorimetric data can be used to determine kinetic parameters and develop a predictive heat flow model, which in turn can be used to simulate required reactor jacket temperature profiles.

Ralph N. Landau and Donna G. Blackmond,
Merck & Co., Inc.

Pharmaceutical and fine chemical reactions are often accompanied by significant heat release, and must therefore be thoroughly understood if they are to be managed successfully on a plant scale. The relatively large heat losses from laboratory-scale apparatus make it difficult (or impossible) to quantify the rate of heat release associated with the chemistry being carried out. The advent of computer-controlled, laboratory-scale calorimeters has provided a useful tool for measuring heat flows accurately (to about 0.1 W/L) with a high degree of reproducibility *(1–4)*. Combining the calorimetric data with kinetic analysis allows the engineer to construct predictive heat flow models that are applicable to vessels of any volume.

This article demonstrates the power of calorimetry as it relates to process design and scale-up of batch and semi-batch processes. It outlines the required methodology of such an analysis, including experimental considerations, development of heat flow models, and subsequent simulations of large-scale plant vessels, ultimately leading to achievable and safe operating conditions.

Experimental concerns

The equipment used for this type of study should be capable of two very important functions: accurate simulation of the chemical process under investigation, and collection of sufficient accurate and precise data. These capabilities are not found in conventional laboratory glassware.

The reaction system should mimic the geometry, mixing conditions, and other features of the industrial process to the greatest extent possible. The system must also contain sufficient mass to properly simulate any heterogeneities found in the full-scale system.

Computer-controlled reaction systems represent the only manageable approach to simulating industrial processes properly. Furthermore, they offer the ability to collect more than sufficient amounts of data, automatically.

Batch reactions

The key to using heat flow techniques in modeling chemical processes lies in the relationship between the rate of heat evolution and the rate of reaction. In the heat flow technique discussed here, an energy balance is performed with limits at the reactor's outer surface. At any given time, for a batch process, the instantaneous heat flow due to reaction, q_r, is given by:

$$q_r = UA(T_r - T_j) + mC_p(dT_r/dt) \qquad (1)$$

A calibration experiment imposing known q_r values on the system (discussed in the sidebar on p. 44) is used to solve for the unknown quantities U and C_p.

For a simple reaction the heat flow is the product of the reaction rate, the specific heat of reaction, and the volume:

$$q_r = V_{rxn}(dC/dt)\Delta H_{rxn} \qquad (2)$$

Generally, for i multiple reactions

occurring simultaneously, Eq. 2 may be written as:

$$q_r = V_{rxn} \sum_i \left(\frac{dC}{dt} \right)_i \Delta H_{rxn,i} \qquad (3)$$

The implication of Eq. 2 is that embedded within heat flow data is information about reaction kinetics.

The choice of kinetic model may vary significantly depending on the ultimate use of the kinetic and thermodynamic parameters obtained. For strictly heat-transfer-related scale-up, often a first-order rate expression is sufficient, since the global heat flow, and not the individual species concentrations, is the focus.

In other studies, intrinsic kinetic parameters are desired, or specific thermodynamic properties (heat of reaction, heat of adsorption, and so on) for series or parallel reactions are needed. In these cases, and for other complex reaction systems, a more descriptive and physically accurate model (often with more adjustable constants) may be required.

For illustrative purposes, and because the focus of this article is on the scale-up of heat transfer, the technique will be demonstrated for a first-order reaction. The general approach for more complex systems will be outlined later.

Considering a single, first-order reaction, a pseudo-first-order model for the reaction can be derived as follows. While this may appear rather intuitive, especially for first-order reactions, unfortunately, in practice, many engineers don't make the connection between concentration and heat flow.

The kinetics are represented by:

$$dC/dt = -kC \qquad (4a)$$

$$\ln(C/C_0) = -kt \qquad (4b)$$

It follows that the heat flow is related to the reaction rate by:

$$q_r = -kC\Delta H_{rxn} V_{rxn} \qquad (5a)$$

$$q_{r,0} = -kC_0 \Delta H_{rxn} V_{rxn} \qquad (5b)$$

The final relationship is then:

$$\ln(q_r/q_{r,0}) = \ln(C/C_0) = -kt \qquad (6)$$

where C_0 and $q_{r,0}$ correspond to the concentration and heat flow values at the beginning of the reaction.

The relationship depicted in Eq. 6 allows determination of a pseudo-first-order rate constant from heat flow data. For predictive purposes, it would be useful to be able to model the variation in rate constants with temperature. This is done using Eq. 7, the well-known Arrhenius expression:

$$k = A_0 e^{-E_a/RT} \qquad (5)$$

The heat of reaction is easily obtained as the integral of the instantaneous heat flow:

$$\Delta H_{rxn} = \frac{\int_0^{t_f} q_r(t)dt}{n_{rxn}} \qquad (6)$$

A series of isothermal experiments over a suitable temperature range provides this kinetic information.

It is also possible to collect non-isothermal data that ultimately lead to the same type of information, and in some cases in a single experiment. In these cases, the Arrhenius parameters are fitted directly. In either case, with the Arrhenius parameters and the heat of reaction, the dynamic heat flow as described in Eq. 2 can be predicted for any imposed reactor temperature profile. Of course, this analysis requires that the chemical pathways do not vary significantly over the temperature range.

Whether investigating a first-order reaction or more complex systems, it is

Calibration procedure

A large part of the method described in this article relies on the use of heat flow data to complete the material and energy balance for the reacting system. Therefore, a brief discussion of the underlying principle may be helpful.

Reaction calorimeters analyze the raw data (temperatures and masses) by either the heat flow technique (heat flowing through the reactor wall) or the heat balance technique (sensible heat change of the cooling or heating medium).

The figure below shows the temperature profiles for the jacket and reactor contents during a typical calibration experiment.

There are two distinct parts to the calibration because Eq. 1 has two unknowns.

The first part of the experiment involves a short ramping of the jacket temperature. During this period, $q_r = 0$ since there is no heat flow.

The second part of the calibration involves a precision heating probe providing a known heat flow. The calorimeter is instructed to maintain the desired isothermal temperature (using the jacket to remove the heat flow), so $dT_r/dt = 0$ and q_r is equal to the power rating of the probe.

From these two experiments, the heat-transfer coefficient and the heat capacity may then be determined.

Normally, this procedure is carried out before initiating the reaction and after the reaction is complete to account for changes in these properties with extent of reaction. Using these two calibration endpoints, the energy balance shown in Eq. 1 is evaluated by appropriately varying the properties U and C_p over the course of the reaction.

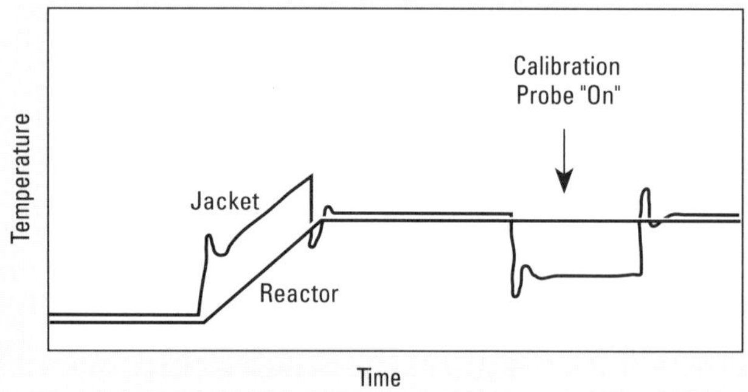

■ ***Temperature profiles for the jacket and reactor contents during a calibration experiment.***

	A	B	C	D	E
1	Time	Temp	k	Conc	Heat Flow
2	(min)	(C)	(min^{-1})	(M)	(W/L)
4	Raw Data	Raw Data	$=Ae^{-Ea/RT}$	$=C_{n-1}(1-k\Delta t)$	$=(\Delta C/\Delta t)^{*}\Delta H_{rxn}$
6	0	25	0.0345	1.000	48.10
7	1	26	0.0380	0.964	50.52
8	2	27	0.0418	0.925	53.57
9	3	28	0.0459	0.885	56.56
10	4	29	0.0504	0.842	59.43
11	5	30	0.0554	0.798	62.12
12	6	31	0.0607	0.751	64.58
13	7	32	0.0666	0.704	66.72
14	8	33	0.0730	0.654	68.49
15	9	34	0.0799	0.604	69.79
16	10	35	0.0875	0.554	70.67
	•	•	•	•	•
	•	•	•	•	•
	•	•	•	•	•

■ *Figure 1. Numerical integration on a spreadsheet; row 4 contains the formulas.*

interesting to note that the heat flow and kinetics parameters acquired, and all of the subsequent insight for potentially complex and energetic reactions, are obtained from two simple temperature measurements of the reactor and the jacket in a noninvasive *in situ* experiment. These parameters can now be used to develop a predictive heat flow model, and eventually may be applied to other reactor vessels and operating conditions.

Heat flow modeling

Once the Arrhenius parameters and the heat of reaction have been determined, the rate of heat evolution can be estimated. The instantaneous heat flow may be calculated by combining the rate constant determined from Eq. 7 with Eqs. 6 and 2. Assuming that the experimental temperature profiles predict the experimental heat flows allows for a check on the model. Figure 1 depicts the numerical procedure carried out on a spreadsheet, where E_a = 17 kcal/mol, $A_0 = 10^{11}$/min, and ΔH_{rxn} = 20 kcal/mol.

The heat flow data produced in column E of Figure 1 may be directly compared to raw experimental data to check the model. Assuming the model is acceptable, modeling of full-scale vessels is now possible.

Note that this type of numerical method requires that the step size (Δt) be small enough to allow dC/dt to be calculated as $\Delta C/\Delta t$. The investigator is encouraged to try progressively smaller step sizes for Δt (while maintaining all other parameters constant) until no differences in heat flow profiles are observed.

Plant modeling

The ultimate use of the kinetic model based on laboratory data is to predict vessel performance in the plant. The ability to predict such behavior depends primarily on estimates of the intrinsic properties of the plant vessel, namely, the overall heat-transfer coefficient. The influences of changes in reactor geometry, agitation, and mass transfer when moving from the laboratory to the plant are assumed here to be minor relative to heat-transfer issues. Laboratory reactors can generally provide better mixing and, therefore, the observed reaction rate will be equal to or higher than that expected in the plant. Thus, this approach, based upon laboratory rate data, represents a worst-case scenario.

Observations of the proposed plant vessel indicate its heat-transfer capabilities. Consider a charge to the vessel that contains solvent and as many other constituents as possible without initiating the reaction. By simply measuring the best consistent cooling rate while maintaining a certain ΔT across the jacket, the heat-transfer coefficient can be estimated. Equation 1 can be solved for UA, and in this case, q_r = 0. From this, an effective UA_{plant} can be obtained by using the heat capacity determined during the laboratory calibrations. This UA_{plant} can be used to simulate several scenarios in the plant, including the desired operating conditions, alternative temperature programming options, and semi-batch operations.

The relative heat-transfer ability of the laboratory and the pilot-plant vessels can now be compared. To compare the plant value to the lab-derived UA, both must be divided by their respective mass charges, yielding $(UA/m)_{lab}$ and $(UA/m)_{plant}$. The ratio of lab-to-plant heat-transfer capabilities can often be 50 or more. Therefore, it is clear that while laboratory studies may indicate ease in handling heat removal (perhaps a 3°C ΔT across the jacket), often larger vessels will present significant challenges (if the ratio is 50 as mentioned above, the ΔT would be 150°C in the plant) when operating at similar conditions.

Using the heat flow model developed earlier, estimates of the required plant jacket temperature profile may be made by rearranging Eq. 1 as follows:

$$T_j = T_r - \frac{q_r - m_r C_{p,r}\left(\frac{dT_r}{dt}\right)}{UA} \qquad (9)$$

By casting the energy balance in terms of required jacket temperature profiles (an easily measurable quanti-

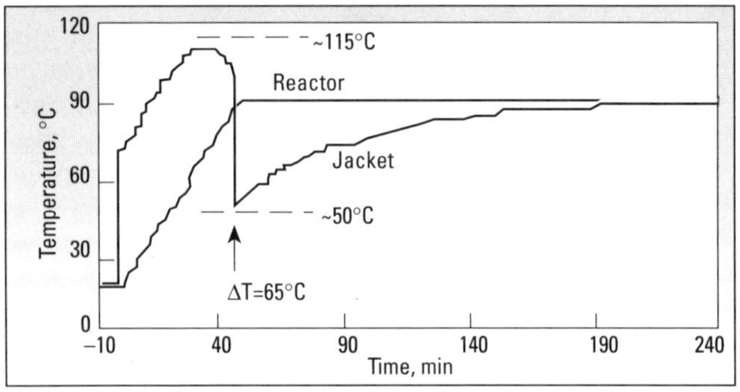

■ *Figure 2. Jacket temperature profile required to achieve reactor temperature ramping of 1.5°C/min from 20°C to 90°C, followed by isothermal reaction at 90°C.*

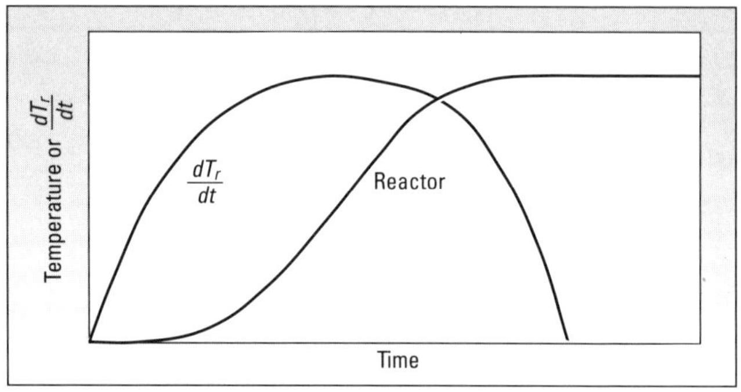

■ *Figure 3. A parabolic dTr/dt curve results in a smooth approach to the desired upper set point for the reactor temperature.*

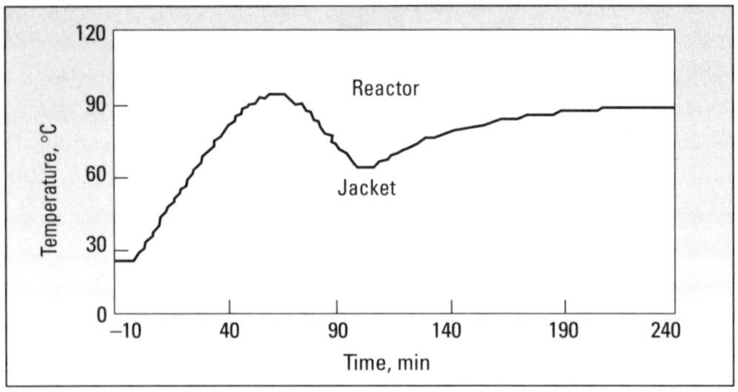

■ *Figure 4. Required jacket temperature profile if the time taken to reach 90°C is 100 min and a parabolic dTr/dt constraint is imposed.*

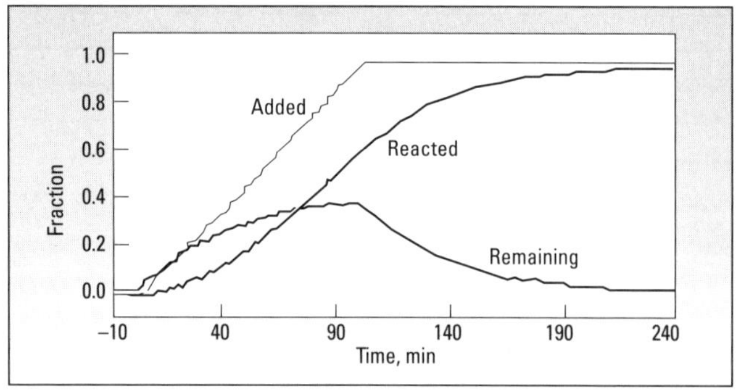

■ *Figure 5. Fractional amount of limiting reagent added, reacted, and remaining in the reactor for a 100-min addition period and reaction at 90°C.*

ty), engineers and operators can readily assess the feasibility of a particular scenario in the plant.

Consider a batch process that begins at ambient conditions, ramps the reactor temperature at a constant rate of 1.5°C/min to 90°C, then undergoes an isothermal reaction at 90°C. Figure 2 shows the required jacket temperature profile to achieve such conditions. (The parameters used for this simulation were $E_a = 21$ kcal/mol, $A_0 = 10^{11}$ /min, $\Delta H_{rxn} = 30$ kcal/mol, and $UA/m = 2$ (W/K)/kg; $\rho_{solution}$ was assumed to be 1 kg/L, and C_p to be 4,000 J/kg•°C.) This calculation is easily accomplished by extending the spreadsheet in Figure 1 (with the appropriate constants and raw data) by adding a column that determines the righthand side of Eq. 9 and calculating dT_r/dt as $\Delta T_r/\Delta t$.

Practically, the discontinuity in the slope of the T_r curve, dT_r/dt, at the moment the reactor temperature reaches the desired set point (90°C) and during the initiation of the ramp causes extreme difficulty, as displayed in the jacket temperature response curve. The jacket must respond by dropping approximately 65°C from 115°C to 50°C in just a few minutes.

The solution is to consider a more realistic reactor temperature profile. Initially, dT_r/dt will be equal to 0, then as the jacket temperature rises, dT_r/dt will slowly increase. As the temperature approaches the upper set point, it would be desirable for dT_r/dt to tend smoothly toward 0 once again to avoid the abrupt discontinuity at the end of the ramp. Mathematically, this is shown as:

$$\left(\frac{dT_r}{dt}\right)_{t=0} = 0 \tag{10a}$$

$$\left(\frac{dT_r}{dt}\right)_{T_r=90°C} = 0 \tag{10b}$$

Equations 10a and 10b suggest the use of a parabola to represent the variation in the derivatives with time, since the curve must pass through 0 at both ends of the ramping period and has a maximum. Using an expression such as Eq. 11b for obtaining a reactor temperature profile will significantly reduce the stiffness of the required jacket temperature response.

$$\frac{dT_r}{dt} = at^2 + bt \tag{11a}$$

$$T_r = \frac{at^3}{3} + \frac{bt^2}{2} + c \tag{11b}$$

Because the derivative is 0 at $t = 0$, there is no zero-order parameter in the expression for dT_r/dt. The initial temperature is assumed to be 20°C, so the constant of integration (c in this case) will be equal to 20.

Figure 3 illustrates this approach qualitatively. During the temperature ramp, the value of dT_r/dt reaches a maximum exactly half way to the desired upper set point for the reactor temperature. There are no sudden changes in dT_r/dt as there were in the previous examples. This will provide the safest mode for a batch operation. The variable that determines the constants a and b in Eqs. 11a and 11b is the time taken to reach 90°C. Under plant conditions, several linear ramps would probably be executed to simulate the parabolic dT_r/dt constraint.

Figure 4 shows the resulting required jacket temperature profile if the time taken to reach 90°C is 100 min. The relationship for the reactor temperature profile is provided by Eq. 11b with $a = -0.00042$, and $b = 0.042$. (The same values for E_a, A_0, ΔH_{rxn}, and UA/m were used as in the previous example.)

Clearly, this approach yields a more manageable jacket temperature profile, with a 50-min period to cool from 94°C to 63°C (approximately 0.6°C/min). Additionally, while it is not explicitly plotted, the conversion is evident by the temperature difference between the jacket and the reactor, and is roughly complete at 200 min. This is essentially the same reaction time required by the case presented in Figure 3, which ramped directly to 90°C at 1.5°C/min but resulted in an unrealistic jacket temperature profile. Therefore, productivity is affected only minimally, while the intrinsic safety has been significantly enhanced.

Each case will require testing various heat-up times to find a manageable and safe profile for the plant. By using a spreadsheet, simply changing one cell entry (the time to reach 90°C) allows for a rapid simulation of the expected plant performance.

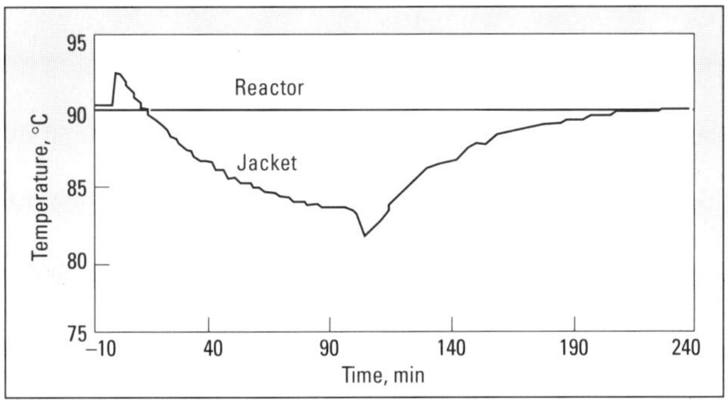

■ ***Figure 6. Required jacket temperature profile for the semi-batch process with a 100-min addition period and reaction at 90°C.***

Semi-batch reactions

Another feasible mode of operation for exothermic reactions is semi-batch *(5–6)*, which can be carried out to satisfy both safety and productivity concerns. In general, semi-batch operations are intrinsically safer than their fully batch counterparts due to greater control of the chemical potential of the material in the reactor. When operating in a fully batch mode, the entire chemical potential of the reaction is present from the outset and may require careful temperature control to maintain an acceptable level of risk during the operation. Unless material handling is extremely difficult or hazardous, semi-batch operations will provide an additional margin of safety because the chemical potential is added to the reactor as needed and can be stopped at any moment simply by stopping the feed.

The procedure for this particular case is to add all of the components to the reactor except the limiting reagent; the latter is mixed with an appropriate amount of the solvent to form a slurry that can be pumped. The reactor is heated to 90°C by any means (no reaction will be occurring yet), then the slurry is slowly added to the reactor. Because the slurry is at ambient conditions, it provides sensible cooling to the reactor, thus reducing the cooling requirements. Equation 10 is the energy balance obtained when operating in a semi-batch mode:

$$q_r = UA\left(T_r - T_j\right) + m_r C_{p,r}\frac{dT_r}{dt} + \frac{dm_d}{dt}C_{p,d}\left(T_r - T_d\right) \qquad (12)$$

Equation 12 differs from Eq. 1 in that it contains a term to account for the temperature difference between the feed and the reactor contents. Generally, if it is desired to maintain the reactor temperature constant (as it is in this case), the second term will disappear.

Nomenclature

- a = third-order coefficient in temperature expression, °C/s³
- A = area available for heat transfer, m²
- A_0 = pre-exponential factor in Arrhenius expression, min⁻¹
- b = second-order coefficient in temperature expression, °C/s²
- c = zero-order coefficient in temperature expression, °C
- C = concentration, moles/L
- C_0 = concentration at $t = 0$, moles/L
- C_p = heat capacity of reactor contents, J/kg•K
- $C_{p,d}$ = heat capacity of added material, J/kg•°K
- $C_{p,r}$ = heat capacity of reactor contents, J/kg•K
- dC/dt = general rate of reaction, (moles/L)/min
- dm_d/dt = addition rate of added material, kg/s
- dT_r/dt = time derivative of reactor contents temperature, °C/s
- E_a = activation energy in Arrhenius expression, kcal/mole
- DH_{rxn} = molar heat of reaction, kcal/mole
- k = pseudo-first-order rate constant, min⁻¹
- m = mass of reactor contents, kg
- m_r = mass of reactor contents, kg
- n_{rxn} = amount of reactant converted, moles
- q_r = instantaneous heat flow, W
- $q_{r,0}$ = instantaneous heat flow at $t = 0$, W
- R = gas law constant = 1.987 ¥ 10^{-3} kcal/mole•K
- t = time, min
- t_f = final time for reaction, min
- T = absolute temperature, K
- T_d = temperature of added material, °C
- T_j = temperature of jacket contents, °C
- T_r = temperature of reactor contents, °C
- U = overall heat-transfer coefficient, W/m²•K
- UA_{plant} = overall heat-transfer coefficient × wetted area for pilot plant, W/K
- V_{rxn} = volume of reaction mass, L

Greek Letters

- $r_{solution}$ = density of solution, kg/L

Figure 5 shows the fractional amount of limiting reagent added, reacted, and remaining (unreacted) in the reactor for a 100-min addition period and reaction at 90°C (again using the same values for E_a, A_0, ΔH_{rxn}, and UA/m). The slurry feed was assumed to be 10% of the total reactor charge (that is needed to evaluate the third term in Eq. 12).

Focusing on the amount in the reactor at any given time, one can see that the highest concentration is about 0.40 (the fraction remaining to react), which corresponds to the end of the addition period. This is less than the 0.58 remaining when the reactor first reaches 90°C in the best-case scenario for a fully batch mode of operation as discussed earlier. This translates into a reduced workload for the jacket, since the reaction rate will be reduced.

Figure 6 depicts the required jacket temperature profile for the semi-batch process. This profile is considerably less demanding than any of the batch profiles, starting at 93°C, dropping slowly over 100 min to 80°C, then slowly returning to 90°C. The slight (roughly 3°C) but rapid changes in the required jacket temperature at the beginning and end of the addition period are due to the same mathematical effect discussed earlier, but in this case are caused by the sudden changes in dm_d/dt. The same approach could be used, utilizing a parabolic constraint on dm_d/dt, to yield a smoother transition. However, because this profile is not exceedingly demanding, further improvements may be unwarranted.

Note that roughly when the addition period is complete, the fraction remaining to react (Figure 5) follows a standard first-order decay. This behavior should be expected at *any* time that the feed is halted. Therefore, during normal operations, if it appears that the reaction is beginning to run away, an adverse situation may be averted by simply stopping the feed. The heat flow will decay, control will be regained, and the addition can be resumed later. In essence, running in a semi-batch mode will provide two leverages against a runaway reaction — the jacket temperature and the feed rate.

Complex systems

The discussion thus far has addressed the use of heat flow data for simple first-order systems. In complex reaction systems, a first-order model will not properly describe the temporal variation observed. Higher-order and multiple-step kinetic models do not produce a simple relationship between q_r/q_{r0} and C/C_0 as depicted in Eq. 6.

However, the numerical approach described in Figure 1 can be applied to the raw heat flow data using Eq. 3. The mathematical model that simultaneously describes the material and energy balances for any complex reaction system will contain several (often too many) adjustable constants, each of which has physical significance. The spreadsheet can, given a set of those constants, simulate the variation of all measurable mass and energy temporal events that are explicitly or implicitly described by the model. Therefore, in theory, by adjusting, or "regressing," these parameters, a minimum discrepancy between the simulation and the raw data can be obtained. Many spreadsheets now have the ability to search for these constants automatically, given a reasonable initial estimate.

After estimating the parameters in this way, one can simulate plant reactor conditions as described for the first-order model, using the complex model in its place.

Influences due to changes in reactor geometry, agitation, and mass transfer when moving from the laboratory to the plant were initially assumed to be minor relative to heat-transfer issues. This restriction may be relaxed considerably if the physical model for the complex system accounts for any potential issues that appear to be significant. CEP

Literature Cited

1. **Regenass, W.,** "Thermal and Kinetic Design Data from a Bench-Scale Heat Flow Calorimeter," ACS Symposium Series, **65**, pp. 37–42 (1978).
2. **Karlsen, L. G., H. Soeberg, and J. Villadsen,** "Optimal Data Acquisition for Heat Flow Calorimetry," *Thermochim. Acta,* **72**, pp. 83–89 (1984).
3. **Litz, W.,** "The Thermokinetic Reactor TKR and its Possible Applications in Chemical Research and Engineering," *J. Thermal Anal.,* **27**, p. 215 (1983).
4. **Landau, R. N., and L. R. Williams,** "Reaction Calorimetry: A Powerful Tool," *Chem. Eng. Progress,* **87**(12), pp. 65-69 (Dec. 1991).
5. **Mills, P. L., P. A. Ramachandran, and R. V. Chaudari**, "Multiphase Reaction Engineering for Fine Chemicals and Pharmaceuticals," Reviews in Chemical Engineering, 8(1–2), pp. 5-176 (1992).
6. **Paul, E. L.,** " Design of Reaction Systems for Specialty Organic Chemicals," *Chem. Eng. Science,* **43**, pp. 1773–1782 (1988).
7. **Landau, R. N., D. G. Blackmond, and H. H. Tung,** "A Calorimetric Investigation of an Exothermic Reaction: Kinetics and Heat Flow Modeling," *I&EC Research,* **33**(4), pp. 814–820 (1994).

Acknowledgments

The authors would like to recognize the support of Dr. H. H. Tung of Merck Research Laboratories, Rahway, NJ. This overview is based on a more complete experimental study available in the literature *(7).*

R. N. LANDAU is a technical operations fellow at Merck & Co., Inc., Whitehouse Station, NJ (908/730-1402; Fax: 908/730-1414). Currently, he manages the project work being carried out in the Reaction Engineering Laboratory (REL), whose primary mission is to conduct integrated process development, process safety, and optimization studies using advanced technologies for both emerging and existing processes. Previously, he was responsible for process safety studies in the Process Safety Laboratory. He holds a BS from the New Jersey Institute of Technology and a PhD from the Univ. of Delaware, both in chemical engineering. He is a member of AIChE.

D. G. BLACKMOND is associate director of the Technical Operations Dept. at Merck & Co., where she conducts research in the area of heterogeneous catalysis. Previously, she was an associate professor in chemical engineering at the Univ. of Pittsburgh. She was named a Presidential Young Investigator by the National Science Foundation in 1986. She currently serves on the editorial board of the Journal of Catalysis and is a director of the Organic Reactions Catalysis Society. She has authored over 35 journal articles and presented more than 85 technical papers, including invited lectures in seven countries. She earned a BS in chemical engineering from the Univ. of Pittsburgh and a PhD in chemical engineering from Carnegie Mellon Univ. She is a member of AIChE.

Organize Your Resources to Improve Maintenance Operations

Follow this method to a safer, successful and cost-saving maintenance program.

Timothy E. Dowdle and Brian R. Goedke,
Quantum Chemical Co.

Within every company, there are valuable resources that can be used to improve equipment reliability and safety, and reduce maintenance costs. These resources are the technical knowledge and skills of employees.

How does a company pull together all of its best maintenance techniques and ideas from its various locations? How does a company establish a multi-site maintenance culture?

Maintenance is the largest controllable cost at most chemical process industries plants. It is vital that this expenditure be controlled and directed to maximize onstream time while minimizing costs.

At Quantum Chemical Co., a maintenance technology center (MTC) was established to foster use of modern maintenance management techniques throughout the company. The MTC is not a physical location, but rather it consists of groups of persons who have expertise in various maintenance technologies.

There has been a great deal of technological improvement in the maintenance field during the last few years. Advanced monitoring equipment has allowed a move from hours-based maintenance to condition-based maintenance. In traditional hours-based maintenance, equipment is routinely disassembled for preventive maintenance. Now, thanks to vibration analysis, oil analysis, thermographic surveys, and hypercompressor plunger monitoring, among others, most equipment is overhauled based on its condition. These techniques allow for the pinpointing of specific problems prior to shutting equipment down for repair.

Goals

The objectives of the MTC are to improve equipment reliability, increase unit onstream time, and reduce maintenance costs. These goals are accomplished through:

- Benchmarking company plants to each other and to plants at outside companies. This is typically done through contacts made at technical meetings and conferences. The process usually starts out informally and may become a formal procedure.
- Establishing uniform maintenance procedures and practices throughout the company.
- Determining weak areas in manufacturing units and recommending programs and projects to take them to a state-of-the-art status.
- Working with the plant superintendents and business teams to focus resources on the projects which promise the highest return on investment.

The first sidebar summarizes the activities of the MTC program.

MTC membership

The MTC is composed of a management team of five members. The team includes each of the maintenance managers at our five major sites. In addition, separate teams, reporting to the MTC, focus on reciprocating, rotating and stationary equipment, electrical and instrumentation, condition monitoring, planning, scheduling, and stores.

Members of the MTC report to the vice president of manufacturing technology. Maintenance team members are chosen from maintenance personnel at each of our plants. Team members are highly qualified, recognized experts in their specific field. The captains of the maintenance teams report to a member of the MTC management team, who sponsors the team.

This type of organizational structure promotes open communication and regular contact with top management. The close working relationship with management is important to the success of the MTC.

Procedures for establishing maintenance program improvement

- ✔ Establish benchmark and write model
- ✔ Issue models to evaluate lines
- ✔ Benchmark all operating lines
- ✔ Use "best" to set standards

- ✔ List opportunities
- ✔ Evaluate and prioritize
- ✔ From model, list all opportunities for improvement
- ✔ Evaluate and prioritize based on projects with greatest impact

- ✔ Define programs for implementation
- ✔ Scope projects
- ✔ Define resource requirements:
  - — Labor
  - — Expenses
  - — Capital
- ✔ Set time line targets

- ✔ Implementation
- ✔ Obtain technology board approval
- ✔ Implement project
- ✔ Post audit to assure success
- ✔ Track progress for continuous improvement

Maintenance technology projects

From their discussions, MTC members develop a list of potential projects. The project proposals are submitted to the manufacturing technology center board for consideration and approval. The board consists of plant managers, the vice president of manufacturing, and the director of engineering. The board meets with the MTC once a year. When the board approves a project, it authorizes the labor expenses required.

The MTC team is responsible for overseeing the implementation of the approved projects. It conducts audits to assure that the expected results are achieved.

There are currently 20 MTC projects in various stages of implementation. These projects include such tasks as automated instrumentation calibration, turnaround planning and execution software, electric motor reliability, and oil mist lubrication. These projects have projected savings of about $20 million per year, upon completion. Three of these projects have already saved in excess of $3.8 million. These projects are improving condition monitoring at the smaller facilities, reducing lubrication costs by standardizing and optimizing lubricants, and, at one site, improving the on-stream time of polyethylene reactors.

A major benefit of the MTC is that members champion the particular innovation or proven technique that their team has targeted. When an improved method is discovered, it is quickly applied division-wide, thus ensuring rapid realization of the project's benefits.

Strong management commitment is essential to the success of the MTC concept. Management support ensures that approved projects are completed. The allocation of personnel, money, and outside resources to MTC projects manifests management's strong commitment to the success of these projects.

Maintenance excellence model

One successful MTC project has been the implementation of the Quantum maintenance excellence model. The model was developed to ensure a consistent approach to management of maintenance within the company.

There are over 30 sections in the current model. These sections cover the entire range of maintenance activities including safety, training, planning and scheduling, machinery alignment, and instrument testing, among others. As an example, the lubrication model is presented in the second sidebar.

The model sections were developed by Quantum maintenance professionals with representation from all major manufacturing locations. The model sections represent the "best of the best" technology and maintenance management techniques. The sections are reviewed annually and updated to ensure that new technology and industry requirements are evaluated and incorporated.

A model's sections are used to evaluate the maintenance functions strengths and weaknesses at the plants through an audit process. The audit feedback process helps to define departmental objectives and necessary resources for continuous improvement. Audits are done once a year.

The audits are done by team members along with a senior maintenance person from another company location. Results of the audits, including a written report and summary of recommendations, are given to the department managers and plant managers.

Meetings

The MTC meets every 2-3 months to review current and new projects, and to discuss changes and recent innovations in the maintenance area and to discuss unique maintenance

Lubrication program model

Score

- **Establish, document, and maintain lubrication specifications.**
  Yes = 5 points; No = 0 points ____________

- **Document all changes to specifications.**
  Yes = 5 points; No = 0 points ____________

- **Label equipment pieces with lubricant type required.**
  Yes = 5 points; No = 0 points ____________

- **Establish a plant-wide (centralized) lubrication crew.**
  Yes = 10 points; No = 0 points ____________

- **Use monthly oil sampling for wear-particle analysis to identify:**
  — *Normal wear*
  — *Abnormal wear*
  — *Severity of wear*
  — *Part(s) generating wear particles*
  Yes = 5 points; No = 0 points ____________

- **Use quarterly oil sampling for physical properties tests consisting of:**
  — *Viscosity*
  — *Color*
  — *Water content*
  — *Oxidation*
  — *Total acid number*
  — *Cleanliness (particle count)*
  Analysis quality based on comments above:
  Yes = 15 points; No = 0 points. ____________

- **Coordinate with vibration analysis group to determine equipment condition and issue reports to maintenance and production.**
  Yes = 5 points; No = 0 points ____________

- **Provide proper storage for lubricants to prevent contamination and degradation of the lubricant and its container. Indoors is preferable, outdoors in horizontal racks and covered is acceptable.**
  Yes = 5 points; No = 0 points ____________

- **Use designated computer program to establish routes and schedules.**
  Yes = 10 points; No = 0 points ____________

Total possible score = 65 points
Total score = ____________

Note: A numerical score is used so that a team can measure its progress from year to year; it is expected that the score will increase each year.

problems encountered at the various sites. To give members firsthand knowledge of each plant's specific equipment and processes, the meetings are rotated among the different Quantum locations. The MTC members also hold regular telephone conference calls to minimize travel costs while ensuring good communication and followup.

Other benefits

In addition to the significant cost savings, other benefits achieved with the MTC concept include:

- Innovative manufacturing practices are rapidly implemented company-wide.
- Duplication of resources is minimized.
- Multiple plant perspectives are considered in developing the best practices to use company-wide.
- Acceptance by all users is established during the improvement review procedure.
- Communication channels are established among plant personnel.

The success of the MTC concept also can be seen in the sense of professionalism among the maintenance personnel at the Quantum plants. CEP

T. E. DOWDLE general manager for Quantum Chemical Company's Morris and Tuscola, IL, plants (815/942-7201; Fax: 815/942-7200). He has been with Quantum since 1980. He holds a BS in mechanical engineering from the University of Illinois in Champaign.

B. R. GOEDKE is plant manager for Quantum Chemical Company's Tuscola, IL, plant (217/253-3311; Fax: 217/253-2906). He also serves as chairperson of Quantum's MTC. He has been with Quantum since 1975. Prior to becoming plant manager at Tuscola in 1993, he was maintenance manager for Quantum's Morris, IL plant. He is a graduate of Illinois Institute of Technology with a BS in electrical engineering.

Use a Quality Management Plan for Engineering Design

This system minimizes risks and costs, and facilitates process design, plant construction, and startup. Controlling the flow and quality of information is crucial.

William J. Colt,
Morrison Knudsen Corp.

This focus explores the design process in three phases: the initial stage of design implementation; the detailed design; and construction, checkout, and startup.

INITIAL DESIGN IMPLEMENTATION PHASE

As with a manufactured product, quality management of engineering design begins with "raw materials" control. Here, information is equivalent to raw materials. Engineering design services, whether provided by an internal unit or contracted outside, receive input from many information sources. To implement a quality program, the design project engineer must employ a quality management plan (QMP) that treats these sources as raw materials suppliers.

The front-end engineering portion of the QMP addresses the design activities that occur during the scope-of-work development (Figure 1). Obviously, these activities are critical in that they form the basis for the project cost estimate, and ultimately, for the approval or disapproval of project funding.

The success of the project initiation portion of the quality plan, and often of the overall project, is predicated on early interface with the "raw materials" suppliers. Therefore, a primary goal of the QMP is to provide the design team with a management-backed tool that outlines the requirements of this early interface. The plan should also emphasize the team's expectations regarding the quality and completeness of input data.

On the other hand, if the early interface is not accomplished, the QMP should facilitate recognition of any poorly defined areas in the scope-of-work, so that appropriate risk management can be exercised. This philosophy must be implemented as a responsibility to the customer's goal. No customer appreciates being informed well into the detailed design that the scope-of-work was incomplete, and therefore, the schedule has slipped and the budget will be exceeded.

Scope-of-work documents

The largest contributor to design-related rework is a poorly defined scope-of-work. During the detailed design phase this problem manifests itself as a productivity issue, resulting in schedule slippage, cost overruns, and field rework. The average cost of rework on industrial projects exceeds 12% of total installed cost. Design deviations are responsible for 80% of this rework.

Figure 2 provides a punch list (check list) that defines a minimum scope-of-work. A quality audit of the scope-of-work begins with the project objectives. The focus of the QMP is to ensure that these objectives are clearly and comprehensively defined, and that the design team's role in meeting them is understood.

Key project objectives include plant capacity, product quality (*e.g.*, product specifications), schedule, and cost. Other objectives include utilization of new technology, safety, maintainability, technology protection, provisions for future expansion, startup considerations, and maintaining continuing operations.

For the chemical process industries (CPI), the primary scope item requiring

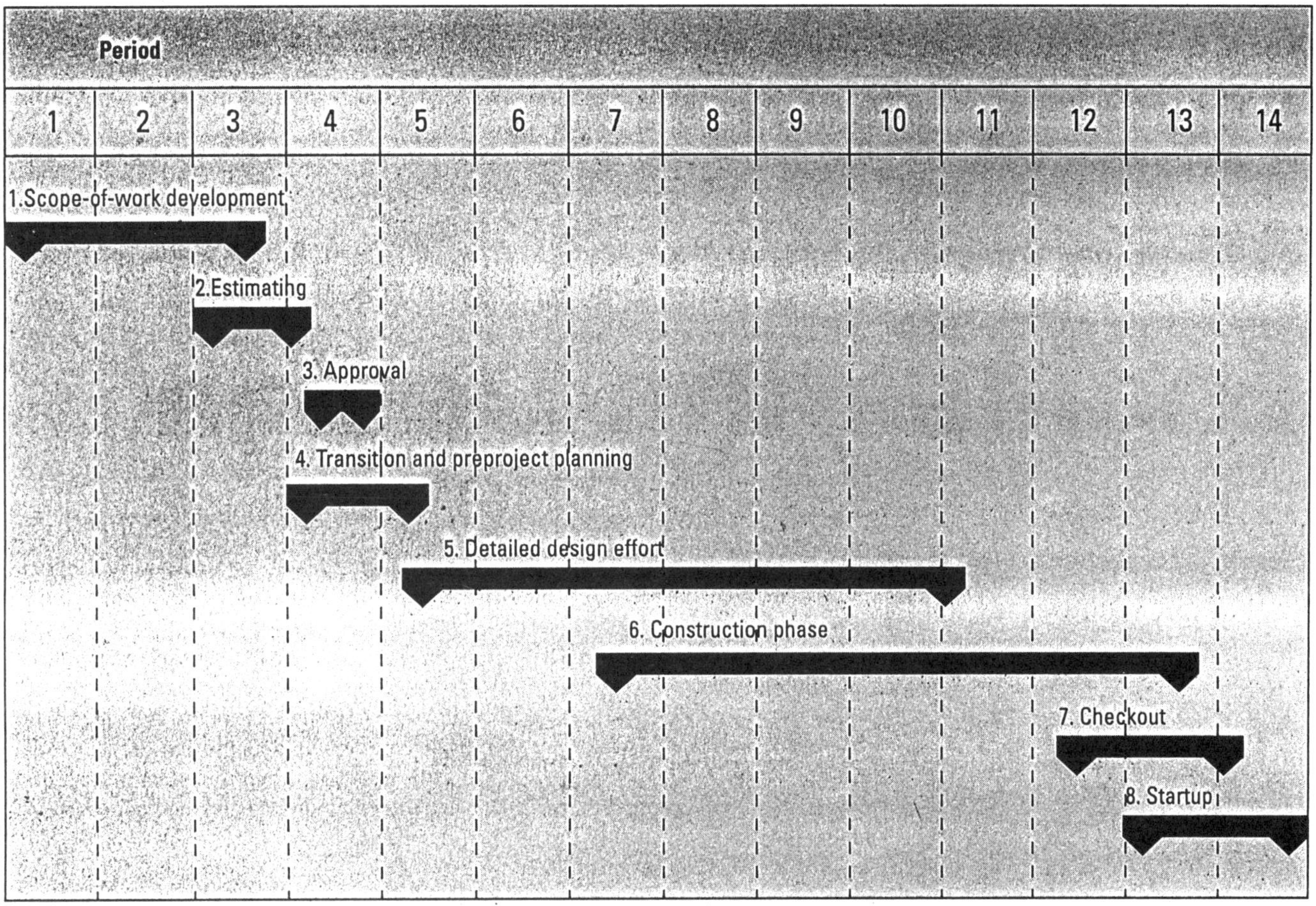

■ *Figure 1. Typical project schedule. Quality management is a concern throughout the life of a project.*

quality control is the process data. Complete, good data should provide the required foundation for a quality detailed design. Data requirements include:

1. Verified material and energy balances;

2. Verified operating and design conditions for each piece of equipment;

3. Identified materials of construction for each piece of equipment;

4. Pipe specifications (including definition of criteria for lines requiring pipe flexibility analysis);

5. Accountability for large scale-up considerations (especially important for separation processes and heat exchange);

6. Appropriate selection of equipment configurations (especially important where exotic materials are involved or where scaleup is significant); and

7. Requirements for expandability, including guidelines on preinvestment costs.

Regulations and site-specific conditions

Environmental requirements and site-specific conditions are often neglected, resulting in scope breakdown. With today's regulatory requirements and complex permitting procedures, the environmental scope takes on a significant position in the overall project definition.

In fact, no process scope can be complete without a thorough evaluation of federal and state regulatory and permitting requirements. Permitting procedure time aside, proper environmental scoping can take up to 18 months or more for a complex chemical facility. These environmental factors can be compounded for an international project in which foreign regulations and practices vary drastically from those in the U.S. Any QMP not addressing the completeness of the environmental scope is at risk of not meeting its objectives.

Accounting for site conditions ranges from using existing pipe racks to the removal of abandoned subsurface foundations. While one sounds trivial and the other complex, both can have profound effects on plant design.

Other site considerations of major importance are: electrical power supply and distribution, real estate allowance (in an existing plant), existing plant right-of-ways, demolition and removal requirements, unique plant specifications, utilization of existing utility systems, soil conditions (contamination), preferred vendors, existing plant operating practices (*e.g.,* control systems philosophy), and local building codes.

Again, the QMP must address the broad issue of site conditions, ensuring that expectations regarding the input data are understood. Also, a contingency plan should be in place to deal with any limited site knowledge.

One approach to scope quality management is to have a member of the customer's team designated as scope manager. It is the scope manager's responsibility to ensure that all inputs meet the same set of quality expectations, and that they are consistent with the project objectives. Figure 3 is a model organization chart for a quality-oriented design team showing the scope manager's position. Dotted lines indicate positions of owner's representatives — such job functions normally would not be part of the contractor's design team.

Safety, technology, and constructability

Three issues which the QMP must address during the front-end engineering phase are safety (including electrical classification), constructability, and technology. These issues require a holistic view of the facility and should draw on all available expert resources for review and input.

Representatives from safety/industrial hygiene, construction, operations, maintenance, R&D, and design should participate in preliminary reviews of these issues. When design and construction are contracted, it is imperative that contractor's representatives be included in these reviews. The QMP must emphasize the need to conduct comprehensive reviews in a timely manner.

Constructability requires additional planning if different design and construction contractors are employed. In this case, either the owner or the design contractor may have to supply construction resources. Where possible, use construction contractor resources for constructability input. Construction input into early project planning is a key to success regarding construction issues and should include cost estimates, schedules, procurement, and contracting strategy.

Technology utilization is most often addressed as a scoping activity because of the implication for change if it is implemented during the detailed design phase. However, more projects are incorporating technology utilization plans into the initial phase of detailed design. These plans are geared toward taking advantage

Figure 2. Punch list defines minimum scope-of-work.

General
1. Project objectives (prioritized)
2. Defined roles and responsibilities
3. Responsibility split between customer and designer
4. Identification of pertinent existing site conditions
5. Definition of any demolition requirements
6. Customer specific project procedures
7. Milestone schedule
8. Procurement strategies
9. Bid package strategies
10. Cost estimate or authorization basis
11. Environmental requirements
12. Unique considerations, such as location, meteorological factors, or logistics constraints

Process
1. Process data — material and energy balances
2. Preliminary piping and instrument diagrams (P&IDs)
3. Defined process hazards review (PHR) procedures and schedule
4. Defined waste streams and processing requirements

Mechanical
1. Equipment arrangement drawings
2. A complete equipment list
3. Equipment data sheets
4. Draft specifications for all major equipment
5. Heating, ventilating, air conditioning, and cooling (HVAC) requirements
6. Fire protection scope-of-work
7. A complete drawing and specification list

Architectural, civil, and structural
1. Codes and standards survey
2. Seismic design factors
3. Geotechnical (soils) report
4. Scope-of-work descriptions for all buildings and structures
5. Plot plans
6. Plot plan markups for civil estimating, such as excavation and back fill requirements, road definition, and contaminated soils removal
7. Complete drawing and specification lists

Electrical
1. Preliminary single-line diagrams
2. Complete equipment list with sizing
3. Preliminary specifications for major equipment
4. Heat tracing scope-of-work
5. Plot plan and equipment arrangement mark-ups defining requirements for cable tray, grounding, and lightning protection
6. Panel list with sizing
7. Hazardous area classification study
8. Complete motor list with horsepower
9. Lighting scope-of-work
10. Complete drawing and specification lists
11. Communications, security and alarms scope-of-works

Piping
1. Plot plan and equipment arrangement markups for pipe bridges and major routing alleys
2. Completed pipe specifications
3. Scope-of-work for pipe flexibility design
4. Existing site tie-in list
5. Modeling scope-of-work (computer vs. hard model basis)
6. Complete drawing list

Control/instrumentation
1. Preliminary specifications for all major items
2. Preliminary instrument list
3. Preliminary input/output (I/O) counts for control system sizing
4. Control panel list and sizing
5. Complete drawing and specification list

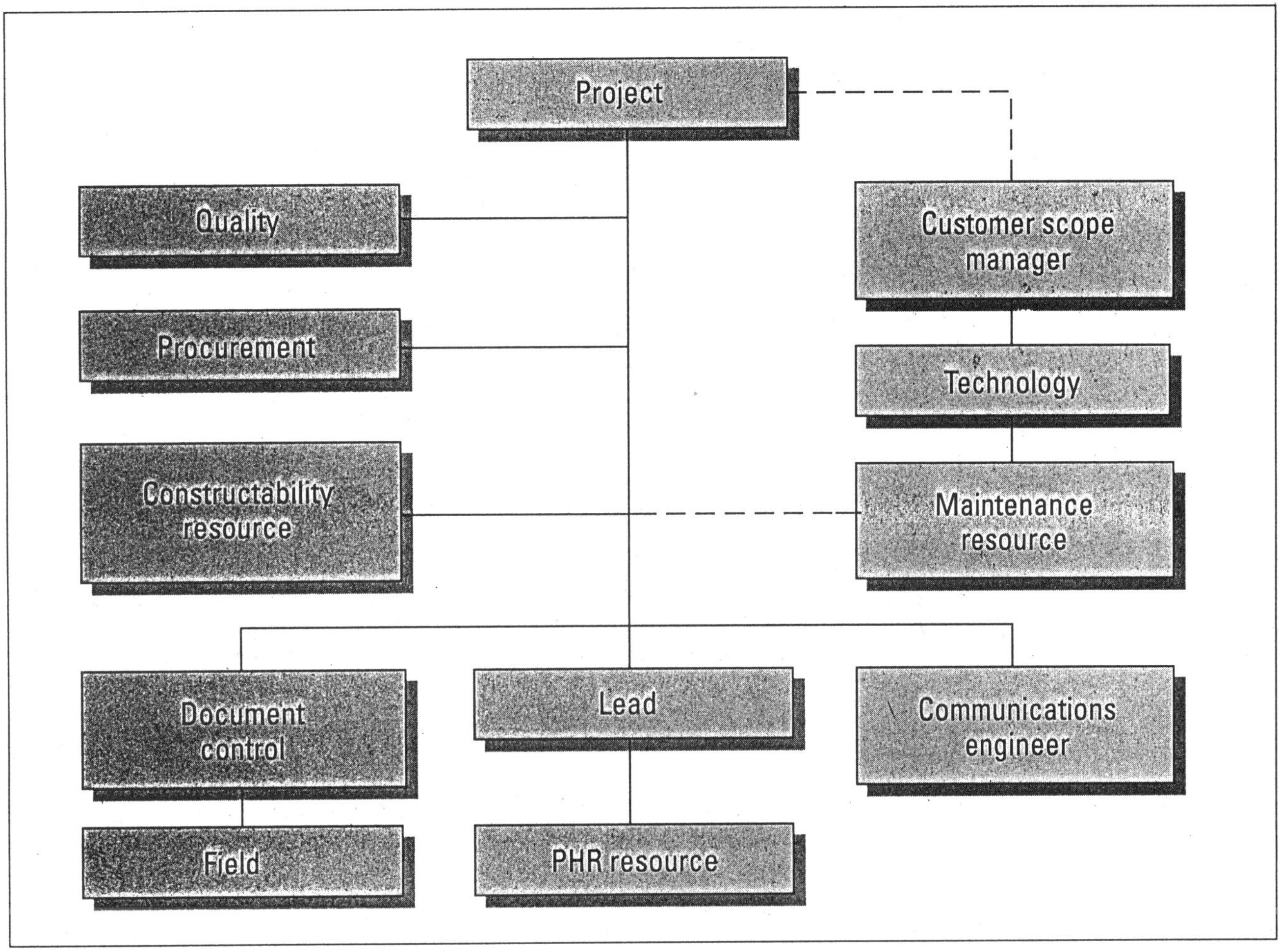

■ *Figure 3. Model organizational chart. The quality manager plays a key role during the project.*

of the collected resources of the newly formed detail design team. However, if this approach is permitted, the quality plan must supply strict economic guidelines with regard to implementing any new technology.

Procurement

A key nontechnical issue is the project's procurement strategy. Procurement protocols — strategies and preferred suppliers — can cause design uncertainties and quality problems if they are not understood clearly. This is a significant factor for major electrical equipment, the control system, and fabricated mechanical equipment.

While any procurement plan is a living document, the timing of its development must be such that it provides adequate detail to allow initial design efforts to proceed, especially for equipment with long lead times and major capital equipment purchases. Ideally, the expectation is that a detailed procurement plan will be prepared as part of the original scope-of-work documents.

Barriers to total quality during front-end engineering

The influence of poor scope definition on project performance is often accepted as the result of market influences. That is, the market demand for the product or the competitive position of the business unit may limit the scoping effort.

Engineers must realize early on that the time to perform the scoping is limited. To do a proper job in a limited time, additional personnel may be needed. However, if the scoping effort remains limited, then the design project engineer must perform a thorough risk analysis to assign contingencies to material quantities, schedules, and budgets. The potential effects on procurement and construction also should be considered, with representatives from those groups participating in the risk analysis.

Equating the lowest-cost facility with the lowest-cost engineering budget is a misconception that often hinders the design approach. Design procedures must continue to improve as to quality. That is, design something correctly the first time. However, project engineers and quality managers need to demonstrate that a tangible economic reward results from design procedures intended to instill quality.

The implementation of the design QMP is usually the design project engineer's responsibility. However, this setup poses a bottleneck: While quality management is the project engineer's responsibility, rarely will that individual have the time to do the job. This problem calls for a full-time quality manager reporting to project engineering. On most projects, this position should remain full-time until design is approximately 60% complete. This allows all engineering disciplines to be well underway and for the establishment of proper interfaces with field activities. Also, this provides time to address major procedural or technical changes so that their effects on quality management are known. The quality manager's role is shown in Figure 3.

DETAILED DESIGN PHASE

Now that the scope-of-work is firm, quality management during the detailed design phase focuses on communications and change management:

Communications — Very few design errors result from technical incompetence. Rather, most result from poor project communications. Communications problems can take many forms: not understanding project objectives, not understanding an end user's need, not using a drawing's current revision, not informing all disciplines of a change, and others.

Therefore, in the detailed design, a primary goal of the QMP is to communicate project objectives and goals, and to facilitate good communications for the ongoing design effort. That is, the plan should address procedures required for documenting, disseminating, and monitoring information; for providing feedback; and for initiating corrective action on all project activities. These quality activities establish the foundation for successful construction and startup.

Management of change — Change management is not only the process of resisting or avoiding change, but includes ensuring that inevitable changes are communicated and implemented properly, and in a timely fashion. The QMP must address those design processes that result in changes and must provide a procedure to manage its impact.

Establishing quality goals

Corporate quality policies are often stated in broad terms, to provide a holistic view of the company's quality goals and reinforce management support. However, for the detailed design phase, individual project teams need to quantify measures of quality. For example:

• Drawings issued on schedule for construction should be on time >98%; and

• Overall piping rework as a result of all issues should be <10%.

While many measurable items (*e.g.*, piping rework) are applicable to process industry projects, actual quantity targets must be adjusted to fit the characteristics of a specific project and the experience of the project team. Also, targets must be specific. For instance, in the above example, <10% of what?: linear feet, total installed cost, line count? Does this figure include approved field change requests?

Some project characteristics that can influence quality standards include: new vs. existing technology, grassroots vs. existing sites, equipment and piping density, firmness of the scope-of-work, domestic vs. international location, physical space — tight vs. open (heavily influenced if existing buildings are being used), makeup of the project team, inherent schedule pressure, and project objective (cost- vs. schedule-driven).

Project objectives must be stated with some practical guidelines. For

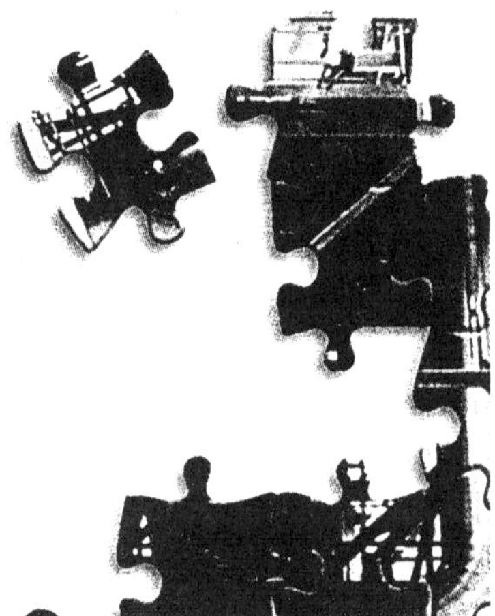

Figure 4. Typical ranges of performance data.

Project activity/error	Frequency range, %
Drawings issued on schedule for construction	50–95
Piping rework from all categories	5–20
Incorrect pipe isometrics	2–10
Pipe fabrication errors	2–10
Other pipe fitup problems	1–5
Instrument errors all categories	1–5
Instrument design/spec errors	1–5
Instrument location problems	1–2
Vendor-supplied package instrument errors	1–2
Power and control wiring errors	Less than 2
Structural design inadequate	0% allowed*
Structural overdesigns	1–5
Structural interferences	0–2†
Equipment specs requiring major changes	Up to 15
Specifications requiring any correcting	Up to 95‡

* No one would ever admit to an inadequate structural design!
† Goal is 0%.
‡ Errors in specifications increase as the scope of the specification increases. That is, a spec for a turnkey building or packaged processing unit is more susceptible to errors and omissions than a spec for an individual piece of equipment.

example, if a project is schedule-driven a statement can be made about the project's critical path and about the application of premium time to reduce the overall duration. If the project is cost-driven, are any cost savings acceptable or is there a minimum return required?

The delineation of these guidelines is more complicated then is presented here. A simple statement that the project is cost- or schedule-driven does not provide the necessary direction. The project team needs to be given specific guidelines regarding the appropriate means to meet the objectives. Note that project objectives can be driven by other forces, such as technology, safety, and maintaining ongoing operations, among others.

Establishing specific quality goals, while not difficult, does take some thought and time. To instill ownership, those individuals (teams or disciplines) whose performance is being measured must establish the goals.

Management mandates are viewed either with skepticism or as unrealistic targets. Management's goals in promoting quality standards should be to: promote a team effort, establish objective measures of performance, highlight and document performance, award outstanding performance, instill a sense of accomplishment and pride, and facilitate continuous improvement of the design process through evaluation and upgrades to design procedures.

One difficulty with regard to establishing quality goals is the lack of published historical data. Design quality, as demonstrated by construction and startup performance, is most often stated in broad terms, and is rarely categorized regarding the source of the quality problem (*e.g.*, design, construction, procurement, or fabrication). Therefore, project teams must rely on their experience in establishing quality goals. A historical range of performance data is given in Figure 4.

Performance measurement

A monitoring/feedback system is needed to validate design performance. This system should be established as a project procedure in the QMP and be administered by the project quality manager for the life of the project. The system should address:

- Description or nature of the quality issue (unsafe, not maintainable, inaccessible, improper application);
- Phase of the project in which the error was discovered (design, procurement, construction, testing, startup);
- Nature of the error (design, vendor, constructor);
- Engineering disciplines requiring action;
- Recommended corrective action;
- Cost; and
- Schedule impact on individual activities, and on turnover (mechanical completion).

Documenting quality

A quality action form, Figure 5, provides a means for documenting quality issues. Use of the form by all project team members should be encouraged early in the life of a project so that it becomes part of the project's culture. Using this type of procedure early in the design phase has three distinct benefits:

1. It helps to identify problem areas or error patterns early on, thereby allowing corrective action to be taken.

2. It promotes corrective action during the design phase, when the cost and schedule impacts are the least significant. For example, an improperly specified control valve identified during design can be corrected without adversely affecting project cost or schedule. The same error left uncovered until installation will cost the purchase price of the incorrect valve. Left undetected until startup, the cost impact could be tremendous, as startup could be delayed, resulting in lost sales revenue.

3. It provides a mechanism for recognition of an employee's contribution toward project quality.

The completed form should be routed so that the project team is aware of the issue, so that items such as cost and schedule are properly accounted for, and to ensure that corrective action is taken. A secondary benefit of using such a form is that it provides an excel-

Figure 5. Quality action form.

Initiated by: ____________________

Problem description*:

____

Construction action required:

____

Engineering action required by:

____

Quality assurance followup required:

____

Cost impact:
Estimate ________
Actual ________

Schedule impact:
Estimate ________
Actual ________

*Note: Attach sketches for clarity and to facilitate cost estimate.

NO. ______

Routing:
______ Project manager
______ Construction manager
______ Procurement manager
______ Other

Hard copy to:
______ Project file
______ Quality assurance
______ Construction engineer
______ Cost engineer
______ Scheduler
______ Engineering disciplines

Error occurred in:
______ Design
______ Procurement
______ Construction
______ Vendor
______ Other

Project status:
______ Scoping
______ Design
______ Procurement
______ Construction
______ Checkout
______ Startup

lent means for establishing a historical database for future reference.

Posting results

Finally, actual performance must be compared to the goals established by the project team. Results should be reviewed on an ongoing basis at project meetings, and can be used as a basis for the project evaluation.

Results should be posted such that the entire project team is aware of them. This can be accomplished by a monthly report, or by posting graphs or charts on a project bulletin board. Some indicators, such as the percentage of drawings issued on schedule, can be graphically enhanced to boldly demonstrate team performance on an ongoing basis, thereby serving as a team motivator.

Project communications

The most critical aspect influencing project success is how well project team members communicate with each other. Good communications require a consistent, common-sense effort, coupled with a keen awareness of the needs of others. Some fundamental approaches to good project-wide communications:

1. Establish a field document control center. A source of construction error is the accidental use of outdated design documents. While this is a construction management issue, it often relates to design procedures for issuing documents.

The best practice is for engineering to issue documents to the field document control center. Field distribution is thus managed locally, with revisions issued to the appropriate field personnel. Also, this reduces the total quantity of documents issued, saving money.

The best policy concerning outdated documents is to destroy them. If it is necessary to save them for historical reasons, keep the documents at a different location, segregated from current versions.

The position of document control clerk often becomes full-time. Design management should consider providing this service, thereby maintaining a stronger link between the field and the design office.

2. Establish engineering document control. Design errors can often be linked to poor document control within the engineering office. By using centralized distribution, the project will benefit from consistency in procedure, clarity of purpose, and on-schedule performance.

The document control center serves as the reference library for the project team. That is, the center maintains all of the current documents referenced by the various engineering disciplines. Often, each discipline attempts to maintain its own reference file of the other discipline's documents.

While this is more convenient and expedient for the individual engineering disciplines, it creates two problems: First, it is expensive to issue multiple quantities of each document so that everybody has a copy; second, individual files often do not have the most recent revision of a document.

3. Establish standard communications meetings. Holding meetings to discuss design issues aids the design effort. Meetings are held to review drawings, meet with vendors, review options, and serve other purposes. However, there are three types of meetings that provide a holistic view of a design effort's progress and facilitate enhanced communications:

The *leads meeting* should be held weekly and include the project manager, project engineers, discipline lead engineers, and the quality manager. The meeting's primary focus is to review those issues that are constraining the design effort. The meeting provides an opportunity for each lead engineer to state his or her priorities and design problems, and to ensure that construction-information needs are being addressed on a priority basis. This meeting also should provide a forum for discussing project procedures and customer relations.

The *schedule meeting* should be held every two to four weeks, depending on schedule criticality or performance. The makeup of this meeting is the same as for the leads meeting, but also includes the project scheduler.

Holding this meeting allows schedule constraints to be identified early, prior to their becoming critical. Work-around strategies can be developed to keep the project moving forward. Also, resource management can be exercised if delays are identified. This is especially important if construction is in progress.

While the project's critical path is the primary focus of this meeting, all areas or activities should be reviewed periodically. Those activities that are not on the critical path at this time have a tendency to become on it at a latter date. Do not wait until schedule performance is a problem before initiating this review meeting; start conducting it at the onset of the design effort.

The *procurement meeting* focuses on the status of the procurement cycle: specification writing, vendor inquiries, bid tabulation, purchase order placement, receipt of vendor drawings, fabrication releases, and delivery. It applies to all items being purchased, including mechanical equipment, electrical equipment, and instrumentation. Due to delivery timing and design information requirements, the emphasis of the meeting usually shifts from mechanical to electrical, and then to instrumentation over the life of the project.

The continuous improvement view on communications

A progressive vision for enhanced project communications is based on the concept that communications should be as important as technical competence. This concept is based on the fact that without quality communications, technical responsibilities cannot be executed to their fullest potential. Therefore, by elevating the importance of good communications, the quality of technical output improves.

How can this be applied to the daily design effort? One approach is to assign a lead communications engineer to each project. The analogy can be made to the assignment of lead discipline engineers to ensure technical quality. The communications lead would report to the lead project engineer (refer to Figure 3) and have the following responsibilities:

- Ensure adherence to project procedures;
- Upgrade and provide training as required for project procedures;
- Facilitate proper documentation of the various project meetings (but not for the process hazards reviews discussed later on);
- Maintain project punch lists;
- Ensure timely followup;
- Carry over key communications insights from project to project; and
- Recognize individuals who contribute to project quality through outstanding communications practices.

Change management

While a formal procedure is followed to document and approve design changes, this procedure does not provide real change management. That is, the procedure is insensitive to the way in which a single, inevitable change must be managed so that it has the least impact on the project. The design team must attack change management from three perspectives:

- First, management must encourage completing, early in the life of the detailed design efforts, those design activities that can result in change or correction to the scope-of-work documents.
- Second, project management must resist additional (late) change because it often results in increased cost and schedule delays.
- Third, engineers must realize that change poses the single biggest communications problem within a project team, and can represent a major source of design error.

Early design activities that promote change include process hazards reviews (PHRs), constructability reviews, and reviews of preliminary drawings, especially piping and instrument diagrams (P&IDs). Most changes resulting from these reviews either provide a viable operating system or eliminate an unsafe condition. By conducting these reviews as early as possible, the impact of changes on cost, the schedule, and communications can be minimized.

After these preliminary design adjustments are made, change must be resisted. Resistance is best employed by insisting that the initiator provide justification for the change. Justi-

Figure 6. Change warning form.

Change Warning — **No.** ________

Title: ______________________________

Distribution: ______
Proj. Engr. ______
Process ______
Mechanical ______
Civil ______
Structural ______
Electrical ______
System Cont. ______
Cost Engr ______
Scheduler ______

Description of potential change*: ______________________________

Estimated cost: ________________

Schedule impact: ________________

Engineering disciplines affected: ___ Process ___ Mechanical ___ Civil ___ Structural ___ Electrical ___ System control

Action required before initiating formal change: ______________________________

Approval: Proceed with formal change notice _____
Hold : Proceed with action required _________
Not approved: Take no action ________________

***Note: Attach sketches to aid description.**

fication usually is an economic advantage, or an unsafe or inoperable condition. In calculating the cost of a change, include an expense for the disruption of the ongoing design effort. That is, most changes have an intangible cost brought about by the disruption of work needed to study, prepare, and implement them.

The most important aspect of change management is communicating the change. Many design errors result from one engineering discipline initiating a change and not informing the others. The other disciplines may waste time pursuing a design based on out-dated information. If informed early on, other disciplines may catch the change before completing any substantial design. The later during design that a change occurs, the more important are timely and effective communications.

Issuing a change warning sheet (Figure 6) is a quick way to notify the project team that a change is occurring. The sheet is not the formal change document. The warning contains a brief description of the change and the reason why it is needed. Economics is not included.

The warning informs all project disciplines that a change is imminent, allowing them to put activities on hold, if appropriate. A warning can be issued by any project team member, and it should be issued as soon as a potential change is identified. All change warning sheets must be followed up by either a formal change document or a cancellation.

Another effective way of managing and communicating change is maintaining a master copy of all drawings. The classic example for the CPI is project P&IDs.

The master copy is a working copy that documents all of the latest corrections to the process. For control, only the lead process engineer makes the corrections on the master copy. Other disciplines must refer to the master copy to ensure that they are designing to the latest information.

Keeping a log book is another effective means of documenting and communicating changes. Any change to a P&ID is entered by the process engineer responsible. The entry consists of a revision number, the drawing(s) affected, and a description of the change, including a sketch, when necessary. Each day, or as required, a copy of the new log entries can be provided to each design discipline, with a copy faxed to the field for construction use.

A critical and difficult aspect of change management is recognizing when the frequency or nature of ongoing changes represents more than isolated issues, but rather a breakdown in either the project scope or the design process.

If this happens, it is often appropriate to interrupt the design effort until a solution or even a rescoping is concluded. This is a critical decision needed to avoid unproductive design and possibly a costly false start of construction. This situation may apply to a sub-area and not to the entire project.

Hazards management

PHRs are one of the most valuable procedures embraced by design in recent years. In its fundamental form, the PHR helps to identify unsafe operating conditions. More importantly, however, PHRs provide the mechanism for design engineers to gain insight into operations, and for operations personnel to provide their input and knowledge toward the safety of a future facility. Through PHR participation, operations personnel gain the confidence that the facility is being designed with their safety in mind.

PHRs can double the amount of instrumentation and completely change the design conditions for major process equipment. If this happens late in the design schedule, the project is doomed to be late and over budget. While formal PHR procedures are beyond the scope of this article, some key management techniques enhance the contributions that PHRs make to quality:

1. Assign a PHR resource person. This person need not be familiar with the project, but should be well versed in PHR techniques and be a good meetings leader. Primary responsibilities include establishing a PHR schedule, leading the PHR meetings, providing PHR meeting documentation, and ensuring that action items from the PHRs are followed up.

2. Schedule the PHRs as early as possible. PHRs promote change, so get these reviews done as soon as possible.

3. Allow ample time for PHRs. The PHR procedure is mentally taxing. A rule of thumb is not to exceed four hours of PHR meetings in a day. The remainder of the work day should be used for documentation and followup.

4. Allow time in the overall engineering schedule to account for PHR changes. Provide a milestone in the engineering schedule to track that followups occur within a reasonable time.

5. Break the project down into discrete areas for PHR review to expedite the overall effort. Assign additional personnel to conduct the sub-area PHRs. Each PHR team should include operations, safety, and design personnel. (For information on process control system reviews, see *CEP*, Oct. 1994, p. 32.)

6. Standardize the format for documenting the PHR meetings. Such a format should be used during meetings so that outside efforts can focus on resolution of problems rather than on clerical tasks. Standardization also allows others to pick up the documentation task easily, and makes retrieval and followup simple. The format for the documentation should emphasize the communication of any PHR change to the design team.

Constructability

Constructability improves the overall project quality by considering construction requirements early in the design phase. Savings in field labor and material are additional benefits realized through constructability reviews (CRs). CRs held during front-end engineering should have defined most of the major construction strategies, mitigating the impact of CRs during the detailed design. However, CRs can promote change, and therefore they must be scheduled early in the life of the design phase. CRs held during the early stages of detailed design should focus on the following:

1. Construction requirements which affect design and the procurement strategy. Examples include:

- Equipment installation sequences which affect the required on-site dates.
- Equipment field dress requirements (that is, ladders, platforms, insulation, and sometimes piping and instruments) that affect the on-site requirement.
- Storage and maintenance requirements for materials received in the field.
- Crane limitations for installing heavy items.

2. Strategies that impact on contracting. Examples include:

- The use of lump sum contracts, and bid-package and work-package definitions. A work package is a job for a specific unit of the plant, such as electrical work for a defined part of the plant. A bid package combines several work packages.
- The detailed construction sequence, identifying the flow of construction work. That is, for a facility with multiple areas, what is the order in which the areas will be built.
- Identifying the field conditions under which the contractors will have to perform the work, including the availability of utilities (such as electric power and water).

3. Construction techniques that may impact on the detailed design approach. Examples include:

- The use of prefabricated buildings.
- Poured-in-place elevated concrete structures.
- Electrical practices such as plug-in designs, wiring harnesses, preinstalled cable, conduit, tray, fixtures, and others.
- Standardizing certain concrete forms and structural steel sizes.
- The use of modular assemblies.

Procurement

Procurement is included as an engineering quality issue because engineering's responsibility is more than just issuing purchase requisitions. The performance of the procurement effort depends on a cooperative planning effort among engineering, construction, and purchasing to ensure that deliveries meet site installation schedules. This scheduling effort must be done at the onset and will, therefore, be based on a preliminary erection sequence.

Schedules can be developed on a unit basis. However, given the early stage at which this must be developed, it is often appropriate to block out the schedule by process sub-areas. Block scheduling is almost always appropriate for instrumentation and field material. Unit scheduling is more appropriate for (large) mechanical and electrical components. Schedules are developed by working backwards from the installation date, and accounting for the time required for the following activities:

1. Field dress requirements such as painting, insulation, and installing ladders or platforms;

2. Shipping (critical for international jobs);

3. Final shop tests or inspections;

4. Fabrication;

5. Vendor drawing approval;

6. Vendor drawing preparation and submittal;

7. Bid analysis and award;

8. Vendor bidding cycle; and

9. Specifications writing.

Good, comprehensive specifications provide the foundation for a consistent bid basis and a smooth fabrication cycle. Allowing nebulous or incomplete specs to be bid makes the bid analysis and the vendor selection a guessing game. It also adds time to the bidding process in that clarifications will be required, and then when bids are delineated each vendor will have a bid based on different assumptions, adding additional time to "standardize" assumptions for all bidders.

Incomplete specs ultimately lead to alterations and cost additions during fabrication. There are no sure-fire approaches to ensuring high quality spec writing. The most critical contribution comes from the authoring engineer and his or her attention to detail. Two simple techniques can be applied to help provide the foundation for a good spec. These are:

1. Use standard specs. For example, if there is a pump spec that has been successfully used in the past, make it a standard to which spec authors can turn for a good starting point. Beware though: Standards must be maintained and upgraded as part of any continuous improvement program. Any changes to codes or upgrades to wording applied on a project-specific spec must be captured and edited on the standard copy.

2. Have each spec checked/reviewed by someone other than its author. Having a different person check the spec helps to avoid errors and omissions, and helps the clarity of wording.

An additional aspect included under procurement is the management of vendor drawings. Vendor drawings are necessary for the accurate completion of all design tasks, especially for the completion of structural, piping, and electrical design.

A computerized tracking system should document the receipt and processing of vendor drawings. While this can be a major clerical task, accurately tracking vendor drawings' progress is an important management requirement. Tracking identifies missing or incomplete documents, as well as highlights problems with in-house turnaround. Since most vendor drawings go through an approval process and are often returned to the vendor with comments, tracking the receipt of final (or certified) drawings should be included in the tracking system. Receipt of final drawings is confirmation that comments have been appropriately considered, and that the design is consistent with the equipment being supplied.

The project evaluation system

The project evaluation system is a management tool being used with increasing frequency. Evaluation takes place at team meetings held periodically throughout the life of the project. During the meeting the project leadership reflects on its successes and shortcomings.

The project evaluation system is a management tool being used with increasing frequency. Evaluation takes place at team meetings held periodically throughout the life of the project.

The focus of this effort is twofold. First, it identifies team strengths and ways in which to maximize their benefits (*e.g.,* Is there a good procedure in one area that can be applied to another to improve performance?). Second, it identifies team weaknesses and allows corrective actions to be developed. This review process is a great communications tool. Improvement actions are often highly tangible, with results being seen immediately.

While this process is usually geared toward a specific project, often improvement ideas are generated which can make a contribution to the company's overall continuous improvement program.

Barriers to total quality during detailed design

Barriers exist which can inhibit the flow and understanding of information. That is, communications are often restricted by procedures and territorial issues. Some engineers feel they communicate well with other disciplines because they issue their specifications and drawings and attend all meetings. Design processes and procedures must emphasize that everything generated is a communications tool.

Engineers must be trained better to focus on the content of design documents and what is required to effectively communicate information to their "customers." The lack of established quality goals can lead to either under-performance or the establishment of unrealistic standards. The lack of established goals is reinforced by the lack of monitoring and feedback mechanisms.

The type of procedure implied by a quality action form often will be viewed as an unnecessary overhead expense.

Personnel constraints are often a barrier to a project's overall quality attainment. Another barrier is having persons assigned to key positions (such as quality manager and communications engineers) who see these assignments as additional work. Often, these roles are approached as tasks rather than as a commitment to their underlying philosophy of total quality and improvement.

Finally, neither procedures nor a QMP will ensure a quality product. Only the commitment and pride of the individual project team members will provide a quality product.

CONSTRUCTION, CHECKOUT, AND STARTUP

With the project focus turning to the field, design personnel can still play a critical role in contributing to the overall quality of the project.

The opportunity to have design personnel support field activities (construction, checkout, and startup) is often not taken advantage of. To promote this opportunity, project management must embrace a holistic view of quality in terms of customer service and future benefits. However, there are tangible benefits for all parties involved, regardless of position (customer or contractor) or role (engineering, construction, or operations).

The engineering team gains the advantage of real-world experience for its designers. This experience provides a common sense regarding "people engineering" which the design engineer will rarely gain through office tenure alone. More importantly though, involving design personnel in field activities enhances their sense of ownership of the project. Encouraging ownership through the startup of the plant can enhance individual performance over the life of the project.

Construction reaps the benefit of having an on-line troubleshooter who is intimately familiar with the design. This not only improves communications and turnaround on problems, but also allows design and construction to thoroughly interact on problems, enhancing the identification of options and the selection of optimum solutions.

During checkout and startup there is a significant advantage to involving an engineer with knowledge gained from months on the design effort. In some cases, the design engineer will know more about a specific unit operation than operations personnel. This often occurs on large projects where operations personnel are stretched thin, or for large vendor-supplied systems which can take months to develop during the design effort. Also, rarely is there a startup in which some extra help cannot be used.

The QMP needs to address these field support positions by defining resource requirements, roles, and responsibilities. While this portion of the QMP requires updating as the project proceeds, establishing generic roles for design personnel early in a project's life allows for resource management and the necessary budgeting. Again, outlining this role in the QMP early in the project's life can enhance individual commitment to the project.

Without being specific, the QMP should address any training requirements needed prior to an individual's transfer to the field. This should include any plant-specific safety training.

Supporting construction activities

In the following discussion, it is helpful to classify the phases of construction activity by engineering discipline. While this oversimplifies the construction effort, it, nonetheless, provides a good basis for identifying design support needs:

- Phase 1. Civil and structural.
- Phase 2. Primary electrical.
- Phase 3. Mechanical (equipment).
- Phase 4. Piping.
- Phase 5. Secondary electrical, and instrumentation.

The type of field design engineer (FDE) required varies with the construction phase. Usually one engineer versed in both civil and structural design can be assigned for Phase 1 support. This assignment starts with the beginning of field activity and runs for the first 30 to 50% of construction. This position should be considered a full-time assignment during this phase.

Phase 2 requires an electrical engineer with an understanding of the primary distribution system from the electrical utility to the main plant substation, as well as any equipment encountered between the substation and the low-voltage motor-control centers. This work overlaps Phase 1 and can occur during anywhere from 10 to 50% of the construction duration. This assignment is usually not long in duration. However, it may require multiple, shorter field assignments.

Phase 3 support is often managed by the traditional construction engineer or mechanical superintendent. Field design support is usually limited, but may be required for complex vendor-supplied systems. If an FDE is required, his or her term will occur between the 20 to 60% construction milestones. The length of the field assignment for this mechanical engineering role can vary from one week to eight months.

The piping phase of the construction effort is the most traditional application for field design support. The assignment can start as early as at the 30% construction-completed point, and it often will last through the end of plant checkout. Also, the piping phase may require more than one FDE.

The electrical and instrumentation phase starts around the 60% construction milestone. It is typical to cover this role with one resource person: a design engineer with an understanding of both disciplines. This assignment also usually lasts through plant startup.

There are other construction situations during which the support of a design specialist is beneficial. These assignments are usually short-lived and occur at specific times during the construction effort. Some examples include the installation of:

- Fire protection systems;
- Waste treatment systems;
- Large-scale utility systems, such as boilers;
- Alarm and communications networks;
- Distributed control system hardware; and
- Electrical pole work.

The role of field design engineers in supporting construction activities is primarily two-fold. Their primary goal is to keep construction going smoothly. That is, if there are any questions regarding installation, the FDE provides immediate clarification. This is especially important during Phases 1 to 3; the installation of concrete, structural steel, and major equipment (mechanical and electrical). This is a time when construction personnel will rarely have

Further Reading

American Society of Civil Engineers, "Quality in the Constructed Project; A Guideline for Owners, Designers and Constructors," Vol. 1, ASCE, New York (1988).

Business Roundtable, "Integrating Construction Resources and Technology into Engineering," Report B-1, Business Roundtable, New York (Aug. 1982).

Construction Industry Institute, "Scope Definition and Control," CII, Austin, TX (June 1986).

Construction Industry Institute, "Input Variables Impacting Design Effectiveness," Publication 8-2, CII, Austin, TX (July 1987).

Construction Industry Institute, "Costs of Quality Deviations in Design and Construction," Publication 10-1, CII, Austin, TX (Jan. 1989).

Harlin, Sewell, "Early Construction Input Improves Efficiency for Project Engineering," *Pulp and Paper,* pp. 116–118 (Sept. 1988).

Lakewood Publications, *Total Quality Newsletter,* monthly, Lakewood Publications, Minneapolis, MN.

Matthews, Michael F., and James L. Burati, "Quality Management Organizations and Techniques," report to Construction Industry Institute, Source Document 51, CII, Austin, TX (Aug., 1989).

Nimmo, Ian, "Extend HAZOP to Computer-Control Systems," *Chem. Eng. Progress,* **90** (10), p. 32 (October 1994).

a couple of days to request an analysis from the engineering home office.

The second task for the FDE is performing ongoing design to accommodate either a constructability issue or a design error. Examples of these situations include:

- Pipe routing interferences;
- Minor equipment fabrication errors such as a nozzle misorientation; and
- Interdisciplinary design information that does not agree.

Supporting checkout and startup

The checkout phase of a project provides the transition from construction to plant operations. The FDE provides an efficient interface with construction, ensuring that punch list items are documented and accounted for. The checkout efforts are focused on validating component and subsystem operation so that costly startup delays are avoided.

During checkout, the need for field design engineers will usually be limited to process (chemical) engineers, and an electrical and instrumentation resource(s) person. Piping may still have a role in punch list cleanup. The individuals involved in the checkout effort are the best candidates for startup support.

The process FDE has primary responsibility for documenting all as-built conditions. Walking down the piping and back checking it against the P&IDs is the time-proven method for "as-builting" a plant. It is also the most effective means of concluding the punch list from a piping perspective. This task can alleviate a lot of frustration during system runs.

As-builting should include performing a reconciliation against operating and safety procedures, especially for interlocks and process safety hardware such as relief valves. Also, it will almost always be necessary to audit any as-built conditions against the PHR.

The electrical and instrument FDE will have the following additional responsibilities during checkout:

- Verify valve operation;
- Conduct cable continuity tests;
- Verify breaker operation; and
- Audit calibrations (these are often performed by the instrument contractor).

Another important task for the electrical and instrument FDE is to ensure that individual components are properly identified (tagged). This is a safety issue for electrical components. The FDE must make sure that motor control centers and breakers identify the equipment being powered.

System runs, usually performed with water, are the pinnacle of the checkout effort. These allow performance verification of the mechanical components and the control system, including interlocks, alarms, and controller set points and parameters.

Once the entire facility has been checked and punched out, it is ready for startup. Specific startup procedures are beyond the scope of the QMP, and will normally be provided by operations personnel. However, the QMP should establish the foundation for design's role to support the operating team during startup, completing the ownership imperative necessary for a commitment to quality.

Performance feedback

The field design engineers are in an excellent position to provide performance feedback to the engineering home office. Again, this is most easily accomplished through the use of the quality action form (Figure 5).

Barriers to total quality during construction, checkout, and startup

The obvious objection to assigning design engineering personnel to field tasks is cost. It is not cheap sending personnel into the field, and this cost is even higher for international assignments. However, the costs of delayed construction and startup can be astronomical.

As an example, assume a $100-million dollar plant with a nominal three year payback. This plant should be generating in excess of $90,000 per day in positive cash flow. The field support described above would cost somewhere between $300,000 and $700,000 for payroll and expenses for the entire effort (assuming an 18-month construction/turnover duration and a domestic project). In other words, if the efforts of the field design personnel can be shown to improve the turnover schedule, then its cost is recouped in a manner of days. CEP

W. J. COLT is a staff project engineer with Morrison Knudsen Corp., Cleveland, OH (216/523-2021; Fax: 216/523-5922). He has 16 years of experience in operations and design, including 10 in project engineering. Colt has been active in the Cleveland section of AIChE since 1983 and has served as its chair twice. He received his BChE from the University of Dayton and his MBA from Cleveland State University.

Use Utility Functions to Select Equipment

Deciding which factors to consider and their relative importance provides a rational basis for making choices.

Joel B. Christian,
Osram Sylvania Inc.

If you are like many engineers, you are called on every day to make decisions, often between complex alternatives. A utility function, also called an objective function, can help you choose the best option for your particular situation. In this article, I will show you how to set up and apply utility functions to select capital equipment. The technique, however, can be used almost anytime that you must make a decision. For instance, it can be applied to risk assessment, system failure networks, and a wide variety of engineering systems analyses.

For utility functions to work, you need to decide two things:

1. What attributes are important to you?

2. How important is each attribute to you?

Then, you must identify your alternatives, and evaluate each against the selected attributes and their specific weightings. With all this in hand, the utility can be calculated.

The attributes can be extended to include probabilities of outcomes using risk numbers or results of fault trees and event trees. Costs of each outcome also can be used. (Indeed, this method was used in recommending a site for the U.S. nuclear waste repository.)

For capital equipment, for instance, the utility to you could be:

- cost;
- time for delivery;
- size;
- capacity;
- vendor's reputation; or
- a probability.

The utility function often is criticized as a way to justify a decision that's already been made. To avoid this rap, you have to be honest with yourself upfront. While it might be easiest to choose based on strictly objective technical attributes, in selecting a lot of capital equipment "soft" attributes, such as the value of unique options, a manufacturer's reputation, and the amount of technical support provided, are important factors that must be considered. So, clearly assessing these (ideally achieving a consensus on them among design team members) before you consider alternatives is critical.

A simple example

Let's start with a familiar decision — buying a vehicle — and limit the field to four types. First, we must list important attributes. Right away, you can see the first requirement — *you must decide what is important to you.* I've picked the attributes for each alternative based on what's important to me, and summarized them as column headings in Table 1. *Any* attribute that is important to you is appropriate. Some, such as horsepower, are easy to give a clear, specific value, while others, such as "sportiness," necessarily are subjective. But the important thing here is to decide *what* is important to *you*.

Next, we need to decide *how important is each attribute to you*? This is a weight that you assign to each attribute. Again, this is your decision. I recommend giving cost equal weight with technical considerations — this seems to work well in most purchase decisions. Let's use the following weights:

Price — 50% (Can we afford it?)
Power — 10% (Will it have zip?)

Sportiness — 10% (Will it be flashy?)

Pull the camper? — 10% (Can we use it on vacation?)

Now, each attribute can be normalized and weighed to give a total score using the equation:

Utility_x = [(least desirable - X)/(least desirable - most desirable)]W

where X is the value for the attribute of an alternative, and W is the weight for the attribute. So, for instance, the score for power for the small pickup = [(80 – 90)/(80 -120)](10) = 2.5.

The results for our example are summarized in Table 2. Using our criteria, the small pickup has the most utility, with a score of 68.5, and would be chosen.

Sensitivity analysis

How do we know that we've made the right decision? How can we be reasonably certain the weights that we've assigned to each attribute are valid? A sensitivity analysis can provide insights. (Performing such an analysis has become much simpler thanks to computer spreadsheets.)

First, pick an attribute, and adjust its weight to the point where another alternative beats the one originally selected. A small range indicates a sensitive parameter, while a large range shows an insensitive parameter. You can use this to concentrate on key attributes to be certain that you've put enough emphasis on them. Consider Table 3, which gives the ranges for each attribute to still select the pickup truck. In this case, a large range of weights can be applied to each alternative to still select the truck. This illustrates how well the technique works. By exercising reasonable care in setting up criteria, the selection of the alternative with the most utility is insensitive to minor judgment errors.

Table 1. Our key vehicle attributes.

Vehicle	Price, $	Seats	Power, hp	Sportiness*	Pull the Camper?
Small pickup	10,000	2	90	6	Yes
Minivan	18,000	7	90	0	Yes
Sedan	12,000	5	80	4	No
Sport coupe	19,000	2	120	10	Yes

* Rating from 0 to 10, with 10 being best.

Table 2. How our vehicle alternatives score on a weighted basis.

Vehicle	Price	Seats	Power	Sportiness	Pull the Camper?	Total Points
(Weight)	(50%)	(20%)	(10%)	(10%)	(10%)	
Small pickup	50	0	2.5	6	10	68.5
Minivan	5.5	20	2.5	0	10	38
Sedan	39	12	0	4	0	55
Sport coupe	0	0	10	10	10	30

Table 3. Changes in attributes required for another option to be favored.

Attribute	Weight,%	Weight range to keep current selection, %
Price	50	(-11 to ∞)
Seats	20	(-∞ to 43)
Power	10	(-44 to 61)
Sportiness	10	(-41 to 106)
Pull the Camper?	10	(-4 to ∞)

An industrial example

Here's a more difficult example. I was involved in selecting clarifiers for a major wastewater-treatment project; this involved choosing from a wide range of companies with varying levels of experience manufacturing clarifiers. In the final analysis, ten alternatives were quoted. These were difficult to evaluate — because technical features as well as prices varied widely, and softer issues, such as a manufacturer's reputation and technical support, also were important.

The utility function was used to determine which alternative offered the best value. In this case, the size of the unit was specified, but the torque capability of the rake drive was left for each manufacturer to determine, based on the waste character and samples provided. The other attributes evaluated were price, support, and corporate stature. Table 4 summarizes the details. There is no clear winner by inspection.

Finding which vendor offers the maximum value, therefore, involved weighing the attributes with their perceived importance. Table 5 shows the final result. This particular analysis selected Vendor E, an average-priced bidder with moderate reputation, good support, and the second highest torque rating. (This was consistent with equipment selection trends for the entire project.) Naturally, changing the attributes or the weights may have changed the outcome, but the weights selected probably accurately reflected the equipment selection philosophy at the time.

Another useful way to look at the same data is to plot the features vs. cost. In this way, you can see the close contenders by inspection. Figure 1 shows cost as a function of utility for the clarifiers.

(A similar analysis often is done in marketing studies, where marketers work utility functions backward, to try to determine the desirable and undesirable attributes for their products — with the goal of then directing resources to those areas that have the most impact on customers.)

A clear choice

The utility function is a simple and useful tool to aid decision-making. It can help you assess a large number of hard facts and soft issues. Try it any time that you have to make a choice based on more than one attribute. Just remember that *you* still have to decide:

1. What attributes are important to you?

2. How important is each attribute to you?

It is worth the effort, because it leads to clear, defensible decisions. CEP

J. B. CHRISTIAN is an advanced research engineer at Osram Sylvania Inc., Towanda, PA (717/268–5345; fax: 717/268–5350). He has 14 years of experience in project management, chemical and environmental process design, operation, and management. He received a BS from the State University of New York at Buffalo and an ME from Cornell University, both in chemical engineering. He is a Diplomate of the American Academy of Environmental Engineers, a registered Professional Engineer in New York, California, and Pennsylvania, and a member of AIChE.

Further Reading

Covello, V. T., "Decision Analysis and Risk Management Decision-Making: Issues and Methods," *Risk Analysis,* **7,** pp. 131–139 (1987).

Keeney, R. L., "An Analysis of the Portfolio of Sites to Characterize for Selecting a Nuclear Repository," *Risk Analysis,* **7,** pp. 195–218 (1987).

Table 4. Rating clarifier alternatives.

Vendor	Company Rating	Support Rating	Torque, lb-ft.	Price, $
A	3	10	45,000	595,000
B	3	3	49,100	626,000
C	3	3	53,920	634,000
D	3	2	31,000	540,000
E	2	7	50,000	515,000
F	2	0	33,120	601,000
G	1	1	34,500	478,000
H	3	2	45,000	575,000
I	1	0	40,000	500,000
J	2	7	59,000	698,400

Table 5. How clarifiers from different vendors compare.

Vendor (Weight)	Company Rating (5%)	Support Rating (5%)	Torque Rating (40%)	Price Rating (50%)	Total Rating, %
A	0.05	0.05	0.20	0.23	53.5
B	0.05	0.02	0.26	0.16	48.8
C	0.05	0.02	0.33	0.15	53.9
D	0.05	0.01	0.00	0.36	41.9
E	0.03	0.04	0.27	0.42	74.7
F	0.03	0.00	0.03	0.22	27.6
G	0.00	0.01	0.05	0.50	55.5
H	0.05	0.01	0.20	0.28	54.0
I	0.00	0.00	0.13	0.45	57.9
J	0.03	0.04	0.40	0.00	46.0

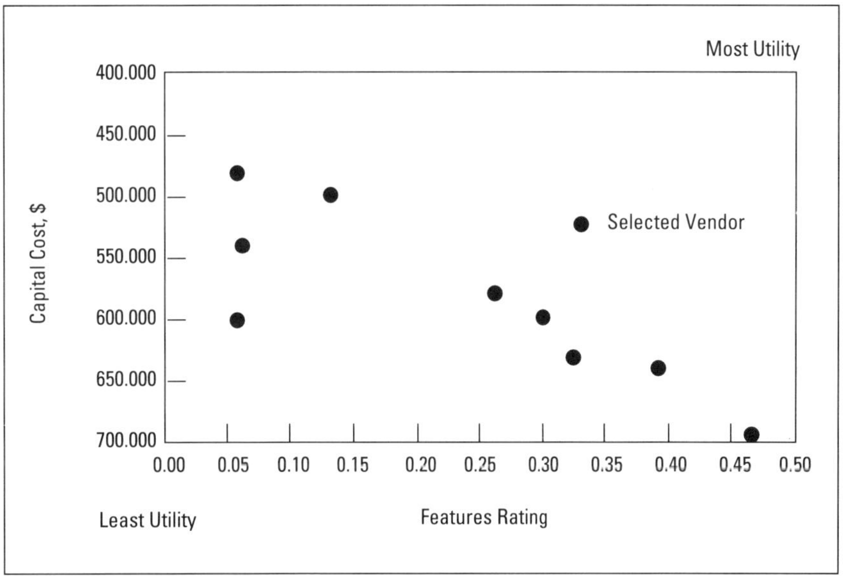

■ ***Figure 1. Cost vs. utility for clarifier alternatives.***

Optimize Your Plant's Compressed Air System

This is a matter of minimizing yearly power costs by selecting the most cost-effective compressor, and managing air distribution and use in the most economical way.

Hank Van Ormer,
Air Power U.S.A., Inc.

Let us define an optimized compressed air system as one in which the overall cost of compressed air is at its lowest effective level per unit of product, not necessarily one that uses the least amount of air. The optimized system should use the least amount of air efficiently.

Using air efficiently is important, because air is basically an inefficient means for transferring energy. For example, to produce 1 hp of power in an air motor requires approximately 30 scfm of inlet air compressed to 90 psig. Producing this compressed air in a relatively efficient manner requires about 6–7 bhp at the compressor shaft. With a 90% efficient motor, this translates into almost 7–8 hp of electric power required to deliver 1 hp in compressed air power.

Managing the cost of compressed air

Compressed air is used for many reasons in the process plant — including safety, cleanliness, repeatability, and control. However, efficient energy transfer is not one of these reasons. Thus, it is to any compressed air user's benefit to continually monitor, audit, and manage the basic efficiency of this expensive but highly productive utility.

Effective management of the cost of compressed air generally has to do with managing the supply, waste, and demand of the air. These are, of course, all connected and a program must deal with all three areas on a continuing basis to be effective and generate permanent positive results.

Power cost is a reflection of the cost of compressed air. There are many ways to define this cost, including initial cost, depreciation, and maintenance cost. All of these are a mixture of facts, opinions, assumptions, and estimates. Costs can vary, depending on the method of their accounting.

Power cost, on the other hand, is easily identifiable and any savings (or extra expense) reflects immediately on the bottom line. For electric motors, the power cost is given by:

$$\text{Power cost} = \text{hp} \times 0.746 \times \text{hours} \times \text{electric rate/motor efficiency}$$

This formula yields accurate estimates and comparisons. Of course, items such as power factor and demand charges must be taken into account.

As an example of electric power cost, consider a 100-hp compressor, operating for 6,000 h/yr, with a motor efficiency of 90% at a rate of \$0.07/kW-h. The yearly cost is \$34,800. A 1,000-hp compressor would have a yearly power cost of \$350,000. For a 100-hp compressor, the initial purchase price of a lubricated machine is \$35,000 to \$50,000, depending on the type.

In short, the power cost of operation in a heavy-duty cycle, full load can equal or exceed the initial cost of the unit every year. A positive variance of 15–25% in power cost can be a significant savings in operations, and this savings is obtainable.

Table 1. Ratings of compressors.

Type	Lubricated Duty Cycle, %	Lubricated Power Consumption Index*	Lubricated Specific Power, bhp/100 cfm*	Nonlubricated Duty Cycle, %	Nonlubricated Power Consumption Index*	Nonlubricated Specific Power, bhp/100 cfm*
Single-stage, single-acting reciprocating (water-cooled)	70	130	25	70	140	27
Two-stage, single-acting reciprocating (air-cooled)	75	120	23	75	125	24
Single-stage, double-acting reciprocating (water-cooled)	75	120	23	75	125	24
Two-stage, double-acting reciprocating (water-cooled)	100	100	19	100	105	20
Single-stage, rotary-vane (oil-cooled)	100	120	23	—	—	—
Single-stage, rotary-screw (oil-cooled)	100	120	23	100†	168†	32†
Two-stage, rotary-screw (water-cooled)	100	100	19	100	115	22
Two-stage, rotary-lobe (water- and air-cooled)	—	—	—	100	140	27
Two-stage centrifugal	—	—	—	100	136	26

Notes:
*At 100 psig discharge pressure. †Water-cooled.

This table does not include any nonlubricated dynamic compressors, such as the multistage centrifugal. Centrifugals range from 100 to several thousand hp. In the smaller sizes (100–450 hp), their performance is in the range of the single- and two-stage oil-cooled rotaries. Sizes 500 hp and larger are more efficient. As the size increases, the power efficiency becomes the highest available. Dynamic compressor performance varies significantly with ambient and cooling-water conditions.

Use of a compressed air survey or audit

Often the problem is not in selecting new equipment but in deciding what to do with an older air system. The problems themselves are not quite so identifiable.

Before embarking on a compressed-air cost management program, it is a good idea to do a compressed air survey or audit. The objectives of this survey or audit should be customized to the individual needs — but would address such things as:

Compressed air supply management

1. See if the primary air compressors are applied correctly, that is, load, type of air, installation, type of compressor, pressure range, and type and application of capacity controls.

2. Determine whether or not the compressed-air treatment devices (such as aftercoolers, dryers, and filters) used with the compressor are applied correctly, that is, minimum pressure drop, necessary performance, and adequate life.

3. Identify what opportunities exist for expansion if demand grows. Find what opportunities there are for economy if demand lessens or changes in some other way.

Waste management

1. Identify areas of wasted compressed air power that can be controlled in the air distribution system and in using tools.

2. Develop continuing plans to implement downstream/operator oriented leak and pressure controls. (Leak management should identify the cost of the leaks.)

3. Develop methods to measure the air flow to various departments and areas of usage for follow-up programs.

4. Identify areas of excess pressure loss and propose corrective actions.

Demand side management

This is one of the most overlooked areas of opportunity in managing the cost of a compressed air system:

1. Identify areas in the plant and the lowest possible pressures they can operate at with optimum production.

2. Identify areas in the system that are running off unregulated air with a resultant pressure spike. Actual air flow in volume (cfm) is directly propositional to pressure on any fixed demand — extra pressure means extra air.

3. Identify the minimum optimum flow and pressure required to perform each job.

Supply side management: optimize system efficiency

Primary equipment — As in the case of many decisions in business, the selection and application of an air system often involves many compromises. Only the users can accurately define the importance of such things as:

- Quality of maintenance available;
- Matching sensitivity of equipment to operating conditions;
- Availability of cooling medium — air or water;
- Initial cost vs. long-term savings;
- Required duty cycle;
- Anticipated future growth; and
- Required quality of air, such as lubricated or nonlubricated.

Types of air compressors — Probably no area is foggier than this one. Let us look at the basic types of compressors and using 100 as most efficient, rate them as a percentage of full load for 100 psig service.

There are many reasons why a compressed air system should be powered by a nonlubricated compressor, but initial cost, maintenance cost, and power are not among them. With the exception of the large-horsepower, multistage centrifugal, a nonlubricated unit can generally be said to cost more to purchase and maintain, and be less power-efficient than a similar lubricated unit. Usually, the nonlubricated will be 3–6% less efficient at best.

Table 1 rates the various types of lubricated air compressors. A two-stage double-acting, reciprocating, water-cooled, lubricated machine is given a power consumption index of 100. The unit has a specific power of 19 bhp/100 cfm. The power consumption index measures the relative bhp at the compressor shaft required to produce the compressed air for 100 psig discharge pressure. Depending on the conditions and application, these figures may vary, but probably not more than 10%. Table 1 shows that:

1. The type of compressor selected has a direct effect on power cost.

2. Nonlubricated units of similar design will always have a higher power cost.

3. Select the air compressor based on the capacity and pressure required — not on horsepower. Selection by horsepower can result in a compressor that is too small or too large.

Certainly many other things enter into compressor selection such as:

- Initial price;
- Maintenance skill level required;
- Maintenance cost;
- Total installed system price;
- Air or water cooling needed;
- Quality of delivered air;
- Pressure requirement; and
- Driver preference.

But, as far as power cost, the plant engineer should have all the power requirement facts from the supplier for the anticipated operating conditions.

Unloading and power cost

All types of plant compressors have two most efficient ways of running — fully loaded and off. The best way from a power standpoint to unload an air compressor is to shut it off. Anything else is a compromise.

When designing an air system, consider the flexibility of using what we would call base load units. Table 1 refers to the duty cycle, which is the percentage the unit can run fully loaded without any adverse effect on its per hour maintenance cost.

All the rotaries and the two-stage double-acting water-cooled unit have duty cycles of 100%. This is mainly because the cooling in these units is so effective that they do not normally develop hot spots and do not require regular cooling off periods.

The use of rotary units as the base load can be very effective, particularly if there are determinable departmental requirements — such as bead blasting or process air. In this case, the rotary can serve a single unit or department — using its unique installation flexibility, running either on or off.

Example — A sandblasting department has four ½-in. nozzles, with a capacity of 330 cfm and a pressure of 100 psig. From an annual power cost, the user will require around 300 hp of compressed air. Table 2 presents yearly power costs for three different options. Note that the type of control and its application very directly affect power cost.

Each type of compressor has its limitations by design as to what can be accomplished with a capacity control. In this example, the standard 75 hp class rotary provides a neat and economical solution. However, often the situation does not fit into a neat solution like this one. In selecting a machine get all the facts needed to do it right.

Motor efficiency

An often overlooked ingredient in the power cost formula controlled by the air compressor package is the motor efficiency, which has a very direct impact on the actual power cost.

The compressor manufacturer or packager selects the motor. Units can be supplied without a motor and the user can add one. However, more and more air compressors are being sold as packages. In such cases, the motor may not be evaluated by the plant engineer.

Motors are touted as being efficient. A salesperson does not say, "I'd like to

Table 2. Comparison of three 1,320-cfm systems.

	One 1,320-cfm, Two-Stage, Double-Acting, Three-Step Unload (Water-Cooled)		Four 335-cfm, Single-Stage, Oil-Cooled Rotary (Air-Cooled)		One 1,320-cfm, Single-Stage, Oil-Cooled Rotary (Air-Cooled)	
	Bhp Required	Power Cost, $/yr	Bhp Required	Power Cost, $/yr	Bhp Required	Power Cost, $/yr
Full Load	265	92,220	330	114,840	330	114,840
¾ Load	206	71,688	247	86,130	264	91,872
½ Load	146	50,721	165	57,420	198	68,904
¼ Load	88	30,450	83	28,710	132	45,936

Note: Cost is based on 6,000 h/yr, 90% motor efficiency, and $0.07/kW-h.

sell you this inefficient motor that will cost you more dollars to run." Yet if you were to check all the standard commercial packages of 50 hp units today, the motor efficiency will range from 87 to 94% at full load. This is a 9% difference in power cost.

Users these days request high-efficiency motors from compressor suppliers, but no mention is often made of how high is high. Be sure of what you are getting. At least get the full-load motor efficiency to see where you are. If you want an optional high efficiency motor, be sure to specify the efficiency, particularly if you are going to buy the package.

Many manufacturers supply standard motors on their premium industrial-grade air compressors with full load efficiencies from 90 to 92%. Be sure the gain in efficiency when specifying a high efficiency motor is worth the premium price.

Pressure losses producing pressure not used

No matter what the type of compressor, the basic horsepower required is determined by the capacity (volume of intake air flow compressed, acfm) and the pressure (amount of stored power, psig).

Sometimes engineers forget that every pound of pressure increase requires more bhp to produce it. So, if you do not need it or use it — do not have it.

Every pound of pressure increase in air compressors (100–150 psig range) requires about ½% increase in power. Therefore, a 10 psig change in pressure can save or cost 5% in bhp required. Per the formula given before, a 5% change in a 100 hp unit is 5 hp = $1,740/yr.

There are many causes of pressure losses in a plant air system, and some are a necessity. Some are bad breaks that cannot be repaired now. However, the plant engineer should be aware of what this costs in the power bill.

If some parts of the plant require high air pressure, then the whole plant may not have to be at the higher pressure. If a department that uses a small volume of air at higher pressures is requiring the total system to run higher pressure, consider installing a small departmental air compressor.

If the central system pressure is being raised to overcome system pressure drop, check the piping, filters, dryers, and heaters. Consider chokes, and add appropriate receivers or departmental air compressors at appropriate points to feed end-of-the-line areas.

Air-treatment auxiliaries

Today's requirements for high-quality air — dry and clean — have deluged the market with filters and air dryers. Although filters and dryers are required, their pressure losses often are neglected during specification.

Generally speaking, filters and dryers can be made smaller and cheaper, but with a higher pressure drop. This is not a good tradeoff because of the potential losses in wasted power through increased pressure drop.

Too often the user specifies a filter by size, not by cfm at a pressure drop. For example, specification may call for a 1-in. filter, of which there are many 1-in. filters on the market. Some have small elements and 1-in. openings. These will have a high pressure drop and a short element life.

Check out compressed air filters by cfm recommended, comparing pressure loss or pressure drop when new and when wet. See what pressure drop change is required.

Also, check out dryers. Evaluate the inlet air temperature, cfm handled, pressure loss, and dew point achieved.

For regenerative pressure-swing desiccant dryers consider the cost of purge air and the inherent savings in purge control.

Potential savings with purge control

Field usage has shown that with an efficient purge control on a regenerative desiccant dryer the purge load can be reduced from 100% of flow to 10–25% of purge flow during the year.

Consider a heatless/pressure-swing dryer. Flow is 1,000 cfm, purge rate is 15%, and power is $0.05/kW-h. Purge air is at 150 cfm and is supplied from a 35 hp motor. The purge is operated at 24% of the cycle. Power = 35 × 0.746/0.90 = 29 kW.

Estimated power cost at full purge flow = 29 × 0.05 × 8,760 = $12,700 annual power cost; estimated power cost at 24% purge flow = 6.96 × 0.05 × 8,760 = $3,048; and potential annual power cost savings = $9,600.

Similar results, although not as dramatic, can be accomplished with purge control or other types of desiccant dryers. Consideration should also be given to reviewing external heat, blower purge, heat of compression, and other terms, depending on conditions.

Even with membrane dryers now becoming popular for small point-of-use work, do not forget the cost of the sweep air.

Waste management

Effective leak control in an air system cannot be overstated. At 90 psig a ¼-in. leak will pass about 95 cfm of air — or about 21 bhp with no work performed. At $ 0.06/kW-h for 8,760 hours per year with a 90% efficiency = $9,150 in power cost.

Fixing leaks is fairly obvious, but removing condensate with open drains or automatic drains is not always obvious.

There are basically three categories of automatic drain valves:

1. *Level-operated mechanically activated* — These do not waste air but are prone to clogging and require continual maintenance to assure operation. These work best in a "power house situation," where continuing regular attention is part of the system.

2. *Dual-timer electronic* — These use an electronic timer to control the number of times per hour they open and the duration of the opening. The theory is that the cycles are adjusted as required to fit the current volume of condensate to be sure to drain fully and minimize the open time without water. This wastes compressed air. The reality is that the cycles either do not get reset from the original factory settings (which causes condensate buildup in the summer) or they get set wide open and not closed down later in cooler weather thus wasting more air.

Consider: The usual factory setting is 20 min with a 10 s duration. A volume of 2,500 scfm of compressed air generates about 63 gal/d in average weather or 2.63 gal/h. Each 20-min cycle will have 0.875 gallons to discharge. This will drain out of a ¼-in. valve at 100 psig in approximately 2.73 s. Compressed air will then blow for 7.27 s each cycle. The 3 cycles an hour will equal 36% of a minute, and this will waste 37.4 cfm of compressed air/h. At $0.06/kW-h power this costs about $3,600/yr per valve. A ½-in. valve under the same conditions would waste 149.4 cfm, significantly more.

3. *Level-operated electronic* — There are a number of different types of these, which receive the signal to open from the condensate high level and the signal to close from the condensate low level. These waste no air and, from a power cost standpoint, are the best selection. Their reliability is many times greater than that of the level-operated mechanical. Many people feel that these drain valves also have a higher degree of reliability than the pneumatic/mechanical valves — particularly when equipped with a positive-opening motorized ball valve.

Pressure spikes and flow control

An even less obvious potential savings in compressed air usage comes from controlling excessive system pressure spikes. These not only automatically cause increased leakage at each leak location but also create an artificial demand throughout the entire system. This causes higher air usage and generally gives no increase in production.

An intermediate controller with proper storage volume can be used to deal with pressure spikes. This controller separates the air demand from the air supply by a pressure differential to allow a smooth uninterrupted flow of air.

The intermediate controller is installed downstream of the cleanup and drying equipment and upstream of the piping distribution system. The intermediate controller creates storage by introducing a controlled pressure differential between the upstream receivers and the distribution piping. This isolates the compressors from the demand side. Peak demands are handled using the reserved stored air energy, allowing more offload time on the primary compressors.

Today's requirements for high-quality air — dry and clean — have deluged the market with filters and air dryers.

The intermediate controller eliminates the artificial demand and its associated wasted over pressuring. This allows the air system to operate at a predetermined correct minimum pressure. Saving in air usage when correctly applied at the right conditions can range from 15-30%.

In summary, there are many areas in any compressed air system where the astute, knowledgeable designer can negatively or positively affect operating costs, particularly power cost.

Operating power cost is only part of the evaluation process in the selection, application, and installation of compressed air equipment — but it is a significant and often overlooked element.

Power cost expense is a very significant measure of relative operating expense because the positive or negative variance moves directly to the bottom line.

Air conservation and leak control management

To effectively conserve air and manage leaks in the plant, a continuing economical program must be in place. Generally speaking, the most effective programs are those that involve the production supervisors and operators in a positive manner working with the maintenance personnel. Some suggestions:

- Identify the annual power cost per hour per cfm. This identifies the annual power cost for any air use.
- Set up continuing leak inspection by maintenance personnel. Each primary sector or plant should be inspected once a month or at least every 6 weeks. A record should be kept of these findings and of the results. A leak can be qualified and measured as to estimated flow and a measured cost assigned to it.
- Create programs in which production people (particularly the operators and supervisors) are positively motivated to identify and repair leaks. Establish and monitor continuing air conservation programs.

One method is to monitor air flow to each responsible sector (perhaps using a recording or nonrecording flowmeters) and identify the air usage as a measurable part of the operating expense of that area. This usually works best when combined with an effective in-house training and awareness program for all those involved . CEP

H. VAN ORMER is the founder of Air Power U.S.A., Inc., Pickerington, OH (614/497-4001; Fax: 614/497-1750). Currently, Van Ormer develops short- and long-term plans to minimize air demand and optimize power costs. He has been associated with the compressed air and air compressor businesses for over 35 years. He has worked with several manufacturers in marketing, product development, training, and other areas, in addition to working in the fabrication and distribution of compressed-air equipment all over the world. Van Ormer holds a BS in economics and physics from Gettysburg College and is a member of the Association of Energy Engineers and the Association of Plant Engineers.

The Importance of Storage, Transfer, and Collection

Hoppers, pneumatic conveyors, standpipes, and gas/solids separators are common in the chemical process industries, but key factors for their successful operation often are underappreciated.

Ted M. Knowlton,
Institute of Gas Technology

John W. Carson,
Jenike & Johanson, Inc.

George E. Klinzing,
University of Pittsburgh

Wen-Ching Yang,
Westinghouse Electric Corp.

Handling bulk solid materials is one of the least understood and, at the same time, one of the most important areas associated with solids processing plants. A study by Merrow *(1)* showed that two of the most common types of problems encountered in solids processing plants were difficulty in transporting solids and in handling fines. Although this is an area primarily ignored by mainstream chemical engineers, a small group of researchers is starting to make inroads in understanding how to design solids transport systems. Progress in this area occurs in small evolutionary steps instead of large revolutionary leaps. The various sections below on hopper design and flow, pneumatic transport, standpipes, and solids separation provide an overall introduction to the area for the uninitiated. It is hoped that the information presented here will increase the level of awareness of the importance of these areas, as well as decrease the level of ignorance about solids handling.

Storage and flow of solids from hoppers

Industries as varied as chemical processing, food preparation, pharmaceuticals, power/utilities, metals, and consumer products, to name just a few, handle millions of pounds of materials every year. Bins, silos, and hoppers used to store these materials vary in capacity from a few pounds of material (such as a press feed hopper used in a pharmaceutical facility) to multi-thousand-ton capacity vessels (such as train-loading silos at a coal preparation plant). The vast quantities of bulk solids that are handled would suggest that this area is well understood by most engineers. Unfortunately, it is not.

Startup delays and ongoing inefficiencies are common in solids processing plants. The most prevalent cause of these problems is the improper design of bulk solids handling equipment — specifically, bins, hoppers, and feeders. Problems are also common in processing vessels that are used to purge, dry, heat, cool, or condition materials.

The Rand Corporation conducted a six-year study of 40 solids processing plants in the U.S. and Canada. Their findings *(1, 2)* reveal that 80% of these plants experienced solids handling problems. The study also found that these plants were slow in coming on-line, with an average startup time for some plants approaching 18 months. Once startup began, poor performance continued to plague operation with performance between 40% and 50% of design.

What are some of the common flow problems in hoppers and silos?

• *No flow.* Lack of discharge can be attributed to the formation of a stable arch over the outlet, or a stable vertical cavity called a rathole *(3)*. Sledge hammers or vibrators, while sometimes useful, cause flow to be erratic at best. In addition, they may compact the material, compounding the flow problem instead of solving it.

• *Segregation.* Many materials experience separation of fine and coarse particles *(4)*. Such segregation can significantly compromise the quality of the final product, and the efficiency of the process. This can lead to a considerable expense when product that doesn't measure up to quality control standards must be discarded or reprocessed.

• *Flooding.* The collapse of a rathole in a bin that contains fine powder can cause the powder to exhibit fluid-like characteristics. The result is uncontrollable flow of material, loss of product, clouds of dust, and a production nightmare *(5)*.

• *Structural failure.* Each year, over 1,000 silos, bins, and hoppers fail in North America alone. Most of these failures could have been prevented if the designer had properly considered the loads imposed by the bulk solid being stored *(6)*.

• *Nonuniform flow in solids processing vessels.* Contact-bed/moving-bed processing vessels are used to dry, heat, purge, or condition bulk solids. The lack of uniform and reliable flow of the bulk solid can drastically undermine the efficiency and cost-effectiveness of a process.

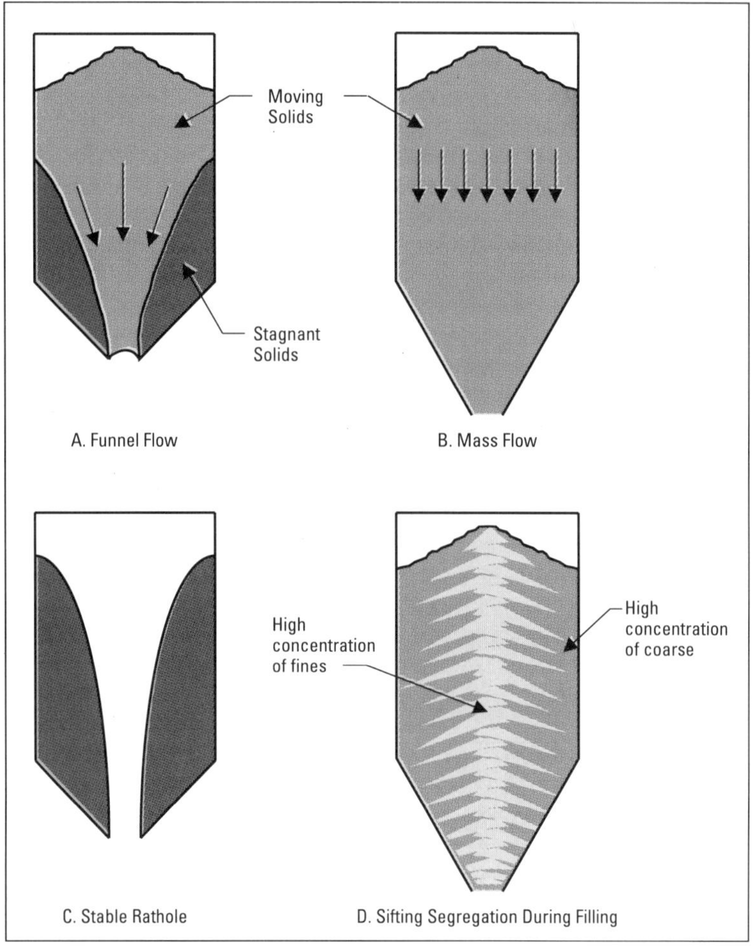

■ ***Figure 1. Types of flow patterns in hoppers.***

Material errors

The design of bins, hoppers, and feeders has never been given the attention that it deserves. Many people still wrongly assume that all materials with the same generic name (for instance, terephthalic acid, sodium carbonate, or even coal) have the same flow characteristics. Nothing could be further from the truth. Different grades and types of a material often exhibit tremendous differences in flowability. Changes in moisture content, particle size, temperature, and so on can also significantly affect flowability.

Techniques such as specifying a material's angle of repose, angle of spatula, and other similar parameters are also ineffective, because the resulting values bear no relationship to the design parameters needed to ensure reliable flow. Likewise, trying to model bulk solids as fluids also leads to a dead end. Flowing bulk solids generate shear stresses and are able to maintain those stresses when their flow rate is changed dramatically, even to the point that they stop flowing. This obviously is in contrast to the behavior exhibited by fluids. It is improper to consider bulk solids as having viscosity since almost all bulk solids have flow properties that are flow-rate independent.

This field was changed dramatically starting in the mid 1950s by the pioneering work of Dr. Andrew Jenike. His approach was to model bulk solids using the principles of continuum mechanics. The resulting comprehensive theory *(7)* describing the flow of bulk solids has been refined over the years, but even today it is recognized worldwide as the only scientific approach to bulk solids flow.

The impact of flow patterns

The type of flow pattern that develops in a bin, silo, or hopper significantly affects how this equipment performs. With a funnel flow pattern (Figure 1A), some of the bulk solid remains stationary, while the rest moves. If the material being stored is subject to caking, spoilage, or oxidation, this first-in/last-out sequence is unacceptable. Ratholing (Figure 1C) is likely, and fine powders will have a tendency to flood through the outlet when a rathole collapses. Funnel flow designs can be cost-effective, but only if segregation is unimportant, and the bulk solid is relatively coarse, free-flowing, and nondegrading.

The alternative to funnel flow is mass flow (Figure 1B). Here, relying only on gravity, particles flow evenly along the hopper walls with no stagnant regions. Ratholing is eliminated and a first-in/first-out flow sequence is established. In addition, mass flow can prevent segregation and ensure uniform flow through the outlet.

Key measures of flowability

To assess flowability, three key parameters should be measured: cohesive strength, wall friction, and compressibility.

Cohesive strength. Using a direct shear tester, a sample of the bulk solid is consolidated by applying both compressive loads and shear (so as to simulate flowing conditions in a bin). Once the sample is consolidated, its strength is measured by shearing it to failure. The resulting cohesive strength vs. consolidating pressure relationship is called a flow function.

Variables that can affect the cohesiveness of a bulk solid are:

- moisture content;
- particle size and shape;
- temperature — both absolute temperature and temperature changes;
- storage time at rest; and
- chemical additives.

Once a material's flow function has been determined, minimum dimensions necessary to prevent flow stoppages due to arching and ratholing can be calculated. This not only determines the minimum size of the hopper outlet, but also the size of the feeder used to control flow *(8)*.

Wall friction. Expressed as either a wall-friction angle or coefficient of sliding friction, this can be measured by sliding a sample of the bulk solid across a stationary surface representative of the vessel wall. The ratio of the shear force to cause sliding of the bulk solid to the applied load acting normal to the plate surface is the coefficient of sliding friction. The arc tangent of this coefficient is the angle of wall friction.

Variables that can affect the friction values of a bulk solid are similar to those that affect cohesiveness.

Some uses of wall friction data are:

- Mass-flow hopper design. Using values of both wall-friction angle as well as effective angle of internal friction, along with design charts given by Jenike *(7)*, one can determine allowable hopper angles for mass flow. The lower the wall-friction angle, the less steep the hopper wall can be to still provide mass flow.
- Bin loads. Wall-friction properties also are used in the evaluation of loads that are applied to the walls of a bin. In both the cylinder and hopper sections of a bin, the shear and normal loads that are applied to the bin walls are strongly influenced by the friction that develops between the bulk solid and the walls *(9)*.
- Sliding on chutes. Beyond the impact point (described below), the acceleration of particles on a chute is directly related to the difference between the chute angle and the bulk solid's wall-friction angle. As long as the chute is steeper than the wall-friction angle, particles will accelerate. Otherwise, they will slow down and eventually block the chute *(10)*.

Compressibility. A solid's bulk density cannot be expressed by just one or two values, such as loose or compacted bulk densities. Instead, bulk density varies continuously as a function of the consolidating pressure that is acting on the material. For many bulk solids, this density/pressure relationship can be expressed as a straight line on a log-log plot, the slope of which is called the material's compressibility.

Compressibility data can be used to determine:

- Minimum dimensions to overcome arching and ratholing. In order to calculate these values, the bulk density of the material at the hopper outlet must be estimated.
- Mass-flow hopper angles. Bulk density values at various points in a hopper are used to calculate the pressures acting on a hopper wall. After running a wall friction test, the wall-friction angle is determined at these pressures, and used to calculate limiting mass-flow hopper angles.
- Bin loads. A solid's bulk density directly affects the pressures and shear stresses acting on a bin wall. By knowing how bulk density changes with pressure, these values can be determined more accurately.
- Feeder loads. Determination of the loads that act on a feeder or gate is also important. These loads depend, in part, on the bulk density of the solid at the point where the feeder or gate is located.

Additional flow tests

Other tests that are often used to characterize the flow properties of bulk solids include:

Permeability. The permeability of a bulk solid can be measured by passing air or another gas through a column of solids that has been compacted to a certain bulk density. The pressure across the bed is regulated to maintain a laminar flow condition, while recording the rate at which the air or gas passes through the material. The resulting value of permeability, which is a strong function of the solid's bulk density, has dimensions of velocity, and is inversely proportional to the viscosity of the gas.

Permeability values can be used to estimate settlement times and critical steady-state solids flow rates through various orifice sizes. These values are also useful in designing processing vessels where gas is introduced to purge, dry, heat, cool, or condition materials.

Segregation tendency. Mixtures of solid particles can separate or segregate while they are being handled. This often results in costly quality-control problems — as well as wasted raw materials, lost production, increased maintenance, and capital costs.

Five primary mechanisms are responsible for most particle-segregation problems *(4)*. By far the most common mechanism, though, is called sifting (Figure 1D). The potential for this occurring can be determined by running a sifting segregation test. After forming a pile of bulk solid under controlled conditions, the pile surface is sampled from top to bottom. The samples then are evaluated for particle-size distribution or chemical composition in order to determine the degree of segregation that has occurred.

Segregation can be minimized by changing the characteristics of the bulk solid (for example, moisture, or particle-size distribution), altering the process by which the bulk solids are handled, or redesigning the equipment used to handle the bulk solid.

Sliding at impact points. Two key factors in chute design are the chute angle, and the smoothness of the point of impact. Too shallow or too rough a surface will prevent cleanoff. The required minimum slope can be measured by placing a ring containing the bulk solid on a wall-surface coupon, applying a predetermined vertical load (to simulate impact), removing the weight, and then raising the plate to determine the angle at which the particles begin to slide *(10)*.

Abrasiveness. Flowing solids can wear the surfaces of bins and hoppers as well as chutes and feeders. In the design of this equipment, the ability to predict wear life can minimize cost, downtime, and maintenance.

A patented tester that can simulate the conditions of wear caused by the flow of a bulk solid now has been developed *(11)*. A supply of fresh bulk solids is continually presented to the wear surface. Solids pressures exerted on a wear surface sample are controlled and measured, as well as the relative velocity between the bulk solid and the wear material. Shear stress exerted on the wearing material is continuously measured, so that changes in the coefficient of friction between the solid and the wear surface can be observed. The resulting data can be used to quantitatively estimate the life of a wear liner and other bulk-solid contact surfaces.

The payoff

Uniform, reliable flow of bulk solids can allow the production of quality products with a minimum of waste, extend the life of a plant, control dust and noise, and maximize a plant's productivity and output.

By conducting laboratory tests and utilizing experts with experience in applying solids flow data, plant start-up delays that can impact schedule and cost can be eliminated.

With existing facilities that present daily bulk-solids flow problems, solids flow testing and analysis can save many hours of expensive downtime and thousands of dollars — thus moving from the quagmire of quick-fix solutions that don't deliver, to proven engineering solutions that work every time.

Dilute- and dense-phase conveying

Pneumatic transport is the art and science of transporting solids in a pipe via a gaseous medium, usually air, under either positive or negative pressures. Since the inception of this art in the late 1880s, the technology has been widely employed in a broad range of industries. The design of pneumatic transport systems also has gradually developed from an art into a science. It is fair to say that pneumatic transport systems for standard applications, where the solids do not change their characteristics during transport, can now be designed reliably — though not necessarily economically. Uncertainties still exist for specialized applications such as high-temperature and high-pressure operation, transport of highly cohesive powders, long-distance transport, and applications requiring branching. For cases where the pneumatic transport line is employed as a transport reactor, the solids flow pattern in the transport line is hardly understood at all.

Most of the focus in early development was on dilute-phase conveying. "Dense phase" is a term that did not appear until about 1974, when a pulsed-piston conveying system was introduced into the market. Since that time, the word "dense" has been used to describe a variety of different systems — therefore, it often is difficult for engineers to understand each other when speaking of dense-phase systems. There is no sharp transition point that marks the beginning of dense-phase and the end of dilute-phase transport. Thus, the word "dense" should always be carefully qualified whenever it is used.

Pneumatic transport will continue to be an important solids-handling technique in the future because of its simplicity, effectiveness, and economics. Continuous improvement is expected in the areas of instrumentation, hardware, and the specialized applications mentioned above.

Advantages and disadvantages

As with any technology, pneumatic transport of solids poses certain tradeoffs.

On the plus side, pneumatic transport systems:

- are simple to design and economical to maintain;
- can be designed to operate as closed systems;
- are adaptable for automation and remote control; and
- are space savers, because piping takes up little space, and can be arranged to avoid obstructions.

On the minus side, such systems:

- are limited to relatively nonfriable materials, due to potential attrition and bend erosion during high-velocity, dilute-phase transport (Dense-phase transport can be used, however, to minimize these problems); and
- generally involve high energy costs. The advantages of pneumatic transport generally outweigh the disadvantages in most industrial applications.

Types of dilute-phase conveying systems

There are many ways to classify the different types of dilute-phase conveying systems. The classification adopted here is based on whether the operating pressure is above or below ambient. The systems operated below ambient pressure are called vacuum or negative-pressure systems, while those operated above ambient pressure are called positive-pressure systems. A third type of system is a combined positive-negative pressure system where the front end is operated at positive pressure, and the back end is operated at negative pressure. Another, special case is the closed-loop system, where the transport gas is recycled.

Vacuum (negative-pressure) systems. These systems (also called "pull" systems) are particularly suitable for handling radioactive and toxic materials. Any leakage will be into the conveying pipe. Vacuum systems are also the most economical if the applications involve multiple feeding points and one common collection point. They commonly are employed for unloading dry bulk material from railroad cars, barges, and ships. The disadvantage of vacuum systems is that the maximum pressure drop is set by ambient pressure. Thus, the application is limited in both distance (usually to about 400 m) and capacity.

Positive-pressure systems. Such systems (also called "push" systems) are the most suitable for distributing materials from a single source to multiple receivers. They are the most extensively used systems in industry. Positive-pressure systems allow for longer distance and greater capacity transport than vacuum systems. For dry bulk material, the maximum conveying distance for a low-pressure system (<1 barg) is approximately 450 meters; for a medium-pressure system (1 barg < P < 3 barg), it is about 1,200 meters; and for a high-pressure system (> 3 barg), solids can be conveyed up to 3,000 meters.

Combined positive-negative pressure systems. The combination of positive and negative pressure in a single push-pull or pull-push system provides a versatile and flexible conveying system. This system can be applied efficiently with multiple feed points and multiple discharge points, and thus is employed extensively in many industries.

Closed-loop systems. These are special cases in pneumatic transport. They are selected when a special conveying gas other than air is used, and the gas must be recycled. For example, food products with moisture sensitivity often are conveyed with carbon dioxide, and nitrogen is used to convey combustible materials in closed-loop systems. Toxic, hazardous, radioactive, and explosive materials are prime candidates for closed-loop conveying systems.

Types of dense-phase conveying systems

Two-phase flow. Starting with a system operating in horizontal dilute-phase flow, more and more solids can be added at the same conveying velocity and pressure to obtain a more highly loaded system. When this occurs, the degree of segregation in horizontal flow becomes more pronounced. At a particular point, a distinct, slower-moving layer of solids on the bottom of the pipe with a faster, more lightly loaded flowing above it is observed. This type of flow is termed two-phase flow. Certain materials, such as fly ash and other aeratable materials, are conveyed easily in this mode.

Extrusion flow. For some materials, when relatively little transport gas is used, it is possible to load the flow so much that essentially the entire pipe is filled with a high concentration of material. This type of flow is called extrusion flow. Feeding such a system often is done by pressurizing a blow tank, and forcing the solids out as in an extrusion process. Extrusion flow generally occurs with finer powders (less than 100 microns in size) that are conveyed over short distances; flexible tubing is often employed as the conduit.

Plug flow. A fixed quantity of material can be injected on a timed sequence to form discrete plugs of solids to facilitate transport. For self-forming-plug material, regulated solid and gas flow can be achieved without a timed sequence. Plastic pellets are easily conveyed in this manner at low velocities — reducing particle attrition and pipe erosion. Air often is injected along the pipeline to provide a fluidized, dense plug, or to break up long plugs that generate large pressure drops.

Gasless transport. A new type of dense-phase transport developed by Molerus *(12)* uses no transport gas to move the solids. Appropriate application of directional vibration to the transport line induces the solids to flow.

In addition to the easily recognizable transport flow patterns discussed above, other flow patterns exist. Immature dune flow, dune flow, degenerated dune flow, immature plug flow, degenerated plug flow, and ripple flow have been observed in horizontal pipes *(13)*. It also should be noted that for long-distance conveying, a system may start out in the dense-phase regime, and end up in the dilute-phase regime as the gas expands and the transport velocity increases with system length. Illustrations of the various flow patterns encountered in horizontal and vertical pneumatic transport are given in Figure 2.

Research areas

To enhance the economics and reliability of dilute-phase and dense-phase pneumatic transport systems, research has been concentrated in the following critical areas.

Choking and saltation. Choking is a flow-transition phenomenon in vertical dilute-phase lines. Saltation is the corresponding phenomenon in horizontal, dilute-phase transport lines. When the transport velocity is decreased in vertical dilute-phase transport lines at a constant solids-conveying rate, a velocity will eventually be reached where the solids can no longer be transported. The flow then collapses into solids slugs spanning the entire cross-section of the pipe. This is a classical example of choking in dilute-phase pneumatic conveying. Actually, if the pipe is large or the solids are very small, classical choking does not occur. The

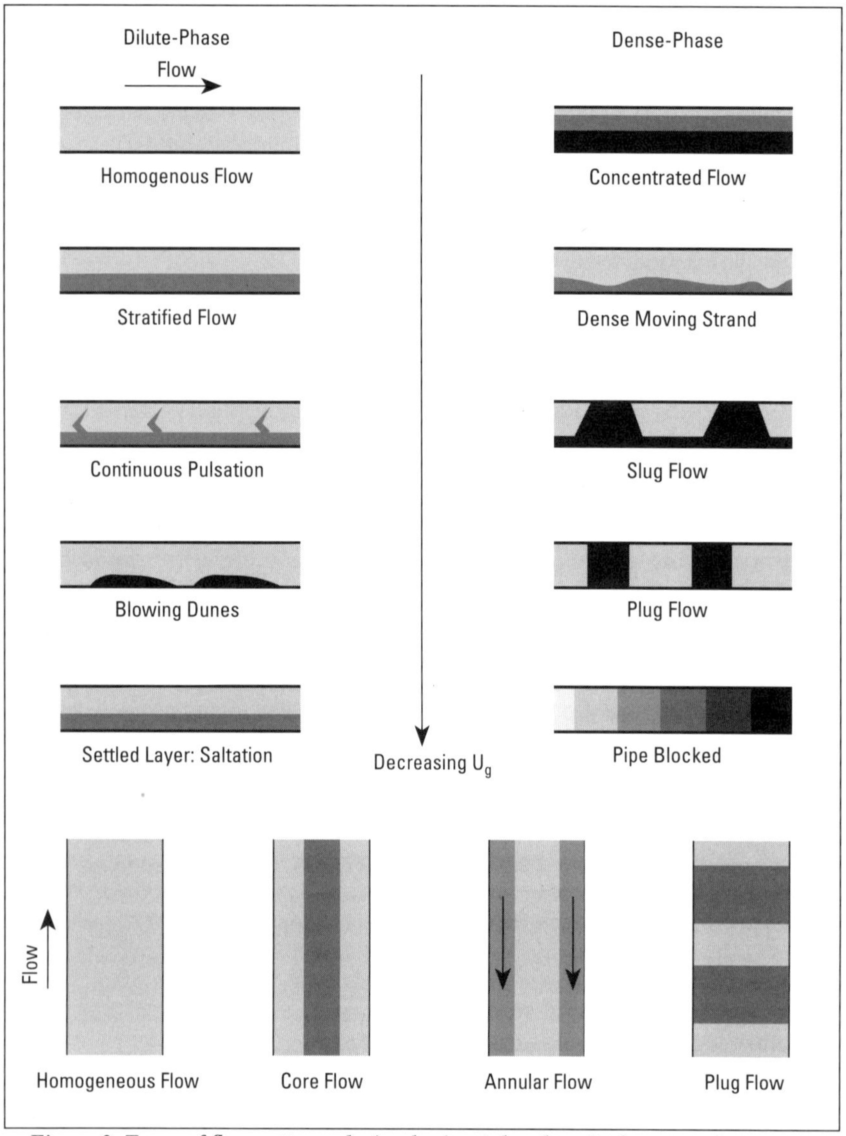

Figure 2. Types of flow patterns during horizontal and vertical pneumatic transport.

flow pattern in the transporting line near the "choking" point (when classical choking does not occur) has a core-annular structure with high-velocity, dilute-phase transport in the core of the pipe, and a dense, annular downflow near the wall *(14)*. This type of structure is a characteristic of "fast-fluidized beds." For vertical flows, choking marks the boundary between dilute-phase and dense-phase flow. Bi and coworkers *(15)* recently subdivided choking into three different types.

In horizontal dilute-phase transport, continuing to decrease the transport velocity while maintaining the solids flow rate constant eventually will produce a phenomenon called saltation — where solids start to become deposited on the bottom of the pipe because of gravity. The amount of saltation generally reaches an equilibrium, and conveying will continue. A further decrease in the conveying velocity will cause the line to convert to dense-phase transport. An extensive review of saltation can be found in Jones and Leung *(16)*.

Flow-regime definitions. Knowing what flow regime the solids are in is much more important in dense-phase conveying than in dilute-phase conveying. The pressure drop in horizontal dense-phase conveying is extremely dependent on the type of flow mode in the pipe. Dense-phase design procedures strongly depend on the type of flow regime in which the solids operate. What flow regime will exist in the pipeline in turn depends on the physical properties of the solids. For example, the conveying of a plug of sand is extremely difficult, while plastic pellets and cohesive Geldart Group C powders form plugs nicely. Plastic pellets form and reform into plugs, while a cohesive material must have good consolidation characteristics to maintain the integrity of the plug throughout the transport line.

Attrition and erosion. Particle attrition and pipe erosion are areas of concern in dilute-phase transport. There is virtually no particle attrition information in the open literature, however, even through particle attrition can be a serious problem for some applications. Particulate Solid Research, Inc. (PSRI), an industrial consortium with 28 industrial members, is conducting a systematic investigation on particle attrition in dilute-phase pneumatic transport lines. PSRI found that attrition rates in bends rise with increasing gas velocity and increasing Hardgrove Grindability Index (HGI) of the transported materials, and drop with increasing solids loading and fines content. They also found that blind tees cause more attrition than long-radius elbows and vortice ells.

Conversely, there are many studies on pipe erosion, because failure due to pipe erosion is common in plants utilizing high-velocity, dilute-phase pneumatic transport. The single most important parameter in pipe erosion is the particle velocity. The erosion rate is proportional to the nth power of the particle velocity, where n ranges from 2 for ductile pipe to 6 for brittle materials. For general industrial applications, $n = 3$ is satisfactory *(17)*.

Bends and branching. Bends are important links in a pneumatic transport line. Solids are slowed down or stopped at bends, and then have to be reaccelerated in a new direction. This increases the pressure drop in the system. Pipe erosion and particle attrition also occur in a bend because of particle impacts on its wall. Because of particle impact at bends, bends are the locations of most erosion failures and plugging. Principal bend types are: long- and short-radius elbows, dead-end tees, and commercially available vortice ells. No one single design is a panacea for general applications.

In many applications, splitting of a single conveying stream into multiple lines requiring equal flows is required. This kind of branching is difficult to achieve simply by design alone. Continuous monitoring of solids and gas flow rates for control is necessary. Development of adequate instrumentation to continuously measure solids flow rates needs to occur before reliable splitting of flows is assured.

Feeding and separation. Pneumatic conveying in pipes is more than just design of pipes and bends. The solids have to be metered into the pipelines and separated from the conveying gas. Thus, the study of solids flow in hoppers, solid feeding devices, and solids separation equipment also is important because they are critical components in the pneumatic transport circuit.

Instrumentation. Instrumentation in pneumatic conveying systems should be as simple and nonintrusive as possible. An up-to-date discussion of instrumentation development in pneumatic transport applications was presented by Marcus and coworkers *(17).* Particulate solids impacting on sensors can destroy them in a very short time—especially when the solids are abrasive or when gas velocities are high. The future focus for instrumentation development in pneumatic conveying is on the accurate, in-line determination of solids flow rate for measurement and control. Electrostatic, acoustic, light, capacitance, microwave, and nuclear instruments are being studied. Measurement of solids flow rates in commercial systems remains a challenging problem, however, and no large-scale, reliable meter is commercially available.

Scale-up. Scale-up is the most ignored subject in pneumatic conveying. The creation of comparable flow patterns in different pipe diameters is not usually the primary concern during design, because generally there is no reaction occurring in the transport line. Therefore, the primary concerns are power consumption for a specific material transport rate, along with particle attrition and pipe erosion. Existing correlations are sufficient to provide this design information. It is well documented, however, that systems larger than 15 cm appear to operate differently than smaller-diameter systems. Above 15 cm, there appears to be a leveling off of the effect of diameter, with the wall region becoming less important *(18, 19, 20).* With the increasing use of transport reactors, such as circulating fluidized beds for industrial applications, how to reliably scale up from small to large reactors has become a central concern. Systematic scale-up technique based on geometric similarity, dynamic similarity, or flow similarity *(21, 22, 23)* have recently been developed, but none have yet been shown to be reliable.

Standpipes

A standpipe is a device used to transfer solids from a region of low pressure to a region of higher pressure. This is exactly the reverse of what occurs in pneumatic conveying, where solids are transferred from high pressure to lower pressure.

Standpipes have been used in solids processing plants for over 50 years. They were first developed in the 1940s by a research team from Jersey Standard that was working on injecting transferred cracking catalyst back into a fluidized bed on a unit to produce gasoline during World War II *(24).*

Most standpipes are oriented vertically. Angled standpipes and combinations of angled and vertical standpipes (called hybrid standpipes), however, also are used to transfer solids against a pressure gradient.

A standpipe generally operates in one of two basic flow regimes: packed-bed flow or fluidized-bed flow. Standpipes also can operate in a dilute phase regime called streaming flow, but this type of flow is generally not desirable, because the pressure buildup possible is reduced.

In packed-bed flow (often called moving-bed flow), the solids in the standpipe are not fluidized, and flow "en bloc" down the standpipe. In this regime, a flow condition called stick-slip flow can occur, wherein the solids appear to quickly start and stop during their flow down the standpipe. This type of flow can cause the standpipe to vibrate excessively, and a loud "chattering" often can be heard when solids are flowing in the stick-slip mode. This flow regime does not seem to occur at low solids flow rates and at extremely high solids flow rates. It should be avoided, but no method exists to predict when it will occur.

When possible, it is usually best to operate standpipes in the fluidized-bed mode. It affords higher solids throughputs, because of the reduced friction of the solids on the wall of the standpipe. In addition, the highest pressure buildup possible occurs in this mode. It is critical, however, to operate the standpipe correctly in the fluidized-bed mode, or significant problems will occur. Large bubbles that can rise up the standpipe will hinder the flowing solids to such an extent that the desired solids flow cannot be achieved. Bubbles also reduce the fluidized density in a standpipe, which then limits the pressure "head" that the standpipe can develop. Thus, a longer standpipe than normal will be required if many bubbles are present.

Theoretically, the best way to operate a fluidized-bed standpipe is to operate it in a bubbleless mode. Practically, this is hard to achieve. Therefore, a standpipe operating in the fluidized-bed mode should operate in such a manner so as to minimize the size of the bubbles.

Solids larger than about 100 microns start to bubble immediately after the minimum fluidization velocity is achieved. Therefore, operation in the fluidized mode without bubbles is impossible with these solids. A standpipe operating with solids of this size should be operated, however, at a velocity as close to minimum fluidization conditions as possible, while still ensuring that the solids will remain fluidized. This will minimize the size of the bubbles in the standpipe, and result in optimum standpipe performance.

Solids smaller than 100 microns in diameter do not start to bubble immediately after the minimum fluidization velocity is achieved. Instead, they remain bubbleless for a certain velocity range above the minimum fluidization velocity. The velocity where bubbles finally start to appear in the fluidized bed is called the minimum bubbling velocity. Because of this property, standpipes operating with these small solids have an operating "window" where they can be fluidized and still remain bubbleless. Therefore, the best way to operate a standpipe transferring small solids is to have the gas velocity in the standpipe near the minimum bubbling velocity.

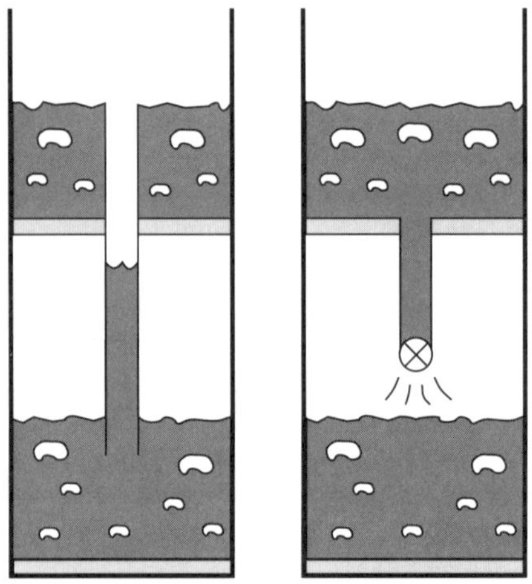

■ ***Figure 3. Overflow and underflow standpipes.***

Cohesive solids clump together and bridge easily, and are very hard to flow in standpipes. One technique that has proven effective with these solids is to periodically pulse aeration gas into the standpipe to prevent particle bridges from forming.

Types of standpipes

Two basic types of standpipe configurations are used to transfer solids: the overflow standpipe and the underflow standpipe. (See Figure 3.) The overflow standpipe is so named because solids overflow from the top of the fluidized bed into the standpipe. (A special case of the overflow standpipe is the cyclone dipleg.) In contrast, solids are introduced into the underflow standpipe from the bottom of a fluidized bed or hopper. There are several important differences between underflow and overflow standpipes as described below.

Overflow standpipes have several distinctive features:

- The bottom of an overflow standpipe must be immersed in a fluidized bed. If not, the standpipe will empty, no pressure seal will form, and solids will not transfer through the standpipe.
- The overflow standpipe operates "automatically." This means that the solids enter and discharge through the standpipe at the same rate without need of a valve.
- This type of standpipe normally operates in the fluidized-bed mode. It is possible to operate an overflow standpipe in the packed-bed mode, but it is difficult to control.
- A disadvantage of the overflow standpipe is that it cannot discharge large solids or agglomerates which may collect on the bottom of the fluidized bed.

Underflow standpipes have the following characteristics:

- Because solids enter from the bottom of the bed, it is necessary to have a valve at the bottom of an underflow standpipe in order to prevent solids in the bed from dumping. This valve also must be used to control the solids flow rate through the standpipe. Therefore, an underflow standpipe cannot operate "automatically."
- The underflow standpipe can operate in either the packed-bed or fluidized-bed mode. Generally, solids greater than 100 microns in diameter will operate in the packed-bed mode,while solids smaller than 100 microns in diameter will operate in the fluidized-bed mode.
- An underflow standpipe does not require immersion in the bed to which it is discharging. If desired, the solids can be discharged into the freeboard instead of into the fluidized bed.
- The underflow standpipe is able to discharge agglomerates that may build up or collect on the bottom of the fluidized bed.

The most common types of standpipes are:

1. a cyclone dipleg (an overflow standpipe used to discharge collected solids from a cyclone back into a fluidized bed). This is used extensively in fluidized catalytic cracking (FCC), fluidized-bed combustion, and gasification processes, and anywhere cyclones are used to collect solids; and

2. an underflow standpipe operating with solids less than 100 microns in diameter in the fluidized-bed regime. This finds service widely in the FCC industry, but is used in any process in which a catalyst is recycled.

Because it is sometimes desired to transfer solids horizontally and vertically at the same time, angled and hybrid standpipes are employed. These are not as efficient at transferring solids against a pressure gradient as vertical standpipes. The gas and solids separate in an angled standpipe, with the gas channeling via bubbles along the upper part of the standpipe while the solids flow along the bottom. When a vertical section is used in conjunction with an angled section to form a hybrid standpipe, slugs often form in the vertical section of the standpipe — blocking solids flow. In general, it is better to use a "true" angled standpipe or a vertical standpipe to transfer solids than a hybrid standpipe.

Standpipe aeration

The primary purpose of adding aeration to a standpipe is to maintain the solids in a fluidized state near minimum fluidization. The reason that aeration must be added to most standpipes is that the interstitial gas between the solids is compressed as it moves down the standpipe and becomes pressurized. As the gas volume in the standpipe "shrinks," the solids move closer together. This can occur to such an extent that the flow regime changes from fluidized-bed to packed-bed flow. Therefore, aeration gas is added to the standpipe to prevent this regime transition from occurring. In general, for solids less than 100 microns in diameter, it is better to distribute the aeration gas equally at several distribution points along the standpipe. For materials with a particle size greater than 100 microns, aeration is best added at the bottom of the standpipe.

During the last 15 years, the importance of correct standpipe design on the operation of solids processing plants finally has been recognized. Unfortunately, too many standpipes still are designed based on "art" instead of science. Although much progress has been made in understanding standpipe operation and design, reliable models to predict aeration locations and amounts for vertical and angled standpipes operating with both fine and large particles are not yet available — and are badly needed. Because more processes are trying to utilize the very fine, cohesive solids, techniques to transfer these cohesive materials reliably in standpipes need to be determined.

Gas/solids separation

Separating solids from a gas stream in commercial processes is very important, indeed often critical. Fines in the exit gas stream in a fluidized-bed process generally must be collected and returned to the bed in order to ensure high conversion efficiencies. Emission standards also may dictate that the solids in the gas exit stream be below a certain concentration. Separation of solids from a gas is accomplished using many different devices. The most common of these devices are:

- cyclones;
- filters;
- scrubbers: and
- electrostatic precipitators.

The selection of the most-appropriate separation device depends on the size of the solids to be collected, the degree of separation desired, pressure drop considerations, reliability, and the operating temperature of the system. Each device has advantages and disadvantages. Two of the most critical factors for choosing a collector are its ability to capture fine particles, and temperature applicability. How each of the collectors mentioned above compare in these two categories is shown in Table 1.

Table 1. Particle collection ranges and maximum temperature limits for solids-collection devices.

Collector Type	Particle Collection Requirement, microns	Range of Temperature Applicability, °C
Cyclone	> 15	> 1,000
Electrostatic precipitator	> 0.1	< 400
Filter		
Fabric	> 0.1	< 250
Ceramic	> 0.1	> 1,000
Scrubber	> 1	> 1000

Cyclones

Cyclones are by far the most common type of gas/solids separation device used in industrial processes. Cyclones have no moving parts, are inexpensive compared to other collection devices, can be used at high temperatures, produce a dry product, have a low energy consumption, and are extremely reliable. Their primary disadvantage is that they have a relatively low collection efficiency for particles below about 15 microns. Even when particle separation below 15 microns is required, a cyclone is generally placed upstream of the final separation device, such as a bag filter or an electrostatic precipitator, to reduce the loading to that device.

Cyclones separate particles from a gas by subjecting the gas/solids stream to a high centrifugal force. This is accomplished by having the solids flow into the cyclone tangentially. Because the particles are much denser than the gas, they are thrown outward to the wall. They flow downward in a thin layer along the wall in a helical path, and are collected at the bottom of the cyclone. The inlet gas stream flows downward in an annular vortex, reverses itself, creates an upward inner vortex in the center of the cyclone, and then exits through the top of the cyclone. This type of cyclone is called a "reverse-flow" cyclone, and is by far the most common type of cyclone.

Cyclone diameters range in size from less than 0.05 m (for use in small laboratory equipment) to 10 m (for cyclones used in circulating fluidized-bed coal combustors). Solid loadings in the inlet gas stream range from 0.1 kg/m^3 to >50 kg/m^3. Typical gas inlet velocities for cyclones are 15–35 m/s. Velocities lower than this result in lower cyclone efficiencies, while higher velocities can cause excessive erosion.

In order to remove as much of the dust from the gas stream as possible, cyclones often are placed in series with each other. This increases the overall collection efficiency relative to a single cyclone, but at the expense of increased pressure drop. For example, in FCC regenerators, three stages of cyclones generally are used to minimize particulates in the exit stream. Collection efficiencies of greater than 99.99% can be achieved in this manner.

The smaller the diameter of a cyclone, the greater the centrifugal force generated, and the better its collection efficiency for fine particles. Cyclones can generate centrifugal forces between 75 and 1,000 times the force of gravity, depending on the diameter of the cyclone. Because the centrifugal forces in a cyclone are many times the force of gravity, the orientation of a cyclone is immaterial. To try to improve collection efficiency, many small cyclones in parallel (called multiclones) often are used to replace one large cyclone. This type of arrangement offers increased collection efficiency at the expense of greater complexity. A disadvantage of parallel cyclones is that equal gas distribution to each cyclone can be difficult to achieve. When gas maldistribution occurs, collection efficiency is reduced, and the advantages of parallel cyclones are greatly negated.

Experience and theory have shown that there are certain relationships among a cyclone's dimensions that should be observed for efficient cyclone performance *(25)*. These dimensions generally are reported as a function of the diameter of the cyclone. There are several different standard cyclone "designs" in which the cyclone dimensions vary slightly from one to another. One of the more popular standard cyclone designs is called the Stairmand design. The dimensions of a Stairmand cyclone for high-efficiency operation are shown in Figure 4.

Because of the relatively high gas velocities at the inlet of cyclones, particle attrition can be significant. In general, particle attrition is at least equivalent to that produced at the grid in fluidized beds. Erosion in cyclones goes hand in hand with particle attrition, and occurs primarily where the particles first impact the cyclone wall. Erosion also will occur at the bottoms of cyclones too short to accommodate the length of the naturally-occurring vortex.

Research needs

One of the earliest cyclones used in an industrial application was patented in the U.S. in 1885 by the Knickerbocker Co. Much information about cyclones subsequently was gained by the extensive research on this device in the 1930s and 1940s. Because of this extensive work, design procedures for cyclones operating at low pressures, at moderate sizes, and at relatively low solid loadings are generally very good. In the last few years, however, cyclones have been operated at higher solids loadings and system pressures than before. In addition, much larger cyclones than before have been used. Systematic data on how these parameters affect cyclone operation are not available — and are badly needed to enable optimum cyclone design for these services. A few workers have started to look into these areas, but much more information is required.

A reliable computer design model for cyclones also is needed. Such a model capable of predicting the effect of geometry changes and system parameters on cyclone efficiency could result in "designer cyclones," optimized for the specific operating conditions of a particular process. Increasing cyclone efficiencies by using computer models to achieve the particulate reduction required to meet emission standards for a process (instead of using expensive electrostatic precipitators for final particulate removal) would greatly simplify collection equipment, and simultaneously reduce costs.

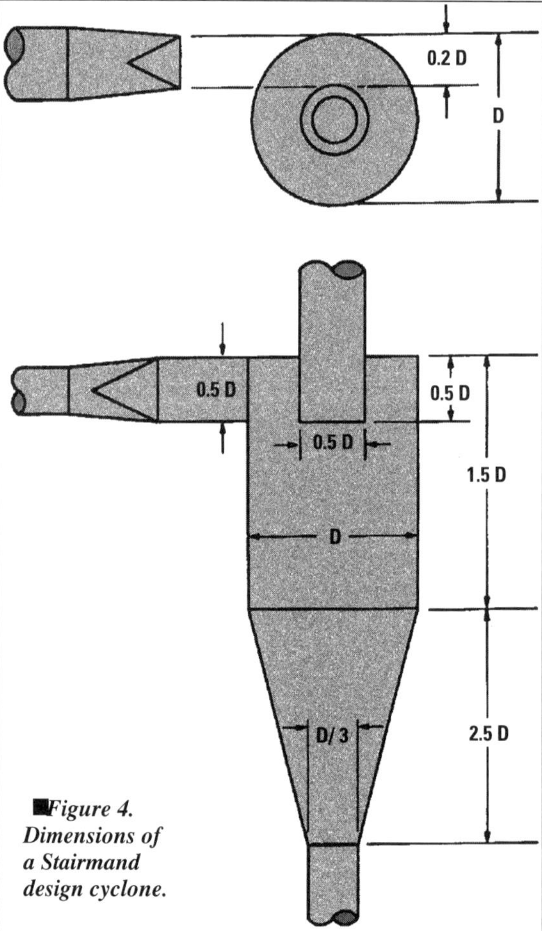

Figure 4. Dimensions of a Stairmand design cyclone.

Filters

In many processes, the final particulate-removal device is a fabric filter (often called a bag filter). These filters can capture particles much smaller in diameter than a cyclone, although a cyclone is nearly always used to remove most of the solids before the exit stream encounters the bag filter.

A bag filter generally consists of a porous fabric in the shape of a cylinder. Many of these bags are placed in a matrix so that their total area will result in a low gas velocity through the bags. This results in a low pressure drop through the filter. Gas velocities through the bag are of the order of 0.005–0.02 m/s. Particulate loadings to these filters generally lie in the range of 0.2–250 g/m^3.

The collected particulates build up on the surface of the bags and cause a gradual increase in the pressure drop through the filter. After a certain limiting pressure drop is reached, the bags are cleaned by pulsing gas back through the filter to remove the solids buildup on the bag.

Ceramic candle filters now are being used to filter fine particulates from hot exit-gas streams. One application of these filters is in the exit line from a coal gasifier in a combined-cycle system. This hot (>1,000°C) gas stream is to be sent to a turbine to produce electrical power. Most turbine blades, however, are adversely affected when particles larger than about 5 microns pass through the turbine.

A ceramic candle filter is composed of multiple porous ceramic cylinders (candles) that filter the gas as it passes through them. Because the candles are ceramic, the filters can withstand very high temperatures.

These filters are also backflushed when the pressure drop across them becomes too high.

The ceramic candle filter is a very promising type of filter. Problems with the filter are breakage due to mechanical and thermal stresses due to the relatively brittle nature of the ceramic. These filters are being used in a variety of hot gas-cleaning applications now. Information gleaned from these duties undoubtedly will enable even better, more reliable candle filters to be designed in the future.

Scrubbers

Particulates are often scrubbed from a gas stream by spraying a jet of a liquid, usually water, into the stream. The particles are intercepted by the droplets of the water spray, and are removed from the scrubber in a slurry.

Although a scrubber is more efficient than a cyclone, it has one definite disadvantage — the collected solids are wet. If this cannot be tolerated, the solids then must be separated from the liquid. Thus, the gas/solids collection problem is replaced by a liquid/solids problem.

Electrostatic precipitators

Electrostatic precipitators separate solids (as well as liquids) from a gas stream by passing the stream though a strong, high-voltage field produced between two electrodes of opposite polarity. The field causes the particles to be charged so that they migrate towards the collecting electrode. They are generally removed from the electrode by periodic rapping.

The advantage of an electrostatic precipitator is that it can collect solids of very small size in a dry form. Electrostatic precipitators, however, generally are large, and capital costs are very high. This is because collection efficiency is proportional to the area of the collecting electrodes, which can become substantial. Operating costs are low, though, and may justify the use of an electrostatic precipitator instead of a bag filter *(26)*. CEP

T. M. KNOWLTON is associate director of fluidization research at the Institute of Gas Technology, Chicago, and is also technical director of Particulate Solid Research Inc., Chicago (312/890–6408; Fax: 312/890–6460). For over 20 years, he has conducted research into the areas of high-temperature and high-pressure fluidization, solids transport, and nonmechanical valves. He is the author of over 50 technical articles in these fields, and teaches the AIChE Continuing Education course in Fluid-Bed Technology, as well as the PSRI Fluidization course. He received BS, MS, and PhD degrees in chemical engineering from Iowa State University. He is a member of AIChE.

J. W. CARSON is president of Jenike & Johanson, Inc., North Billerica, MA, (508/392–0300; Fax: 508/393–9980) a firm that he joined in 1970. He has been active in research, consulting, and management, and has written over 50 technical papers and articles in the areas of bin loads, flow of fine powders, vibration, and using computers to analyze solids flow. He lectures extensively on the topic of storage and flow of solids, and devotes much of his time to consulting with clients. He received a BS in mechanical engineering from Northeastern University, and a PhD from Massachusetts Institute of Technology. He is a member of AIChE.

G. E. KLINZING is a professor of chemical engineering at the University of Pittsburgh (412/624–9814; Fax: 412/624–1108), where he has conducted research in solids processing and pneumatic conveying for over 30 years. He is a regular lecturer in the AIChE Continuing Education Series in this area. His current research interests are flow measurement and control of solids processing systems. He received a BS in chemical engineering from the University of Pittsburgh, and MS and PhD degrees from Carnegie Mellon University.

W.-C. YANG is an advisory engineer at the Science and Technology Center of the Westinghouse Electric Corporation, Pittsburgh (412/256–2207; Fax: 412/256–1222) He received BS, MS, and PhD degrees from the National Taiwan University, the University of California at Berkeley, and Carnegie Mellon University, respectively. He has authored more than 75 technical articles, as well as two chapters on pneumatic transport, fluidization, and fluid-particle systems in the "Encyclopedia of Fluid Mechanics." He is programming chair and secretary of the Particle Technology Forum. A Fellow of AIChE, he is the 1993 recipient of the Fluidized Processes Recognition Award presented by the Particle Technology Forum of AIChE.

Literature Cited

1. **Merrow, E. W.,** "Linking R&D to Problems Experienced in Solids Processing," *Chem. Eng. Progress,* **81** (5) p. 14 (May, 1985).
2. **Merrow, E. W.,** "Estimating Startup Times for Solids-Processing Plants," *Chem. Eng.,* **89** (10) p. 89 (Oct. 1988).
3. **Marinelli, J., and J. W. Carson,** "Solve Solids Flow Problems in Bins, Hoppers, and Feeders," *Chem. Eng. Progress,* **88** (5), p. 22 (May 1992).
4. **Carson, J. W., T. A. Royal, and D. J. Goodwill,** "Understanding and Eliminating Particle Segregation Problems," *Bulk Solids Handling,* **6**, p. 139 (Feb. 1986).
5. **Royal, T. A., and J. W. Carson,** "How to Avoid Flooding in Powder Handling Systems," *Powder Handling and Proc.,* 1/93, p. 63 (Mar. 1993).
6. **Jenkyn, R. T., and D. J. Goodwill,** "Silo Failures: Lessons to be Learned," *Eng. Digest,* p. 17 (Sept. 1987).
7. **Jenike, A. W.,** "Storage and Flow of Solids," Bulletin No. 123, University of Utah Engineering Experiment Station, Salt Lake City, UT (Nov. 1964).
8. **Marinelli, J., and J. W. Carson,** "Use Screw Feeders Effectively," *Chem. Eng. Progress,* **88** (12) p. 47 (Dec. 1992).
9. **Carson, J. W., and R. T. Jenkyn,** "Load Development and Structural Considerations in Silo Design," presented at Reliable Flow of Particulate Solids II, Oslo, Norway (Aug. 1993).
10. **Stuart-Dick, D., and T. A. Royal,** "Design Principles for Chutes to Handle Bulk Solids," *Bulk Solids Handling,* p. 447 (Sept. 1992).
11. **Johanson, J. R., and T. A. Royal,** "Measuring and Use of Wear Properties for Predicting Life of Bulk Materials Handling Equipment," *Bulk Solids Handling,* **2**, p. 517 (1982).
12. **Molerus, O., and W. Siebenhaar,** "Vibration Induced Pneumatic Conveying," presented at Pneumatech 4, Glasgow, Scotland (June 1990).
13. **Wen, C. Y., and H. P. Simon,** "Flow Characteristics in Horizontal Fluidized Solids Transport," *AIChE J.,* **5** (2), p. 263 (June 1959).
14. **Yang, W. C.,** "Criteria for Choking in Vertical Pneumatic Conveying Lines," *Powder Tech.,* **35**, p. 143 (1983).
15. **Bi, H. T., J. R. Grace, and J. X. Zhu,** "On Types of Choking in Vertical Pneumatic Systems," *Intl. J. Multiphase Flow,* **19**, p. 1,077 (1993).
16. **Jones, P. J., and L. S. Leung,** "A Comparison of Correlations for Saltation Velocity in Horizontal Pneumatic Conveying," *Ind. Eng. Chem. Proc. Des. Dev.,* **17**, p. 571 (1978).
17. **Marcus, R. D., L. S. Leung, G. E. Klinzing, and F. Rizk,** "Pneumatic Conveying of Solids," Chapman and Hall, London (1990).
18. **Wypych, P. W., and P. C. Arnold,** "On Improving Scale-Up Procedures for Pneumatic Conveying Design," *Powder Tech.,* **50**, p. 281 (1987).
19. **Marcus, R. D., and J. D. Mohlmann,** "Parameters Influencing the Pneumatic Conveying of Large Rocks," Proceedings of Pneumatech 2, Univ. of Kent, Canterbury, U. K., p. 230 (1984).
20. **Cabrejos, F. M. J., and G.E. Klinzing,** "Pickup and Saltation Mechanism of Solids Particles in Horizontal Pneumatic Transport," *Powder Tech.,* in press.
21. **Glicksman, L. R., D. Westphalen, K. Woloshun, T. Ebert, K. Roth, M. Lints, C. M. Brereton, and J. R. Grace,** "Experimental Scale Models of Circulating Fluidized Bed Combustors," *Proceedings,* 1991 International Conference on Fluidized Bed Combustion, ASME, New York, p. 1,169 (1991).
22. **Chang, H., and M. Louge,** "Hydrodynamic Scaleup of Circulating Fluidized Beds," *Proceedings,* 1991 International Conference on Fluidized Bed Combustion, ASME, New York, p. 1,215 (1991).
23. **Horio, M., H. Ishii, Y. Kobukai, and N. Yamanishi,** "A Scaling Law for Circulating Fluidized Beds," *J. Chem. Eng. Japan,* **22**, p. 587 (1989).
24. **Campbell, D. L., H. Z. Martin, and C. W. Tyson,** U. S. Patent 2,451,803 (1948).
25. "Gas Fluidization Technology," D. Geldart, ed., John Wiley, London, p. 206 (1986).
26. **Svarovsky, L.,** "Solid-Gas Separation," Elsevier, Amsterdam (1981).

Use Pinch Analysis to Knock Down Capital Costs and Emissions

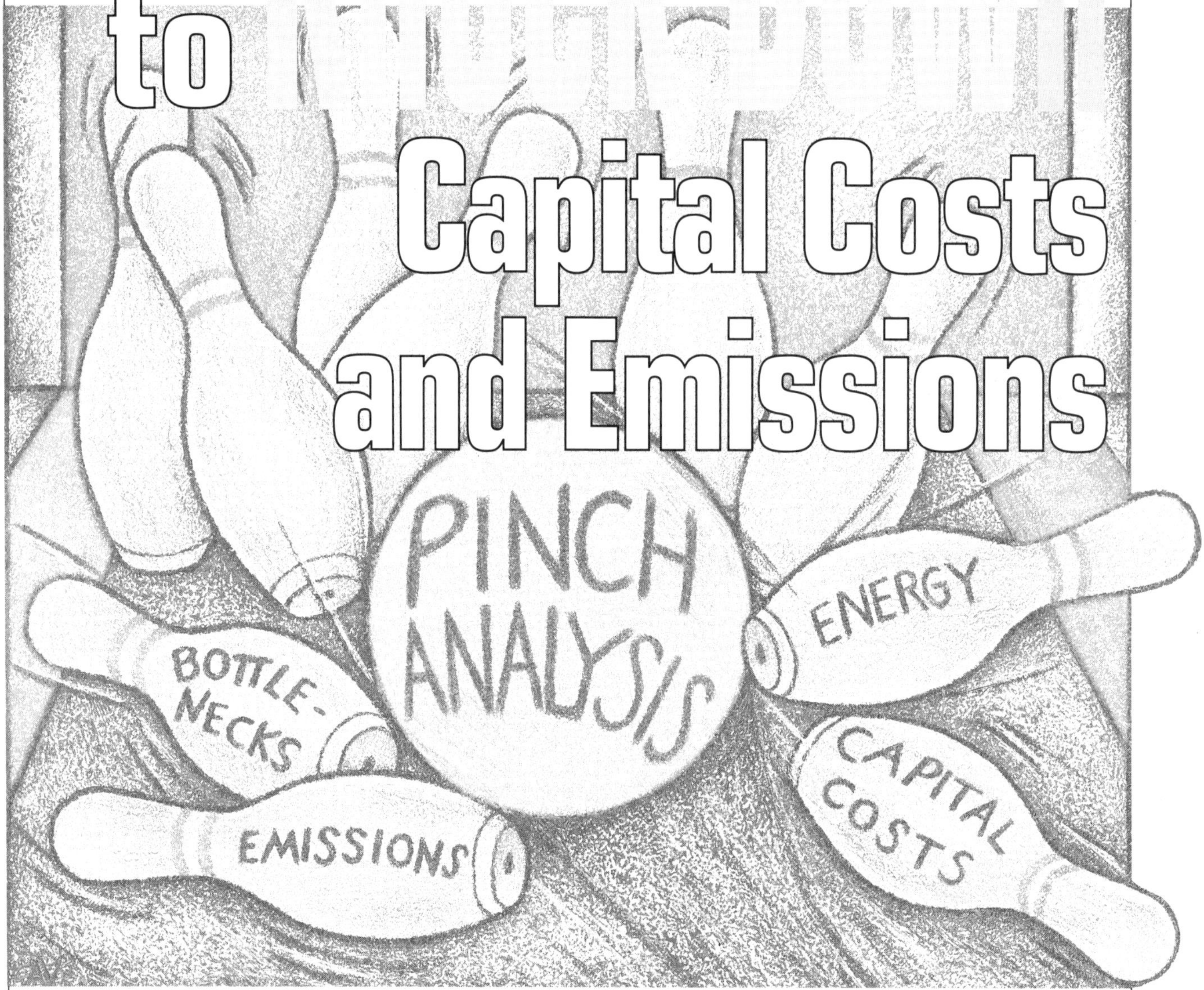

More than ten years ago, against the background of the then-current energy crisis, Pinch Technology emerged as a tool for the design of heat exchanger networks (HENs). Its key contribution was to give the engineer simple concepts to use interactively. The approach differed from the black-box computer methods proposed at the time, because the engineer could stay in control. Applying the concepts of Pinch Technology, the engineer could use simple targets, and take account of operability, plant layout, safety, and so on, to drive the design toward integrated solutions that were not only thermally efficient (the input of Pinch Technology), but also industrially acceptable (the engineer's input).

Since then, the world has moved on. Contrary to all expectations, the cost of energy in real terms has come down. Nevertheless, the importance of continuous improvement in process design has grown due to three major trends: the response of the chemical process industries (CPI) to environmental pressures; the need for ever better capital cost efficiency; and the general drive for "quality" in the broadest sense These three trends have touched on the life of the professional in all corners of the CPI, including process design. Against this background, Pinch Technology has evolved. Its scope has broadened and its application has changed.

The scope today includes separation, waste-removal, and utility systems, as well as total sites. The focus often is on nonenergy objectives such as capital cost, operability, and emissions. The technology, in fact, has evolved into a general approach for process design upstream from simulation and is nearly as comprehensive.

The mode of application of Pinch Technology also has been broadened. A major application has become targeting rather than design. It often is used for process analysis and therefore increasingly is referred to as Pinch *Analysis* rather than *Technology*. Morgan of M. W. Kellogg explains that today's experienced practitioner uses Pinch Analysis to scope and screen major design options early, prior to simulation, with significant beneficial effects both "...for the process design and for the design process..." *(1)*.

Pinch Analysis ten years ago was a specialized tool, used only in certain circumstances. Today, it is a general tool. In line with the CPI's drive for quality, its application has become routine in major corporations. It simply is good practice. It virtually guarantees consistency in process integration. It has become an integral part of the state-of-the-art design environment dealing with integrated processes, utility systems, and total sites, and aiming at lower emissions and reduced investment. Often, its application has little to do with energy costs.

Pinch Analysis has evolved during the last decade from a specialized tool for heat recovery into a broad-based methodology for reducing capital costs, emissions, and energy consumption, spanning process design and total site planning.

This article gives an up-to-date overview. A summary of basic concepts, with emphasis on those which are less well known than others, is followed by a description of eight recent developments. Some of these developments sharpen the focus of the established concepts, while others widen the scope. All of them follow the established pattern for Pinch Analysis of setting targets prior to design. The eight developments are:

1. Pressure Drop Optimization;
2. Multiple-Base-Case Design;
3. Distillation Column Profiles;
4. Low-Temperature Process Design;
5. Batch Process Integration;
6. Water Pinch;
7. Total Site Integration; and
8. Emissions Targeting.

To complete the overview, the article provides a comprehensive summary of previous literature. Theory and applications are covered, and an historical perspective is given. The relative roles of Pinch Analysis and other methods for process integration are discussed.

Here, we concentrate on the essence not the detail, which can be found in the cited references. Case studies are described in outline. Practical applications are discussed, including the aspect of "making it happen." Some organizations simply rely on technical experts — and fail — whereas other firms integrate the methodology into the design process — and succeed. Successful users typically address several questions: Where does the new activity fit into the organization? What tools (software packages) are required? At what stages in a project is Pinch Analysis best used? What data are required? Who needs what training? As is so often the case, success depends on good management as well as on good technology.

Key established principles

Figures 1 and 2 relate to the three concepts probably best known in Pinch Analysis, the "Composite Curves" *(2)*, the "Grid Diagram" *(3)*, and the "Pinch" *(4)*. All three concepts relate to a *given economic level* of heat recovery that needs to be optimized (as we will discuss).

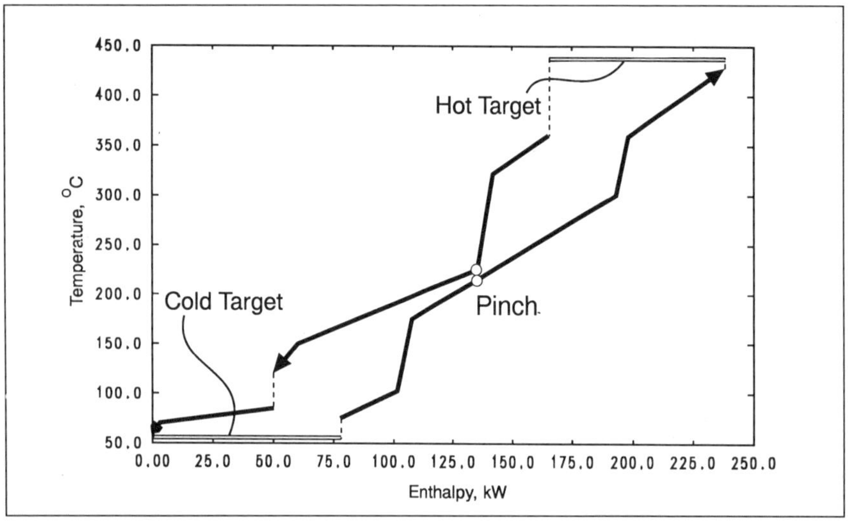

Figure 1. The best established concept of Pinch Analysis is that of Composite Curves, which set energy targets ahead of design and locate the Pinch point. They also are used to visualize heat transfer (sources, sinks, driving forces) in the overall context.

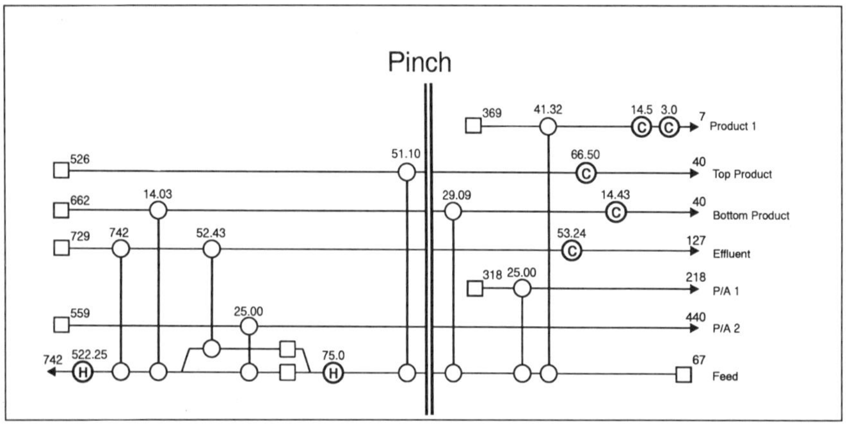

Figure 2. The Grid Diagram of streams is laid out prior to design. Hot streams run from left to right at the top, cold streams run countercurrent at the bottom. The position of streams does not change; however, the design changes. The Pinch is easily located in the grid.

The Composite Curves are constructed from stream data derived from a process heat-and-material balance. The Composite Curves allow the designer to predict hot and cold utility targets ahead of design, to understand driving forces for heat transfer, and to locate the heat recovery Pinch.

The Grid Diagram enables the designer to develop heat recovery networks using the "Pinch Design Method" *(5)*, a mixture of methodology and engineering judgment.

The Pinch principle states that no design can achieve the targeted utility requirements if there is any "cross-Pinch" heat transfer from above the Pinch to below the Pinch:

Actual = XP + Target

where XP = cross-Pinch heat transfer.

If $XP = 0$, then the actual heat requirement equals the target. The Pinch principle allows the designer to keep the level of heat recovery optimized all through the evolution of even complex designs simply by making sure that cross-Pinch heat transfer is avoided.

Figure 3 illustrates probably the next best known concepts, the "Grand Composite Curve," and "Appropriate Placement."

Grand Composite Curves *(6)* are based on the same process stream data as Composite Curves, and highlight the process/utility interface. They help the designer to select different utility sources and sinks.

Appropriate Placement holds true for any unit operation and any process. Figure 3 shows examples of a heat pump and a heat engine *(6)*. Appropriate Placement for heat pumps is across Pinches *(7)*. Appropriate Placement for heat engines is above or below Pinches. Two turbine stages (1 and 2) above the Process Pinch consume boiler steam, deliver process steam — both high pressure and medium pressure — to the mains, and generate shaftwork. (We will use VHP, HP, MP, and LP for very high-, high-, medium-, and low-pressure steam, respectively, from this point.) Given that without a turbine the process requires heat in any case, the shaftwork generated is equal to the additional boiler steam consumed. Stages 1 and 2 are *appropriately placed.* Essentially, shaftwork is generated from steam one-to-one.

Stage 3 below the Process Pinch is also *appropriately placed.* It generates work from process heat that otherwise goes to cooling.

Finally, the heat pump is *appropriately placed.* It receives heat from the process below the Pinch and supplies heat above the Pinch.

The concept of Appropriate Placement applies to heat engines *(7)*, heat pumps *(6)*, distillation columns *(8)*, evaporators *(9)*, furnaces *(10)*, and to any other unit operation, be it process or utility, that can be represented in terms of heat sources and sinks. It gives the designer powerful guidelines for the design of heat engines, heat pumps, distillation columns, evaporators, and the like in the context of integrated processes.

Figure 4 relates to the concept of "SuperTargeting" *(11)*. It is possible to predict from the Composite Curves

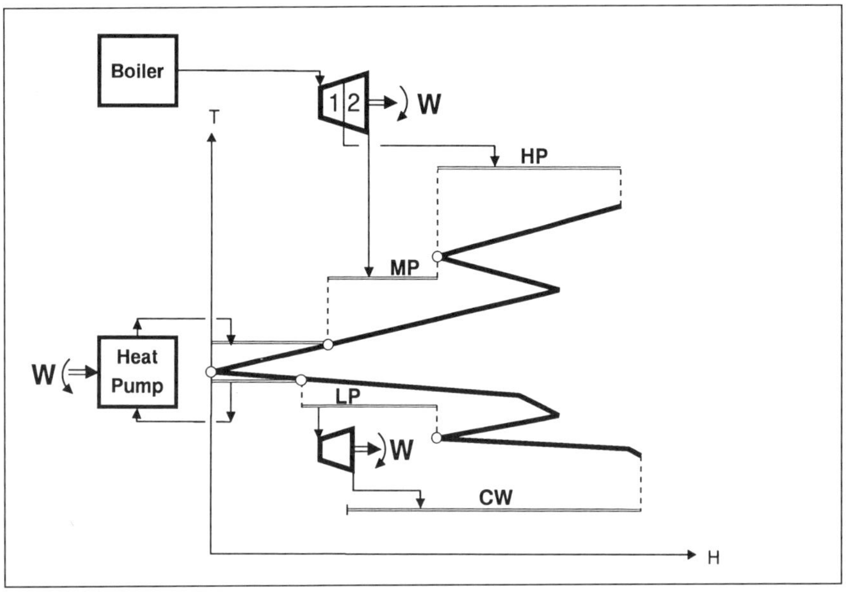

Figure 3. The Grand Composite Curve helps to select utilities. The concept of Appropriate Placement assesses the integration of individual process or utility units in the overall context. The Appropriate Placement of heat pumps is across the Pinch; that of heat engines is above or below the Pinch.

the overall target surface area for any stream data. It further is possible to predict the target capital cost that is necessary for the target area. Based on target energy cost and capital cost, the target overall cost is predicted for any given ΔT_{min} (the minimum allowable temperature difference for heat exchange). This enables the designer to optimize ΔT_{min} *prior to design.* In other words, Pinch Analysis does not just address energy cost, but takes into account capital cost. It guides the overall optimization of *energy and capital cost* of integrated systems at the targeting stage.

Figure 5 relates to "Process Changes" as discussed in the literature *(6)* — namely, changes to the basic process heat-and-material balance, but not necessarily to the chemistry of the process. The so-called "Onion Diagram" is used to represent the hierarchy of chemical process design. The start is from reactors converting feeds to products. Once the reactor design is settled, the separation needs are known (for example, to separate unconverted feeds from products), and the separators can be designed. Once the separator design is settled, mass flows, vapor/liquid equilibrium, temperatures, pressures, and flow rates are known, and the HEN can be designed. Prior to designing the HEN, the basic heat-and-material balance is defined, and the Composite Curves representing the base-case process can be constructed.

The relevant concept for process changes is the "Plus/Minus Principle" *(12)*, which essentially is a generalization of Appropriate Placement. It uses the Composite Curves to identify process changes by going backwards into the onion. The example shown in Figure 5 relates to the change of pressure in a distillation column. If column pressure is reduced, then both condenser and reboil temperature decrease. The condenser does not cross the Pinch. Therefore, there is both a minus and a plus in the hot stream enthalpy balance below the Pinch. The effects cancel out. There is no net change in enthalpy for hot streams below the Pinch. The reboiler, by contrast, originally placed above the Pinch, crosses the Pinch as a result of the process modification. There is a minus above and a plus below the Pinch in the cold-stream enthalpy balance, which results in both hot and cold utility targets being reduced. The targeted saving equals the distillation reboil duty. Essentially, the distillation column now could be run on integrated heat flow, or zero utility *(8)*.

The Plus/Minus Principle is a general formulation for desirable process changes (relating to distillation, reaction, and so on) that gives the designer guidelines as to what changes are beneficial for energy (as in Figure 5) or capital costs *(13)*.

Figure 6 shows a "Balanced Composite Curve" *(13)* and the corresponding "Balanced Grid" *(13)*. Balanced Composite Curves represent process and utility streams in one construction. Utilities are integrated with process streams so that the enthalpy balance is closed. Balanced Composite Curves

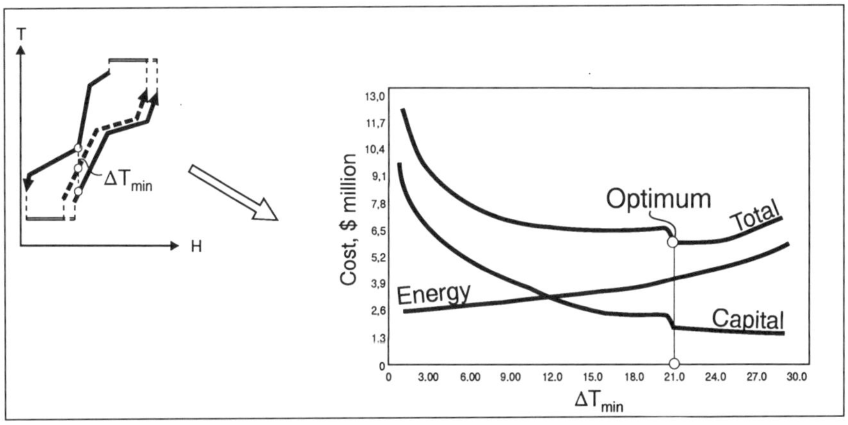

Figure 4. Both energy and capital cost targets are set from the Composite Curves. SuperTargeting implies the setting of energy and capital cost targets prior to design, over a range of ΔT_{min} to identify the optimum combination of both targets.

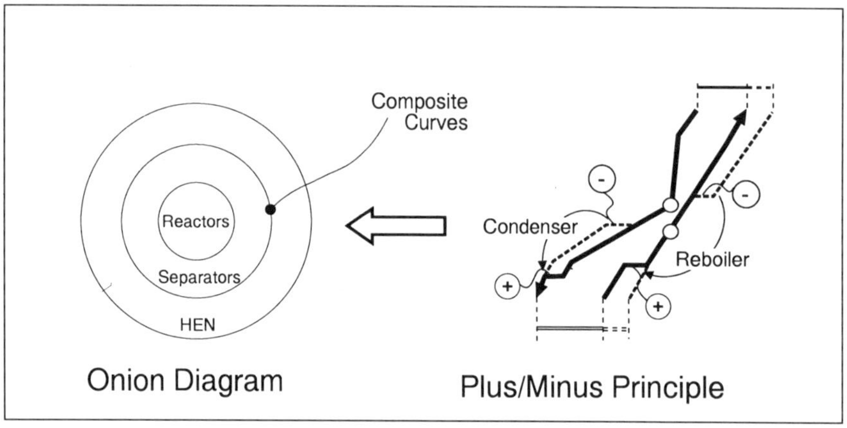

■ ***Figure 5. Composite Curves are based on the reactor/separation scheme of the base-case process. Composite Curves therefore can be used to modify the reactor/separation scheme. This task is guided by the Plus/Minus Principle.***

are desirable for a clear visualization of overall heat transfer (utility/process and process/process) and of "Utility Pinches." The Balanced Grid consists of process streams *and* utilities as in the Balanced Composite Curves. Process changes and optimized values for ΔT_{min} as obtained through Super-Targeting usually are incorporated in both the Balanced Composite Curves and the Balanced Grid.

The stream and utility data used in Figure 6 are based on the example given in Figure 3. Note that the upper-temperature Utility Pinch is caused by and is located at the utility (MP steam), but the lower-temperature Utility Pinch is caused by utility (LP steam), and is located at the start of process stream No. 5. The concept of Utility Pinches generally is less well understood than that of the Process Pinch. A full discussion is provided elsewhere *(13)*.

Figures 7 and 8 demonstrate the usefulness of the Balanced Grid. Figure 7 shows two competing designs for a HEN in a simple Grid (without utilities). Both designs feature fired-heater coil duties of 300 units (the target, see *(13)*). There is a difference, however. Process temperatures in the furnace coils are different, resulting in different flue-gas temperatures and, therefore, different fired-heater efficiencies. Even though both designs achieve the targeted heater duty, fuel consumption differs significantly (93% vs. 100%). In Figure 8, the Balanced Grid is laid out prior to design, including the flue-gas (which is targeted both for heat duty and stack temperature, from 1,500°C to 160°C). Normal use of network design procedures now identifies process/process exchangers as well as fired heater duties in one coherent approach. The resulting design achieves both the targeted heat duty, and the targeted minimum flue-gas temperature. Fuel consumption is reduced further to 81%.

The most common mistake made even by otherwise experienced users of Pinch Analysis is the consideration

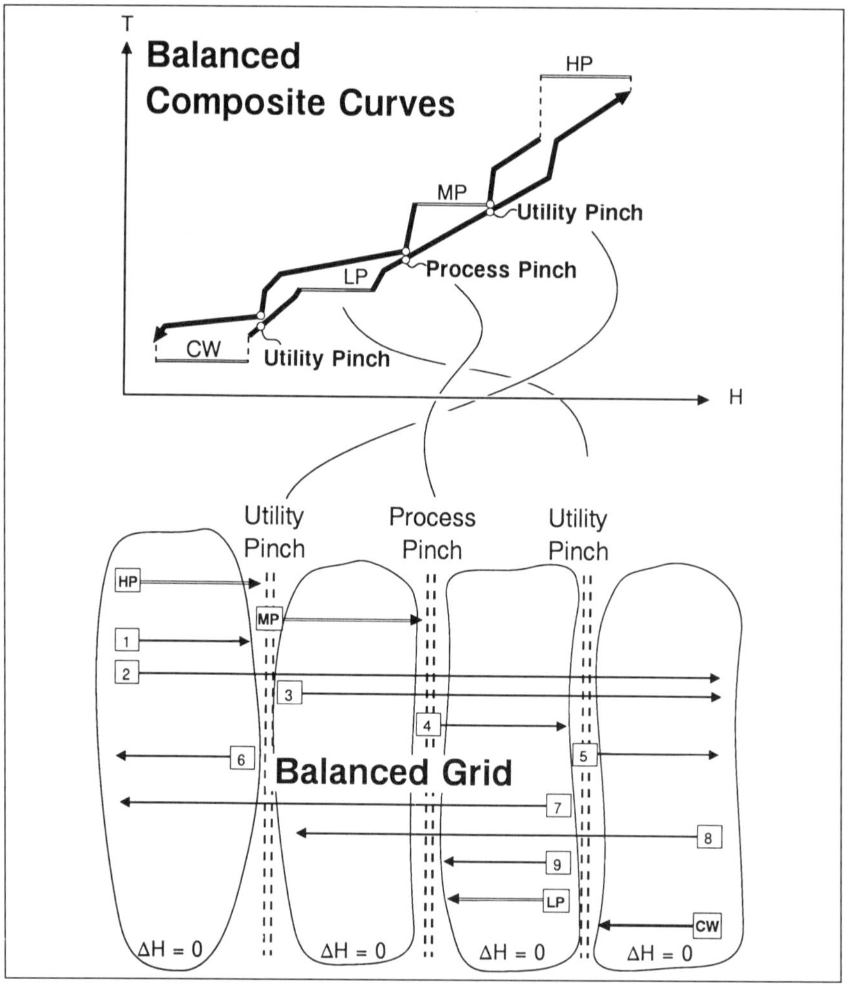

■ ***Figure 6. Balanced Composite Curves represent process streams and utilities such that the enthalpy balance is closed. They help locate Utility Pinches and the visualization of overall heat transfer process-to-process and process-to-utility. The Balanced Grid is based on the Balanced Composite Curves. Design carried out in the Balanced Grid gives equal and simultaneous consideration to process-to-process and utility-to-process units.***

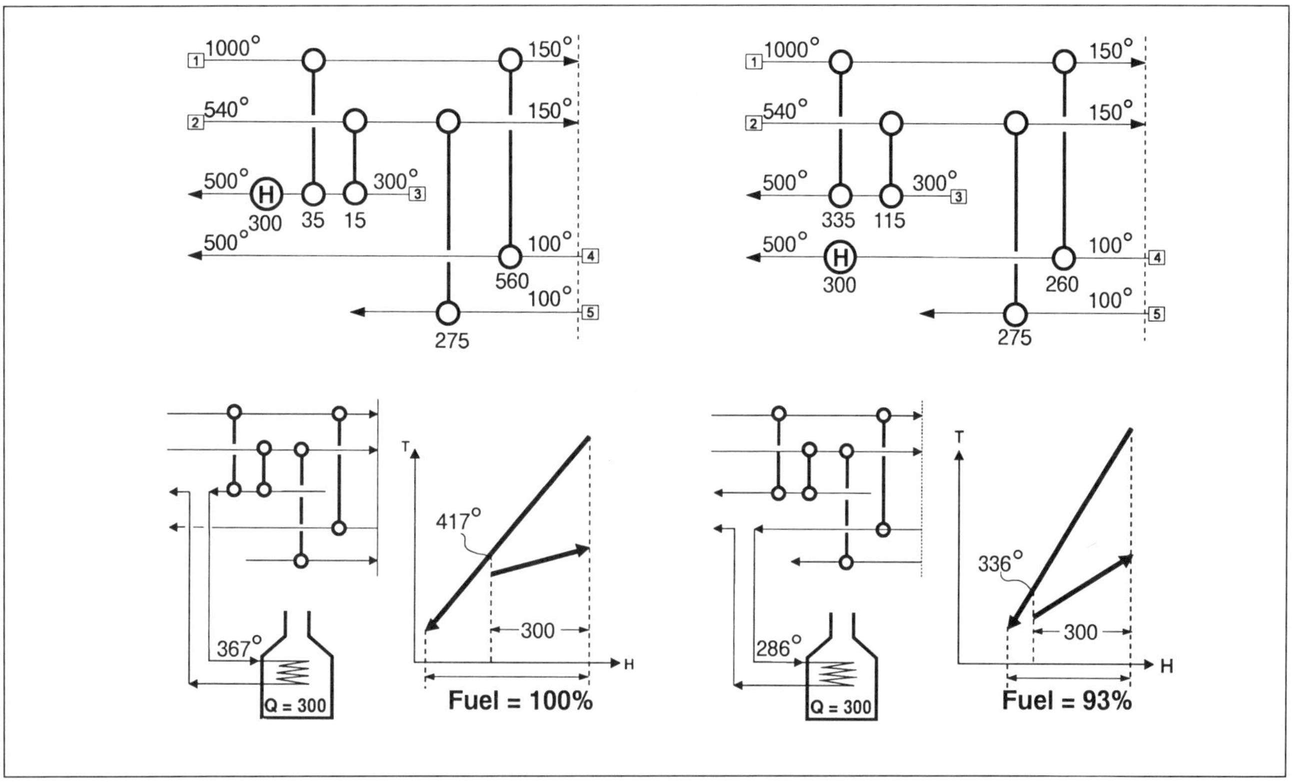

Figure 7. In this example, two HEN designs are shown, both achieving 300 units of heat input (the energy target). However, the location of the heater is different in each design. Therefore, the process temperature at the heater input is different, resulting in a different furnace efficiency and fuel consumption.

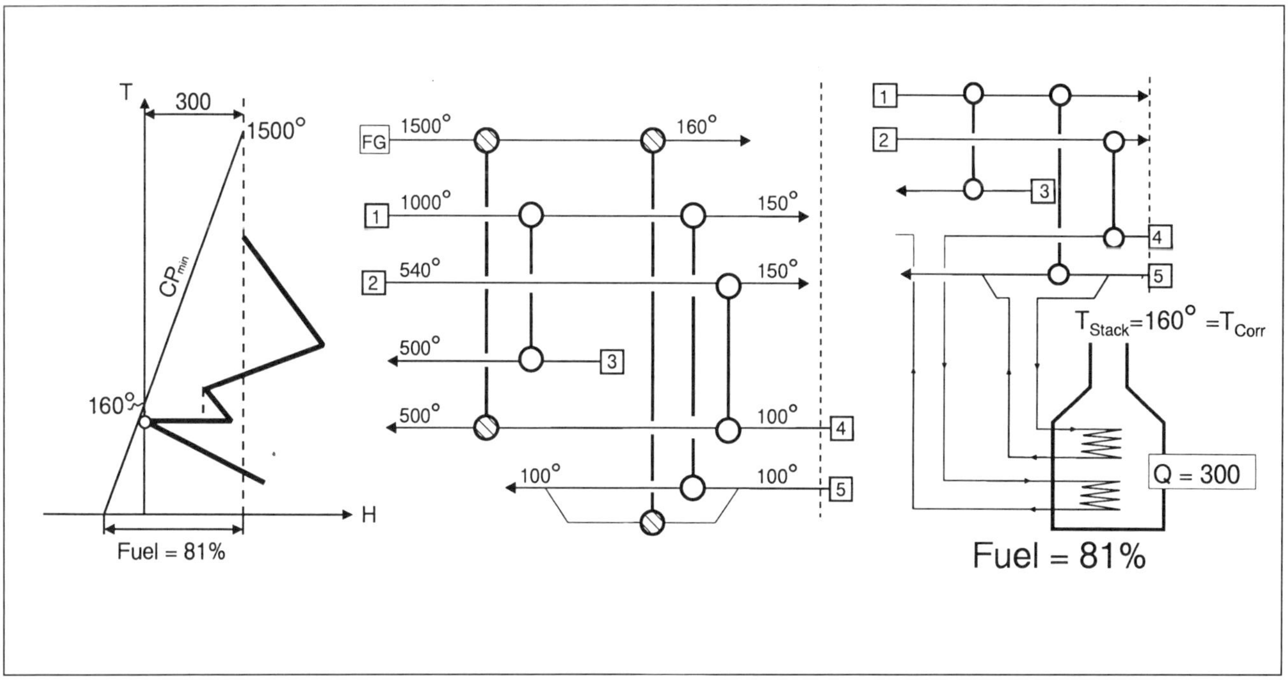

Figure 8. The design from Figure 7 should be carried out in the Balanced Grid, based on the minimum flue-gas flow rate. This ensures that the fuel target is reached. Once the design is completed, the Balanced Grid representation is easily changed into a more conventional drawing that shows the intended use of the process furnace.

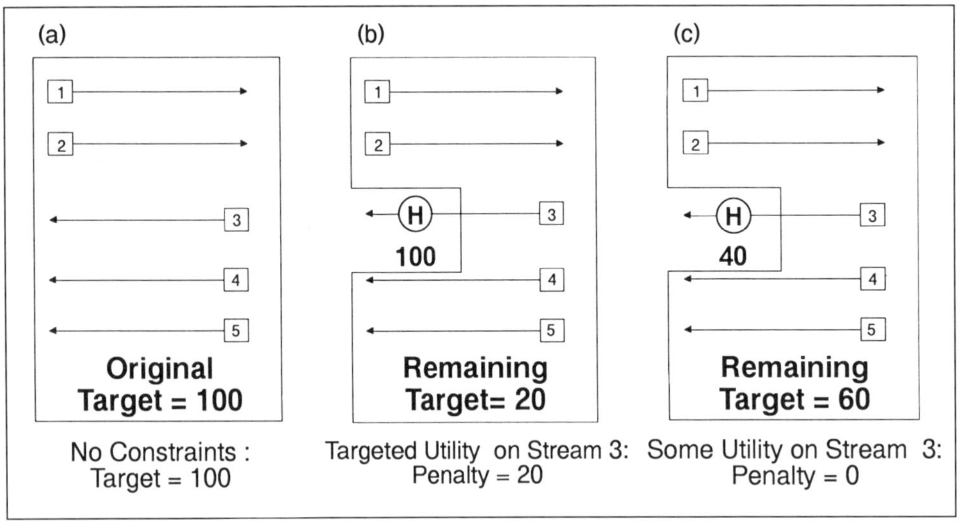

Figure 9. The target for the unconstrained problem is 100 units. If this target heat duty is placed on stream No. 3 (for operability reasons), then RPA reveals that there is now a target of 20 units for the remaining problem. Thus, overall consumption must be at least 120 units — a penalty of 20 units. If only 40 units of utility are used on stream No. 3, there is a target of 60 units for the remaining problem. Thus, 40 units can be freely used on stream No. 3 with no penalty.

of utility choices *after* network design — as was done in Figure 7. Design should be carried out based on Balanced Composites and in the Balanced Grid. This ensures that it is carried out *after* proper targeting of utilities (both for heat duties *and* temperatures), consideration of process changes, and optimization. The Balanced Composite Curves and the Balanced Grid also are less well understood than most other basic concepts. They are, however, of key importance.

Figure 9 illustrates "Remaining Problem Analysis" (RPA) *(13)*, which is the least well-known of the basic concepts. RPA considers remaining stream data after a design decision. Design decisions eliminate stream data step-by-step, and RPA retargets based on what is left. Assume that the simple energy target for a given stream set is 100 units, and further assume that there is an operability constraint: a startup heater is necessary on stream No. 3. To save equipment, the designer wishes to place a steady-state utility heater with the targeted heat load (that is, 100 units) on stream No. 3 where the heater also can be used for startup. The impact of such a decision is tested using RPA (center). The energy target for the "Remaining Problem" is 20 units. This means that the heater on stream No. 3 is well located for startup but not for thermal integration; if all 100 units of steady-state heat load are placed on stream No. 3, an overall penalty of 20 units is incurred. A compromise must be explored (right). For example, it is possible to place 40 units of utility on stream No. 3 (for startup *and* steady-state load) without penalty.

RPA can be carried out with respect to both energy- *and* capital-cost targets. It is a powerful evaluation tool that allows the designer to explore the impact of key design decisions early. Decisions such as "a startup heater on stream No. 3, with a heat duty of 40 units" are settled upfront, and form the basis for subsequent design work.

The overall approach

Figure 10 summarizes traditional Pinch Analysis. The start is from the base-case heat-and-material balance representing reactors and separators. The Grand Composite Curves are constructed, and an initial assessment is made of utilities. SuperTargets are set and process changes are examined. Utility options are tested again. RPA is used to evaluate the impact of constraints. Capital costs are traded off against energy costs. Based on these targeting exercises, decisions are made regarding key process design options (for instance, reactor and separator design), key utility scenarios (such as flue-gas vs. steam), key process constraints (like startup heaters), and major capital-cost tradeoffs. At the end of this iterative targeting procedure, the designer will have converged on a basic design strategy — and will have selected the correct broad scenario and initialized the design near-optimally. The design of the HEN now follows (in the Balanced Grid), as does the design of the main process items and utility units.

Figure 10 makes it once more appropriate to reflect on the terms Pinch *Technology* vs. *Analysis*. When "Pinch" first emerged, the Grid Diagram and HEN design were the key contributions. Targeting was secondary. *By contrast, the key contribution today is the broad scoping and screening of process and utility scenarios at the targeting stage.* It therefore would appear indeed more appropriate today to refer to "Pinch" as an analysis tool rather than a technology. This explains why workers in the field and users alike increasingly use the term Pinch Analysis in preference to Pinch Technology. This article uses Pinch Analysis consistently, except where it is not appropriate in the historical context.

Finally, the procedure shown in Figure 10 clearly supports what was stated before: Pinch Analysis is no longer a tool solely for the design of HENs. Rather, it has evolved into a methodology for general process integration addressing capital cost as much as energy cost.

Most generally, whatever our objective is in design (capital cost

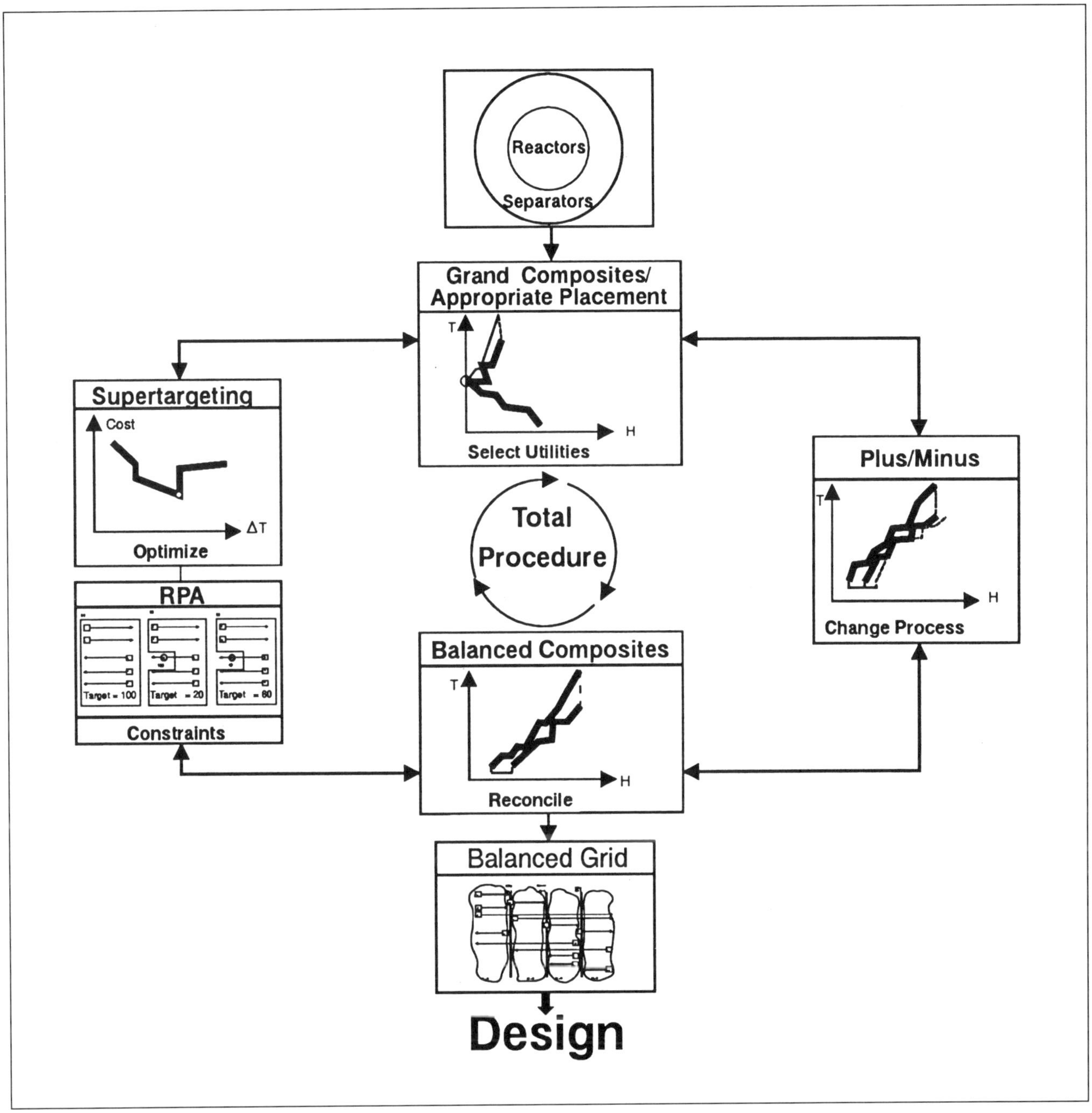

■ *Figure 10. Pinch Analysis is all about the scoping and screening of design options in targeting mode prior to design. Utility selection, capital cost optimization, the impact of constraints, and changes to the process/reactor/separation system all are evaluated and put into context. The Balanced Composite Curves and Balanced Grid represent an optimized set of targets. Design only proceeds after this initialization.*

reduction, operability, etc.), we start from a heat and material balance. Pinch Analysis essentially helps us to better understand the energy component of the heat and material balance. This better understanding, in turn, improves both "...the process design and ...the design process..." *(1)*.

A multitude of principles and concepts

This article can give a description of only a small number of the 60 or so principles and concepts of Pinch Analysis. Table 1, though, gives a near-complete summary. Not all of these concepts are covered in the literature quoted so far. Most of the remaining concepts are covered either in Linnhoff March's "Foundation Training Course" *(13)* or in the "MS Course in Process Integration" of the University of Manchester Institute of Science and Technology (UMIST) *(14)*. In addi-

Table 1. Principles, tools, and design rules of Pinch Analysis — and references for more information.

Principles and Tools	Rules
Basic HEN Targeting and Design	
Composite Curves *(2,6)*	Pinch Design Method *(5)*
Problem Table *(3)*	Stream Splitting Rules *(13)*
Threshold Plot *(4)*	Mixing Rules *(14)*
Bath Area Target *(11)*	HEN Evolution *(6,13)*
Delta *T* Contributions *(11)*	Topology Traps *(11)*
Euler and Number of Units Target *(3)*	Retrofit Design *(14)*
Grid Diagram *(3)*	
Pinch Principle *(4)*	
CP-Matrix *(13)*	
Crisscross Principle *(11)*	
Driving Force Plot(s) *(11,14)*	
Remaining Problem Analysis *(13)*	
Loops and Paths *(3)*	
SuperTargeting *(11)*	
Delta *p* Targeting and SuperTargeting *(15)*	
Retrofit Targeting (constant α) *(16)*	
Area Matrix *(17)*	
Utility Targeting and Design	
Grand Composite Curve *(6)*	Furnace Integration *(10)*
Utility Composite and Utility Pinches *(13)*	Heat Engine Placement *(7)*
Balanced Composites and Balanced Grid *(13)*	Heat Pump Placement *(6,7)*
Balanced Grand Composite Curves *(18)*	Multiple Utility Optimization *(23)*
Total Site Profiles (*T–H*) *(19)*	Low *T* Process Design *(24)*
Cooling Water Targets *(14)*	
Exergy and Pinch: The η_c–*H* Plot *(20)*	
Exergy Grand Composites *(21)*	
Low *T* ShaftworkTarget *(21)*	
Total Site Profiles (η_c–*H*) *(19)*	
Power Cycle Targeting *(22)*	
Advanced HEN Design	
Rigorous Area Targets *(25)*	Constrained HEN Design *(37)*
Constrained HEN Targeting *(26)*	Condensing Steam Cycle Power Block Design *(38)*
Area Matrix Retrofit Targets *(17)*	
Area Integrity Matrix *(27)*	
Area Cost Targets for Different Materials of Construction *(28)*	
Number of Shells Targets *(29)*	
Downstream Paths *(30)*	
Sensitivity Tables *(31)*	
Multiple-Base-Case Targets *(32)*	
Repiping and Rerouting Targets *(32)*	
Nonconvexity *(33)*	
Resiliency Index *(33)*	
Time-Slice Targets *(34)*	
Cascade Analysis *(35)*	
Batch Utility Curves *(36)*	
Process Design	
The Onion Model *(6)*	Data Extraction Rules *(13)*
Keep Hot Streams Hot and Cold Streams Cold *(6)*	Appropriate Placement of Distillation Columns (8)
Plus/Minus Principle *(12)*	Appropriate Placement of Evaporators (9)
Column Grand Composite Curves *(39)*	Wastewater System Designs Rules (40)
Column Composite Curves *(39)*	
Wastewater Targets *(40)*	

tion, Table 1 includes further references *(15–40)*.

The reader wishing to understand the essence rather than the detail should consult previous overview articles from ICI *(41,42,43,44)*, the Energy Technology Support Unit (ETSU) of Harwell Laboratory *(45)*, BASF *(46,47)*, Union Carbide *(48,49)*, UMIST *(50,51)*, Linnhoff March *(52)*, M. W. Kellogg *(1)*, and Mitsubishi Kasei *(53)*. These articles summarized developments up to their time, and reported on their impact and implications. They offer the most convenient access, available to date, to the basics of Pinch Analysis.

The remainder of this article outlines eight selected areas of recent progress. The discussion centers on application, scope, and impact. Appropriate references are given for details of methodology. Five of the developments originate from the Department of Process Integration at UMIST. For the remaining three developments, there are references from other sources as well. Clearly, UMIST has played a dominant role in developing the concepts of Pinch Analysis over the years. It, however, is promising that some of the more recent developments originate from other centers as well.

Pressure Drop Optimization

Consider the revamp of a crude-oil preheat train. Furnaces and pumps often are the limiting factor for revamping or debottlenecking. The challenge usually is to make the unit more fuel-efficient and to debottleneck the furnace by adding and reconfiguring heat-transfer surface area *while at the same time avoiding the installation of new pumps*. In other words, revamping or debottlenecking objectives need to fit in with pressure drop limits set by existing pumps. This is a complicated design problem.

Traditional Pinch Technology ignored pressure drop. The heat recovery scheme was designed in the context of temperature and heat load considerations. Pressure drop was an afterthought. It became apparent, however, during the first practical applications of Pinch Analysis that pressure drop could not be treated in this fashion. The thermal design settled heat exchangers at given sizes. Subsequently, optimization of heat-exchanger surface area against pressure drop might double certain exchangers in size, while others became much smaller. This clearly invalidated any optimization that had

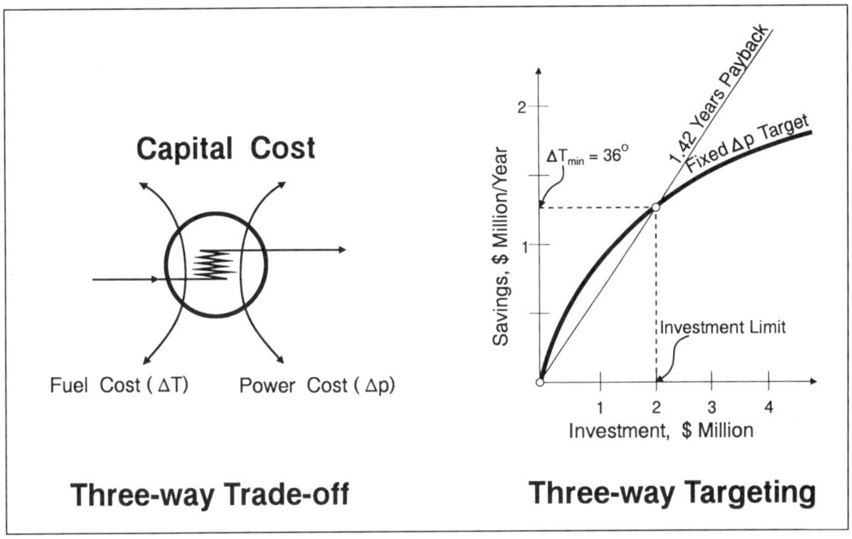

■ *Figure 11. Surface area in heat exchangers is optimized both against temperature difference, that is, fuel, and pressure drop, that is, power. There is a three-way tradeoff. Pressure Drop Targeting considers this three-way tradeoff either by allowing pressure drop to be optimized (new design) or by accepting given pressure drop limits (retrofit).*

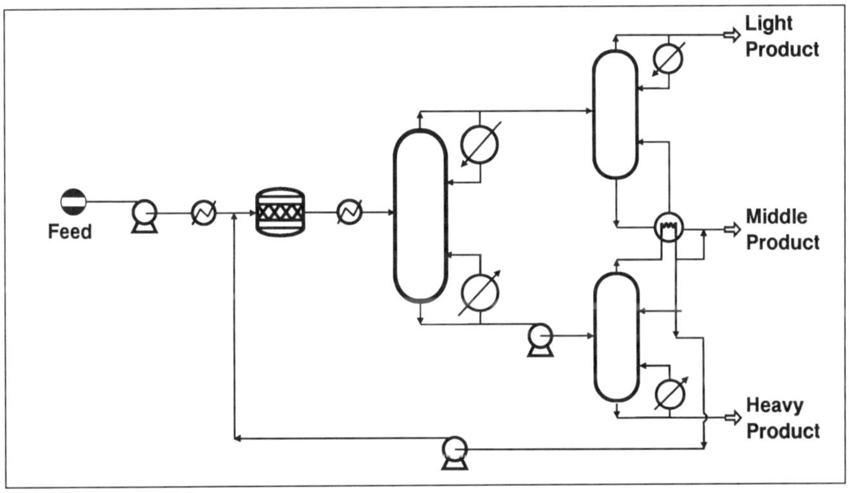

■ *Figure 12. In this example, condenser/reboiler integration may have beneficial or detrimental implications on operability. If catalyst deactivates and reactor product composition changes, the integration may prove detrimental. If overall throughput increases, and the utilities become limiting, the integration may prove beneficial.*

taken place during the initial design. Worse, the design might exceed available pressure-drop limits, rendering it infeasible. The conclusion soon was reached that the optimization of heat-exchanger surface area vs. thermal energy is inextricably linked with the optimization of heat-exchanger surface area vs. pressure drop. There is a three-way tradeoff.

This problem led, after several years of work, to the concepts of "Pressure Drop Targeting" and "Three-Dimensional SuperTargeting" (see Figure 11) *(15)*. Three-Dimensional SuperTargeting addresses pressure-drop-related costs or pressure-drop limits for hot and cold streams. Targeting then is carried out consistent with the costs of (or limits on) pressure drop, fuel, and heat-exchanger surface area. Prior to design, streams are set at optimized pressure drop, or at the available pressure-drop limit, and utilities are set at the optimized level of thermal recovery. Heat exchangers are placed in this context. If an attempt is made to optimize a design so initialized for fuel, capital cost, or pressure drop, little benefit is found. Pressure Drop Targeting and Three-Dimensional SuperTargeting are important contributions to making Pinch Analysis more practical, particularly in the context of oil refinery applications. Practical projects include debottlenecking, retrofitting for different feeds or products, and energy revamps, in that order.

Multiple-Base-Case Design

Another subject that proved to be important during the first practical applications of Pinch Analysis was that of flexible designs, or design for multiple-base cases. Few designs, if any, are operated only under design base-case conditions. Processes need to operate efficiently, reliably, and safely for different capacities, product specifications, and feedstocks, as well as with fresh or spent catalyst, varying ambient temperatures, clean and fouled equipment, and so on. The first industrial users of Pinch Analysis were sceptical as to the flexibility of integrated designs. The common thinking was that integration would lead to operability problems.

Initially, this hurdle was overcome only by hard work. Integrated structures were evolved for the base case, and operability was checked in the traditional way, that is, through simulation, modifications to the design, more simulation, and so forth. Experience quickly showed, however, that there was a relationship — but not necessarily a conflict — between integration and operability. Indeed, in some cases, integration proved beneficial for operability. In other cases, though, integration was detrimental.

Consider Figure 12. A product separation/recycle system consists of three distillation columns. The condenser of the heavy products column is integrated with the reboiler of the light products column. Now, consider two potential changes to operating conditions: first, a change of catalyst performance and, therefore, reactor

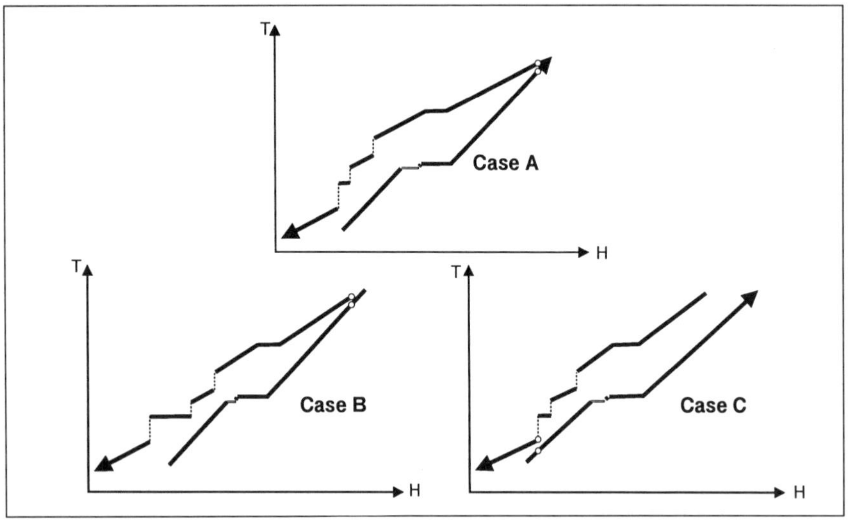

■ *Figure 13. Composite Curves make it obvious why good compatibility is expected between Operating Cases A and B, but poor compatibility is expected with Case C.*

product composition; and, second, a change of feed flow rate. If the catalyst deteriorates and reactor product composition changes, the proposed integration may prove detrimental to operability. The load on one column may increase while that on the other column decreases, and the condenser/reboiler integration is likely to bottleneck the overall unit. If, by contrast, the overall feed flow rate changes, the proposed integration may prove beneficial. For a given variation in overall process capacity, the integration results in a *smaller* variation in utility loads and, in a case where the utility system is limiting, this may prove beneficial to flexibility in overall process capacity.

The confirmed experience today is that integrated systems can be more operable than their less-integrated counterparts. To take advantage of this opportunity, operability must be taken into account early on. The approach taken in Pinch Analysis is to include operability objectives in targeting and during the development of the integrated structure.

An example of this approach already has been discussed in the context of RPA. In Figure 9, RPA is used to settle the issue of a startup heater prior to design.

Figure 13 depicts another example, showing different operating cases in terms of their Composite Curves. It immediately becomes clear that a design that operates Case A also is likely to operate Case B with few additional features, but that more significant changes are required to operate Cases A and C, or Cases B and C in one design.

The consideration of operability at the targeting stage has proved very successful. In addition, there has been work addressing the design of flexible structures. Much of this work has stood the test of industrial application. The main result usually is a reduction in capital cost used to "buy" operability. Key references are articles by Calandranis and Stephanopoulos *(33)*, Colberg and coworkers *(54)*, Floudas and Grossman *(55)*, Kotjabasakis and Linnhoff *(30,31,56)*, and Saboo and coworkers *(25,57)*.

Distillation Column Profiles

Moving on to subjects relating to the design of the basic process rather than to the HEN, a breakthrough was achieved recently with regard to distillation. Consider Figure 14. On the left, the traditional Pinch Technology view of distillation is shown. A so-called "box" in temperature-enthalpy coordinates describes the reboiler as a heat sink and the condenser as a heat source. This is contrasted on the right with a graph, not unlike a Grand Composite Curve, between reboil and condensing temperatures called a "Column Profile" *(39)*. The Pinch point of the Column Profile is located at the column feed. Column Profiles are used like Grand Composite Curves. They indicate at what temperature heat needs to be supplied and rejected up and down the column. Not all heat needs to be provided at reboil temperature.

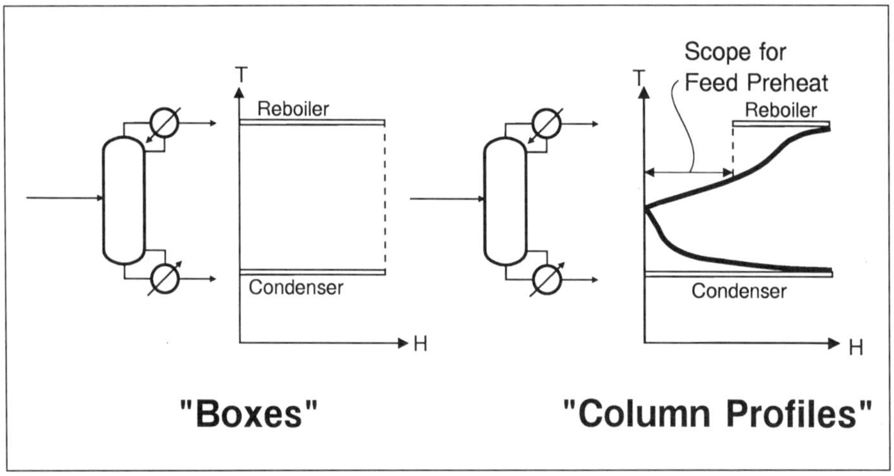

■ *Figure 14. Traditionally, Pinch Analysis described distillation through the thermodynamic footprint of the condenser and the reboiler in the so-called box representation. Now, Column Profiles are used instead. In the example, the Column Profile of a petrochemical distillation suggests that additional feed preheat leads to a significant reduction in reboil duty.*

Some can be supplied at lower temperature. Likewise, not all heat needs to be removed at condensing temperature. Partial heat removal at higher temperatures may be appropriate.

Graphs such as Column Profiles are not new. They have been detailed previously in the literature *(58,59)*. In previous work, however, such graphs were described theoretically. They realistically could not be computed for anything but ideal binary mixtures. Their computation for real columns was nearly impossible. The significant progress made recently is that Column Profiles now are generated with reasonable accuracy from a single converged tray-by-tray simulation even for multicomponent nonideal mixtures *(39)*.

The graph in Figure 14 is based on a real case. There is scope for significant heat supply directly above the feed point. Such an extreme situation usually does not call for an intermediate reboiler. Rather, it is a strong indication that there is scope for feed preheating, which, in this case, should result in a reduction in reboil duty more or less on a one-to-one basis.

Figure 15 shows the construction of a "Column Grand Composite Curve" and "Column Composite Curves" in a tray-by-tray fashion. The Column Composite Curves describe vapor and liquid traveling up and down the column, and depict available driving forces. They are a strong clue to column capital costs. As with standard Pinch Analysis, Column Composite Curves allow the designer to include the consideration of capital costs (reflux ratio and number of trays) hand-in-hand with technical feasibility.

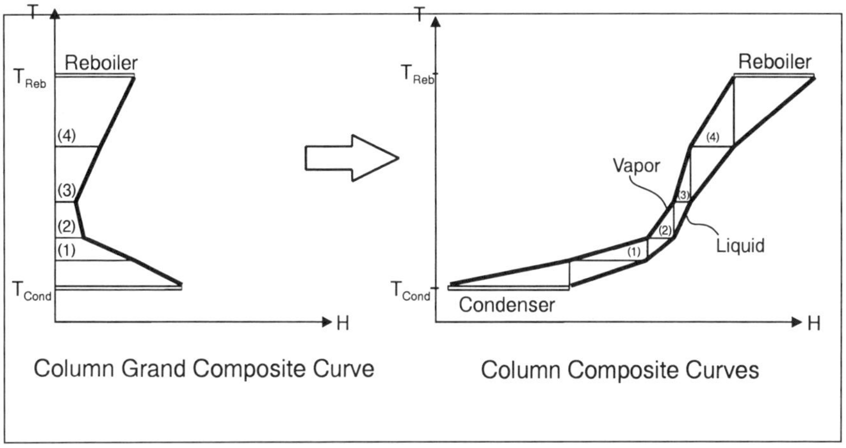

■ *Figure 15. Column Grand Composite Curves and Column Composite Curves have a relationship similar to the established Grand Composite Curves and Composite Curves. Column Grand Composite Curves help assess the use of external heat sources and sinks. Column Composite Curves help interpret column internal processes, driving forces, and capital costs.*

Finally, Figure 16 contrasts the traditional approach based on boxes with Column Profiles, in the context of an integrated process design. The example is based on three columns, A, B, and C. The overlap between boxes indicates that for integration, the pressure in Column A should be increased, and the pressure in Column C should be reduced. Alternatively, a side condenser or a side reboiler might be contemplated for Column B. Consideration of the same problem in terms of Column Profiles makes it apparent that a side reboiler in Column B would enable integration to take place between Columns A and B without a pressure change in Column A, and a side reboiler in Column C would help integration between Columns B and C. Clearly, the conclusions are different — with the latter being the valid ones.

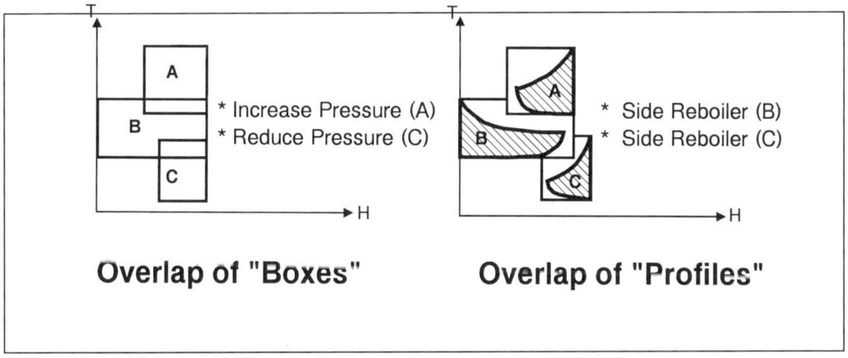

■ *Figure 16. An integration problem is assessed using the traditional box representation and the more recent concept of Column Profiles. The conclusions are different. Column Profiles offer a better assessment.*

Column Profiles have proved to be a powerful tool over the past two years in their first industrial applications, both in the optimization of individual columns and in the reconfiguration of integrated distillation schemes. There is work left to be done, however, to further develop these concepts. This work should prove rewarding. Distillation is without doubt the single most important aspect of process design for energy efficiency. Moreover, distillation columns represent heavy investment. Designs for the future either must avoid or improve distillation. Column Profiles help the designer with both objectives.

Low-Temperature Process Design

Many important processes involve low-temperature separations. For example, the ethylene production process (Figure 17) extends significantly below ambient temperature, with several major distillations carried out across and below ambient. There are many heat sources and heat sinks

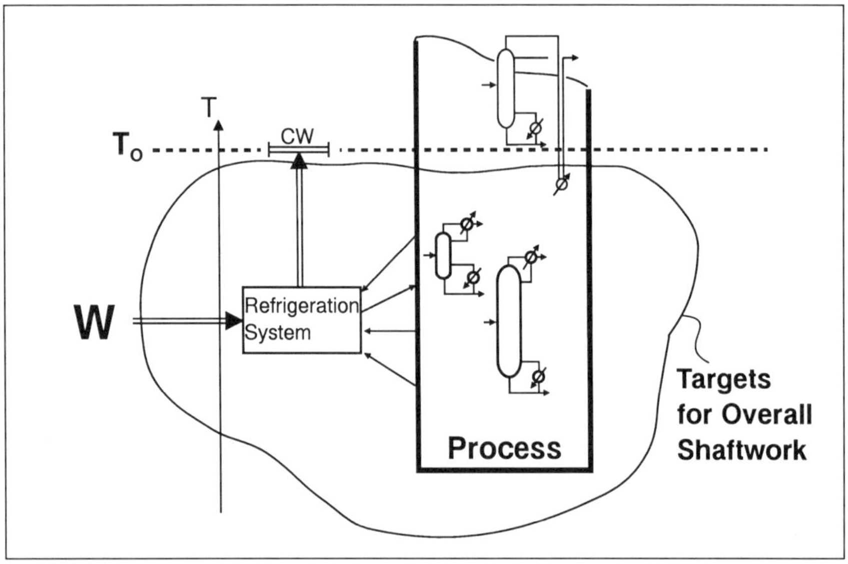

■ ***Figure 17. Low-temperature separation processes, such as encountered in ethylene production, bring together the design of low-temperature process units, complex heat-recovery networks, and multilevel refrigeration systems. There are only two external utilities: shaftwork and cooling water.***

below ambient, with numerous integration opportunities. A complex refrigeration system supplies and removes heat to and from the process below ambient. The refrigeration system usually consists of two cycles (ethylene and propylene), and operates at several levels. There are only two utilities: cooling water and shaftwork.

The overall design of such systems is complex. There are issues both of process and refrigeration-system design. Finally, there is integration between the process and the refrigeration system. It is a simultaneous design problem of low-temperature distillation, other process operations, process heat recovery, and the refrigeration system.

Standard Pinch Analysis could help in this context only to a limited extent. The usual approach was to use Pinch Analysis to set process heat-load targets, to identify refrigeration levels and loads from these targets, and to design the refrigeration system subsequently. Any change in process design, however, then had to be translated into new refrigeration levels and loads, and the design of the refrigeration system needed to be updated accordingly. The ultimate effect of changes in the process on capital costs and power consumption was therefore evaluated by means of a somewhat cumbersome procedure.

Recent work has led to shaftwork targets for overall low-temperature systems *(21)*. Consider Figure 18. A process Grand Composite Curve is shown below ambient with the Carnot Factor as vertical axis instead of temperature. Due to this substitution, the area in the construction represents exergy *(20)*. Specifically, the cross-hatched area between the so-called "Exergy Grand Composite Curve" (EGCC) and the refrigeration levels represents exergy loss. Consequently, a change in refrigeration system design as shown in Figure 18 (the example relates to the introduction of an additional level) is assessed in terms of the consequent reduction of exergy supplied by the refrigeration system. The reduction in overall shaftwork is:

$$(W_A - W_B) = (1/\eta_{ex})(\alpha)$$

where η_{ex} (the exergetic efficiency of the refrigeration system) is approximately constant. In other words, it is possible to completely bypass the design of the refrigeration system in targeting for shaftwork.

The approach extends to process changes, and its use is demonstrated in Figures 19 and 20 *(24)*. Figure 19 shows how the EGCC is built up from base-case process components. The example relates to a distillation process. Note that the engineer is free at this stage to make any assumptions with respect to intermediate reboilers, condensers, and other design details. An inner and an outer EGCC are developed. The inner EGCC is based on Column Profiles, and the area surrounded by it represents — approximately — the reversible exergy requirement of the overall process. The outer EGCC repre-

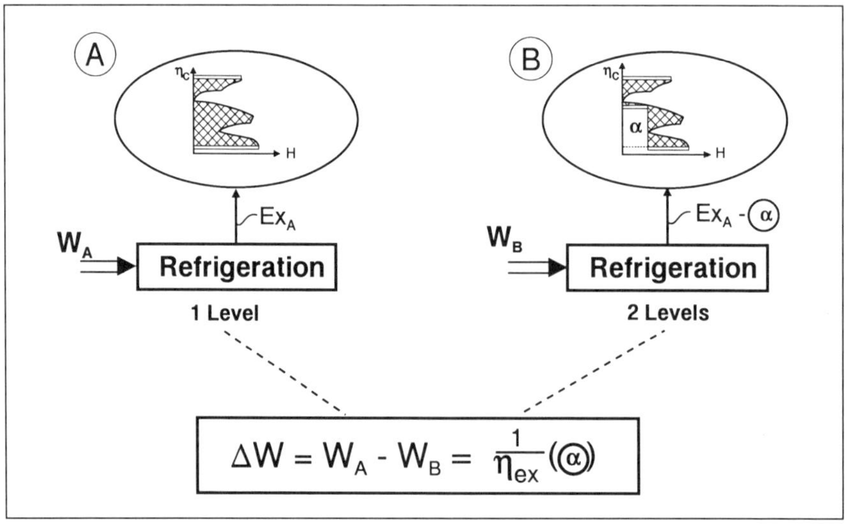

■ ***Figure 18. The change from a single-level refrigeration system design to a two-level system saves shaftwork proportional to area in the η_c vs. H plot.***

■ *Figure 19. There are two stages in the application of Pinch Analysis to the design of low-temperature distillation-based processes, such as ethylene production. First, each distillation column is optimized standalone in terms of capital cost vs. exergy to define a base case. Then, two Exergy Grand Composite Curves (EGCCs) are built up — an inner one based on Column Profiles, and an outer one representing the base-case reboil and condensing duties.*

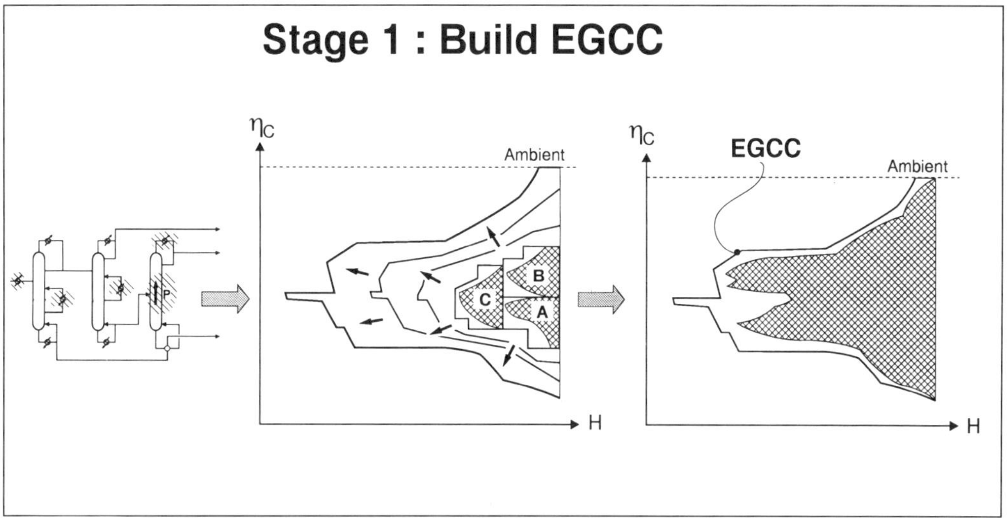

sents the actual exergy requirement of the base-case process.

In Figure 20, the refrigeration levels are added. The area between the outer EGCC and the refrigeration levels represents the exergy loss in the heat recovery system. Also in Figure 20, two process manipulations are implemented. First, the intermediate condenser for Column A is eliminated. In the isolated base-case design for Column A, the intermediate condenser reduces exergy loss. In the overall context, however, this saving leads to no more than a corresponding increase in heat recovery losses; there is no effect on overall refrigeration requirements.

The second process manipulation changes the overall EGCC in several subtle ways and one obvious way: changes are made to several reboil and condensing pressures so that the horizontal "spike" that creates the overall Process Pinch in Figure 19 is "sucked in." This change clearly has a dramatic effect on overall refrigeration requirements.

The engineer uses the construction shown in Figures 19 and 20 to assess process changes, the HEN design, and the refrigeration system design in one coherent approach *(24)*. The combined benefit of distillation column profiles and low-temperature shaftwork targets in the design of low-temperature distillation-based processes (such as ethylene) has been significant. Results achieved with these techniques offer significant improvements over those attained using the previously established principles of Pinch Analysis *(1,60)*. Typical objectives pursued are debottlenecking, yield optimization, and energy efficiency.

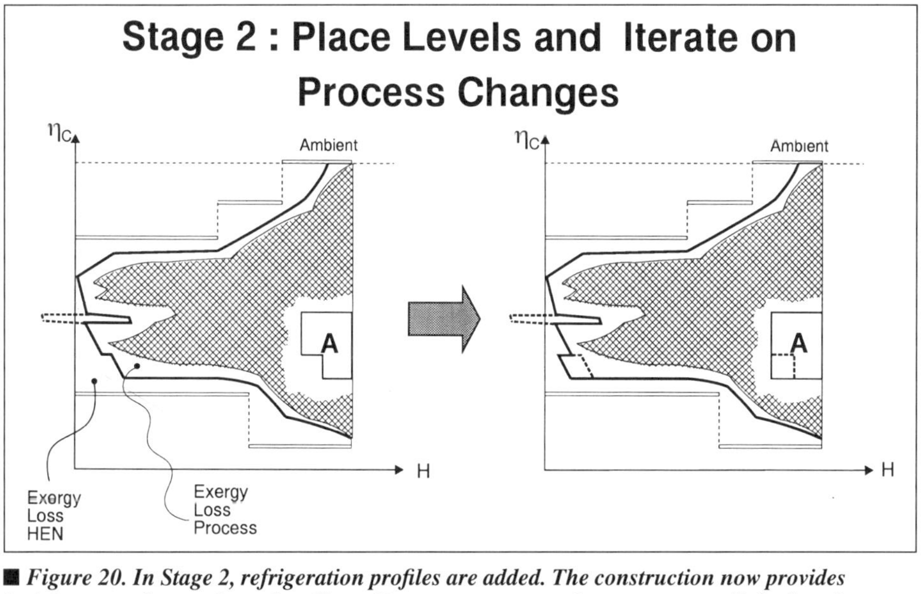

■ *Figure 20. In Stage 2, refrigeration profiles are added. The construction now provides insights, at a glance, about the effect of innermost process changes on overall shaftwork required. For example, the spike from Figure 19 has been eliminated. Likewise, the intermediate condenser for Column A has been deleted, as it leads to no overall benefit.*

Batch Process Integration

A batch production line can be dedicated or multipurpose, and batch processes can be cyclic and random. Dedicated and cyclic production often is found in the food and beverage industry. Multipurpose and random production is used in the manufacture of glue and resins, pharmaceuticals, and other low-volume specialty products. The concept of integration appears more obvious for dedicated and cyclic production than multipurpose and random production. Nevertheless, experience suggests that the integration of batch processes can be worthwhile even in the least likely environment. In a summary report issued by ETSU in Britain, the results of 26 practical applications of Pinch Analysis across various sectors of industry were published *(45)*.

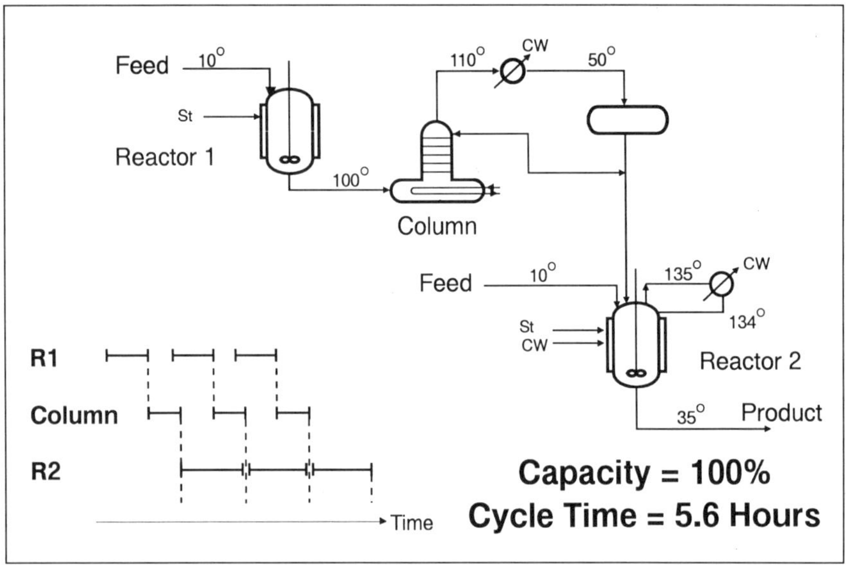

■ ***Figure 21. Time-event chart and process sketch of a simple batch operation. Reactor No. 2 is limiting (36).***

Eighteen of these projects related to batch or partial batch production, ranging from dedicated/cyclic to multipurpose/random. Some of the most worthwhile results actually were achieved in multipurpose/random batch environments.

The results usually come from better energy management — though energy cost reduction frequently is not the key motivation. Instead, efforts often are aimed primarily at debottlenecking, improved yields, more consistent product quality, less rework, reduced product cycle times, and the like.

Batch processes usually are constrained by a combination of the following four factors:

1. material flow (for instance, waiting for the next charge);

2. heat flow (such as waiting to reach temperature);

3. equipment capacity (for example, waiting for the next empty vessel); and

4. labor (like waiting for next shift).

These factors typically are interlinked and heat flow plays a dominant role for feasibility. Material flow may be controlled by the rate of heating the charge or cooling the product. Equipment capacity may be regulated by reactor residence time, which, in turn, is controlled by heat transfer. The quality of separation may be limited by cooling temperature limits, and so on. Careful energy management usually results in cost savings for capacity (debottlenecking), yield, product quality, and energy in that order *(45)*.

So, the obvious question is: How does the engineer take account of the time dimension? There are various concepts and procedures such as "Time Average Models," "Time Slice Models," and "Time Event Charts" *(34)*. There is "Cascade Analysis" *(35)*, and there are "Batch Utility Curves" *(36)*. An overall summary of today's state-of-the-art procedures was compiled recently *(61)*.

A simple example of successful batch-process integration is given in Figures 21 and 22 *(36)*. Figure 21 shows the flowsheet of a specialty process operated in cycles, but in the context of overall random production and in multipurpose reactor vessels. Figure 21 also provides the Time-Event Chart for the process in question. The cycle time is 5.6 h, and Reactor No. 2 is limiting. Figure 22 shows the flowsheet and the Time Event Chart for the project proposal. Reactor No. 1 has been debottlenecked through external feed heating while charging occurs. Reactor No. 2 has been debottlenecked through external feed heating *and* product cooling, both steps being integrated for heat recovery but decoupled timewise through the introduction of a simple storage tank. Cycle time is reduced to 3.25 hours. Capacity is increased from 100% to 172%.

As stated above, process integration may not be an obvious concept in the context of batch processing. There is, however, a significant body of commer-

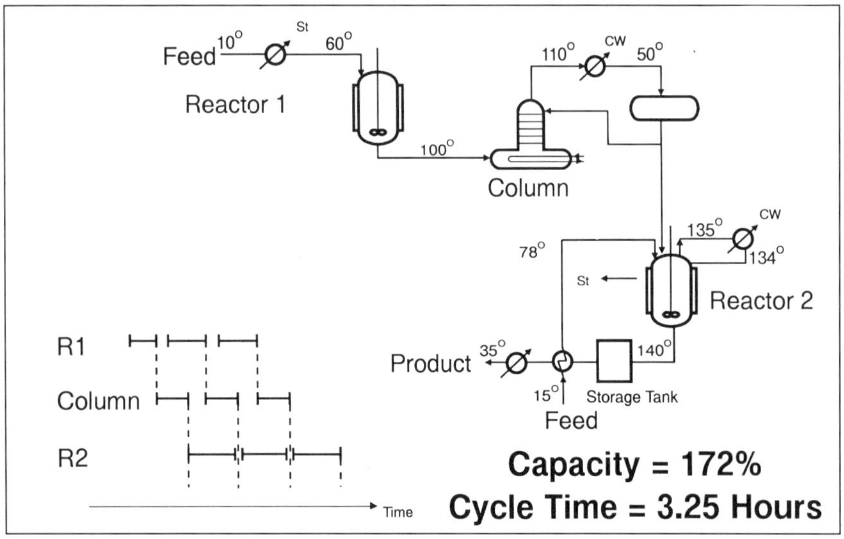

■ ***Figure 22. Project proposal for the system from Figure 21. Both reactors have been debottlenecked through external heat transfer (charge and discharge). Preheating and product cooling on Reactor No. 2 are integrated, but decoupled timewise through a storage tank. Overall system capacity is increased by 72% (36).***

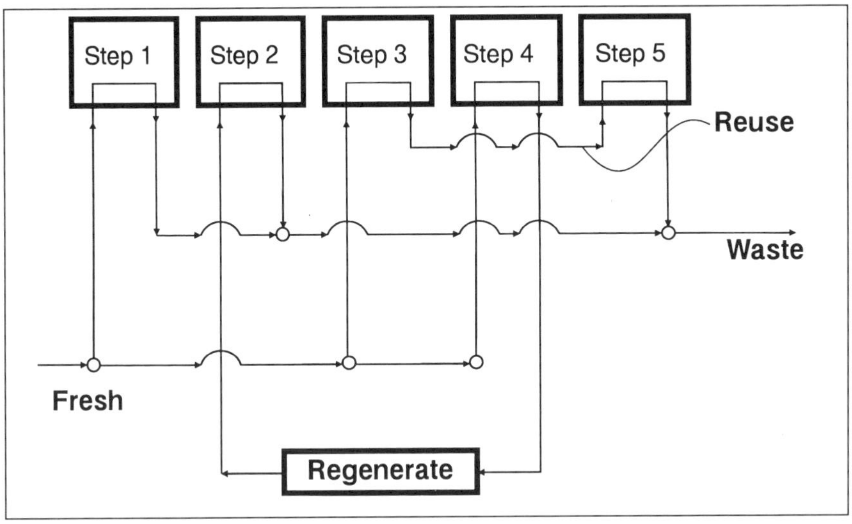

Figure 23. Schematic showing the logistics of wastewater system design. Water is either used, reused, or regenerated and reused before it goes to waste. Reuse and regeneration lead to a reduction in overall water flow rate. The correct structure of wastewater systems incorporating reuse and regeneration sets a complex integration task.

cial applications experience and, surprisingly, cost savings as a percentage of total operating costs often are greater than those found in continuous processes. I believe the key reason is that, in many batch process environments, there simply has never before been a *methodical* search for process integration opportunities.

Water Pinch

Water and wastewater minimization are generic problems of integrated process design. Most (waste) water systems consist of a number of steps that relate to different parts of the process and are linked by water. Water may be supplied as fresh water or may originate from the process. An example based on a simplified schematic involving five process steps is shown in Figure 23. Each step requires water, for filter cake washing, steam stripping, extraction, and so on. Fresh water is fed to Step 1 and spent water is piped from Step 1 to waste. Likewise, fresh water is supplied to Step 3. Spent water from Step 3, however, is used in Step 5 before it is passed to waste. In other words, water is reused. Reuse results in a reduction of overall water flow rate. Finally, Steps 4 and 2 are linked through regeneration, which is recycle with partial treatment prior to reuse. This also results in a reduction of the overall water flow rate.

Integrated (waste)water-system design is a complex problem. There may be a large number of process steps, and water can be reused or regenerated in all manner of possible configurations. As a result of reuse and regeneration, the overall water flow rate through the system is reduced. In practical applications, however, there are constraints. For example, water from one step may not be used in another step — because, for instance, of the danger of process contamination. Even given these constraints, identifying the correct steps for reuse and regeneration can be a near impossible task. At the same time, reduction of water flow rate is important. A reduction in flow rate decreases both fresh-water and effluent-treatment costs. Energy costs (pumping), capital costs (piping), and environmental load factors all usually improve. The incentives for water minimization can be strong.

Figure 24 highlights a relevant recent development in Pinch Analysis. Concentration profiles are shown for three process steps and for the water

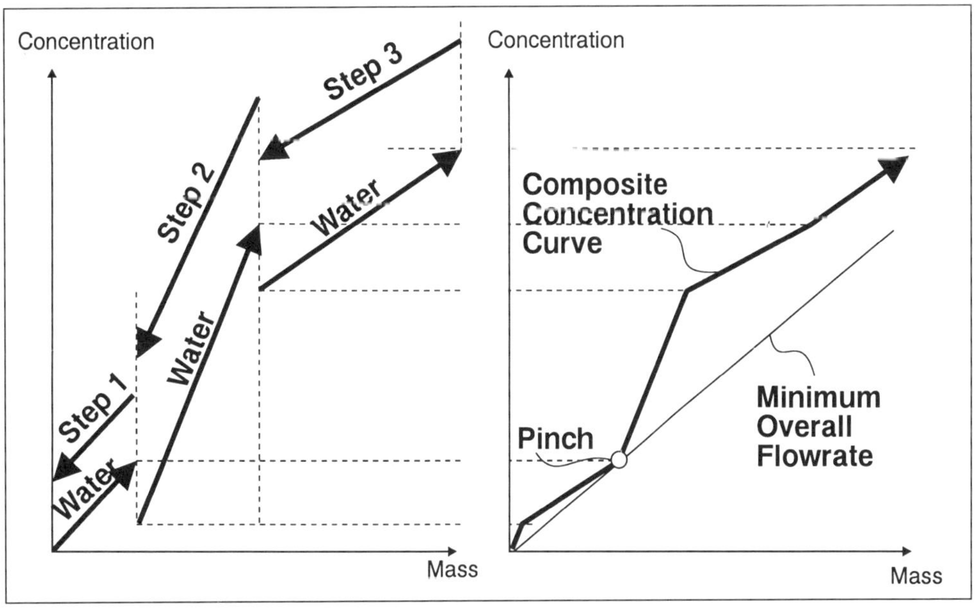

Figure 24. Each process step requiring water is represented in the coordinates of concentration and mass (of the contaminant). The construction of the Composite Concentration Curve is analogous to that of the established Composite Curves for temperature-enthalpy. The steepest possible tangent on the Composite Concentration Curve represents the minimum feasible water flow rate for the overall system. This sets the (waste)water flow rate target.

required in each step. Next, the water concentration profiles from the individual process steps are combined in a so-called "Composite Concentration Curve." The rules of construction are analogous to those for Composite Curves in temperature enthalpy terms in traditional Pinch Analysis *(62)*. Finally, a straight line is drawn through the extreme point of the concentration profile (the "Pinch"). Clearly, a straight line of a given slope represents a specific overall water flow rate. A straight line of steeper slope represents a greater concentration change for a given amount of picked-up contaminant, that is, a lower water flow rate. The slope of the line obtained in Figure 24 determines the minimum feasible overall water flow rate for any system. There is a minimum water target *(40)*.

Again, a target is set prior to design. The minimum feasible water flow rate *of any system* can be predicted. However, it can go further *(62)*. The Pinch Point in Figure 24 is significant. There are rules — analogous to those for HENs — for the design of the integrated water system. For example, there must be no concentration mixing across the Pinch. A methodology follows for the design of integrated process water systems that achieve minimum water flow rates.

The concepts of "Water Pinch" recently have been extended to multicomponent systems, (that is, systems with more than one contaminant), and to the optimal placement of regeneration units *(40)*. Current development aims to extend the concepts further to the optimization of tradeoffs between energy/capital/water costs/emissions, and to the design of process wastewater systems in the context of the overall plant. Distributed-effluent water treatment can be beneficial if there is process waste from certain processes that jeopardizes central treatment. The emerging technology addresses the task of optimized water use within processes simultaneously with either distributed or centralized treatment in the plant as a whole.

Total Site Integration

Total Site Integration goes back to the roots of Pinch Analysis, namely, to the optimization of heat and power. It, however, extends beyond the boundaries of the individual process. It addresses the task of designing each process and the utility infrastructure optimally in the context of the overall plant.

A recent development is the concept of "Total Site Profiles" *(19)*. Figures 25 and 26 describe the concept by way of a case study *(63)*.

Figure 25 illustrates an existing plant with six processes (left) and a project proposal (right). In the existing plant, there is a HP mains that receives steam from a central cogeneration set. There is a MP mains that recovers heat from certain processes and supplies heat to other processes, and so on. The proposed expansion plan is based on a new process to be built on-site that will consume 70 units of MP steam. That steam is to be supplied via a new turbine that links into the existing VHP and MP sys-

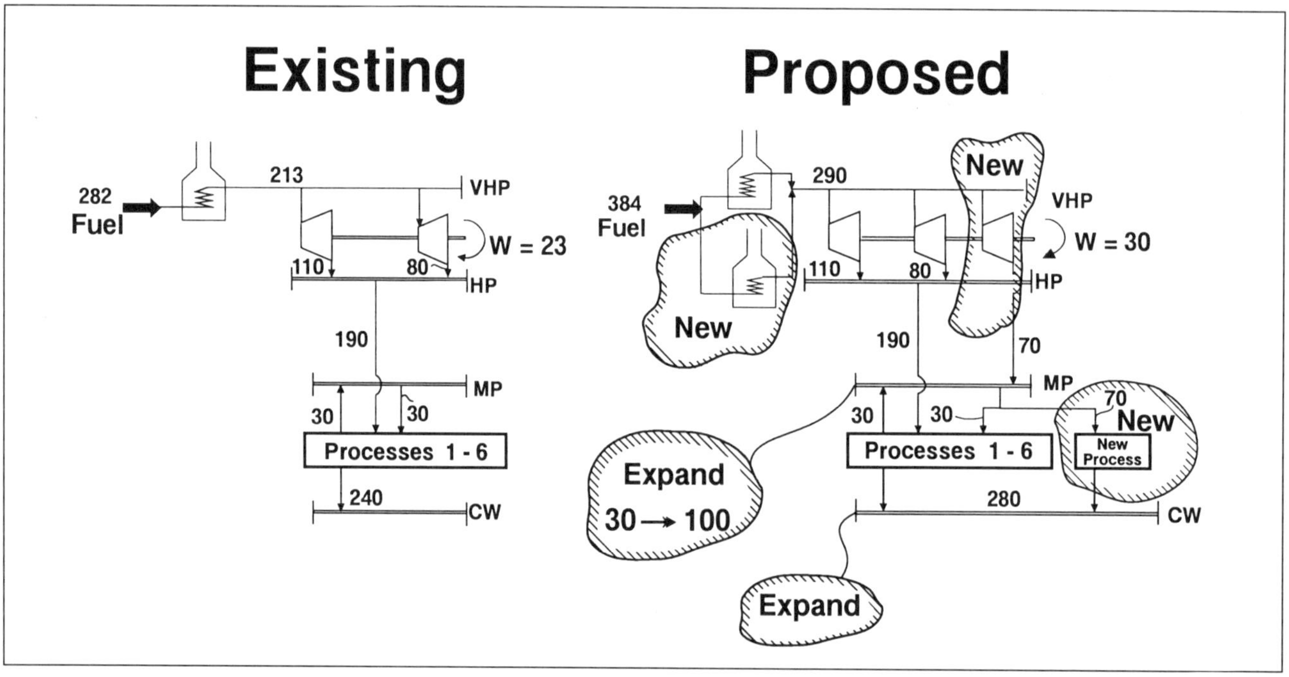

■ *Figure 25. There are six processes on the existing site, and a conventional energy system incorporating cogeneration and two steam distribution levels (HP and MP). The proposed expansion is centered around a new process to be built on-site. Substantial investment is required in the infrastructure. The project requires additional cooling capacity, expansion of the MP steam mains, a new turbine stage (VHP/MP), and an additional furnace.*

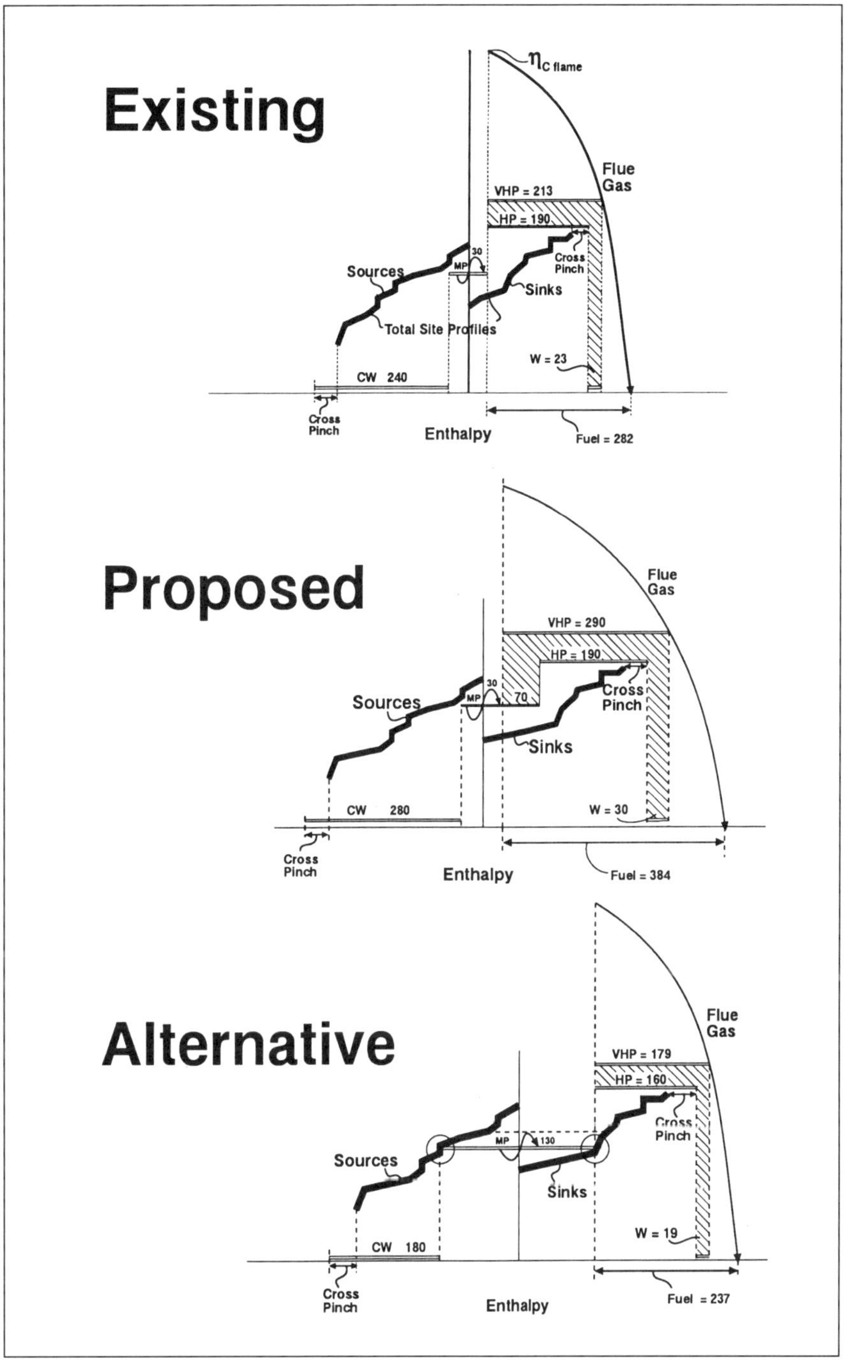

Figure 26. Total Site Profiles representing the existing site (top), the proposed expansion (center), and an alternative (bottom). The profiles show that investment in infrastructure becomes minimal if the pressure can be reduced in the MP steam-distribution level.

tems. To generate the additional VHP steam required, a new boiler is planned. Also, an expansion (to 280 units from 240) of the cooling water system is necessary.

Figure 26 shows the same situation in terms of Total Site Profiles. The Profiles are made up from the Grand Composite Curves of the processes on-site *(19)*. There is a sink profile and a source profile. The vertical axis is the Carnot Factor, the horizontal axis is enthalpy. Based on thermodynamic arguments similar to those used in Figures 18–20, making the Carnot Factor the vertical axis means that the shaded area in Figure 26 is proportional to shaftwork cogenerated on-site.

There is a one-to-one relationship between the utility systems in Figure 25 and the construction based on Total Site Profiles in Figure 26. For the existing situation, there are 282 units of fuel consumed, 23 units of power cogenerated, 213 units of VHP steam generated, 190 units of HP steam, 30 units of MP steam, and 240 units of cooling water. The Total Site Profiles are a thermodynamic representation of the energy flows on-site, given the existing processes and infrastructure.

The second set of Total Site Profiles in Figure 26 relates to the proposed expansion plan. Again, there is a one-to-one relationship between the utility system sketch and the Total Site Profiles. There are 384 units of fuel, 30 units of cogeneration, 290 units of VHP steam, 190 units of HP steam, 100 units of MP steam, and 280 units of cooling water.

The Total Site Profiles, however, reveal additional information. The pressure level of MP steam clearly is not appropriate. As a result of the expansion, the process profiles have changed significantly. Importantly,

Table 2. Comparison of conventional and alternative plans.

	Proposed (Fig. 25)	Alternative (Fig. 27)
Total Utilities Operating Cost	100%	81%
Capital Cost Implications	Expand medium-pressure steam	Shift medium-pressure steam pressure and expand
	New turbine	Adapt processes to suit
	New boiler	
	Expand cooling	

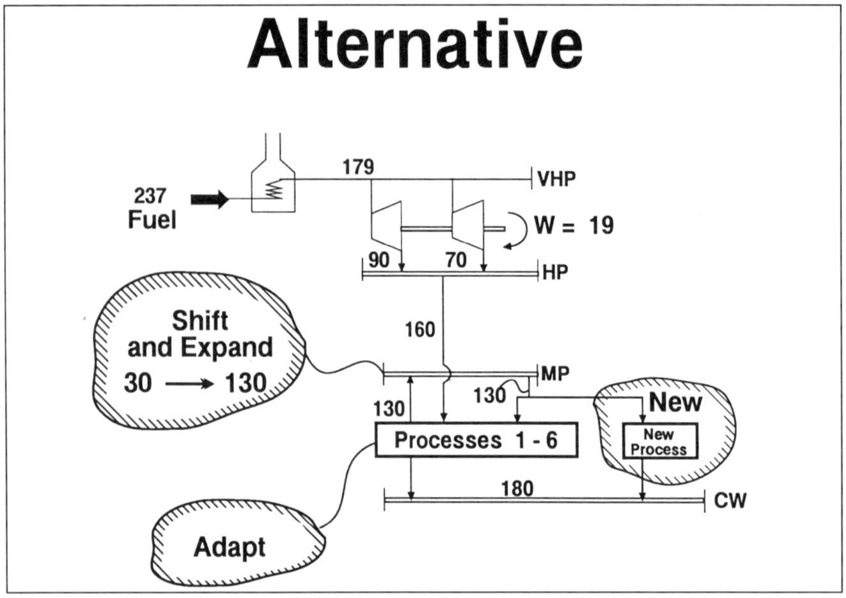

Figure 27. The alternative project as developed in Figure 26. Energy costs and CO_2 emissions are significantly reduced — as is capital cost.

the distance has increased between steam levels and most of the *existing* process sinks. It is apparent that MP steam does not exploit the driving forces available between the steam levels and the process sink curve. The experienced user of Total Site Pinch Analysis recognizes a clear signal to reduce the pressure in the MP steam mains.

This idea is implemented in the third set of Total Site Profiles in Figure 26. The pressure of the MP steam mains is reduced such that the amount of steam recovered from processes into the MP steam mains equals the amount of heat supplied. The MP steam mains is in balance, and its load is maximized. Only a small pressure shift is necessary. The results are dramatic. The amount of VHP steam that must be generated is reduced significantly.

The alternative plan leads to a 19% energy savings compared to the conventional expansion plan. (See Table 2.)

Surprisingly, there also is a significant reduction in capital costs *(63)*. Whereas the conventional expansion plan requires a new turbine, a new boiler, and an expansion in the cooling system, the alternative plan does not need any such investment in infrastructure. Instead, there is moderate investment in the processes so as to enable the pressure of the MP mains to be downgraded (see Figure 27). Once again, Pinch Analysis has led to the *simultaneous* reduction of energy cost, emissions, and capital costs.

Recently, Total Site Integration has been developed further *(64)*. Figure 28 depicts so-called "Site Composites Curves" and two corresponding "Site Utility Grand Composites" for another example. The Total Site Composites show the Total Site Profiles overlapping, not unlike Process Composite Curves. The important difference is that heat transfer only can take place indirectly — through the steam mains — and *not* directly from the hot to the cold composite. Therefore, the "Total Site Pinch" occurs *not* where the profiles touch in the extreme (for example, at ΔT_{min} = 0), but where there is no overlap between utilities.

White areas between the Site Composites (marked in Figure 28 via vertical outlined arrows) represent exergy loss in heat transfer between processes on-site and utilities. Shaded areas represent potential for cogeneration.

The Site Utility Grand Composites relate to the Site Composites in the usual way. They allow the engineer to determine, at a glance, how various drives can be placed to exploit

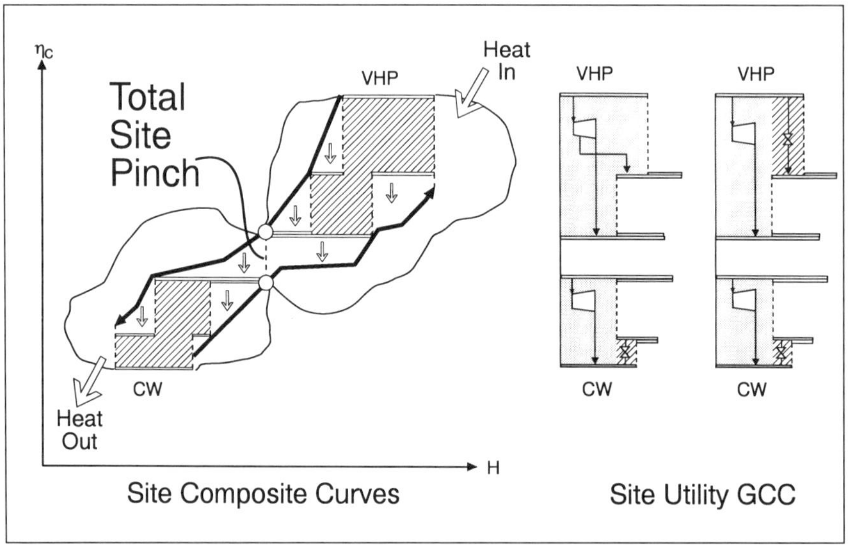

Figure 28. The Site Composite Curves and Site Utility Grand Composite Curve as recently introduced (64). The Total Site Pinch is located at the zero-overlap point between utilities. White areas represent exergy losses; crosshatched areas, the potential for cogeneration on-site; and gray areas, the shaft power cogenerated. Existing or suggested steam system configurations easily are assessed in the Site Utility Grand Composites.

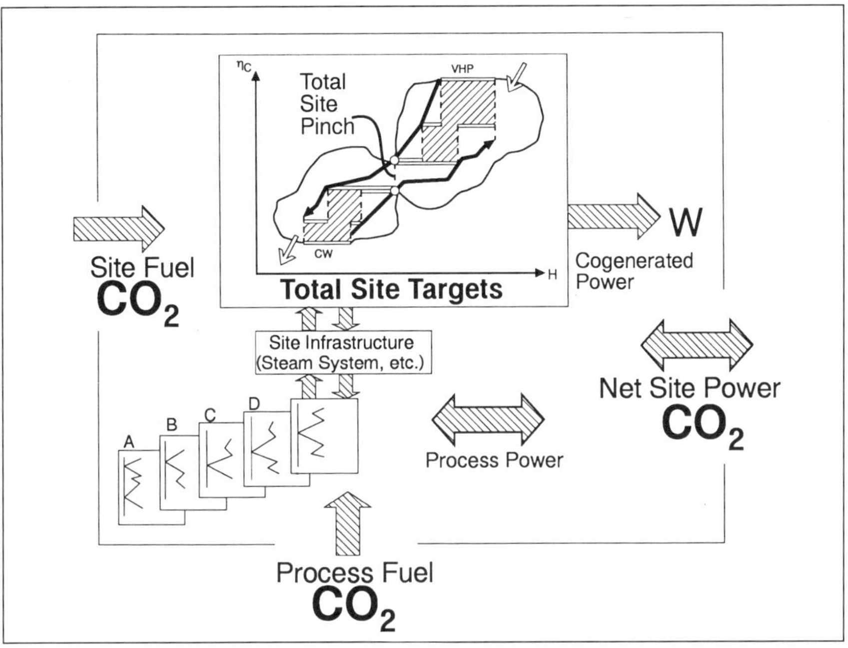

Figure 29. Total Site targeting allows data describing the processes and the utility systems to be brought together in one coherent approach. Design changes, planning scenarios, capacity increases, and so on, immediately are understood in terms of energy cost and fuel-related emissions, regardless of fuel mix, power balance, or the like. Importantly, the approach is independent of any cost of steam. Steam flow is strictly an internal parameter.

the site's potential for cogeneration.

Reference 64 expands on this discussion to explain how process modifications are interpreted in the Total Site context. The Plus/Minus Principle is applied with respect to the Total Site Pinch. Process changes are divided into three categories:

1. those leading to site fuel savings;

2. those leading to additional cogeneration on-site at the expense of an additional fuel requirement equal to the work cogenerated; and

3. those leading to an additional fuel requirement, compensated for by additional condensing power generation.

Another point of interest is that Total Site Integration as described here removes the age-old problem of steam pricing. Steam is strictly an intermediate parameter. The implications of any imaginable design or planning step are easily understood in terms of fuel and power at the plant gate. This enables the planner/designer either to abandon the concept of "cost of steam" or to set the costs so as to *truly* represent those at the plant gate.

Total Site Integration has been a major breakthrough in Pinch Analysis. Know-how has developed quickly, building on the considerable applications experience gained in individual process integration over ten+ years. Successful industrial applications already have been cited in the literature *(53,65,66,67)*, with Linnhoff March alone reporting the completion of well over 30 projects in more than ten countries in less than three years *(68)*.

Emissions Targeting

Emissions reduction and Emissions Targeting have a process dimension and a total site dimension. The process dimension is easily understood. Lower process emissions usually require sharper separations that, in turn, require additional energy. Therefore, the reduction in process emissions needs to be assessed relative to the increase in fuel-related emissions. This dilemma increasingly is recognized both by designers and legislators *(69,70)*. There have been instances in the past decade where regulations requiring excessively low ppm limits on process emissions have led to additional fuel consumption such that *overall* emissions deteriorated. It clearly is necessary to assess the overall picture. Pinch Analysis targeting concepts have an obvious contribution to make in this context. They allow designers, planners, and legislators to come to a rational assessment of tradeoffs between process-related and fuel-related emissions and to agree on achievable targets *(71,72)*.

Over the past several years, emissions have taken on a total site dimension in many industrial situations. The so-called "Bubble Concept" often guides planners and regulators in setting limits for emissions from either a company or a production site. Figure 29 relates to "Global Emissions" (in this case, CO_2) for a general site in the CPI. Total-Site Pinch Analysis easily incorporates aspects such as changes in production mix and capacities, investments in the infrastructure, and investments in the processes. The analysis brings together these and other aspects in one coherent approach and gives targets for:

1. central site combustion;

2. total-site electric power import or export; and

3. decentralized combustion on-site.

Global CO_2 emissions follow, and can be targeted as a function of investment as shown, for example, in Figure 30.

Emissions Targeting is specifically discussed elsewhere *(69,70)*. Pinch Analysis can help predict both long- and short-term targets for any process and site. The procedures involved are uncomplicated, accurate, and practical.

Pinch Analysis and other process integration methodologies

So what is different about Pinch Analysis? Why have other methods for process integration found less application in practice? This is a question usually asked "after hours" at chemical engineering conferences. Especially, what is the difference between Pinch Analysis, or the *ther-*

Table 3: Overview of industrial applications of Pinch Analysis.

Technique	Use							
	Capital Cost Reduction	Energy Cost Reduction	Debottlenecking	Emissions Reduction	Yield Improvement	Operability Improvement	Retrofit	New Design
Basic HEN Concepts	✔	✔	✔			✔	✔	✔
Pressure Drop Optimization	✔		✔			✔	✔	✔
Multiple-Base-Case Design	✔	✔	✔			✔	✔	✔
Distillation Column Profiles	✔	✔	✔	✔	✔		✔	✔
Low-Temperature Process Design	✔	✔	✔		✔		✔	✔
Batch Process Integration	✔	✔	✔	✔	✔	✔	✔	✔
Water Pinch	✔	✔	✔	✔		✔	✔	✔
Total Site Integration	✔	✔	✔	✔		✔	✔	✔
Emissions Targeting				✔			✔	✔

modynamic approach, and the *mathematical programming* approach? This question deserves discussion.

Many areas of engineering design combine physical insights with mathematical methods. For example, fluid dynamics gives fundamentals, numerical methods allow application of these fundamentals in the design of practical flow systems. Physical insights are the engineer's "eye" while numerical methods are the "muscle." It is possible to argue that the maturity of any given area of engineering design methodology is measured by the degree to which *both* physical insights *and* numerical techniques are in use.

Process Integration is a relatively new area for engineering design methodology. Until recently, engineers developed conceptual process designs by experience and intuition without using methodology. In my opinion, many of the so-called mathematical methods simply introduced muscle before there was an eye. Pinch Analysis contributed the eye. In this, Pinch Analysis was not alone. For example, "Exergy Analysis" *(73)* and the heuristic approach recently summarized by Douglas *(74)* attempt to do likewise. But process design methodology, as a professional discipline, still was relatively little developed, and there were few insights at the time. This is why the impact of Pinch Analysis was significant.

It is inevitable that the distinction between *mathematical* and *thermodynamic* will blur over time. Gundersen and Naess *(75)* discussed work combining insights with numerical methods six years ago. Notable examples are articles from Westerberg *(76)*, Floudas *(77)*, and Grossmann and coworkers *(78)*. Likewise, there is a growing body of powerful software supporting the more recent developments in Pinch Analysis discussed in this article. Numerical methods are used for increased power and speed of data handling, the optimization of targets, the automatic definition of key design patterns, and so forth. Pinch Analysis, by

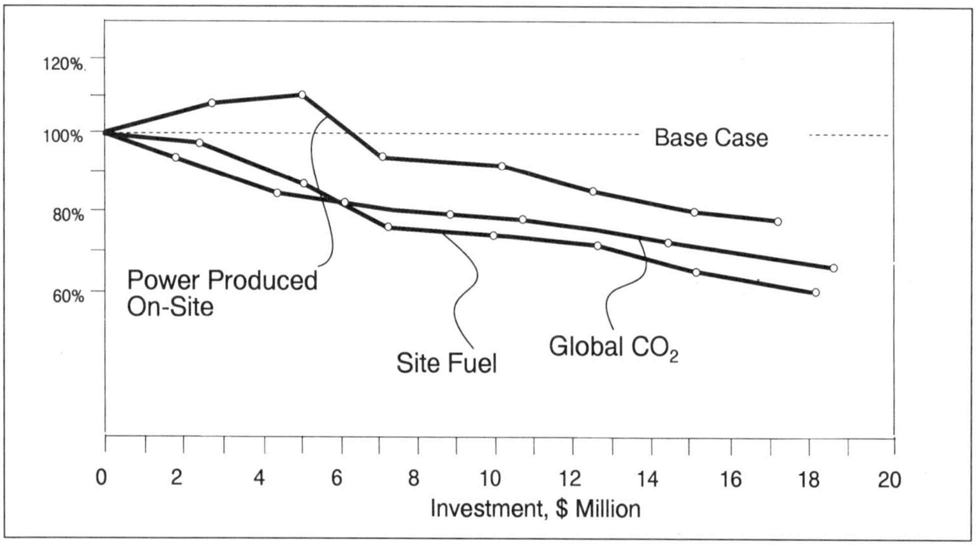

Figure 30. Targets are set as a function of investment for any specific site. In this example, based on a recent case study, power produced on-site initially increases in spite of reduced site fuel consumption. Usually, cogeneration potential is expected to decrease as less fuel is burned. Experience has shown that better level matching between processes and utilities can lead to an increase in the amount of power produced per unit heat.

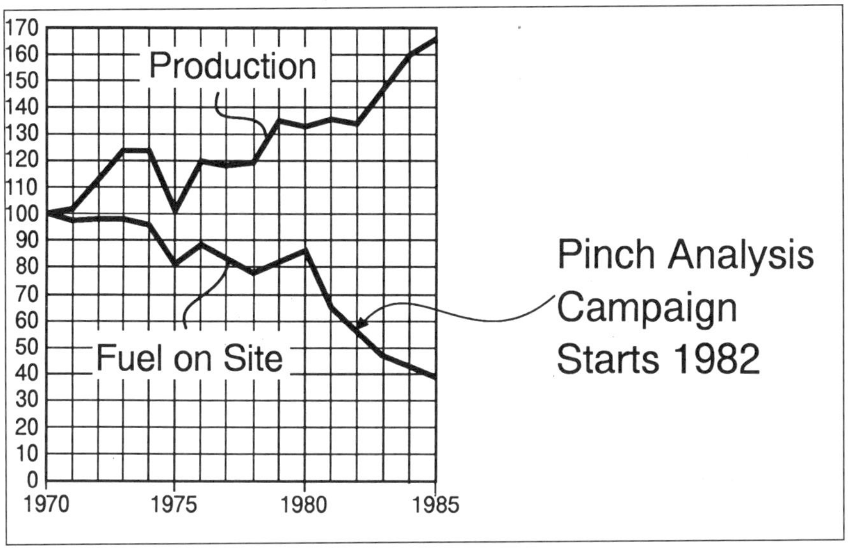

■ ***Figure 31. Historical data for BASF's site at Ludwigshafen, Germany. From 1970–1982, the firm used conventional approaches to reduce energy consumption. Between 1979 and 1982, production became limited by the energy infrastructure. Between 1982 and 1985, Pinch Analysis was relied on heavily in identifying and implementing numerous short-payback retrofit projects (46).***

contrast, provides the eye. There is emerging a whole family of design methodologies, many of which are software based, that will provide the process design engineer with a practical, mature, and powerful working environment.

An industrial record to build on

There are still those who regard Pinch Analysis, or Pinch Technology, as a tool for specialists addressing energy efficiency in the (grassroots) design of HENs. Table 3 should dispel that view. It shows which of the concepts discussed here are routinely applied industrially for what objectives and in what context — retrofit or new design. It underscores that Pinch Analysis is used for capital cost reduction, energy cost reduction, emissions reduction, operability improvement, and yield improvement — in that order — both for retrofit and new design.

Probably the most common and also the most rewarding application of Pinch Analysis occurs with expansion projects — be they involving design of new units or retrofitting (for instance, debottlenecking) in the process itself or the infrastructure (such as boiler replacement). It generally leads to capital cost savings coupled with other benefits. The project described in Figures 25–27 and summarized in Table 2 is a case in point.

In addition, so-called "opportunity" projects, having little or no capital cost, usually are identified in the context of expansion. For instance, a new steam line at different steam pressure may not be economic as an energy revamp on its own, but, if the expansion takes place anyway, there may be the opportunity to adopt a new specification and *not* replace like with like. Thus, Pinch Analysis can be used over time to modernize entire production sites with little capital investment, by using a "roadmap" in successive expansion projects that takes advantage of each opportunity in turn.

A possible reason for the still widespread perception that Pinch Analysis is a narrow tool may stem from the fact that industrial applications were widely reported early in its history. The first publicly reported applications — by ICI — averaged energy savings approaching 30% in processes previously thought optimized *(43)*. Applications reported by Union Carbide a few years later showed even better results, mainly due to progress in the understanding of how to effect process changes *(48)*. BASF reported completing over 150 projects and achieving energy, debottlenecking, and environmental benefits. Figure 31 illustrates the development of production and total fuel consumption over time on BASF's site in Ludwigshafen, Germany *(46)*. Taking 1970 as reference, production in 1980 stood at approximately 135% and total fuel consumption at approximately 85%. It is evident from the graph that production was closely tracking fuel efficiency — fuel-firing capacity was the site's bottleneck. Planning permission for the construction of a new power station was withheld and production stabilized in the early 1980s. BASF reacted with a determined campaign to improve energy efficiency, reducing total fuel consumption to approximately 55% by 1982. Then, in 1982, the firm started a site-wide campaign using Pinch Analysis to identify additional short-payback retrofit projects. The results are shown in Figure 31. By 1985, production had reached approximately 160% while energy consumption had dropped to approximately 45%. This implies a 32% improvement in *specific* (that is, per unit production) energy consumption relative to 1982 ((0.45/1.60)/(0.55/1.35) = 0.68). Not all of this improvement was due to the Pinch campaign, but it was credited with approximately 25% *(46)*.

There have been many public testimonials both from operating companies and contractors over the years reporting on the developing breadth of the technology, on applications, and on the results achieved. In all, projects have been reported in over 30 countries and throughout the last decade.

Few applications, if any, have had the impact of BASF's campaign. And while it is true that no one could expect the same impact today (process design has moved on generally), it also is true that an essential ingredient in BASF's success often is overlooked: the firm did not just rely on training experts. Rather, BASF managed a campaign. It

B. LINNHOFF is Head of the Department of Process Integration at the University of Manchester Institute of Science and Technology, Manchester, U.K. (0114461/200–4380; Fax: 0114461/236–7439), and Chairman of Linnhoff March International, Houston (713/787–6861; Fax: 713/787–6865), a company specializing in the development and transfer of process design technology. Helped by fifteen years of experience in process design, both in academe and industry, he has developed Pinch Analysis and pioneered its use. He holds an MS in mechanical engineering from Eidgenoessische Technische Hochschule, Zurich, and a PhD in chemical engineering from Leeds University. He is a Member of AIChE.

committed to adopting — corporately — a new technology. It thought the matter through and provided the budgets, the infrastructure, the software, the training, and so on. It determined how Pinch Analysis fitted into the design and project organizations inside BASF, and how projects could be "sold" internally and implemented. It also carefully managed the innovation process. In my opinion, this committed and comprehensive approach is the key to success.

Morgan of M. W. Kellogg makes a powerful case for the thoughtful integration of Pinch Analysis, or any other technique for process integration, into the design process *(1)*. Consider process simulation. In many ways, process integration and process simulation are similar. Only 10–20 years ago, most companies left process simulation to a few specialists using in-house software on *ad hoc* projects. Most companies came nowhere near exploiting the potential of the technology. Today, while many companies still do not fully exploit the potential, there at least is growing awareness that managers, not just technical specialists, should think through the information flow, the organization of project work, the windows in time to take advantage of "opportunities" (see above), the spreading of general awareness, the outsourcing of software, and other aspects, so as to make the most of simulation. Process integration is 10–20 years behind. In most firms, it still is regarded as an area solely for specialists. Those few companies that already have integrated the technology into their organization have derived by far a greater benefit.

The role of software in this discussion is especially important. Many people still think of process-integration software packages (for instance, software generating Composite Curves, Grand Composite Curves, Total Site Profiles, and Grid Diagrams) as simple one-off tools that might as well be sourced from universities or be written in-house. In reality, process integration is an integral part of process design, alongside simulation, and process integration software should be integrated into the process-design software environment with the same consideration given to user interfaces, file handling, data transfers, and so on, as is customary for

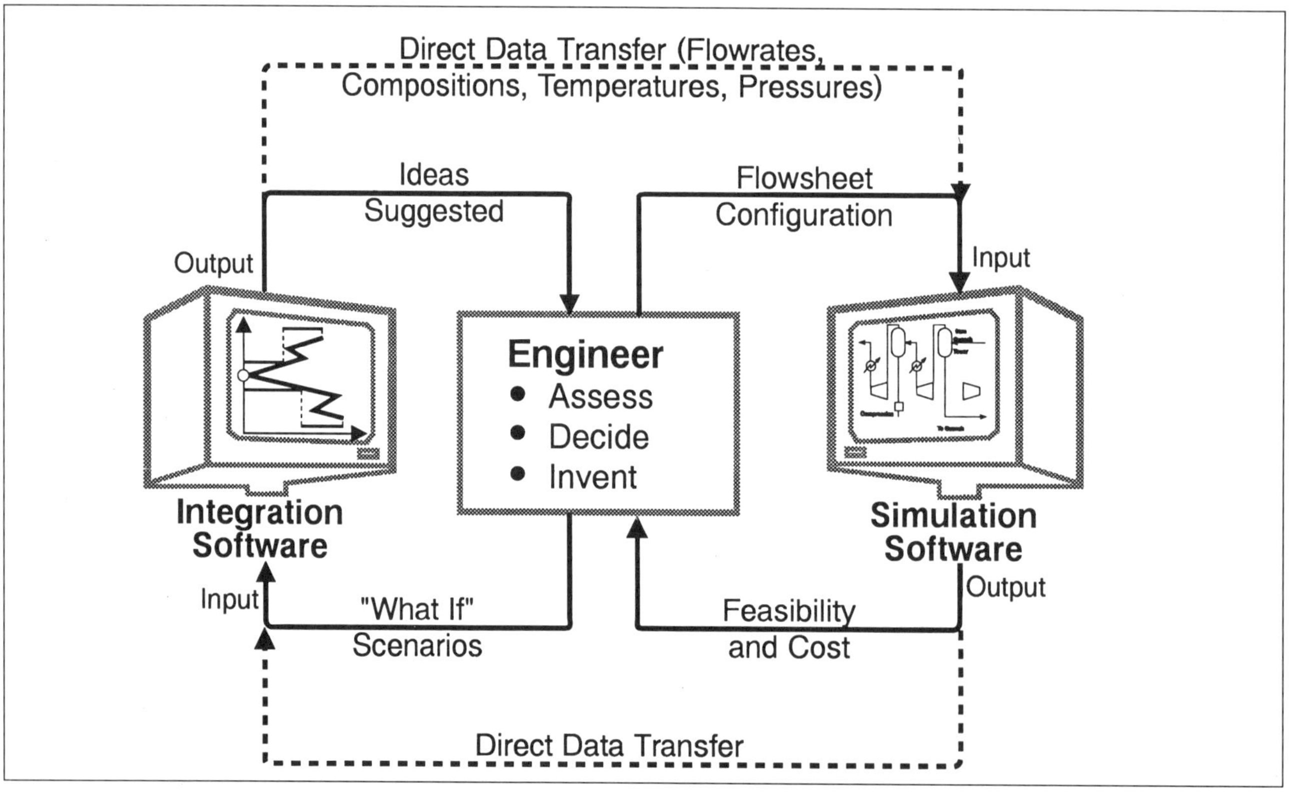

Figure 32. Process Integration software needs to be incorporated into the suite of process design software, alongside simulators, databases, and so on, with careful attention given both to the user interface and to interfaces with other software.

simulation packages, databases, and the like.

Consider Figure 32. The engineer still drives the design process — describing the unit operations, defining the flowsheet structure, and using simulation software to check on feasibility and costs. The engineer then uses process integration software to consider "what if" scenarios. What if the distillation pressure is changed? What if reactor conversion is increased? What if air preheat is used? What if an additional steam level is introduced? Pinch Analysis software helps the engineer to identify promising options, to establish outline feasibility and costs, and to recognize key problems and characteristics for each scenario. Only the most promising scenarios then are examined further through simulation for a more detailed evaluation. For time efficiency, there is a need for interfaces between the Pinch Analysis software, the databases, and the process simulator. Given the help or hindrance effect of suitable software in today's design environment, any corporate plan to adopt process integration and Pinch Analysis should include a realistic assessment not just of software capabilities, but also of user-friendliness and flexibility of interfaces.

> ***In many ways, process integration and process simulation are similar... However, process integration is 10-20 years behind. In most firms, it still is regarded as an area solely for specialists. Those few companies that already have integrated the technology into their organization have devrived by far a greater benefit.***

To conclude, many papers and articles detail the industrial application and impact of Pinch Analysis (for instance, Refs. 43,45,46,48,53,79). These and others establish a track record and testimonials that should not be ignored. Also, they are representative of results achieved in different industries and countries over time. Whereas many of the early applications were reported from English-speaking countries, it seems that exploitation of the technology in Germany and, particularly, Japan is edging ahead now. This comes back to the importance of the corporate approach. Japanese companies with their emphasis on long-term thinking and proper planning generally have adopted Pinch Analysis later than their U.S. and European counterparts, but they are doing it with more consistent success. Major contractors such as Japanese Gas Corp. *(80)* and large operating companies such as Mitsubishi Kasei *(53)* have followed BASF's example. In many cases, they started from a higher baseline — as the Japanese petrochemical and oil refining industries always have been technology and efficiency focused. The results nevertheless have been significant enough for the prestigious Japanese newspaper *Nikkei* to report on Pinch Analysis three times in less than a year, and to describe it as being "strategically important" for Japan *(81,82,83)*.

For those who wish to get started, Table 1 and the list of references should prove helpful. Commercial software is available from Aspen Technology *(84)*, Linnhoff March *(85)*, Britain's National Engineering Laboratory *(86)*, Simulation Sciences *(87)*, and others. Training is available primarily from Linnhoff March *(13)* and UMIST *(14)*. CEP

Literature Cited

1. **Morgan, S.,** "Use Process Integration to Improve Process Designs and the Design Process," *Chem. Eng. Progress,* **88** (9), pp. 62–68 (Sept. 1992).
2. **Huang, F., and R. V. Elshout,** "Optimizing the Heat Recovery of Crude Units," *Chem. Eng. Progress,* **72** (7), pp. 68–74 (July 1976).
3. **Linnhoff, B., and J. R. Flower,** "Synthesis of Heat Exchanger Networks: Part I: Systematic Generation of Energy Optimal Networks," *AIChE J.,* **24** (4), pp. 633–642 (Apr. 1978). "Part II: Evolutionary Generation of Networks with Various Criteria of Optimality," *AIChE J.,* **24** (4), pp. 642–654 (Apr. 1978).
4. **Linnhoff, B., D. R. Mason, and I. Wardle,** "Understanding Heat Exchanger Networks," *Comp. & Chem. Eng.,* **3**, pp. 295–302 (1979).
5. **Linnhoff, B., and E. Hindmarsh,** "The Pinch Design Method of Heat Exchanger Networks," *Chem. Eng. Sc.,* **38** (5), pp. 745–763 (1983).
6. **Linnhoff, B., D. W. Townsend, D. Boland, G. F. Hewitt, B. E. A. Thomas, A. R. Guy, and R. H. Marsland,** "User Guide on Process Integration for the Efficient Use of Energy," IChemE, Rugby, U.K. (1982).
7. **Townsend, D. W., and B. Linnhoff,** "Heat and Power Networks in Process Design, Part I: Criteria for Placement of Heat Engines and Heat Pumps in Process Networks," *AIChE J.,* **29** (5), pp. 742–748 (May 1983). "Part II: Design Procedure for Equipment Selection and Process Matching," *AIChE J.,* **29** (5), pp. 748–771 (May 1983).
8. **Linnhoff, B., H. Dunford, and R. Smith,** "Heat Integration of Distillation Columns into Overall Processes," *Chem. Eng. Sc.,* **38** (8), pp. 1175–1188 (1983).
9. **Smith, R., and P. S. Jones,** "The Optimal Design of Integrated Evaporation Systems," *Heat Recov. Sys. & CHP,* **10** (4), pp. 341–368 (1990).
10. **Linnhoff, B., and J. de Leur,** "Appropriate Placement of Furnaces in the Integrated Process," presented at IChemE Symposium Understanding Process Integration II, UMIST, Manchester, U.K. (Mar. 1988).
11. **Linnhoff, B., and S. Ahmad,** "Cost Optimum Heat Exchanger Networks, Part 1: Minimum Energy and Capital using Simple Models for Capital Cost," *Comp. & Chem. Eng.,* **14** (7), pp. 729–750 (1990). **Ahmad, S., B. Linnhoff, and R. Smith,** "Part 2: Targets and Design for Detailed Capital Cost Models," *Comp. & Chem. Eng.,* **14** (7), pp. 751–767 (1990).
12. **Linnhoff, B., and S. Parker,** "Heat Exchanger Networks with Process Modifications," presented at IChemE Annual Research Meeting, Bath, U.K. (Apr. 1984).
13. **Linnhoff, B., et al.,** "Pinch Analysis," Foundation Training Course, Linnhoff March Ltd., Knutsford, U.K. (1986).
14. "MS Course in Process Integration," Dept. of Process Integration, UMIST, Manchester, U.K.
15. **Polley, G. T., M. H. Panjeh Shahi, and F. O. Jegede,** "Pressure Drop Considerations in the Retrofit of Heat Exchanger Networks," *Trans. of IChemE,* **68**, Part A, pp. 211–220 (1990).
16. **Tjoe, T. N., and B. Linnhoff,** "Using Pinch Technology for Process Retrofit," *Chem. Eng.,* pp. 47–60 (Apr. 28, 1986).
17. **Shokoya, C. G.,** "Retrofit of Heat Exchanger Networks for Debottlenecking and Energy Savings," PhD Thesis, UMIST, Manchester, U.K. (Jan. 1992).
18. **Hall, S. G.,** "Targeting for Multiple Utilities in Pinch Technology," PhD Thesis, UMIST, Manchester, U.K. (Nov. 1989).
19. **Dhole, V. R., and B. Linnhoff,** "Total Site Targets for Fuel, Co-generation, Emissions, and Cooling," *Comp. & Chem. Eng.,* **17** suppl. pp. s101–s109 (1993).
20. **Linnhoff, B.,** "Pinch Technology for the Synthesis of Optimal Heat and Power Systems," *Trans. of the ASME, J. of Energy Res. Tech.,* **111** (3), pp. 137–147 (Sept. 1989).
21. **Linnhoff, B., and V. R. Dhole,** "Shaftwork Targets for Low Temperature Process Design," *Chem. Eng. Sci.,* **47** (8), pp. 2,081–2,091 (1992).
22. **Dhole, V. R., and J. P. Zheng,** "Applying Combined Pinch and Exergy Analysis to Closed Cycle Gas Turbine System Design," presented at ASME Cogen Turbo Power, The 7th Congress and Exposition, Bournemouth, U.K. (Sept. 1993).
23. **Parker, S. J.,** "Supertargeting for Multiple Utilities," PhD Thesis, UMIST, Manchester, U.K. (Feb. 1989).
24. **Dhole, V. R., and B. Linnhoff,** "Overall Design of Subambient Plants," *Comp. and Chem. Eng.,* **18**, suppl. pp. s105–s111 (1994).
25. **Saboo, A. K., M. Morari, and R. D. Colberg,** "RESHEX — An Interactive Software Package for the Synthesis and Analysis of Resilient Heat Exchanger Networks, Part I: Program Description and Application," *Comp. & Chem. Eng.,* **10** (6), pp. 577–589 (1986). "Part II: Discussion of Area Targeting and Network Synthesis Algorithms," *Comp. & Chem. Eng.,* **10** (6), pp. 591–599 (1986).
26. **Cerda, J., A. W. Westerberg, D. Mason, and B. Linnhoff,** "Minimum Utility Usage in Constrained Heat Exchanger Networks — A Transportation Problem," *Chem. Eng. Sc.,* **38** (3), pp. 373–387 (1983).
27. **Ahmad, S., and D. C. W. Hui,** "Heat Recovery Between Areas of Integrity," *Comp. & Chem. Eng.,* **15** (12), pp. 809-832 (1991).
28. **Hall, S. G., S. Ahmad, and R. Smith,** "Capital Cost Targets for Heat Exchanger Networks Comprising Mixed Materials of Construction, Pressure Ratings and Exchanger Types," *Comp. & Chem. Eng.,* **14** (3), pp. 319–335 (1990).
29. **Ahmad, S., and R. Smith,** "Targets and Design for Minimum Number of Shells in Heat Exchanger Networks," *Chem. Eng. Res. & Des.,* **67**, pp. 481–494 (Sept. 1989).
30. **Linnhoff, B., and E. Kotjabasakis,** "Process Optimization: Downstream Paths for Operable Process Design," *Chem. Eng. Progress,* **82** (5), pp. 23–28 (May 1986).
31. **Kotjabasakis, E., and B. Linnhoff,** "Sensitivity Tables for the Design of Flexible Processes, Part 1: How Much Contingency in Heat Exchanger Networks is Cost-Effective?," *Chem. Eng. Res. & Des.,* **64**, pp. 197–211 (1986).
32. **Jones, P. S.,** "Targeting and Design for Heat Exchanger Networks Under Multiple Base Case Operation," PhD Thesis, UMIST, Manchester, U.K. (Oct. 1991).
33. **Calandranis, J., and G. Stephanopoulos,** "Structural Operability Analysis of Heat Exchanger Networks," *Chem. Eng. Res. & Des.,* **64**, pp. 347–364, (Sept. 1986).
34. **Obeng, E. D. A., and G. J. Ashton,** "On Pinch Technology Based Procedures for the Design of Batch Processes," *Chem. Eng. Res. & Des.,* **66**, pp. 255–259 (1988).
35. **Kemp, I. C., and A. W. Deakin,** "The Cascade Analysis for Energy and Process Integration of Batch Processes," *Chem. Eng. Res. & Des.,* **67**, pp. 495–525 (1989).
36. **Gremouti, I. D.,** "Integration of Batch Processes for Energy Savings and Debottlenecking," MS Thesis, UMIST, Manchester, U.K. (Aug. 1991).
37. **O'Young, L.,** "Constrained Heat Exchanger Networks: Targeting and Design," PhD Thesis, UMIST, Manchester, U.K. (Jan. 1989).
38. **Linnhoff, B., and F. J. Alanis,** "A Systems' Approach Based on Pinch Technology to Commercial Power Station Design," presented at ASME Winter Annual Meeting, San Francisco (Dec. 1989).
39. **Dhole, V. R., and B. Linnhoff,** "Distillation Column Targets," *Comp. & Chem. Eng.,* **17** (5/6), pp. 549–560 (1993).
40. **Wang, Y. P., and R. Smith,** "Wastewater Minimization." *Chem. Eng. Sci.* **49** (7), pp. 981–1,006 (1994).
41. **Boland, D., and B. Linnhoff,** "The Preliminary Design of Networks for Heat Exchange by Systematic Methods," *The Chem. Eng.,* pp. 9–15 (Apr. 1979).
42. **Linnhoff, B., and J. A. Turner,** "Simple Concepts in Process Synthesis Give Energy Savings and Elegant Designs," *The Chem. Eng.,* pp. 742–746 (Dec. 1980).
43. **Linnhoff, B., and J. A. Turner,** "Heat-Recovery Networks: New Insights Yield Big Savings," *Chem. Eng.,* pp. 56–70 (Nov. 2, 1981).
44. **Townsend, D. W., and B. Linnhoff,** "Designing Total Energy Systems by

Literature Cited

Systematic Methods," *The Chem. Eng.*, pp. 91–97 (Mar. 1982).

45. **Brown, K. J.,** "Process Integration Initiative" (review of the process integration initiatives funded under the Energy Efficiency R&D Programme), Energy Technology Support Unit, Harwell Laboratory, Didcot, U.K. (July 1989).
46. **Körner, H.,** "Optimal Use of Energy in the Chemical Industry," *Chem. Ing. Tech.*, **60** (7), pp. 511–518 (1988).
47. **Linnhoff, B., and W. Lenz,** "Thermal Integration and Process Optimization," *Chem. Ing. Tech.*, **59**, pp. 851–857 (Nov. 1987).
48. **Linnhoff, B., and D. R. Vredeveld,** "Pinch Technology Has Come of Age," *Chem. Eng. Progress*, **80** (7), pp. 33–40 (July 1984).
49. **Linnhoff, B., and W. D. Witherell,** "Pinch Technology Guides Retrofit," *Oil & Gas J.*, pp. 54–65 (Apr. 7, 1986).
50. **Linnhoff, B., and G. T. Polley,** "Stepping Beyond the Pinch," *The Chem. Eng.*, pp. 25–32 (Feb. 1988).
51. **Smith, R., and B. Linnhoff,** "The Design of Separators in the Context of Overall Processes," *Chem. Eng. Res. & Des.*, **66**, pp. 195–228 (May 1988).
52. **Linnhoff, B., G. T. Polley, and V. Sahdev,** "General Process Improvements Through Pinch Technology," *Chem. Eng. Progress*, **84** (6), pp. 51–58 (June 1988).
53. **Natori, Y.,** "Managing the Implementation of Pinch Technology in a Large Company," presented at the IEA Workshop on Process Integration, Gothenburg, Sweden (Jan. 1992).
54. **Colberg, R. D., M. Morari, and D. W. Townsend,** "A Resilience Target for Heat Exchanger Network Synthesis," *Comp. & Chem. Eng.*, **13** (7), pp. 821–837 (1989).
55. **Floudas, C. A., and I. E. Grossman,** "Synthesis of Flexible Heat Exchanger Networks for Multiperiod Operation," *Comp. & Chem. Eng.*, **11** (2), pp. 123–142 (1987).
56. **Kotjabasakis, E., and B. Linnhoff,** "Better System Design Reduces Heat-Exchanger Fouling Costs," *Oil & Gas J.*, pp. 49–56 (Sept. 1987).
57. **Saboo, A. K., M. Morari, and D. C. Woodcock,** "Design of Resilient Processing Plants: VIII. A Resilience Index for Heat Exchanger Networks," *Chem. Eng. Sci.*, **40** (8), pp. 1,553–1,565 (1985).
58. **Kaibel, G.,** "Gestaltung destillativer Trennungen unter Einbeziehung thermodynamischer Gesichtspunkte," PhD Thesis, Technical University, Munich, Germany (1987).
59. **Fonyo, Z.,** "Thermodynamic Analysis of Rectification: 1. Reversible Model of Rectification," *Intl. Chem. Eng.*, **14**, pp. 18–27 (1974).
60. **Trivedi, K. K., K. H. Pang, H. R. Klavers, D. L. O'Young, and B. Linnhoff,** "Integrated Ethylene Process Design using Pinch Technology," presented at AIChE Meeting, Atlanta (Apr. 1994).
61. **Klemes, J., et al.,** "Design and Operation of Energy Efficient Batch Processes," final report — Contract No. JOU-43/00, Commission of the European Communities, Brussels (1994).
62. **El-Halwagi, M. M., and V. Manousiouthakis,** "Synthesis of Mass-Exchange Networks," *AIChE J.*, **35** (8), pp. 1,233–1,244 (1989).
63. **Linnhoff, B., and V. R. Dhole,** "Targeting for CO_2 Emissions for Total Sites," *Chem. Ing. Tech.*, **16**, pp. 256–259 (1993).
64. **Raissi, K.,** "Total Site Integration" PhD Thesis, UMIST, Manchester, U.K. (July 1994).
65. **Snoek, J., and T. N. Tjoe,** "Process Integration Experience in a Large Company," presented at the IEA Workshop on Process Integration, Gothenburg, Sweden (Jan. 1992).
66. **Davison, R.,** "Pinch Helps Energy Efficiency Drive," special report, *Eur. Chem. News* (July 27, 1992).
67. **Spriggs, H. D.,** "Experiences in Total Site Studies," presented at the IEA Workshop on Process Integration, Gothenburg, Sweden (Jan. 1992).
68. **Rudman, A., et al.,** "Experience from Total Site Integration," Linnhoff March International, Knutsford, U.K. (1994).
69. **Rossiter, A. P., and H. D. Spriggs,** "Expanded Pinch Analysis Procedure for Pollution Prevention at Amoco's Yorktown, Va., Refinery," report prepared by Linnhoff March Inc. for Amoco/U.S. EPA Pollution Prevention Workshop, Williamsburg, VA (Mar. 1991).
70. **Rossiter, A. P., H. D. Spriggs, and H. Klee, Jr.,** "Apply Process Integration to Waste Minimization," *Chem. Eng. Progress*, **89** (1), pp. 30–36 (Jan. 1993).
71. **Smith, R., E. A. Petela, and H. D. Spriggs,** "Minimization of Environmental Emissions through Improved Process Integration," *Heat Recov. Sys. & CHP*, **10** (4), pp. 329–339 (1990).
72. **Smith, R., and E. A. Petela,** "Waste Minimisation in the Process Industries," "1: The Problem," *The Chem. Eng.*, pp. 24–25 (Oct. 1991); "2: Reactors," *The Chem. Eng.*, pp. 17–23 (Dec. 1991); "3: Separation and Recycle Systems," *The Chem. Eng.*, pp. 24–28 (Feb. 1992); "4: Process Operations," *The Chem. Eng.*, pp. 21–23 (Apr. 1992); "5: Utility Waste," *The Chem. Eng.*, pp. 32–35 (July 1992).
73. **Rant, Z.,** "Exergie, ein neues Wort Für technische Arbeitsfähigkeit," *Forschungs Ingenieur-Wesen*, **22**, p. 36 (Jan. 1956).
74. **Douglas, J. M.,** "Conceptual Design of Chemical Processes," McGraw Hill, New York (1988).
75. **Gundersen, T., and L. Naess,** "The Synthesis of Cost Optimal Heat Exchanger Networks — An Industrial Review of the State of the Art," *Comp. & Chem. Eng.*, **12** (6), pp. 503–530 (1988).
76. **Cerda, J., and A. W. Westerberg,** "Synthesizing Heat Exchanger Networks having Restricted Stream/Stream Matches using Transportation Problem Formulations," *Chem. Eng. Sc.*, **38** (10), pp. 1,723–1,740 (1983).
77. **Floudas, C. A., A. R. Ciric, and I. E. Grossman,** "Automatic Synthesis of Optimum Heat Exchanger Network Configurations," *AIChE J.*, **32** (2), pp. 276–290 (Feb. 1986).
78. **Papoulias, S. A., and I. E. Grossmann,** "A Structural Optimization Approach in Process Synthesis — II, Heat Recovery Networks," *Comp. & Chem. Eng.*, **7** (6), pp. 707–721 (1983).
79. **Steinmetz, F. J., and M. O. Chaney,** "Total Plant Process Energy Integration," *Chem. Eng. Progress*, **81** (7), pp. 27–32 (July 1985).
80. **Obata, K., and H. Shibuya,** "The Challenge of Minimum Energy Plant Design," presented at Ascope 93, Bangkok (Nov. 1993).
81. "Imported Design Technology From British Company," *Nikkei*, p. 11 (Dec. 3, 1992).
82. "Growing European Business related to the Environment," *Nikkei*, p. 9 (Mar. 25, 1993).
83. "Selling Energy Saving Technology to Japan," *Nikkei*, p. 14 (June 21, 1993).
84. "ADVENT" pinch technology software, Aspen Technology, Inc., Cambridge, MA.
85. "SUPERTARGET" pinch analysis software, Linnhoff March International, Houston.
86. "HEATNET" pinch technology software, National Engineering Laboratory, East Kilbride, Scotland.
87. "HEXTRAN" software for heat-exchanger-network design and simulation, Simulation Sciences, Brea, CA.

Acknowledgments

This manuscript took a lot of organizing. Three people deserve thanks: Lil Falkensson for keeping the pressure up, and Vrunda Dhole and Olive Ireland for persevering. How did they put up with me? As far as the contents are concerned, I have worked with several organizations and teams over the years, and many other people have contributed to the development of Pinch Analysis. Special thanks are due to my colleagues, students, sponsors, and customers in UMIST, Linnhoff March, and other companies and universities worldwide. Major contributions were made by my PhD students and coworkers in the Linnhoff March Technology Centre: Su Ahmad, Francisco Alanis, Vikas Dhole, Stephen Hall, David Hui, Manolis Kotjabasakis, Lionel O'Young, Stuart Parker, Kalliopi Raissi, Tina Shokoya, David Tjoe, Bill Townsend, and Dale Williams. They all deserve credit.

INDEX

F

Q

R

S

T